# ENCYCLOPAEDIC DICTIONARY OF P...

*Volumes 1–9*

# ENCYCLOPAEDIC DICTIONARY
# OF PHYSICS

## SUPPLEMENTARY VOLUME 1

SUPPLEMENTARY VOLUME 1

# ENCYCLOPAEDIC
# DICTIONARY OF PHYSICS

GENERAL, NUCLEAR, SOLID STATE, MOLECULAR
CHEMICAL, METAL AND VACUUM PHYSICS
ASTRONOMY, GEOPHYSICS, BIOPHYSICS
AND RELATED SUBJECTS

EDITOR-IN-CHIEF

## J. THEWLIS

ATOMIC ENERGY RESEARCH ESTABLISHMENT

HARWELL

*Associate Editors*

**R. C. GLASS**
LONDON

**W. J. STERN**
LONDON

**A. R. MEETHAM**
TEDDINGTON

## PERGAMON PRESS

OXFORD · LONDON · EDINBURGH · NEW YORK
TORONTO · PARIS · FRANKFURT

Pergamon Press Ltd., Headington Hill Hall, Oxford
4 & 5 Fitzroy Square, London W.1

Pergamon Press (Scotland) Ltd., 2 & 3 Teviot Place, Edinburgh 1

Pergamon Press Inc., 44–01 21st Street, Long Island City, New York 11101

Pergamon of Canada, Ltd., 6 Adelaide Street East, Toronto, Ontario

Pergamon Press S.A.R.L., 24 rue des Écoles, Paris 5ᵉ

Pergamon Press GmbH, Kaiserstrasse 75, Frankfurt-am-Main

First published 1966

Library of Congress Catalog Card No. 65–28743

# ARTICLES CONTAINED IN THIS VOLUME

# LIST OF CONTRIBUTORS TO THIS VOLUME

ABSON W. (*Harwell*)
ADDERLEY E. E. (*Sydney, Australia*)
AITKEN M. J. (*Oxford*)
ANDERSON A. R. (*Harwell*)
APPLETON A. S. (*Liverpool*)
ASSENHEIM H. (*London*)
AVEYARD S. (*Harwell*)

BAKER J. M. (*California*)
BARNES R. S. (*Harwell*)
BARRON T. H. K. (*Bristol*)
BAUMAN R. P. (*New York*)
BENSON F. A. (*Sheffield*)
BEYER N. S. (*Argonne*)
BONNOR W. B. (*London*)
BOTTOMLEY S. C. (*London*)
BRADSELL R. H. (*Teddington*)
BRANDON D. G. (*Geneva*)
BRAY W. J. (*London*)
BROOKER G. A. (*Oxford*)

CALDIROLA P. (*Milan*)
CHERNICK C. L. (*Argonne*)
CHIU H. Y. (*New York*)
CLAPHAM P. B. (*Teddington*)
CLAYDEN W. A. (*Sevenoaks, Kent*)
COEKIN J. A. (*Nigeria*)
COSSUTTA D. (*Surrey*)
COWAN J. D. (*London*)
CRASTON J. L. (*Harwell*)

DARBYSHIRE J. (*Anglesey*)
DAVIES R. D. (*Manchester*)
DIXON J. A. (*Warrington, Lancs.*)
DOWNTON F. (*Leicester*)
DUMMER G. W. A. (*Malvern, Worcs.*)
ENTWISTLE K. M. (*Manchester*)

FLETCHER J. G. (*California*)
FRANK R. C. (*Illinois*)
FROST B. R. T. (*Harwell*)
FULLER W. (*London*)

GIRDLER R. W. (*Durham*)
GOLDSMID H. J. (*Wembley, Middx.*)
GORE J. E. B. (*Berkeley, Glos.*)
GRIDGEMAN N. T. (*Ottawa*)
GROSS W. (*California*)

HALL I. M. (*Manchester*)
HEADING J. (*Southampton*)
HIGSON G. R. (*Isleworth, Middx.*)
HILL D. W. (*London*)
HILL J. F. (*Harwell*)
HOEPPNER C. (*Florida*)
HOLISTER G. S. (*Swansea*)
HOWE E. D. (*California*)
HOWIE A. (*Cambridge*)
HOWIE A. J. (*Glasgow*)

JENSCH A. (*Jena*)

KAPLAN L. (*California*)
KELL R. C. (*Wembley, Middx.*)
KING-HELE D. G. (*Farnborough, Hants.*)

LIEBHAFSKY H. (*New York*)
LOWELL J. (*Oxford*)
LUCK G. A. (*Erith, Kent*)
LYONS L. E. (*Brisbane*)

MARTIN D. G. (*Harwell*)
McCLAIN E. P. (*Washington*)
McCUTCHEON C. W. (*Maryland*)
McMASTER R. C. (*Ohio*)
MELLOR M. (*New Hampshire*)

MENDENHALL R. M. (*Chicago*)
MORAY N. (*Sheffield*)

NAMIAS J. (*Washington*)
NAYLER J. L. (*Claygate, Surrey*)
NEWKIRK J. B. (*New York*)
NOAKES M. (*Harwell*)
NORTON F. J. (*New York*)

O'BRIEN J. F. (*Maryland*)
OWEN R. B. (*Harwell*)

PARKE S. (*Sheffield*)
PARKS P. C. (*Southampton*)
PARROTT J. E. (*Cardiff*)
PEARSON J. (*California*)
PEIRSON D. H. (*Harwell*)
PIERCE O. R. (*Michigan*)
POOLE D. M. (*Harwell*)
POSTLE L. J. (*Manchester*)
PRICE M. S. T. (*Winfrith Heath*)

RANBY P. W. (*London*)
REED T. B. (*M. I. T.*)
RENKEN C. J. (*Argonne*)
RICHARDSON J. M. (*Washington*)
RILEY B. (*Harwell*)
ROBERTS A. (*Argonne*)
ROBERTS A. C. (*Harwell*)
ROBERTS D. H. (*Towcester, Northants*)
ROGERS G. T. (*Wantage*)
ROSS D. S. (*Glasgow*)
ROWELL P. M. (*Nottingham*)

SAWYER J. S. (*Bracknell, Berks*)
SAYERS J. B. (*Harwell*)
SCHOFIELD B. H. (*Waltham, Mass.*)

SELWOOD P. W. (*California*)
SHARPE R. S. (*Harwell*)
SINGER J. R. (*California*)
SKIDMORE I. C. (*Aldermaston*)
SMITH E. K. (*Boulder, Colorado*)
SPIERS F. W. (*Leeds*)
STANG L. G. (*Brookhaven*)
STUART P. R. (*Teddington*)
SWIFT-HOOK D. T. (*Southampton*)

TAYLOR C. A. (*Manchester*)
TOSI M. P. (*Argonne*)
TROTMAN-DICKENSON A. F. (*Aberystwyth*)
TROUP G. (*Florence*)
TUCKER G. D. (*Birmingham*)

VEIS G. (*Athens*)

WELSBY V. G. (*Birmingham*)
WESTON D. E. (*Teddington*)
WILLMORE A. P. (*London*)

WILSON B. (*Towcester, Northants*)
WINTER D. F. T. (*Sevenoaks, Kent*)
WOOD J. A. (*Chicago*)
WOODWARD P. M. (*Malvern, Worcs.*)
WRIGHT P. A. (*Warrington, Lancs.*)

ZWORYKIN V. K. (*Princeton, N.J.*)

# FOREWORD

THE reception of the Encyclopaedic Dictionary of Physics has been very gratifying. With a few trivial exceptions the criticisms made have been constructive and will, I am sure, help us to establish and maintain the usefulness of the Supplementary Volumes, of which this is the first, and to continue to offer an authoritative, unified and modern account of the present state of knowledge in physics and those related branches of science that fall within the scope of the work.

Of its very nature, a work of the size and scope of the Encyclopaedic Dictionary of Physics can never reach ultimate completion. However, by issuing a continuous series of supplementary volumes, we shall strive to keep it as up to date and comprehensive as we can (having regard to the inevitable time lapse between writing and publication), and as free from errors as may be.

The volumes in this series are intended to form part of a unified whole, and are numbered accordingly. They are designed to deal with new topics in physics and related subjects, new development in topics previously covered and topics which have been left out of earlier volumes for various reasons. They will also contain survey articles covering particularly important fields falling within the scope of the Dictionary.

The contents of these volumes will be arranged alphabetically, as in the previous volumes. Articles will be reasonably short and will be signed. Cross references to other articles will be incorporated as necessary, and bibliographies will be included as a guide to further study. Each volume will have its own index, prepared on the same generous scale as before; and, in addition, it is intended to issue a cumulative index every five years. Errata and addenda lists will be published, referring to the original Encyclopaedic Dictionary of Physics and to those supplementary volumes which will already have been published.

In preparing the Supplementary Volumes regard will be had to the changing emphasis in many branches of physics—the invasion of the biological sciences by physics, the possibilities opened out by the increasing use of computers in all branches of science and technology, the ever-increasing scope of theoretical physics, the progress in high energy physics, the emergence of new instrumental techniques etc.; and, at the same time the authors of previously published articles will be given the opportunity of bringing those articles up to date. Naturally there are many articles for which this will not be necessary, and it is certainly not intended that new articles shall be written if there is no need for them. In short, it is our intention to produce a series of volumes in which the high standard already achieved in the Encyclopaedic Dictionary of Physics is fully maintained.

I should like, once again, to express my gratitude to the publishers, and particularly to Mr. Robert Maxwell, for their constant support; to thank my Associate Editors, Dr. A. R. Meetham, Mr. R. C. Glass and Mr. W. J. Stern, for all the help they have given; to pay tribute to Mr. S. Crimmin of the London Office of Pergamon Press for his indispensable work behind the scenes; and to acknowledge the debt I owe to my wife who, as always, has been of invaluable assistance in a variety of ways.

J. THEWLIS
*Editor-in-Chief*

# ADDENDA: VOLUMES I-VII

### Volume I

p. 367, Col. I, line 9. *Delete*: When such a projectile... and following five lines. *Insert*: One of the hardest problems is that of re-entry. When such a projectile re-enters the denser air of the Earth's atmosphere it is travelling so fast that unless it can be slowed the aerodynamic heating will burn it up. On the other hand to retard in the rarefied outer atmosphere is very difficult. Despite this, the problem has been solved and several astronauts brought safely back to earth.

p. 442, article Biorheology. *Add* to bibliography: Copley A. L. and Stainsby G. (Eds.) (1960). *The Rheology of Blood and other Biological Systems*, Oxford: Pergamon Press.

p. 548, Col. I. *Replace* book quoted in Bibliography by: Hyde C. G. and Jones M. W. (1960) *Gas Calorimetry*, London: Benn.

### Volume II

p. 271, article Deformation, plane. *Add* after last sentence: The deformation is called *plane*, when $e_z$, $\gamma_{yz}$ and $\gamma_{xz}$ (or an analogous triple) vanish.

p. 411, article Dilatancy. *Add* immediately above figure; The term "rheopexy" is often wrongly used to describe these phenomena.

### Volume III

p. 241. *Replace* Tables 1 and 2 by Tables 1 and 2 below.

*Table 1: Overall collection efficiency on standard test dust*

|  | Overall Efficiency (%) | Efficiency (%) | | |
|---|---|---|---|---|
|  |  | 5 microns | 2 microns | 1 micron |
| Medium-Efficiency Cyclone | 65.3 | 27 | 14 | 8 |
| High-Efficiency Cyclone | 84.2 | 73 | 46 | 27 |
| Low Pressure-Drop Cellular Cyclone | 74.2 | 42 | 21 | 13 |
| Tubular Cyclone | 93.8 | 89 | 77 | 40 |
| Irrigated Cyclone | 91.0 | 87 | 60 | 42 |
| Electrostatic Precipitator | 99.0 | 99 | 96 | 86 |
| Irrigated Electrostatic Precipitator | 99.0 | 98 | 97 | 92 |
| Low-Velocity Fabric Filter | 99.8 | > 99 | > 99 | 99 |
| Shaker-type Fabric Filter | 99.7 | > 99 | > 99 | 99 |
| Reverse-jet Fabric Filter | 99.9 | > 99 | > 99 | 99 |
| Spray Tower | 96.3 | 94 | 87 | 55 |
| Irrigated-Target Scrubber | 97.9 | 97 | 92 | 80 |
| Self-induced Spray Deduster | 93.6 | 94 | 83 | 48 |
| Disintegrator | 98.5 | 98 | 95 | 91 |
| Venturi-Scrubber | 99.7 | 100 | > 99 | 97 |

*Table 2: Approximate costs of various dedusting systems treating 60,000 ft³/min of dusty gases at 68°F*

| Equipment | Capital Cost (£*) | | Total Running Cost (£/annum) | Capital Charges (£/annum) | Total Cost, including Capital Charges | |
|---|---|---|---|---|---|---|
|  | Total | per ft³/min capacity |  |  | (£/annum) | d/1000 ft³ |
| Medium-Efficiency Cyclones | 7900 | 0.13 | 1750 | 790 | 2540 | 0.021 |
| High-Efficiency Cyclones | 15300 | 0.25 | 2320 | 1530 | 3850 | 0.032 |
| Low Pressure-Drop Cellular Cyclones | 13500 | 0.22 | 720 | 1350 | 2070 | 0.017 |
| Tubular Cyclones | 16600 | 0.28 | 2040 | 1660 | 3700 | 0.031 |
| Irrigated Cyclones | 18800 | 0.31 | 2850 | 1880 | 4730 | 0.039 |
| Electrostatic Precipitator | 74000 | 1.23 | 1400 | 7400 | 8800 | 0.073 |
| Irrigated Electrostatic Precipitator | 95000 | 1.58 | 1980 | 9500 | 11480 | 0.096 |
| Low-Velocity Fabric Filter | 48000 | 0.80 | 4690 | 4800 | 9490 | 0.079 |
| Shaker type Fabric Filter | 52000 | 0.87 | 5070 | 5200 | 10270 | 0.086 |
| Reverse-jet Fabric Filter | 92000 | 1.53 | 8580 | 9200 | 17780 | 0.148 |
| Spray Tower | 44000 | 0.73 | 6000 | 4400 | 10400 | 0.087 |
| Irrigated-Target Scrubber | 26000 | 0.43 | 3770 | 2600 | 6370 | 0.053 |
| Self-induced Spray Deduster | 21000 | 0.35 | 3140 | 2100 | 5240 | 0.044 |
| Disintegrator | 43000 | 0.72 | 23760 | 4300 | 28060 | 0.234 |
| Venturi-Scrubber | 37000 | 0.62 | 12100 | 3700 | 15800 | 0.132 |

### Volume III

p. 851, article Insulation, temperature limits of, Col. II, line 7. After Commission, *delete*: full-stop and "This document represents." *Insert*: and British Standard 2757. These documents represent ...

### Volume V

p. 430, Col. I *Add* to Bibliography: Hamer F. M. (1964) *The Cyanine Dyes and Related Compounds*, New York: Interscience.

p. 548, Col. II, sixteen lines up. *Add*: In 1963, photo-electric polarimeters employing other electro-optical modulators began to appear. The birefringence induced by an electrostatic field in some crystals (Pockels effect) can be converted into the equivalent of an optical rotation by a quarter-wave plate.

p. 589, article Porosity in metals, line 4. After "examination" *delete*: full-stop and *insert*: and for industrial uses by X-ray, gamma-ray, ultrasonic and liquid penetrant methods (non-destructive testing).

### Volume VI

p. 316, article Rheology, line 14. *After*: flow of matter. *Add*: Moreover, since flow is a type of deformation, the inclusion of both terms in the definition is really a tautology.

p. 317, line 14. *For*: must be expressed *read*: are generally expressed.

p. 317, line 18. *After*: components. *Add*: (A. S. Lodge has proposed an alternative and perhaps better classification involving generalized coordinates and matrices in the place of tensors.)
At end of article, *add*: Just as classical thermodynamics has been used to account for the phenomena of rubber-like elasticity, so the new science of irreversible thermodynamics would seem to be especially suited to the study of living systems.
To bibliography *add*: Lodge A. S. (1964) *Elastic Liquids*, New York: Academic Press; Eirich F. R. (1956, 1958, 1960) *Rheology; Theory and Applications*, Vols. 1–3, New York; Academic Press.

p. 318, article Rheopexy. At end of article, *add*; This term has recently been wrongly but widely used to describe properties of non-thixotropic systems.

p. 515, article Soap manufacture, Col. II, end of second paragraph. *Add*: often under pressure.

p. 516, Col. I line 18. *For*: are *read*: used to be.

p. 516, Col. I, nineteen lines up. *For*: more recently, various extrusion processes *Read*: adiabatic cooling followed by extrusion from an evacuated chamber.

p. 540, article Solids, liquids and complex bodies, eleven lines up. *For*: in 1957 *read*: in modern times.

### Volume VII

p. 387, Col. II. *Delete*: Grothendieck etc. in Bibliography and *insert*: Kelley J. L. *et al.* (1963) *Linear Topological Spaces*, Princeton: The University Press: Robertson A. P. and Robertson W. J. (1964) *Topological Vector Spaces*, Cambridge: The University Press.

p. 635, article Viscoelastic gases. *Add* to bibliography: Foux A. and Reiner M. (1964) *Second-order Effects in Elasticity, Plasticity and Fluid Dynamics*, Oxford: Pergamon Press.

p. 686, article Wave breaking, Col. II. *Replace* first paragraph by: The physical condition usually responsible for breaking is that the surface curvature at the crest tends to increase without limit. Also, in a progressive wave the horizontal particle velocity tends to exceed that of the wave itself, and in a standing wave the downward vertical acceleration tends to exceed $g$. The mathematical theory associated with breakers is largely concerned with the highest wave which can exist without producing these, or other equivalent conditions. Some of the main results are given below: *Replace* last sentence of second paragraph by: While the standing wave has crest acceleration $g$, the Stokes $120°$ crest has only $0 \cdot 5g$.

p. 746, article Weather forecast. *Add* after last sentence: Since this was written considerable advances have been made in long-range forecasting: see, for example "Weather forecasting, long-range" and "Weather forecasting, numerical" in the present volume.

# ERRATA: VOLUMES I-VII

## Volume I

p. VIII, line 7. *For*: depth *read* debt.

p. 1, Col. I, Line 1, *For*: Abbé *read* Abbe.

p. 2, Col. II. Rotate Fig. 4 through 90°.

p. 72, Col. I, line 11. *For* $5 \times 10^{12}$ ncm$^{-2}$/sec$^{-1}$ *read* $5 \times 10^{12}$ ncm$^{-2}$ sec$^{-1}$.

p. 73, Col. I. See also: *for* Antenna *read*: Various articles beginning "Antenna".

p. 114, Col. I, line 42 *should read*: i.e. ratio of total scattered energy to incident energy.

p. 128, Col. II. Beginning of second paragraph. *For*: principle *read*: principal.

p. 135, Col. II, line 6. *For*: and *read*: to an. Equation 2, *for*: $A$ *read*: $\sqrt{A}$.

p. 145, Col. I, line 17. *For*: of *read*: $\varphi$.

p. 145, Col. I, Fig. 2. I $(\omega L - 1/\omega C)$ should be to left of vertical axis.

p. 145, Col. II. Interchange captions of Figs. 3 and 4.

p. 146, Col. II, lines 31 and 33. *For*: i–k *read* k–i.

p. 156, Col. I, end of first equation. *For*: COO *read*: COO$^-$.

p. 187, Col. I, line 2. *For*: unvariant *read*: invariant.

p. 200, article Anode resistance, first and last lines. *For*: $\pi_a$ *read*: $r_a$.

p. 220, Col. I, line 40. *For*: 1 (a) *read*: 1 (b).

p. 243, Col. I. See also: *For*: Spectrograph grating *read*: Spectrograph, concave grating. *For*: Ampére (unit) *read*: Ampère.

p. 359, *under* Balance, single pan *add* See: balance.

p. 359, Col. II. Second paragraph from bottom. *This should read*: *Fundamental parameters*. The internal diameter of the tube, the calibre, provides a natural unit of length. Bore lengths are then a few tens of calibres. If the shot weight is $W$ lb and the calibre $d$ in., then $W/d^3$ differs from 0·5 by rarely more than a factor of two. Low values … .

p. 414, article Bilinear form: *add* rotations *after last word*.

p. 446, article Black nucleus, *for*: nucles *read*: nucleus; *for*: $\pi(R + \lambda)$ *read*: $\pi(R + \lambda)^2$.

p. 467, Col. I, equation 2 *should read*

$$\frac{\partial f}{\partial t} + (V \cdot \nabla) f + \frac{1}{m} (F \cdot \nabla_v) = \left(\frac{\partial f}{\partial t}\right)_{\text{coll.}}$$

p. 467, Col. I, two lines up. *For*: $\Delta_v$ *read* $\nabla_v$.

p. 479, article Borrmann effect. *For*: dank *read*: dark.

p. 495, Fig. 2. *Interchange* (b) and (c).

p. 496, Col. II, table. *For*: of *read*: or.

p. 497, Col. I. *For*: G. L. T. Bailey *read*; G. L. J. Bailey.

p. 536, Col. I. Figure to be turned through 90°.

p. 556, Col. II, Equation in centre. *For*: $\left(1 + \dfrac{d'}{ba}\right)$ *read* $\left(1 + \dfrac{d'}{6a}\right)$.

p. 592, Col. I, line 38. *Insert* $10^{-8}$ *before* dyne.

p. 686, Col. II, line 13, *For*: sec $x$ *read*: sec$^2$ $x$.

p. 722, Col. I. The denominator in equation 3 should read: $\gamma(I_\varrho^{(1)}/I_\varrho^{(2)}) + \gamma(I_\varrho^{(2)} + I_\varrho^{(1)})$; *Add* to equation 5: $= \Gamma_{12}(\tau)/\gamma I_1 I_2)$; equations 6 and three lines above it replace $\gamma_{12}(\tau)$ by $\Gamma_{12}(\tau)$.
Col. II, line 5. *Delete*: are two points on a subsequent wave front. *Insert*: are the two conjugate image points associated with an optical system.

p. 776, Col. II, three lines up. The equation should read

$$F(V, J) = B_v J (J + 1) + D_v J^2 (J + 1)^2.$$

## Volume II

p. 40, Col. II, line 7. *For*: gamma *read*: gramme.

p. 56, article Conjugate planes, line 2. *For*: from *read*: of an.

p. 242, Col. I, article Curie scale of temperature, line 9. *For*: $T^*_{H=0} = C^*_{H=0}$.

p. 242, Col. II. *Delete* article Curl; *insert*: *Curl*. If $S$ is a plane area containing the point P and has normal $\boldsymbol{n}$, we define curl $v$ at the point P by stating that its component in the direction $\boldsymbol{n}$ is the limit of

$$\frac{1}{S} \int_{\Gamma} \boldsymbol{v} \mathrm{d}\boldsymbol{s}$$

as $S \to 0$ in such a way that the perimeter of $\Gamma$, the boundary of $S$, tends to zero. d$\boldsymbol{s}$ is the element of arc along $\Gamma$.
In cartesian coordinates curl $v$ has components

$$\frac{\partial v_z}{\partial y} - \frac{\partial v_y}{\partial z}, \quad \frac{\partial v_x}{\partial z} - \frac{\partial v_z}{\partial x}, \quad \frac{\partial v_y}{\partial x} - \frac{\partial v_x}{\partial y}$$

where $\boldsymbol{v}$ has components $(v_x, v_y, v_z)$ in a rectangular frame of reference $O(x, y, z)$.

p. 246, Col. I, article Current efficiency, electrochemical, line 3. *For*: 96,500 *read*: 96,490.

p. 470, Col. II, lines 12 and 13. *For*: $3 \cdot 57 Z_1 Z_2 \text{Å}$, $Z_1$ and $Z_2 \ldots$ *read*: $3 \cdot 57 z_1 z_2 \text{Å}$, $z_1$ and $z_2 \ldots$

p. 513, Col. I, line 16 *should read*: … of ferrous to ferric compounds and the reduction of ceric to cerous.

p. 673, Col. I, line 16. *For*: 16 *read*: $10^6$.

p. 680, Col. II, Interchange diagrams of Figs. 1 and 2.

### Volume III

p. 17,   article Etalon, line 4. *For*: 40 *read*: $\frac{1}{0}$.

p. 91,   Col. I, line 29. *For*: $Fe_{0.5}^+$ *read*: $Fe_{0.5}^{3+}$.

p. 91,   Col. I, seven lines up. *Insert* $4\pi M$ *after* magnetization.

p. 92,   Col. I, line 1. *For*: 5000 *read*: 10,000.

p. 92,   Col. I, line 4. *For*: 100 *read*: 10.

p. 92,   Col. I, Fig. 1. Left hand ordinate *should read*: $\mu'$.

p. 94,   Col. I, line II. *For*: 1600 *read*: 1800.

p. 94,   Col. I, line 19. *For*: 30 *read*: 20.

p. 127, Col. I, line 15. *For*: $a + a \sqrt{-1}$ *read*: $a + b \sqrt{-1}$.

p. 400, Col. I, four lines up. *For*: 273·16 *read* 273·15.

p. 469, Col. II, equation in Gladstone and Dale law should read $\dfrac{n-1}{\varrho} = $ constant.

p. 473, Col. I, line 3. *For*: $-3 \times 10^{-6}$ *read*: $3 \times 10^{-6}$

pp. 484 and 485. Interchange diagrams for Figs. 2 and 3.

p. 575, Col. I, equation 2. *For*: $\neq$ *read*: $=$.

p. 575, Col. I, four lines after equation 2. *For* $Tu$ *read*: $T_u$.

p. 575, Col. II, first equation. *For* $P = Po(d/D)^x$ *read*: $P = P_0(d/D)^x$.

p. 575, Col. II, seven lines up. *For*: $P = 4W/\pi d2$ *read*: $P = 4W/\pi d^2$.

p. 576, Col. I, line 1. *For* $Y = b\varepsilon^x$ *read*: $Y = be^x$.

p. 577, Col. I, line 17. *For*: $M = kH$ *read*: $M = kH^{1/3}$.

p. 577, Col. II, line 8. *For*: $Ca_{10}(Po_4)_6F_2$ *read*: $Ca_{10}(PO_4)_6\ F_2$.

p. 792, article Illumination, cosine law of, line 3. *For*: $I(a)$ *read*: $I(0)$.

### Volume IV

p. 2,   article Intermolecular potential. *For*: charged *read*: charges.

p. 14,   Col. II, eleven lines up. *For*: replased *read*: replaced.

p. 14,   Col. II, sixteen lines up. *For*: reacting *read*: reaction.

p. 44,   Fig. 4 Ignore numbers on ordinate axis.

p. 44,   Fig. 5. *For*: Famada *read*: Yamada.

p. 46,   article Ionic strength. *For*: $\frac{1}{2} M_i z_i^2$ *read*: $\frac{1}{2} \Sigma M_i z_i^2$ and *For*: $\frac{1}{2} m_i z_i^2$ *read*: $\frac{1}{2} \Sigma m_i z_i^2$.

p. 158, Col. II, seven lines up. *For*: their mean distances *read*: the cube of their mean distances.

p. 196, Col. II, equation 2. *For*: $Al(q_1, q_2, \ldots q_n, t)$ *read*: $A_1(q_1, q_2, \ldots q_n, t)$.

p. 197, Col. II, equation 8. *For*: $\dfrac{d}{dt}\left(\dfrac{\partial L}{\partial q_i}\right)$ *read*: $\dfrac{d}{dt}\left(\dfrac{\partial L}{\partial \dot{q}_i}\right)$.

p. 459, Col. II, article Magnetization of rocks, line 27. *For*: magnetite *read*: maghaemite.

p. 493, Col. I, line 7. *For*: $(10^{-2} - 10^{-6} \text{ mm}^2\text{Hg})$ *read*: $(10^{-2} - 10^{-6} \text{ mm Hg})$.

p. 574, article Meso-ionic compounds. Formula should read:

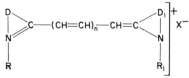

p. 575, Col. I. Formula (I) should read:

p. 700, Col. II, article Molecular effusion. *For*: $\dfrac{S\sqrt{(kt)}}{2\pi m}$ *read*: $S\sqrt{\left(\dfrac{kt}{2\pi m}\right)}$.

p. 771, Fig. 7. The acknowledgement should read: (Crown copyright. Reproduced by permission of the Controller, H. M. Stationery Office.)

### Volume V

p. 32,   Col. I, line 14. *For*: $f_0 = 10/h$ read: $f_0^2 = 10/h$.

p. 100, Col. I, last line. *For*: spare *read*: space.

p. 200, Col. II, nine lines up. *Insert* (3) at right of equation.

p. 223, article Optical parts, use of Canada balsam for cementing, line 7. *For*: caproate *read*: caprate.

p. 429, Col. II. Second formula should read:

p. 547, Col. II, line 1. *For*: flat *read*: diametral.

p. 548, Col. II, line 15. *For*: Ericson *read*: Ericsson.

p. 516, article Planck's constant, last line. *For*: $10^{27}$ *read*: $10^{-27}$.

p. 655, Col. I, line 7. *For*: $\theta l$ *read*: $\delta l_i$.

p. 655, Col. II, seventeen lines up. *For*: their *read*: either.

p. 657, Col. I, lines 7 and 14 up. *For*: $\dfrac{\lambda}{x}$ *read*: $\dfrac{\lambda}{X}$.

p. 731, Col. I, line 23. *For*: matric *read*: matrix.

### Volume VI

p. 234, article Reduced stress. *For*: $\sigma_{ij} = \sigma_{ij} - \sigma\delta_{ij}$ *read*: $\sigma'_{ij} = \sigma_{ij} - \sigma\delta_{ij}$; *for*: hydraulic *read*: hydrostatic.

p. 297, Fig. 2.

For      Read

p. 323, Col. II, line 12. *For*: imply $a = \bar{c}$ *read*: imply $\bar{a} = \bar{c}$.

p. 500, article Skewness, line 27. *For*: $3\sigma$ *read*: $3\bar{\sigma}$; line 28, *for*: $\sigma$ is *read*: $\bar{\sigma}$ is. Line 31, *for*: $|\sigma|$ *read*: $|\bar{\sigma}|$.

p. 500, Col. II, line 1. *For*: $\gamma_1(= \sqrt{\beta},)$ *read*: $\gamma_1(= \sqrt{\beta_1})$.

p. 500, Col. II, line 7. *For*: recently *read*: recently (1959).

p. 537, article Solar X rays, line 10. *For*: stotm *read*: storm; line 12, *for*: fall-out *read*: fade-out.

p. 694, Col. II. Lines 21, 22 and 32. *For*: $^{4}H^{13/2}$ *read*: $^{4}H_{13/2}$.

p. 851, Col. II, line 33. *For*: $f(x) = (\sigma \sqrt{2\pi})^{-1} \exp \ldots$ *read*: $f(x) = \sigma \sqrt{(2\pi)} \exp \ldots$

p. 853, Col. I, nineteen lines up. *For*:

$$t = (\bar{x}_1 - \bar{x}_2) \left| s\left(\frac{1}{n_1} + \frac{1}{n_2}\right) \right.$$

*read*: $t = (\bar{x}_1 - \bar{x}_2) \left| s\left(\frac{1}{n_1} + \frac{1}{n_2}\right)^2 \right.$.

*Volume VII*

p. 48, Col. I, line 21. *For*: geometric *read*: geomagnetic.

p. 356, Col. II, line 17. *For* $\dfrac{Ga^2}{E}$ *read* $\dfrac{GE}{a^2}$.

p. 387, Col. I. second para. line 3. *For*: $\|\alpha\|$ *read*: $|\alpha|$. Line 4, *for*: $\|y\|$ *read*: $\|y\|$.

p. 396, article Torricellian vacuum. *For*: less than $10^{-6}$ *read*: of order of $10^{-4}$.

p. 676, Col. I., article Water, density and expansion of. Figures in last lines should read: $-0{\cdot}06427 \times 10^{-3}$, $8{\cdot}5053 \times 10^{-6}$, $-6{\cdot}79 \times 10^{-8}$.

p. 767. Col. I. See also; *For*: Dynamic instability, dynamic *read*: Instability, dynamic.

p. 860, Col. II, line 6. *For*: class oscillators *read*: class of oscillators.

**ACOUSTIC ARRAYS, DIRECTIONAL.** A directional array is an arrangement of acoustic sources or receivers which is designed to have directional properties. The term is sometimes extended to include certain electrical signal-processing operations which influence the directional properties. An example is the "multiplicative" array.

The response of the array is expressed by a *directional function*. For a transmitting array, this relates the strength of the acoustic signal at a distant receiving point to the bearing angle as the array is rotated and, for a receiving array, it expresses the electrical output amplitude from a fixed distant source as the array is rotated.

For an array of identical sending or receiving transducer "elements", the directional response is the product of two factors, one denoting the intrinsic directional properties of each element and the other the response of a hypothetical array of point sources or receivers.

The following discussion of directivity refers to measurements made in the far-field (or Fraunhofer region), i.e. at distances large compared with $l^2/\lambda$ where $l$ is the length of the array and $\lambda$ is the wavelength. Generally similar results may be obtained at distances less than this but the directional patterns tend to become more and more distorted as the near-field (or Fresnel region) is approached.

The simplest array consists of $n$ point elements, transmitting an unmodulated carrier wave and spaced equally at a distance $d$ apart. The acoustic path lengths to a receiver in the far-field, measured from any pair of adjacent elements, will differ by $d \sin \theta$, where $\theta$ is the angular bearing of the receiver relative to the perpendicular axis of the array. This causes corresponding phase differences so that the total received signal can be expressed in exponential form as

$$\sum_{r=1}^{n} a_r \exp 2jrp \cdot \exp j\omega t$$

where $$p = \pi d \sin \theta / \lambda.$$

The $a$'s are complex coefficients denoting the amplitudes and relative phases of the signals fed to the array elements. The normalized directional function is of the form

$$|D(p)| = \frac{|\sum a_r \exp 2jrp|}{|\sum a_r|} \qquad (1)$$

In the particular case of a *uniform line array*, with all the elements fed in phase and at equal amplitudes, this reduces to

$$|D(p)| = \left| \frac{\sin np}{n \sin p} \right|. \qquad (2)$$

The same equations apply also to a receiving array where the $a$'s are regarded as "sensitivity coefficients" denoting the relative amplifications and phase shifts applied to the electrical outputs of the elements before they are added together.

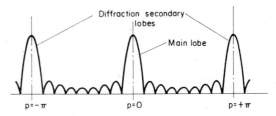

*Fig. 1. Directional pattern for uniform line array*
$$\text{when } n = 9; \quad |D(p)| = \left| \frac{\sin 9p}{9 \sin p} \right|.$$

Figure 1 shows a typical form of $|D(p)|$ for a uniform line array. In addition to the main lobe at $p = 0$, "diffraction secondary" lobes appear at intervals of $\pi$ along the $p$-axis.

The amplitudes of the smaller lobes of the pattern can be reduced, at the expense of some widening of the main lobe, by feeding all the elements in phase but causing the amplitudes of the coefficients to fall progressively from the centre towards the ends of the array. This process is called "tapering" or "shading".

An important property of a line array is that if the phases of all the element feeds are changed by an amount which is proportional to distance measured along the array, the directional pattern remains unchanged in shape but is shifted along the axis of $p$ by an amount $\frac{1}{2}\varphi$ where $\varphi$ is the phase shift between adjacent elements. This gives a method of rotating the directional response of a stationary array by purely electrical means and forms the basis of electronic sector-scanning techniques.

*Generalized theory.* The far-field directional pattern of any acoustic transmitting array is, in fact, the diffraction pattern corresponding to a particular distribution of source amplitudes and phases over some reference surface. The effective area of the surface, over which the source amplitude is significant, is called the aperture of the array. For a line array the aperture is merely a straight line. Let a complex function $Q(\sin \theta)$ denote the amplitude and phase of the signal received at a point in the far-field and let another complex function $P(2\pi y/\lambda)$ denote the amplitude and phase of the transmitted signal at any point whose distance along the straight-line aperture is $y$ from its centre. Then it can be shown that $P$ and $Q$ are a pair of Fourier transforms. This fundamental relationship applies also when the array is used for reception; $P$ is then a "sensitivity function" describing the amplitude and phase adjustments made to signals from each part of the array before addition takes place.

Application of the sampling theorem leads to the conclusion that practically all the useful directional information contained in the signals received by a given aperture can be extracted in the form of samples taken at intervals not less than $\lambda/2$ along the aperture. Although, theoretically, it is possible, by reducing the sampling interval to less than $\lambda/2$, to produce a "super-directive" array which behaves as though its effective aperture length were greater than its real length, such arrays are rarely realizable in practice. Any appreciable amount of superdirectivity is usually accompanied by an intolerable worsening of the signal-to-noise performance of the system and also by unattainable limits of accuracy for the element sensitivities.

The practical "resolving power" of an array is ultimately limited by its aperture size in wave-lengths, even if noise can be neglected. (It is instructive to compare this with the well-known corresponding result in the theory of optical telescopes and microscopes.)

*Multiplicative (or correlation) arrays.* For reception only, some advantages can be obtained by splitting a line array into two parts and then multiplying together the two signal voltages obtained from them. This process halves the width of the main lobe of the directional pattern and greatly reduces the size of all

except the first pair of subsidiary lobes. The importance of the latter is also reduced because they are reversed in sign relative to the main lobe. The directional function for a uniform multiplicative array with $m$ elements in one group and $n - m$ in the other is

$$D(p) = \left(\frac{\sin mp}{m \sin p}\right)\left(\frac{\sin (m-n)\,p}{(m-n)\,p}\right) \cdot \cos np. \qquad (3)$$

The narrowing of the main beam does not necessarily mean that the resolving power of the system has been increased because unwanted cross-multiplication effects are also introduced. In practical acoustic systems however, these unwanted effects often tend to average out so that multiplication does possess real advantages. Figure 2 shows the typical form of the directional function given by equation 3.

*Arrays of other shapes.* A development of the line array consists of a stack of $m$ uniform, $n$-element arrays, forming a plane matrix with $m \times n$ identical elements. The directional function is obtained by multiplying together the functions for one row and one column of the matrix, the diffraction pattern forming a two-dimension "pencil" beam. The pattern can be modified by "tapering" the element amplitudes and also can be deflected by suitable adjustments of the relative phases of the feeds to the elements.

A whole series of curved arrays can be developed by defining the function $P$ over an arc of a circle instead of a straight line. The additional complexity of the design computation, compared with that for the straight-line array, may be justified in some cases. One advantage is the relative ease with which the directional pattern of a circular array can be deflected over very wide sector angles without distortion because the pattern is expressed directly in terms of the bearing angle $\theta$, rather than $\sin\theta$.

*Wide-band arrays.* For the "narrow-band" arrays so far mentioned, the directional pattern is a fixed function of the wave-length. It is sometimes desirable to have patterns which have beam-widths which remain practically constant over a given frequency range. This can be done, for example, by using delay lines which introduce suitable real time delays into the element feeds instead of phase shifts.

In another type of wide-band array, a directional pattern comparable with that of a multi-element array can be obtained, with an array having only two receiving elements, by using a "multi-frequency" carrier wave containing components which are equally spaced in the frequency spectrum.

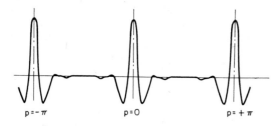

*Fig. 2. Directional pattern for uniform multiplicative array when $n = 9$ and $m = 4$*

$$D(p) = \frac{\sin 4p}{4 \sin p} \cdot \frac{\sin 5p}{5 \sin p} \cos 9p.$$

*Bibliography*

KOCK W. E. and STONE J. L. (1958) *Proc. I.R.E.* **46**, 499.
SCHELKUNOFF S. A. (1943) *Bell System Techn. Journal.* **22**, 80.

TORALDO DI FRANCIA G. (1956) *Trans. Inst. Radio Engrs.* (AP–4), 473.
TUCKER D. G. (1957) *Nature* **180**, 496.
WELSBY V. G. and TUCKER D. G. (1959) *J. Brit. I.R.E.* **19**, 369.
WOODWARD P. M. and LAWSON J. D. (1948) *J. Inst. Elect. Engrs.* **95**, 363.

V. G. WELSBY

**ACOUSTIC NOISE IN THE SEA.** There are many causes of this, and it forms a serious limitation to the performance of sonar and other underwater detection systems. One form of underwater acoustic noise is a form of thermal-agitation noise which gives an electrical output from an electro-acoustic transducer exactly corresponding to that which would arise if the radiation resistance caused by the loading effect of the water on the transducer face were an ordinary electrical resistance at the temperature of the water. Other sources of noise are usually dominant at low frequencies, and are due, for example, to waves breaking; to fish and other marine animals, such as porpoises and shrimps; to the rolling of shingle under the action of currents; to ship's propellers, to dockyard work; etc. Although the overall effect of this noise is usually to give a continuous frequency spectrum, yet the spectrum density is found to fall at a rate of about 6 dB per doubling of frequency. This kind of noise is therefore usually unimportant in sonar systems when the operating frequency is above say 200 kc/s. In low-frequency passive detection systems it is the main limitation on range of detection.

*Bibliography*

KNUDSEN V. O. *et al.* (1948) *J. Marine Res.* **7**, 410.

D. G. TUCKER

**AERODYNAMIC RANGES.** An aerodynamic range is a device for observing the motion of a projectile in free flight. The projectile may be a shell fired from a normal gun, for instance, or may be a scale model of an aircraft or missile, launched from a smooth-bore gun using a sabot to hold the model in the desired attitude and to protect it during launch. Such a sabot is designed to be discarded after leaving the muzzle of the gun, allowing the model to fly down the range.

Depending on the application, projectiles may be launched at representative speeds up to 8 km/sec (the orbital velocity for vehicles which will re-enter the Earth's atmosphere), these high speeds being reached by the use of light gas guns. During the flight down the range, measurements of velocity and attitude are made which enable aerodynamic data such as drag, lift and oscillatory behaviour to be derived. By using an enclosed range, it is possible to simulate flight at altitudes other than sea-level. At very high simulated altitudes, however, the aerodynamic forces become so small that measurement is difficult; for instance, in the 1000 ft ranges in the U.S.A. a typical projectile would be 3 cm diameter and the maximum simulated altitude about 50–60 km. (It is necessary for the projectile to complete more than one oscillation during its flight if moment coefficients are to be derived from the measurements.)

The velocity of the projectile in a range may be measured by placing foil or wire screens so that the projectile makes or breaks an electrical circuit as it passes through them. The screens may damage the projectile or affect its motion, and do not operate correctly at high speeds when the air round the projectile is ionized. The projectile may also interrupt or reflect a light beam to actuate a photocell. Any of these methods may be used in conjunction with a spark photographing the projectile to obtain a more accurate position measurement, and all depend on some form of electronic timing device to measure the interval between two stations. Another method of velocity measurement uses the doppler shift of a microwave signal reflected from the projectile as it approaches an antenna.

To obtain projectile drag, it is necessary to observe the change in velocity over at least two intervals, and so the accuracy of velocity measurement if this quantity is required must be high. For instance, over a 4 m base length at 4 km/sec, a velocity measurement is accurate to 1 per cent if the distance is measured to 2 cm and the time to 5 microseconds. However, the aerodynamic drag might only change the velocity by 40 m/sec between successive measuring stations, and to obtain the drag coefficient to 1 per cent it would be necessary to measure to 0·2 mm and 0·05 microsecond.

Oscillatory motion may be observed by photography at successive stations, using orthogonal pictures from which the angle and direction of projectile deviation may be deduced. The same result may be obtained, at any rate with heavy projectiles at low speeds, by using yaw cards, thin cards which are penetrated by the projectile and leave a silhouette. From sufficient yaw cards or photographs it is possible to deduce all the required aerodynamic data.

With any such system, however, the analysis is very complicated, and data is only available at discreet stations. To overcome both these difficulties, telemetry of such quantities as projectile yaw and deceleration has been developed. Since many of the projectiles have to withstand very high acceleration during launch, however, it is difficult to combine measuring sensitivity with ruggedness, and telemetry has not been used extensively in range work. A system by which projectile orientation is detected by magnetic or electromagnetic interaction with coils mounted in the range shows more promise.

An aerodynamic range has some advantages over a wind tunnel, because it is possible to launch projectiles into still air, with none of the problems of support interference which arise in a tunnel. In addition, the velocity of the projectile can be made to agree with full-scale, as well as the Mach number, and this becomes particularly important when the air affected by the projectile becomes dissociated or ionized. On the other hand, the projectile is moving and apart from some limited information which may be telemetered, it is not possible to obtain measurements apart from photographic observations. The aerodynamic range and wind tunnel are thus complementary, the range being more suitable for observing overall behaviour, e.g. aerodynamic forces, and the tunnel providing more opportunity for detailed measurements, e.g. pressure distribution.

D. F. T. Winter

**ANIMAL JOINTS, LUBRICATION OF.** The slipperinesss of animal joints is a very old phenomenon, but its emergence as a puzzle is recent. It came, and could only have come, after the performance limitations of conventional bearings were established experimentally and theoretically—and animal joints were found to exceed these limitations.

Joints vary enormously in design. In some there is no bearing at all in the ordinary sense of the word. A flexible block (an intervertebral disk for example) provides limited freedom of motion. Some, like the "sutures" joining the skull bones allow little or no relative motion. But the rest, however much they differ in appearance, all follow the same principle.

The bone ends form a conventional-looking bearing with rubbing surfaces made of articular cartilage. The necessary constraints to hold the joint together and eliminate unwanted degrees of freedom are provided by a system of flexible but relatively inextensible ligaments. Loads upon the joint are borne by tension in the ligaments and compression in the rubbing surfaces.

These rubbing surfaces slide over each other with a friction coefficient which has been reported variously from 0·007 to 0·02. This is markedly lower than the 0·05 of PTFE against PTFE, or the 0·03 of an ice skate on ice. Yet the bearing, crudely illustrated minus ligaments in Fig. 1, looks very ordinary. The rubbing surfaces are immersed in synovial fluid, a liquid which is about 98% water, and they generally do not mate well with each other. The cartilage is 1–2 mm thick in man (less in smaller animals) and, while it is reported to be locally very smooth ($\pm 125$ Å over a length of 5000 Å), it has longer wave-length lumpiness which is easily visible to the naked eye. To the touch it is sluggishly elastic, and might be mistaken for plasticized polyvinyl chloride were it not that it feels damp—with a clamminess which ordinary drying with a towel will not remove.

Even before the first measurements of joint friction were made, M. A. MacConaill suggested that joints enjoyed hydrodynamic lubrication. And it is true that, under the right conditions, hydrodynamic lubrication can achieve friction coefficients as low as those in joints. But one objection seems insurmountable. In the hydrodynamic process, the motion of the rubbing surfaces drags the lubricant, by means of its viscosity, into the load-bearing region where it forces the surfaces apart.

If there is no motion there is no dragging in of lubricant, and the surfaces will remain in contact. A hydrodynamic bearing is poorly lubricated at low speed, and not at all when at rest.

Joint friction remains low right down to zero speed. This seems to dispose of the hydrodynamic explanation, which is almost a pity, because the rheological properties of synovial fluid at first appear to make it an ideal hydrodynamic lubricant. Its content of *hyaluronic acid*, one of the body's many *mucins*, renders it strongly non-Newtonian with a viscosity which rises sharply with decreasing shear rate. In a hydrodynamic bearing such a lubricant would be thin at high speed and viscous at low speed, an ideal characteristic.

The trouble with this beautiful picture is that the rise in viscosity occurs at shear rates so low (most of the rise occurs below 20 sec$^{-1}$) that they would occur only in bearings with very large clearance moving very slowly, conditions which imply a loading order of magnitude lower than is found in animals and machinery. Under conventional loadings, synovial fluid is a nearly Newtonian liquid of low viscosity. This is confirmed by the failure of synovial fluid to show any unusual lubricating properties in man-made bearings.

Because joints failed to show the characteristics expected of hydrodynamic lubrication, J. Charnley concluded that boundary lubrication must be occurring. Either the cartilage–cartilage system, or one made

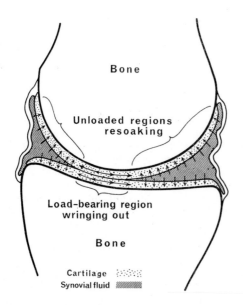

*Fig. 1. A joint, minus ligaments and tendons, showing the long, narrow path traversed by liquid as it is wrung out of the cartilages, and the much easier path available when it returns.*

of cartilages "plated" with synovial fluid, seemed to be extremely slippery.

A *priori* there can be no objection to this view because it follows the sound logic of describing a material in terms of its measured properties. Of course, one is naturally cautious about conceding to nature such a complete victory over friction by direct assault. In what follows it will be seen that there is another objection.

More recently it has been suggested that joints enjoy a form of lubrication previously unknown to engineering. This has been called *weeping lubrication*, and is best understood by the following argument. The only man-made bearing which can equal cartilage in its low friction at all rubbing speeds is the hydrostatic bearing. Here the load is carried by fluid supplied at high pressure by a pump. There are no pumps in animal joints, but the structure of cartilage and the intermittent character of the load it carries combine to produce the same effect.

Cartilage is a spongy, liquid-soaked material—hence the persistent dampness of its surface. The skeleton of this sponge is very deformable (its Young's modulus is about $6 \times 10^6$ dynes/cm$^2$); so deformable that by itself it would be unable to support the load it carries in service. But any flattening of the skeleton requires that the liquid within it be expelled, and, as Fig. 1 shows, the available excape route is long and thin (the bone ends are nearly impervious). Further, the sponge skeleton is very fine grained. The resulting high flow resistance allows flattening to occur only very slowly.

When two cartilages are pressed together their skeletons are, at first, almost undeformed. This, combined with their small stiffness, means that the force they exert on each other is small. So most of the load is carried by hydrostatic pressure in the liquid; and it is carried frictionlessly. The only source of friction is the light contact between opposing high-spots on the sponge skeletons.

Eventually the liquid will wring out, the cartilages will squash down, and their skeletons will take over the load carrying. Friction will rise. This is easily demonstrated in experiments, but unlikely to occur in animals. The poor mating of the rubbing surfaces of a joint means that as long as the animal moves (and move he will unless extremely ill), the load will keep shifting from place to place in the bearing. Eventually every region of the cartilage will get its chance to expand and soak up a fresh charge of liquid.

The pressure difference which draws the liquid back into the cartilage is produced by the feeble stiffness of the sponge skeleton and is much less than that which occurs during the expulsion phase. Nevertheless, in the competition between resorbing and wring-out the latter does not win. This is because the path open to the returning liquid is broad and the distance it needs to travel is short (see Fig. 1), while the wring-out path is thin and long.

To confirm that cartilage enjoys weeping lubrication, its friction coefficient was measured as a function of time. The measurement was against smooth glass rather than against another piece of cartilage. The latter experiment is much more difficult, because then neither surface is flat so the normal (in the sense of perpendicular) and tangential directions vary from place to place. Furthermore, a cartilage–cartilage experiment measures deformation losses as well as surface friction because the cartilages dent each other and the dents move as one cartilage is slid over the other.

A further advantage of the cartilage-glass system is that wring-out is faster so the experiments take less time. The fit of the two surfaces is much worse than in an animal joint. This makes the loaded region smaller, so it contains less liquid, while the wring-out path is shorter and thus has lower flow resistance.

Strictly these experiments relate only to the cartilage-glass system, but as this has a friction coefficient

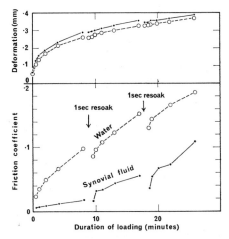

*Fig. 2. As the liquid is wrung out the cartilage the deformation rises (the cartilage gets thinner) and its friction against glass increases. Short periods of resoaking produce only transient reductions in friction.*

as low as the measured value of cartilage–cartilage friction, it is clear that some remarkable process is occurring, and it is fairly safe to assume that it is either the same process or one closely related.

The data is shown in Fig. 2. The spherical upper end of a pig humerus with a 5lb load (giving about the same pressure as that occurring in the pig) at first shows a low coefficient of friction which rises with the passage of time. At the same time, the cartilage deformation increases; it gets thinner as the liquid wrings out. This is what a weeping bearing ought to do. If one waits until wring-out is completed (not shown) the friction coefficient rises to 0·4, an entirely ordinary value. This is hard to explain on the theory that the remarkable properties of cartilage result from boundary lubrication because a boundary lubricant does not normally lose its effectiveness with the passage of time.

If one unrealistically ignores the porosity of the cartilage, one might imagine that the initial low friction was provided by a thick film of liquid trapped between the rubbing surfaces. The wringing-out of this liquid would be retarded by its viscosity so the bearing might enjoy a form of hydrodynamic lubrication. With the wringing out of this film, the friction would rise.

That this is not the explanation was shown by separating the rubbing surfaces for one second. This will restore any hydrodynamic film, so on hydrodynamic theory the friction should fall to the low value found at the beginning of the experiment. In fact, only a slight dip occurred in the friction record. Clearly the friction depends on the amount of liquid in the cartilage rather than upon free liquid on its surface.

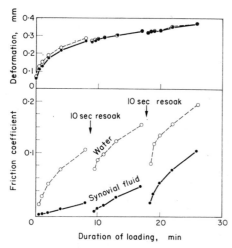

*Fig. 3. Longer resoaking periods produce a longer-lasting reduction in friction and a noticeable swelling of the cartilage.*

The slight dip in friction occurs because any re-soak, however short, saturates the outermost layer of the cartilage, and this lowers the friction until the imbibed liquid is redistributed throughout the depth of the cartilage. In Fig. 3, the re-soak was increased to 10 seconds, the dip in the friction coefficient is much more pronounced, and now enough water is imbibed so that the swelling of the cartilage can be seen in the deformation record.

These experiments also show that synovial fluid is a good lubricant for cartilage. As long suspected, its lubricating property seems to reside in its mucin content. Blood serum, which is much like synovial fluid, but with the mucin left out, lubricates not nearly so well. Saliva lubricates cartilage and it also contains one or more mucins.

That cartilage can be lubricated is no surprise. The skeletons of the cartilages press only lightly against each other (or against the glass plate), but they do press, so it is entirely to be expected that a lubricant should reduce friction. Mucins are not the only lubricants for cartilage. The cationic detergent, cetavlon, also lubricates, but then soaps are well known as lubricants.

Synovial fluid, by contrast, has never shown any marked lubricating ability in man-made bearings. But note again Figs. 2 and 3. They show that even for cartilage, synovial fluid begins to lose its lubricating ability as wring-out proceeds. Other experiments in which wring-out was allowed to proceed much further, showed that both synovial fluid and saliva eventually ceased to lubricate any better than water. When the cartilage is wrung out, all the load is carried by the skeletons and it is clear that the mucins cannot hold them apart under the resulting 5 to 10 atm pressure.

This completes the pragmatic explanation of why synovial fluid fails to lubricate man-made bearings. It has insufficient viscosity at high shear rate to provide hydrodynamic lubrication and insufficient strength (in some manner not clearly defined) to provide boundary lubrication for the highly loaded high spots which carry the load in a slow-moving man-made bearing.

One would not expect mucin solutions to be satisfactory high pressure lubricants. The mucin molecules are long, flexible polymer chains with molecular weights in the millions. These coil and flail about at random, occupying a volume much larger than that of the polymer chains themselves. Whatever form of lubrication such an inherently dilute material may provide at low loads, it is reasonable to expect that it would fail at loads sufficient to flatten the polymer molecules.

This lubricating ability of mucins under low loads is the major outstanding puzzle, but there are also other places where we are ignorant. The holes in the cartilage skeleton have never been seen. From measurements of flow resistance and of the total volumes of the holes it has been deduced that their diameter is of the order of 60 Å. This would be big enough to be seen easily with an electron microscope were cartilage a good subject for examination, which, unfortunately, it is not.

Whatever may be the precise size and arrangement of the holes, they perform their necessary function of holding a large amount of liquid and simultaneously presenting a large resistance to liquid flow. These characteristics, combined with the high deformability of the sponge skeleton, explain why cartilage is almost the world's only weeping bearing. Few other porous materials retain liquid long enough under load for its effect to be noticed.

The holes are sufficiently small so that, even allowing for a large factor of ignorance, it seems unlikely that mucin molecules will willingly enter them. Even if the chain diameter is low enough to permit them to be threaded endwise into the holes one would expect that the accompanying reduction in entropy would prohibit their entry. So in being imbibed by the cartilage the synovial fluid probably has its large molecules filtered out, but this does not render it less able to carry load by hydrostatic pressure. What happens to the mucin molecules is therefore not of importance to weeping lubrication, but rather to the next, unwritten chapter about why mucins make a weeping bearing better.

*See also*: Lubrication, weeping.

*Bibliography*

BARNETT C. H. *et al.* (1961) *Synovial Joints*, London: Longmans, Green.
CHARNLEY J. (1959) *Symposium on Biomechanics*, London: The Institution of Mechanical Engineers.
McCUTCHEN C. W. (1962) *Wear*, **5**, 1.
McCUTCHEN C. W. (1961) *New Scientist* **15**, 412.

C. W. McCUTCHEN

**ARCHEOLOGY, PHYSICS IN.** The applications of physics in archaeology can be functionally classified as *finding*, *dating* and *analysis*. Projects in the second category are not solely "aids to archaeology" but may yield fundamental information unobtainable except through archaeological evidence and material (for example, the past history of the geomagnetic secular variation). In all categories there are many *possible* methods; this article will be restricted to those actually in use or showing high promise at the time of writing (1964).

### 1. Field Surveying

*1.1 Magnetic location.* Closely-spaced measurements of the geomagnetic field strength, usually with the magnetic detector one foot above ground level, can be used to locate such features as:

  (i) Iron weapons and tools,
  (ii) Kilns, furnaces, ovens and hearths,
  (iii) Pits and ditches,
      and, in special circumstances,
  (iv) Walls, foundations and empty tombs.

A magnetic disturbance arises from kilns and furnaces, etc., because of the *thermoremanent magnetism* acquired by clay when it cools down from firing. The crude clay used in the building of these structures may contain about 5 per cent of iron oxide, and the permanent magnetization acquired in this way (more fully explained in 2·2 following) can attain remarkably high values—up to $5 \times 10^{-2}$ emu/g for grey reduced clay and up to $5 \times 10^{-3}$ emu/g for red oxidized clay. Pits and ditches show up magnetically because of the higher susceptibility of the filling (up to $10^{-3}$ emu/g for rich black soil) compared to that of the sub-soil or rock (less than $10^{-5}$ emu/g except for igneous rock) into which the feature has been cut. The filling may be top-soil that has silted in or a pit may have been used for rubbish or as a latrine. The enhanced susceptibility arises because of conversion of the iron oxide from the

very weakly ferrimagnetic *haematite* ($\alpha$—$Fe_2O_3$) to the more strongly ferrimagnetic *maghemite* ($\gamma$—$Fe_2O_3$). This takes place via the strongly ferrimagnetic *magnetite* ($Fe_3O_4$) and is favoured by high humus content and burning (as for example in ground clearance).

Walls and tombs are detectable as a *reverse* anomaly when they are substantial enough and buried in a strata of sufficiently high susceptibility (e.g. Mediterranean "terra-rossa", top-soil in ironstone districts, alluvial deposits of volcanic origin, volcanic tufa). Walls built of brick or volcanic rock may be detectable because of their thermoremanent magnetism.

The standard instrument used for magnetic detection is the proton magnetometer. This has a sensitivity of 1 gamma ($10^{-5}$ oersted, or 79·6 amp-metre$^{-1}$) and this is adequate since random variations from point to point often amount to 5 gamma at the usual height of 1 foot. Except on magnetic storm days, difficulties due to the diurnal variation can be avoided by taking all the readings within a standard area (50 ft by 50 ft) as quickly as possible (100 measurements would normally be made in such an area in less than 10 minutes). Since the archaeological features show up as abnormal readings the actual value of the general level does not matter. The proton magnetometer can be transistorized and made to weigh only 10 kilogram including accumulators.

Two other instruments are used: the *proton gradiometer* and the *fluxgate gradiometer*. Both of these eliminate the diurnal variation and can therefore be used on magnetic storm days, and also in the proximity of d.c. electric trains and trams. Two detectors are used, one at either end of a staff which is held approximately vertical. In the simplest form of proton gradiometer, the "Max Bleep", the output consists of the amplified sum of the two free-precession signals. If the detectors are in identical field strengths the frequencies of the two signals are identical and a steady note is heard for the relaxation time of the detector used. If the fields are different then "beats" are produced.

*1.2. Resistivity surveying.* The resistivity of soil and rocks depends on the water content, electricity being conducted by the process of electrolysis. Hard compact rocks like granite are very poor conductors, while the more porous limestones are much better, though still poor by comparison with soil, sand and clay. As with magnetic anomalies, the ease of detection of archaeological features depends on the degree of contrast between the material of the feature and the strata surrounding it. Low resistivity anomalies are typically produced by soil-filled pits and ditches, tombs and caverns, and high resistivity anomalies by stone walls, foundations and roads intruding into soil.

The usual technique for the measurement of the resistivity utilizes four steel probes which are inserted into the ground along a straight line. A current is passed through the ground between the two outer probes and the voltage developed between the two inner probes is used as a measure of the resistivity of the ground contained in a hemisphere of a radius of the same order as the probe separation. The use of separate probes for current and voltage reduces effects due to contact resistance between probes and soil. To avoid the effects of contact voltages, natural earth currents and probe polarization, it is essential to use alternating current. The most commonly employed probe separation (the *Wenner configuration*) is equidistant and roughly equal to the expected depth of feature. Sometimes it is preferable to use the *Schlumberger configuration* in which the distance between the two inner probes is small compared to that between the outer ones. In the *Palmer configuration* each inner and outer form has a closely-spaced pair which are widely separated from each other.

Early work was carried out using a Megger earth tester. In this the driving voltage (40 c/s) is produced with a manual generator and rectifying contacts allow the effective inner-probe resistance to be read directly on a meter. Subsequent instruments use null-balancing techniques (in which the inner-probe voltage is balanced against voltage-drop produced by the outer-probe current when it passes through a resistance in the instrument) thereby completely eliminating the effects of contact resistance. A transistorized version can be made small enough to be held in the palm of the hand.

The resistivity technique is more painstaking to apply than magnetic location. In addition a period of rain tends to weaken low resistance anomalies and a period of dryness weakens high ones. On the other hand the resistivity method can reach to greater depths and it is more use in the detection of walls and roads etc.

### 2. Age Determination

*2.1. Radiocarbon dating.* Cosmic-ray neutrons in the upper atmosphere transmute nitrogen-14 to carbon-14 at a global rate of about 7·5 kg per year. Carbon-14 is radioactive and has a half-life of 5730 ($\pm$40) years so that the equilibrium value is 60 metric tons. After formation carbon-14 atoms are oxidized to heavy carbon dioxide. These mix in with the ordinary $CO_2$ of the atmosphere and enter plant-life by photosynthesis, and the oceans as dissolved carbonate and bicarbonate by an exchange reaction. The former eventually returns to the atmosphere by decomposition, some of it via animal life. The atmosphere, biosphere, and the organic and inorganic matter in the oceans form the constituent parts of the carbon exchange reservoir between which the carbon atoms circulate fairly rapidly. The 60 metric tons of carbon-14 are distributed almost uniformly through the reservoir, which is estimated to contain $40 \times 10^{12}$ metric tons of ordinary carbon-12. Consequently, the ratio of carbon-14 atoms to carbon-12 atoms is 1 to $0·8 \times 10^{12}$.

Radioactive decay of a carbon-14 atom is accompanied by the emission of a beta particle (maximum energy, 160 keV) and the carbon-14 concentration is determined by measuring the specific radioactivity: in living wood, in ocean carbonate, in atmospheric carbon dioxide, etc., this is about 15 disintegrations per minute per gramme of carbon. When the wood dies, or the carbonate is deposited, the carbon atoms no longer exchange with those in the reservoir and the specific radioactivity decreases exponentially, by 1 per cent every 83 years (corresponding to the half-life of 5730 years); this is the basis of the age determination. The most suitable samples are those for which on death there is least likelihood of further exchange of carbon with the reservoir; they include wood, charcoal, peat, leaves, nuts, hair, skin and leather. Paper, cloth, dentine, charred bone can also be dated, as well as coarse-grained sea-shell. The amount of carbon required varies between 1 and 100 grammes depending on the type of sample and the measurement technique.

The most common technique employed is to convert the carbon into carbon dioxide, methane, or acetylene and use this as the gas in a proportional counter having a volume of several litres and comprehensively shielded against background radiation. Synthesis of the carbon into an organic liquid scintillator is also used. Such systems can determine ages back to 50,000 years (corresponding to a specific radioactivity of only 0·03 disintegrations per min. per gramme). A further 20,000 years can be added to this by isotopic enrichment prior to measurement, using thermal diffusion columns. For such old samples the possibility of contamination by 'modern' carbon, e.g. by seepage in of humic liquids while still buried, becomes of tremendous importance.

The most critical assumption of the technique is that the radiocarbon concentration in the exchange reservoir has always been the same. This implies both a constant production rate of carbon-14 and a constant size to the reservoir, averaged over thousands of years. The former implies a constant cosmic-ray intensity and one necessary condition for this is a constant geomagnetic field; evidence about this is obtainable through archeomagnetism (see 2.2.). The size of the reservoir is mainly determined by the ocean carbonates (accounting for $35 \times 10^{12}$ out of the $40 \times 10^{12}$ tons); a significant change in the amount of water locked up in glaciers (as during the Ice Ages) would alter this. Also a change in the average ocean temperature would alter the carbonate concentration, and therefore the total amount. The most convincing method of checking these assumptions is by measuring 'known-age' samples. Wood from Egyptian tombs is one source of these: back to 2000 B.C. the Egyptian calendar provided fairly reliable dates for these but in the initial period 2000 B.C. to 3000 B.C. the chronology is more controversial. Annual growth rings from giant sequoia trees and bristle-cone pines (back to 1700 B.C.) are another important source: it has been established that the carbon in such a ring remains fixed and does not exchange with adjacent rings. However, the tree-ring dates are not infallible, for occasionally some trees add more than one ring per year.

Back to the earliest 'known-age' check at 3000 B.C. the discrepancies do not exceed 6 per cent. This is using a half-life of 5730 years which was adopted in 1962 as the most reliable value from laboratory measurements of the disintegration rate. Previously 5570 years had been used and the discrepancies against 'known-age' checks had been greater. The contrary is the case with another type of 'known-age' check—against varve chronology (annual layers of sediment deposited in lakes formed by melting glaciers) which extends back to 8000 B.C.

Careful checks over the last two thousand years have revealed short term fluctuations in the past radiocarbon level, amounting to ±1 or 2 per cent. Between 1860 and 1954 there has been a decrease of about 2 per cent due to the burning of 'fossil-fuels' such as oil and coal which release appreciable quantities of 'old' carbon (in which the radiocarbon has decayed away). Since 1954 the radiocarbon content of the atmosphere has risen dramatically due to hydrogen bomb explosions: the time delays observed between the increases due to a given explosion has provided useful information about mixing rates between different parts of the carbon exchange reservoir.

Radiocarbon dates are sometimes quoted as so many years B.P. Although this stands for 'Before Present' it is an accepted convention that 'Present' is taken as A.D. 1950.

*2.2. Magnetic dating ('archaeomagnetism').* In well-baked clay the iron oxide is in ferrimagnetic form, either as magnetite ($Fe_3O_4$) or haematite ($\alpha$-$Fe_2O_3$). On heating, the magnetic domains are able to re-orientate (at a temperature $T_B$—the blocking temperature—which is dependent on the size, shape and material of the grains) due to thermal agitation and on cooling the clay has a hard permanent magnetization (*thermo-remanent magnetism*—TRM) in the same direction as the geomagnetic field. Thus if the clay has remained *in situ* subsequently, the ancient direction of the geomagnetic field can be found by measuring the direction of magnetization in extracted samples on which the

directions of true North and the horizontal have been carefully marked before disturbance.

The ancient direction in London since 1580 is known from the recorded observations of scientists on suspended magnets. This 'directly-known' curve is shown in the figure together with some directions inferred from TRM measurements on structures for which archaeological evidence provides some indication of the date. The most useful structures are pottery kilns but the ancient direction can also be obtained from tile kilns, furnaces, ovens and hearths. In the first place the results provide interesting geophysical information about the secular variation—from the figure it is clearly unpredictable, contrary to what was thought at the end of the nineteenth century, but in agreement with present day geophysical ideas about the cause of the secular variation. Secondly, once the details of the variation have been established from archaeological structures of known date the process can be reversed and the TRM direction found in a structure can be used to determine a 'magnetic date'. The basic reference curve must, however, be established for each region of, say five hundred miles across since the secular variation is different in different parts of the world.

The foregoing paragraphs refer to samples extracted from structures. Pots and vases also carry a magnetic moment. If the glaze or decoration is sufficiently sophisticated it may be justifiable to assume that the pot must have been baked on its base on a floor that was within a few degrees of horizontal. In this case it is possible to determine the ancient angle of Inclination (I). Sometimes this is also possible with ancient bricks and tiles.

The ancient intensity of the geomagnetic field can be found by the technique of *re-*magnetization. The magnetic moment of baked clay is proportional to the field intensity in which it was baked, thus

$$\frac{F}{F_0} = \frac{M}{M_0}$$

where $F$ is the present day field intensity,

   $F_0$ is the ancient field intensity,

   $M_0$ is the moment found in the clay
       (i.e. the TRM from the ancient firing),

   $M$ is the moment acquired when the clay is re-fired in the present day field.

In practice the process is often invalidated by the occurrence of mineralogical changes when the brick is re-fired. These can be detected by re-magnetizing in steps of temperature. The ratio of ancient moment lost to new moment gained should be the same for each temperature interval. This is because all those magnetic domains having blocking temperatures ($T_B$) below the refiring temperature are remagnetized and all those having $T_B$ above the refiring temperature are unaffected, and the same was the case in the original firing. Additionally, if the original firing did not reach the highest blocking temperature of the domains

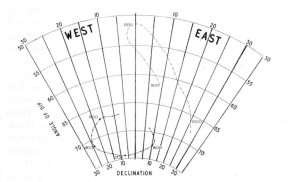

present (675°C, the Curie point of haematite), then this will be revealed in the refiring, as also will be the occurrence of any secondary firings to a lower temperature than the original one but in a different field.

*2.3. Thermoluminescent dating.* If a small amount of ground-up ancient pottery is heated rapidly (in about half-a-minute) to 500°C, a small amount of light (detectable with a sensitive photomultiplier) is emitted in the range 250–500°C. This is in excess of the ordinary thermal ('black-body') radiation, as may be established by a second heating when only the latter will be observed. The excess light on the first heating is produced by the release of electrons trapped in metastable states in the constituent minerals of the clay. The number of trapped electrons is proportional to the radiation dose received since the original firing of the pot, and therefore the excess light (or thermoluminescence) is proportional to the age. The radiation dose is from the natural radioactive impurities in the clay of the pot (mainly uranium, thorium and potassium-40) and the dose-rate for each piece of pot dated must be estimated by determining the concentration of these impurities. In addition the susceptibility to the acquisition of thermoluminescence is measured by exposing each piece to a standard artificial radioactive source. The 'specific glow', defined as

$$\frac{natural\ glow}{(artificial\ glow) \times (specific\ radioactivity)}$$

should then be proportional to the age independent of different mineral constituents between pots.

At the time of writing (1964), the principle of the method has been established but the eventual degree of precision is uncertain.

### 3. Physical Methods of Chemical Analysis

The objectives served by analytical programmes may be broadly classified as: (i) information about ancient techniques (e.g. in metallurgy); (ii) dating information through a knowledge of the history of techniques; and (iii) information about place of manufacture because of geological variation in the impurity contents of raw materials. In general useful results are only obtainable with large scale programmes involving several hundred specimens. In consequence rapid non-destructive or partially non-destructive methods are at a premium.

*3.1. Density determinations.* This technique for determining the fineness of gold objects was first used by Archimedes. Uncertainty in the density of the pure metal itself limits the accuracy obtainable (likely error, 1%) even assuming the alloying constituent is known. If it is unknown the method becomes semi-quantitative except for objects of high gold content; the uncertainty is less than 2 per cent as long as the gold content exceeds 95 per cent.

*3.2. Emission spectrometry.* A few milligrams of the specimen are introduced into an electric arc or spark and the spectrum of the emitted light is dispersed with a prism (usually quartz) and recorded on a photographic plate. Standardization with known samples is essential and these must resemble the actual composition of the specimen as closely as possible. The detection sensitivity varies from element to element; in many cases it is a few parts per million.

One important application has been to Bronze Age weapons and tools. Likely geographic locations for the sources of metal used in various categories of artifact in prehistoric Europe have been determined. The technique has also produced useful results from the analysis of pottery. For example, in the period 1400–1200 B.C., pottery made in Crete shows a distinctively higher content of magnesium and chromium compared to pottery from mainland Mycenae. Consequently by examining pottery found around the shores of the Mediterranean it is possible to gauge the relative importance of these two exporting centres.

*3.3. X-ray fluorescence spectrometry.* The characteristic secondary radiations emitted when the inner electron shells of an atom are excited by a primary X-ray source (also, excitation can be by a beam of electrons) can be used to measure the concentrations of all elements except those of too low an atomic number (say $Z = 11$ and below). Only a thin layer (0·01–0·1 mm) of specimen is analysed and consequently the most important application is the examination of glazes on pottery, particularly Chinese porcelain because of the method's non-destructiveness. As with the emission spectrometer standardization is essential. The limits of detection are between 0·1 per cent and 0·001 per cent. It is not a satisfactory technique for the analysis of coins and metal objects because of corrosion and surface enrichment effects.

*3.4. Neutron activation analysis.* The nuclei of many elements can be made radioactive by neutron bombardment. In most cases the subsequent decay is accompanied by a gamma ray of characteristic energy which can be used to identify and measure the concentration of the element responsible. For archaeological purposes the particular value of the technique is that it can be entirely non-destructive, yet at the same time give an analysis that refers to a substantial volume of the object concerned. It is particularly suitable for the analysis of precious coins. The main drawback is that without chemical separation after irradiation the resolution between different elements is comparatively poor and consequently the method is of limited value when many different elements are present.

*3.5. Beta-ray back-scatter measurements.* The backscattering of beta-rays from a surface is strongly dependent on the atomic numbers of the constituent elements. Based on this fact a simple transistorized portable detector and radioisotope source (e.g. thallium-204) can be used to estimate the lead content of glazes on pottery and glass. The lower limit of detection is a few per cent but the question to be answered is usually whether or not there has been deliberate addition of lead. For instance, lead was used in pottery glazes in Mesopotamia in the seventeenth century B.C. but it did not appear in Egyptian glazes until much later.

*See also:* Proton magnetometer.

*Bibliography*

AITKEN M. J. (1961) *Physics and Archaeology*, New York: Interscience.
AITKEN M. J. (1962) *Contemporary Physics* **3**, 161 and 333.
BLACKETT P. M. S. *Rock Magnetism*, New York: Interscience.
BOWEN R. N. C. (1958) *The Exploration of Time*, London: Newnes.
BROTHWELL D. and HIGGS E. (Eds.) (1963) *Science in Archaeology*, London: Thames & Hudson.
LIBBY W. F. (1955) (2nd Edn) *Radiocarbon Dating*, Chicago: The University Press.
PYDDOKE E. (Ed.) (1963) *The Scientist and Archeology*, London: Phoenix House.

M. J. AITKEN

**ASTRONOMICAL OBSERVATIONS FROM SPACE-CRAFT.** Astronomical observations of several different kinds can be carried out above the denser layers of the Earth's atmosphere; these include the direct sampling of the interplanetary medium and of planetary atmospheres and surfaces by experiments on spacecraft. However, astronomical observations in the more conventional sense of studies of the electromagnetic waves emitted by extraterrestrial objects may also benefit from the use of a spacecraft. This is true even in the visible part of the spectrum because of the elimination of atmospheric scintillation, the rapidly-varying component of atmospheric refraction which causes a serious degradation of image quality in the best telescopes, and also because of the elimination of scattered or emitted light from the night sky.

The use of a spacecraft, however, offers its greatest advantages in the broad spectral regions, shown diagrammatically in Fig. 1., where atmospheric absorption is high at the Earth's surface. The entire ultra-violet and X-ray spectrum at wave-lengths shorter than 3000 Å is absorbed in this way, since these radiations are capable of ionizing or dissociating various constituents of the atmosphere. Some atmospheric "windows" exist in the infra-red, between the strong absorption bands of, primarily, water vapour and carbon dioxide. Finally, after the broad transparent range of short radio waves, the ionosphere becomes reflecting at about 30 metres and radiation of longer wave-length is therefore not transmitted to the ground. Figure 1 shows, as a function of wave-length, the altitude at which light is reduced in intensity to $1/\varepsilon$ of its value outside the atmosphere. If we take this as the minimum altitude for an observing platform, it will be seen that this varies from 50 to 1000 km or more.

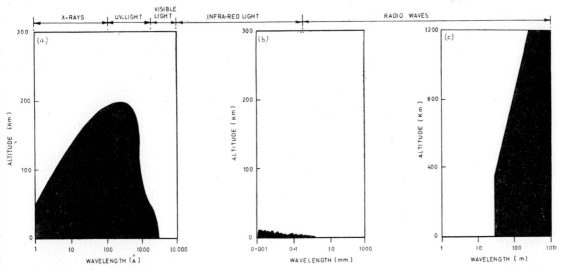

*Fig. 1 (a), (b) and (c). The three sections show the depth in the atmosphere to which extraterrestrial radiation penetrates. No attempt has been made to reproduce the detailed structure of the actual curves which are anyway subject to considerable variation with time of day and meteorological and solar conditions, but only to indicate their general form. The two main atmospheric windows in the visible and radio spectra are shown, but there are others particularly in narrow bands in the infra-red. Note the changes of scale between the three sections of the diagram.*

Spacecraft observations of the Sun can be made without significant interference from absorption or scattering in the interplanetary medium. Radiation from more distant sources may be subject to considerable modification by absorption in interstellar hydrogen and other gases, and must also compete with an extensive glow in the night sky believed to arise from the scattering of sunlight in the hydrogen geo-corona. In particular, strong absorption is to be expected between about 20 Å and the Lyman-alpha line of hydrogen at 1216 Å (Strom 1961, Cook 1963). It is possible that even in this range strong emission lines of the nearest stars will prove to be detectable, in which case a considerable improvement in the knowledge of the interstellar medium will result from a comparison of the spectra of similar stars at varying distances. For example, molecular hydrogen, at present undetectable, may be detected by its absorption between 1000 and 1100 Å. In general, however, it is believed that studies of stellar and galactic emissions will not be possible between 20 and 1216 Å.

### Technique of Observation at Ultra-violet and X-Ray Wave-lengths

The simplest method of observation from a spacecraft is to use an unstabilized rocket so that its normal roll and precession motion will cause the field of view of the instrument to scan a portion of the sky. The instrument may consist of little more than a simple photoelectric photometer with filters, wave-length-sensitive detectors or an objective prism. Such a simple arrangement is very limited in scope, because it is not possible to control the area of sky under observation, only a poor wave-length resolution is obtainable, and the sensitivity is so low that only the brightest objects can be detected because the observing time is short (being governed by the time taken for the star to cross the field of view) and the light-collecting area small.

In a more elaborate experiment, the following problems must be solved. First the chosen object must be acquired. Then the instrument must be pointed towards it with an accuracy which lies typically in the range 1 arc min to 1 arc sec or less. Next, means must be provided to select the desired spectrum interval, and finally the selected photons must be registered. We shall discuss the solution of these problems separately.

(a) *Spacecraft stabilization.* A spacecraft approaches closely the ideal of a rigid body falling freely in inertial space and so possesses a high degree of inherent stability, particularly when spinning. However, both moving parts within the spacecraft and also perturbing torques of external origin will cause the optical axis to move at an unacceptably large rate of the order of 0·05–1 arc sec/second (the former rate applying to a satellite with some gyroscopic stability). The principal perturbing torques arise from the interaction of the magnetic moment of the spacecraft with the geomagnetic field, from the interaction of the geogravitational field with a spacecraft whose mass distribution is not spherically symmetric, from aerodynamic forces and from solar radiation pressure. Counteracting these forces requires active stabilization of the spacecraft, correcting torques being derived from gas jets, the interaction of current loops with the geomagnetic field, from accelerating flywheels ('inertia wheels') or from a combination of some or all these.

Initial acquisition of a chosen star or small region of the sky necessitates the establishment of known reference directions. Small star tracker telescopes directed at standard stars may be used for this; at least four will be required if occultation by the Earth is considered but more may be included for redundancy. They may also control the generation of the correcting torques, though if a stabilization to much better than 1 arc min is required it will probably be necessary to derive error signals within the instruments from the chosen object itself or from a bright "finder" star close to it. This system is that used in the Orbiting Astronomical Observatories (Scott 1962) of the United States, which are able to accommodate an optical system 1 m in diameter and point it at a chosen star with an accuracy of about 0·1 arc sec.

(b) *Photon detection.* Although on account of their unrivalled capacity for storing information photographic emulsions have been widely used in rockets to record the solar spectrum, their poor quantum efficiency and the difficulties of in-flight processing of the emulsion make them less satisfactory for observations of stars from a satellite. Accordingly photoelectric cells or photon counters have generally been used. These can be made, by a suitable choice of window and photocathode or gas filling, to respond only to a limited band of wave-lengths and so can be used without a dispersing optical system if high wave-length resolution is not required; even if dispersion is present it is often useful to arrange that the detector does not respond to wave-lengths greater than the longest it is desired to observe, thus eliminating the response to visible light which has been scattered in the instrument.

Below about 80 Å, where thin organic films of materials such as melinex become transparent, ionization chambers and proportional counters can be made (Kreplin 1964). These will function as broad-band detectors sensitive typically to a wave-length range from $\lambda$ to about $2\lambda$. Below 20 Å a much improved wave-length resolution can be obtained by using their proportional properties, the electric charge produced by an incident photon being proportional to its quantum energy (Pounds and Willmore (1963). Their quantum efficiency may exceed 50 per cent, though when used as a broad-band detector the average value will be rather lower.

Between 80 Å and the short-wave cut-off of lithium fluoride at 1150 Å, windowless photomultipliers must be used. The photocathode and secondary emitting surfaces must evidently be so chosen that exposure to the atmosphere prior to the launch does not change their sensitivity. On account of their stability and general robustness, photomultipliers employing crossed electric and magnetic fields to transport the electron avalanche have been popular for solar observations, and it would be anticipated that if stellar observations can be made in this portion of the spectrum they will also be useful here. With a tungsten photocathode, the quantum efficiency is about a few per cent.

At wave-lengths exceeding 1150 Å, windowed photomultipliers and ionization chambers once again become feasible and a band-pass response is achievable. The quantum efficiency can exceed 10 per cent over a portion of the spectrum. Instead of a UV-sensitive photocathode, a fluorescent wave-length shifter such as sodium salicylate may be employed in conjunction with a conventional photomultiplier sensitive to visible wave-lengths though the advantage of insensitivity to longer-wave-length scattered light is thereby lost.

In this portion of the spectrum, imaging tubes of various kinds become feasible. Their advantage is that whilst a photocell cannot record the light from more than one line at a time and must be slowly scanned through the spectrum behind an exit slit, an image tube can record and store a substantial portion of the spectrum at once. The spectrum can then be scanned at convenient intervals which might for very faint objects be as long as several hours. The addition of image intensification enables low light thresholds (less than 100 photoelectrons per image point) to be obtained, whilst the use of electric or magnetic deflexion to stabilize the final image may eliminate the need for fine stabilization of the spacecraft.

(c) *Optical instruments.* The reflectivity of the usual reflecting materials diminishes at shorter wave-lengths than visible light. Thus whilst freshly-deposited aluminium films have a reflectivity at 1216 Å of 38 per cent, with atmospheric exposure this rapidly falls owing to the formation of an oxide film to 25 per cent (Berning *et al.* 1960). Since most instruments involve several reflections a large loss of sensitivity is entailed. Special coatings can be used to reduce this—for example an evaporated coating of 250 Å of magnesium fluoride on freshly-deposited aluminium has a reflectance of 80 per cent at 1216 Å and is stable against ageing.

At still shorter wave-lengths this method is ineffective and it is necessary to use reflecting surfaces—mirror or grating—at grazing incidence angles to maintain a reasonably high reflectivity. For mirrors this is satisfactory down to X-ray wave-lengths, and reflectors in the form of deep metal parabolas have been designed to give a reflectivity of 30 per cent or more down to wave-lengths of a few Ångstrom (Giacconi 1960).

These points apart, spacecraft instruments have for the most part employed well-established optical

designs, large incidence angles being employed at wave-lengths above 1000 Å, and grazing incidence at shorter wave–lengths.

*Stellar Ultra-violet Spectra*

The hotter stars, i.e. those of spectral classes ABO, radiate the greater part of their energy in the ultra-violet. At the present time, the temperature and energy output of these stars are estimated from calculations of model atmospheres whose emission matches that of the star in the visible. The emission in the ultra-violet, however, depends much more sensitively on the temperature and the structure of the star than does that in the visible, so that our knowledge of these properties (together with the total energy emitted) will be much improved by observations in the ultra-violet.

Several rocket observations have so far been made at wave-lengths down to 1216 Å. These have used an

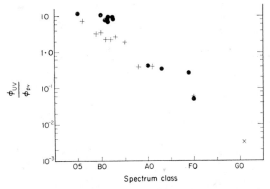

*Fig. 2. The ratio of the flux emitted by a star at a wave-length of approximately 2000 Å to that emitted at 5390 Å is shown as a function of spectral class; + indicates the experimental values obtained by Alexander et al. whilst those marked · are obtained from model atmosphere calculations. X refers to the Sun.*

objective prism spectrophotometer (Boggers 1961; Stecher and Milligan 1961) or broad-band detectors (Alexander *et al.* 1963). The results (see, for example, Fig. 2) have generally shown that the ultra-violet emission of O and B stars has been overestimated theoretically by a factor of around two. A detailed interpretation of this observation is at present lacking.

The strong ultra-violet absorption of the interstellar gas makes the comparison of the spectra of otherwise similar stars at varying distances from the Earth a potentially powerful tool for studying the composition and distribution of the interstellar medium. In this connexion the possible detection of molecular hydrogen by its absorption between 1000 and 1100 Å has already been mentioned. Such experiments are planned but have not yet (1964) been conducted.

## X-Radiation From the Night Sky

Although theoretical estimates of the X-ray emission to be expected from a wide variety of stellar systems have been made, the strength of these still remains very uncertain. This reflects the fact that the X-ray emission often depends on quantities which are very poorly known so that the measurements are of great interest.

The solar X-ray spectrum—though much brighter than would be expected for a 6000°K black body on account of the coronal emission which has a colour temperature of $\sim 10^{6}$°K—is still so weak that at a distance of a few light-years the Sun would be undetectable. Some stars, however, are expected to have much denser and hotter coronae than the Sun and a correspondingly enhanced X-ray emission which may be observable (de Jager and Neven 1961). Moreover, the total emission of X rays in the galaxy by this process may produce general galactic background of detectable intensity. During a flare, the solar X-ray emission is strongly enhanced, particularly at the shorter wave-lengths which are most likely to penetrate the interstellar medium and so be observable. The emission may increase by several orders of magnitude. Consequently flare stars, of which several are known to exist, may be powerful X-ray emitters, particularly if they are also strong radio emitters (e.g. UV Ceti).

The close association of X-ray and radio wave emission in the Sun leads to the supposition that many discrete sources of radio-emission may also be X-ray sources. If the ratio of the two emissions is assumed to be the same as in the Sun, several of the strongest radio sources (e.g. the Crab Nebula) should be detectable.

The quasi-stellar radio sources, about which not a great deal is known, may be strong X-ray emitters, particularly as the great jets seen in them may generate non-thermal radiation. Neutron stars, if these indeed exist, may also emit X rays in detectable quantities.

Some preliminary rocket surveys of the night sky have already been made using collimated photon counters as detectors (Giacconi et al. 1962; Gursky et al. 1963; Bowyer et al. 1964). These give rather low sensitivity and angular resolution. The measurements have been made from unstabilized rockets so that the counter scans a wide area of the sky because of the rocket motion. The wave-lengths used have been in the range 1–8 Å. Complete agreement does not exist between different measurements, but the following results are suggested:

(i) the presence of a source near the galactic centre with an intensity of $\sim 10$ photons/cm$^2$ sec,

(ii) the presence of two weaker sources, one in Scorpio and the other near the Crab Nebula,

(iii) the existence of a general galactic background which has not been established to be of extraterrestrial origin but which could, in fact, arise from the combined emission of a large number of unresolved sources.

## Gamma-Radiation from the Night Sky

The occurrence of nuclear processes in stars and in the interaction of cosmic rays with interstellar material might be expected to result in the emission of detectable γ-ray fluxes. Attempts have been made to detect these using counter telescopes with the result that fluxes possibly of stellar origin of about $5 \cdot 5 \times 10^{-4}$/cm$^2$ sec sterad and more have been observed near 100 MeV (Clark and Kraushaar 1963). There is no clear indication of an intensity variation over the celestial sphere.

## Long-wave Radio Emission from the Galaxy

Measurements of extraterrestrial radio emission at wave-lengths beyond the ionospheric cut-off at approximately 30 m are complicated by the need for a large aerial system. Even if only a half-wave electrical dipole is used a length of 150 m is required at 1 Mc/s. Such a system gives very little information on the direction of the source of the radiation, though some information about this can be obtained (in a satellite experiment) from the occultation of approximately one celestial hemisphere by the Earth and, at suitable altitudes, from the fact that the ionospheric cut-off wave-length decreases from the zenith towards the horison so that at any wave-length radiation from a limited cone only is accepted (Jennison 1961). The size of the cone can be calculated if sufficient ionospheric information is available.

Experiments to date have been conducted between 0·5 and 4 Mc/s in rockets and satellites, using simple dipole aerials. It has generally been found difficult to make an adequate correction for the effects of ionospheric refraction, but the results suggest a sharp fall in the radio noise level below 2 Mc/s, which has been attributed to interstellar absorption by ionized hydrogen (Haddock et al. 1963).

## Bibliography

ALEXANDER J. D. H. et al. (1963) *Space Research III* (Ed. W. Priester) Amsterdam: North-Holland.

BERNING P. H. et al. (1960) *J. Opt. Soc. Amer.* **50**, 586.

BOGGESS A. (1961) *Ap. J.,* **66**, 293.

BOWYER S. et al. (1964) *Nature* **201**, 1307.

CLARK G. W. and KRAUSHAAR W. L. (1963) *Space Research III* (Ed. W. Priester) Amsterdam: North-Holland.

COOK A. H. (1963) *J. Roy. Astron. Soc.* **4**, 203.

DE JAGER C. and NEVEN L. (1961) *Mem. Soc. Roy. Sci. Liege* **4**, 522.

GIACCONI R. (1960) *J. Geophys. Res.* **65**, 773.

GIACCONI R. et al. (1962) *Phys. Rev. Lett.* **9**, 439.

GURSKY H. et al. (1963) *Phys. Rev. Lett.* **11**, 530.

HADDOCK F. T. et al. (1963) *Ap. J.* **68**, 75, and *Space Research IV* (in press).

JENNISON R. C. (1961) *J. Brit. I.R.E.* **22**, 205.

KREPLIN R. (1964) *Space Research V* (in press).

POUNDS K. A. and WILLMORE A. P. (1963) in *Space Research III* (Ed. W. Priester) Amsterdam: North-Holland.

SCOTT W. H. Jr. (1963) in *Space Age Astronomy* (Eds. Deutsch A. S. and Klemper W. B.) New York: Academic Press.

STECHER T. P. and MILLIGAN J. E. (1961) *Ap. J.* **66**, 296.

STROM S. E. and STROM K. M. (1961) *Publ. Astron. Soc. Pacific* **73**, 430.

A. P. WILLMORE

**AUTORADIOGRAPHY OF METALS AND IN-ORGANIC SOLIDS.** The principles of autoradiography and the techniques used with biological materials have been described in Autoradiography in Biology and Medicine (see Bibliography). It is a technique for highly localized analysis, applicable mainly to elements that can be labelled with a radioisotope that emits $\alpha$ particles or low energy $\beta$ particles. An autoradiograph shows directly the distribution of an element in a solid though in many applications, particularly to biological systems, the chemical binding of an isotopic-ally labelled element, for example carbon, can be chosen so that the distribution of molecular species is studied. The value of the technique is directly related to the highest spatial resolution and sensitivity of which it is capable.

Close contact between the specimen and the auto-radiographic emulsion is the most important factor determining resolution. With biological materials the greatest emphasis has, therefore, been placed on stripping film and liquid emulsion techniques since these give the closest and most reproducible contact (Taylor 1960; Kopriwa 1962).

Somewhat different techniques must be used for determining distributions, in relation to microstructu-ral features, of elements present at low concentrations in metals, alloys and inorganic solids. Techniques which involve wet emulsions in contact with specimens are of limited application for these materials because of their greater tendency to react chemically with emulsion. A protective layer on the specimen is almost always necessary; a plastic film one micron thick is about the thinnest that is sufficiently protective to allow *in situ* development, and this limits resolution to 2–3 microns. The difficulty of preventing reaction, particularly during development, while retaining resolution makes dry contact autoradiography of considerable value with reactive solids.

*Contact autoradiography.* The specimen is placed in good mechanical contact with an emulsion and the two are separated after exposure. The method requires thin parallel-sided specimens that are flat and have highly polished faces. As most autoradiographic emulsions are unsupercoated and therefore easily pressure marked, particularly by slip, a jig for bring-ing the specimen and emulsion together without slip greatly reduces such marking. Resolution comparable

with stripping film technique can be obtained but this can vary with each exposure and from point to point over a specimen due to varying contact. Resolution has therefore to be assessed for individual autoradiographs by exposing, whenever possible, suitable resolution test pieces with the specimen.

A different approach is to use a dry, gelatin-free, replicating film (Rogers 1963). This gives many of the advantages of stripping film.

*Replicating film technique.* This uses a film consisting of a single layer of sensitive silver bromide crystals at the surface of a plastic film. The film is softened in an organic solvent and applied to the specimen. As the solvent evaporates the film conforms to the surface and, when hard, forms a replica of the surface topo-graphy. After exposure the film is stripped from the specimen and developed. As the plastic is unaffected by water, the final autoradiograph retains a replica of the specimen surface. Even with metals as reactive as magnesium a supercoat about $0 \cdot 3 \, \mu$ thick is protective under these conditions. Resolution is better than $2 \, \mu$. The replica of the surface on the autoradiograph assists interpretation by providing a means for realigning the autoradiograph with the original surface with a precision of a few microns. A number of autoradio-graphs can be taken from one specimen and a flat specimen surface is not required.

This technique may have advantages for the auto-radiography of biological materials containing water soluble labelled compounds as the specimen never comes in contact with aqueous media.

*Electron microscopic autoradiography.* Techniques giving autoradiographs that can be examined with the electron microscope have been developed (Pelc 1961; Silk 1961; Bachmann 1964) in attempts to increase resolution. Thin tissue sections are mounted on electron microscope specimen grids and covered with a single layer of silver halide grains. After exposure and development the autoradiograph is examined in contact with the specimen. If the latent image is developed physically (Bachmann 1961) the emulsion has the potential to resolve $0 \cdot 1$–$0 \cdot 2$ but a distribution of radioactive material showing this resolution has not so far been demonstrated. There is little gain in resolution when the latent image is chemically develop-ed as the silver grains appear as tangled filaments covering areas with linear dimensions of the order of a micron. However, these filaments are easily distinguish-able from other foreign particles, an advantage when examining autoradiographs from very weakly radio-active specimens.

*Sensitivity.* The resolution required determines the maximum thickness of specimen that can be used and the thickness of emulsion within which an image can show that resolution. (When using radionuclides that emit low-energy particles the range of the particles in the specimen may limit the thickness that contributes to the autoradiograph and hence determine the resolution.) The amount of an element that can be detected is then mainly determined by the specific activity with which it can be labelled in the specimen, the initial background in the emulsion and its rate of increase with time, and the exposure time. When the specimen contains features which are clearly identifiable microscopically (e.g. regions of a separate phase, containing the element under investigation) and when the autoradiograph can be examined in precise registry with the specimen, greater sensitivity can be obtained without undue labour. Only this situation will be considered.

Consider an autoradiograph taken using an emulsion of thickness $2a$ separated by a gap of width $0·2a$ from a specimen of thickness $2a$ in which the features can be approximately represented as cylinders $2a$ in diameter with their axes perpendicular to the emulsion. If $a = 10^{-4}$ cm the autoradiograph will have a resolution of about 2 microns. When the cylinders are uniformly labelled with $d$ disintegrations $sec^{-1} cm^{-3}$, the mean intensity of radiation, neglecting absorption, in the emulsion along the cylinder axis is $0·13ad$ particles $sec^{-1} cm^{-2}$. The mean intensity over the volume of emulsion directly above the feature is never less than 85 per cent of this value (Lamerton 1954). Hence if each particle produces $n$ developable grains in passing through the emulsion the rate of increase of signal above the feature is $0·11nad$ grains $sec^{-1} cm^{-2}$. If the feature, density $\varrho$ g/cm³, atomic weight $A$, contains $P$ atomic per cent of trace element labelled with a specific activity $S$ curie/g atom then

$$d = 3·7 \times 10^{10} \frac{\varrho}{A} \frac{P}{100} S.$$

The signal registered above the feature after an exposure of $T$ days is $3·5 \times 10^{12} \dfrac{na\varrho PS_0}{A\lambda} (1 - e^{-\lambda T})$ grains $cm^{-2}$ where $\lambda$ day$^{-1}$ is the decay constant of the label and $S_0$ is the specific activity at the start of the exposure. The background after $T$ days will be $B_0 + bT$ where $B_0$ is the initial background and $b$ its rate of increase. Useful information can be obtained with signal-to-background ratios of 5 but it is more convenient to work with ratios greater than 10. Hence a measure of sensitivity is given by

$$10(B_0 + bT) = 3·5 \times 10^{12} \frac{na\varrho PS_0}{A\lambda} (1 - e^{-\lambda T}).$$

and

$$P = 2·9 \times 10^{-12} \frac{A\lambda(B_0 + bT)}{na\varrho S_0(1 - e^{-\lambda T})} \text{ atomic per cent}.$$

Values of $B_0$ and $b$ and signal-to-background ratios can vary considerably with experimental conditions but $B_0 = 5 \times 10^5$ grains $cm^{-2}$, $b = 0·2 \times 10^5$ grains day$^{-1} cm^{-2}$ have been reported (Kopriwa 1962). By exposing at 4°C in dry air or $CO_2$ latent image fading can be greatly reduced (Ray 1962) so that signal and background increase linearly with time for exposures of 80 days and more (Kopriwa 1962). For isotopes

emitting $\beta$ particles with energies below about 0·4 MeV, $n$ can be of the order of unity while maintaining low rates of growth of background (Herz 1959). For a range of metals $A/\varrho \sim 10$. Orders of magnitude for $P$ for some elements of metallurgical interest calculated for 2 micron diameter cylinders $a = 10^{-4}$ cm and 100 days exposure are given in the table. $S_0$ for each element has been taken as 0·1 of the maximum specific activity that is commercially available.

| Radio-element | $S_0$ curie/g atom | P atomic % for 100 days exposure |
|---|---|---|
| ³H | $2·6 \times 10^3$ | $2·8 \times 10^{-6}$ * |
| ¹⁴C | 4 | $1·8 \times 10^{-3}$ |
| ⁶³Ni | 29 | $2·5 \times 10^{-4}$ |
| ³⁶Cl | $2 \times 10^{-3}$ | 3·6 |

\* This figure is somewhat too small as, with tritium, absorption in the feature cannot be neglected.

*Specimen preparation.* The introduction of a suitable radioisotope into the material often presents the greatest problem in applying autoradiographic methods. Neutron irradiation of a material in the state to be examined generally produces a large number of radioactive species. An autoradiograph shows only a distribution of radioactivity and this is weighted in favour of radionuclides of short half-life that decay by the emission of low energy charged particles. Those elements that have high neutron capture cross-sections for the production of such nuclides and are most localized will contribute most to an autoradiograph. Methods for distinguishing the constituents of mixtures of low energy particle emitters are inadequate and have very low spatial resolution. The degree of localization of elements is generally unknown. Some differentiation of radioactivities by half-lives may be possible by a suitable choice of irradiation time and by taking a series of autoradiographs over a period after irradiation but there is generally considerable uncertainty in element distributions obtained from neutron irradiated materials. Whenever possible the label should be introduced by a preparation that starts with chemically and radiochemically pure radioisotopes. Because, for high sensitivity, labelling at high specific activity is required, preparations need to be carried out on a small scale to keep the total amounts of radioactive materials involved within manageable proportions. Normal methods can often be modified but an additional complication is the need to allow for radioactive decay particularly when slow time dependent steps, such as age hardening, are involved.

To obtain autoradiographs showing a high resolution that is constant across a specimen, it must be of a uniform thickness, comparable with or smaller than the resolution required, unless the effective thickness is determined by the range of particles in the specimen.

For contact autoradiography the specimen must also be flat to a similar order. The surface on which the autoradiograph is taken must be polished to a standard similar to that required for metallographic examination, the reverse surface being finished to a surface roughness small compared with the specimen thickness. Several machining and grinding techniques have been developed for obtaining specimens with parallel faces down to a few microns thick and the surface conditions can usually be achieved by modifications of conventional techniques for the preparation of specimens for metallography. The possibility that the radioactive species may move during polishing must be considered particularly when the tracer could escape into, or exchange with air. Tritium close to a free surface may escape as gas or exchange with water vapour. As tritium $\beta$ particles have only a very small range in solids such a loss could lead to an apparent total absence of tritium.

*Relationship to the electron probe microanalyser.* Autoradiography can be considered as complementary to the electron probe microanalyser as a technique for highly localized elemental analysis. Both techniques can have a spatial resolution down to about $1\,\mu$. The probe analyser will detect elements of atomic number greater than about 12 when present in concentrations greater than about 1 per cent.

As the table shows radioisotopes of a number of elements are available at specific activities which, provided they can be introduced into the specimen, make autoradiography the more sensitive method. It provides a technique for studying some elements with atomic numbers below 12 where the probe analyser is insensitive. The low energy of tritium $\beta$ particles (18 keV) and the high specific activity at which it is available makes autoradiography specially suitable for the study of hydrogen distributions. Carbon, using $^{14}C$, and boron distributions, making use of the $^{10}B\,(n,\,\alpha)\,^{7}Li$ reaction, can also be studied.

The probe analyser has the advantage that specially labelled material is not required and so analysis can be made directly on specimens from other investigations and production materials. The analyser is also generally more rapid since the time to prepare radioactively labelled materials must be included in considering autoradiography.

*Artefacts.* Physically and chemically produced artefacts are liable to occur in all autoradiographic techniques. Most forms of artefact produce images with sharper outlines than true autoradiographs and such images should be regarded as suspect.

Pressure artefacts can result from specimen-emulsion slip and asperities on the specimen. They may also arise from stresses introduced into emulsion during coating, drying, and processing; frequently these are associated with sharp changes in contour on the specimen or in the emulsion. If optical fluorescence of parts of a specimen occurs this will produce a spurious image.

Chemical interactions can produce directly visible attack, developability desensitization and, in principle, sensitization, though no example of the latter is known to the writer. Attack is generally immediately apparent as massive deposits of silver associated with etching or corrosion of the specimen. When a feature of the specimen, e.g. grain boundaries, makes emulsion developable, labelled and unlabelled specimens will give similar images. The density of these is often unexpectedly high and does not vary with exposure and decay rate of the label in the expected manner. Grain free areas associated with features on the specimen suggest that these features are causing desensitization.

As there are many causes and types of artefact caution is necessary in the interpretation of autoradiographs. Whenever possible several autoradiographs from a labelled specimen should be compared with ones from unlabelled specimens and the causes of all reproducible features ascertained.

*Applications.* Autoradiography will show how a radioactive component, such as plutonium, is distributed in an alloy in which it is present at low concentrations. The technique is specially suited to showing whether hydrogen, carbon, and boron are components of phases identified by optical metallography.

It is of unique value when used in conjunction with radioactive tracers. In the life sciences it is well established as a technique for locating molecular species and the sites of biosynthetic reactions in cells. For structure determinations large labelled molecules are degraded to smaller molecular products which are separated on paper chromatograms or by paper electrophoresis. Autoradiographs of the paper show immediately where labelled products have separated. This may make possible great savings in the subsequent analytical work required. In the physical sciences the examination of self-diffusion couples shows the contributions of grain boundary and bulk diffusion to the overall diffusion rate. Its use is essential in tracer investigations to establish the source of a trace constituent found at identifiable regions in a solid.

*Bibliography*

BACHMANN L. and SALPETER M. M. (1964) *Naturwissenschaften* **51**, 237.
BOND V. P. (1961) in *Encyclopaedic Dictionary of Physics* (Ed. J. Thewlis) **1**, 340.
HERZ R. H. (1959) *Laboratory Investigation* 8, No. 1, 71.
KOPRIWA B. M. and LE BLOND C. P. (1962) *J. Histochemistry and Cytochemistry* **10**, 269.
LAMERTON L. F. and HARRISS E. B. (1954) *J. Phot. Sci.* **2**, 135.
PELC S. R. *et al.* (1961) *Expt. Cell Research* **24**, 192.
RAY R. C. and STEVENS G. W. W. (1953) *Brit. J. Radiol.* **26**, 362.
ROGERS G. T. and HUGHES J. D. H. (1963) *Nature* **199**, 566.
SILK M. H. *et al.* (1961) *J. Biophys. Biochem. Cytol.* **10**, 577.
TAYLOR J. H. (1960) *Advances in Biological and Medical Physics* **7**, 107.

G. T. ROGERS

# B

**BOUNDARY SCATTERING OF PHONONS.** In low temperature experiments on heat transport in insulating crystals, it is found that for small enough samples the quantity heat flow divided by temperature gradient is no longer proportional to the cross-sectional area but decreases at a faster rate. Strictly speaking this prohibits the use of the term thermal conductivity in such situations but it is generally more convenient to still use the expression and to regard the thermal conductivity as size dependent. The reduction of the thermal conductivity in conductors of small cross-section is due to the scattering of the entities responsible for lattice heat conduction, the *phonons*, at the boundary surfaces of the conductor. It was first observed by de Haas and Biermasz in 1938 and has been widely investigated since that time.

Similar effects are observed in electrical conduction in thin specimens, where it is due to boundary scattering of electrons. However, generally the effect is observable in the lattice heat conduction case in larger specimens than is possible for the electrical conduction case.

Boundary scattering of phonons manifests itself in another, more subtle, way. In certain materials, most conspicuously in the semiconductors germanium and silicon, the thermoelectric coefficients are enhanced by dragging effects between electrons and phonons. For example, in the Peltier effect the electron current will, in appropriate circumstances, drag a phonon current along with it and the phonon contribution to the Peltier coefficient may be larger than that due to purely electronic effects. However, at low enough temperatures the Peltier coefficient is found to be size dependent in the same way as the lattice thermal conductivity and for the same reason. This relatively unusual effect was first seen by Geballe and Hull in 1954.

In large samples the lattice thermal conductivity can be written as

$$\varkappa \cong \tfrac{1}{3}CVl. \tag{1}$$

where $C$ is the specific heat, $V$ the phonon velocity and $l$ the mean free path. This very approximate equation is derived by analogy with the similar equation for the thermal conductivity of a gas. The mean free path is limited by phonon scattering mechanisms due to anharmonic three-phonon processes and lattice defects. The thermal conductivity becomes size

dependent when the dimensions of the specimen become similar to $l$. Since $l$ generally increases rapidly as the temperature is lowered, the effect is observable only at low temperatures. It is possible to define a boundary scattering mean free path $L$, equal to the diameter of a cylindrical specimen, which can be added reciprocally to the mean free path due to other scattering to give a total mean free path. An approximate result can be obtained by inserting this in (1) but since $l$ depends on frequency this equation should really be replaced by a summation of such terms, one for each normal mode of vibration of the lattice.

In reasonably pure specimens the total mean free path can become approximately equal to $L$ at low temperatures. Since $C \propto T^3$ in this situation we should find that $\varkappa \propto T^3$ also. Approximate agreement with the expected dependence on size and temperature was found by de Haas and Biermasz in quartz and potassium chloride, by Berman, Simon and Ziman in diamond and by Berman, Foster and Ziman in sapphire. Results by Carruthers, Cochran and Mendelssohn on a very pure sample of germanium show quite exact agreement between 0·2 and 2°K corresponding to three decades of thermal conductivity. It should be remembered however that quite different results are found in less pure specimens.

The experiments of de Haas and Biermasz stimulated a theoretical investigation by Casimir. He assumed that the scattering of the phonons at the boundary surface was completely diffuse, i.e. after scattering the phonons were in equilibrium at the local temperature of the boundary at which they were scattered. This situation has a strong analogy with that prevailing in the theory of black body radiation. He showed that for a circular cylinder $L$ was exactly equal to the diameter whilst for a square cross-section of edge $d$, $L = 1\cdot12d$. Similar results are obtainable by solving the Boltzmann transport equation with this boundary condition.

The interpretation of the results on diamond and sapphire obtained by Berman *et al.* required some elaboration of the theory. It was found necessary to allow for the possibility of specular reflection of phonons and for the finite length of the sample. Specular reflection alone would not produce any additional thermal resistance and any proportion of specular reflection will tend to increase $L$. If $p$ is the fraction reflected, assumed independent of wave-length and

18

angle of incidence, they found that at the lowest temperatures $p = 0.4$ could be obtained by polishing the surface. However, $p$ appeared to vary with temperature as $T^{-1/2}$ between $2°$ and $6°$K. This increase in specularity with decreasing temperature is probably due to the increasing importance of very long wave-length phonons since contrary to the assumption made initially $p$ must presumably increase with phonon wave-length.

Berman, Foster and Ziman have considered two approaches to the problem of constructing a microscopic theory of boundary scattering of phonons. In the first they consider a "uniformly rough surface" where topography is described by a Gaussian distribution function. Such a surface gives rise to a scattered beam varying slowly with angle together with a narrow beam corresponding to specular reflection. For normal incidence

$$p(\eta, \lambda) = e^{-8\pi^2\eta^2/\lambda^2}. \qquad (2)$$

where $\eta$ is the asperity parameter defining surface roughness and $\lambda$ is the wave-length of the phonon. The form of $p(\eta, \lambda)$ gave the effective $p$ a temperature dependence of $T^{-3}$ instead of $T^{-1/2}$. This discrepancy was removed in their second approach using a non-uniformly rough surface divided into small regions of uniform roughness. This gave the correct temperature dependence and reasonable values for the dimensions of the regions within which the roughness was uniform.

At somewhat higher temperatures the rate of increase of thermal conductivity is much less than $T^3$ and eventually it passes through a maximum and starts to decrease again. This is largely due to point defects which can arise due to isotopic mass differences even in chemically very pure specimens. It was pointed out by Herring that there should be a perceptible size effect even at quite high temperatures because the long wave-length phonons are scattered so feebly by the point defects. It can be shown that if $\varkappa_0$ is the bulk thermal conductivity one actually has

$$\varkappa = \varkappa_0 - \frac{A}{L^{1/2}}. \qquad (3)$$

where $A$ is a parameter depending on the other phonon scattering mechanisms and varying approximately as $T^{-9/2}$. This effect is essentially neglected by treatments using only equation 1.

In order to measure this rather small effect Geballe and Hull used a differential method with two parallel germanium samples of different cross-sectional area. They found that in fact $\varkappa_0 - \varkappa \propto T^{-5}$ and that the magnitude of the effect was approximately that predicted by Herring.

The phonon drag contribution to the Seebeck coefficient, $Q$, can be written as

$$Q_p \cong -\frac{l_v}{\mu_L T}. \qquad (4)$$

where $\mu_L$ is the electron mobility due to electron–phonon interactrons. At room temperature in germanium $Q_p^{\varpi}$ is quite small but as the temperature decreases it increases approximately as $T^{-3}$ being limited mainly by three phonon processes. Eventually the rate of increase slows down and $Q_p$ decreases again below about $35°$K. For smaller samples $Q_p$ was smaller in magnitude and its maximum occurred at a higher temperature. Theoretically $Q_p$ should be proportional to $T^{1/2}$ where $l \cong L$; but measurements have not yet been made at temperatures low enough to observe this. However there seems no doubt that a size effect of the same kind as that found in thermal conductivity is also at work here.

*Bibliography*

BERMAN R. *et al.* (1953) *Proc. Roy. Soc.* A **220**, 171.
BERMAN R. *et al.* (1955) *Proc. Roy. Soc.* A **231**, 130.
CARRUTHERS J. A. *et al.* (1962) *Cryogenics* **2**, 160.
CARRUTHERS P. (1961) *Rev. Mod. Phys.* **33**, 92.
CASIMIR H. B. G. (1938) *Physica* **5**, 495.
DRABBLE J. R. and GOLDSMID H. J. (1961) *Thermal Conduction in Semiconductors*, Oxford: Pergamon Press.
GEBALLE T. H. and HULL G. W. (1954) *Phys. Rev.* **94**, 1134.
GEBALLE T. H. and HULL G. W. (1955) *Conf. Phys. Basses. Temp.* Paris, 460.
DE HAAS W. J. and BIERMASZ T. (1938) *Physica* **5**, 47.
HERRING C. (1954) *Phys. Rev.* **95**, 954.
HERRING C. (1954) *Phys. Rev.* **96**, 1163.
HERRING C. (1958) *Halbleiter und Phosphore*, Brunswick: Vieveg.
KLEMENS P. G. (1958) *Solid State Physics* **9**, 1.
ZIMAN J. M. (1960) *Electrons and Phonons*, Oxford: Clarendon Press.

J. E. PARROTT

**BRAGG CUT-OFF WAVE-LENGTH.** The coherent elastic scattering of X rays and slow neutrons in crystals occurs at particular angles according to the Bragg relationship $\lambda = 2d \sin \theta$; $\lambda$ is the X ray or neutron wave-length, $2\theta$ the scattering angle and $d$ the interplanar spacing for a particular Bragg reflection. This spacing assumes a number of discrete values, the largest of which we denote by $d_{max}$. If $\lambda > 2d_{max}$ the Bragg relationship is not satisfied at any angle, so that no reflections occur. $2d_{max}$ is, therefore, the Bragg cut-off wave-length.

D. G. MARTIN

**BROADBAND ELECTROMAGNETIC TESTING METHODS.** Broadband electromagnetic tests can be considered an extension or generalization of the nondestructive sinusoidal eddy current test methods which are extensively used for sorting and quality control of metallic parts, and broadband methods are in principle useful in the solution of the same type of problems. While the overwhelming preponderance of electromagnetic or eddy current testing (the terms are somewhat interchangeable in this context) has been accomplished with equipment which uses test coils or probes carrying a current varying sinusoidally in time, it has also been recognized that the use of current of more complex time variation might provide at least theoretical advantages over the single frequency sinusoidal current. The term "broadband" arises from a consideration of the Fourier analysis of the exciting current wave form. Any time varying function which is repetitive can be analysed into the sum of a series of sinusoidal functions of the form

$$f(t) = a_0 + \sum_{n=1}^{\infty} (a_n \cos nx + b_n \sin nx).$$

provided the function $f(t)$ fulfills certain mathematical conditions. Except in certain special cases this series is infinite; in any case, electronic circuits which are to amplify wave forms other than single frequency sinusoids are required to have some finite bandwidth. If the wave form is some type of a pulse an amplifier that is to faithfully amplify it must have among other characteristics a relatively wide bandwidth.

It is interesting to note that apparently the first reported experiments with what is now termed electromagnetic testing, performed by D. E. Hughes in 1879 actually used a type of pulsed current generated by the ticks of a clock falling upon a microphone. This current was used in connexion with a bridge arrangement to compare various metallic specimens with each other. However, practically all of the pioneer developmental work undertaken by Förster in Germany and Farrow in the United States in the 1930's, as well as the advances that followed have used single frequency sinusoidal current. The revival of interest in broadband electromagnetic test methods was apparently initiated with the work of Waidelich in 1955. Since then other workers in the field have produced developments to the point where the method is useful for numerous important test applications.

*Diffusion of pulsed currents into conductors.* As migth be expected, more of the theoretical problems that are of interest in eddy current testing have been solved for the single frequency sinusoidal case than have been solved for the more complicated broadband case. The problem of a sinusoidally varying plane wave impinging normally upon a plane conductor, which produces the familiar formula for nominal depth of penetration $\delta = \dfrac{1}{(\pi f \mu \sigma)^{1/2}}$, has been solved, however, for various pulse wave forms. With the magnetic field parallel to the surface of the plane infinite conductor, $\bar{H} = H_y \bar{j}$, the current density, $\bar{J} = J_x(z, t)\,\bar{i}$ will vary with depth inside the conductor. Maxwell's equations reduce to

$$\frac{\partial^2 H_y}{\partial Z^2} = \frac{1}{h^2}\frac{H_y}{\partial t} \quad \text{and} \quad \frac{\partial^2 J_x}{\partial Z^2} = \frac{1}{h^2}\frac{\partial J_x}{\partial t}.$$

neglecting displacement current.

Where
$$h = \frac{1}{(\mu\sigma)^{1/2}}.$$

Following the results of an analysis by Vallese, the Laplace transform of the current $J(z, s)$ is in the general case.

$$J(z, s) = \frac{(s)^{1/2}}{h} I(s) \exp\left[-\frac{z}{h}(s)^{1/2}\right].$$

where $I(s)$ is the transform of $i(t)$, the total current in the conductor for a strip of unit width in the direction of the $y$ axis. In the sinusoidal case, the skin depth $\delta$ is the depth at which the current density drops to $1/e$ of its value at the surface. The equivalent skin depth in the transient case does not appear so neatly. As a first approximation, the equivalent skin depth for a square wave function $i(t) = I_0\,[U(t) - U(t - T_0)]$ is given by $z' = h(2T_0)^{1/2}$; for a function $I_0(t)^{-1/2}$ $[U(t) - U(t - T_0)]$, $z' = h(6T_0)^{1/2}$; and for the function $i(t) = I_0 e^{-\alpha t}$, $z' = h(2T_0)^{1/2}(1 + 2\alpha T_0)$. Inversion of $J(z, s)$ increases rapidly in difficulty if more complicated current wave forms are considered. The current $i(t)$ is the total current in the conductor for a strip of unit width in the direction of the $y$ axis, which is the same as the magnetic field intensity just outside the conductor. The metal test specimen itself influences the field at its surface, so even if practical field sources for broadband electromagnetic testing emitted plane waves, which they do not, the current waveform flowing in such a source would not be duplicated in the metal.

Some of the disadvantages of the broadband method are already apparent from the foregoing discussion. But it is certain that information exists in a field containing a broad band of frequencies that is not contained in any given single frequency sinusoidal field being affected by the same conductor. The extraction of this additional information is one of the central problems of the broadband electromagnetic test method. One of the simplest means of extraction is amplitude sampling in the time domain. As an example, the figure shows the voltage induced in a small pickup as a result of the field outside a good conductor. The conductor is, in this case, a thin-

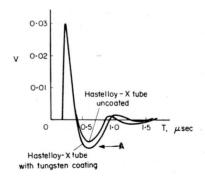

Hastelloy – X tube
uncoated

Hastelloy-X tube
with tungsten coating

walled tube of Hastelloy-X (50% Ni, 19% Fe, 9% Mo, 20% Cr) ($\varrho = 118.3\mu$ ohm-cm) of outer diameter $4.98$ mm × $0.38$ mm wall. Part of the tube is coated on the inner surface with $0.1$ mm of tungsten ($\varrho = 5.65\mu$ ohm-cm). If the intent is to measure the tungsten thickness, the obvious area in which to sample is in the vicinity of the point marked "A". The overall amplitude of the reflected field is sensitive to variations in distance between field source and specimen, but information about conditions deeper inside the specimen appears later in the reflected field pulse. This is due to the relatively slow rate of diffusion of current into good conductors. Simple pulse amplitude sampling in the time domain has been found to be a satisfactory means of information extraction in many practical test problems, particularly the testing of tubes and pipes for defects and the non-destructive inspection of the bond between two metals. The full potential of the broadband method, however, is not realized unless more sophisticated methods are used to analyse the field reflected from the metal. Methods of signal analysis have been developed which allow the separation of a number of parameters of the test specimen which can all vary simultaneously. The separation of up to 4 simultaneously varying test specimen parameters has been demonstrated experimentally. The calculations and adjustments necessary to accomplish this feat are necessarily complex, but such a task seems to be beyond the capability of any existing single frequency test method.

Most of the broadband test systems which have been constructed to date have placed the test coils or probes which are used to induce and detect the flow of current in the test specimen in the same geometrical and electrical arrangements familiar in sinusoidal eddy current test equipment, but broadband electromagnetic test equipment which drives the field source with a short current pulse has the advantage of allowing great flexibility in the test coil or probe arrangement which is used to induce and detect the flow of current in the test specimen. The total time of current flow may be typically one one-thousandth of the time between the repetition of cycles which means that the peak pulse power applied to the field coil and, consequently, the field about it, can be relatively high compared to continuous wave equipment. These strong fields have made possible the development and use of devices in which a field generating coil is positioned in an enclosure made from a high conductivity material, near an aperture through which some of the flux passes. Typical aperture sizes are in the range $1.02$–$1.65$ mm in diameter, so the field near the aperture has a small cross-sectional area which of course increases as the distance from the aperture increases. But if the test specimen is positioned close to the aperture—within the probe-to-specimen distances normally used in eddy current testing—the test equipment which uses such a field source can be capable of a high degree of surface resolution. A small pickup coil is used to detect the field reflected from the test specimen, and by suitable design of the enclosure containing the aperture, and proper positioning of this pickup coil, a condition can be obtained in which no voltage is developed unless a specimen is near the aperture. The voltage developed across the pickup when a test specimen is placed near the aperture is then a function of the field reflected from the metal surface and interior. The advantages of such devices besides surface resolution include some simplification in application compared to some electromagnetic test equipment because there are no bridges or equivalent devices to balance. This type of equipment is also quite stable because changes in the impedance of neither the field coil nor the pickup coil as a result of ageing or ambient temperature changes have any significant effect on the reflected field.

There are a number of disadvantages in broadband electromagnetic test methods as compared to their narrowband counterparts. These include:

1. The unavoidable increase in noise voltage imposed on the desirable test signal as a result of the wide bandwidths inherent in the method.

2. The loss in maximum possible inspection speeds if the time between cycles is long compared to the duration of current flow in the field generating coil.

3. The loss of the relatively simple impedance plane concept so useful in sinusoidal eddy current theory and practice.

*Bibliography*

HUGHES D. E. (1879) *Phil. Mag.*, Series 5, 50.

LIBBY H. L. and COX C. W. (Jan. 1961) Broadband Electromagnetic, Testing Methods Part 2, Signal Analysis. HW–67,639, Hanford Atomic Products Operation, Richland, Washington.

RENKEN C. J. (1964) *A Pulsed Electromagnetic Test System Applied to the Inspection of Thin Walled Tubing*, ANL–6728, Argonne National Laboratory, Argonne, Illinois.

VALLESE L. M. (Feb. 1954) *J. Appl. Phys.* **25**, No. 2.

WAIDELICH D. L. (1955) *Proc. National Electronic Conference* **10**.

WISE S. (1961) in *Encyclopaedic Dictionary of Physics* (J. Thewlis Ed.) **4**, 437, Oxford: Pergamon Press.

C. J. RENKEN

# C

**CHEMICAL RECORDER.** This is a device which records electrical signals on paper through chemical action produced by the flow of current in the chemical with which the paper is impregnated. Typically potassium iodide is used in slightly damp paper, and will mark by the release of iodine with signals of a few volts; dry types are also available commercially (e.g. Teledeltos paper) in which rather higher voltages (e.g. 50 V) are needed. One of the main uses of such an instrument is in sonar systems, where the motion of a stylus across the paper represents a time-base and the density of a mark indicates the strength of a received signal; its position in the stylus traverse indicates the range of the echoing object. The paper is moved slowly in a direction at right angles to the stylus traverse, so that successive traverses are recorded side-by-side. This leads to greatly improved detection of weak signals. Very fast stylus movement is now available, speeds of 10 m/sec having been reliably achieved.

*Bibliography*

GRIFFITHS J. W. R. and MORGAN I. G. (1956) *J. Soc. Instrum. Techn.* **8**, 62.

D. G. TUCKER

**CHONDRULES.** About 85 per cent of meteorites seen to fall to Earth are stones containing small (0·5–3 mm) spheroidal structures called *chondrules*. These stones, the *chondrites*, are debris from broken planets (probably the asteroids), and are believed still to be in or near the same state that the rocky matter of the inner planets initially assumed when the solar system was formed.

Chondrules make up over 50 per cent of the volume of some chondrites. They are less abundant in most chondrites, but these have obviously been recrystallized during a high-temperature stage in their history, resulting in a homogenization of their textures that must have erased most of the chondrule structures originally present.

Chondrules are variable in chemical composition, but their average composition is similar to that of chondrites as a whole, approximately as follows (relative atomic abundances of the fifteen most abundant elements):

| | | | | | |
|----|-----|----|-----|----|------|
| O  | 344 | Al | 7·9 | Mn | 0·56 |
| Si | 100 | Ca | 5·5 | P  | 0·45 |
| Mg | 93  | Na | 4·5 | K  | 0·33 |
| Fe | 71  | Ni | 3·6 | Ti | 0·23 |
| S  | 10  | Cr | 0·77| Co | 0·12 |

These elements appear in the chondrules in the form of several minerals, principally olivine, $(Mg, Fe)_2SiO_4$; pyroxenes, $(Mg, Fe, Ca)SiO_3$; metallic nickel–iron; troilite, $FeS$; and feldspar, $NaAlSi_3O_8$-$CaAl_2Si_2O_8$. Chondrites contain random mixtures of chondrules displaying a variety of internal textures and containing various proportions of the above minerals.

Textures of the chondrules and minerals they contain are analogous in most respects to those of terrestrial volcanic rocks. Some contain glass, and some are attached to and indented by other chondrules, showing they behaved plastically at one time. Evidently the chondrules were once dispersed droplets of molten lava, which cooled and crystallized rapidly, retaining their spheroidal shapes. These hardened droplets were thoroughly mixed, then they aggregated, along with fine-grained material of similar composition, to form a chondritic agglomerate.

A number of different processes have been proposed that might have produced dispersed lava droplets. Most writers have advocated secondary processes, i.e. processes that operated on or in the parent meteorite asteroids after their formation, such as explosive vulcanism, or high-velocity collisions between asteroids, with fragments melted by the energy of impact. A more intriguing possibility is that they were primary, that the first material to condense when the solar system was formed included lava droplets, and that the solid particles or "planetesimals" from which the inner planets accreted consisted largely of chondrules.

The concept of chondrules as primary condensations is supported by the fact that the chondrules in unrecrystallized chondrites contain very little iron in an oxidized state, although they frequently contain metallic iron. They are in a reduced state, and their small content of ferrous iron is approximately that which silicate droplets would contain, if in equilibrium with gas of solar composition at approximately $2000°K$ (based on thermochemical calculations). The view is widely held today that the solar system evolved from a hydrogen-rich gas nebula, and it is tempting to suppose the chondrules condensed as liquid droplets in cooling regions of the primordial nebula. However, the solar or cosmic abundance of metallic elements is so low relative to hydrogen and helium that only at rather high total gas pressures (> 100 atmospheres) would the triple points of iron metal and of silicate mixtures be exceeded in such a gas system. None of the theoretical models for stars, protostars or the solar nebula developed to date contain regions of sufficiently high pressure juxtaposed with moderate temperature (approximately 2000°K) where iron and silicate liquids would be thermodynamically stable, so if the chondrules are indeed

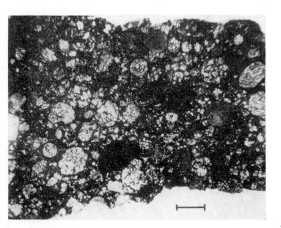

*Thin-section of chondrite Bishunpur, viewed by transmitted light, showing chondrules. Scale bar, 1 mm.*

primary solid particles, they must have been created under special circumstances not yet understood.

Recent studies of the mass spectra of noble gases, especially xenon, extracted at high temperatures from meteorites and from isolated chondrules, have revealed isotopic anomalies that appear to support the concept of primary chondrules. A number of chondrites contain an excess of $^{129}Xe$ relative to terrestrial xenon. This is attributed to decay of $^{129}I$ (half-life, $16\cdot4 \times 10^6$ years) which was created by nucleosynthetic processes operating prior to or during the formation of the solar system, and which was subsequently incorporated in the meteorites. Merrihue (1963) has found anomalous $^{129}Xe$ is 2·9 times more abundant in chondrules of the chondrite Bruderheim than in the chondrite as a whole. Melting would have driven $^{129}Xe$ out of the chondrules, so their liquefaction must have occurred before substantial $^{129}I$ decay had taken place, i.e. within a few tens of millions of years after cessation of nucleosynthesis. This condition would have been met more easily by primary chondrules than by chondrules that were not generated until after planets accreted and heated up internally.

Bruderheim chondrules also contain anomalously high abundances of $^{124}Xe$ and $^{126}Xe$. These isotopes are thought to be spallation products of charged-particle irradiation of heavy elements. They can be most satisfactorily accounted for in the chondrules if the latter remained for a time dispersed in space, exposed to protons and alpha particles emitted by an active sun or protosun.

*Bibliography*

MERRIHUE C. M. (1963) *J. Geophys. Res.* **68**, 325.
REYNOLDS J. H. (1963) *J. Geophys. Res.* **68**, 2939.
TSCHERMAK G. (1885) *Die mikroskopische Beschaffenheit der Meteoriten*, Stuttgart: E. Schweizerbart'sche Ver. (translated re-edition in *Smithsonian Contrib. Astrophys.* 4, No. 6, 1964).
WOOD, J. A. (1963) *Icarus* **2**, 152.

J. A. WOOD

## COLD NEUTRON SCATTERING FROM DEFECTS.

There are a number of processes by which neutrons may be removed from a slow neutron beam when it is made to pass through a crystal. Amongst these are absorption, and also incoherent, inelastic and magnetic scattering effects. For simplicity we shall consider initially only diamagnetic materials, so that the magnetic scattering may be neglected.

If the crystal is structurally perfect, the only other additional scattering process is that due to Bragg reflections. This is an elastic scattering process which occurs only at specific scattering angles. If, however, the crystal is not structurally perfect there will also be some diffuse elastic scattering which will occur at all angles. The observation of this diffuse scattering may be used therefore as a technique to estimate the nature of imperfections on an atomic scale. Since its principal application to date has been in the study of radiation damage we will illustrate the general principles of defect scattering by means of this particular application.

Because the intensity of this diffuse scattering is so much smaller than that due to Bragg scattering, the latter must be eliminated in any experiment. This is achieved either by using a suitably oriented single crystal as the sample or, more generally, by using cold neutrons whose wave-lengths are longer than the Bragg cut-off wave-length.

Let us now consider the scattering intensity due to interstitial atoms or vacancies in various stages of aggregation. For a single interstitial atom or vacancy in a crystal the scattered intensity is mathematically equivalent to a void of the same dimensions as the crystal together with an atom of the element corresponding to the defect and located at its position; thus the defect scattering is isotropic. For a number of such defects randomly arranged in the crystal the scattered intensity is simply enhanced proportionately. When defects are aggregated the scattering is no longer isotropic owing to interference effects between the scattered waves from each point defect comprising a particular cluster. It is now a calculable function of the scattering angle and neutron wave-length, the details of which will depend on the nature of the cluster.

In principle, therefore, it is possible to estimate the state of aggregation of defects from neutron scattering measurements. Also the absolute defect concentrations may be determined, since the only other parameter in the calculated cross-section is the bound atom coherent scattering cross-section, which is a well known, experimentally determined, figure. In radiation damage studies this technique is limited by the indistinguishability of vacancies and interstitial atoms, and also by the difficulties which can exist when more than one type of defect is present in the material. In addition, owing to the comparatively low intensities of cold neutron beams that are available, defects whose largest dimension is greater than about 50 Å are often difficult to observe by this technique.

In practice calculations of the scattering cross-section for particular defect configurations are more complex than has been suggested. Firstly, in non-metallic materials defects may form paramagnetic centres by trapping electrons, and these will provide an additional magnetic contribution to the defect scattering. However, only under particular experimental conditions is this effect likely to be significant. Secondly, atoms in the neighbourhood of defects will be displaced from their lattice sites, and so make a contribution to the defect scattering. In many cases this contribution is appreciable, and therefore must always be considered when such calculations are performed.

Experimental considerations are governed by the smallness of the scattering cross-section to be measured. Thus, for a material containing 1 per cent of point defects, the total cross-section is only of the order of 50 millibarns. As a result most experimenters have measured transmission cross-sections as a function of neutron wave-length rather than the scattering intensity directly as a function of angle owing to its smallness in comparison with reactor backgrounds. Also other extraneous scattering processes must be reduced to a minimum, and this limits observations to materials containing elements with favourable nuclear properties, such as small absorption and incoherent scattering cross-sections. Other favourable features in the choice of suitable materials are those possessing a comparatively low inelastic scattering cross-section and short Bragg cut-off wave-length.

So far, investigations have been made on the following materials, all of which had been irradiated in a reactor in the temperature region 30–100°C: graphite, BeO, $\alpha$-$Al_2O_3$, and $SiO_2$. The results have shown that in conjunction with other techniques they can make a useful contribution to our understanding of the detailed nature of radiation damage.

This technique can clearly be applied to other types of defect as well as radiation damage. For example, Low and Collins have shown that it may be used to estimate up to a distance of several ångströms the spatial distribution of the magnetic moment disturbance around an impurity in a ferromagnet. Again, it should be possible in principle to examine atomic relaxations around impurity atoms in dilute solid solutions; also dislocation distributions in material, and in particular, in ferromagnets.

*See also*: Bragg cut off wavelength.

*Bibliography*

Low G. E. E. and Collins M. F. (1963) *J. Appl. Phys.* **34**, 1195.
Martin D. G. (1964) *The Interaction of Radiation with Solids* (R. Strumane *et al.* Eds.) (North-Holland: Amsterdam).
Thewlis J. (Ed.) (1962) *Encyclopaedic Dictionary of Physics* 4, 814, **5**, 2, 3. Oxford: Pergamon Press.

D. G. Martin

**COLD WORK, INTERNAL ENERGY OF.** Plastic deformation of annealed metals at low temperatures introduces lattice defects, such as dislocations, vacancies and stacking faults and in alloys may change the degree of order. These structural changes represent an increase of energy of the metal which will be stored indefinitely if the temperature is kept sufficiently low. As the temperature rises the defects become mobile and anneal out of the crystal lattice, which in consequence reverts to its pre-deformed state. The energy associated with the lattice defects appears as heat during annealing. At high temperatures the defects will be sufficiently mobile to anneal out during the deformation process and no stored energy remains after deformation ceases. Plastic deformation carried out under these conditions is described as hot working; in contrast, at lower temperatures, where defects generated by plastic flow persist after deformation is complete, the deformation is referred to as cold working. This causes the structure and the physical properties such as hardness and electrical resistivity to be permanently altered at the temperature of deformation.

There is no sharply defined temperature separating hot and cold working processes. The most persistent defects that have the most marked effect on the energy and strength of metals are dislocations. These are drastically reduced in density during recrystallization, when new grains are nucleated and grow by a process of boundary migration that sweeps up dislocations in the path of the advancing boundaries. For this reason recrystallization is an effective, but not unique, process for the removal of the effects of cold work and, since it can easily be detected experimentally, the temperature at which recrystallization begins is often taken as an approximate indication of the transition from cold to hot working.

Elimination of defects created by cold work can be detected below the recrystallization temperature; these processes are called recovery.

Mechanical work must be done on a metal that is deforming plastically; during the process heat is evolved but careful measurements reveal that the amount of heat evolved is less than the thermal equivalent of the mechanical work done. The difference is the energy stored in the metal. Experimental difficulties arise because the stored energy rarely exceeds 10 per cent of the work done during deformation. For this reason efforts have been made to develop direct methods for the measurement of the stored energy either by measuring the heat evolved during annealing or by comparing the heats of solution of annealed and deformed specimens. ·

*Thermodynamic considerations.* The change of internal energy $\Delta E$ produced by cold working is given by the First Law of Thermodynamics:

$$\Delta E = W - Q$$

where $W$ is the work done on the body and $Q$ is the quantity of heat generated. Measurement of $W$ and $Q$ allow the stored energy, which is equal to $\Delta E$, to be determined.

Annealing causes the stored energy to be transformed into heat. The heat generated is the change of enthalpy $\Delta H$ which is related to the internal energy by

$$\Delta H = \Delta E + P \cdot \Delta V$$

where $\Delta V$ is the permanent volume change that takes place and $P$ is the hydrostatic pressure on the specimen. The product $P \Delta V$ is always small so that

$$\Delta H = \Delta E$$

and negligible error arises if the change of enthalpy is set equal to the stored energy.

The change of free energy $\Delta G$ produced by cold working is

$$\Delta G = \Delta H - T \Delta S,$$

where $\Delta S$ is the change of entropy. At normal pressures we can again set

$$\Delta H = \Delta E$$

so that $$\Delta G = \Delta E - T \cdot \Delta S.$$

No known technique exists for the experimental determination of $\Delta S$ and the best estimates are calculated from a knowledge of the nature of the lattice defects that are introduced by cold work. We now consider the entropy contribution from the two important defects, dislocations and vacancies.

Cottrell has shown that the major part of the free energy associated with dislocations, about 5 eV per atom plane, resides in the elastic strain energy of the associated strain field, which is an internal energy term. A configurational entropy term arises because the dislocations can be arranged in many ways in the

crystal. An upper limit is achieved if the dislocation is perfectly flexible, which in practice is not the case, and here in a lattice of coordination number 12 there is an entropy contribution of $-\mathbf{k}T \ln 6$ which is less than $-0{\cdot}05$ eV at room temperature. In addition a vibrational entropy term arises because the dilational strain round a dislocation will change the local interatomic forces and modify the atomic vibrations. This effect is estimated to account for an entropy contribution of about $-3\mathbf{k}T$, a little more than the configurational entropy. The total entropy change appears however to be at most about 2 per cent of the free energy at room temperature and therefore it can be assumed to a close approximation that the dislocation contribution to the free energy is equal to the internal energy term.

Analysis of the contribution made by vacancies to the free energy of cold worked metals predicts that the entropy term can be about one fifth of the free energy at 200°K. However, the total free energy contribution from vacancies is at least an order of magnitude less than that contributed by dislocations.

*Measurement of stored energy.* The stored energy in cold worked metals seldom exceeds 1 cal/cm³ which is about 1 per cent of the latent peat of fusion. The small thermal changes equivalent to this amount of energy make exacting demands on calorimetric techniques.

(*i*) *Measurement by difference.* This method of measuring the stored energy, used particularly successfully by Taylor and Quinney, is based on the thermodynamic relation for the stored energy

$$E_{\mathrm{s}} = W - Q.$$

Taylor and Quinney plastically deformed tubular specimens in torsion and measured the work done $W$ by the area under the curve relating torque and twist for the tube. The heat evolved during plastic deformation, $Q$, was measured about one second after deformation was complete either from the increase of the surface temperature of the specimen or by dropping the specimen quickly into a calorimeter. Care was taken to adjust the experimental conditions so that heat loss by conduction from the ends of the specimen was small.

A correction must be made if the applied strain has a dilational component as for example in simple tension, since an elastic strain of this kind produces a change of temperature through the thermoelastic effect. An increment of tensile stress $\Delta \sigma$ applied adiabatically to a solid of positive coefficient of thermal expansion $\alpha$ at absolute temperature $T$ and specific heat per unit volume at constant stress $C_\sigma$ produces a fall of temperature

$$\Delta T = -\frac{\alpha T}{C_\sigma} \cdot \Delta \sigma,$$

and therefore contributes $-\alpha T \Delta \sigma$ to the internal energy. In pure torsion no thermoelastic temperature change is generated and therefore no correction was necessary on this account to Taylor and Quinney's measurements.

*(ii) Annealing methods.* Annealing methods have produced the major part of the reliable published data on stored energy. Thermal effects produced by the transformation of stored energy into heat are measured in a specimen that is either heated steadily through the temperature range in which the stored energy is released or is maintained isothermally at a selected temperature in this range. Isothermal methods give valuable data on the kinetics of energy release but may fail to detect energy liberated during heating to the annealing temperature. It is important to avoid spurious energy changes caused by surface effects, for example errors have been traced to the dissociation of surface oxide and to the evaporation of a condensed film of moisture from the specimen surface.

All annealing techniques compare the thermal behaviour of a cold worked specimen with that of an annealed specimen of identical shape and size. Several methods have relied on the assumption, which can be shown to be invalid, that in successive heating runs the thermal resistance between a specimen and its surrounding furnace enclosure remains constant. Quinney and Taylor developed a method that did not require this condition to be preserved. They made cylindrical specimens from deformed material, fitted an internal heater and placed the specimen on insulating supports of small cross-sectional area in an evacuated enclosure. The specimen was electrically heated by a constant current and the temperature of the enclosure surrounding the specimen was continuously adjusted to be equal to the specimen surface temperature. These experimental provisions aimed to reduce heat losses from the specimen by conduction, convection and radiation to such a low value that it could be assumed that the rise of specimen temperature depended only on the energy supplied by the heater and the release of energy stored in the metal. The energy release was obtained from the difference between the final temperature attained in successive runs with the same current passing through the specimen heater. The first run released all the stored energy and therefore the second run was that for an annealed specimen.

A development by Clarebrough and his colleagues is currently the most advanced technique in this class, and aimed to dispense with the need in Quinney and Taylor's method to determine the specific heat-temperature curve for the annealed material in a separate experiment. Clarebrough *et al.* used two cylindrical specimens of identical size, one annealed and one cold worked; these were placed side by side on thermally insulating supports in an evacuated enclosure. Both specimens had internal heaters and the input of electrical energy to these was automatically controlled by saturable reactors to keep the two specimens and the wall of the enclosure all at the same temperature

whilst the enclosure temperature increased steadily at about 6°C/min. The imposed experimental conditions prevent heat transfer either between the specimens or between the specimens and the enclosure and the integrated difference between the input of electrical energy to the two specimens is the energy of cold work released by annealing. The difference between the rate of energy supply to the specimens was measured directly on a specially-constructed differential wattmeter.

Isothermal annealing methods were placed on a firm foundation by the work of Borelius and his school in Sweden. Gordon has followed up this work. In his equipment the cold-worked specimen is placed in an enclosure the walls of which are maintained at a constant temperature to within $10^{-4}$°C by a vapour thermostat. The enclosure is evacuated and the specimen is supported in such a way as to achieve a high thermal resistance between it and the enclosure walls. In an isothermal anneal the specimen generates heat as the stored energy is released and a temperature difference $\Delta T$, at most 0·1°C, develops between the specimen and the enclosure wall. $\Delta T$ is measured as a function of time by a sensitive thermopile and $\Delta T$ is later converted into rate of heat evolution by an experiment in which known amounts of heat are generated in the specimen by the Peltier effect at a junction of known thermoelectric power carrying a measured current.

*(iii) Solution methods.* The principle of the solution method is to determine the difference between the heats of solution of deformed and annealed samples. It is essential to select a solvent in which the heat of solution of the annealed material is low since this increases the accuracy of the determination of the small heat change that is equivalent to the stored energy of cold work. The success of the method developed by Bever and Ticknor rests on the choice of a liquid metal solvent. They found that the temperature change following the addition of annealed 75 wt.% gold—25 wt.% silver alloy at 0°C to liquid tin at 240°C was very small because the heat of solution liberated was just that required to raise the temperature of the alloy to 240°C. Under these circumstances there is a marked difference between the temperature change of the liquid tin bath following the addition of equal weights of deformed and annealed alloy. It is from this temperature difference and the thermal capacity of the calorimeter and contents that the energy of cold work is determined.

*Interpretation of stored energy measurements.* Some of the published values of stored energy are of doubtful validity since the experimental techniques used to obtain them are unsound, but a general pattern is emerging from the small volume of reliable data.

It is clear that the amount of stored energy reaches a saturation value at high strains that correspond closely to the degree of plastic deformation beyond which little further increase of yield stress due to work hardening takes place. Figure 1, for example, shows results of measurements by Taylor and Quinney for copper deformed plastically in torsion. The graph relates the ratio of the increment of stored energy $\Delta E_s$ to the increment of mechanical work done on the specimen $\Delta E_w$ as a function of torsional plastic strain. The sharp fall in this ratio at a strain of 1·2 corresponds to the point at which the stored energy is beginning to approach saturation; Taylor and Quinney observed that the rate of work hardening had also fallen to a low value at this strain.

The relation between work hardening and stored energy is significant in giving a clue to the mechanism of energy storage. It is now thought that the increase of yield stress brought about by plastic flow arises

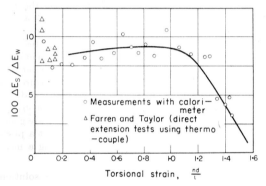

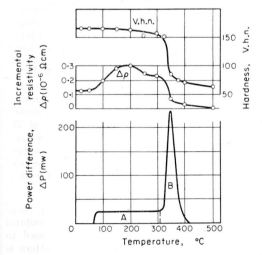

Fig. 1. *Ratio of the incremental values of stored and expended energy as a function of strain of copper in torsion. (Taylor and Quinney.)*

Fig. 2. *Power difference, representing release of stored energy from cold-worked arsenic-bearing copper, as a function of annealing temperature. Incremental resistivity and hardness also shown.* (Clarebrough, Hargreaves and West.)

from the multiplication of dislocations that interact to form a complex network that resists further dislocation generation and glide. The connexion between work hardening and stored energy leads to the expectation that the increased dislocation density generated by plastic flow might be an important source of stored energy. That this is so is confirmed by a study of the pattern of stored energy release during heating a cold worked specimen. Figure 2 shows that the rate of energy release from cold-worked arsenical copper has a prominent peak B coinciding with the temperature that produces a sharp drop of indentation hardness. Metallographic examination confirms that these changes accompany recrystallization and justifies ascribing the release of stored energy producing peak B to the reduction of dislocation density caused by the gross modification of the grain structure. Further confirmation of this deduction comes from the isothermal annealing experiments of Gordon. He found that the increase with time of the released recrystallization energy followed a similar curve to that relating percentage recrystallization with time deduced from micro-indentation hardness measurements.

If all the internal energy of cold work is assumed to reside in the stress fields of the dislocations in the metal, it is possible to make estimates of dislocation density that turn out to be in agreement with determinations by other methods. Approximate estimates of the energy of unit length of dislocation line using an elastic continuum analysis give about $10^{-4}$ erg/cm so if all of this energy is that of the dislocation stress-fields a dislocation density of about $10^{12}$ cm/cm$^3$ must exist which is the right order of magnitude. More precise analysis which has not yet been carried out demands a detailed knowledge of the dislocation structure of the metal, since the dislocation energy is affected by interaction with adjacent dislocations.

Figure 2 reveals that a significant amount of stored energy can be released at temperatures below that at which recrystallization begins. This energy evolution arises from recovery processes and appears to be more prominent in less pure metals. One mechanism in this class is the redistribution of dislocations into vertical walls, termed polygonization, that results in a configuration that has a lower energy than the same dislocation population more randomly distributed. Other recovery processes that release energy arise from the elimination of vacancies generated by cold work. These point defects are mobile at room temperature in pure metals. Henderson and Koehler deformed pure copper at $-185°C$ and detected peaks in the curve relating rate of energy release with temperature for this material at $-90°C$ and $-25°C$. They suggest that the lower temperature peak results from the annihilation of pairs of vacancies and the peak at $-25°C$ corresponds to the annealing out of the less mobile single vacancies. These contributions from vacancies in part explain the observation that more energy can be stored in metals deformed at low temperatures where vacancies are immobile. Leach *et al.*

measured a stored energy of 242 cal/g-atom in drillings of a copper–gold alloy machined at $-195°C$ whereas drillings from a specimen at room temperature stored only 82 cal/g-atom. Measurement of energy release at elevated temperatures on specimens that have been stored for some time at room temperature after deformation will miss much of the energy associated with point defects that will have annealed out prior to the test. In impure metals and alloys interaction between point defects and solute atoms may retain vacancies in solute-vacancy pairs. These dissociate at elevated temperatures and the annihilation of the released vacancies may contribute to energy release during recovery.

Other structural changes induced by cold work and capable of storing energy are the generation of faults in the regular sequence of stacking of the close-packed planes of atoms, particularly in face-centred cubic structures, and changes in the degree of order in solid solutions when slip displaces adjacent sheets of atoms relative to each other. X-ray measurement of stacking fault density from the displacement of diffraction peaks produced by cold work in 70/30 brass lead to the result that one plane in 40 may be faulted; if the stacking-fault energy is about 40 erg/cm² then the total energy stored in stacking faults could be 10 cal/g-atom which represents a significant contribution to the stored energy of cold work.

Long-range internal stresses, such as arise from macroscopically inhomogeneous plastic flow, make only a small contribution to the stored energy.

Tabulated stored energy values published in the literature up to 1958 are presented in a comprehensive review of the internal energy of cold work by A. L. Titchener and M. B. Bever in Progress in Metal Physics (1958).

*Bibliography*

COTTRELL A. H. (1964) *The Mechanical Properties of Matter*, New York: Wiley.

TITCHENER A. L. and BEVER M. B. (1958) in *Progress in Metal Physics* **7**, 247.

K. M. ENTWISTLE

**COLLECTIVE PARAMAGNETISM.** Certain unique magnetic properties may be exhibited by an assembly of ferromagnetic particles so small that each particle consists of but one magnetic domain. These unique properties include that the magnetization of such an assembly is equilibrated by thermal energy at the temperature of observation, and that this occurs in a time which is short compared to the duration of observation. The magnetization of assemblies of this kind, in its relation to field and temperature, is thus comparable with that of paramagnetic atoms and molecules, except that the magnetic moment per particle may be several orders larger. This pheno-

menon is known by various names, of which collective paramagnetism and *superparamagnetism* are two.

The observed magnetization, $M$, of an assembly of particles, each of which possesses a permanent moment $\mu$, is given by the *Langevin equation*:

$$\frac{M}{M_s} = \coth\frac{\mu H}{kT} - \frac{kT}{\mu H}, \tag{1}$$

where $M_s$ is the saturation magnetization, $H$ is the field, $T$ the absolute temperature, and $k$ the Boltzmann constant. For a ferromagnetic particle of spontaneous magnetization $I_{sp}$ and volume $v$, $\mu = I_{sp}v$, whence it follows that:

$$M = I_{sp}V\left(\coth\frac{I_{sp}vH}{kT} - \frac{kT}{I_{sp}vH}\right). \tag{2}$$

$V$ being the total volume of the sample.
For an assembly of non-uniform particles such that

$$\int_0^\infty f(v)\,dv = 1$$

the magnetization becomes:

$$M = I_{sp}V\int_0^\infty\left[\coth\left(\frac{I_{sp}vH}{kT}\right) - \frac{kT}{I_{sp}vH}\right]f(v)\,dv \tag{3}$$

provided that $I_{sp}$ is independent of $v$, as appears to be the case, except as noted below. It must then follow that magnetizations obtained at various values of $H$ and $T$ should superimpose if plotted against $H/T$. This effect has been demonstrated and, together with the absence of remanence, it forms a convenient diagnostic test for collective paramagnetism. Substances which have been shown to behave in this fashion include the metals nickel, cobalt, and iron, the oxides $Fe_3O_4$ and $\alpha$-$Fe_2O_3$, and a few alloys, all in particle sizes ranging down from about 100 Å diameter.

Collective paramagnetism may arise through physical rotation of the individual particles, and this kind of behaviour has been known for a long time. In 1949 L. Néel (whose name is closely connected with the whole development) showed that a single domain particle, if sufficiently small, could undergo a change in the direction of magnetization without physical rotation. Néel also derived the law of approach to thermal equilibrium for a sample magnetized to saturation along the easiest axis. Thus the remanent magnetization, $M_r$, at time $t$ has decayed exponentially as follows:

$$M_r = M_s\exp(-t/\tau) \tag{4}$$

where $\tau$ is the relaxation time, and:

$$1/\tau = f_0\exp(-Kv/kT) \tag{5}$$

where $f_0$ is a frequency factor and $K$ is the anisotropy energy per unit volume.

The quantity $K$ arises from the crystal anisotropy, from the particle shape, and from other causes. For a spherical particle of iron of radius 125 Å, at room temperature $M_r/M_s$ will become 0·01 in about 6 min. The rate of decay is strongly dependent on particle shape and volume and it also varies with temperature. The original impetus to studies of this kind was based on the observation that certain rock deposits have remained magnetized for geological times; their study has become an important branch of geophysics. Similarly, small single domain, elongated particles of iron are available commercially as permanent magnets. Our chief concern in collective paramagnetism is, however, with those systems in which the decay time is short.

Several further questions arise, even in connexion with those assemblies exhibiting true collective paramagnetism. One of these questions is whether the spontaneous magnetization is the same in a very small particle as it is in massive metal. The answer appears to be that it is so, at least down to the lowest diameter thus far measured, namely about 7 Å. The second question is whether the Curie temperature is dependent on particle size. The experimental evidence shows reasonably well that a slight diminution does occur. A third question is the possible influence of particle clumping. Such clumping apparently does occur and it probably has an influence on estimates of the individual particle size determined magnetically as described in the following paragraphs.

Applications of collective paramagnetism have been made to particle size determination in various alloy systems, and in some classes of catalytically active solids. Of the several methods for doing this one will be described briefly.

If all particles in an assembly had identical moments the volume, $v$, of a particle would be simply $\mu/I_{sp}$, and would readily be obtained from equation 2. For a distribution of particle sizes one may use equation 3. In a plot of $M$ vs. $H/T$ the initial slope of the magnetization curve will be more influenced by the larger, more readily magnetized, particles. The approach to saturation will be more influenced by the smaller particles. From the low field measurement we may obtain:

$$\frac{\bar{v}^2}{v} = \frac{3kT}{I_{sp}H}\left(\frac{M}{M_s}\right) \qquad (6)$$

and from the high field:

$$v = \frac{kT}{I_{sp}H} \cdot \frac{1}{1-(M/M_s)} \cdot \qquad (7)$$

This, and related methods, in several refinements have been usefully applied to problems of particle size determination in, for instance, precipitates of

cobalt in a copper matrix, and in nickel hydrogenation catalyst systems. The particle diameters to which the method has been applied range from about 15 to 80 Å for nickel and somewhat smaller for cobalt and iron.

Another application of collective paramagnetism is to problems of adsorption, especially of reactive vapour molecules on the surfaces of catalytically active solids, of which nickel is one. A monolayer of, say, hydrogen adsorbed on an assembly of nickel particles exhibiting collective paramagnetism may lower the apparent moment of the particles by 10 to 20 per cent in the case of typical catalyst preparations. This effect has proved to be useful in gaining information concerning the mode of surface bonding between adsorbent and adsorbate for a large number of systems of interest in the area of heterogeneous catalysis.

An effect somewhat related to collective paramagnetism may be observed in small particles of antiferromagnetics. This is briefly described under Superantiferromagnetism.

*See also:* Superantiferromagnetism.

*Bibliography*

JACOBS I. S. and BEAN C. P. (1963) in *Magnetism*, (Eds. RADO G. T. and SUHL H.) New York: Academic Press.

SELWOOD P. W. (1962) *Adsorption and Collective Paramagnetism*, New York: Academic Press.

P. W. SELWOOD

**COLOUR RADIOGRAPHY.** What is commonly meant by colour radiography is the technique of producing a colour display of alterations experienced by penetrating radiation (usually X or gamma), as it passes through an object. As in conventional black and white radiography, the alteration of primary interest is that of intensity change. A colour radiograph is not intended to show an object's natural colour as might be perceived by reflected visible light.

Before discussing the techniques used to produce colour radiographs, a simplified description of the visual attributes of colour will be presented. The description should help to show the potential value of colour radiographs and perhaps justify developmental work in colour radiography. One way of diagrammatically showing the visual attributes of colour is that depicted in Fig. 1. One visual attribute of colour is its degree of brightness, and is indicated on the vertical axis in the diagram. Position on the axis is intended to represent various values of colour brightness perceived by the eye. It ranges from black at the lower end to white at the upper, with varying shades of grey in between. The information contained on a conventional black and white radiograph is restricted to this single dimension. In other words, when studying a conventional radiograph, only differences in grey may be interpreted as alterations of the X-ray beam. A second attribute of colour perception, and the one learned early in childhood, is difference in hue. It is simply the condition of the object which we perceive as either red, blue, green, etc. Hue is designated in the diagram by a particular angle of rotation made about the brightness axis. It is measured in some isobrightness plane

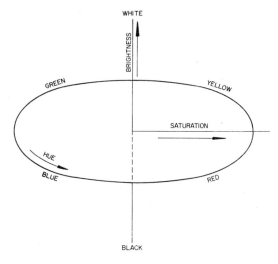

*Fig. 1. Visual attributes of colour.*

which is perpendicular to the brightness axis. A third attribute of colour perception is that of saturation. Saturation is that condition perceived as a redder red, a greener green, etc. It is presented on the diagram as a radial distance from the brightness axis. These three stimuli, brightness, hue and saturation should each contribute information about an object radiographed in colour. As mentioned earlier, conventional black and white radiographs display information only by brightness differences. The possibility that a colour radiograph is capable of containing more information due to the addition of the two parameters hue and saturation gives justification to the study of the production of colour radiographs.

Several different methods of producing colour radiographs have been attempted which have met with some success. The methods involving multiple exposures are interesting. Donovan produced colour radiographs by simultaneously projecting three films of an object through blue, green and red filters. He thus built-up a composite radiograph in colour. The three films were exposed at low, medium and high X-ray tube voltages. He was attempting to develop a method analogous to that of using different colours or wave-lengths of visible light to produce a composite colour photograph. The analogy, however, was not a good one, since the greater penetrability of the high kilovoltage exposure made it responsible to some extent, for every tone on the film. Another type of multiple exposure technique has been done by Bryce. It con-

sists of first preparing both negative and positive plates from a conventional black and white radiograph. These are then enlarged in succession on a single piece of photographic colour printing paper using different coloured filters. In this fashion, he prepared a two colour print. By using one of the plates twice for two different colours, a third colour was added. He reports that exact registration of the images was difficult, but that the result was not unpleasant.

It is interesting to note that television equipment has been used to produce colour radiographs. One technique reported by Fisher and Gershon-Cohen involved the scanning of a conventional black and white radiograph with a flying spot scanner. Light transmitted through the film was converted into electrical voltages using a photomultiplier tube. Depending on the amplitude of these voltages, they were each, in turn, applied to either the red, green or blue control electrodes of a three-gun colour television picture tube and a colour image of a radiograph was produced.

One of the most successful methods of producing colour radiographs having good hue differences in-

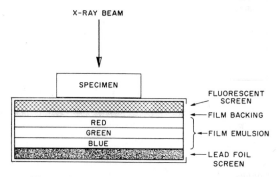

*Fig. 2. An exposure technique for producing a colour radiograph.*

volves the use of conventional multilayer colour film. The technique to be described below is that used by Beyer. He has adapted portions of the methods reported by Bryce and Blais. Figure 2 is a diagram of one arrangement used to expose the colour film. A multilayer colour film having the blue, green and red sensitive layers is shown sandwiched between a fluorescent intensifying screen and a lead screen. The relative positions of the screens shown in the figure may be reversed, but the positions shown are preferred. Conventional radiographic fine detail calcium tungstate screens have been used primarily; however, screens of other phosphors are under study by Beyer. After the film has been exposed to the beam of radiation, it is processed by techniques prescribed by the manufacturer except for one additional step. The additional step amounts to an interruption of the processing after partial development in the first developer. At this point, the film is exposed to coloured light and then normal processing is completed. Exposures made with either red or green lights have been the most satisfactory. It is believed that the coloured light exposure prints on one of the lower layers an image of the relatively heavy silver image which has been produced in the outer emulsion (blue sensitive) after partial development. The combination of the various hues and degrees of brightness and saturation produced in each layer by the screen fluorescence, X-ray beam, and coloured light exposure combine to form the resultant differences on the colour radiograph.

Figure 3 is a diagrammatic representation of a colour radiograph of four uranium foils which was made on multilayer film, as described above. The square image was from a foil 0·001″ thick and the circular

COLOR RADIOGRAPH OF STACKED
URANIUM FOILS

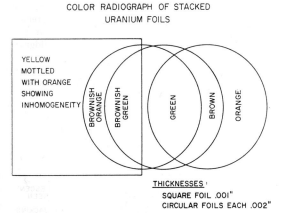

*Fig. 3. Colour radiograph of stacked uranium foils.*

images were each from foils 0·002″ thick. By a staggered piling arrangement, images of 0·001″, 0·002″, 0·003″, 0·004″, 0·005″ and 0·006″ thicknesses were obtained. The large hue difference between 0·001″ and 0·003″ thickness is interesting to note, and the hue difference between 0·002″ and 0·003″ is also significant. The radiograph was made by exposing the film in the type of cassette arrangement shown in Fig. 2. For this case, a conventional medium speed calcium tungstate fluorescent screen was used. During the processing of this exposed film, the first development was interrupted approximately midway and the film was flashed to green light. Typical of this technique is the predominence of the hue of the coloured light flash exposure in areas of low X-ray exposure. The same technique was used to produce colour radiographs of an aluminium step wedge and a 1/2″ aluminium weld. The 1/4″, 1/2″ and 3/4″ steps were displayed as yellow, tan and dark brown respectively. A 0·001″ tungsten wire placed on the wedge was clearly visible as a tan hue. Lack of penetration in the weld appeared as a yellow hue and tungsten inclusions were dark brown tending toward green, but the definition

was not as good as can be attained with a top quality black and white radiograph. Two per cent radiographic sensitivity was displayed on the weld plate image by using a standard ASME type penetrameter having a specified thickness of 0·010″ (i.e. 2% of 0·500″), and holes of 0·020″, 0·030″ and 0·040″ diameter.

A slightly different technique than described above was used by Beyer to produce colour radiographs of cracks. Compounds which have a predominant hue in their fluorescence were mixed with conventional penetrating solutions and the mixture applied to cracked surfaces. Phosphors with acceptable characteristics (i.e. small and uniform grain size, moderate persistence, etc.) and which were readily available were found to be those produced for colour television tubes. Using a P-22 phosphor (red zinc phosphate), cracks were penetrated and filled on one side of the test specimen. This side of the specimen was then placed in direct contact with the emulsion of the colour film. This was, of course, done in a darkroom environment and placed in a light-tight envelope for exposure to the X-ray beam. The film was processed in the conventional manner. There was no flashing to coloured light. The cracks had a red appearance on the resulting radiograph. The solid portions of the specimen were dark blue and other cracks (interior or opposite surface) were lighter blue. By penetrating the cracks on the opposite surface with a yellow fluorescing phosphor such as P-7 (Zn Cd S : Cu), a second radiograph was produced which displayed its surface cracks as a yellow hue. This procedure is shown diagrammatically in Fig. 4. A black and white radiograph displays all the cracks as grey images and does not lend itself to easy differentiation of the surface upon which they are located. Crack location is simpler when a colour radiograph of the type described above is used. A surface crack that becomes an interior crack can be located by noting a crack image that changes from surface hue (for example, red) to the light blue colour of an interior crack. Of course, purely interior cracks can also be located easily by their hue difference.

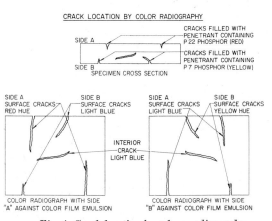

*Fig. 4. Crack location by colour radiography.*

In most cases, the production of colour radiographs is difficult and time consuming. At the present stage of development, the usefulness of colour radiographs produced by any of the above methods is limited when compared to good quality black and white radiographs. However, if developmental work continues, it may be possible to realize at least some of the potential value of colour radiography.

*Bibliography*

BEYER N. S. (1961) *Argonne National Laboratory Report* 6515, 203.

BLAIS J. M. and SCHWERIN A. K. (1955) *Radiography* **21**, 254.

BRYCE A. (1955) *Brit. J. Radiol.* **28**, 552.

DONOVAN G. E. (1951) *Lancet* **1**, 882.

FISHER J. F. and GERSHON-COHEN J. (1958) *Amer. J. Roentgenol.* **79**, 342.

N. S. BEYER

**COLOUR TELEVISION.** The instantaneous transmission of scenes or moving pictures by electrical means, providing a rendition of colour as well as brightness differences.

*Colour television systems.* All colour television systems proposed so far rely on the fact that any colour can be reproduced or closely approximated by the addition of light of three suitably chosen primary colours (red, green, and blue). The addition may take place, e.g. by the projection in superposition of the three lights on a white reflecting screen, by the emission of light of the three colours from contiguous luminous elements so small that they are not individually resolved by the human eye or by the successive presentation of the three lights to the eye within a period smaller than the time of persistence of vision. All these techniques have found application in colour reproduction in television.

The colour television system which is simplest in concept is the so-called field-sequential system (Fig. 1). In principle, it can use a standard black-and-white television chain provided that the frame and line frequencies (for equal resolution and freedom from flicker) are increased by a factor of three. The only additions required for conversion to colour are colour filter disks synchronized with the field deflexion which rotate in front of the camera tube target and the viewing tube screen. In effect, the television system transmits the red component of the scene, the green component, and the blue component in temporal succession and these are fused by the viewer through persistence of vision.

The simple system described, which formed the basis of regulations governing colour broadcasting in the United States for a brief period (1950–53), had two principal drawbacks. These are (1) that, for the broadcast bandwidth of a standard black-and-white television channel, it provides a degraded picture, and (2) that it is incompatible in the sense that field-sequential colour broadcasts cannot be utilized for monochrome reception by unmodified black-and-white television receivers. A colour television system which overcame both difficulties, i.e. could utilize standard monochrome channels without appreciable sacrifice of resolution and freedom from flicker and was fully compatible, was pioneered by the Radio Corporation of America. Optimal standards for such a system were worked out and recommended for adoption to the U.S. Federal Communications Commission by both RCA and the National Television System Committee (NTSC) of the American Radio Industry in 1953. These standards now form the basis of colour television broadcasting in the United States and Japan. The NTSC System, modified for operation with 625 lines/sec and 50 fields/sec, and two systems (SECAM and PAL) which differ from the NTSC System in certain respects have been considered (April 1965) as the basis of a unified colour television system in the European Broadcasting Area.

The NTSC System accomplishes its objective of transmitting the colour information along with the brightness information in the picture within a bandwidth normally required for transmitting the brightness information alone by an adaptation of the system to the limitations of the human eye. Specifically, the system utilizes the following properties of the eye:

1. The resolution of the eye for variations in chrominance (i.e. different colours of equal brightness or luminance) is less than for variations in luminance.
2. The resolution of the eye for colour variations in the range from green to its complementary colour purple is less than that for colour variation in the range from orange to its complementary colour cyan.
3. The threshold for flicker perception for chrominance differences lies at lower frequencies than for luminance differences.
4. The threshold for flicker perception for small-area luminance (or chrominance) variations lies at lower frequencies than for large-area variations.

In the NTSC System (Fig. 2) the transmitted colour picture signal is the sum of a monochrome signal with a frequency spread equal to that of a black-and-white television picture signal and a chrominance signal carrying the colour information in the picture. Both are usually derived from three camera signals $E_R$,

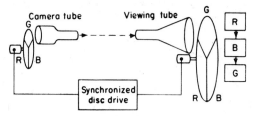

*Fig. 1. Field-sequential colour television system.*

$E_G$, $E_B$, the spectral sensitivity and optical filters in the camera channels being chosen so that intensities $E_R$, $E_G$, $E_B$ of the red, green, and blue primary sources in the display reproduce, in combination, the colour and relative brightness in the original image. The display primaries are assumed to have the CIE colour coordinates

$$\text{R} \quad x = 0\cdot67 \quad y = 0\cdot33$$
$$\text{G} \quad x = 0\cdot21 \quad y = 0\cdot71$$
$$\text{B} \quad x = 0\cdot14 \quad y = 0\cdot08.$$

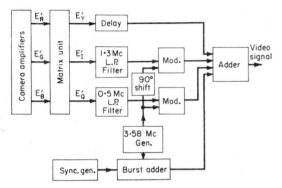

Fig. 2. Generation of NTSC video signal.

The monochrome signal then has the form

$$E_Y = 0\cdot30E_R + 0\cdot59E_G + 0\cdot11E_B.$$

The coefficients represent the relative luminous efficiencies of the display primaries when the relative scales for measuring $E_R$, $E_G$, $E_B$ have been chosen so that equal values of these three quantities reproduce so-called C-illuminant white, with the CIE coordinates

$$x = 0\cdot310 \quad y = 0\cdot316.$$

The indicated monochrome signal is a measure of the luminance distribution of the picture. In practice the signals $E_R$, $E_G$, $E_B$ are predistorted to compensate for the non-linear relation between beam current and signal in the reproducer, and these signals (indicated by primes) are employed to generate the monochrome signal $E'_Y = 0\cdot30\ E'_R + 0\cdot59\ E'_G + 0\cdot11\ E'_B$. This practice is advantageous in minimizing the visibility of noise in the picture and simplifies receiver construction. The assumed contrast gain or $\gamma$ of the picture reproducer is $2\cdot2$.

The chrominance signal is a subcarrier with a frequency which is an odd integer multiple of half the line frequency modulated in phase and amplitude by the colour information. More specifically, the chrominance signal $S$ is given by

$$S = E'_Q \sin(\omega_s t + 33°) + E'_I \cos(\omega_s t + 33°)$$

where

$$E'_Q = 0\cdot41(E'_B - E'_Y) + 0\cdot48(E'_R - E'_Y)$$
$$E'_I = -0\cdot27(E'_B - E'_Y) + 0\cdot74(E'_R - E'_Y)$$

and $\omega_s/2\pi$ is the frequency of the colour subcarrier, equal to $445/2$ times the line frequency or approximately $3\cdot58$ mc/s in the system adopted in the United States and Japan. $E_Q$ represents the departure of the colour of the picture element being scanned from white, measured along the green-purple axis on the colour triangle, $E_I$ that measured along the orange-cyan axis perpendicular thereto. The amplitude of the chrominance signal $S$ is a measure of the purity of the colour (i.e. the closeness with which it approaches the nearest spectrally pure colour), while the phase determines the hue or dominant wave-length of the colour. Before modulation of the subcarrier, the $E'_Q$ signal is passed through a low-pass filter attenuating it beyond $500$ kc/s and the $E'_I$ signal is passed through a filter attenuating it beyond $1\cdot3$ mc/s. Finally, the monochrome and chrominance signals and synchronizing pulses are added and applied to the transmitter modulator. The synchronizing signal itself is modified by the addition of a "colour burst" consisting of about 9 cycles of the colour subcarrier, which is utilized in the receiver to establish a reference phase for the detection of the colour information. The composite transmitted video signal itself is attenuated beyond $4\cdot2$ mc/s, so that the higher-frequency components of the orange-cyan signal are transmitted by single sideband (Fig. 3).

In the colour receiver (Fig. 4), the detected video signal is applied to the synchronization separator, which also serves to channel the colour burst to the local oscillator for colour demodulation; to a bandpass filter separating out the chrominance signal; and to the monochrome video amplifier and delay circuit. The local oscillator operating at subcarrier frequency is phase-controlled by the colour burst signals and its output is applied to the Q and I demodulators in the phase appropriate to recover the signals $E'_Q$ and $E'_I$,

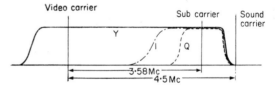

Fig. 3. Distribution of chrominance signal components.

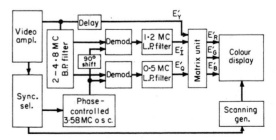

Fig. 4. Schematic diagram of colour television receiver for NTSC signal.

respectively. These signals are passed through 500-kc/s and 1·2-mc/s low-pass filters and combined, finally, with the monochrome signal in a matrix network. This network derives the three colour signals $E'_R$, $E'_G$, $E'_B$ which are utilized in the colour display device for picture reproduction.

The standards for the chrominance signals have been prescribed so as to minimize spurious effects in the reproduced picture arising from interference between the monochrome and chrominance channels. Thus, for a stationary picture, the luminance signal can be expressed as a Fourier series with frequency components which are integer multiples of the frame frequency. The colour subcarrier, on the other hand, has a frequency equal to a half-integer multiple of the line frequency and, hence, of the frame frequency, since the line frequency is an odd multiple of the frame frequency. The modulated colour carrier, or chrominance signal, similarly has only terms with frequencies which are a half-multiple of the frame frequency. Accordingly, the chrominance signal reverses sign in successive frames. Consequently the chrominance signal added to the luminance signal superposes, in a black-and-white display as well as in a colour display, a fine grained pattern (corresponding to frequency components in excess of 2 mc/s) which reverses polarity in successive frames and would be completely invisible (in view of the low threshold frequency of the eye for small-area flicker) except for receiver non-linearities. In similar fashion, high-frequency luminance components, demodulated in the colour demodulator along with the chrominance signal, give rise to changes in hue which reverse in sign in successive frames and tend to balance out. Again receiver non-linearity makes resulting colour errors (e.g. at edges in the picture) more readily visible. Full colour reproduction is provided only for relatively coarse detail, corresponding to picture signal frequencies up to 500 kc/s, where both sidebands of the chrominance signal are transmitted. For finer detail, (up to 1·2 mc/s) the colour variation is reproduced as a variation along the orange-cyan axis; this is justified by the fact that, in this range of detail, any colour variations can be adequately matched by variations along the orange-cyan axis. Single-side-band transmission results in the generation of spurious components in the Q-channel, which are, however, eliminated by the low-pass filter in this channel. The finest detail is reproduced simply as its variation in luminance, which alone is perceived by the eye. It is thus evident that the NTSC System permits:

1. The reproduction of black-and-white detail from a colour transmission on a colour or black-and-white receiver without any reduction in resolution and without the superposition of spurious signals.

2. The reproduction of coloured portions of the picture in true colour on a colour receiver and in black-and-white on a black-and-white receiver. In both instances chrominance-luminance channel interference results here in spurious patterns of minimal visibility.

3. The reproduction of pictures from monochrome transmissions in black-and-white on a colour receiver; the chrominance channel is automatically inactivated in this case.

An alternative to the NTSC System proposed recently by the Compagnie Francaise de Television is the SECAM (Sequentiel Avec Memoire) system. It dispenses with the necessity for synchronous detection of the colour subcarrier by modulating it by a single signal, namely $E'_R - E'_y$ or $E'_B - E'_y$ in alternate lines. Delay lines and switching circuits in the receiver make each of these signals serve for two successive lines. The combination of the colour difference signals with the monochrome signal to generate the three colour signals to be applied to the colour display is similar to that in a NTSC system receiver. The primary advantage claimed for the SECAM system is greater immunity to multipath interference which, in the NTSC system, might distort the colour burst and, consequently, reduce colour fidelity through shifts in the reference phase. The employment of frequency modulation of the chrominance subcarrier may reduce, furthermore, the influence of amplitude distortion in the transmission channel.

Still another system, PAL (Phase Alternation Line), proposed still more recently by Telefunken in West Germany, differs from the NTSC System in reversing the polarity of the I-signal in successive lines. In a simpler version of the PAL receiver, a synchronized switching system reverses once more the polarity of the I-signal in alternate lines, restoring the original signal. Errors in the reference phase produce here errors in hue which deviate from the correct hue in opposite direction in successive lines and thus tend to cancel out. In a more elaborate version of the PAL receiver, the chrominance signal is applied to a 64-microsecond (= one line period) delay line and the output of the delay line is added to the original signal to yield the Q-signal and subtracted from it to yield the I-signal. Particularly high immunity to interference is claimed for a receiver of this type. The selection of the subcarrier in the PAL system also differs slightly from that in the NTSC system and is designed to minimize the visibility of residual spurious patterns in the system.

Both SECAM and PAL share with the NTSC system compatibility with monochrome television and a bandwidth requirement comparable to that of monochrome television. All can provide excellent pictures under favourable conditions of transmission. The relative acceptability of the three systems depends largely on relative immunity to various types of interference and defects in the transmission channel on the one hand and relative economy of receiver construction on the other. Of the three systems, only the NTSC System has been tested by extensive practical experience.

*Colour television cameras.* Colour television cameras have in the past utilized, in general, three camera tubes which, individually, generate the picture signals for the red, green and blue component images. For live pickup, the camera tubes have in general been image orthicons; for film pickup, where no limitation is placed on picture illumination, vidicons. Light from the camera objective is split up by *dichroic mirrors* into its red, green, and blue components and directed onto the three camera tube targets. Commonly a relay lens, imaging the image plane of the objective onto the camera tube targets, is employed to provide the optical distance necessary for accommodating the dichroic mirror system.

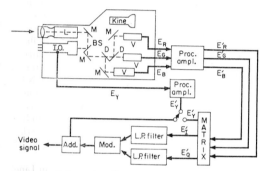

*Fig. 5. Schematic diagram of colour television camera with separate luminance channel. L: zoom lens (focal length range 4–100 cm). M: mirror. BS: neutral beam splitter. I.O.: image orthicon. D: dichroic mirror. V: vidicon.*

Good resolution is obtained with cameras of this type only if the three optical images and the corresponding scanning patterns in the three tubes are accurately registered and identical in the three channels. Refinements in optical, mechanical, and electrical design have achieved the realization and maintenance of this condition over long periods of time. However, the demands on precision of adjustment can't be materially reduced if the monochrome signal carrying the luminance information is obtained from a single camera tube, which may be, e.g. a 4–1/2 inch image orthicon tube (Fig. 5). The chrominance signal is then derived from three 1-inch vidicon tubes, for which the registry requirements become materially less stringent than in the three-tube camera, since the chrominance signal merely serves to "tint" the high-definition image provided by the image orthicon. In the camera sketched in Fig. 5, a Rank-Taylor Hobson Varotal III zoom lens provides a wide range of focal lengths and a long back focus to accommodate the beam-splitting system.

*Colour television displays.* In the simplest colour television display, the three signals are applied to individual cathode-ray tubes or kinescopes with red, green, and blue phosphor screens, respectively, and

the three images are then superposed optically. This technique is employed for large-screen projection colour television. The individual images are projected on the viewing screen, e.g. by spherical mirrors with a Schmidt correction plate to compensate the spherical aberration of the mirror. Home receivers, on the other hand, quite generally employ a single viewing tube. The screen here consists of a mosaic of red, green, and blue-emitting phosphor elements so small in area that they are not individually resolved.

In the shadow-mask colour kinescope (Fig. 6) employed exclusively in home colour television sets manufactured in the United States as well as in most sets in Japan—the two areas of the world where colour is broadcast regularly—there are three electron guns, modulated by the red, green, and blue picture signals. The beams emitted by the three guns converge toward a common spot on the screen and fall on a "shadow mask" mounted a small distance from the screen. In a 21″ tube, the shadow-mask is perforated by over 350,000 apertures in a hexagonal array. Accurately registered with every one of the shadow-mask apertures, there is a trio of red, green, and blue-emitting phosphor dots so oriented that the portion of the "red" beam passing through the mask apertures strikes only the red phosphor dots, the "green" beam the green phosphor dots, etc.

In the shadow-mask kinescope, a major portion of the beam electrons strikes the mask and hence does not contribute to the generation of the picture of the screen. In post-deflexion focused (PDF) colour tubes (commonly called Lawrence tubes after the American physicist E. O. Lawrence), this loss is largely avoided by replacing the shadow-mask by a grill of fine wires. A potential difference 2 to 3 times as great as the accelerating potential of the electrons arriving at the grill is applied between the grill and the screen

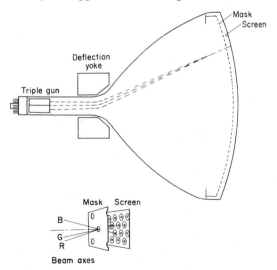

*Fig. 6. Schematic diagram of RCA shadow-mask colour viewing tube.*

This focuses the electrons incident on the space between two grill wires into a narrow line on the screen,
midway between the shadow projection of the wires,
and, incidentally, results in contrast loss from secondary-electron bombardment of the screen. The screen
itself consists of an array of red, green, and blue
phosphor lines. A tube with post-deflexion focusing
may be employed with three guns, like the shadow-
mask kinescope, or with a single gun. In the latter
case, alternate wires of the grill are connected together
and a deflexion voltage with the frequency of the
colour subcarrier (3·58 mc/s) is applied between
them to displace the electron line from the central phosphor line of a trio to the phosphor lines on either side
of it. Such single-gun operation with limited duty
cycle reduces the picture brightness by a large factor.
The output signal of the receiver is, of course, adapted
to this "element-sequential" mode of operation.

Numerous other colour display devices have been
proposed and constructed experimentally. In some

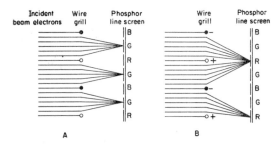

*Fig. 7. Focusing and colour selection in the
single-beam post-deflexion focusing tube. A: grill
wires at uniform potential. B: alternate grill wires
at different potential. In both cases average potential difference between grill wires and cathode is
$\frac{1}{4}$ potential difference between screen and cathode.*

of these, employing a phosphor line screen like the
post-deflexion focused tube described above, the
mask or grill is omitted altogether and indexing strips
provide secondary-emission or ultra-violet signals in
response to beam bombardment. These signals,
through feedback circuits, control the beam deflexion
so as to maintain it in phase with the colour signal.
An even more radical departure from the usual display
is represented by the "banana tube" system. Here,
horizontal scan is provided along a single trio of red,
green, and blue phosphor strips. This is viewed with
an appropriately shaped mirror through a rotating
cylindrical-lens drum so as to stretch the line trio
vertically into a picture with the customary 3:4
aspect ratio. Additional experimental display systems
are described in one of the references (Zworykin and
Morton (1954)).

*Bibliography*

CARNT P. S. and TOWNSEND G. B. (1961) *Colour Tele-
vision*, London: Illiffe.
Colour Television (1963) *Industrial Electronics* (G.B.)
**1**, 637.
SCHAGEN P. (1961) *Proc. Inst. El. Engrs.* **108 B**, 1.
ZWORYKIN V. K. and MORTON G. A. (1954) *Television*,
New York: Wiley.

V. K. ZWORYKIN and E. G. RAMBERG

**COMMUNICATION-SATELLITE SYSTEMS.** *1. Introduction.* Communication satellite systems present
a fascinating complex of scientific, technological, economic and administrative problems involving nego-
tiation and agreement on the international plane, since
such systems by their very nature require the close
cooperation of many countries.

The need for international agreement arises, for
example, in the problem of finding frequency space
for communication-satellite systems. That part of
the radio spectrum—1000 to 10,000 Mc/s—technically
suitable for such systems is already widely used for
terrestrial radio services such as high-power radar,
tropospheric-scatter and radio-relay systems. It is
only by the shared use of certain frequency bands by
communication-satellite systems and low-power radio-
relay systems under suitable controlled conditions that
sufficient frequency space has been found on an internationally agreed basis as, for example, by the Extraordinary Administrative Radio Conference of the
International Telecommunication Union (I.T.U.) held
in Geneva, October 1963. Without such agreement the
extensive use of communication-satellite systems
would be virtually impossible.

Since communication-satellite systems will, with
submarine cable systems and inland radio-relay and
line systems, form part of the world communication
network, they should conform with internationally
agreed standards of transmission, e.g. in respect of the
quality of transmission and freedom from noise and
interference, if conversation between telephone subscribers is to be satisfactory. Conformity with internationally agreed standards is also necessary for the
transmission of telegraphy, facsimile and television
signals.

Agreement on the technical characteristics of
communication-satellite systems is clearly necessary
between countries participating directly in the use of
such systems by the establishment of earth-stations.
Certain of the characteristics, and the siting of earth
and radio-relay stations, must also be agreed by neighbouring countries whose radio services might otherwise
suffer interference. The technical characteristics of
communication-satellite systems, with other radio
systems, are a matter for study by the International
Radio Consultative Committee of the I.T.U. The
technical Recommendations and Reports of the
C.C.I.R. greatly facilitate the orderly and efficient use

of the radio spectrum and the establishment of agreed transmission standards.

In all these matters in which international agreements are necessary, considerable progress has been made during recent years.

Furthermore, the large amount of test data and useful experience gained since mid-1962 with the experimental communication-satellites, Telstar, Relay and Syncom launched by the U.S. National Aeronautics and Space Administration has provided a solid basis on which to build the design of future operational systems. In this work the British Post Office satellite communication earth-station at Goonhilly, Cornwall, the American Telephone & Telegraph Company station at Andover (Maine) and the French Post, Telephone & Telegraph Administration station at Pleumeur Bodou (Brittany) have played significant roles. These stations have since been joined by others near Fucino (Italy), Raisting (Federal Republic of Germany), Rio de Janiero (Brazil) and Tokyo (Japan).

Design studies for possible world-wide systems have been made by a number of organizations in the U.S.A. and Europe, and in the United Kingdom by the Post Office and the Ministry of Aviation. The technical factors on which such studies are based are surveyed below.

*2. Technical aspects.* The design of a communication-satellite system is determined in the first instance by the operational requirements and the performance objectives. These in turn influence the choice of orbit, type and numbers of satellites, frequencies and modulation methods and other technical characteristics. Ultimately, however, the choice between alternative design of systems meeting the specified requirements must be based on economic factors, including the costs of providing and maintaining the system, since commercial viability is essential for a civil system.

The operational requirements for a world-wide system are likely to call for large traffic capacity, e.g. a thousand or more telephone channels and perhaps a television channel, in each satellite. The total capacity should be capable of subdivision, e.g. into blocks of 24 or 120 telephone channels, to meet in a flexible manner the needs of both low- and high-capacity earth stations. The telephone channels should be available on a 24 hours per day basis, and be suitable for telegraphy, facsimile and data transmission as an alternative to speech.

The transmission performance objectives for multi-channel telephony and television signals transmitted via communication-satellite systems have been defined on a similar basis to those used for other long-distance transmission systems such as submarine repeatered cables, coaxial cables and radio-relay systems so that these different transmission media may be used interchangeably.

The possible orbits include circular and elliptic orbits in the equatorial plane, and in polar and inclined planes Fig. 1. The heights of greatest interest for

communication purposes range from about 5000 to 36,000 km, the corresponding orbital periods ranging from about 4 to 24 hr. The "synchronous" circular orbit at a height of 36,000 km and with a period of 24 hr is of special interest since, if such a satelliet moves from West to East in the equatorial plane, it appears to be stationary relative to an observer on Earth. This is the basis of the communication-satellite system first proposed by A. C. Clarke in 1945, in which he envisaged three such satellites providing world-wide coverage except for the polar regions, Fig. 2. The successful launching by N.A.S.A. of the synchronous satellite Syncom 2, Fig. 3, on the 26th July, 1963 is a remarkable demonstration of the validity of Clarke's proposals which were made

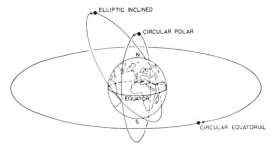

*Fig. 1. Typical communication satellite orbits.*

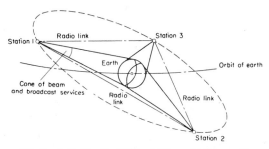

*Fig. 2. A. C. Clark's 1945 satellite system proposal.*

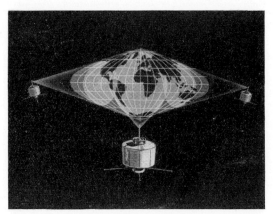

*Fig. 3. Syncom satellite.*

some twelve years before the first satellite was placed in relatively low orbit by the U.S.S.R. in October 1957. Syncom 2 is not, however, a stationary satellite, since the orbit is inclined to the equatorial plane.

Sub-synchronous satellites have periods corresponding to an integral fraction of 24 hr, e.g. 8 hr at 14,000 km height. At this height some 10 or 12 satellites would be required to provide world-wide coverage, except for the polar regions.

In general, orbits above some 10,000 km height are preferred since they need fewer satellites for a given coverage. However, more powerful launchers are required and the longer transmission delay from earth station to satellite and back to earth station may be a disadvantage for public telephony.

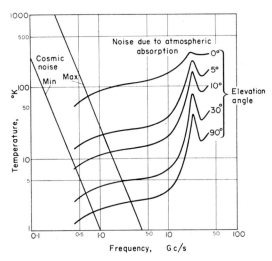

*Fig. 4. Aerial noise temperature due to cosmic noise and atmospheric absorption.*

The satellites may be either active, i.e. with receiving and transmitting equipment as in Telstar, Relay and Syncom, or passive reflectors of radio waves as in the case of the metallised-balloon satellite Echo. The concensus of opinion confirmed by the successful tests with Telstar and Relay is that active, rather than passive, satellites are required for a high-capacity system meeting the internationally agreed performance objectives. The satellites may be "random" or "station-keeping"; in the latter case, they are controlled so as to maintain the same relative positions as they move round the orbit, whereas in a random system the relative positions vary continuously. Random systems require large numbers of satellites and present major problems of control if used to provide world-wide networks. Satellites may also be "attitude-stabilised" in which case one axis of the satellites points continuously at the centre of the Earth, or "spin-stabilized", the direction of the spin axis remaining substantially fixed in space.

The optimum frequency range for satellite communications lies between about 1 and 10 Gc/s (i.e. 1000 and 10,000 Mc/s), although higher frequencies than 10 Gc/s are not excluded. The lower limit is set by cosmic noise, i.e. radio noise from the star galaxy, Fig. 4. The upper limit is determined by radio noise due to absorption in the Earth's atmosphere and rain noise, both of which increase with frequency. The choice of frequency range is critical in view of the very low levels of signals received from satellites and the consequent need to minimize noise as far as possible. By international agreement a number of frequency bands have been allocated to communication-satellite systems, nearly all of them on shared basis with line-of-sight radio-relay systems, e.g. 3700 to 4200 Mc/s (transmission from satellites) and 5925 to 6425 Mc/s (transmission from earth stations). Frequency sharing is only possible without mutual interference if limits are set to the power radiated by the transmitters in satellites, earth stations and radio-relay systems. The power limits and the co-ordination procedure to be employed in the siting of earth stations to avoid interference to and from radio-relay systems have also been agreed internationally. The frequency bands allocated to communication-satellite systems, the conditions under which they may be used, the co-ordination procedures to be used in the siting of earth stations and procedures for the international registration of frequencies to be used at earth stations and by satellites, are given in the "Final Acts of the Extraordinary Administrative Radio Conference to Allocate Frequency Bands for Space Radiocommunication Purposes (Geneva, 1963)".

As a result of the limitation on satellite transmitter power and the long transmission path, the signals received at earth stations may be only of the order of a micro-microwatt ($10^{-12}$ W), even when earth-station aerials some 85 ft (26 m) in diameter are employed, such as the one at the Post Office communication-satellite earth station at Goonhilly, Fig. 5. Thus, it is

*Fig. 5. Goonhilly aerial.*

necessary to use exceptionally low-noise receivers employing, for example, a liquid-helium cooled maser or cooled parametric pre-amplifiers to prevent the weak received signal from being swamped by noise. With such receivers overall system noise temperatures of the order of 50°K at the zenith can be achieved consistently.

The choice of modulation method must take into account a number of factors. For example, the low received signal levels may make it desirable to use modulation methods such as wide-deviation frequency modulation in which the wider bandwidth is used to obtain an improvement in the signal-to-noise ratio in the telephone and television channels. The tests with Telstar and Relay have demonstrated the practicability of this approach; however, alternative methods

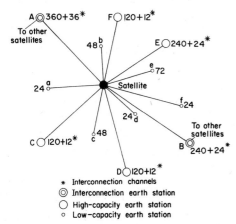

* Interconnection channels
◎ Interconnection earth station
○ High-capacity earth station
∘ Low-capacity earth station

*Fig. 6. Example of use of multiple-access satellite by high and low-capacity earth stations.*

such as pulse-code vestigial-sideband phase modulation and single-sideband amplitude modulation are also under consideration for multi-channel telephony. The modulation method should also permit "multi-station access" to satellites, i.e. enable several high- and low-capacity earth stations to communicate with each other via a satellite as shown for example in Fig. 6.

The design of the communication satellite itself poses a number of problems, in addition to those involved in the design of the communications package. These arise from the inevitable weight and size limitations and the constraints imposed by the space environment, and include the achievement of a satisfactory thermal balance and the provision of long-life power supplies, e.g. from solar cells. The success of the Telstar, Relay and Syncom satellites indicates that considerable progress has been made towards the solution of these problems, and in the precise achievement of the desired orbits.

*3. Synchronous and sub-synchronous orbit communication-satellite systems.* Both synchronous and sub-

synchronous orbits may find application for communication-satellite systems.

The synchronous equatorial orbit at 36,000 km height with its nominally stationary satellites considerably simplifies the tracking problem and enables coverage to be provided between about 75°N and 75°S latitude with only three satellites (or six if standby satellites are provided). However, the transmission delay of about 270 milliseconds (one-way) on each satellite link may present difficulties for telephone subscribers if very long inland circuits were connected to the satellite link, or two such links were connected in tandem (540 milliseconds delay) to provide circuits half way round the world. Thus synchronous orbit systems would appear to be best adapted to providing extensive 'regional' rather than full 'world-wide' coverage.

One possible approach to a sub-synchronous orbit system using 12 satellites in a circular equatorial orbit at 14,000 km height (8 hr orbit) is shown in the table.

*Equatorial Orbit Sub-Synchronous Satellite Communication System*

**Orbit**

| | |
|---|---|
| Type: | Circular, equatorial. |
| Height: | 8610 s. miles (14,000 km) |
| Period: | 8 hr (actual), 12 hr (apparent) |

**Satellites**

| | |
|---|---|
| Type: | Active, station-keeping, attitude-stabilized. |
| Number: | 12 |
| Capacity: | 1200 + 360 telephone channels (TP) One television channel (TV) |
| Access: | Up to 10 high-capacity ground stations (TP + TV) Up to 10 low-capacity ground stations (TP) |
| Power: | 3 × 2 W (telephony), 2 W (television) |
| Aerials: | 14 db gain, 43° beamwidth |

*Frequencies and Modulation*

| | |
|---|---|
| Ground-to Satellite: | 6000 Mc/s Band: FM/Multiple Carriers |
| Satellite-to-Ground: | 4000 Mc/s Band: Wide-Deviation FM |

*Earth Stations*

| | High-Capacity | Low-Capacity |
|---|---|---|
| Aerials: | 85 ft (25 m) | 30 ft (9 m) |
| Power: | n × 0·5 kW | m × 0·5 kW |
| Capacity: | n × 120 telephone channels. 1 television channel | m × 24 telephone channels. |

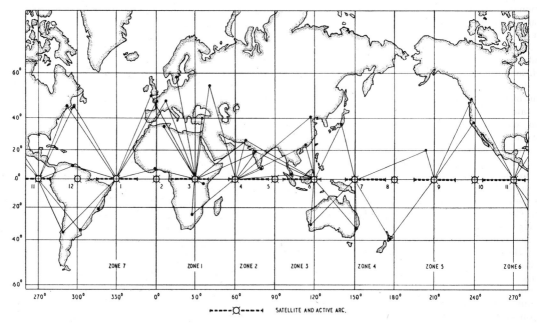

*Fig. 7. Representative links in a world-wide system using 12 satellites in an equatorial 14,000 km (8 hr) orbit.*

A sub-synchronous equatorial system at 14,000 km height could provide world-wide coverage between latitudes of 60°N and 60°S. All the satellites would follow the same track as seen from a given earth station and would re-appear at the same local times each day. The transmission delay on each satellite link would be about 120 milliseconds (one-way), thus permitting extension over long inland circuits and the connexion of two satellite links in tandem (240 milliseconds delay) to provide circuits half way round the world. However, additional aerials are needed at earth stations (compared with a synchronous orbit) to avoid breaks in transmission when switching from satellite-to-satellite.

It is to be noted that the use of satellite attitude-stabilization would permit relatively simple directional aerials to be employed on satellites, thereby reducing the satellite power supply and weight.

Figure 7 shows representative links in a world-wide communication-satellite system using 12 satellites in a sub-synchronous 14,000 km (8 hr) circular equatorial orbit. In this system the world is divided into seven overlapping zones; all the earth stations in a given zone use the same satellite at the same time as it traverses the "active arc" appropriate to that zone. Such a transit occupies one hour, a second satellite appearing at the beginning of the active arc as the first satellite leaves it, thus enabling continuity of transmission to be maintained.

An example of a satellite communication equipment, suitable for synchronous and sub-synchronous orbits, with facilities for multiple-access by high- and low-

capacity earth stations and providing high- to low-capacity interconnexion in the satellite, is shown in simplified schematic form in Fig. 8. The example shown is based on the use of transistorized microminiature baseband and intermediate-frequency circuits, and travelling-wave tubes for the output stages; alternatively solid-state varactor output stages could be used.

*4. Conclusion.* Much detailed study and experimental work, and international discussion—because the essence of international telecommunications is inter-

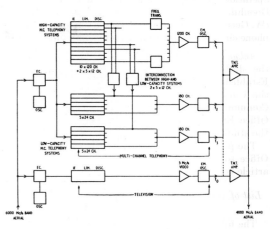

*Fig. 8. Simplified schematic of satellite communications equipment.*

national agreement—will be required before the design of a world-wide communication-satellite system can be established. A major step forward was made in August, 1964 when the U.S.A., U.K., Canada, Australia, Japan and seven countries of Europe (in addition to the U.K.) signed agreements providing for the design, development, construction and establishment of the initial phases of the "space segment", i.e. the satellites themselves and the ground stations for their control, of a global communication-satellite system as a cooperative international enterprise. The telecommunications earth stations are to be owned by the countries, or groups of countries, in which they are located. Control of the space segment will be exercised by an International Committee comprising representatives of the participating countries, the U.S. Communication Satellite Corporation acting as manager for the space system, pursuant to the general policies of the International Committee.

In this brief survey it has only been possible to outline some of the technical problems involved in the design of a system. Nevertheless, the encouraging results of tests using the Telstar, Relay and Syncom satellites provide a basis for the belief that communication-satellites are capable of providing a high-capacity flexible system which would be complementary to other intercontinental transmission systems such as submarine repeatered cables, in providing an improved and expanded world-wide communication network.

### The Early Bird Satellite

The communication satellite "Early Bird" (HS 303), designed and constructed by the Hughes Aircraft Company, was successfully launched by the U.S. National Aeronautics and Space Aministration for the Communication Satellite Corporation on 6th April, 1965. A synchronous orbit very close to the equatorial plane of the Earth was achieved, the satellite then being substantially stationary as seen from the Earth. Commercial operation via the earth stations at Andover (US), Goonhilly (UK), Pleumeur Bodou (France) Raisting (W. Germany) commenced in June 1965, some 240 telephone circuits being provided across the North Atlantic.

*Acknowledgements.* The author's thanks are due to the staff of the Space Department, Royal Aircraft Establishment, Ministry of Aviation, to members of the United Kingdom C.C.I.R. Study Group IV, the Commonwealth team, and his colleagues in the Post Office Engineering Department, who collaborated in the studies outlined in the article.

The permission of the Engineer-in-Chief of the Post Office to make use of information contained in this article is gratefully acknowledged.

### List of Definitions Relating to Communication-satellite Systems

The following definitions relating to communication-satellite systems were adopted in the 'Final Acts of the Extraordinary Administrative Radio Conference to Allocate Frequency Bands for Space Radiocommunication Purposes' (Geneva, 1963).

#### Space Systems, Services and Stations

*Space service.* A radiocommunication service:
—between earth stations and space stations,
—or between space stations,
—or between earth stations when the signals are retransmitted by space stations, or transmitted by reflection from objects in space, excluding reflection or scattering by the ionosphere or within the Earth's atmosphere.

*Earth station.* A station in the space service located either on the Earth's surface, including on board a ship, or on board an aircraft.

*Space station.* A station in the space service located on an object which is beyond, is intended to go beyond, or has been beyond, the major portion of the Earth's atmosphere.

*Space system.* Any group of co-operating earth and space stations, providing a given space service and which, in certain cases, may use objects in space for the reflection of the radiocommunication signals.

*Communication-satellite service.* A space service:
—between Earth stations, when using active or passive satellites for the exchange of communications of the fixed or mobile service, or
—between an earth station and stations on active satellites for the exchange of communications of the mobile service, with a view to their re-transmission to or from stations in the mobile service.

*Communication-satellite earth station.* An earth station in the communication-satellite service.

*Communication-satellite space station.* A space station in the communication-satellite service, on an Earth satellite.

*Active satellite.* An Earth satellite carrying a station intended to transmit or re-transmit radiocommunication signals.

*Passive satellite.* An Earth satellite intended to transmit radiocommunication signals by reflection.

*Satellite system.* Any group of co-operating stations providing a given space service and including one or more active or passive satellites.

#### Space, Orbits and Types of Objects in Space

*Orbit.* The path in space described by the centre of mass of a satellite or other object in space.

*Angle of inclination of an orbit.* The acute angle between the plane containing an orbit and the plane of the Earth's equator.

*Period of an object in space.* The time elapsing between two consecutive passages of an object in space through the same point on its closed orbit.

*Altitude of the apogee.* Altitude above the surface of the Earth of the point on a closed orbit where a satellite is at its maximum distance from the centre of the Earth.

*Altitude of the perigee.* Altitude above the surface of the Earth of the point on a closed orbit where a satellite is at its minimum distance from the centre of the Earth.

*Stationary satellite.* A satellite, the circular orbit of which lies in the plane of the Earth's equator and which turns about the polar axis of the Earth in the same direction and with the same period as those of the Earth's rotation.

*Bibliography*

BENTLEY R. M. (1964) *The Significance of Synchronous Satellites, New Scientist,* No. 386, 9th April.

BRAY W. J. (1964) *Technical Aspects of the Design of Communication-Satellite Systems, Proc. I.E.E.* **3**, No. 4, April.

BRAY W. J. and TAYLOR F. J. D. (1962) *The Post Office Satellite System Ground Station at Goonhilly Downs, British Communications & Electronics* **9**, No. 8, 574, August.

CLARKE A. C. (1945) *Extra-Terrestrial Relays, Wireless World* **51**, 305.

HUSBAND H. C. (1962) *The 85ft. Steerable Dish Aerial at Goonhilly Downs, British Communications and Electronics* **9**, No. 8, 584.

KIESLING J. D. (1964) *The NASA Relay I Experimental Communication Satellite, R.C.A. Review* **25**, No. 2, 232, June.

PIERCE J. R. (1955) *Orbital Radio-Relays, Jet Propulsion,* April.

PIERCE J. R. and KOMPFNER R. (1959) *Transoceanic Communication by Means of Satellites, Proc. Inst. Rad. Eng.* **47**, 372.

The Andover Station for Project Telstar (1962), *British Communication and Electronics* **9**, No. 8, 580.

The Telstar Experiment (1963) *Bell System Technical Journal* **42**, No. 4, Parts 1, 2 and 3, July.

W. J. BRAY

**COMMUNICATION THEORY.** The mathematical theory of communication shares with physics one basic concept, that of *entropy*. It was Boltzmann who first pointed out that entropy may be interpreted as "missing information" and it was R. V. L. Hartley, a communication engineer, who first drew explicit attention (in 1928) to the fact that the logarithm of the number of possible "states" of a communication channel provides a measure of its information-handling capacity. However, it was not until twenty years later that the theory was developed by C. E. Shannon to the point where the full significance of probabilities became apparent. Once this step had been taken, the mathematical ground was prepared for theorems of

the greatest significance to the engineer and of more than passing interest to the physicist.

The simplest example of communication to illustrate basic principles is a stream of symbols, each chosen from the same alphabet with certain probabilities for each symbol. For simplicity, assume that each choice is independent of all the others, so eliminating questions of transition probabilities. Let us, then, consider a language with four letters in the alphabet, having the following probabilities or statistical frequencies,

$$A \quad \tfrac{1}{2},$$
$$B \quad \tfrac{1}{4},$$
$$C \quad \tfrac{1}{8},$$
$$D \quad \tfrac{1}{8}.$$

A typical message in such a "language" might have the following appearance

ABBAAADABACCDAAB

Here there are only 16 symbols, but the strongest and simplest theorems apply to infinitely long streams in which statistical fluctuations are negligible. On the principle that the commonest symbols should be the easiest to write, let us now convert the above stream into code using the substitutions

$$A = 0$$
$$B = 10$$
$$C = 110$$
$$D = 111$$

giving

0101000011101001101101110010

The reader will easily verify that the original stream can be recovered uniquely in spite of the differing lengths of the encoded letters. Applying a further coding by taking pairs of digits, using (say) the substitutions

$$00 = A$$
$$01 = B$$
$$10 = C$$
$$11 = D$$

we obtain a new stream in the original alphabet:

BBAADCCBCDBDAC

A remarkable effect may now be observed. What originally took 16 letters to say has now been said in only 14 letters or 28 binary digits. The original stream was capable of being 'compressed' in such a way that, on average, each letter could be represented without loss of information by 1·75 binary digits. With this simple example, due to Shannon, it can be proved that no further compression is possible. As each of the original symbols carried only $1\tfrac{3}{4}$ "bits of information", not 2 as might at first have been supposed, we may say that one eighth of the original symbolism was redundant. The reader will see intuitively that, if the original probabilities had all been equal, no tricks of compression would succeed, and each letter would indeed have been worth 2 bits. Or, to take another simple case, if the probabilities of A and B had been each a half, and those of C and D each zero, then each letter would have been worth only one bit.

The above ideas can be mathematically formalized by associating with each selection a quantity of information

$$I = -\log_2 p \qquad \text{(bits)} \qquad (1)$$

where $p$ is the probability of that selection. This a more general way of saying that selecting one symbol out of $n$ equally probable symbols represents a quantity of information

$$I = \log_2 n \qquad \text{(bits)} \qquad (2)$$

Using formula (1), the average information content per symbol is the weighted mean

$$\bar{I} = - \sum_i p_i \log_2 p_i \qquad (3)$$

where $p_i$ is the probability of choosing the $i$th symbol of the set. Compare this with the formula for the entropy of a physical system,

$$S = -k \sum_i p_i \ln p_i \qquad (4)$$

where $k$ is Boltzmann's constant and $p_i$ represents the probability of finding the system in its $i$th state.

Logarithms to the base 2 are not, of course, compulsory in communication theory. They merely give answers in 'bits' where natural logarithms would give them in 'natural units' of information, the two measures differing only by a constant factor. The logarithms in this article are written 'log' when the choice of base is immaterial in such a sense.

A suggestive way of deriving (3) is to consider the limiting process

$$\bar{I} = \lim_{t \to \infty} \frac{\log n(t)}{t} \qquad (5)$$

for a stream of $t$ symbols, where $n(t)$ is the total number of sequences of length $t$. Then, ignoring fluctuations and considering all possible combinations in which symbols may occur, the application of Stirling's asymptotic formula for factorials quickly leads to (3). The process is exactly that used in statistical mechanics.

When comparing (3) and (4), one is at first faced with a seeming contradiction, which has in the past confused many scientists unnecessarily. It would appear that, units apart, the same formula is being used in physics as a measure of disorder, or ignorance, and in communication theory as a measure of positive information. But one should, as originally pointed out by D. K. C. MacDonald, describe (3) as a measure of *a priori* ignorance in terms of *a priori* probabilities. This ignorance is entirely removed once a definite message has been selected, and hence provides a measure of its information content. The final condition, when the message is known, is one of complete certainty and zero entropy. The *a posteriori* probabilities are all noughts or one, and would make every term in the entropy formula zero.

Communication channels are, in practice, subject to interference by random fluctuations which, in an otherwise perfect system, would be produced by thermal noise. The results in the foregoing paragraphs have been generalized by C. E. Shannon to deal with this more interesting situation. We now require the following probabilities for a noisy communication channel:

$p(i)$ = probability of $i$th symbol being transmitted,
$p(j)$ = probability of $j$th symbol being received,
$p_i(j)$ = conditional probability of $j$th symbol being received when the $i$th was transmitted,
$p_j(i)$ = conditional probability of $i$th symbol having been transmitted if the $j$th is received,
$p(i, j)$ = joint probability of $i$th symbol being transmitted and the $j$th received.

The information rate for a noisy channel may be arrived at, intuitively, by considering the effect of the arrival of one symbol at the receiving end. The ignorance removed is

$$
\begin{aligned}
I &= \text{(initial ignorance)} - \text{(final ignorance)} \\
&= -\log p(i) - (-\log p_j(i)) \\
&= \log \frac{p_j(i)}{p(i)} \qquad (6)
\end{aligned}
$$

Averaging over all $i$ and $j$ we obtain the average information communicated per symbol

$$\bar{I} = \sum_i \sum_j p(i, j) \log \frac{p_j(i)}{p(i)} \qquad (7)$$

By the laws of probability, we have

$$p(i, j) = p(i)\, p_i(j) = p(j)\, p_j(i) \qquad (8)$$

and using this, (7) may be written symmetrically in $i$ and $j$, thus

$$\bar{I} = \sum_i \sum_j p(i, j) \log \frac{p(i, j)}{p(i)\, p(j)}. \qquad (9)$$

This is a generalization of (3), reducing to it when $i$ is always equal to $j$, but reducing to zero if $i$ and $j$ are statistically independent. The information rate, $R$, for a communication channel is simply

$$R = m\bar{I} \qquad (10)$$

where $m$ is the number of symbols per unit time. All these formulae would merely be hollow definitions if it were not for a coding theorem, also due to Shannon, which shows that it is possible to use noisy channels for almost error-free communication. This is in spite of the fact that the individual transmitted symbols $i$ are not always equal to the received symbols $j$. Shannon proves that a source of (say) binary digits can be communicated with an arbitrarily small proportion of errors by suitable encoding into a sequence of the transmitted symbols and decoding from the resulting (imperfectly corresponding) sequence of received symbols, provided that the information rate of the source is less than that of the channel. The coding may be difficult, as it cannot generally be done digit by digit or symbol by symbol. Long sequences must be done at a time, and a delay is thereby introduced, but the rate $R$ can always be approached.

An important result for the engineer and applied physicist is obtained when (9) is generalized to continuous signals. The sums become integrals and the sampling theorem of waveform analysis is invoked to replace the notion of symbols per second by amplitudes or 'signal values' per second. It is well known that a frequency band of width $W$ can transmit $2W$ independent signal values per second. The information rate for a channel of bandwidth $W$ is then

$$R = 2W \iint p(x, y) \log \frac{p(x, y)}{p(x)\,p(y)} \, \mathrm{d}x \, \mathrm{d}y \qquad (11)$$

where $x$ and $y$ denote transmitted and received signal values. The units of $R$ are bits/sec if the logarithm is base 2. This expression can be simplified when the received signal has merely suffered the addition of a random fluctuation, i.e.

$$y = x + n, \qquad (12)$$

and the integral may readily be evaluated if $n$ is "white Gaussian", $p(x)$ is suitably chosen and $n$ is statistically independent of $x$. For best use of the channel, $p(x)$ should be chosen to maximize $R$, though unless constraints are introduced, $R$ would be infinite. With a constraint on mean signal power

$$P = \overline{x^2} \qquad (13)$$

it can be shown that $x$ (and hence $y$ also) is Gaussian. Writing the mean noise power as

$$N = \overline{n^2} \qquad (14)$$

the expression (11) yields the maximum information rate, or *channel capacity*

$$C = R_{\mathrm{max}} = W \log \left( 1 + \frac{P}{N} \right). \qquad (15)$$

This is probably the best known and most widely used result of the theory; it had been less rigorously anticipated by W. G. Tuller before his untimely death in an air accident.

If the noise is purely thermal, we have classically

$$N = W \cdot kT \qquad (16)$$

where $kT$ has its usual meaning. The expression (15) increases with $W$, and as $W$ tends to infinity, it asymptotically approaches the limit

$$C = P/kT \qquad (17)$$

natural units of information per unit time. It follows that, on average, a signal of energy $E$ cannot yield, in the presence of undular thermal noise, a quantity of information greater than $E/kT$ natural units.

The further elaborations and recent advances in the subject are of interest more to the engineer and mathematician than the physicist, being mainly concerned with coding. The source is generally not a quasi-random stream of binary digits and its statistics will not generally happen to correspond to those of the optimized channel. The order and decoder required for purposes of statistical matching would in general be quite complex electronic machines. A summary of fairly recent advances has been given in a book by R. M. Fano.

*Bibliography*

BRILLOUIN L. (1956) *Science and Information Theory*, New York: Academic Press.

FANO R. M. (1961) *Transmission of Information: a statistical Theory of Communication*, New York: Wiley.

SHANNON C. E. and WEAVER W. (1949) *The Mathematical Theory of Communication*, Urbana: Illinois University Press.

P. M. WOODWARD

**CONGESTION.** Physical systems having a random element sometimes behave temporarily as if they were overloaded, despite the fact that they are working within their capacity, because of the operation of chance. Thus in a queueing system with random service and/or input of demands, long queues may build up despite the fact that, on the average, demands are dealt with more quickly than they arrive. Congestion theory studies the magnitude, duration and other aspects of such overloads.

*See also:* Monte Carlo methods.

A. J. HOWIE

**CONTINENTAL DRIFT, INITIATION OF.** The idea that the continents were once part of one or more large landmasses probably goes back to Francis Bacon (1561–1626) and Alexander von Humboldt (1767–1835). Later, the German meteorologist Alfred Wegener made a careful study of many geological and climatological aspects of the continents and constructed a series of maps showing the possible distribution of land and sea in past geological eras. He first published his researches in 1912. Since this time, the theory that earlier continents broke up and drifted apart has largely been out of favour until the advent of palaeomagnetism. This has strongly supported continental drift indicating that the continents have moved relative to one another since the time rocks acquired their magnetization.

The early workers were impressed by the geometrical fit of the continents especially of those on either side of the Atlantic Ocean. Bullard and his co-workers have now made a careful study of the fit of the continents using an electronic computer. With a "least squares" criterion they find that the best fit is obtained

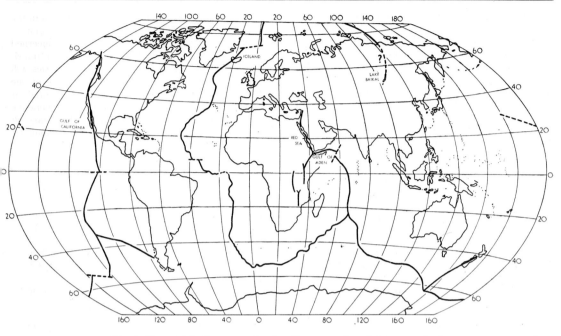

*Fig. 1. The world rift system showing the areas where it intersects land.*

for the 500 fathom (914 m) line near the continental edge. The mean square misfit for South America and Africa is found to be 0·8° when the recent features of the Niger Delta and Walvis Ridge are excluded. When North and South America are compared with Greenland, Europa and Africa the mean square misfit is only 2°. This is probably the best direct evidence for considering these continents were once part of a large single landmass.

The main phase of continental drift is generally considered to have started 200 to 300 million years ago, i.e. late in the history of the Earth (age of Earth $= 4·6 \times 10'$ yr). Various movements of crustal blocks (e.g. along the San Andreas fault of California of the order of 1 to 5 cm/yr) suggest that continental movement may be costinuing. So far, there have been no geodetic measurements which confirm that an actual separation of the continental crust is in progress today. The Red Sea and Gulf of California are two possible locations where this might be tried.

The seismicity of the Earth provides further evidence of present day crustal movements. In 1954, Professor Rothé noticed that the rift system of Africa including the Red Sea and Gulf of Aden is associated with shallow seismicity (earthquake focal depth $\leq 70$ km) and the narrow region of shallow seismicity continues into the Indian Ocean following the Carlsberg and mid-Indian ridges and around Africa to join with the mid-Atlantic rift. The inference is that the belt of shallow seismicity locates the presence of a rift or fracture zone. Two years later, Ewing and Heezen extended this idea and mapped the fracture zone for

the whole world. The location of this zone is shown in Fig. 1. It is seen that the rift zone, in general, is found in the ocean floor but occasionally intersects the continents. These regions are the East African rift system, the Gulf of California, and the Lake Baikal area of Siberia. Iceland lies astride the mid-Atlantic rift but it is unlikely that the Island is composed of continental crust. In these regions, there is considerable geophysical evidence for crystal separation and it is likely that the whole of the world rift system is a region of crystal extension with varying amounts of shear.

*1. The Red Sea and Gulf of Aden.* In recent years, it has confirmed that the crust beneath the centres of the Red Sea and Gulf of Aden is much closer to oceanic than continental. The extents of these regions of new oceanic crust are shown in Fig. 2.

This was suspected from the gravity anomalies. Von Triulzi (1898, 1901), Vening–Meinesz (1934) and Harrison (1957) have shown that the centre of the Red Sea is characterized by large positive Bouguer anomalies of 100 to 150 mgals. The anomaly has been interpreted by Girdler (1958) as being due to basic rocks intruded into a crack in the down faulted Arabian-Nubian Shield.

The centre of the Red Sea has an axial trough reaching depths of 2 km. Seismic refraction measurements by C. L. Drake in this area show the presence of high velocity (6·7 to 7·4 km/s) material a few kilometres below the sea floor. The velocities are appropriate for basic, intrusive rocks.

Total intensity magnetic field measurements have been made by C. L. Drake and by T. D. Allan. These show that the deep, axial trough is characterized by large magnetic anomalies of amplitude 1200 gammas (1 gamma = $10^{-5}$ oersted). Similar anomalies are often found over the mid-oceanic rifts.

The nature of the topography, the gravity and magnetic anomalies and the seismic refraction results

of the order of 100 mgal (Harrison and Mathier). The anomaly occurs over deep water and indicates that the deep part of the Gulf is underlain by denser rocks than on either side. Seismic refraction measurements for this area show the presence of a higher velocity layer of 6·3–6·7 km/s at shallow depths of 4–5 km below sea level. The Gulf is therefore another example of new oceanic crust occurring between continental crust. The extent of the area of oceanic crust is indicated in Fig. 3.

The San Andreas fault continues north from the Gulf of California and this has a measurable strike-slip motion. The motion is in the sense shown in Fig. 3 of several centimetres per year. It seems that the Baja Peninsula has rifted away from the mainland in a relatively northward direction resulting in the formation of a strip of new oceanic crust along the centre of the southern part of the Gulf of California.

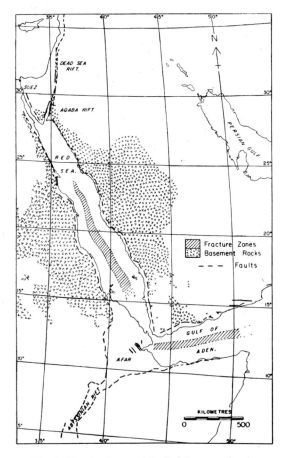

*Fig. 2. Structural map of the Red Sea area showing the extent of new oceanic crust (fracture zones) deduced from geophysical evidence.*

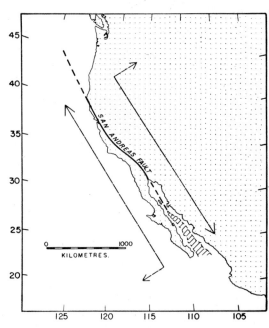

*Fig. 3. The Gulf of California showing the extent of new oceanic crust and its relation to the San Andreas fault.*

all indicate the presence of quasi-oceanic rather than continental crust beneath the centre of the Red Sea and suggest a recent breaking apart of the Afro-Arabian landmass.

*2. The Gulf of California.* A branch of the world rift system (Fig. 1) also runs into the Gulf of California via the seismically active East Pacific Rise. Geophysical results show that the structure of the Gulf is very similar to that of the Red Sea and Gulf of Aden.

South of latitude 28°N, the axis of the Gulf is associated with a positive Bouguer gravity anomaly

*3. The Lake Baikal area.* Heezen and Ewing (1961) have plotted the epicentres of shallow earthquakes in the Arctic Ocean and have shown that the mid-Atlantic rift continues across the Arctic to meet Siberia near the mouth of the River Lena. It is possible that the seismically active zone continues to Lake Baikal. Little information is available for this area but Beloussov has indicated that Lake Baikal may be somewhat similar in structure to the Red Sea.

*4. Iceland.* Although Iceland is not composed of continental crust, it does provide an example of crustal drift. The mid-Atlantic rift crosses Iceland from south-west to north-east (Fig. 1) and the central valley is observed to be growing wider at a rate of 3·6 m/1000 years for every kilometre of its width (Bernauer quoted by Heezen 1960). The expansion is thought to be due to the intrusion of basaltic material along a series of fissures. Båth (1960) made a series of seismic field refraction measurements and obtained a crustal structure for the centre of Iceland consisting of 2·1 km of material 3·7 km/s (lava and ash), 15·7 km of material 6·7 km/s and about 10 km of material 7·4 km/s. The latter two velocities are considered to be appropriate for two varieties of basalt. No velocity appropriate to granitic crust was found.

*Some general characteristics of the world rift system.* Geophysical survey work over the past decade has shown that the rifts have one or more of a number of characteristics. In the ocean floor the rift zone is associated with a broad rise or ridge. Several heat flow measurements have been made and values several times the normal value for the Earth of $1·1 \times 10^{-6}$ cals/cm²/s have been found along narrow belts associated with the rift zone or crest of the rise. Such high values have been found in the Gulf of California, along the crest of the East Pacific Rise, along the mid-Atlantic rift and in the centres of the Gulf of Aden and Red Sea. The rift zone is frequently but not always associated with gravity and magnetic anomalies and seismic velocities around 7 km/s are usually found at shallow depths. These velocities suggest material intermediate between the normal oceanic "basaltic" layer and that of the upper mantle. Olivine basaltic rocks have been found in several localities along the rift system. The high heat flow and seismicity of the rifts suggest that they are of recent origin and still in process of formation.

*Possible mechanisms for the origin of the world rift system and continental break-up.* The geophysical evidence taken together suggests that the rifts are widening and that they are due to tensional forces. There are two possible ways of explaining these world wide zones of extension—either the Earth is expanding or convective motion is occurring in the mantle. There is no sound physical reason for rejecting the former but the latter hypothesis has the advantage of explaining several other features of the Earth's surface. In particular, there are many compressional features and these are thought by many workers to be a consequence of *converging* mantle convection currents. These features include fold mountains (e.g. the Alps and Andes), the island arcs, deep sea trenches and andesitic volcanism. Further, fault plane solutions of earthquakes indicate that 70 per cent of all motions are transcurrent and it is easier to explain the predominance of shear with the convection hypothesis than with the expanding Earth hypothesis.

At present, it is common to consider mechanism whereby convection currents are rising under the

oceans and descending under the continents. In this way, an explanation is obtained for the mid-oceanic rifts and the drifting apart of the continents. The fact that the rift system intersects Afro-Arabia and western north America indicates that various stages of rift formation and continental drift may be observed. Figure 4 shows a possible mechanism for the early stages of rift formation and continental break-up. It seems likely that the Red Sea and Gulf of Aden and the Gulf of California may be consequences of new regions of uprising mantle convection.

Any theory concerning mantle convection must take into account that the pattern of convection must change with time. This is because relics of ancient fold mountains and rifts are found to be distributed differently in the geological past from today. Runcorn has suggested that the convection pattern changes as a consequence of the growth of the Earth's core and at

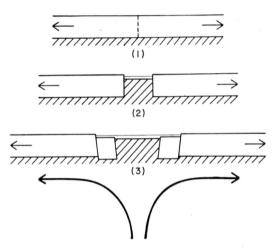

*Fig. 4. The formation of a rift structure such as the Red Sea with the formation of new oceanic crust, the crustal extension being a consequence of uprising, diverging mantle convection.*

each changeover, continental drift occurs. Chandrasekhar and Vening Meinesz have shown that the complexity of the convection is a function of the thickness of the mantle. Hence, for the present core radius of $0·55 R$ the convective motion excited at marginal stability shows very strong third, fourth and fifth order harmonics. The present day distribution of the seismically active rift zones tends to support a recent changeover from the fourth to fifth order mantle convection.

This has probably been the most successful attempt so far at an explanation of continental drift. Even if this theory proves to be incorrect, some theory involving a change in mantle convection pattern seems to be the most satisfactory way of explaining the initiation of continental drift.

*Bibliography*

Båth M. (1962) *Crustal Structure in Iceland & Surrounding Ocean*, ICSU Rev. 4, 127.

Bullard E. C. (1964) *Continental Drift, Quart, J. Geol. Soc. Lond.* 120, 1.

Drake C. L. and Girdler R. W. (1964) *A Geophysical Study of the Red Sea, Geophys. J. Roy. Astron. Soc.* (in press).

Girdler R. W. (1963) *Geophysical Studies of Rift Valleys, Phys. & Chem. Earth* 5, 122.

Hill M. N. (Ed.) (1963) *The Sea* 3, New York: Interscience.

Runcorn S. K. (Ed.) (1962) *Continental Drift*, New York: Academic Press.

Thewlis J. (Ed.) (1961) *Bouguer anomaly*, in *Encyclopaedic Dictionary of Physics*, Vol. 3, Oxford: Pergamon Press.

Van Andel and Shor G. C. (Ed.) (1964) *Marine Geology of the Gulf of California*, Mem. Amer. Assoc. Pet. Geol. No. 3.

Wegener A. (1915) *Die Entstehung der Kontinente und Ozeane*, Vieweg.

R. W. Girdler

**CORNER FRACTURE.** The term corner fracture is used in the explosive metal working field to define a fracture surface produced by the interaction of two or more waves of tensile stress. While fractures of this type appear in impulsively loaded metal systems having a wide range of configurations, the dynamics of the fracture process remains essentially the same.

The formation of corner fractures is illustrated in the figure which shows the cross-section of a long metal cylinder of square outer configuration internally loaded by means of a detonating explosive. When the

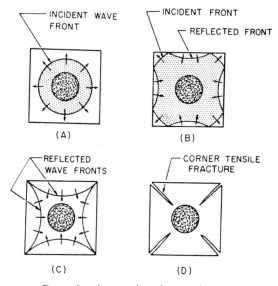

*Dynamics of generation of corner fractures.*

explosive charge is detonated inside the cylinder, a compressive, divergent, sharp-fronted longitudinal disturbance is created which propagates outward through the cylinder wall. When the front of this disturbance reaches the outer surface, release of tension waves begin to propagate back into the stressed region due to the reflection process. As the fronts of these release waves meet, high tensile stresses exist momentarily along the diagonals of the cylinder section. If these localized stress values are of sufficient magnitude, fracture surfaces (corner fractures) will be produced along the planes of stress wave interaction.

J. Pearson

**COSMIC NEUTRINOS AND THEIR DETECTION.** Although it was quite obvious from the basic work of Bethe and Critchfield on hydrogen reactions in stars that stellar neutrinos were produced abundantly in nature, only recently have attempts been made to detect them and other cosmic neutrinos. One of the obvious reasons for this is the smallness of the cross-section in neutrino interactions (the cross-section is $\sim 10^{-44}$ cm$^2$ at 1 meV neutrino energy). In lead, the mean free path of 1 meV neutrino is one light year ($\sim 10^{18}$ cm). Recent developments in the theoretical and experimental aspects of the weak interaction have made it possible to study cosmic neutrinos, including stellar neutrinos.

*Neutrino interactions.* The present theory of weak interactions originated from Feynman and Gell-Mann. They postulated that the weak interactions were caused by the interaction of a current $J$ with itself. The current has the following form:

$$J = \sqrt{g}[(\Psi_e \gamma_\alpha (1 + \gamma_5) \, \Psi_{\nu_e}) + (\Psi_p \gamma_\alpha (1 + \gamma_5) \, \Psi_n) +$$
$$+ (\Psi_\mu \gamma_\alpha (1 + \gamma_5) \, \Psi_{\nu_\mu}) + \text{strange particles}] \qquad (1)$$

where $\psi_A$ is the wave function for a particle $A$, $\gamma_\alpha (\alpha = 1, 2, 3, 4)$ are the Dirac Matrices and $\gamma_5 = i \gamma_0 \gamma_1 \gamma_2 \gamma_3$, $g$ is the weak interaction coupling constant and numerically $g m_p^2 = (1 \cdot 01 \pm 0 \cdot 01) \times 10^{-5}$ ($m_p$ is the mass of proton). The neutrino $\nu_e$ associated with electrons is distinguished from that associated with the $\mu$-meson ($\nu_\mu$). That $\nu_e \neq \nu_\mu$ has been demonstrated in a recent experiment by Danby *et al.*

We may abbreviate $J$ as follows:

$$J = (e\nu_e) + (pn) + (\mu\nu_\mu) + \cdots \qquad (2)$$

where $e$, $\nu_e$, ... now stand symbolically for particles or anti-particles. The weak interaction Hamiltonian is then given by

$$JJ^* = (e\nu_e)(\bar{e}\bar{\nu}_e) + (pn)(\bar{e}\bar{\nu}_e) + \cdots \qquad (3)$$

Each term in $JJ^*$ now gives rise to a reaction consistent with all conservation laws (charge, lepton number, energy, momentum, etc.). For example, some of the allowable reactions of the first two terms in equation 3 are

$$e\pm + \nu_e, \bar{\nu}_e \to e\pm + \nu_e, \bar{\nu}_e \quad \text{(Scattering of neutrinos by electrons)}$$

$$e^- + p \to n + \nu_e \quad \text{(electron capture)}$$

$$n \to p + e^- + \bar{\nu} \quad \text{(neutron decay)} \qquad (4)$$

At present an experiment at CERN (and a similar one at Brookhaven National Laboratory) is under way to test a new hypothesis in weak interactions (on the existence of an intermediate boson) from which the validity of the theory of Feynman and Gell-Mann could be established. Preliminary results support their theory.

*Neutrino production associated with stellar nuclear processes.* Neutrinos are produced through the $(e\nu_e)(pn)$ interaction of equation 3 in hydrogen reactions in which helium is built up. There are two energy production cycles, the proton–proton reaction is important in stars less massive than the Sun, and the carbon cycle is more important in the other case. In all cases the temperature required is around $1–3 \times 10^7 °K$. The proton–proton reaction chain is

$$p + p \to D + e^+ + \nu_e$$
$$D + p \to {}^3He + \gamma$$
$$^3He + {}^3He \to {}^4He + 2p. \qquad (5)$$

As helium is built up in the centre, the following bi-cycle may take place:

$$^3He + {}^4He \to {}^7Be + \gamma$$

$$^7Be + e^- \to {}^7Li + \nu_e \quad (E_\nu = 0.861 \text{ meV}, \quad 0.383 \text{ meV})$$
$$\text{Line Spectra}$$

$$^7Li + p \to {}^8Be + \gamma$$

$$^8Be \to 2{}^4He \qquad (6)$$

or:

$$^7Be + p \to {}^8B + \gamma$$

$$^8B \to {}^8Be + e^+ + \nu_e \quad (E_\nu{}^{(max)} = 14.06 \text{ meV})$$
$$\text{Continuous spectra}$$

$$^8Be \to 2{}^4He \qquad (7)$$

The last reaction is of particular interest to neutrino astronomers since these neutrinos have higher energy and should be easier to detect.

The carbon cycle is:

$$^{12}C + p \to {}^{13}N + \gamma$$

$$^{13}N \to {}^{13}C + e^+ + \nu_e \quad (E_\nu{}^{(max)} = 1.2 \text{ meV})$$
$$\text{Continuous Spectra}$$

$$^{13}C + p \to {}^{14}N + \gamma$$

$$^{14}N + p \to {}^{15}O + \gamma$$

$$^{15}O \to {}^{15}N + e^+ + \nu_e \quad (E_\nu{}^{(max)} = 1.74 \text{ meV})$$
$$\text{Continuous Spectra}$$

$$^{15}N + p \to {}^{12}C + {}^4He \qquad (8)$$

Note $^{12}C$ acts as a catalyst.

The total energy release in building up one helium nucleus from four protons is 26·7 meV, about 0·52 meV to 1·7 meV are in the form of neutrinos. Hence we can say generally in hydrogen burning stars about 2–8 per cent of energy radiated is in the form of neutrinos.

Table 1 lists the flux intensity expected on the Earth.

Very small amounts of neutrinos are produced in association with nuclear energy generations in later stages of stellar evolution. Almost all neutrinos produced at a later stage come from direct production processes.

*Direct neutrino production processes.* Of all astrophysical processes for neutrino production, the more

Table 1. Solar neutrino fluxes on Earth $(\nu/cm^2 sec)*$

| | $\nu_{pp}$ | $\nu_{Be}$ | | $\nu_B$ | $\nu_N$ | $\nu_O$ |
|---|---|---|---|---|---|---|
| Spectrum type | continuous | line | | continuous | continuous | continuous |
| Maximum energy (meV) | 0·42 | 0·383 12% | 0·861 88% | 14·06 | 1·20 | 1·74 |
| Mean energy (meV) | 0·26 | — | — | 7·25 | 0·710 | 1 |
| Flux | $(5·3 \pm 0·6) \times 10^{10}$ | $1·5 \times 10^9$ | $(1·2 \pm 0·5) \times 10^{10}$ | $(2·5 \pm 1) \times 10^7$ | $(1 \pm 0·5) \times 10^9$ | $(1 \pm 0·5) \times 10^9$ |

$\nu_{ppe}$ neutrino from $p + p + e^- \to D + \nu$, etc.
* Sears (1964).

important ones are:

URCA process

$$e^- + (Z, A) \rightarrow (Z-1, A) + \nu_e \quad \text{beta decay}$$
$$(Z-1, A) \rightarrow (Z, A) + e^- + \bar{\nu}_e \quad \text{interaction} \qquad (9)$$

Photo neutrino process

$$e^- + \gamma \rightarrow e^- + \bar{\nu}_e + \nu_e \qquad (10)$$

Annihilation process

$$\gamma + \gamma \rightleftharpoons e^- + e^+ \rightarrow \bar{\nu}_e + \nu_e \qquad (11)$$

$(e\nu_e)(e\nu_e)$ interaction

Plasma process

$$\gamma(\text{plasmon}) \rightarrow \bar{\nu}_e + \nu_e \qquad (12)$$

The Urca process is at present of historical importance, since it was the first one to be considered. In the Urca process beta decay and inverse beta decay reactions occur alternatively, depleting the kinetic energy of the electrons. This process requires a minimum electron energy roughly equal to the beta decay energy. Since the beta decay energy $E_b$ of most stable nuclei is of the order of a few meV, and the kinetic temperature of the electrons hardly exceeds a few tenths of meV, the Boltzmann factor $\exp(-E_b/kT)$, which is a measure of the fraction of the number of electrons with energy $>E_b$, is rather small and this process is not very important. Moreover, elements with small $E_b$ are not stable against photodisintegration at the temperature when the calculated energy conversion rate becomes large.

In the photoneutrino process the energy of a photon is converted into a pair of neutrinos. This is quite analogous to the Compton scattering process. In the annihilation process the photons are in equilibrium with electron pairs (at a temperature $\sim 6 \times 10^9\,°K$)

and the electron pairs can annihilate to form neutrinos via the $(e\nu_e)(e\nu_e)$ interaction.

Ordinarily a free photon cannot decay into two neutrinos because of the conservation laws. Inside an electron gas, because of interactions, photons are not free. The relation between the frequency $\omega$ of a photon and its wave number vector $\boldsymbol{k}$ is similar to that for a particle with a mass $\hbar\omega_0$.

$$\hbar^2\omega^2 = \hbar^2\omega_0^2 + k^2c^2$$

whereas

$$\omega_0^2 = \frac{4}{\hbar^3} e^2 p_F^3 \frac{1}{\pi E_F}$$

and $p_F$ is the Fermi momentum for electrons. $E_F$ is the Fermi energy including the rest energy $mc^2$. Hence inside an electron gas a photon can decay into two neutrinos. The plasma process is especially important in dense stars (such as white dwarfs and neutron stars).

Detailed calculation of the energy conversion rate on these processes have been performed by Chiu, Stabler, Ritus, Adams, Ruderman, and Woo. The detailed computation is quite complicated. We here list the asymptotic formulae for certain limiting cases.

*Nucleosynthesis beyond hydrogen burning. Supernova.* As hydrogen becomes exhausted at the centre of a star, a dense helium core develops and the star becomes a red giant, with a vast envelope. Because of gravitational contraction, the temperature at the centre increases, and at $T \sim 10^8\,°K$ helium begins to react via the $3\alpha \rightarrow {}^{12}C$ reaction.

At a temperature of around $4 \times 10^9\,°K$, nuclear reaction rates become quite rapid, as a consequence all elements come to statistical equilibrium and the most favoured element is ${}^{56}Fe$. At a temperature of

*Table 2. Summary of photo neutrino and annihilation energy loss rates:*
($\varrho$ in g/cm³, $T_n = T/10^n$)

| Temperature and density region: | Photo neutrino loss (ergs/g sec) | Pair annihilation loss (ergs/g sec) | Plasma process (ergs/g sec) |
|---|---|---|---|
| Non-relativistic, non-degenerate $\epsilon_F \lesssim kT \ll mc^2$ | $\dfrac{10^8}{\mu_e} T_9^8$ | $\dfrac{4 \cdot 8 \times 10^{18}}{\varrho} T_9^3\, e^{-2mc^2/kT}$ | |
| Extreme relativistic, non-degenerate $mc^2 \ll kT,\ \epsilon_F \lesssim kT$ | $\dfrac{2 \cdot 5 \times 10^{14}}{\mu_e} T_{10}^8\, (\log_{10} T_{10} + 1 \cdot 6)$ | $\dfrac{4 \cdot 3 \times 10^{24}}{\varrho} T_{10}^9$ | $1.5 \times 10^{22} \left(\dfrac{\hbar\omega_0}{mc^2}\right)^{7 \cdot 5}$ $\odot \left(\dfrac{mc^2}{kT}\right)^{-1 \cdot 5} \exp\left(-\dfrac{\hbar\omega_0}{kT}\right),\ \dfrac{\hbar\omega_0}{kT} \gg 1$ |
| Non-relativistic, extreme degenerate $kT \ll \epsilon_F < mc^2$ | $1 \cdot 5 \times 10^2\, T_8 \left(\dfrac{1}{\mu_e}\right)^{2/3} \dfrac{1}{\varrho^{1/3}}$ | $4 \cdot 5 \times 10^6\, T_8^{3/2}\, \mu_e \exp\left[\dfrac{-(2mc^2 + \epsilon_F)}{kT}\right]$ | |
| Extreme relativistic, extreme degenerate $kT,\ mc^2 \ll \epsilon_F$ | $\dfrac{6 \cdot 3 \times 10^6}{\mu_e}\, (1 + 5 T_9^3)\, T_9^7 \left(\dfrac{mc^2}{\epsilon_F}\right)^3$ | $\dfrac{1 \cdot 4 \times 10^{11}}{\mu_e}\, T_9^4 \left(\dfrac{\epsilon_F}{mc^2}\right)^2 \exp(-\epsilon_F/kT)$ | $2.96 \times 10^{22}$ $\odot \left(\dfrac{\hbar\omega_0}{mc^2}\right)^6 \left(\dfrac{mc^2}{kT}\right)^{-3},\ \dfrac{\hbar\omega_0}{kT} \ll 1$ |

$\epsilon_F$ = Fermi energy excluding the rest mass of the electron.

around $8 \times 10^9$ °K, the equilibrium configuration changes to $^4$He. This change is endothermic. Also, the neutrino processes we discussed previously also dissipate stellar energy quickly. With these two endothermic reactions the star contracts rapidly and collapses. The result is a supernova.

In the figure we plot the neutrino luminosity $L$ of a star as a function of its central temperature. At $T > 5 \times 10^8$ °K, the energetics of a star are entirely governed by neutrinos. Overall, roughly 25 per cent of stellar energy is released in the form of 1 meV neutrinos.

In Table 3 we list the neutrino history of a star.

As a star collapses to become a supernova, more high energy neutrinos of energy $\gtrsim 100$ meV (e and $\mu$ type) may be emitted. Rough estimates indicate that energy up to $10^{52}$ ergs or more may be dissipated in the form of $\mu$-neutrinos. These neutrinos may be detected by using gigantic neutrino detectors to monitor future supernova explosions in our galaxy.

*Cosmic-ray secondary neutrinos.* When primary cosmis rays hit the upper atmosphere, nuclear reactions take place in which $\pi$-mesons are created. These $\pi$-mesons subsequently decay into $\mu$-mesons and neutrinos ($\mu$-type). Because $\pi$-mesons produced in the horizontal direction travel for a longer time, thus reducing the probability to interact strongly with the atmosphere, cosmic ray secondary neutrinos (and $\mu$-mesons) show an anisotropy, favouring the horizontal direction.

The flux of these neutrinos is the same as cosmic ray $\mu$-meson flux. The flux is too low to be detected directly.

However, a detector embedded in Earth may detect secondary $\mu$-mesons produced by neutrino reactions. These $\mu$-mesons also favour the horizontal direction.

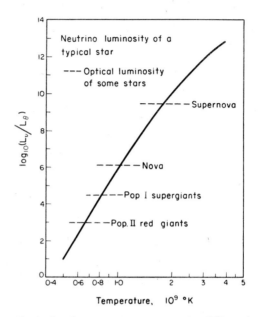

$L_\odot$ *is the solar energy output rate = 4 × $10^{33}$ ergs/sec*

*Table 3. Neutrino history of a star*

*The Average life time is taken to be 3 × $10^9$ years, the total optical energy output is taken to be 3 × $10^{50}$ ergs*

| Stage | $\langle E_\nu \rangle$ (meV) | Total Energy (ergs) | Duration (years) |
|---|---|---|---|
| Main sequence (Hydrogen burning) (Beta process) | 0·26 | $10^{49}$ | $3 \times 10^9$ |
| | 0·8 | $4 \times 10^{49}$ | $3 \times 10^9$ |
| | 7·0 | $4 \times 10^{44}$ | $3 \times 10^9$ |
| Red giant (Helium burning and plasma process) | 10 keV | $10^{47}$ | $10^8$ |
| Late stages (presupernova, photoneutrino and pair annihilation processes) | 100 keV − 1 meV | $10^{50}$ | $10^4$ very rapidly |
| | URCA process 2·58 meV | | |
| Supernova explosion and collapse | 100 meV | up to $10^{54}$? | a fraction of a second? |
| | Very uncertain | Numbers vary, depending on who speculates | |
| White dwarfs (plasma process) | 1 − 10 keV | $\sim 10^{48}$ | $10^8 \sim 10^9$ |

*Table 4. Cosmic neutrino fluxes*

| $\langle E_\nu \rangle$ | Flux on Earth ($\nu/\text{cm}^2$ sec) | Energy density (eV/cm$^3$) | Detectability | Solar back-ground flux |
|---|---|---|---|---|
| 1–10 keV | $10^5 \to 7$ | $10^{-3} \to 10^{-1}$ | no at present | none |
| 0·26 meV | $10^6$ | 1 | no at present | $5 \times 10^{10}$ |
| 0·8 meV | up to $10^6$ | 1 | yes, but difficult | $10^{10}$ |
| 1 meV ($\nu_e$ and $\bar{\nu}_e$) | vary rapidly with time | 1 on the average | yes, but difficult | $10^{10}$ |
| 7 meV | up to $10^2$ | $10^{-3}$ | Yes, but solar background must be eliminated (going to Pluto?) | $10^7$ |
| 100 meV | $\sim 10^{-4}$ | ? | yes | none |
| >1 BeV, cosmic secondaries | same as $\mu$ flux | not meaningful to define | yes | none |

Plans are being made to detect these underground secondary $\mu$-mesons.

Table 4 lists cosmic neutrino fluxes for different ranges.

*Solar neutrinos and their detection.* The flux of solar neutrinos is big enough for terrestrial detection at the present status of neutrinology. Pontecorvo has suggested that the reaction

$$^{37}\text{Cl} + \nu_e \to {}^{37}\text{A} + e^- \quad \text{(Threshold 0·8 meV)}$$

$$e^- + {}^{37}\text{A} \to {}^{37}\text{Cl} + \bar{\nu}_e \quad \text{(Half life 30 days)}$$

may be used to detect solar neutrinos. The $^{37}\text{A}$ atom may be chemically separated from $^{37}\text{Cl}$ (prepared as a liquid carbon tetrachloride $CCl_4$) by using carriers ($^4\text{He}$), afterwards $^{37}\text{A}$ is separated from $^4\text{He}$ by absorption at low temperature ($\sim 20°\text{K}$). The half-life of $^{37}\text{A}$ limits the integration time to around 30 days. This method has been put into practice by Raymond Davis. A total amount of $0·5 \times 10^6$ litres of $CCl_4$ will be needed, and the daily production rate of $^{37}\text{A}$ is around 10 atoms. (The natural abundance of $^{37}\text{C}$ is 25 per cent.)

J. N. Bahcall pointed out that transitions occurring between the ground state of $^{37}\text{Cl}$, ($J = 3/2 +$, $T = 3/2$) and the excited state of $^{37}\text{A}$ $J = 3/2 +$, $T = 3/2$ (5·1 meV) is superallowed and a cross-section at 10 meV neutrino energy of around $0·8 \times 10^{-42}$ cm$^2$ may be expected. The size of the cross-section is a large determinary factor for the feasibility of solar neutrino astronomy.

*Conclusion.* Stellar neutrinos are produced abundantly and in the future neutrino astronomy may be expected to play an important role in our study of the cosmos.

*Bibliography*

ADAMS B. *et al.* (1963) *Phys. Rev.* **129**, 1383.
BAHCALL J. N. (1964) *Phys. Rev. Letters* **12**, 300.
CHIU H.-Y. and MORRISON P. (1960) *Phys. Rev. Letters* **5**, 573.
CHIU H.-Y. and STABLER R. (1961) *Phys. Rev.* **122**, 1317.
DANBY G. *et al.* (1962) *Phys. Rev. Letters* **9**, 36.
DAVIS R., jr. (1964) *Phys. Rev. Letters* **12**, 302.
FEYNMAN R. P. and GELL-MANN M. (1958) *Phys. Rev.* **109**, 193.
GAMOW G. and SCHONBERG M. (1941) *Phys. Rev.* **59**, 539.
PONTECORVO B. (1956) *J. Exp. Theoret. Phys.* (*U.S.S.R.*) **37**, 1751. (Translation: *Soviet Phys.—J.E.T.P.* **10**, 1236, (1960).)
SALPETER E. E. *Stellar Energy Sources.*
SEARS R. L. (1964) *Astrophys. J.* (147, 471).

H.-Y. CHIU

**CREEP, IRRADIATION-INDUCED.** Creep is usually defined as the change in strain of a material under constant stress. Another form of the creep phenomenon is stress relaxation, which is defined as the change in stress at constant strain. The creep behaviour of a number of materials has been studied during neutron irradiation, under both constant stress and constant strain conditions.

Neutrons may cause radiation damage in crystalline materials in two ways:

(a) Directly by knocking atoms out of place in the lattice creating vacant sites and interstitial atoms.

(b) Indirectly by producing changes in atomic nuclei and thereby creating impurity atoms.

Although neutrons of all speeds are scattered by nuclei, only the fast ones freshly created from nuclear fission produce radiation damage by scattering because only these are capable of giving a large recoil energy to the struck nucleus. The atoms displaced by collisions with fast neutrons will acquire an energy much larger than that required to move them from their lattice site and so the displaced atoms can displace further atoms by colliding with them. Each displaced atom is expected to give up about one half of its energy to its immediate neighbours as atomic vibrations. These vibrations cause a short lived (about $10^{-12}$ sec) localized increase in temperature which may be as high as 4000°C.

New atoms may be created by thermal neutron capture and subsequent radioactive decay. For example, most metals and alloys contain boron either as a trace impurity or an alloying element. The $^{10}B$ isotope in natural boron reacts with a thermal neutron as follows:

$$^{10}_{5}B + ^{1}_{0}n \rightarrow ^{7}_{3}Li + ^{4}_{2}He + 2 \cdot 8 \text{ MeV}$$

and the presence of lithium and helium in the material may lead to changes in properties. Also the kinetic energy liberated in the reaction will be dissipated locally in the metal or alloy and will therefore displace a large number of atoms in addition to those displaced by the energetic neutrons.

Thus the creep behaviour of a material may be affected by neutron irradiation in the following ways:

(a) The presence of vacant lattice sites and interstitial atoms will increase the hardness and strength of the material but will decrease the ductility.

(b) In the small high temperature region created when an atom is displaced constitutional changes may occur, in which alloying elements and impurities are redistributed. Plastic flow is also expected in these high temperature regions and consequently changes in creep behaviour may occur.

(c) Internal stresses caused by the thermal gradients around the localized high temperature regions may also affect the creep properties.

(d) Chemical changes may occur in the material due to transmutations by thermal neutrons. Chemical changes are more common in nuclear fuel than in structural material since, in the fuel, impurity atoms are introduced into the lattice as fission fragments.

Although creep is generally considered to be a high temperature property, exposure of some materials to neutrons significantly lowers the temperature at which creep occurs. In-pile creep measurements are difficult and expensive and generally specimens have been withdrawn from the reactor and examined after various irradiation doses.

An increase in the creep rate of alpha uranium is observed during neutron irradiation at about 80°C and may be accounted for by internal stresses caused by irradiation growth. The rate of creep strain observed, $\dot{\varepsilon}_e$, is given by the equation,

$$\dot{\varepsilon}_e \simeq (\sigma/\sigma_y) \, \dot{\varepsilon}_g$$

where $\sigma$ is the externally applied stress, $\sigma_y$ is the yield stress and $\dot{\varepsilon}_g$ is the rate of radiation growth.

It is also found, rather surprisingly on the growth model, that some of the creep strain is recovered after irradiation. Neutron irradiation does not appear to affect the creep behaviour of uranium at temperatures above about 450°C, which is the highest temperature for irradiation growth to occur.

Stress relaxation of (18% Cr, 8% Ni)-type stainless steel occurs during neutron irradiation to doses of $9 \times 10^{20}$ fast neutrons per square centimetre at a temperature below 100°C. The creep of this steel without irradiation is negligible at temperatures below 400°C. The relaxation that occurs may be as high as 7 kg/cm². A slight increase in the creep rate under constant load of a similar steel has also been observed.

Creep experiments on zirconium made at 260°C show that irradiation in a flux of $3 \times 10^{12}$ n cm$^{-2}$ sec$^{-1}$ produces a large decrease in the creep rate. There is no significant effect of irradiation on the creep behaviour of aluminium at 350°C, constantan at 300°C or nickel at 700°C. In-pile creep experiments on creep resisting alloys working at temperatures near the upper limit of their useful range do not appear to have been made. Since both thermal spikes and diffusion due to the migration of point defects may alter the state of precipitation in such alloys their creep resistance may be significantly lowered.

Neutron irradiation enhances creep in graphite; it now appears well established that during irradiation a loaded graphite specimen deforms more than is expected from the radiation induced changes in Young's modulus. Although the amount of creep deformation observed depends on the applied stress and the irradiation does it is independent of temperature, in the range 200–650°C.

*Bibliography*

JOSEPH J. W. (1959) *A.E.C. Research and Development Report*, DP 369.

KONOBEEVSKII S. T. (1960) *Atomnaya Energija* **9**, 194.

NIGHTINGALE R. E. (Ed.) (1962) *Nuclear Graphite*, New York: Academic Press.

ROBERTS A. C. and COTTRELL A. H. (1956) *Phil. Mag.* **1**, 711.

A. C. ROBERTS

**CRYOPUMPING.** *1. Introduction.* If a surface, at a temperature low enough to condense the gas present, is introduced into an enclosure, the resultant pressure in the enclosure will be that of the vapour pressure of the condensate. As the condensate temperature is lowered the vapour pressure is further reduced to the triple point of the gas and freezing occurs. A slight decrease in temperature will then result in a rapid decrease in vapour pressure. This is essentially the principle of cryopumping and in its simplest form has been in operation for many years as a cold trap used to remove unwanted vapours in conventional vacuum systems.

Recently large scale applications have beome possible with the availability of very low temperature refrigerants on a commercial scale and cryopumping is being increasingly used to operate low density wind tunnels, space simulation chambers and similar devices where very large pumping speeds are required at low pressures.

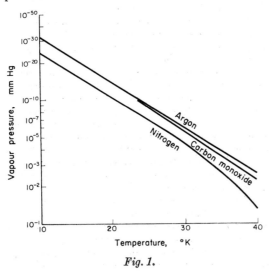

*Fig. 1.*

Figure 1 shows the vapour pressure of several common gases as a function of temperature. It may be seen that a surface temperature of 20°K is sufficiently low to produce a pressure of $10^{-10}$ mm Hg with pure nitrogen. However for gases like air, nitrogen, oxygen and argon the practical useful range of a cryopump appears to be in the range of $10^{-2}$–$10^{-6}$ mm Hg. At pressures below $10^{-4}$ mm Hg and pumping speeds higher than about $10^5$ l./sec it appears that the power and space requirements of conventional pumping systems become orders of magnitude larger than cryopump requirements.

Some fundamental considerations concerning the design and operation of cryopumps are given below in more detail.

*2. The solid–gas interface problem.* In operation, the gas freezes on to the cold surface as a solid layer and if the gas flow is considerable the pump will not be a continuous device because the pumping will cease when the heat transfer through the layer becomes too small to maintain the gas-solid interface at a sufficiently low temperature. Thus the running time will depend upon the flow rate and condensing surface area. This condensate layer problem has been analysed as a one-dimensional heat conduction problem with a

moving boundary and a typical result shows that a mass flow of 1 g/sec of nitrogen can be maintained for 10 hr at a pressure of $10^{-3}$ mm Hg with a condensing surface area of 6 m². One problem with this type of calculation is that the thermal conductivity of the frozen condensate is not known accurately since it depends upon how the condensate is deposited. This in turn depends upon the distribution of temperature near the interface and the deposits may be "snowy" with poor conductivity or "icy" with good conducitvity.

*3. Removal of non-condensable gases.* Most commercially available gases contain traces of neon, hydrogen and helium (normal atmospheric air contains about 0·16 per cent) which will not be condensed by a cryopump and have to be removed in some other manner. This is normally accomplished with a diffusion pump placed in series with the cryopump. The quantity of non-condensable gases to be removed can be reduced by using commercially pure gases in the system and reducing leaks to a minimum. Another mechanism also exists for the removal of non-condensable gases. This occurs if the flow rate is sufficiently high when the non-condensable gases are entrained by the condensable gases during the condensation process and trapped in the condensate. This phenomenon has been observed in several laboratories but has not yet been evaluated quantitatively.

*4. Condensing surface and radiation shield geometry.* Apart from considerations involving the running time discussed in Section 2 the area of the condensing surface must be large enough to allow the random flow on to the surface to equal the mass flow ihto the vacuum system. Assuming free molecule flow the condensing surface area, $A$, is given by

$$m \sim C\varrho\bar{c}A/4$$

where $m$ is the mass flow, $\varrho$ is the density of the gas, $\bar{c}$ is the mean molecular speed and $C$ is the "sticking" coefficient, defined as the ratio of moleculeswhich strike the surface and freeze on to the surface to the total striking the surface. Values of the "sticking" coefficient have not yet been obtained systematically but for practical cases may be taken to be near unity.

The gain of heat by radiation to the condenser surface can be an order of magnitude larger than the heat load due to the sensible and latent heat of the condensed gas. It is therefore necessary to provide some form of radiation shielding which can consist of several layers of aluminium foil but is more effective if cooled with liquid nitrogen. The radiation shielding is designed to maximize the pumping speed and minimize the thermal radiation load. A number of possible geometries suitable for wind tunnels or space simulation chambers which are used or have been proposed are shown in Fig. 2. These shielding arrangements ensure that the optical path between the condensing surface and the vaouum system is zero.

The condenser not only has to remove the latent heat but also the sensible heat of the gas in the system which in the case of a wind tunnel employing some form of heater can be an order of magnitude larger than the latent heat. A precooler is consequently commonly employed to remove as much of the sensible heat as possible more economically than the very low temperature condensing surface. The precooler is frequently combined with the radiation shield in one device such as a baffle composed of liquid nitrogen cooled chevrons (Fig. 2(a) and (d)). The disadvantage of such a system is that it severely restricts the pumping speed. For example in free molecule flow with an optimum configuration of chevrons only 28 per cent

of molecules striking the chevrons are able to pass through to the condensing surface and a baffle area of 20 m² is required per 10⁶ l./sec pumping speed. Thus although a cryopump is essentially a constant mass flow device with pumping speed inversely proportional to pressure, as the pressure is lowered the pumping speed will ultimately be restricted by the conductance of the circuit connecting the system to the pump. A typical mass flow characteristic for a low density wind tunnel operated with a 1000 W 20°K refrigerator is shown in Fig. 3.

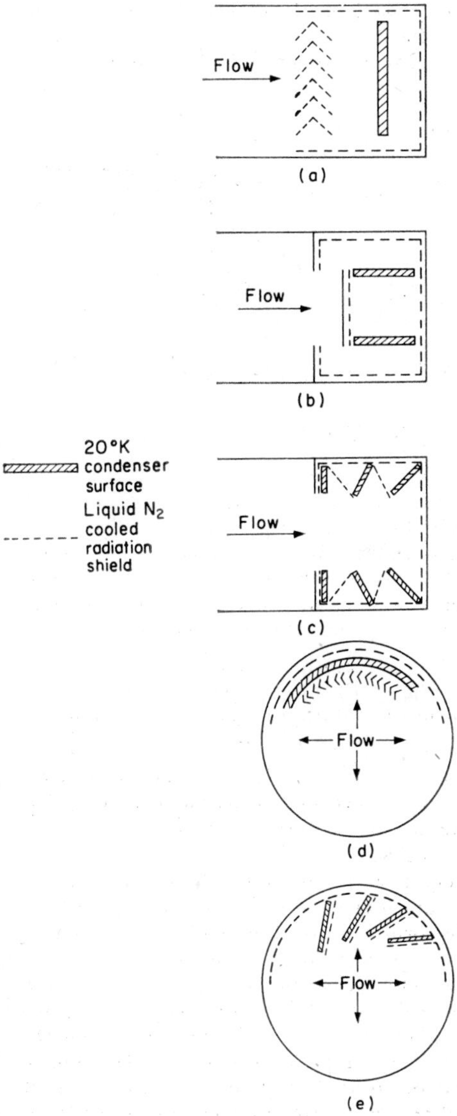

*5. Refrigeration systems.* For most practical purposes a condensing surface temperature of ∼20°K is required to pump the common gases and a refrigeration power of 280 W is required to freeze 1 g/sec of nitrogen after it has been cooled to 100°K by a precooler. Radiation losses approximately double the power requirements.

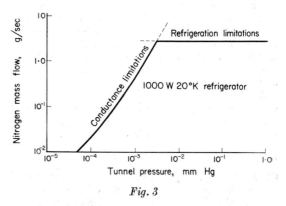

*Fig. 3*

For a small laboratory installation refrigeration can be most conveniently obtained by boiling liquid hydrogen or liquid helium in an open cycle device, the choice between the two gases depending upon local availability and the ultimate vacuum required. For large scale applications closed cycle operation appears to be most economical although the capital cost of a commercial refrigerator to produce one kW of power at 20°K is considerable. With closed cycle operation it is also possible to use neon as a refrigerant. The relative merits of various systems which have been considered are discussed below.

Neon has the advantage that it can be liquefied by a refrigerator using compressors and the Joule-Thompson process and thus the capital costs are less than the alternative systems. However, the liquefaction temperature is 27·6°K at atmospheric pressure and whilst this is sufficiently low to produce pressures ∼10⁻³ mm Hg with the common gases, in a practical application the running time is more limited than with the other systems because the liquefaction temperature is not low enough to allow sufficient heat transfer through a thick condensed layer. A slight improvement could be obtained by boiling liquid neon at reduced pressures

Flow

( a )

Flow

( b )

20°K condenser surface

Liquid N₂ cooled radiation shield

Flow

( c )

←Flow→

( d )

←Flow→

( e )

*Fig. 2.*

but the possibility of air leaks into the neon with subsequent contamination would rule out this method for practical equipment. The possibility of loss or contamination of the neon cannot be tolerated because the cost of the neon required to charge the plant represents about 10 per cent of the total capital cost.

Liquid helium is not normally warranted because of its low latent heat and because operating temperatures are not usually required below 15°K. Cold helium is obtained by the use of precision work engines which are used to extract heat from the helium after precooling with liquid nitrogen. The work engines are required because the inversion point for the Joule-Thompson process is 20°K. Whilst the work engines are simple in principle, considerable development work was required before helium refrigerators became commercially available, and capital costs are still considerable.

Hydrogen may be liquefied by the Joule-Thompson process after precooling with liquid nitrogen and by boiling at a reduced pressure it is possible to achieve a temperature of 15°K. The direct capital cost of a liquid hydrogen refrigerator may be only half that of a comparable helium refrigerator but once the extra costs of safety precautions have been included it does not appear that there is much to choose between hydrogen and helium refrigeration.

At the time of writing whilst several large scale cryopumps are operational there are still a number of features which have not yet been investigated thoroughly and the science is still in its infancy.

*Bibliography*

CHUAN R. *et al.* (1961) *The cryopump, first and second generation: A report on the operational history of the University of Southern California Cryopump and the design of a new facility*, Second annual symposium on space vacuum simulation, Cambridge, Mass.: Arthur D. Little Inc.

KLIPPING G. and MASCHER W. (1962) *Vakuum-Technik* **11** (3), 81.

SCOTT R. B. (1959) *Cryogenic Engineering*, Princeton: Van Nostrand.

SIMONET C. (1963) *Le Vide* **18** (105), 311.

W. A. CLAYDEN

**CRYOTRONS FOR COMPUTERS.** In 1956 D. A. Buck first described a superconductive switching element which he named the cryotron. It consisted of a small solenoid of fine insulated niobium wire, up the centre of which ran a fine tantalum wire. When the element was cooled in liquid helium to about 4·2°K both wires, being superconducting, were resistance free. However, on passing a current through the niobium coil, the *control*, a magnetic field was developed which was sufficient to send the inner tantalum wire, the *gate*, into the resistive state. On switching off the control current the gate returned to the superconductive state. Throughout this sequence the niobium control remained superconducting, niobium having a

higher critical field and critical temperature, $T_c$, than tantalum. Because the cryotron can be used to steer current from one superconductive circuit to another superconductive circuit in parallel, it can be used to perform binary logic and is therefore of considerable interest to computer engineers. As will be shown later the cryotron can be used to make a computer storage cell and this is an even greater attraction.

The wire-wound cryotron was far too slow in its operation to be of immediate use, the high control inductance and low gate resistance leading to $L/R$ circuit time constants of the order of a millisecond, However, in thin film or "planar" form the cryotron is much faster, and has the added attraction that it should be very cheap to fabricate in one sequence of vacuum depositions thousands of already interconnected elements.

A typical planar cryotron is shown in Fig. 1.

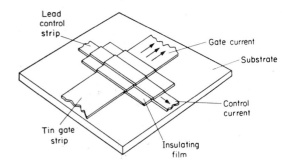

*Fig. 1. Typical crossed-film cryotron.*

Because thin films of tantalum and niobium are extremely difficult to prepare, tin ($T_c = 3\cdot73$°K) and lead ($T_c = 7\cdot29$°K) are used, immersed in liquid helium boiling under reduced pressure (approx. 2/3 atmosphere, i.e. approx. 3·5°K). Typical dimensions of the crossed strip thin film cryotron are:

| | | |
|---|---|---|
| tin gate | width 0·6 mm | (600 $\mu$) |
| | thickness 5000 Å | (0·5 $\mu$) |
| lead control | width 0·1 mm | (100 $\mu$) |
| | thickness 7000 Å | (0·7 $\mu$) |

Typical operating currents are in the range 200–800 mA. Indium ($T_c = 3\cdot34$°K) is sometimes used as a gate material, in which case the cryotron must be operated at about 3·2°K.

The cryotron possesses the important feature of current gain if the control current necessary to send the gate normal in the absence of gate current is less than the "self-switching" gate current necessary to send the gate normal in the absence of control current. Current gain, which is a temperature dependent quantity, is necessary if it is required to connect the gate of one cryotron, A, in series with the control of another, B, so that with increasing current B is switched before A switches itself. In theory the current gain of a cryotron should equal the quotient of the width of the gate divided by the width of the control. In practice, the current gain of thin film cryotrons is usually about half of this figure, this reduction being mainly caused by the inferior superconductive properties of the diffuse edges of the vacuum deposited gate strips.

A typical control characteristic of a crossed film cryotron is shown in Fig. 2. For control currents, $I_c$, and gate currents, $I_g$, corresponding to points within the closed area, the gate is superconducting, Outside the closed area it is resistive.

The current gain is given by $\dfrac{I_{go}}{I_{co}}$.

A weak magnetic field will not penetrate a superconducting material, surface currents being induced instead. Superconducting thin films can therefore be used as "ground planes" to shield one plane of cryo-

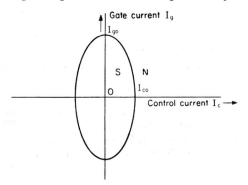

Fig. 2. Control characteristic of crossed-film cryotron.

trons from another. Superconductive "ground planes" deposited between the cryotron crossing and the substrate have two other useful functions besides that of screening. The confinement of the magnetic field under the gate strip results in a gate current which is more evenly distributed across the width of the gate, with a consequent improvement in current gain. The ground plane also leads to a considerable reduction in circuit inductance and consequently a marked improvement in circuit operating speed. Further reduction in the $L/R$ circuit time constant should be possible through the use of alloy gates, for alloys show a considerably higher low-temperature resistivity than that of either of their components when pure.

An alternative form of cryotron, the "in-line" or parallel strip cryotron, is shown in Fig. 3. The control and gate are superimposed and because the gate is now long and narrow, the gate resistance is much higher than that of the crossed strip cryotron. Consequently the circuit $L/R$ time constants are much shorter.

The control characteristic of a typical in-line cryotron is shown in Fig. 4.

It will be noted that the current gain, $\dfrac{I_{go}}{I_{co}}$, is now less than unity and an additional superimposed con-

trol strip providing bias is necessary to overcome this disadvantage. The operation of the in-line cryotron is affected by whether the control and gate currents are in parallel or antiparallel and this assymetry may prove useful.

In-line cryotron circuit should possess $L/R$ time constants of about 10 ns., in which case the operating speed would probably be limited by heat transfer rates, for heat is developed in the gate when it is resistive and before the current flowing through it has had

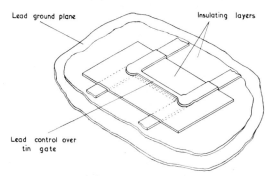

Fig. 3. In-line cryotron.

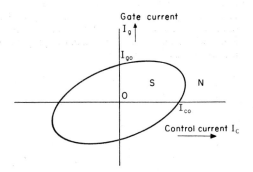

Fig. 4. Control characteristic of an in-line crotron.

time to be completely diverted to an alternative path. Furthermore, the latent heat of the superconductive transition has to be exchanged between the gate and the substrate each time switching occurs. For these reasons it is desirable to use substrates of good low-temperature thermal conductivity, such as aluminium or crystalline quartz.

Since superconductors are completely resistance free, once a current is set up in a superconducting loop circuit, it will continue circulating until resistance is somehow introduced into the circuit. Persistent currents can therefore be used in the store of a computer, the presence of a persistent current indicating the storage of a binary digit. The basic circuit used for storage is shown in Fig. 5.

The stages of operation are as follows:

(a) Current $I$ divides according to the relative inductances of the two sides of the loop, so that no flux passes through the loop. Cryotron B remains superconducting.

(b) A "write" current $i$ is applied through the control of cryotron A. The gate of cryotron A becomes resistive and all the current $I$ flows through the control of cryotron B, whose gate becomes resistive. Flux now passes through the loop.

(c) $i$ is switched off, A becomes superconducting again but, since flux is now trapped in the loop, $I$ continues to flow down the right hand side only.

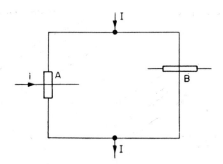

*Fig. 5. Cryotron storage element.*

(d) $I$ is switched off. Flux is still trapped in the loop so a persistent current is forced to flow round it. Cryotron B shows this circulation without interfering with it, so that the cell has the property of non-destructive readout.

(e) To erase, switch $i$ on again. The persistent current dies down to zero, the stored magnetic energy being converted into ohmic heating of the gate of cryotron A. Cryotron B becomes superconducting again.

The problems of operating a computer at very low temperatures are not as great as might be imagined, and the main obstacle to exploiting the unique properties of cryotrons is lack of reproducibility of the superconductive properties of thin films. It would seem necessary to deposit the films from the vapour in a vacuum system in which the pressure of residual gases is less than $10^{-7}$ torr, for oxygen, carbon dioxide and water vapour have a marked influence on the mode of growth of thin films of tin, indium and lead. Another major problem in the fabrication of cryotrons is the making of thin insulating films which are free of pinholes and able to withstand the severe stress of *thermal cycling*, between liquid helium and room temperatures, for example. Thin polymer insulating films have been developed for this purpose.

*See also:* Superconducting alloys.

*Bibliography*

BREMER J. W. (1962) *Superconductive Devices*, New York: McGraw-Hill.
WALKER P. A. (1963) *Cryotrons and Cryotron Circuits— A Review, J. Brit. I.R.E.* **25**, 387.

P. R. STUART

# D

**DEPLETED URANIUM.** *Origin.* Depleted uranium is a by-product of the nuclear industry and arises from two sources. In its natural state uranium consists essentially of two isotopes, U 238 and the fissionable U 235, the relative quantities being 99·3% U 238 and 0·7% U 235. If the percentage of U 235 is reduced by separating the isotopes in a diffusion plant, or by irradiating the natural uranium in a reactor, the resultant material (which contains a higher percentage of U 238) is known as depleted uranium. It usually has about half the U 235 content of natural uranium. However, chemically and physically it is identical with natural uranium.

The depleted uranium can be obtained in metallic form or as compounds, usually oxides, and various uses have been found for it.

*Properties.* The most significant physical property of depleted uranium is its high density. The theoretical density of the metal is 19·04 g/cm³, but varies between 18·75 g/cm³ and 19·0 g/cm³, depending on the method of working the metal. The usual method of producing metal components is by casting which gives a minimum density of about 18·90 g/cm³.

The metal melts at 1131°C and below this temperature exists in three allotropic forms—alpha, beta and gamma. The alpha phase is stable up to 665°C and the transition from beta to gamma occurs at 771°C.

Alpha uranium is anisotropic and its mechanical properties are affected by crystal orientation, but the properties of cast metal are sufficiently good for cast components to have wide industrial applications.

Because uranium has a high affinity for oxygen at elevated temperatures most fabrication processes have to be carried out either under vacuum or in an inert atmosphere, which enables the material to be fabricated by casting, rolling, pressing, extrusion, etc.

Uranium may be machined by using conventional workshop practices, but certain precautions have to be observed, since, when finely divided, uranium is pyrophoric, and an extra supply of coolant or lubricant is necessary to avoid ignition of the chips and turnings. In general, heavy cuts and high tool speeds are recommended when machining uranium.

Welding the metal is possible but is hindered by the rapid oxidation of the metal, and soldering and brazing is even more difficult.

The metal can be electroplated with the majority of conventional protective metals such as nickel, cadmium, copper, chromium, etc.

*Uses.* Depleted uranium can be used in several ways depending upon whether the applications make use of its nuclear, physical or chemical properties.

As the material is largely U 238, it is a good neutron absorber and is used in thermal reactors for flux flattening, and also as a blanket material in fast breeder reactors, where, by absorbing a neutron, U 238 leads to the production of plutonium.

The majority of non-nuclear applications of depleted uranium make use of its high density, which offers important advantages in the manifactire of the following:

1. Radiological shielding for equipment using gamma isotopes and high energy X rays. Gamma-ray attenuation depends primarily on the density of the shielding medium and uranium can offer distinct advantages over lead shielding despite its higher cost. Using uranium only approximately half the thickness of shielding is required. This fact can usually reduce considerably the size and weight of shielding required, especially in the manufacture of cobalt-60 radiotherapy equipment and portable radiography cameras used for non-destructive testing. In these applications up to 50 per cent saving in weight can be achieved.

2. Transport flasks and containers for the shipping of isotopes and irradiated fuel elements. In this case the use of depleted uranium gives savings in volume and weight, thus attracting much lower freight and shipping rates.

3. Depleted uranium has been used as counterweights and mass balance weights in aircraft, where a dense metal is of great value for use in small, confined spaces.

4. Other uses for which the high density of uranium appeals to design engineers is in engine crankshafts, high-speed rotors for gyro compasses and high inertia flywheels.

5. Because of the high atomic number of uranium it is a good target material in an X ray tube giving off hard X rays of short wave-length; although its rather low melting point is a disadvantage.

6. The high density and superior mechanical properties of depleted uranium are useful in the manufacture of conventional ammunition.

Several other uses of depleted uranium metal depend upon its chemical properties, and its applications in the ferrous and non-ferrous metals industry have been investigated.

Uranium additions to steel have not proved as successful as was originally hoped for and, although certain advantages were discovered, they were of no economic value over cheaper alloying elements.

However, in the manufacture of copper from high lead content ores and also in the fabrication of high lead and bismuth alloys the use of uranium has met with some success. The uranium removes the lead and bismuth impurities from the copper, and so renders possible the hot working of the copper and its alloys.

Uranium is also used as a 'getter' or 'scavenger' for the purification of gases. At elevated temperatures small pieces of uranium will absorb oxygen, hydrogen, water vapour, carbon monoxide and carbon dioxide; purification systems based on a uranium 'scavenger' are now used where it is possible to recycle inert gases, and such systems are very economic.

Since the metal has a high positive electrode potential of about $1 \cdot 2$ volts a use in cathodic protection has been suggested, but the advantages over magnesium and zinc are marginal.

Uranium forms a large number of compounds with many unique properties. As it is a very reactive metal, it forms compounds with most non-metals, the most common being the series of oxides where uranium exhibits several valencies. Three oxides are of importance, namely, uranium trioxide ($UO_3$), uranium octoxide ($U_3O_8$), and uranium dioxide ($UO_2$), each having a different colour (orange, green-black and brown respectively).

By far the best known non-nuclear use of uranium compounds in large quantities is in the ceramic industry. Uranium oxides in ceramic glazes can be fired to give a variety of brilliant colours. Additionally, the brown $UO_2$ is stable and maintains its strength at high temperatures and is utilized for the manufacture of refractory ceramics.

The addition of uranium oxides to glass imparts fluorescent qualities and other colours of glass can be made by uranium additions.

Uranium forms compounds with a wide variety of colours, and these form the basis for pigments which have found large-scale usage in the pottery and paint industries.

Limited experimental work in the past has shown that depleted uranium could be used as a catalyst. This work resulted from the position of uranium in the periodic classification in the same group as chromium, molybdenum and tungsten, but it was found that uranium had no superiority over other elements.

Recently, however, as a result of further research several new possible catalytic reactions have been observed. The large-scale use of uranium catalysts is a distinct possibility and several patents have now been filed on the application of uranium.

A further use for the uranium oxides is in the manufacture of high density concretes for specialized shielding applications such as in the manufacture of transport containers for irradiated fuel elements. The concretes made during experimental trials have densities ranging from $6 \cdot 0$ to $7 \cdot 0$ g/cm³, 50 per cent increase over barytes concretes (at present widely used for reactor shielding), and have physical and mechanical properties similar to ordinary concrete. Mixes of uranium oxide and hard-setting resins have been made to produce special shielding blocks for use in hospitals for screening X-ray rooms. Such blocks have a slightly higher density than concrete bricks and extremely good properties for building uses.

*Health and safety.* The main objection to the use of depleted uranium is the possible hazard associated with the handling of a slightly radioactive material.

Uranium presents two slight hazards; first, from its residual activity and, secondly, from its chemical toxicity. The residual activity of the material is mainly in the form of beta radiations, which, in the case of handling the metal, could be hazardous if large amounts are in contact with the body for long periods. This hazard is overcome by wearing rubber gloves and limiting the periods in which workers handle the material. The requirements are laid down in Government regulations.

The chemical properties of uranium present a hazard only after it is absorbed into the blood stream. This chemical toxicity is more significant when handling the oxides due to the problem of inhalation of fine oxide particles. Being compounds of a heavy metal the main hazard is similar to any other heavy metal toxicity and precautions similar to those taken for heavy metal compounds, e.g. white lead, are adequate for uranium compounds. Uranium is absorbed only to a limited extent through the skin and its specific activity is quite low in comparison with the radioactive elements.

Both associated hazards can be prevented by good housekeeping and common sense.

*Summary.* Uranium is a new metal in the field of industry and, although a vast amount of knowledge is available on its nuclear properties, knowledge about uranium for industrial applications is scant.

However, more uses for this unique metal are being discovered every year and it is now looked upon as an ordinary metal and not just as a fissionable material.

P. A. Wright

**DICHOTIC PRESENTATION, DICHOTIC LISTENING.** When a message is presented to a listener through only one ear it is monaural. When the same message is presented to both ears (as when listening to the output of a loudspeaker) the message is binaural. When (using headphones) one ear receives one message while the other simultaneously receives a different one, presentation is said to be dichotic.

Occasionally diotic is used as a synonym for binaural, but this use has little to commend it.

N. Moray

**DICHROIC MIRRORS.** Mirrors formed by the deposition on a transparent substrate of alternating layers of high-index and low-index dielectric material with thicknesses equal, e.g. to a quarter wave-length of the radiation for which maximum reflection is to take place. Such mirrors have the property that the wave-length of maximum reflection varies with the direction of incidence of the light and that, for incident white light, the colours of the transmitted and reflected light are complementary. Dichroic mirrors find application as beam splitters in colour cameras and colour television cameras.

V. K. Zworykin and E. G. Ramberg

**DISLOCATIONS, TRANSMISSION ELECTRON MICROSCOPY OF.** Transmission electron microscopy was first used for the identification and study of dislocations in thin films by Bollmann and independently by Hirsch, Horne and Whelan in 1956. Since that time the technique has been considerably developed and several other lattice imperfections, such as stacking faults and various point defect agglomerates, as well as precipitates have now been examined, not only in many of the common metals and alloys but also in a number of ionic and covalent crystals. The electron microscopes usually employed have a double condenser lens enabling a spot not more than a few microns in diameter to be focussed on the specimen and they accelerate the electrons to about 100 kV. Under favourable conditions crystalline films of thicknesses up to a few thousand ångstroms are transparent to electron beams of this energy.

Thin metal films are normally prepared by electropolishing methods. The most convenient starting material is a foil a few tenths of a millimetre in thickness but for some purposes, in particular the study of dislocation arrangements in large single crystals, slices of various orientations are cut from the bulk material by spark erosion or acid sawing techniques and these slices are then electropolished. Thin films of metals and other material can also be prepared by ion bombardment, evaporation, microtonomy or cleavage. In all cases the influence of the specimen preparation technique on the defect structure of the film must be considered, particularly if it is desired to relate the observations to the behaviour of bulk material.

In the observational method most frequently employed, dislocations and other defects are rendered visible in thin foils by a mechanism of Bragg diffraction contrast which is illustrated in Fig. 1. The rays leaving the specimen consist of a transmitted wave

and several waves which have suffered reflection at small angles ($\sim 1°$) from crystal planes as shown. On passing through the objective lens, the rays form a diffraction pattern in the plane AB and an image of the specimen in the plane CD. By focusing the lower lenses of the microscope (not shown in Fig. 1) on these respective planes, diffraction or microscopy conditions are obtained. Under microscopy conditions an objective aperture is usually inserted in the plane AB (see Fig. 1) and permits only the transmitted or possibly only one of the reflected waves to contribute to the micrograph. Image contrast then depends on the intensity of the transmitted or reflected wave used and is influenced by any change in the conditions of Bragg reflection. Figure 1 shows how such a change will occur in any strained region of the lattice, such as near

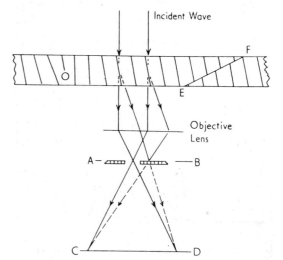

Fig. 1.   The mechanism of Bragg Diffraction Contrast.

a dislocation 0 where the reflecting planes are buckled, or at a stacking fault EF where the planes undergo an abrupt displacement. Dislocations are generally visible as dark lines about one hundred ångstroms in width (e.g. at A in Fig. 2) and stacking faults appear as a set of fringes running parallel to the intersections of the fault with the surfaces of the thin film (e.g. at B in Fig. 2).

In general the study of lattice defects by this method does not utilize the full resolving power of the instrument since the visibility of small defects is usually determined by the magnitude of the local change in the Bragg reflected intensity rather than by any instrumental properties. In an alternative technique, due to Menter, all of the transmitted and reflected waves are allowed to contribute to the final image so that, provided the resolution of the instrument is adequate, the lattice planes of the crystal and any defects of the structure can be directly imaged. We confine further discussion to the diffraction contrast method of studying lattice defects since, possibly because of the less exacting requirements for both the specimen and the instrument, this technique has so far found wider application than the direct resolution method. A good understanding of the diffraction contrast mechanism is essential for interpreting the micrographs, however, since it can give rise to a large number of complicated image phenomena which are not always related in any simple way to the structure of the lattice defects in the thin film. There is now a

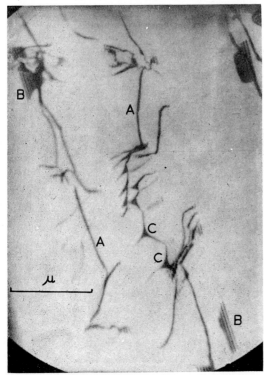

*Fig. 2. Dislocations (A), Stacking faults (B) and extended dislocation nodes (C) in Ni (30%) Cr (30%) Co (40%) alloy. (A. Thölén).*

well developed theory of the image contrast which explains most the experimental observations in terms of the dynamical theory of electron diffraction. The general principle which is most useful in practice is that no image contrast arises if the crystal is oriented to excite a Bragg reflection for which the relevant crystal planes are unaffected by the presence of the defect. The invisibility of various defects under given conditions of Bragg reflection can thus be used as a means of identifying the nature of the strain field present. In the case of screw dislocations for instance, the lattice displacements are all in the direction of the Burgers vector $b$ and hence neither the spacing nor the orientation of crystal planes which contain this vector are affected by the presence of the dislocation which is therefore invisible if the Bragg reflection from

such a set of planes is used to form the image. This invisibility criterion, conveniently expressed by the equation $g \cdot b = 0$ where $g$ is the reciprocal lattice vector normal to the crystal planes considered, enables the Burgers vector of screw dislocations to be determined. The condition only holds approximately for edge dislocations since, in addition to the lattice displacements parallel to $b$, there are in this case displacements normal to the slip plane and these produce a characteristic (usually weak) contrast effect. For detailed studies of this type it is essential to employ a goniometer stage in the electron microscope so that the same area of the specimen can be examined using a number of different Bragg reflections.

With the development of the transmission electron microscope technique a large number of the theoretical predictions about the properties and behaviour of dislocations have been confirmed in great detail. Different dislocations have been identified and their strain fields examined by the methods just described. A large number of dislocation interactions have now been studied in different crystal structures in particular those reactions leading to the formation of dislocation networks and Lomer–Cottrell locks. The dissociation of dislocations into a pair of partial dislocations bounding a stacking fault ribbon has been directly observed in a number of metals and alloys with face-centred-cubic structures. Unless the stacking fault energy is extremely low the width of these extended dislocation ribbons is less than the width of the dislocation image so that the ribbon width cannot be measured directly. However, interactions between extended dislocations can sometimes lead to the formation of hexagonal networks (see Fig. 2) containing extended dislocation nodes. The form of the partial dislocation bounding the nodes is governed by the balance between the dislocation line tension ($\sim \mu b^2 / 2$ where $\mu$ is the shear modulus) and the attractive surface tension due to the stacking fault energy $\gamma$. A simple formula $\gamma = \mu b^2 / 2R$ due to Whelan enables $\gamma$ to be estimated from measurements of $R$, the radius of curvature of the extended dislocation node. Measurements of $\gamma$ have been made in a number of alloys.

The high resolution of the transmission electron microscope technique makes it particularly suitable for the detailed study of the complex dislocation arrangements found in deformed crystals where the dislocation density $\varrho$ may vary from $\sim 10^5$ cm$^{-2}$ to $\sim 10^{12}$ cm$^{-2}$. Simple density counts have indicated that the resolved shear stress is given approximately by the relation $\tau = 0 \cdot 5 \ \mu b \sqrt{\varrho}$. More detailed examination of single crystals suggests that although most of the slip during plastic deformation occurs as a result of dislocation motion on the main glide system, many dislocations on other systems are also present. These studies have also shown that at low deformations in single crystals, large numbers of edge dislocation dipoles (parallel pairs of dislocations of opposite sign) are created. These dipoles are also found in enormous numbers in fatigued metals. So far the correlation of these observations with macroscopic mechanical properties is only in a preliminary stage.

The strain field from isolated point defects such as vacancies or interstitials is too small to be visible in the electron microscope but agglomerates of point defects, formed as a result of a supersaturation of the defects created by quenching or radiation damage, can be observed. Vacancies for instance can form a disk on the close packed planes. Collapse of these disks leads to the creation of small dislocation loops observed in aluminium, nickel and a number of other substances. In quenched gold, the observation of small tetrahedra of stacking faults, a defect whose existence had not been predicted, represented one of the most remarkable successes of the electron microscope technique. These tetrahedra remain stable at temperatures up to 600°C. The theory of the image contrast can be usefully applied to determine both the form and the sense (i.e. vacancy or interstitial nature) of the strain field round agglomerates of point defects or round impurity precipitates, which often give rise to similar effects. Much work remains to be done in correlating the presence of such defects in crystals with their mechanical or electrical properties.

The development of both heating and cooling stages for the microscope, particularly when the goniometer tilting facility can be retained permits the study of a large number of dynamic processes. Thus the motion of dislocations, not only by glide under the stresses set up as a result of the heating effect of the electron beam, but also by climb during the heating of specimens containing dislocation loops has been observed. Recrystallization processes and phase transitions can also be studied directly although it must be borne in mind that in these experiments the thin film may have markedly different properties than the bulk material.

*Bibliography*

BOLLMANN W. (1956) *Phys. Rev.* **103**, 1588.

HIRSCH P. B. (1959) *Metallurgical Reviews* **4** (14), 101.

HIRSCH P. B. *et al.* (1956) *Phil. Mag.* **1**, 677.

HOWIE A. (1961) *Metallurgical Reviews* **6** (24), 467.

THOMAS G. (1962) *Transmission Electron Microscopy of Metals*, New York: Wiley.

THOMAS G. and WASHBURN J. (1963) *Electron Microscopy and Strength of Crystals*, New York: Wiley.

WHELAN M. J. (1959) *Proc. Roy. Soc.* **A 249**, 114.

A. HOWIE

**ÉCHELON, ÉCHELLE AND ÉCHELETTE GRAT-INGS.** The échelon, échelle, and échelette differ generally in methods of manufacture and dimensions of spacings. The transmission échelon, for example, may be constructed by assembling flat plates of glass or silica to form a "staircase" pattern. When light is passed through these steps (vertically through the "stair treads") this is entirely equivalent to a coarse transmission échelette, used on the high order blaze. An échelle is intermediate in coarseness between the échelette (usually 100–80,000 grooves/cm) and the échelon (usually less than 10 steps/cm) and is commonly produced by machining, often followed by a lapping operation. These dispersive gratings should not be confused with the Fabry-Perot étalon, which operates without angular dispersion.

Thewlis J. (Ed.) (1961) *Diffraction grating; Echelle grating*, in *Encyclopaedic Dictionary of Physics*, Vol. 2, Oxford: Pergamon Press.

R. P. BAUMAN

## ELECTROCHEMICAL MACHINING.

### Definitions of Terms

*Cutting forces.* In metal cutting theory, any cutting tool may be considered to experience a reaction from the work which (usually) tends to separate the work and the tool. The forces between tool and work may have the effect therefore of distorting the work material in bending, and also may cause deflexion of the tool holding device. The first effect may cause permanent set in the workpiece, and coupled with the second effect, may lead to inaccuracy and inferior surface finish in the completed work.

*Hydrostatic screw and nut.* The frictional effect of a nut on a leadscrew is of such a nature as to make the arrangement usually of less than 50 per cent efficiency in the transmission of power. The intermittent sliding and stopping action when in use is termed "stick-slip" motion.

By using a specially designed thread form having large flank areas, together with a mating nut which contains small diameter passages for the supply of externally pressurized oil, a leadscrew can have an efficiency of over 95 per cent with no "stick-slip" characteristic being detected. The screw virtually floats on a film of oil, thus preventing metal to metal contact, and eliminating the corresponding frictional effects.

*Recirculating ball screw.* This is a device which minimizes the frictional effects between a screw and nut by means of supporting the nut in a continuous path of ball bearings which fit into the special form of the screw thread. As the screw is rotated, the balls feed forward along the thread in a rolling action, thus eliminating the effects of sliding friction. The balls are trapped within the nut and are returned to the other end of the nut via internal passageways, to continue a further circuit along the thread. Continuous rolling action over any length of thread is thus possible.

*Forging dies.* Items to be produced as forgings are made by forming the hot material by impact between mating halves of a die having the approximate shape of the finished article. Dies are of heat treated die steel, and are costly due to the complexity of the cavities which must be machined into each half.

Conventional techniques of metal machining rely upon physical force for the removal of material by means of edge-cutting tools. Such machining processes suffer severe restrictions when applied to hard, tough or brittle materials, as the heat generated in the cutting process often affects the structure and mechanical properties of the material. Internal stresses may also be induced.

Machine tool research has therefore been directed to the investigation of other scientific principles which could be adapted to the needs of industry in metal removal applications.

Among a number of quite revolutionary techniques which have been developed during the last ten years or so, electrochemical machining (E.C.M.) now holds a prominent place and has good prospects of future development and application.

*Basic principle of electrochemical machining.* The process can be described as the reverse of electroplating, i.e. deplating, and is based upon the laws of electrolysis stated by Michael Faraday in 1834.

It has been defined as the deplating (selected metal removal) of an electrically conductive workpiece to tolerance by electrochemically dissolving the work in a suitable electrically conductive solution.

The process may be best explained by considering the electrolytic cell shown in Fig. 1.

As in electroplating, metal ionizes at the anode and goes into solution in the electrolyte, usually in the form of metallic oxides or hydroxides. Plating at the cathode is prevented, however, using a suitable choice of electrolyte, hydrogen gas being given off. The metal removal or anode cutting which is achieved in this fashion forms the basis of E.C.M.

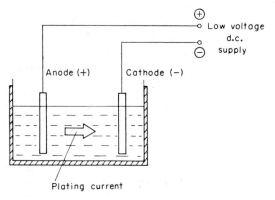

*Fig. 1.*

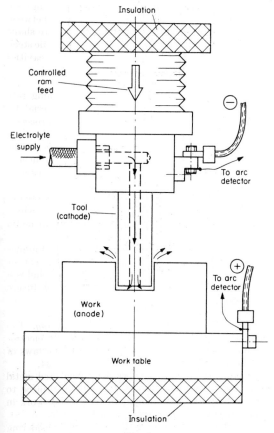

*Fig. 2.*

In order to make the process economic, by achieving high metal removal rates, the gap between the anode and cathode must be much smaller than exists in electroplating processes, (e.g. 0·001 and 0·020 in.). Current densities are many times higher than those used in plating, (e.g. 5000 times).

To maintain the electrochemical action, it is necessary to remove the metallic oxides and hydrogen gas which are formed in the inter-electrode gap. This is effected by pumping the electrolyte through the gap, at high speeds, the contaminated electrolyte being filtered prior to recirculation. A typical arrangement suitable for cavity sinking is shown in Fig. 2.

Rates of flow, liquid pressures and current densities are high. Supply current may be as high as 10,000 amp. In accordance with Faraday's laws, the metal removal rate is directly proportional to the total current passing between the anode and the cathode, directly proportional to the time during which the current passes, and inversely proportional to the valency of the anode (work) material. It is possible therefore to calculate theoretical current flows and metal removal rates.

*Process parameters.* The process parameters inherent in E.C.M. are as follows:

1. Tool design and insulation.
2. Tool material.
3. Tool surface finish.
4. Tool feed rate.
5. Electrolyte flow.
6. Electrolyte composition.
7. Electrolyte temperature.
8. Electrolyte back pressure.
9. Work material, form and microstructure.
10. Current density.
11. Voltage.

These parameters will now be considered in more detail.

*Tool design and insulation.* The electrode tool (cathode) is shaped to the form of the hole or cavity which is to be produced. Allowance is made for the free exhaust of the electrolyte, so that the tool will be

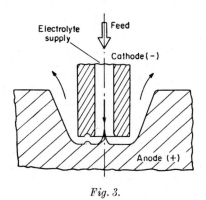

*Fig. 3.*

slightly smaller than the form it produces. The difference between hole and tool size is termed the "overcut".

Figure 3 illustrate this feature, and in addition shows that a non-parallel hole or cavity will result due to the side cutting electrochemical action. The amount of overcut varies from 0·001″ to 0·020″.

Any small form on the lower edge of the tool is reproduced on the hole bottom, the small peak at the centre being due to the path of the electrolyte flow.

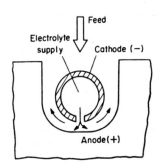

*Fig. 4.*

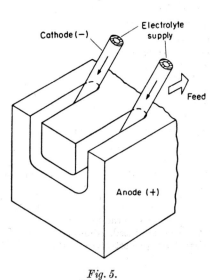

*Fig. 5.*

The cathode may also take the form of a tube having a row of side holes through which the electrolyte is supplied (Fig. 4). This form of tool is used in the operation known as wire cutting (Fig. 5), whereby slugs of costly material can be cut out of a solid piece, thus reducing the amount of scrap.

In an alternative method of suppling the electrolyte known as "edge flow", the electrolyte is constrained between barriers of insulating material (Fig. 6) and flows from one side of the cathode to the other. Turbine blades of aerofoil section are machined by the arrangement shown in Fig. 7.

A further type of operation comparable to a planing operation for machining surfaces longitudinally is available.

The side-cutting effect on tubular electrodes can be prevented by applying insulation on the surface

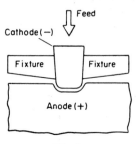

*Fig. 6.*

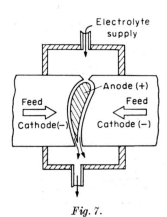

*Fig. 7.*

*Fig. 8.*

where the electrochemical effect is to be prevented. This is equivalent to the "stopping off" process in electroplating (Fig. 8).

Trepanning of holes is effected by insulating both inner and outer surfaces of the electrode with a suitable material, of which there are a great number available (Fig. 9). Porcelain enamels, epoxy resins, nylon and delrin are among the best insulators for this purpose.

*Tool material.* The materials most suitable for use in E.C.M. are copper, brass, stainless steel and titanium, the first two of this group being most popular due to their high conductivity, both electrically and thermally, and their machinability characteristics. Where the tool is slender and greater stiffness is required, stainless steel or titanium are used.

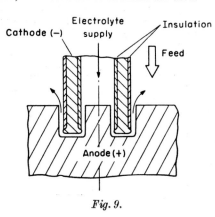

*Fig. 9.*

*Tool surface finish.* The smoothness of the tool (cathode) surfaces is very important, since any surface defect will tend to produce a mirror image pattern on the workpiece.

Any projection which produces a disturbance of the flow of electrolyte may be the cause of a linear defect in the workpiece, due to a starvation of electrolyte downstream, in line with the tool defect.

*Tool feed rate.* The tool feed rate must be accurately maintained in order to provide accuracy in machining. Control systems for the machine slide carrying the cathode are therefore required, and these include a spark detection device.

Rates of feed up to 1·00 in./min have been achieved. The graph in Fig. 10 shows the typical relationship between penetration rate and current density.

*Electrolyte flow.* One of the major difficulties in tool design is that of ensuring that the flow of electrolyte in the inter-electrode gap is as uniform as possible, and that there are no stagnant areas. If such areas exist, there is insufficient electrolyte to sweep away the metallic oxides and hydrogen gas formed by the electrochemical reaction, thus limiting the rate of metal removed in these areas. Flow rates of 17,500 linear ft/min (200 mph) have been used.

Research is being conducted at present into the flow of electrolyte through small channels, and also into the pressure-velocity relationships in the working gap between the electrodes. The effect of electrolyte flow rate on the rate of machining is also being studied.

*Electrolyte composition.* The electrolyte serves to remove the reaction products and the heat generated in the machining area. Practical tests have shown that

no one electrolyte is ideal for all metals. Electrolytes can be acids, bases or salt solutions. The salt solutions may also be acid, basic or neutral. Non-corrosive fluids are preferred so that the machine tool itself is not destroyed.

One of the most common electrolytes used is sodium chloride (2 lb/gallon). Using this fluid, a metallic sludge is formed, which must be removed from the solution. Apart from these disadvantages, it is cheap, and it is easy to maintain at a constant concentration.

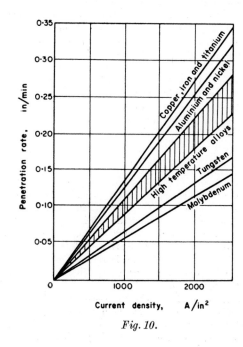

*Fig. 10.*

It is important that concentration of the electrolyte is kept fairly constant (± 1 per cent) as the conductivity of the liquid is changed with varying concentrations, causing difficulty in machining if too weak a solution is used, or conversely crystallization if the solution is too strong.

*Electrolyte temperature.* By increasing the temperature of the electrolyte above room temperature, the conductivity of the electrolyte may be considerably increased. It is most important however, to ensure that the electrolyte does not boil, since if this occurs, the gas bubbles do not conduct current in the working gap between the electrodes, terminating the reaction.

One advantage in using a higher temperature is that less power is required for the machining operation, due to increased conductivity of the fluid.

*Electrolyte back pressure.* In certain instances it has been shown to be desirable to create a back pressure on the electrolyte in the machining area, of a value considerably higher than atmospheric pressure. This has the effect of increasing the accuracy of the hole or cavity being machined, and also reduces or completely eliminates flow lines on the finished work surface. An increase in the conductivity of the electrolyte also results.

*Work material, form and microstructure.* The selection of a suitable electrolyte is related to the type of workpiece material, since better surface finishes may be obtainable by judicious choice of electrolyte. Certain materials, such as cast iron, are reported to be rather unsuitable for E.C.M. applications.

The form of work material is also of interest. Surface finishes obtained on forgings of a specific material are found to be better than those obtained on castings of the same material. In the machining of holes in case-hardened components, the diameter and surface finish produced in the hard casing may be found to be different from those in the softer core.

Considering microstructure, it is suggested that large grain size results in a rougher surface finish than is obtained with fine grain size, and that the presence of chemically insoluble particles results in pitted surfaces.

*Current density.* The current density for any particular application is related to the penetration rate of the tool. For a penetration rate of 1 in./min, a solid square cathode machining an inch square hole requires approximately 10,000 amp, this being equivalent to a current density of 10,000 amp/sq. in. If the penetration rate is reduced to 1/10 of its original value, the current density is reduced in direct proportion.

The current density can be increased until a limiting condition is reached whereby the flow of electrolyte is insufficient to prevent the electrolyte from boiling, when the electrolytic process ceases to function.

Practical current densities quoted for various materials for a penetration rate of 0·1 in./min are given in Table 1, although as suggested earlier the tool feed rates may be as much as seven times the rate on which this table is based.

The current density across the inter-electrode gap between two flat-surfaced electrodes is uniform, and therefore the weight of metal dissolution across the

tool face is also uniform. If, however, as is the more common case, the electrode faces are not parallel (Fig. 11), the current density is increased at the points of minimum inter-electrode space, thus giving preferential dissolution of the anode material. As the minimum working gap is maintained by a control servomechanism, the condition is eventually reached when the electrode faces become sensibly parallel, irrespective of their shape (Fig. 12), the current density then being uniform.

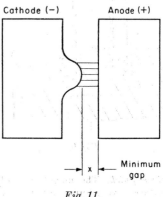

*Fig. 11.*

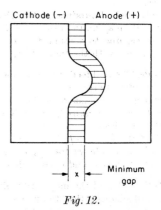

*Fig. 12.*

*Voltage.* Low voltages are employed in all cases of E.C.M., the usual values being below 30 V. It is important, if maximum accuracy in the finished work is desired, that the voltage be kept as constant as possible, maximum deviation figures of $\pm 1$ per cent being quoted for the best possible results.

*Machining characteristics.* Theoretical values of metal removal rates, determined from the electrochemical equivalents of the respective materials are shown in Table 2. This table is based on 100 per cent anode efficiency, but in practice this is never achieved. A conservative figure of 80 per cent is usually taken in estimating production times.

*Table 1*

| Material | Current Density amp/sq. in. |
|---|---|
| Aluminium | 800 |
| Cobalt | 790 |
| Iron | 750 |
| Magnesium | 380 |
| Molybdenum | 1490 |
| Nickel | 790 |
| Titanium | 990 |
| Tungsten | 1620 |

*Table 2. Theoretical cutting speeds of metals assuming 100 per cent anode efficiency*

| Metal | Valence | Density lb/cu. in. | Removal rates at 1000 amp/sq. in. lb/1000 amp hr. | cu. in./min. |
|---|---|---|---|---|
| Aluminium | 3 | 0·098 | 0·74 | 0·126 |
| Antimony | 3 | 0·239 | 3·33 | 0·232 |
|  | 5 |  | 2·00 | 0·139 |
| Arsenic | 3 | 0·207 | 2·05 | 0·165 |
|  | 5 |  | 1·23 | 0·099 |
| Beryllium | 2 | 0·067 | 0·37 | 0·092 |
| Bismuth | 3 | 0·354 | 5·73 | 0·270 |
|  | 5 |  | 3·44 | 0·162 |
| Cadmium | 2 | 0·313 | 4·62 | 0·246 |
| Chromium | 2 | 0·260 | 2·14 | 0·137 |
|  | 3 |  | 1·43 | 0·092 |
|  | 6 |  | 0·71 | 0·046 |
| Cobalt | 2 | 0·322 | 2·42 | 0·125 |
|  | 3 |  | 1·62 | 0·084 |
| Columbium (Niobium) | 3 | 0·310 | 2·55 | 0·132 |
|  | 4 |  | 1·92 | 0·103 |
|  | 5 |  | 1·53 | 0·082 |
| Copper | 1 | 0·324 | 5·22 | 0·268 |
|  | 2 |  | 2·61 | 0·134 |
| Germanium | 4 | 0·192 | 1·49 | 0·129 |
| Gold | 1 | 0·698 | 16·22 | 0·387 |
|  | 3 |  | 5·40 | 0·129 |
| Hafnium | 4 | 0·473 | 3·94 | 0·139 |
| Indium | 1 | 0·264 | 9·43 | 0·595 |
|  | 2 |  | 4·71 | 0·296 |
|  | 3 |  | 3·14 | 0·198 |
| Iridium | 3 | 0·813 | 5·27 | 0·108 |
|  | 4 |  | 3·96 | 0·081 |
| Iron | 2 | 0·284 | 2·30 | 0·135 |
|  | 3 |  | 1·53 | 0·090 |
| Lead | 2 | 0·410 | 8·52 | 0·346 |
|  | 4 |  | 4·26 | 0·173 |
| Magnesium | 2 | 0·063 | 1·00 | 0·265 |
| Manganese | 2 | 0·270 | 2·26 | 0·139 |
|  | 4 |  | 1·13 | 0·070 |
|  | 6 |  | 0·75 | 0·047 |
| Manganese | 7 |  | 0·65 | 0·040 |
| Molybdenum | 3 | 0·369 | 2.63 | 0·119 |
|  | 4 |  | 1·97 | 0·090 |
|  | 6 |  | 1·32 | 0·060 |
| Nickel | 2 | 0·322 | 2·41 | 0·129 |
|  | 3 |  | 1·61 | 0·083 |
| Osmium | 2 | 0·815 | 7·60 | 0·160 |
|  | 3 |  | 5·06 | 0·107 |
|  | 4 |  | 3·80 | 0·080 |
|  | 8 |  | 1·95 | 0·040 |
| Palladium | 2 | 0·434 | 4·38 | 0·168 |
|  | 4 |  | 2·19 | 0·084 |
| Platinum | 2 | 0·775 | 8·02 | 0·173 |
|  | 4 |  | 4·01 | 0·087 |
| Rhenium | 4 | 0·756 | 3·82 | 0·084 |
| Rhodium | 3 | 0·447 | 2·82 | 0·105 |
| Silicon | 4 | 0·084 | 0·58 | 0·114 |
| Silver | 1 | 0·379 | 8·87 | 0·390 |
| Tantalum | 5 | 0·600 | 2·98 | 0·083 |
| Thallium | 1 | 0·428 | 16·80 | 0·654 |
|  | 3 |  | 5·60 | 0·218 |
| Thorium | 4 | 0·421 | 4·76 | 0·188 |
| Tin | 2 | 0·264 | 4·88 | 0·308 |
|  | 4 |  | 2·44 | 0·154 |
| Titanium | 3 | 0·163 | 1·31 | 0·134 |
|  | 4 |  | 0·99 | 0·101 |
| Tungsten | 6 | 0·697 | 2·52 | 0·060 |
|  | 8 |  | 1·89 | 0·045 |
| Uranium | 4 | 0·689 | 4·90 | 0·117 |
|  | 6 |  | 3·27 | 0·078 |
| Vanadium | 3 | 0·220 | 1·40 | 0·106 |
|  | 5 |  | 0·84 | 0·064 |
| Zinc | 2 |  | 2·69 | 0·174 |
| Zirconium | 4 | 0·234 | 1·87 | 0·133 |

Since the electrochemical equivalent can be considered as representing the "machinability" of metals being subjected to E.C.M., it is of interest to note the similarity of many of the values shown in Table 2. The electrochemical equivalents do not change to any marked degree when these metals are alloyed, and some of the toughest of the modern alloys can be machined as easily as aluminium. The rate of metal removal is essentially dependent on the power supplied and is unaffected by the physical properties of the material.

The power requirements for E.C.M. are high compared to conventional machining using edge cutting tools. For lathe turning, 1 hp/in³/min is normal, whereas in E.C.M. 160 hp/in³/min may be required. The major advantage offered by E.C.M. however, is that of the ability to machine tough or hard metals at a much faster rate than by conventional methods.

Dimensional accuracy in the production of holes can be held to ±0·0015 in., while roundness can be maintained within 0·0005 in., and taper to 0·001 in. over any length. Square corners or true radii cannot be produced in the bottom of blind holes, even using a square cornered tool (Fig. 13). The form of curve produced is of approximately exponential form.

Surface finishes vary, depending upon the material, material form and state of hardness. Optimum surface finishes are obtained when machining at the highest rates, this being the converse of conventional machining processes.

Typical values of surface finish are given in Table 3.

*Table 3.*

| Operation | Material | Surface finish (microinches) |
|---|---|---|
| Cavity sinking | Plain carbon steel | 25 |
| Cavity sinking | H. 11 Die steel (hardened) | 25 |
| Planing | Titanium alloy | 10 |
| Planing | Haynes 25 alloy | 5 |
| Face contouring | Plain carbon steel | 10 |

*Machine tool requirements and forms.* The high flow rates of electrolyte through the machining gap between the electrodes require high pumping pressures of the order of 600 psi. The effect is to tend to separate the anode and cathode, hence the practical design of electrochemical machines must provide for extreme rigidity to resist deflexion.

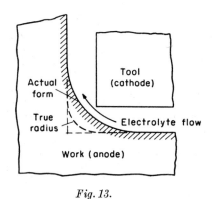

*Fig. 13.*

Stainless steel tables and working areas, piping and pumps, give freedom of choice of electrolyte.

Electrical conductors for attachment to the work material must have adequate current carrying capacity, and the type normally supplied is of flexible braided construction.

Two main designs of electrochemical machines have developed, although the possibility of alternate forms is still evident.

The first is the *cavity sinking machine*, using stationary electrodes in the manner described under the heading of Process Parameters-Tool Design.

The second and more recent development is the *electrolytic turning lathe*, in which the anode (workpiece) rotates on a spindle as in a conventional turning lathe. The cathode is a non-rotating electrode which is carried in the position normally occupied by the tool post. The work spindle is of high conductivity bronze, connected by a heavy duty brush commutator to the positive terminal of a 3000 amp d.c. supply.

Spindle speed is variable, but is usually about 200 r.p.m. for all "turning" work. Work diameters up to 19½ in. may be accepted, and axial movement of the cathode is 4 in., with feed rate up to 0·250 in/min.

This form of machine is ideally suitable for internal or external circular work, such as are illustrated in Fig. 14.

All machines are fitted with enclosures around the work area, with extraction equipment for the removal of the hydrogen gases evolved. Anti-friction slides are essential in the cathode-holding ram, with re-circulating ball screws and nuts, or hydrostatic types, for supplying the feed motion free of stick-slip effects.

*Economics.* The cost of manufacture of complex electrodes precludes the use of E.C.M. for one-off jobs so far as economy is concerned. Simple hole drilling might very well be acceptable on a very low quantity basis. Batches of 10–20 components can be advantageous and for large production quantities, the economic benefits are very substantial.

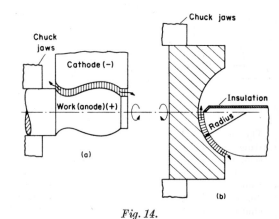

*Fig. 14.*

Savings of between 80 and 98 per cent can be achieved depending upon the complexity of the component being produced, when compared to conventional methods.

Savings in manufacturing time can be as high as 95 per cent.

*Advantages and disadvantages.* The process provides for the stress-free machining of hard, high strength and brittle metals, involving mass metal removal. Multiple operations and complicated shapes present no major difficulties and there are no burrs left on the finished work.

Thin sections of metal may be machined as readily as heavy sections. Accuracies are better than ±0·002″ and surface finishes are good.

On the debit side, factors which must be considered are the cost of extremely complicated tooling, the corrosive nature of some electrolytes, and the very high cost of the capital equipment.

*Other electrochemical machining applications.* The principles of E.C.M. given earlier are incorporated in a number of other applications such as Electrolytic grinding, Electro-polishing, Electro-superfinishing and Electro-deburring, Jet etching, surface marking, etc.

*Electrolytic grinding.* In this process, a modified tool and cutter grinder is used as the basic machine tool (Fig. 15). The spindle carries a metal bonded diamond impregnated wheel (cathode), which is connected to the negative side of the low voltage d.c. supply, via pick-up brushes. The tool (anode) is connected to the positive side of the d.c. supply. Electrolyte is carried into the gap (0·001 in.) between wheel and work, also serving as a coolant.

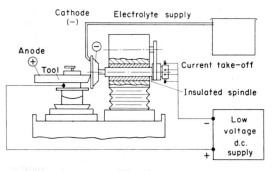

*Fig. 15.*

The abrasive wheel provides some mechanical cutting action, in addition to the electrochemical action, and also scrapes away the E.C.M. products. Thus, the process may be described as mechanically assisted E.C.M., and the rate of metal removal may be higher than the theoretical value for purely electro-chemical dissolution.

Surface finish on tools is consistently less than 10 micro-inches, and does not depend on operator skill. Production rates are about ten times faster than conventional grinding, and wheels are expected to last up to a year. Stellite, H.S.S. tools, Nimonic, Titanium, etc. are easily machined.

*Electropolishing, electro-superfinishing and electro-deburring.* These are probably the first known applications of anodic dissolution, when the anode is separated from the cathode by approximately 10–12 in. The current distribution is sensibly uniform between the two electrodes. Metal removal of a small order occurs, producing surface polishing, and in addition, preferential dissolution of high spots takes place, removing burrs and peaks created by prior machining processes. The shape of the workpiece (anode) remains substantially the same as before treatment, and dimensional accuracy can be maintained by making the necessary allowance for machining before electropolishing.

Electro-superfinishing is used to show up fine cracks and inclusions in high duty components, and

also may be used to remove the surface characteristics produced by previous processes.

*Jet etching.* A jet of electrolyte of between 0·006 and 0·015 in. diameter is used for etching a small circular recess in the face of thin transistor elements. Light is accurately focused on to the sample to ensure generation of electron-hole pairs. The spent solution is sucked away by an aspirator tube.

The surface finish and final form of the etched recess is extremely good.

*Surface marking.* Electrolytic surface marking may be applied to any metal, whether hard or soft, the marking time being only a few seconds.

The electrolyte is carried on a felt pad, which is covered by a stencil of the form which is required on the marked surface. The component is connected to the positive side of a low voltage d.c. supply, and on contact with the negatively connected felt pad, an etched mark is produced.

Stencils are of the standard commercial type, and can be used for trade marks, emblems and similar work.

*Bibliography*

COLE R. R. (May 1961) *Trans. A.S.M.E.*, Series B **83**, 194.

COLE R. R. *Trans. A.S.M.E.*, Paper No. 62–WA–71.

*Elements of Electrochemical Machining*, The Cincinatti Milling Machine Company, Ohio.

Electrochemical Machining of Metals (1963) *Battelle Tech. Review* **12**, No. 1.

GROSS J. A. (Jan. 1963) S.A.E. Paper 618 C, Auto Eng. Congress, January, 1963.

HAGGERTY W. (1963) S.A.E. Paper No. 680 C.

KLEINER W. B. S.A.E. Paper No. 618 D.

HUGHES H. D. (Nov. 1963) Eng. Materials and Design Conference, Paper No. 13, London.

SNAVELY C. A. and GROSS J. A. (1961–62) A.S.T.M.E. Paper, No. SP 62–39.

WILLIAMS L. A. (Jan. 1963) Anocut Engineering Company, *S.A.E. Automotive Engineering Congress,* Detroit, Michigan, Paper 618 B.

D. S. ROSS

**ELECTROLUMINESCENT IMAGE RETAINING PANEL.** When the technical possibilities of electroluminescence were first demonstrated about 15 years ago, it seemed reasonable to expect that this technique would eventually provide the illuminating engineer with a completely new type of lamp with which to solve his problems. The long-established incandescent filament lamp already provided a point source of light; the then more recently established fluorescent lamp provided a linear light source; and now electroluminescent panels, that is, panels which glowed uniformly over their surface, would provide area light sources. Unfortunately, these high hopes have, even now, not materialised and although present-day electroluminescent lamps are many hundreds of times better than the original panels and are proving extremely useful for a number of specialised lighting problems, their use for general room lighting is still not a practical proposition because of their low efficiency.

Other high hopes for electroluminescence have centred around a form of "picture on the wall" television panel. A thin flat panel which could replace the conventional television tube would open up tremendous changes in the television industry, and although techniques by which electroluminescence would be used to achieve these ends have been proposed, again, the hopes have not materialised and the chances of any immediate progress in this direction seem slight. Still another field in which it seemed reasonable to expect electroluminescence to provide a new solution to an existing problem, is that of image intensification, but here again the progress is slow.

*Phosphors for d.c. operation.* One feature common to fluorescent lamps, television tubes and electroluminescent lamps, is the use of a phosphor. These are chemically stable inorganic materials used as finely divided powders to absorb one form of energy and re-emit at least part of it as light.

Commercially available electroluminescent lamps use zinc sulphide phosphors and emit light only when connected to an alternating supply. The vitreous enamel or "ceramic" form of an electroluminescent lamp would be very suitable for the lighting of motor car dash boards except that it involves the additional expense of a convertor to operate from the conventional 12 V battery supply. By making slight changes to the composition of a conventional electroluminescent zinc sulphide phosphor, the ceramic lamp can be constructed so that it emits green light when operated from the normal 240 V 50 c/s supply, but glows an orange yellow when a corresponding d.c. potential is applied.

It was thought that by making even greater chemical changes to the phosphor before using it in this type of construction, it might be possible to develop d.c. electroluminescence to the point where panels could be operated from much lower voltages, perhaps even the 12 V d.c. used in the motor industry. Instead, a phosphor was developed which initially appeared to show d.c. electroluminescence at voltages of about 200, but after switching off the panel it did not re-light when the voltage was re-applied. It soon became apparent that the panel could be made to re-emit light provided it was again exposed to the ordinary room lights, and when these lights were extinguished the panel was again found to be glowing. From this has developed the image retaining panel, but before describing this it is necessary to consider in a little more detail the construction of an ordinary electroluminescent lamp.

*Construction methods.* The conventional form of an electroluminescent lamp is essentially a "luminous capacitor" from which light is emitted when an

alternating field is applied to a thin layer of phosphor particles sandwiched between two flat electrodes, one of which is transparent to light. There are various constructions, but the most robust form employs a metal base plate to which a number of very thin vitreous enamel layers are applied. One of these enamel layers incorporates the phosphor. It is covered with a transparent electrically conducting film, so that this film and the metal base plate function as the two electrodes of the "luminous capacitor." Finally, a transparent overglaze is applied to protect the structure. This vitreous enamel form is called, perhaps rather misleadingly, a "ceramic" electroluminescent lamp, and the construction is shown diagrammatically in Fig. 1.

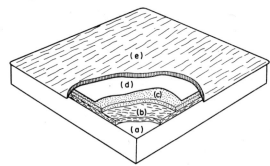

*Fig. 1. Diagrammatic representation of the construction of a "ceramic" electroluminescent lamp. (a) Metal plate (b) Ground coat (c) Phosphor in ceramic layer (d) Transparent conducting film (e) Transparent overglaze.*

The image retaining panel (i.r.p.) is very similar in construction to the ceramic lamp, and consists of a thin metal plate coated on one side with vitreous enamel layers and a transparent electrically conducting film. A special electroluminescent type of phosphor is embedded in one of the enamel layers. If light or X-rays are allowed to fall on the coated surface while a potential difference of 60–120 V d.c. is applied between the front and back electrodes, the plate emits a yellow light where it has been irradiated. When an opaque object is placed between the panel and the exciting source before the panel is irradiated, an image of the object appears on the surface after the source of excitation and the object are removed (Fig. 2).

Provided the d.c. potential is maintained, the glowing image can be retained for 20 or 30 min after irradiation has ceased, but it disappears as soon as the potential is removed or the polarity reversed. If, after a few seconds the potential is re-applied, the panel is again available to receive another image by irradiation. In this way, one of these panels can be re-used thousands of times to retain images for short periods of time.

When light is used to excite an i.r.p., it can be as a very bright short duration flash, such as from a photoflash bulb, or in the form of a much longer exposure at a low level of intensity. All levels of emission brightness from zero to the maximum (about 1 lumen/ sq ft) can be observed over the surface of an i.r.p., giving the image half-tones representative of the density of the original object to the exciting radiation.

*Fig. 2. The image retaining panel (left) with a plastic comb resting on it. When the light is switched off and the comb removed, the image remains "printed" on the panel (right).*

*Electrical properties.* The electrical properties of these panels can be used as a measure of their response to radiation. For example, when a voltage is applied to a panel in the dark, the current taken is about 0·15 mA/sq in and the surface remains unemissive. On exposure to exciting radiation, the conductivity increases.

To prevent damage to the panel the current should not be allowed to exceed about 2 mA/sq in. If the exciting radiation is removed, the current taken by the panel decreases as shown in the conductivity curve of Fig. 3. The light emission from the panel shows a similar rise of excitation with time, and the conductivity curve can therefore be used to indicate the response of the panel to various forms and intensities of excitation.

After irradiation, the brightness can be varied by altering the applied d.c. potential, provided this is done smoothly without any discontinuity. In this way the brightness of the image can be increased somewhat by increasing the voltage, but since the current taken by the panel should not exceed 2 mA/sq in, this usually

means the applied voltage should not be increased beyond about 120 V. Under these conditions the panels remain substantially cool to the touch; actually, the panels can be used at temperatures up to about 80°C, but the image retaining properties decrease at higher temperatures.

*Excitation properties.* Although the i.r.p. is excited by a wide spectrum of electromagnetic radiation there is a pronounced decrease in its response to excitation in that region of the visible spectrum where the panel emits, and it is thought that this is the reason why a glowing image retained on the i.r.p. does not self-excite the remainder of the panel and so the picture quality is retained. The relative response to radiation between 4000 and 9000 Ångström units is shown by curve A of Fig. 4, while curve B shows the spectral energy distribution of the light emitted from the panel.

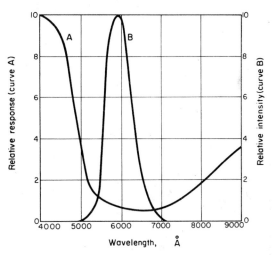

*Fig. 4. Relative response of image retaining panel to radiation between 4000 and 9000 Å (curve A); spectral energy distribution of the light emitted by the panel (curve B).*

These panels also respond to excitation from optical lasers emitting up to $1\cdot2\mu$ and possibly up to $1\cdot6\mu$, and because of this, are proving of use in the setting up of optical lasers. Other forms of excitation can also be used such as gamma-radiation, nuclear particles and high energy electrons. For example, in cathode ray tubes at 10 kV, writing speeds of 10 cm/sec can be recorded and the trace held for periods of half an hour or until the plate is switched off. However, it is their response to X-rays which is of widest interest.

*Excitation by X-rays.* Excitation of an i.r.p. by X-rays can be used to produce a picture which can be viewed and studied after the X-rays have been switched off. This is different from the conventional fluoroscopic screen where the picture is viewed during the actual excitation by X-rays and different again

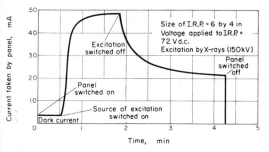

*Fig. 3. Conductivity curve of an image retaining panel while operating at a constant voltage of 72 V d.c.*

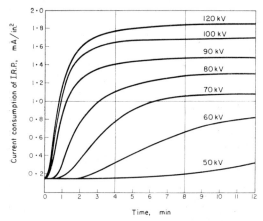

*Fig. 5. Typical response characteristics of an
i.r.p. under X-ray excitation.*

| | |
|---|---|
| *Voltage applied to i.r.p.* | *= 120 V d.c.* |
| *X-ray tube current constant at* | *= 5 mA* |
| *Filter between X-ray source and i.r.p.* | *= 0·5 in Al* |
| *Distance between X-ray source and i.r.p.* | *= 30 cm* |

from the photographic technique where a processing of the film is necessary before the picture can be studied. Consequently, these panels have excited considerable interest from radiographers in both the medical and industrial fields where an image retention time of about 20 minutes is often sufficient for many purposes.

Because of this interest, the response of the panel to X-rays is obviously of prime importance and measurements show that with a constant X-ray tube current, the sensitivity of the panel increases with the tube voltage as shown in the typical curves in Fig. 5. After excitation, the brightness of the image shown by the panel decreases and this is most rapid during the first few seconds after excitation. In practice, this decrease can be reduced by increasing the applied voltage to the panel, always provided the current is not allowed to become excessive. However, this effect of decreasing image brightness especially during excitation by X-rays at lower level than normal, means that the relationship between the X-ray tube current and exposure time during X-ray excitation at a given tube voltage, is not reciprocal, so that the normal photographic technique of doubling the intensity and halving the exposure time cannot be used. The relationship is shown in the curves of Fig. 6.

The X-ray picture obtained by using one of these panels is obviously different from that obtained by using photographic film. The picture from the image retaining panel resembles that seen on a fluorscopic screen. Compared with X-ray film, the picture quality is not yet as good, but is possibly better than that of a fluoroscopic screen. There is reason to believe the picture quality will be improved even more by further developments, and at present the resolution of a typical panel is of the order of 100 line pairs per inch.

In a sense the i.r.p. should be considered as an X-ray viewing device, intermediate between the conventional fluoroscopic screen (which can only be viewed during X-ray excitation, but of course is reusable almost indefinitely) and the permanent record obtained on an X-ray film.

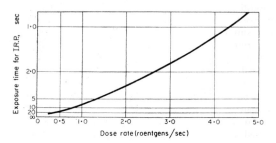

*Fig. 6. Relationship between exposure time and
X-ray dose required for an i.r.p. operating under
optimum conditions (i.e. maximum current
density of 2 mA/sq in.).*

*Possible X-ray applications.* In radiology it is unreasonable at this stage to expect to use the i.r.p. for observing the extremely fine detail detectable by photographic film, but improvements in this direction may result from further phosphor development. However, even in the present form there are a number of possible X-ray applications for these panels. For example, not all industrial radiography is concerned with the detection of fine cracks and there are many cases where correct positioning of components within a sealed container is checked by X-rays. In cases like this the use of the i.r.p. would represent a considerable saving in time and expense by eliminating the photographic process and providing a virtually immediate picture. Another associated advantage is that by providing rapid results, the labelling of components necessary to correlate an X-ray photograph with a specific object in routine testing, can often be eliminated. The X-ray detection of faults during the early stages of the welding of pipelines by using the flexible form of the i.r.p. in place of photography, is another possibility.

The picture quality is also already good enough to be of interest in X-ray examinations during orthopaedic surgery. For example, in orthopaedic surgery of the hip joint it may be necessary to X-ray the patient during the actual course of the operation and since any delay in obtaining the result involves loss of valuable time, it is likely that the i.r.p. will be of considerable value as soon as the X-ray dose which can be used for this purpose has been reduced to a safe medical level. Another advantage in this type of surgery would be that, because the panel is so thin and yet robust, X-ray pictures could be readily taken from different angles. Another application which looks promising is the use of these panels for checking the location of the radiation in radiotherapy treatment.

*Some general features.* Because these panels are excited by light and also the brightness of a glowing image does not exceed about 1 lumen/sq ft, it is necessary to view and inspect the image in a dark room or a light excluding viewing box. It also means that when used for excitation by light, the panel must be adequately shielded from extraneous excitation, for example, by using a cassette. When used for obtaining images from an X-ray source the panel can be covered by a suitable black plastic or paper envelope which is then removed in the dark for viewing purposes.

The panels are sensitive to the polarity of the supply and will only operate if the front electrode is connected to the battery or supply positive. Reversing the polarity does no permanent damage but renders the panel insensitive to radiation while the potential is applied; it also serves to erase an image rapidly. If an i.r.p. is used twice in quick succession, a "ghost" of the first image can be superimposed on the second. To avoid this, the polarity of the applied voltage should be reversed and the panel exposed to light for a few seconds before the second exposure is made with the d.c. voltage now correctly applied. In this way, "ghost" images can be completely avoided. If the panel is left unused for more than a few minutes the "ghost" image will disappear of its own accord.

Although physically robust, moist or damp objects should not be placed directly on the surface of the panel but separated from it by a thin transparent film.

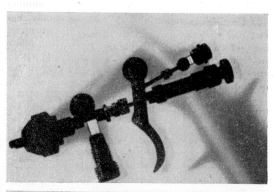

*Fig. 7. Examples of X-ray pictures obtained on an i.r.p. and recorded on 35 mm film. (a) Spray gun head. (b) Grid bias dry battery.*

These panels are fabricated on a steel base plate about 0·02 in. thick and so are fairly rigid. They can, however, be prepared on much thinner metal so that they have a degree of flexibility for use over curved surfaces where the radius does not exceed about 5 in. In such cases the image retaining surface must not be strained into a position of tension.

Because of the way in which these panels are prepared they can be of various shapes and sizes, but for most applications simple rectangles are likely to be of most general application. The sizes in which they can be prepared are limited only by the manufacturing equipment required in applying the very carefully controlled enamel layers, and at present the maximum size normally available is about 10 by 8 in. although narrow strips of 35 or 70 mm width in lengths of 2 ft or more can be specially prepared.

Photographic recording of the image by contact printing directly from the i.r.p. is not very satisfactory, but permanent records on film can readily be obtained with a camera, and some examples of X-ray pictures obtained in this way are shown in Fig. 7.

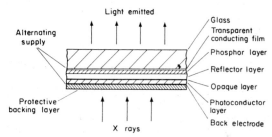

*Fig. 8. Diagrammatic representation of the construction of an X-ray image convertor using electroluminescent and photoconducting materials.*

The i.r.p. should not be confused with image convertors and intensifiers of the type which use multilayer structures of microcrystalline photoconductors such as cadmium sulphide, and conventional electroluminescent sulphides, as diagrammatically represented in Fig. 8. The latter device is intended to intensify an X-ray image, but does not retain the picture for any appreciable time after the X-ray source is removed, while the i.r.p. does not intensify, but retains the image. Some of the physical differences between these devices are listed in Table 1 and it is perhaps worth noting that whereas the intensifier is a logical attempt to combine the two separate processes of photoconduction and electroluminescence, the i.r.p. has developed from the discovery of a phosphor material which combines these properties but does not show normal a.c. or d.c. electroluminescence. Consequently, although the mechanism of image retention may involve two or more process, from the practical point of view the i.r.p. is easier to construct than the intensifier because it involves only one microcrystalline layer.

An enormous amount of work has been done throughout the world in trying to develop electro-

*Table 1. Properties of i.r.p. and image intensifier*

|  | Image retaining panel | PC-EL Image intensifier |
|---|---|---|
| Retention of image after X-ray source removed | 20–30 min | 30–45 sec |
| Irradiate, and view from | Same side | Opposite sides |
| Operating supply | D.c. only; 60–120 V | A.c. only (50 c/s preferred); 600–1000 V |
| Robustness | Unbreakable | Fragile |
| Flexibility | Semi-flexible or rigid | Only rigid |
| Humidity resistance | Very high | Very low |
| Construction | Vitreous enamel layer construction on metal back plate | Organic layers on glass plate |

luminescence towards a preconceived objective. The electroluminescent image retaining panel on the other hand has grown out of the application of an unforeseen property of electroluminescence to produce an equally unexpected effect for which there was at the time of its discovery no conscious need. It is most encouraging that already within the past year the sensitivity and quality of these panels has been very considerably improved.

*Bibliography*

ARDRAN G. M., CROOKS H. E. and RANBY P. W. (1963) *Brit. J. Radiol.* **36**, 927.

British Patent Appln. Nos. 26957 (1959); 34426 (1962); 2907 (1963); 45505 (1963).

HENDERSON A. S. (1962) *New Scientist* **16**, 686.

KAZAN B. Non-destructive Testing **16**, 438; (1958) HENDERSON S. T. (1960) *Phys. Med. Biol.* **4**, 339; and FOWLER J. F. (1960) *Brit. J. Radiol.* **33**, 352.

PAYNE E. C., MAGER E. L. and JEROME C. W. (1950) *Ill. Eng.* **45**, 668.

RANBY P. W. (1963) *Electrical Times* **144**, 341.

YANDO S. (1962) *Proc. IRE* **50**, 445.

P. W. RANBY

**ELECTRON BEAM MELTING.** The melting of metals by an electron beam has been known for some time. M. von Pirani's Patent of 1905 describes an apparatus in which tantalum or other high melting point metals were melted with cathode rays. Progress in beam technology was, for many years, severely limited through the inadequacy of suitable vacuum equipment with high enough pumping speeds, to prevent glow discharges occurring when gas was evolved from the liquid metal; hence the interest in electron beam devices was confined to basic research and equipment on a laboratory scale.

Over the past five years, however, beam melting has been developed to the state where refractory and reactive metals such as tungsten, molybdenum, zirconium and similar metals can be refined and cast quite readily and in some instances the method has proved practical on an industrial scale.

An electron beam apparatus consists of a vacuum chamber and a device for producing electrons and accelerating them towards the work piece (in some cases this includes controlling the direction and focusing of the beam). The vacuum systems used generally are conventional two-stage systems comprising a mechanical fore pump and a high vacuum oil diffusion pump. The degree of vacuum specified varies considerably; proper control can be exercised over the electron beam where the gun operates in a vacuum of $10^{-4}$ torr or less.

There are essentially two types of devices used to produce controlled beams of electrons suitable for melting, the ring cathode and the beam gun.

In the ring cathode system electrons are accelerated from a heated annular cathode surrounding the work piece which is itself the anode. Such an arrangement can be used for zone refining by passing the bar to be refined through the ring at a controlled rate. When used for melting, a consumable electrode is fed through the ring cathode and melted, and the molten metal drips into a retractable hearth, solidifying in bar form, Fig. 1.

The advantages of this system are simplicity of electron beam generation, easy interchangeability of cathode and axial symmetric power distribution. The disadvantages, however, are that the molten pool cannot be heated independently of the consumable electrode (except by the use of two ring beams, which, however, results in the melting plant becoming more complicated). In addition, with the work piece in close proximity to the emitter sudden discharges of gas impurities and/or metal vapours cause glow discharges to occur, which may destroy the cathode. Splashing of metal on to the cathode severely limits its life to a few hours, particularly when melting material in powder or granule form.

These drawbacks can be minimized by the addition of an accelerating anode with a ring-shaped slit for emission of electrons positioned near to the heated cathode. The distance between the feed material and the cathode can then be increased and it becomes possible to melt in a field-free space.

The beam gun is a development to provide high thermal energy similar to those guns used in the cathode ray tube, the X-ray and the electron microscope. The appratus resembles an ordinary diode wherein the heated filament (cathode) emits electrons with an initial velocity which is a function of the filament temperature. The high-velocity electrons are

not effect operation of the gun. The remote mounting of the guns reduces the risk of damage from metal spatter.

The cathode can be heated directly or indirectly, that is, with a heating coil or by bombardment, and lasts longer, up to 100 hours, when the heating is by bombardement. Vacuum conditions restrict the

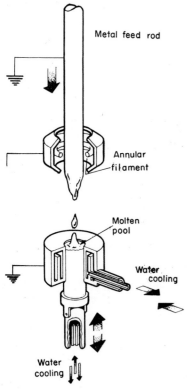

Fig. 1. Drip melt. (Courtesy of W. C. Haraeus Hanau.)

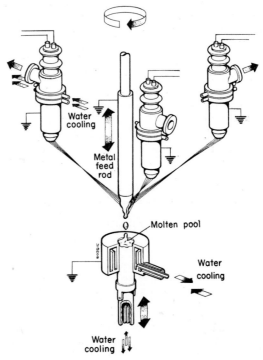

Fig. 2. Drip melt. (Courtesy of W. C. Haraeus Hanau.)

accelerated and projected towards the work piece, which is held positive with respect to the filament. The kinetic energy of the electrons is converted into heat on impact with the material.

By interposing an electric field and/or an electromagnetic lens between the filament (cathode) and the work piece, the electron beam can be collimated and focused on to a particular point; additionally by making one of the lenses the anode the electrons can be made to travel in free flight beyond the lens as a convergent beam, striking the plate or work piece, which can be electrically insulated from the gun. Figure 2 illustrates a gun system. The beam gun system eliminates some of the disadvantages associated with the ring beam. By installing the gun, or a number of guns, in a chamber evacuated independently of the furnace chamber, fluctuating pressures in the latter due to sudden gas discharges from the melt will

choice of material to tantalum or tungsten. Tantalum has the higher emission efficiency but lower resistance to sagging, which places an upper limit on the operating temperature. Tungsten is generally used where the current is passed through the filament to generate electrons. Tantalum emitters are used, however, in the indirect heating method where the curve disk geometry provides mechanical strength.

Electron beam equipment is available with a beam power in the order of 1 kW for machining techniques, 10 kW for welding and up to 250 kW for melting, refining and evaporation process.

Beam melting has several advantages over other thermal sources. By combining several guns there is no upper limit to the amount of energy that can be concentrated in a given area (except space limitations in the chamber). In addition, very high temperatures, together with superheating of the material can be attained under high-vacuum conditions, allowing purification of the melt to take place. Visual observation of the melt is possible, the temperature, melting and cooling rate of which can be controlled.

Electron beam melting and casting equipment employs accelerating potentials up to 30 kilovolts with high beam currents; voltages of this order generate only soft X rays which can be effectively stopped by $\frac{1}{4}$ in. thick mild steel and the chamber provides adequate shielding. A possible source of leakage, however, is through a viewing port but a thickness of lead glass may be added over an ordinary thickness of plate glass to provide necessary shielding.

Higher voltages produce a greater accuracy of focusing and a wider range of power density but weighed against this is the extra shielding protection needed against the X-ray hazard.

An electron beam melting furnace provides a versatile tool in which a number of melting techniques can be carried out:

(a) Floating zone melting where the ends of the rod-shaped material are firmly clamped and the main part of the bar oscillates vertically through the beam zone.

(b) Boat type melting in which the material to be melted is contained in a water-cooled copper boat-shaped crucible, which moves horizontally, the electron beam striking the material vertically, causing a liquid zone to run through the material and affect a zone-refining process.

(c) Drip melting with the feed material introduced into the melt chamber as an ingot rod or in powder or pellet form. As melting proceeds the hearth is retracted to form a rod.

(d) Centrifugal casting with the assembly rotating at speeds up to 700 r.p.m. with the moulds radially located at the periphery of the crucible. Melting is initiated with a small charge in the crucible, thereafter, further material is fed in as granules, Fig. 3.

(e) The technique, producing button melts for metallurgical investigations can be provided by melting a series of alloys at one loading, (Fig. 4) where the material is in the form of powder, compacts or swarf.

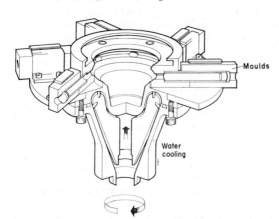

*Fig. 3. Centrifugal casting assembly.*

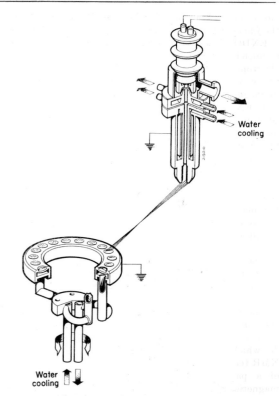

*Fig. 4. Button melt. (Courtesy of W. C. Haraeus Hanau.)*

*Bibliography*

BAKISH R. *Introduction to Electron Beam Technology*, New York: Wiley.
GRUBER H. (1962) *Electron Beam Melting of Metals, Metal Treatment and Drop Forging*, April, May, June.
PIRANI M. VON (1905) German. Pat. Spec. 188, 466.
SMITH H. R., HUNT C. D'A and HANKS C. W. (1958) *A.I.M.M.E., Third Reactive Metals Conf.*, Buffalo, May.
SMITH H. R., HUNT C. D'A and HANKS C. W. (1958) *Third Plansee Seminar*, Reutte. June 22 to 26.
SMITH H. R. (1958) *Vacuum Metallurgy*, New York: Reinhold.

M. L. NOAKES

## ELECTRON NUCLEAR DOUBLE RESONANCE.

Electron nuclear double resonance (*ENDOR*) is a combination of the techniques of electron paramagnetic resonance (*EPR*) and nuclear magnetic resonance (*NMR*). It is a form of radiofrequency spectroscopy. The spectroscopic transition whose frequency is measured is one in which only the quantum state of the nucleus changes, so that it is really a nuclear magnetic resonance; the electron paramagnetic resonance merely serves as a sensitive means of detecting the nuclear resonance.

ENDOR has been used mainly to observe resonance in nuclei which are strongly coupled to the unpaired electrons of paramagnetic centres: this is in contrast to conventional NMR which is usually performed in diamagnetic materials. The information obtained from ENDOR measurements is similar to that obtained from conventional NMR; one can measure nuclear magnetic moments and the magnetic field in which nuclei are situated. In contrast to conventional NMR the magnetic field which the nuclei experience is often quite different from the applied magnetic field because of the internal field $H_e$ set up by the paramagnetic centres. This is particularly so for the nucleus of a paramagnetic ion where the field from the unpaired electrons may be several million gauss. $H_e$ may also be appreciable for the nuclei of the ligands of the paramagnetic centre, and for the nuclei of more distant diamagnetic neighbouring atoms. The measurement of $H_e$ for these nuclei reveals the strength of their interaction with the paramagnetic centre, which in turn gives information about the spread of the paramagnetic wavefunction and its interaction with its surroundings.

It is just the nuclei which experience appreciable $H_e$ which are difficult to observe with the standard NMR techniques. Each nucleus in the neighbourhood of a paramagnetic centre experiences a different magnetic field $H_e$, so that it resonates at a different frequency from the other neighbouring nuclei. In a rigid structure, such as a crystalline solid, the surroundings of the paramagnetic centres in a unit cell are identical to those in all the other unit cells. Hence the number of nuclei contributing to any one nuclear resonance line is comparable to the number of paramagnetic centres. If the sample contains $10^{20}$ centres per cm³, which is the number needed typically to observe a conventional nuclear resonance, there is a strong spin-spin interaction between the paramagnetic centres. This interaction causes rapid fluctuations of the quantum state of the paramagnetic electron spin, and a consequent fluctuation of the magnetic field which the electron exerts on the nucleus. Such a fluctuation broadens the nuclear resonance so much that it is unobservable. To remove this broadening dilute material must be used with about $10^{16}$ centres per cm³, such as is customarily used for EPR; the dilution separates the paramagnetic centres and so reduces their spin-spin interaction. In such dilute material the number of nuclei is too low for detection by standard NMR techniques. ENDOR makes use of the strong EPR signal in such material, and the fact that the strong coupling between the electrons and the nuclei causes the EPR signal strength to change when the NMR is simulteneously excited. The change in the EPR signal enables the NMR to be detected.

The spectroscopic transitions involved in ENDOR are illustrated in Fig. 1 for the simple case of an electron of spin $S = 1/2$ coupled to a nucleus of spin

$I = 1/2$, and magnetic moment $\mu_n$. Suppose that when the electron is in the $S_z = -1/2$ state the field $H_e$ is in the same direction as the applied field $H_0$. The nucleus experiences a field $(H_0 + H_e)$ and the NMR frequency $\nu_1$ is given by:

$$h\nu_1 = 2\mu_n(H_e + H_0).$$

For the $S_z = 1/2$ the orientation of the electron spin in the applied field is reversed, so that the direction of $H_e$ is opposite to $H_0$. The NMR frequency $\nu_2$ is given by:

$$h\nu_2 = 2\mu_n(H_e - H_0).$$

The measurement of both NMR frequencies enables both $\mu_n$ and $H_e$ to be evaluated separately. A measurement of the frequencies of the two EPR lines, which are hyperfine structure components, gives only the product $\mu_n H_e$.

The EPR signal strength of, for example, the transition from level $a$ to level $b$ is proportional to the difference in population of these levels $N_a - N_b$. This population difference is determined by a detailed balance between the transitions induced by the microwave power, which tends to equalise $N_a$ and $N_b$, and the electron spin-lattice relaxation, which tends to maintain thermal equilibrium and the Boltzmann distribution. In addition to a relaxation process which couples levels $a$ and $b$ directly there are others in parallel which proceed through transitions to levels $c$ and $d$ or both, as shown in Fig. 1. The strong excitation of either of the two nuclear resonance transitions effectively short circuits the relaxation process in parallel with that transition. This changes the overall spin-lattice relaxation time between levels $a$ and $b$, and hence changes the magnitude of $N_a - N_b$, and the strength of the EPR signal. The EPR signal is most sensitive to changes in spin-lattice relaxation time when the transition rate induced by the microwave power is about the same as that due to relaxation;

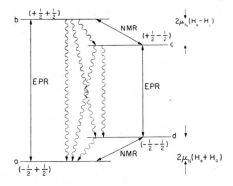

*Fig. 1. Energy-level diagram for coupled electron and nuclear spins ($S = I = \frac{1}{2}$) showing EPR and NMR transitions; the states are labelled ($S_z I_z$). The wavy lines show possible relaxation processes which have different relative magnitudes in different systems.*

this is when the EPR is just beginning to show signs of saturation.

To perform ENDOR experiments one needs a sensitive EPR spectrometer suitably modified to allow the irradiation of the sample with power at the nuclear resonance frequency (often referred to as the second frequency). As the sensitivity is greatest when the EPR is partially saturated it is often necessary to go to low temperatures, so as to achieve partial saturation with a moderate microwave power. Most ENDOR experiments have been performed at temperatures below 20°K, but some have been done at room temperature in systems with long spin-lattice relaxation times. For the greatest ENDOR effect it is necessary also to saturate the nuclear resonance, so that a fair amount of power may be necessary at the second frequency. To drive the nuclear resonance strongly requires a large amplitude of the oscillating magnetic field at the second frequency. Most of the differences in design of apparatus is centred on the problem of how best to introduce this second frequency without materially reducing the sensitivity of the EPR spectrometer. Other differences arise because of the wide frequency range over which experiments can be performed: a resonance from a weakly-coupled nucleus might occur at a few megacycles, while the nucleus of a lanthanon ion might occur at several hundred megacycles. In practice the second frequency has been introduced into the microwave cavity in a variety of ways; coils of one or more turns have been used inside the cavity; coils have also been wound on the outside ot the cavity, the r.f. field leaking into the cavity either through slits or through very thin walls of thickness intermediate between the skin depth at the microwave and second frequencies; also a helical coil carrying the second frequency current has been used as the microwave cavity. Figure 2 illustrates the arrangement which was used by Dr. G. Feher for the first ENDOR experiment in 1955. The emphasis in second-frequency circuitry is always on getting the maximum field at the sample, so that one does not require a high Q coil or a balanced bridge system as one does for conventional NMR. In fact a high Q system has the serious

disadvantage that it is not easy to vary the second frequency when searching for a resonance.

In addition to the variety of experimental arrangements there is also a variety of ways in which the experiment can be done. Typically one can excite EPR with a steady magnetic field and a steady microwave frequency adjusted for the centre of an EPR line, and then search for the nuclear resonance by varying the second frequency. The sensitivity may be increased by modulating the second frequency and detecting the modulation of the EPR signal which is produced when NMR is excited. In some experiments the magnetic field has been modulated to facilitate EPR detection, and in others transient techniques have been used to excite either or both of the resonances. The most suitable technique depends largely upon the electron spin-lattice and the nuclear spin-lattice relaxation times. All of the methods have in common the principle that the excitation of the NMR in some way disturbs the EPR, and this disturbance enables the NMR to be detected.

All ENDOR experiments so far have been done in solids, usually single crystals. The nuclear resonance frequencies are rarely isotropic so that powder specimens would tend to display averaged lines which would be wide and weak; and liquids would be the same for long correlation times, and would show only the isotropic interactions if the correlation time were short.

It has been mentioned that dilute material must be used to reduce the line broadening due to fluctuations of the field $H_e$ produced by interactions between the electron spins. A lower limit to both the ENDOR and the EPR line widths is set by the rapidity of this fluctuation; by the uncertainty principle the line width is related to the mean lifetime of the states. The observed EPR line is often much wider than this because of the variation of static magnetic field from one paramagnetic centre to another which is produced by the dipolar fields of randomly oriented neighbouring nuclei. The EPR line is then a composite of many closely-spaced narrow lines, as illustrated in Fig. 3. These narrow components are called "spin packets". A composite line is called "inhomogeneous", as

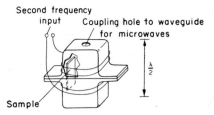

Fig. 2. Sketch of the arrangement for introducing the second frequency used by G. Feher. The microwave cavity is glass with a thin silver coating on the inside. The coil of wire wound outside the cavity carries current at the second frequency, and the r.f. field penetrates the thin walls of the cavity.

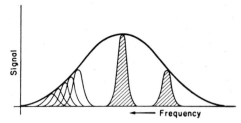

Fig. 3. Some of the overlapping "spin packets" which comprise the EPR line (large envelope) are shown on the left. In a typical ENDOR experiment the central shaded spin packet is saturated and spins are transferred from the other unsaturated shaded spin packet by the NMR, as shown by the arrow.

opposed to a "homogeneous" line where the spin packet is wide enough to swamp unresolved structure, and the whole line is one spin packet. If the EPR line is homogeneous the ENDOR lines have the same width as the EPR lines, but for an inhomogeneous line the ENDOR lines may have the much narrower spin packet width. Experiments have mostly been done with inhomogeneous lines, and the widths observed, usually between 10 and 100 kc/s, are comparable with the line widths of conventional NMR in solids, and are much smaller than the 10 Mc/s typical of conventional EPR.

The many lines which compose an inhomogeneous line are unresolved hyperfine structure due to all the surrounding nuclei. If one particular nucleus changes its quantum state the frequency of the EPR is shifted due to the change in local field at the paramagnetic centre. Thus the spin packet to which the centre contributes is changed. For example the component at the centre of the line in Fig. 3 might be the line c ↔ d in Fig. 1, and the other shaded component might be the line a ↔ b Were it not for the broadening due to all the other nuclei this hyperfine structure would be resolved as two narrow lines. The ENDOR process singles out the two shaded spin packets from the rest of the distribution. The EPR frequency is kept constant so that only the central component of the EPR line is saturated. ENDOR transfers unsaturated paramagnetic centres as shown by the arrow, from the other shaded spin packet to the one at the centre of the line. The contribution of these new centres to the EPR is the ENDOR signal. This example shows how ENDOR methods can be used to measure the resonance of a nucleus even though the hyperfine structure due to that nucleus is not resolved in EPR.

*Applications.* ENDOR has been used to measure nuclear magnetic moments. It does not give as accurate values for nuclear magnetic moments as high resolution NMR because of the limitations set by the linewidth. However, conventional NMR may only be done in a diamagnetic material, and there are some atoms, particularly the lanthanon series, which do not form diamagnetic compounds. For these atoms the ENDOR method, and a triple resonance atomic beam method which is rather similar to ENDOR in principle, are the only ways of obtaining accurate values for the nuclear moments. In addition to the magnetic moments the interaction between the nuclear electric quadrupole moment and the electric field gradient at the nucleus is measured: extraction of the value of the nuclear quadrupole moment requires a calculation of the electric field gradient.

Most of the information which ENDOR gives comes from the measurement of the magnetic field $H_e$ at the nucleus. For most paramagnetic ions with nuclear magnetic moments a hyperfine structure is resolved in EPR measurements. The hyperfine structure gives the product $\mu_n H_e$, so that if $\mu_n$ is known $H_e$ can be found; ENDOR measurements merely give a more accurate value of $H_e$. In most cases the theoretical interpretation of $H_e$ cannot be done with sufficient accuracy to make the additional accuracy of the ENDOR measurements worth while. However, if two isotopes exist which have nuclear magnetic moments, the accuracy of ENDOR measurement is sometimes sufficient for calculation of a hyperfine structure anomaly. The anomaly is defined in terms of different ratios for the hyperfine structures and the nuclear moments of the two isotopes, but it is easily shown to be given by:

$$\Delta = (H_e^I / H_e^{II}) - 1.$$

The anomaly is principally due to unpaired s-electrons, so that its measurement gives information about the amount of unpaired s-electron which is contributing to the field $H_e$. Even in transition metal ions, where the s-electrons are expected to be paired off, there is some unpairing of the closed s-shells due to interaction with the unpaired transition shell electrons. ENDOR measurements of the anomaly enable the s-electron contribution to be evaluated so that one knows what contribution is made to $H_e$ from the other unpaired electrons. A knowledge of this provides some confirmation of the validity of the wave functions which may be used to describe the unpaired electrons, as $H_e$ is proportional to $\langle r^{-3} \rangle$ of the wave functions.

ENDOR also enables one to measure the field $H_e$ at ligand or more distant nucleus. The hyperfine structure due to this field is sometimes resolved in EPR experiments, but very often it is not. The much greater resolution of ENDOR then makes possible measurements of $H_e$ which are not possible using EPR. The field $H_e$ at neighbouring nuclei often has neither the symmetry nor the magnitude of a purely dipolar field from a magnetic moment at the paramagnetic centre. This is because the closed shells of s- and p-electrons on the neighbours may be unpaired due either to covalent bonding or exchange interaction with the unpaired electrons on the paramagnetic centre. ENDOR measures the strength of these interactions. Because of its high resolution it can often be done for several shells of neighbouring ions, and one can discover how the interactions vary with distance from the paramagnetic centre and with position in the crystal structure.

A particularly good example of this is the spectrum of phosphorus donor atoms in silicon. Here the extra electron moves in a large hydrogen-like orbit which overlaps many silicon nuclei. The silicon-29 nuclei enabled ENDOR to be done and the variation of $H_e$ to be measured at various silicon sites up to 11.8 Å from the donor nucleus. From this a map of the wave function of the donor electron was constructed. This work, which was done by G. Feher, is described in his article listed in the bibliography.

*Distant Endor.* For nuclei sufficiently distant from the paramagnetic centre $H_e$ becomes small, and the NMR frequencies approach that for a free nucleus·

In the system illustrated in Fig. 1 the two EPR frequencies are nearly the same. If they lie within the same spin packet they are saturated equally. Simultaneous excitation of the NMR does not then change the EPR signal, and ENDOR of the sort previously described does not work. However, a different mechanism was discovered by Lambe and co-workers; it was called distant ENDOR because it involves nuclei well separated from the paramagnetic centres.

Distant ENDOR makes use of the fact that the small coupling between the nuclei and the electron causes the normally forbidden transitions between levels $a$ and $c$ and between levels $b$ and $d$ in Fig. 1 to become weakly allowed. These transitions are separated from the allowed EPR by the NMR frequency, so that when EPR is saturated at frequency $v_0$ for some paramagnetic centres, a "forbidden" transition is simultaneously saturated for those centres whose allowed EPR is greater or less than $v_0$ by the NMR frequency. Simultaneous excitation of the NMR does not affect the allowed EPR signal strength, but it does change the strength of the "forbidden" transitions.

Both ENDOR processes can occur simultaneously but the conditions for observing them are different. For distant ENDOR the effect is a maximum at the side of a EPR line, and is zero at the centre of the line. Also when the NMR is switched off distant ENDOR recovers with the relaxation time of the nuclear spins; this may be much longer than that of the electrons, which governs the recovery of local ENDOR, and makes the use of modulation techniques difficult for distant ENDOR.

The distant ENDOR does not itself give much useful information, which could not be obtained by conventional NMR (its main advantage is in increased signal strength), but it has been used by Lambe as the basis for a triple resonance experiment. The NMR in a second type of nucleus, whose abundance is low, causes changes in the distant ENDOR of more abundant nuclei due to some form of spin-spin coupling between the two species of nuclei. Such a technique may facilitate the measurement of the properties of nuclei whose abundance is low.

*See also:* Spin packet.

*Bibliography*

FEHER G. (1959) *Electronic Structure of Donors by the Electron Nuclear Double Resonance Technique, Phys. Rev.* **114**, 1219.

LAMBE J. *et al.* (1961) *Mechanics of Double Resonance in Solids, Phys. Rev.* **122**, 1161.

PAKE G. E. *Paramagnetic Resonance,* London: Benjamin.

WOODGATE G. K. and SANDERS P. G. H. (1958) *Measurements of Nuclear Moments in Atomic Beams, Nature* **181**, 1395.

J. M. BAKER

**ELECTRON-PROBE MICROANALYSIS.** Electron-probe microanalysis is a non-destructive technique for the identification and estimation of the component elements in a selected micro-volume at the surface of a solid specimen. An electron gun and lens system produces an electron-probe focused to $\sim 1\ \mu$ diameter at the specimen surface and subject to an accelerating potential usually in the range 5–40 kV. X-ray lines characteristic of the elements in the analysed volume of the sample (Fig. 1) are identified by crystal spectro-

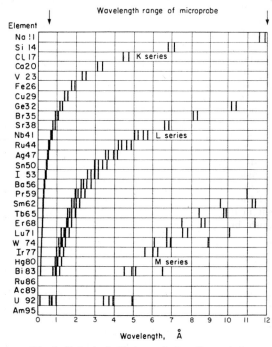

*Fig. 1. Principal X-ray emission lines of the elements.*

metry or by analysis of the pulse height spectrum from an X-ray detector of the proportional type; the measured line intensity can be related to concentration of the corresponding element in the specimen in a manner to be discussed later.

The method was originally investigated by Castaing and his prototype instrument was developed into the CAMECA "Microsonde", shown in Fig. 2, which serves to illustrate the main features of this type of equipment.

Selection of the point of impact of the probe may be by direct viewing with an optical microscope or alternatively by a scanning display as provided on many instruments. Electromagnetic deflexion of the probe, or mechanical specimen movement is arranged so that an area of the specimen is scanned; the intensity of a particular X-ray line modulates a C.R. tube spot moving in synchronism thus building up an "X-ray" picture of the distribution of a selected element over a given area, usually of the order of 10–500 $\mu^2$; the "X-ray picture" may give sufficient information in itself or it may be used to position the probe for point analysis. In addition to the "X-ray picture" the back-scattered or transmitted electrons may be used to give an "electron picture" of the area; in this case the intensity and contrast are due partly to topography and partly to differences in atomic number of the elements in various parts of the specimen.

Many different crystal spectrometer designs are employed on the various instruments utilizing a variety of analysing crystals such as quartz, lithium fluoride, gypsum, mica etc. which generally enable characteristic radiations to be selected for the elements from Na (At. No. 11) upwards; in general little difficulty is experienced in unambiguously identifying the characteristic radiations involved. The X-ray detector may be any of the three main types—Geiger, proportional or scintillation counters; the latter two have the advantage that pulse height discrimination can enhance the peak-to-background ratios obtained from the crystal spectrometers, but they require more complex associated counting circuitry.

Identification of characteristic radiations by pulse height analysis of proportional counter outputs, without the use of a crystal spectrometer can be employed, but in its simple form this technique cannot distinguish between radiations of closely similar wavelengths, e.g. from elements which are near neighbours in the periodic table.

Recently the use of artificial stearate "crystals" has been publicised for the selection of the soft characteristic X rays emitted by elements down to the region of carbon (At. No. 6) and gratings may also be considered for the analysis of the very soft X rays from the low atomic number elements. An alternative technique for analysis of soft X-ray spectra is the proportional counter and pulse height analysis method developed by Dolby in which overlapping pulse height distribution curves are separated by an external network.

Qualitatively the method can identify and indicate the distribution of components with a good sensitivity, but the details of performance depend critically on the equipment, the specimen and the skill with which the appropriate analysing conditions are selected. The size of the analysed region (or the approximate resolution in a scanning picture) may be as small as about 1 micron in diameter and depth, but under commonly used conditions may be several microns. Detection limits depend on the existence of a count rate or accumulated count total from the characteristic line which can be distinguished from the corresponding background, and may range from a few 10's of ppm to a few per cent according to conditions; values below 0·1 per cent are quite common.

Quantitative analysis is normally carried out by referring the measured intensity of a characteristic

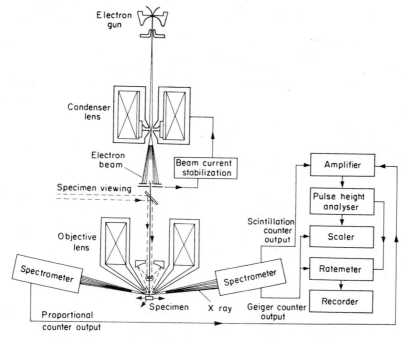

*Fig. 2. Schematic diagram of C.A.M.E.C.A. Microsonde (First model). (By courtesy of Compagnie d'Applications Mécaniques à l'Electroniques au Cinéma et à l'Atomistique and d-mac Ltd.)*

ine of element A in the specimen (IA) to that from a standard of pure A (I(A)) excited and measured under identical conditions of probe current, voltage, spectrometer setting etc. Background effects and counter dead-times if any are allowed for. Conversion of this measured ratio to a concentration figure may then be via calibration curves if available, or via various corrections the details of which are not yet firmly established; these corrections attempt to determine the true generated intensity ratio and to relate this to the weight percentage of the element analysed and in most instances can do this with a reasonably high accuracy.

These corrections arise from the different properties of specimen and standard with respect to both electrons and X rays. Differences of electron back-scatter and retardation result in a departure of the generated intensity from a simple linear dependence on composition, this is generally called the "atomic number effect" since in general, the departure from linearity increases with the atomic number separation of the component elements of the sample. In addition the distribution in depth of the generated X rays will vary from specimen to standard due to these same differences in electron retardation etc. and hence the absorption of the emerging X rays will differ for the two samples even if the mass absorption coefficients are identical; there is thus an atomic number contribution to the X-ray absorption correction which is relatively minor but which is often the cause of confusion.

In general the main correction for X-ray absorption arises from mean mass absorption coefficient differences between specimen and standard; graphical or tabular presentations of absorption functions which take into account absorption coefficients, X-ray take-off angle, probe kilovoltage and, less clearly, sample atomic number are available (Castaing, Birks). Philibert and others have derived formulae by which the correction may be calculated.

In addition to the characteristic X rays excited directly by the electron beam significant enhancement may occur due to fluorescence effects which will result in incorrect analyses if due allowance is not made in the corrections. The more energetic part of the continuum can cause fluorescence but the magnitude of the effect and its variation from specimen to specimen has been shown (Henoc *et al.*) to be small and generally is entirely neglected. In an alloy sample characteristic lines of other elements may cause fluorescence which can be an important effect under certain conditions and methods of calculation of its magnitude are given by various authors (Casting, Wittry, Reed, Duncumb). A special case of fluorescence can occur close to the interface with a second phase and has been considered, by Austin *et al.* and Dils *et al.*

Examples of applications of the technique are widespread through the literature and the reader is referred to the bibliography by Heinrich.

*Bibliography*

BIRKS L. S. (1963) *Chemical Analysis*, Vol. 17, New York: Interscience.
CASTING R. (1960) *Advances in Electronics and Electron Physics* **13**, 317, New York: Academic Press.
DUNCUMB P. (1961) *J. Inst. Metals* **90**, 154.
HEINRICH K. F. J. (1963) *Bibliography on Electron Probe X-ray Microanalysis and Related Subjects* Delaware: Du Pont et Nemours.
POOLE D. M. (1963) *Applied Materials Res.* **2** (1), 3
WITTRY D. B. (1963) *Proc. of A.S.T.M. Symposium X-ray and Electron-Probe Microanalysis*.

## ELECTRON SPIN RESONANCE, APPLICATION OF.

Paramagnetism occurs whenever a system of charge has a resultant angular momentum, Here we are considering the cases where this momentum is of electronic origin, the charge and resulting magnetic moment being due to the electron and its spin about its own axis, and the angular momentum due to its orbit around the nucleus.

Now if this magnetic dipole is placed in an external magnetic field it will orientate itself in discrete directions, depending on the interaction between the electron magnetic moment and the magnetic moment of the nucleus which is being embraced by the electron orbit. The magnitude of this interaction will depend on such factors as the type of atom involved, the phase of the sample, or the solvent used for a liquid sample. In the general case, a nuclear spin of $I$ will give $(2I + 1)$ possible orientations in the applied magnetic field. This splitting of an energy level into two or more components when placed in an external magnetic field is the Zeeman effect, and the technique of electron spin resonance is purely to make electron transitions from the lower level, or ground state, to the higher level. If the two energy levels involved are $E_1$ and $E_2$ then the energy required to produce one transition from $E_1$ to $E_2$ is:

$$E_2 - E_1 = h\nu$$

$h\nu$ being the quantum of energy of frequency $\nu$ and Planck's constant. The distribution of electrons between the two energy states is given (in most cases) the Maxwell–Boltzmann expression. There are, course, relaxation processes involved which allow the transfer of electrons back to the lower energy level. is because the energy difference between the two levels involved in a transition determines the frequency of the radiation emitted or absorbed by the particular atom, molecule or nucleus being investigated that the fundamental problem is to measure the frequency as accurately as possible, this entailing consideration of such questions as line widths, intensities and standards of frequency.

The basic energy equation for resonance is:

$$h\nu = g\beta H$$

$\beta$ being the magnetic moment of the electron, the Bohr magneton, $H$ the applied magnetic field, and $g$ the spectroscopic splitting factor, which is close to 2 for a relatively free spin, i.e. a free radical. It should be noted that in any atom where there is no resultant magnetic moment, due to the cancellation of orbital and spin angular momenta in a closed shell of electrons, the atom is not paramagnetic, but diamagnetic and unsuitable for this method of investigation.

This limitation, together with the high cost of the necessary apparatus may constitute a serious drawback to its general use in industry. Three factors which offset this apparent disadvantage are that; firstly, many materials affect the results of electron spin resonance without actually giving a resonance. One example of this is oxygen, the presence of which rapidly alters the relaxation times of many resonances. Secondly, the very necessity of requiring unpaired electrons may sometimes be turned to an advantage in that results can be obtained from the paramagnetic (unpaired system) independently of the surrounding diamagnetic material which might swamp the required information in a less selective method of measurement. Thirdly, many diamagnetic materials when subjected to radiation lying in the visible ultra-violet or X-ray spectrum, or to nuclear radiation will give an electron spin resonance spectra. In fact nearly all substances, if irradiated long enough and in the proper temperature range will show some electron spin resonance if observed with a sensitive spectrometer, although it should be pointed out that it is not always easy to identify the radicals giving rise to the spectrum.

E.S.R. measurements may be made by placing the sample to be examined in a transmission or reflection type cavity and measuring the change in complex magnetic susceptibility at the condition of resonance. This may be detected as a change in $Q$ of the cavity and monitored by a bridge system, using a hybrid $T$, a circulator or directional coupler in the case of a reflection cavity, or using a transmission cavity. Dispersion or absorption may be detected by monitoring the real or the imaginary components of the complex magnetic susceptibility. The absorption signal is more often studied.

The more usual type of spectrometer incorporates either,

(a) A crystal video system, where by a direct absorption or reflection technique the signal is displayed on a cathode-ray oscilloscope.

(b) A homodyne system which normally includes high frequency modulation of the magnetic field, and a slow sweep through the resonant condition. This will give a first derivative output after phase sensitive detection.

(c) A superheterodyne system, whereby the field is modulated at a low frequency but high frequency detection is employed. The detection frequency is normally obtained by beating or mixing the microwave frequency with the signal from another oscillator.

The occurrence of paramagnetism can be listed as follows:

1. All atoms and molecules having an odd number of electrons, e.g. atomic hydrogen or nitrogen, and the molecule NO, and a few molecules which have an even number of electrons but still have a resultant magnetic moment, e.g. $O_2$.

2. The transition group of elements, i.e. due to partly filled inner electron shells.

3. Free radicals or biradicals.

4. Metals and semiconductors, where the paramagnetism is caused by conduction electrons, and,

5. Compounds whose normal bonds have been modified or broken, say, by irradiation or dislocation. This category includes colour centres, which generally involve electrons or holes trapped in various regions of a crystal lattice, i.e. F or V centres respectively.

Perylene positive ion

←5 gauss→

*Fig. 1.*

Examples of typical traces are shown. Figure 1 is the absorption spectrum of the naphthalene positive ion, and is shown as a first derivative. This illustrates hyperfine splitting due to the interaction of the spinning electron with protons of the parent molecule. Figure 2 shows the first derivative absorption spectrum of acetyl methionine, and is perhaps typical of spectra formed by high energy irradiation of the amino acids and their derivatives. In this case the irradiation breaks a bond in the molecule, and so forms one or more free radicals.

*Free radicals formed by irradiation.* When an organic material is irradiated, the bonds between atoms can be broken. The minimum energy required to break these bonds is determined by the strength of the bond itself. Most bond strengths correspond to wave-lengths in the ultra-violet region and the energy required can often be determined by variation of the wave-length of the incident radiation.

X-ray or $\gamma$-ray irradiation corresponds to much greater energies, and a larger number of bonds can therefore be broken. This usually gives a high concentration of free radicals, but as more different damage mechanisms are possible, and spectra may even overlap, identification is less certain.

Irradiation of inorganic materials (often in the form of single crystals) produces damage centres trapped within the crystal lattice, and may cause the breaking of chemical bonds, or the formation of free atoms, molecules or radicals, which can be stabilised within the crystal.

Often spectra cannot be observed because of fast recombination of these radicals, as the concentration of radicals formed is purely a balance between the

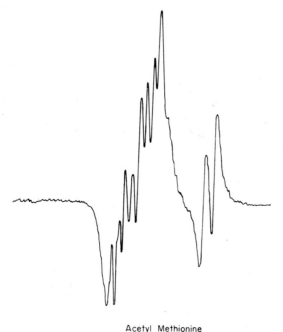

Acetyl Methionine

*Fig. 2.*

rate of formation and the rate of recombination. Sometimes even the relaxation time is too short but these difficulties can generally be overcome by stabilising the radicals at low temperatures.

Another very important act induced by irradiation of hydrogen-containing solids, is the abstraction of the hydrogen atoms. As the excitation energy of the irradiating source is quite often in considerable excess over the binding energy, a small fraction of the hydrogen atoms abstracted may possess very high energy and will display very specific chemical properties. Hydrogen atom reactions and a high level of irradiation in, say, solid hydrocarbons can only be judged by their secondary reaction products, and electron spin resonance is an ideal tool for the analysis of these complex radicals. In frozen aqueous solutions of $H_2SO_4$, $H_3PO_4$ or $HClO_4$, H atoms are quite stable at liquid nitrogen temperatures (77°K) and this provides excellent opportunities for further study.

Secondary reaction radicals can also be formed in solids irradiated by u.v. or high velocity electrons and

evidence has been presented on the existence of these secondary radicals in irradiated PTFE after exposure to oxygen, and in other irradiated polymers.

One other example of hydrogen abstraction lies in the irradiation of frozen (−50°C) cyclohexanol. The intensity distribution and hyperfine structure are consistent with the removal of a hydrogen atom from the carbinol atom.

Free radicals formed by high energy radiation in solids can then be identified by analysis of their e.s.r. spectrum. The radical formation, as we have seen may be due to bond breakage, crystal imperfections, or secondary radicals produced by abstraction of hydrogen atoms. In addition to this, there are small fragments, such as hydrogen atoms which may be trapped in solids at low temperatures.

*Organic free radicals.* In the study of a free radical, much information can be gained about its structure and nature by electron spin resonance techniques. Initially, the stability of the radical is noted. The hyperfine structure is characteristic of its particular chemical group, and the group included in the orbit of the unpaired electron. The number and relative intensity of the hyperfine components will give the spin number $I$ of the interacting proton, or protons and radicals containing $n$ interacting protons will have a hyperfine pattern of $(n + 1)$ lines. Overlapping spectra from different groups can often be separated and additional effects of other non-equivalent protons can be deduced. The $g$ value gives much information, being a measure of the contribution of the spin and orbital motion of the electron to its total angular momentum. Organic free radicals have $g$ values close to 2. Electronic splitting occurs when an atom has more than one unpaired electron associated with it.

A great deal of information about free radicals has been collected and these can be split into about 5 groups:

1. Stable free radicals like diphenyl picryl hydrazyl and similar derivatives.

2. Substituted methyl group radicals like triphenylmethyl and dimesitylmethyl.

3. Aromatic ions in solutions, e.g. naphthalene, perylene, tetracene, anthracene, diphenyl, and phenanthrene.

A detailed study of the unstable intermediates formed in the production of the positive and negative ion aromatic hydrocarbons has been undertaken and full information is available on hyperfine splitting constants.

4. Semiquinones, which are formed during the slow reduction of quinones and related compounds. Examples of these are benzosemiquinone, methyl-substituted benzosemiquinones, ortho- and para-benzosemiquinones, and some naphtho-semiquinones. As can be seen, the semiquinones are ideal insofar as substitution reactions are concerned.

5. Radicals containing sulphur. Measurements have been made on such compounds as thiophenol, thiocresol, thionaphthol and diphenyl disulphide to mention but a few. These sulphur radicals are characterized by $g$ values quite appreciably different from 2.

Much work has been done on irradiated organic compounds, such as paraffin, hydrocarbons and alcohols, alkyl halides, amino acids, acrylic and vinyl polymers and polynuclear hydrocarbons. Biradicals may also be detected and identified.

The use of electron spin resonance has also revealed the presence of free radicals in coal, heat-treated carbon, irradiated graphite, charred carbohydrates and irradiated and unirradiated diamond, and more recently the effect of diphenyl picryl hydrazyl on carbon blacks has thrown more light on the use of carbon blacks as rubber reinforcing agents.

*Free radicals in polymers.* The radicals formed by irradiating small molecules are often interesting as the simplest prototype of radicals to be expected in irradiated polymers where identification is less certain. The usual practise in using e.s.r. as a tool in the study of the radiation chemistry of polymers is to irradiate the polymer and observe the e.s.r. spectrum at 77°K, at which temperature the free radicals can be considered immobile, i.e. trapped. If the sample is then allowed to warm up to room temperature, and the reactions and decay of the free radicals followed, information is given, not only on the radiation chemistry, but also on the solid state reactions taking place. Also it has been found that the e.s.r. pattern depends on the orientation of the polymer chain, and the interpretation of the e.s.r. pattern obtained is often simplified when the polymer is orientated by stretching before irradiation.

Several workers have observed the formation of a 6-line spectrum in polyethylene on irradiation with X rays and observation at 77°K and several types of polyethylene have been studied; in all cases a similar spectrum has been observed. On warming these polymers to room temperature the 6-line spectrum usually disappears leaving a more complex spectrum. Similar results have been obtained with alkathene and terylene.

An 8-line spectrum has been detected in polypropylene using high-energy irradiation at 77°K, and the characteristic decay observed as the sample is warmed to room temperature. Polymethyl methacrylate, polyacrylamide (perspex), polysiloxanes, vinyl monomers, P.T.F.E., polyvinyl chloride also give characteristic spectra, to mention but a few of the materials which have been studied after high-energy irradiation.

The nature and decay characteristics of the free radicals which are produced in Nylon by X-irradiation have been studied, and the results found to be dependent on irradiation temperature, irradiation dose, measurement temperature, storage time, and storage environment. These particular results have been explained by a mechanism which involves radical site migration and an H atom shift.

Bombardment with hydrogen atoms has also helped in gaining knowledge of the identities, concentration and reactions of free radicals and atom intermediaries during the reactions between the gaseous atoms and organic liquids or polymers. Polystyrene has been studied in this way and information has been gained on the mechanisms of radiolytic and photolytic reactions. The effects of oxygen on irradiated polytetrafluroethylene has been studied, and the formation of free radicals in polymers at 77°K during mechanical breakdown.

*Inorganic radicals.* Applications to inorganic chemical research lie mainly in the fields of transition group ions and metallic conductors.

One limitation that applies only to transition group ions is that useful results can usually only be obtained with single crystals. Most spectra are anisotropic, i.e. the $g$ value changes with different orientation of crystal, so in powdered or polycrystalline samples the spectra are smoothed out and resonances may not be detected. There are some exceptions to this, such as $Mn^{2+}$ or $Gd^{3+}$ which are isotropic in many cases.

However, electron spin resonance, when it can be measured, gives a direct and accurate description of the effect of the crystalline environment on the energy levels of the paramagnetic ion. Careful measurement of the resonance spectra enables one to determine the significant "crystal field" parameters. Initial work on inorganic materials was done on ions of the iron group as these were readily available and could be introduced into a great many crystal structures as impurity ions in a host crystal. Work has since been carried out on the rare-earths, palladium and platinum group ions, and more recently on $U^{3+}$, $Cm^{3+}$, $(NpO_2)^{2+}$ and $(PuO_2)^{2+}$ of the higher transition elements.

The usual technique is to dilute the ions magnetically in a diamagnetic host crystal in order to reduce the line-broadening effect of spin-spin interaction, i.e. to reduce the coupling between electron spins. A comprehensive table has been compiled by Orton on some of the impurity ions which have been measured in given host crystals. Two of the transition elements merit particular attention, namely those of manganese and copper. Much work has been done on the $Mn^{2+}$ ion and it has been found that the separation in the $Mn^{2+}$ spectra depends on the nature of the host crystal. More work has been done on the behaviour of unpaired electrons from impurity ions in various fluoride crystals.

The copper ion has been studied in some detail and it has been shown that the two isotopes $^{63}Cu$ and $^{65}Cu$ can be readily detected in a diluted copper potassium sulphate crystal, grown in heavy water to reduce line broadening effects.

Other studies, in brief, include the detection of iron-group impurities in alkali and silver halide crystals, e.g. irradiated silver bromide gives a resonance due to the metallic silver which is formed in the lattice. This mechanism is similar to that of the irradiated lithium hydride resonance, due to the formation of metallic lithium.

Some metallic conductors have been studied and resonances have been observed from lithium, potassium, sodium, beryllium, and caesium. Donor impurities in semiconductors may give a resonance on account of the unpaired electron not required for bonding, and results have been obtained from lithium, phosphorus, arsenic and antimony.

These techniques clearly facilitiate the detection of minute quantities of impurities in a specimen, and a good example of this is its application to vanadium as an impurity in crude oil. Crude oils almost invariably contain trace metals such as vanadium, nickel, copper and iron. Vanadium and nickel are known to be combined at least partly in the form of porphyrin complexes and vanadium porphyrin has in fact been isolated and detected in certain crude oils, and its quantitative detection may be extremely useful as vanadium is known to poison certain catalysts which are used in the refining operations.

It has been known for some time that the oxy-radicals like manganate and chromate can be detected and information is now available on the oxyanions of chlorine, viz. ClO, $ClO_2$ and $ClO_3$.

The radicals $PO_3^{2-}$ and $SO_3^-$ have been studied by Horsfield, some evidence is available for the trapping of $SO_4^-$ radicals and the $NH_3^+$ radical has been identified.

A systematic study of the oxyanions of the non-metals has been carried out and large numbers of simple salts have been exposed to high energy ultra-violet and gamma radiation. By these methods, paramagnetic centres have been produced, and, when the nature of the product is known, considerable information concerning electronic structure has been forthcoming. All of the work undertaken has thrown more light on the molecular structure of the materials being investigated; bond angles and bond lengths can be measured, and it has also been useful in quantitative and qualitative analysis.

*Biological and medical.* Many biological processes are closely connected with free radical reactions, and perhaps a good place to start would be considerations of enzyme interaction. Many years ago it was suggested that many enzyme reactions involved chain reactions in which free radicals acted as an intermediary, and more recently, evidence in favour of this viewpoint has been put forward from the work done on oxidation-reduction systems.

The enzymes catalase, peroxidase, and tyronnase are known to have paramagnetic ions, and the single-electron intermediaries may play an important part in normal biochemical processes. Also of possible application to medical work are the results of Commoner *et al.* who have discovered free radical resonances in several kinds of organic tissue such as liver, brain and

muscle. These may have an important application in the field of cancer research. It has also been shown that free radical concentration is associated with the protein components and that denaturation of the protein destroys the free radical concentration.

Free radicals have been found in germinated and ungerminated seeds, and have been produced by photosynthesis of organic materials such as leaves and barley etc.

Work on carcinogenic activity has been carried out by several workers and it has been suggested that the carcinogenic activity of certain large ring structures may be related to their ability to form negative ion free radicals with mild reducing agents. In contrast to this, non-carcinogenic hydrocarbons require strong reducing agents before such radicals are formed. A high free radical concentration has been found in melanin, which is a pigmentation found in various biological tissues.

Ingram *et al.* have carried out work on free radical concentration in condensed cigarette smoke, and found that some of the free radicals formed disappear on warming the condensate, and it has been noted that most bioassays in tobacco carcinogenis have been carried out using relatively old condensates, i.e. condensates in which the short-lived radicals have already disappeared.

Many early experiments on photosynthesis were carried out on lyophilized materials and free radicals have since been detected in aqueous suspensions of chloroplasts, which are known to be responsible for the essential steps of photosynthesis. The chloroplasts were prepared from tobacco leaves. Very recently work has been done on the formation of free radicals in lyophilized bacterial cells in the presence of oxygen.

Work carried out in the resonance from iron-group ions in the organic compounds phthalocyanine, haemoglobin, myoglobin and their derivatives has thrown more light on the possibility of using electron spin resonance techniques to investigate such compounds in detail; in particular, the haemoglobin of the blood. It is possible that certain blood disorders may be detected in this way.

Irradiation of biological materials is now playing a very important part in the search for an understanding of biological processes. Gordy *et al.* have studied many X-irradiated proteins, viz. cystine, hair, nail, feather etc., and have given suitable explanations of the results. Further studies by Gordy, with the purpose of building up systematic data for X-irradiated proteins, includes work on the polypeptides, complex proteins such as histone, insulin, haemoglobin and albumin, hormones and vitamins such as progesterine, parathyroid, vitamin A, vitamin R and absorbic acid, and nucleic acids such as DNA, RNA, adenosine, cytidine and inosine.

Several other workers have reported the presence of very broad line resonances in unirradiated nucleic acids (DNA and RNA) but the general feeling appears to be that these results are not conclusive and depend to a great extent on chemical treatment.

A series of physical and chemical studies on the factors controlling free radicals in an irradiated biological system has yielded evidence that the radicals formed in cells by high-energy X-irradiation may be responsible for some of the biological effects of irradiation. These experiments were carried out with dried spores of the bacterium Bacillus megaterium. The criteria used were biological in nature, i.e. the ability of the irradiated spore to germinate and to give rise to a microscopically visible colony of cells. The chemical studies give information on the purely chemical aspects of the free radicals produced, whilst physical studies using e.s.r. techniques give information on the type of radicals formed and their physical nature. This problem is perhaps typical of those at present being investigated.

*Physical and solid-state applications.* As far as physical and solid state applications are concerned, a brief survey of some of the fields of interest is given.

The measurement of magnetic susceptibility has already been mentioned, and to this we can add optical absorption, irradiatiation effects, and the calculation of specific heats at low temperatures.

Electron spin resonance effects in solid-state physics cover such measurements as the investigation of the effects of impurities and colour centres in semiconductors, the study of conduction bands and trapped electrons in metals, and the study of radiation damage and property changes in silicon and germanium (e.g. neutron-irradiated silicon, both $p$ and $n$ types).

Cyclotron resonance has been used in the study of semiconductor material and also to study the current carriers in metallic conductors.

The more recent applications in this field are in the design of Masers, but much yet remains to be done in the way of fundamental measurements, e.g. detailed investigation of relaxation lines and their variation with different physical conditions, so that the most suitable type of material can be chosen for Maser operation.

(This article is published by courtesy of Hilger and Watts Ltd.)

*Bibliography*

INGRAM D. J. E. (1959) *J. Brit. I.R.E.* **19**, 357.
INGRAM D. J. E. *Free Radicals as studied by Electron Spin Resonance*, London: Butterworths.
INGRAM D. J. E. (1962) *Research and Development* **5**, 58, Jan.
ORTON J. W. (1959) *Rep. Prog. Phys.*

H. M. ASSENHEIM

**ELEVATED ELECTRON TEMPERATURE.** In a conductor in which the current is carried mainly by the electrons, each electron moves as soon as an e.m.f. is applied, and it is accelerated until it has a collision. Thus all the electrical energy (ohmic heating) goes initially into the electrons which then distribute it throughout the conductor by collisions with the atoms. Because of the large difference in mass, only a small fraction of the energy is exchanged in each elastic collision, although it is randomized upon the first collisions. About $10^4$ elastic collisions will be required to redistribute the initial energy by which time considerably more energy may have been fed into the electrons thereby raising their total energy. In electron–electron collisions the excess energy is shared upon the first collision so that the electrons will rapidly reach an equilibrium Maxwellian distribution amongst themselves, but this may correspond to a much greater total energy, i.e. higher temperature than if they were in equilibrium with the gas. If $\delta m/m_0$ is the fractional energy loss per collision between electrons and atoms (mass $m_0$) then

$$kT_e = kT_0 + (2m_0/3\delta)(J^2 n_e e)$$

where current density is $J$, and electron number density $n_e$.

This effect is important in solid semiconductors and in gaseous conductors or plasmas, as for instance in a fluorescent lamp in which the electron temperature may exceed 15,000°C while the mercury vapour is not much above ambient.

It should be noted that although the electrons have a Maxwellian distribution and may be in a steady state they are not strictly in equilibrium and so the electron temperature does not necessarily obey the usual laws of thermodynamics. It is, however, useful for describing the mean random energy of the electrons.

If ionization in the conductor is caused by electron collisions, the level of ionization will be strongly dependent upon the energies of the electrons, i.e. upon the electron temperature.

This effect may be important in MHD generation for closed cycle applications, since high gas conductivities are possible at gas temperatures considerably lower than those required for thermal ionization. The use of lower gas and MHD duct temperatures may well improve the prospects for closed cycle MHD.

D. T. SWIFT-HOOK

**ENVIRONMENTAL TESTING.** Any device, materia or equipment must be designed and constructed to resist the environment in which it is to be operated. The environments vary widely according to the use of the equipment—it may be a device hooked up on a bench following some fundamental physical research, or may be an electronic equipment operating in an aircraft or missile. In the first case the device may only have to withstand the normal handling it may receive on the bench, whilst in the latter case the environments are severe vibration, shock, temperature extremes, pressure variations, etc.

Whilst environmental testing may be, and is, applied to cars, aircraft, ships, tanks, etc., this section will be confined to electrical and electronic applications.

*Classes of environments.* It is convenient to divide environments into three groups, according to their origin:

(1) Natural environments: e.g. gravity, climatic conditions such as temperature, air pressure, humidity, rain, wind speed, biological and entomological attack.

(2) Environments generated by the equipment or by interaction of equipment with the surrounding medium: e.g. acceleration, vibration, chemical, electrical or kinetic heating, ionization, aerodynamic excitation, etc.

(3) Environments generated by enemy action: e.g. overpressure, earth or water shock, fragments from explosive heads, nuclear radiation, electromagnetic interference, etc.

According to the use to which the equipment is to be put, these environments will exist and their effects must be obviated as far as possible in design. The climatic conditions existing in all parts of the world are detailed, followed by a summary of environmental stresses.

### Climatic Conditions

*Temperate zones.* It is difficult to determine precisely the area of temperate regions because of the gradual merging of the sub-tropical into temperate and the temperate into sub-arctic. However, a general definition is that the temperate zone extends from 40 to 65°, and within these general boundaries deterioration and degradation of electronic equipment and components is not severe. However, some deterioration does take place—some of it man-made—as industrial pollution can have considerable damaging effects on equipments in certain industrial areas; in addition, there are the corrosive effects of a marine or coastal environment.

*Desert regions.* In general, desert regions such as those in North and Central Africa, Arabia, Iran and Central Australia are characterized by high temperatures (and large diurnal variations) and low relative humidities. Air temperatures may range from 60°C (140°F) by day to −10°C (14°F) by night. Diurnal variations of 40°C (72°F) are quite normal to the desert. Solar radiation can be very high and, as a result, ground temperatures in exposed places and the surfaces of equipments and packages exposed to direct radiation from the sun often reach 75°C (167°F). Maximum relative humidities are of the order of 10 per cent, i.e. when the temperature falls during the night. Minimum relative humidities can be as low as 3 per cent and this level has been frequently measured in the Sahara Desert.

One important aspect of the desert areas is that of ultra-violet radiation. Owing to the low atmospheric moisture content, there is little diffusion of the incoming solar radiation, and therefore the percentage of ultra-violet within the solar radiation is relatively high at some 3 per cent of the total radiation, that is the quantity below 3900 Å.

Another condition experienced in the desert is that of dust and sand. Dust may be encountered either as a cloud rising only a few feet above the ground, created by vehicles moving over dry earth or laterite roads, or as a storm created by strong winds of gale force which are a characteristic of all desert areas. Dust particles may be angular, are certainly abrasive, and can be hygroscopic. They vary considerably in size, ranging on the average between 0·005 and 0·02 mm.

Sand may be encountered as a cloud raised by strong winds. Such clouds may extend only a few feet above ground level, because the particles have relatively high mass. Such particles consist largely of grains of quartz varying considerably in size, but having an average diameter of 0·4 mm. On the other hand, at certain times of the year, for instance, the northerly wind blowing across the Sahara, raises sand particles to a height of 10,000 ft which remain suspended for several weeks. In such conditions, visibility is frequently reduced to below 1000 yards. Analysis of this sand/dust shows a deposition of 200–300 particles cm$^{-2}$ hr$^{-1}$, with sizes ranging between 0·5 and 5·0 $\mu$.

*Tropical (hot/wet) regions.* These areas are to be found between the Tropics of Cancer and Capricorn at latitudes 23° north and south of the Equator. The tropics are characterized by sustained high temperatures with small diurnal variations and by high relative humidities. Precipitation is high and the rainfall is usually spread over a large portion of the year. Countries which lie within the tropical belt include West, Central and East Africa, Central America, Malaya, Burma, East Indies, and many islands such as New Guinea.

Air temperatures in the tropics may rise to 40°C (104°F) during the day and rarely fall below about 25°C (77°F) at night. During the heavy part of the rainy season, temperatures may fall down to around 20°C (68°F) for a few nights. The relative humidity is high, rises to over 90 per cent during the night, but falls during the day to between 70 and 80 per cent. Normally, periods of lower humidity coincide with higher air temperatures and frequently the air becomes saturated with water vapour at night (i.e. the relative humidity reaches 100 per cent).

*Maritime climates.* The maritime is largely a mild climate with little of thermal extremes. It includes the Mediterranean coastline and west coast of North America, west coast of Norway and countries such as the British Isles and New Zealand.

These areas do not have identical overall climates, but they can be grouped together because they have in common a lack of temperature extremes. They are regions embracing humidity, rain, fog and salt spray.

*Tropical sea coast.* The climate along the shore of most tropical territories is a combination of fairly high relative humidity with prolonged and sustained high temperatures. In addition to this is the high saline content of the atmosphere which, added to the high relative humidity and high temperature, provides conditions which are conducive to rapid corrosive action on metals.

On most tropical beaches the surf is heavy and continuous. Often a salt cloud is visible when looking along a beach, visibility sometimes being reduced to 1 mile. These salt clouds have been observed to extend roughly 50 yards along the surf line and some 30 feet high.

Another factor affecting degradation at coast sites is that of ultra-violet radiation. With prolonged sunlight and somewhat lower level of relative humidity to that in the rain forest areas, the u.v. content is high but is increased even further by reflected radiation from the sand, which can often be as much as 30 per cent of the direct radiation.

However, decreases in corrosion with distance from the sea is very great, corrosion at only a few miles inland being appreciably lower than in inland rural Britain and indeed not greatly higher than in the dry climate of semi-desert regions.

The preponderating effect of salinity makes it difficult to assess the effects of other ambient conditions. Other things being equal, humidity cannot but assist corrosion, although it is ineffective by itself.

*Sea temperatures in the tropics.* The highest recorded air temperatures in navigable sea areas are 52°C (125°F) in harbour and 38°C (100°F) at sea, the highest sea temperature being 20°C (84°F).

*Polar and arctic climates.* Arctic conditions are characterized by low temperatures which are by no means restricted to the polar regions. Although the weight of water vapour in unit volume of the air at low temperatures may be small, the relative humidity may be high.

In countries such as Siberia, Alaska, Northern Canada, North Eastern Europa, and parts of the Southern Hemisphere, temperatures as low as − 40°C (− 40°F) may often be experienced. In isolated regions − 55°C (− 67°F) is relatively common, but temperatures as low as − 70°C (− 94°F) have been recorded. Seasonal variations of from − 55°C (− 67°F) to 35°C (95°C) have been observed, but the diurnal variation is of the order of 20°C (36°F).

*Sea temperatures in the arctic.* The lowest recorded air temperatures in navigable sea areas are − 40°C (− 40°F) in harbour, and − 30°C (− 22°F) at sea, the lowest sea temperature being 0°C (32°F).

*The upper atmosphere environment.* In considering world environments, high altitudes and, indeed, interstellar space must be included together with the probable impact that it may have on electronic equipment incorporated in missiles and space vehicles. More information is being acquired each month on conditions at high altitudes and it might be advanta-

geous to examine the physical conditions which occur as one ascends from the Earth.

Meteorologists have split up the atmosphere into zones covering specific climatic changes. The first region, which includes the Earth's surface is the troposphere; it extends upwards to approximately 40,000 ft. Next is the stratosphere up to approximately 125,000 ft, the mesosphere to approximately 260,000 ft, and the thermosphere above that.

The ozone layer which absorbs about 5 per cent of total solar radiation lies between about 50,000 and 160,000 ft, with concentration at 56,000 and 82,000 ft. Absorption is mostly in that portion of the electromagnetic spectrum emitting ultra-violet light at short wave-lengths, i.e. below 3000 Å. The precise level of the ozone layer depends to some extent, of course, on latitude and season of the year.

There is also a decrease in the moisture content of air and at about 50,000 ft the air is very dry with relative humidity of some 2–3 per cent.

Air temperature falls about 3°F for every 1000 ft altitude until at about 40,000 ft (top of the troposphere), the average temperature is − 40°F and at 55,000 ft about − 70°F. At this level there is a pause in the temperature fall, and it begins to rise again as the stratosphere is entered.

Wind strength increases with altitude in the troposphere and tends to become westerly no matter what its direction may be at Earth's surface. This is due to the fact that in the upper troposphere there is an extensive low-pressure area centred around the North Pole, extending nearly to the equator. The upper winds swirl counter-clockwise round this polar "low", becoming westerlies. They attain their maximum speed near the tropopause, and then fall off again in the stratosphere. This is known as the "jet stream", a name given to narrow belts of high-speed winds which from time to time alarm aircraft pilots flying near to the base of the stratosphere. These jet streams may be hundreds of miles long, and have velocities of as much as 270 m.p.h. or more. Jet streams are probably most powerful at latitudes around 36° since here the extra impetus given by the Earth's rotation will be at maximum.

Increased altitude in the stratosphere (or isothermal layer as it is sometimes called) brings a slow rise in temperature, until at about 100,000 ft the temperature may be about − 40°F. There are both upward and downward currents in this region, thought not nearly such vigorous ones as those near to the ground. There are also tenuous "mother-of-pearl" clouds at a height of somewhere about 80,000 ft.

Through the use of rockets in researches in the upper atmosphere, considerably more factual data are now being accumulated and results show that between 100,000 ft and 130,000 ft the temperature begins to rise again at nearly the same rate as it falls in the troposphere—approximately 3°F per thousand feet of ascent. In this upper warm region, indications are that the temperature reaches a maximum of between 95°F and 115°F, at from 165,000 to 180,000 ft.

After this peak is passed the temperature declines again to about 50°F at 200,000 ft, and then to a deep minimum of perhaps −120°F at 270,000–300,000 ft. At this height are found the lower boundaries of the aurora and the luminous night clouds. It is not yet fully established whether the clouds are formed of ice crystals or cosmic dust but the amount of water vapour at that height is very small.

The increase in temperature between 100,000 ft and 180,000 ft is probably explained by the strong absorption of ultra-violet energy from the sun at the top of the ozone layer.

Above 300,000 ft there is thought to be a steady rise in temperature again, perhaps through the absorption of solar radiation by monatomic oxygen.

At this height comes the ionosphere, starting with the Kennelly–Heavyside layer (E-layer) at 270,000 ft, and above it the Appleton layer (F-layer), at 490,00 ft.

Higher still, the gas particles are mostly in an atomic state and, finally, there is the outermost shell of the atmosphere called the exosphere, where the particles gradually escape into outer space, almost a perfect vacuum.

### Environmental Stresses

*Vibration.* Vibration is induced in all moving systems, more severe in military equipments and missiles, where rocket motors and aerodynamic forces exert considerable stresses. Amplitudes and frequencies cover a wide range, and in both cases they are highest when considered in relation to guided missiles. Vibration effects are partially amenable to alleviation in that vibration mounts can be designed for affected components to lower their sensitivity. The effectiveness of the vibration mounts are greatly dependent upon how well the forcing amplitudes and frequency spectrum are known.

*Shock.* Shock is induced by sudden application of loads such as transportation, manual or mechanical mishandling, explosive forces, launching and boosting phases of missiles and space vehicles. Here again a wide range of effects is produced. It is possible, by a compromise between mounts for isolation of vibration and shock, to alleviate to some extent the effects of shock. However, it is not possible fully to eliminate shock effects and assemblies and components must be robust enough to withstand a considerable degree of shock.

*Temperature extremes.* The operation of equipment in high desert temperatures and also sub-zero temperatures of the polar regions must be considered in equipment design to avoid over-heating, etc., problems. Even more serious is the requirement for equipments to operate in supersonic aircraft where cooling systems must be adequate to deal with very high temperatures far exceeding those encountered by ground equipment. Finally there are temperature effects associated with rockets and missiles. These may include high temperature burning of flame deflectors due to misalignment during launch, extreme temperatures produced during re-entry, malfunctioning of propellents subjected to extremes. Some of these effects can be alleviated by the use of cooling systems on the ground, and radiation shields in space. Re-entry problems can be largely overcome by providing ablating nose cones.

*High humidty.* High humidity combines with temperature to produce serious and quick deterioration of electrical and mechanical equipment and, also, to affect human efficiency critically. On the other hand lack of a certain level of humidity may cause dehydration and result in the cracking of materials and components. In the case of human efficiency in high humidity, the only feasible approach is to alleviate these conditions,

*Acoustic noise.* This can be serious in the case of jet-powered aircraft and missiles or other systems working in a medium of high acoustic noise content where degradation of components can be considerable, and again can affect the efficiency and reliability of human operators. For sensitive components it is necessary to provide acoustic shielding and human operators must have similar protection.

*Low pressure.* Pressure effects are generally those associated with low pressures encountered at high altitudes, producing corona effects and outgassing of materials. Sealing problems are usually difficult to overcome—in particular where components such as semiconductor integrated circuits must preserve a hermetic seal and a vacuum far better than the surrounding pressure and thus prevent the ingress of gases given off by adjacent materials.

*Ozone.* Ozone effects are mostly linked to rockets and missiles, and other space vehicles, as these effects are limited by altitude to approximately 12–19 miles above the Earth's surface, where infra-red absorption characteristics may prove to be a problem in connexion with satellite surveillance and detection. The effects of ozone can be ignored except in special cases.

*Cosmic radiation.* Cosmic radiation effects will vary considerably with distance from the Earth, and with rate of solar activity. Data collected so far suggests the radiation belt begins about 250 miles above the Earth, reaches maximum intensity at about 6000 miles and disappears at about 40,000 miles. Intensities great enough to give a human the maximum permissible lifetime dosage in about 3 hr of exposure have been recorded. In addition, aircraft and crews flying at great heights may be extremely susceptible to nuclear explosions in the upper atmosphere.

*Acceleration.* Acceleration effects will be felt by some equipments—particularly those employing electromechanical components—throughout many types of operations. The human will also be affected considerably by steady-state acceleration. Ingenuity is called for to ensure that components will withstand the required levels by mounting them in the correct plane so as to minimize the effects.

*Wind, snow, dust, atmospheric contamination and fungus.* These environments are important in isolated cases and each must be considered on its individual merit. In general the exclusion of dust, snow, fungi, and atmospheric contamination depends upon the efficient sealing of the equipment to ensure that no ingress is permitted.

*Salt spray.* Salt spray is an important environment for items operated on or close to the ocean or a surf beach. These include ships, search and anti-submarine aircraft, buoys and beacons, and coastal surveillance radar. This problem is essentially one of protection of materials against corrosion and the prevention of electrolytic action by dissimilar metals.

*Electromagnetic radiation.* Electromagnetism is an important phenomenon to consider wherever electrically operating equipment is used and electrical and magnetic fields generated. The possibility of stray high-frequency fields capable of inducing high voltages in seemingly harmless wires and thereby setting up transients at inconvenient times should always be borne in mind. In many cases, removal of the stray fields by changes in the basic design are advisable.

*Meteorites.* These are, of course, only of interest in relation to the launching of satellites and space vehicles. Meteoritic bombardment will produce important environmental effects if a vehicle enters a dense meteor shower such as often occurs in the Earth's vicinity. Predictions by Soviet Russia as a result of data collected by Sputnik III are that a space vehicle with an area of 100 m² will encounter a meteor body with a mass of 1 g on the average of only once in 14,000 hr of flight, a mass of 0·01 g once every 140 hr, and a mass of 0·001 g once every 10 hr. Consideration of this environment leads to the conclusion that this is a statistical problem. The effects of contact with meteorites must be alleviated, but alleviation means increased weight and possible compromise of test goals, so some balance must be struck to enable a certain probability of success in a particular test to be accepted.

*Weightlessness.* Weightlessness, while definitely a factor in space flight, is almost completely unknown in so far as long term effects are concerned. Short term effects appear encouraging. Most test subjects seem to enjoy weightlessness. Possible difficulties may arise when long time flights are made, and humans try to sleep while weightless, whereupon they may get impressions from their subconscious minds that they are falling. However, it is not known that this will be the case.

*Testing.* In order to ascertain whether equipment will withstand the environments in which it is to be operated, testing must be carried out under simulated conditions. Pre-testing in climatic chambers, vibration machines, etc., will save considerable money and effort in evaluation under actual conditions. Faults which occur in operation may be corrected so that the reliability of the equipment can be considerably improved. It is not always possible to simulate exactly the more difficult environments, such as space, but a series of environmental tests has been issued for most military equipment.

These tests involve the design and construction of testing equipment to simulate high/low temperature, high humidity, low pressure, shock and vibration. Many contractors can now supply this equipment for their use by large firms, but smaller firms may have their equipment tested by special contractors who will carry out environmental tests to any required sequence of tests. The importance of environmental testing at the various stages of development is now generally recognized and there is greater emphasis on combined environmental testing, such as temperature with vibration, and there is a trend towards more accurate simulation of any given environment, such as simulating the flight plan of a modern supersonic aircraft. The widening field of application of electronics and the requirement for greater reliability, means that even more environmental testing is likely to be done in the future.

*Bibliography*

DEF 5133, (Feb. 1963) *Climatic, Shock and Vibration Testing of Service Equipment*, H.M. Stationery Office.

DUMMER G. W. A. and GRIFFIN N. (1962) *Environmental Testing Techniques for Electronics and Materials*, Oxford: Pergamon Press.

GREATHOUSE G. A. and WESSELL C. J. (1954) *Deterioration of Materials, Causes and Preventive Techniques*, New York: Reinhold.

RYCHTERA M. and BARTÁKOVÁ B. (1963) *Tropicproofing Electrical Equipment*, London: Leonard Hill.

G. W. A. DUMMER

**ERGODIC THEORIES.** Two distinct approaches can be followed in constructing the theory of statistical mechanics. The theory, that is, which is to describe average properties of very complicated physical systems (gas, liquid, solid) made up by an enormous number of particles.

The first consists in assuming a certain number of *a priori* probability postulates, the second consists in deriving the laws of statistical mechanics from the laws of pure dynamics of isolated conservative systems. The latter eventually aims to justify the probability definitions assumed in the former, by linking them directly to dynamical laws. Ergodic theory is just that set of theorems which underlie this justification.

*Ergodic theory in classical mechanics.* Discussion about the ergodic problem started as long ago as the time of Boltzmann but it was not until rather recent times that it was given a sufficiently satisfactory solution.

At first it was believed that the justification of the statistical ensemble method could be based on the validity of the following statement, which was known as ergodic hypothesis of Boltzmann: "There exists on every hypersuperface of constant energy $H = $ const. in phase space at least one trajectory which goes through every point of such a hypersurface".

Otherwise stated one made the assumption that the system could go through all the microstates *a priori* consistent with its energy during its time evolution, and this was meant when one said that the system was ergodic.

Plancherel and Rosenthal (1913) though proved that such a hypothesis is not even consistent with the dynamical equations which are satisfied by the particles of the system (and this because Hamilton equations do not allow trajectories with multiple points): this is equivalent to saying that the Lagrangian system that one considers cannot be ergodic in the sense of Boltzmann. It is true nevertheless that it is not necessary, *a priori*, to require such a restrictive condition, as implied by the ergodic hypothesis of Boltzmann, to give a justification of Gibbs method. As a matter of fact the initial condition of a system is always known up to a certain indetermination, consequently in the geometrical representation in phase space a certain region, even though very small, and not a point, will correspond to the initial condition of the system. The point representative of the system will then belong to this region.

During time evolution such a region will move within the layer between the two hypersurfaces $H = E$ and $H = E + \Delta E$, of energy $E$ and $E + \Delta E$ respectively, it will change its shape but its volume will remain constant as follows from Liouville's theorem.

Having this in mind Paul and Tatiana Ehrenfest suggested another, much less restrictive, hypothesis, the so-called "quasi ergodic hypothesis": "There exists, on every hypersurface $H = $ const. in phase space at least one trajectory which fills it densely, that is to say a trajectory which passes at a distance that can be taken arbitrarily small from any given point of the hypersurface"; or equivalently stated: "There exists always a trajectory that crosses two elements of a hypersurface $H = $ const. no matter how small these two elements are chosen".

Fermi in 1923 was able to prove that Lagrangian systems sufficiently complicated and without very special symmetry properties satisfy the quasi ergodic hypothesis, that is to say, they are indeed quasi ergodic systems. It is true though, that nobody was ever able to prove rigorously that the Ehrenfest quasi ergodic hypothesis does indeed allow one to substitute time average with microcanonical average, as it is needed for the justification of Gibbs' method. These first formulations, mathematically unsatisfactory, have, to-day, a purely historical value. Modern ergodic

theory in fact developed following new lines of thought. The first one initiated by Von Neumann, culminated in Birkhoff's theorem (1931) which can be stated as follows: "Let us consider an isolated conservative dynamical system with an arbitrary number of degrees of freedom. The time average (taken over a given dynamical trajectory)

$$\bar{f}(P) = \lim_{T \to \infty} \int_{t_0}^{t_0+T} f(P_t)\, \mathrm{d}t$$

of every (Lebesgue integrable) function $f$ of the point $P_t$, in phase space, representative of the system at the time $t$ exists, is independent of the initial time instant $t_0$ (this result is a consequence purely of the fact that time evolution is canonical), and is still an integrable function and a constant of the motion". From this it follows immediately, then, that if and only if the system is metrically transitive, $\bar{f}(P)$ is a constant on the energy surface $F$ (which is finite in the physically interesting cases) with the exception at most of a set of points of Lebesgue measure zero. That is to say if the energy surface $F$ cannot be decomposed into two sets $F'$ and $F''$, each of positive measure, and invariant with respect to the Hamiltonian group of motion. This means that the time average of $f(P_t)$ is equal to the microcanonical average with the only exception of the above-mentioned exceptional initial phases. Let then $M$ be the time average and $A$ the variable point on $F$, one has:

$$M[f(P_t)] = \frac{\int\limits_{F} f(A)\, \mathrm{d}\sigma}{\int\limits_{E} \mathrm{d}\sigma},$$

where $\mathrm{d}\sigma$ is the microcanonical measure of the surface element of $F$.

The hypothesis of metrical transitivity is essentially different from the old quasi ergodic hypothesis of P. and T. Ehrenfest, since a metrically transitive system is quasi ergodic but not vice versa. It must be said however that no criterion has, as yet, been established to ascertain whether a dynamical system is metrically transitive or not, even though it seems plausible that physical systems sufficiently complicated should have this property.

Another line of approach to classical ergodic theory is the so-called ensemble theory, here the fundamental theorem is Hopf's "mixing theorem" (1953) which states: "if and only if the system is metrically transitive in the space direct sum $\Sigma = \Gamma \oplus \Gamma$, one has:

$$M\left\{ \left[ \int\limits_{F_2} \varrho(A, t)\, \mathrm{d}\sigma - \sigma/\sigma_2 \right]^2 \right\}$$

where $\varrho(A, t)$ is any positive Lebesgue integrable function and $F$ is an arbitrary set of points of $F$. Otherwise stated this theorem asserts that any given distribution $\varrho(A, t)$ is nearly always grossly uniform in the energy shell. Finally let us remark that the frequency of occurrence in practice of any given initial condition cannot evidently be foreseen dynamically. So that the connexion between Birkhoff's theorem and physical reality is established only if one makes some hypothesis on the probability with which the exceptional initial conditions may occur. Obviously the most natural one would be to give a *a priori* probability equal to zero to every set of zero Lebesgue measure. In this connexion it is to be noted also that the ensemble theory approach based on Hopf's theorem automatically gives zero statistical weight to the set of exceptional initial conditions of Birkhoff's theorem. The difficulties met in the classic ergodic approach of Birkhoff and Hopf, the impossibility, that is, of singling out in practice metrically transitive systems have been recently overcome by Khinchin, Truesdell and Morgenstern.

These authors in fact succeeded in justifying (K.T.M. theorem) statistical mechanics only on the basis of the fact that time evolution is canonical and the number of degrees of freedom of the system is very large. They did not make use of any ergodicity condition. K.T.M.'s theorem has in practice the same consequences as Birkhoff's theorem but the set of exceptional initial phases instead of having zero measure has a small measure $\mu$ such that the ratio between $\mu$ and the microcanonical measure of the energy surface approaches zero as the number of degrees of freedom approaches infinity.

More precisely K.T.M.'s theorem states what follows: the relative measure of the set of points for which the following inequality holds true:

$$\left| \frac{M[f(P_t)]}{\int_F f(A)\,d\sigma / \int_F d\sigma} - 1 \right| > Kn^{-1/4}$$

(where $K$ is a positive constant) is a small quantity, of the order of magnitude $n^{-1/4}$, where $n$ is the number of degrees of freedom of the system.

*Quantum ergodic theory.* This was initiated by Von Neumann in 1929. One considers an isolated system enclosed in a finite volume, the Hamiltonian operator is denoted by $H$ and the vector of the system in the Schrödinger representation is denoted by $\Psi(t)$.

Let us assume, for the sake of formal simplicity alone, that the series expansion of $\Psi$ in eigenvectors of $H$ contains only a finite number $S$ of non-zero components, that is to say, that $\Psi$ belongs to a unitary $S$-dimensional space which is called a quantum energy shell. Let

$$\Psi(t) = \sum_{n=i}^{S} \tau_k(t)\,\omega_k$$

be the expansion of $\Psi(t)$ in a given set of basis vectors $\{\omega_k\}$ $(k = 1, ..., s)$. Let us denote these vectors with a double subscript in the following way:

$$\{\omega_{\nu,j}\} \quad (\nu = 1, ..., N; \quad j = 1, ... s_\nu); \quad (\sum_{\nu=1}^{N} s_\nu = S).$$

The linear manifold spanned by the vectors $\omega_{\nu i}, ..., \omega_{\mu, s\nu}$ will be called the $\nu$th cell of the energy shell. If we denote $u_\nu(t)$ the probability that, when a measurement at the time $t$ is made; the system be found in a state represented by a vector belonging to the $\nu$th cell, Von Neumann's fundamental theorem can be stated as follows: "If $H$ does not have degeneracies or resonances (ergodicity condition) and if $s_\mu \gg N$ for every $\nu$, one has:

$$\frac{\mathfrak{M} M\{[u_\nu(t) - s_\nu/S]^2\}}{s_\nu^2/S^2} \ll 1,$$

where $M$ is the time average and $\mathfrak{M}$ is the average over all possible bases $\{\omega_{\nu,j}\}$ of energy shell, all taken as *a priori* equiprobable. Von Neumann's theorem does not have a clear-cut physical significance as pointed out by Fierz (1955). The theorem in fact does not state for which observable the time average of the probability $u_\nu(t)$ is equal to the microcanonical average but only that such an equality is verified for the overwhelming majority of the bases of the energy shell.

It was proved later by Landsberg and Farquhar and by Bocchieri and Loinger that the hypothesis of no degeneracy and no resonances used by Von Neumann in deriving the theorem and indeed even the time evolution of the state-vector actually do not play any role at all. Von Neumann's theorem is simply a consequence of the averaging operation and cannot lead to any real ergodicity condition.

Bocchieri and Loinger successively turned around Von Neumann's point of view and were able overcome the above mentioned difficulties by proving the following inequality:

$$\mathfrak{A}\left\{ M \sum_{\nu=1}^{N} (\omega_\nu(t) - s_\nu/S)^2 \right\} < \frac{1}{S+1} \ll 1 \qquad (1)$$

where $\mathfrak{A}$ is an average over all the initial states of the energy shell, equally weighted. Such a result essentially states that for the overwhelming majority of the initial states $u_\nu(t)$ stays very approximately equal to $s_\nu/S$ and that sizeable deviations can occur only for very short intervals of time. This theorem cannot lead to ergodicity conditions in so far as it does not take into account the fact that a macroscopic measurement on the system precipitates its state vector into a given cell $\nu'$; That is to say by making a measurement one necessarily restricts the initial condition to those corresponding to a manifold with a number $s_\nu$ of dimensions smaller than $S$, these initial conditions have therefore zero weight in an average such as $\mathfrak{A}$.

Prosperi and Scotti were able to show later 1960 (their result was then improved upon by Grossmann (1962)) that an inequality such as (1) is still verified if the average $\mathfrak{A}$ is substituted with an average $\mathfrak{B}$ taken only over the states belonging to the cell $\nu'$ and if a relation of the following type is satisfied:

$$\frac{1}{s_{\nu'}} \sum_{i=1}^{s_\nu} \sum_{j=1}^{s_{\nu'}} \sum_{\varrho=1}^{N} |(\omega_{\nu j}, P_\varrho\,\omega_{\nu' j})|^2 = \frac{s_\nu}{S}\left[1 + 0\left(\frac{1}{N}\right)\right] \qquad (2)$$

where $P_\varrho$ are projections operators on the energy eigenstates ($H = \sum_\varrho E_\varrho P_\varrho$) and $\{\omega_{vi}\}$ is the basis of the energy shell which, now, is supposed to be suitably chosen to represent the macroscopic states of the system.

Relations (2) are ergodicity conditions; a similar result was independently obtained in a slightly different way by Ludwig.

*Master equation and ergodicity conditions.* Another possible way for the construction of statistical mechanics is the one based on the master equation which was suggested for the first time, in quantum machanics, by Pauli (1927):

$$\frac{d\omega_v(t)}{dt} = \sum_{v'} \left( \frac{s_v}{S} w_{vv'} u_{v'}(t) - \frac{s_{v'}}{S} w_{v'v} u_v(t) \right) \quad (3)$$

where $w_{vv'} = w_{v'v} > 0$ is the transition probability per unit time. This equation, which was believed to be of very general validity, has been proved up to now only in very special cases (Van Hove 1955–1957) such as the case of a uniform gas infinitely extended and in the limiting conditions of weak interaction between the molecules, low density, short-range intramolecular forces and the case of a solid with weak interaction between the normal oscillators. Making the hypothesis that, for arbitrarily given $v$ and $v'$, there exists a chain of states $v_1, v_2 \ldots, v_n$ such that

$$v_{vv_1} \neq 0 \quad v_{v_1v_2} \neq 0 \ldots v_{nv'} \neq 0 \quad (4)$$

one can prove (Pauli 1927), (Siegert 1949) that one has the following as a consequence of the master equation:

$$u_v(t) \xrightarrow[\lim t \to +\infty]{} s_v/S. \quad (5)$$

An equation similar to (3) and properties similar to (5) are valid also in classical mechanics (Brout and Prigogine 1956). Employing the same techniques developed for the derivation of (3) one can prove (5) directly under much more general conditions which are satisfied for instance by a gas or a solid not necessarily uniform and with interactions not necessarily weak. One has always to assume though that the system is infinitely extended (Van Hove 1959, Janner 1962 for quantum mechanics; Prigogine and Resibois (1961) for classical mechanics).

Regarding the connexion between the approach just discussed and ergodic theory proper, we remark that relations (4) are essentially "grossgrain transitivity conditions" (i.e. they state that transitions are always possible between any two cells).

Finally it is relevant that one can show (Prosperi 1961) that for systems belonging to the above-mentioned category ergodicity conditions (2) are satisfied.

*Bibliography*

Ergodic Theories (1961) in *Proceedings of the International School of Physics* "Enrico Fermi" Varenna, Italy, New York: Pergamon Press.
Farquhar E. (1964) *Ergodic Theory in Statistical Mechanics*, New York: Interscience.

TER HAAR D. (1954) *Elements of Statistical Mechanics*, New York: Rinehart.
JANCEL R. (1963) *Les fondaments de la Mecanique Statistique classique et Quantique*, Paris: Gauthier-Villars.
KHINCHIN A. I. (1949) *Mathematical Foundations of Statistical Mechanics*, New York: Dover.

P. CALDIROLA

**ETTINGSHAUSEN EFFECT.** When an electric current flows along an electrical conductor in a transverse magnetic field, a temperature gradient appears at right-angles to both the electric current and the magnetic field. This effect, the Ettingshausen effect, was discovered in 1887 by the Austrian scientist Baron von Ettingshausen. It is rather small for metals but can be measured with ease using certain semiconductors and semi-metals.

The definition of the Ettingshausen coefficient $P$ is given in Fig. 1, in which the sign convention adopted

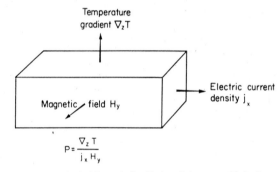

$$P = \frac{\nabla_z T}{j_x H_y}$$

*Fig. 1. Definition of the Ettingshausen coefficient. When $j_x$, $H_y$ and $\Delta T_z$ are positive in the directions shown the Ettingshausen coefficient is negative.*

by most Western authors has been employed. It should be noted that most Russian authors adopt the opposite sign convention. The conditions under which the Ettingshausen coefficient should be measured are that the temperature gradient $\nabla_x T$ in the direction of the primary current flow should be zero and the transverse electric current density $j_z$ and heat current density $w_x$ would also be zero.

The Ettingshausen effect is closely related by thermodynamics to the Nernst effect, just as the Peltier and Seebeck effects are related to one another. The thermodynamic equation

$$\varkappa_i P = Q_i T, \quad (1)$$

when $\varkappa_i$ is the so-called isothermal heat conductivity and $Q_i$ is the isothermal Nernst coefficient, is known as the Bridgman relation. The isothermal heat conductivity is defined with the measured temperature gradient in the $z$ direction and with $\nabla_x T$, $j_x$ and $j_z$ all equal to zero. The isothermal Nernst coefficient is defined with the measured electric field in the $x$ direction and with $\nabla_x T$, $j_x$ and $j_z$ again equal to zero.

As will be explained shortly, the Ettingshausen and isothermal Nernst coefficients can provide useful information about the processes that control the scattering of charge carriers in semiconductors. The Ettingshausen coefficient must be measured in conjunction with the isothermal heat conductivity if it is to be used quantitatively for this purpose and it might be thought that the measurement of the Nernst coefficient would give the required information more directly. However, it is not easy to measure the isothermal Nernst coefficient; generally one has to determine the adiabatic Nernst coefficient, the Righi–Leduc coefficient and the Seebeck coefficient if $Q_i$ is to be specified. Sometimes even the signs of the adiabatic and isothermal Nernst coefficients are different.

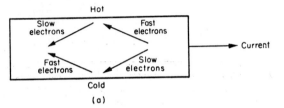

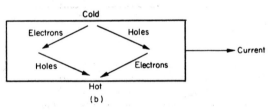

*Fig. 2. Origin of the Ettingshausen effect in (a) an extrinsic conductor and (b) an intrinsic conductor. In (a) it is supposed that the scattering of the electrons increases with energy. The magnetic field is directed towards the reader.*

The origin of the Ettingshausen coefficient can be explained with reference to Fig. 2. If the current carriers all have the same sign, in an extrinsic semiconductor, the situation is shown in Fig. 2(a). The electrons in an $n$-type semiconductor (or holes in a $p$-type semiconductor) all move in the same longitudinal direction under the influence of an electric field. Thus, on the application of the magnetic field they tend to move in the same transverse direction, so building up a Hall field. In equilibrium the Hall field prevents any net flow of charge in the tranverse direction and, if the charge carriers all had the same energy, there would be no tendency for heat to flow in the transverse direction. However, in practice, the electrons have a range of energies and their relaxation time between scattering events is usually energy-dependent. If the relaxation time falls as the energy rises, the fast electrons drift with the Hall field while the slow electrons drift against this field. It is this

different behaviour of the fast and slow electrons that is responsible for the appearance of the Ettingshausen temperature gradient. If it is supposed that the relaxation time $\tau$ depends on energy $E$ according to a law of the form $\tau \infty E^{\lambda}$, the Ettingshausen coefficient has the same sign as the scattering constant $\lambda$. The sign of the Ettingshausen coefficient does not depend on the sign of the charge carriers.

The Ettingshausen coefficient for an extrinsic semiconductor is given by the approximate relation

$$ P \cong \frac{1}{e z_i} \frac{\mu}{1 + \mu^2 H^2} \left[ \frac{\langle \tau^2 E \rangle}{\langle \tau^2 \rangle} - \frac{\langle \tau E \rangle}{\langle \tau \rangle} \right] \quad (2) $$

where the angular brackets denote average values for all energies, $e$ is the electronic charge and $\mu$ is the mobility of the charge carriers. Equation 2 shows that the Ettingshausen coefficient tends towards zero in high magnetic fields ($\mu H \gg 1$). For a non-degenerate semiconductor in a low magnetic field ($\mu H \ll 1$), equation 2 reduces to

$$ P = \lambda \frac{kT}{e} \frac{\mu}{z_i}. \quad (3) $$

In an intrinsic semiconductor or semi-metal there are equal numbers of electrons and positive holes that move in opposite longitudinal directions when an electric field is applied. However, on applying the magnetic field both the electrons and holes tend to move in the same transverse direction as shown in Fig. 2(b). If the drift speed is the same for the two types of carrier, there is not net current flow in the transverse direction and the Hall field is zero. On the other hand, the flow of heat can be considerable since the electron-hole pairs are generated near one face of the sample and recombine, liberating heat, at the opposite face. This bipolar effect leads to a positive Ettingshausen coefficient of considerably greater magnitude than the negative Ettingshausen coefficient in an extrinsic conductor.

When the mobilities of the electrons and holes are equal the Ettingshausen coefficient is given by

$$ P = \frac{\mu}{2 e z_i} \left[ E_g + 2 \frac{\langle \tau^2 E \rangle}{\langle \tau^2 \rangle} \right] \quad (4) $$

where $E_g$ is the energy gap ($E_g$ is negative for a semi-metal and positive for a semiconductor). The bipolar contribution to the Ettingshausen coefficient does not fall to zero in a high magnetic field but remains more or less constant as the field is increased. It is, therefore, possible to observe very large transverse temperature gradients due to the Ettingshausen effects in intrinsic conductors at high magnetic field strengths.

O'Brien and Wallace suggested the use of the Ettingshausen effect as a means of refrigeration in 1958, while Delves, in 1962, drew attention to the fact that reasonably efficient Ettingshausen refrigerators would require intrinsic rather than extrinsic conductors. Since then, very promising results have been found in experimental work on bismuth and its alloys with antimony. For example, Kooi and his colleagues have observed a lowering of temperature below 150°K of 36° using a rectangular bar of $Ni_{97}Sb_3$; a lowering of temperature of 54°K was obtained with a specially-shaped bar of the same alloy. The best performance is achieved when the product $\mu H$ is appreciably larger than unity. Thus, if $H$ is restricted to, say, 1 weber/metre² (i.e. 10 kilogauss) the mobilities of the charge carriers, electrons and holes, must be rather greater than 1 metre²/V sec. Such high mobilities are usually found only at low temperatures so Ettingshausen refrigeration will probably be regarded as a useful technique primarily in low-temperature research. The application of the Nernst effect in the generation of electricity from heat is probably impracticable because of the same requirement that the product $\mu H$ be appreciably greater than unity.

*Bibliography*

CAMPBELL L. L. (1923) *Galvanomagnetic and Thermomagnetic Effects*, London: Longmans Green.
DELVES R. T. (1962) *Brit. J. Appl. Phys.* **13**, 440.
KOOI C. F. *et al.* (1963) *J. Appl. Phys.* **34**, 1735.
LINDSAY P. A. and SIMS G. (1963) in *Encyclopaedic Dictionary of Physics* (Ed. J. Thewlis) **7**, 298, Oxford: Pergamon Press.
O'BRIEN B. J. and WALLACE C. S. (1958) *J. Appl. Phys.* **29**, 1010.
PUTLEY E. H. (1961) *The Hall Effect and Related Phenomena*, London: Butterworths.
STANDLEY K. J. (1962) in *Encyclopaedic Dictionary of Physics* (Ed. J. THEWLIS) **4**, 796; **6**, 321, Oxford: Pergamon Press.
TSIDIL'KOVSKII I. M. (1962) *Thermomagnetic Effects in Semiconductors*, London: Infosearch.
URE R. W. (1963) *Proc. I.E.E.E.* **51**, 699.

H. J. GOLDSMID

**EXPLOSIVE WORKING OF METALS.** The explosive working of metals includes those operations which involve the change of shape of a metal part; the displacement, removal or joining of metals; cutting and shearing; and changes in the engineering and metallurgical properties of a material all through the use of explosive energy. The workpiece material in such operations might be in bulk, plate, sheet, or powder form. Both detonating and deflagrating explosives are utilized as energy sources, with the industrial use of detonating explosives predominating.

Historically, explosives have been used sporadically for the last seventy-five years to work metals in relatively crude commercial applications. However, starting about 1955 the requirements of the aircraft and missiles industry led to the rapid development and growth of metal working techniques requiring the more sophisticated use of explosives. The need for complex and unusual designs, the application of new

materials, and problems peculiar to the short-run production of parts were all related to this development.

Explosive metal working operations can be divided into two main groups, depending on the position of the explosive charge relative to the workpiece. The first group consists of those applications where the charge is located some distance from the workpiece and the energy is transmitted through an intervening medium such as water, air, or oil. These are classed as *standoff operations* and include the forming, sizing, deep drawing, embossing, and flanging of metal parts. All such operations are frequently classed together under the general term *explosive forming*. The second

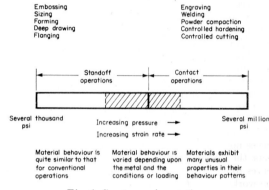

Fig. 1. Spectrum of operations.

group includes applications where the explosive charge is placed in intimate contact with the workpiece. These are known as *contact operations*, and include controlled cutting, controlled hardening, welding, cladding, high density compaction and engraving.

Conceptually, explosive metal working applications can be placed in a spectrum of operations as a function of strain rate or pressure as shown in Fig. 1. The working pressures over this spectrum vary from several thousand psi to several million psi. Most operations are performed in the microsecond to millisecond range, with the contact type having the shortest times and the greatest pressures. At the low pressure end of the spectrum the problems encountered in metal processing are not too dissimilar to those of conventional operations. However, at the high pressure end materials tend to exhibit many extreme behaviour patterns including stress-wave-induced fractures, extensive plastic deformation, and severe work hardening. In the region between, no single generalization of behaviour holds, with the pattern depending on the metal and the specific conditions of the loading. Representative operations for both the stand-off and contact areas are described below.

A typical system for a commercial stand-off operation consists of four basic parts: (1) an explosive charge, (2) an energy transmittal medium, usually water, (3) a die assembly, and (4) a workpiece or preform. Auxiliary equipment might include forming tanks, air compressors, vacuum pumps, and heavy duty handling equipment. The system may be open or closed with major differences existing in the design of the die, the manner in which the die is supported, and the method of containing the water or other medium. Both types normally use only the female part of the die. Due to basic design differences in these two general types of systems, the size of the part that can be worked in a closed system is limited, while with the open system there is essentially no size limitation. Detonating explosives are normally used with an open system; detonating or deflagrating explosives in the closed types.

The physical forms of explosives successfully used in these operations cover almost the entire available range, and include detonating fuse, plastic sheet, plastic bulk, cast solid, pressed solid, powdered, gelled, and liquid explosives. The charge configuration may be a hollow tube, rod, solid cylinder, sphere, ring, sheet, mat, and others, with the shape of the charge being related to the geometry of the end product. For closed systems, special gun powder cartridges have been developed to produce a specific nominal pressure in the forming cavity based on the cavity volume.

The medium between the charge and workpiece serves not only to transmit the pressure, but also affects the delivered impulse and time of operation, serves to distribute the pressure more uniformly over the workpiece, and occasionally is of value in temperature control of the preform. The extensive use of water in stand-off operations is related to its excellent ability to transmit pressure, ease of handling, and economy. Air is the next most commonly used medium, with salts, gelatin, etc., employed only on a restrictive basis for special operations.

The requirements for open system die materials are determined by the type and thickness of the metal to be worked, the configuration and tolerances of the finished part, and the number of items to be produced. Kirksite and plastic-faced dies have been used for light forming requirements; tool steels, and ductile iron for medium requirements; and high impact strength steels for severe services. Some of the larger dies in current use weigh in excess of 44 tons. Closed system dies are usually made from the same materials as are used for dies in comparable conventional forming operations.

Die design details vary with the type of operation. The cavity forming and sizing of parts requires air removal from the die through the use of either a vacuum or an adequate system of venting. Hold-down rings are employed in the drawing of parts to prevent wrinkling and to provide control over the stretch-draw ratio of the piece. Underwater open system dies require a sealing method to keep water from the die cavity. Closed system dies require a gas-tight arrangement, especially when using a low explosive. Parts of simple design are normally produced by means of a single explosive charge, while those of complex design may

require several firing operations and a series of stages in the die.

Figures 2 and 3 are representative of open system methods for forming parts with a detonating explosive and a water medium. The method of Fig. 2 requires a large, water-filled tank where the charge weight, stand-off distance, and other design parameters are related

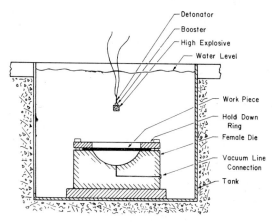

*Fig. 2. Open system explosive forming in water tank.*

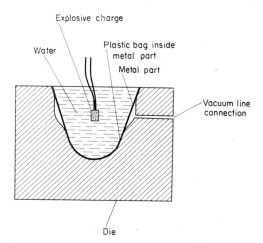

*Fig. 3. Explosive forming with self-contained open system.*

to the size, thickness, configuration, depth of draw, and properties of the workpiece. The forming of the part results mainly from the action of the initial pressure pulse produced by the explosive detonation. In some systems additional working may be obtained from the bubble effect which results from the behaviour of the detonation products. In Fig. 3 the system is self-contained with the explosive charge and water all located within the preform of the workpiece. Explosive weights in such operations vary from a few ounces to several pounds depending on the part to be produced.

Figure 4 shows a closed system method for bulge forming cylindrical parts. The energy source is a powder cartridge inserted and fired at the top of the die enclosure. In this system the workpiece contains air and the bulging is performed by the expanding gaseous products of the powder charge. Closely related are the *cartridge-hydraulic systems* where the explosive force actuates a ram which, in turn, forces a hydraulic fluid to perform such operations as bulging, blanking, shearing, and cupping.

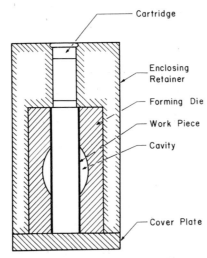

Cartridge

Enclosing Retainer

Forming Die

Work Piece

Cavity

Cover Plate

*Fig. 4. Closed system forming with a powder cartridge.*

Parts currently produced in commercial stand-off operations range in weight from a few ounces to several tons; their major dimensions extend from a fraction of an inch to over twenty feet; and thicknesses vary from a few thousandths of an inch to about six inches. Items have been produced in almost the complete range of present day metals. The unusual and more complicated shape is the rule rather than the exception, for it is in this area that explosive stand-off operations have demonstrated their commercial capabilities.

The effects of a contact operation on a metal body are frequently of an extreme nature. Extensive plastic deformation, distortion, work hardening, fracturing, and occasionally localized phase transformations in the structure of the material may be expected. Many commercial applications are designed around these behaviour patterns and use them to advantage as in the explosive hardening, cutting, welding, and compaction of metals.

When an explosive charge is detonated in contact with a metal specimen work hardening in the region of the loading can be expected, and changes in the engineering properties of the metal will result. The hardness value and the depth of work hardening are dependent on the initial properties of the metal, and on the nature of the impulse delivered by the load. The major industrial use of *explosive hardening* to date has been with parts of austenitic manganese steel which are subject to severe impact and abrasion, such as railroad frogs, rock crusher jaws, grinding mills, and power shovel buckets.

Commercial explosive hardening is normally accomplished by detonating a thin layer of a plastic sheet explosive in contact with the surface to be hardened. This surface may be only a small restricted portion of the part, or an extensive area. Explosive hardening increases the BHN of the metal both at the surface and in depth. In doing so, it increases the yield and tensile strength so that the piece offers more resistance to impact type wear and has less tendency to deform.

The *explosive cutting* of metals can be divided into different categories, depending upon the dynamics of the cutting process. Three such categories of cutting are: (1) with contact demolition charges, (2) with shaped charges, and (3) with explosively induced stress waves. All methods have commercial application.

To cut metal with a contact demolition charge, the explosive is detonated in direct contact with the part. The cutting action is a shearing process which results from the large contact pressure generated at the metal-explosive interface. The commercial uses of this oldest and most commonly used method of explosive cutting originated from military demolition requirements.

To perform cutting or drilling operations with a shaped charge, an explosive system with a metallic liner is placed a short distance from the workpiece. Under the action of the detonating explosive the metal liner collapses and is projected as a jet of high velocity fragments. The cutting or drilling action results from the high velocity impingement of the jet fragments against the workpiece. Shallow line cuts, or deep, small-diameter holes can be made depending on the geometry of the charge. This technique was developed for ordnance requirements in World War II and is now finding numerous commercial applications.

To cut with stress waves, a relatively small amount of explosive is placed in contact with the part. When the explosive is detonated, stress waves which follow theoretically predictable paths are generated in the body. The cutting process results from the interaction of stress waves at predetermined locations within the part. Fractures produced in this way are specifically categorized by such terms as spalling or *scabbing* and *corner fractures*, depending upon the dynamics of the stress wave interaction by which they are formed.

*Explosive welding* and *explosive cladding* are operations in which detonating explosives are used to bring metal surfaces together at high pressures and high relative velocities in such a manner as to produce severe but localized plastic flow at the interacting surfaces, resulting in a high strength bond between the parts. Current industrial interest includes the cladding of large plates, lining the inner surfaces of tubular parts, joining billets of dissimilar metals prior to rolling, butt-welding pipes, and the use of metal-explosive devices for performing unusual welding operations in remote locations. Bimetallic applications appear to be most numerous. Many of the heavy-duty type welds are performed with contact charges, whereas explosive cladding is usually the result of a short stand-off operation with water or some other material used as a buffer. In general, the strength in shear across a good welded surface is comparable to the strength of the parent metal, or better.

A wide variety of metals are compatible to the explosive welding process in both similar and dissimilar combinations. Sheets of several different metals can be welded into a multilayer composite part in one operation. Many of the welds are performed with a vacuum between the interacting surfaces, others are performed in air.

For the high density compaction of powders, a variety of expendable, *explosively-activated presses* have been developed to impart high-intensity, short-duration loads. Much of the current interest in the *explosive compaction of powders* stems from the need for high density parts of powdered materials for use in missile and atomic energy applications. Press design is based on the geometry of the pressed part and the desired pressures, with most presses using either explosively-driven pistons or a "squeeze" effect derived from semi-implosion concepts. Pressures developed in such presses normally range from several hundred thousand psi to several million psi.

The *explosive engravement* of metal surfaces utilizes a low density stencil between the workpiece and the explosive charge. Paper, cardboard, lace and similar materials are appropriate for use as stencils. The engravement process is the result of differential pressure focusing on the metal surface based on the design of the stencil. Adequate control can be exercised over engravement depth in the range of about one thousandth to ten thousandths of an inch. Commercially, explosive engraving serves as a low cost method for producing intricate designs on prototype parts.

*Bibliography*

Cook M. A. (1958) *The Science of High Explosives*, New York: Reinhold.
Dove T. E. (1961) *in Encyclopaedic Dictionary of Physics* (J. Thewlis Ed.) **3**, 46, Oxford: Pergamon Press.
Kellner K. (1961) *in Encyclopaedic Dictionary of Physics* (J. Thewlis Ed.) **3**, 49, Oxford: Pergamon Press.
Kolsky H. (1953) *Stress Waves in Solids*, Oxford: The University Press.
Mercer D. M. A. (1961) *Explosive wave or Shock*, *Encyclopaedic Dictionary of Physics* **3**, 50, Oxford: Pergamon Press.
Rinehart J. S. and Pearson J. (1954) *Behavior of Metals Under Impulse Loads*, Cleveland: American Society for Metals.
Rinehart J. S. and Pearson J. (1963) *Explosive Working of Metals*, London: Pergamon Press.
Shewmon P. G. and Zackay V. F. (1961) *Response of Metals to High Velocity Deformation*, New York: Interscience.

J. Pearson

# F

**FALL-OUT FROM NUCLEAR EXPLOSIONS.** The radioactive debris from a nuclear weapon consists of fission products, unexpended fissile material such as uranium-235 and plutonium-239, and activation products—the result of the capture of excess neutrons in the weapon material or in the environment of the explosion. This vaporized mixture is forced upwards into the atmosphere by the tremendous release of energy in the explosion. As the cloud rises and cools, the radioactive products condense to form particles of solid debris; it is this particulate material together perhaps with gaseous tritium (from fusion reactions) and carbon-14 (from the neutron reaction with atmospheric nitrogen) that later constitutes fall-out. Nuclear radiations are emitted as the radioactive atoms disintegrate. The life-time of the many products varies from fractions of a second to many years so that the radioactivity of a sample of the mixture decays with time in a complex manner.

The first nuclear explosion occurred in 1945. During 1954 to 1958 the rate of weapon testing was continuously high: after 1958 there were no substantial explosions until the autumn of 1961 when a massive programme of testing started that, by the end of 1962, had roughly doubled the amount of fission produced.

The radioactive fall-out produced by nuclear weapons should be recognized as an artificial addition to the natural fall-out on to the Earth's surface. The natural fall-out arises from the continuous production of radioactive isotopes in the atmosphere by cosmic rays and also from the decay products of radon emanation from the ground.

*Local fall-out* consists principally of the larger particles that are deposited in the vicinity of the explosion. The fraction that falls out locally, within a few hours, varies from virtually none, for nuclear explosions high above the ground, to practically all, for ground level explosions. The local deposition has not materially affected the world-wide pattern.

*Tropospheric fall-out* refers to that part of the debris remaining in the troposphere after the cloud has stabilized. The troposphere is the layer of air extending from the Earth's surface to about 30,000 feet in polar regions and to about 55,000 feet in the tropics. In this layer, which contains most of the world's weather (i.e. cloud and precipitation), circulation is more marked than north-south motion so that tropospheric debris is removed within a few weeks by rain, in the general latitude of the testing site. Because of the short residence, tropospheric debris is dominated by the short-lived fission products such as barium-140 (half-life 12·8 days), zirconium-95 (65 days), strontium-89 (50 days) and iodine-131 (8 days). The latter has particular biological significance because of the potential hazard to the human thyroid.

*Stratospheric fall-out.* The stratosphere is the region of the atmosphere lying above the troposphere in which vertical motions are inhibited by a zero or small positive temperature gradient. Apart from gravitational settling of the larger particles ($> 10$ micron) the motion of stratospheric debris downwards towards and through the tropopause will be slow. The bulk of the radioactivity from nuclear weapons is derived from large explosions in the megaton range; these are sufficiently energetic to insert their debris into the stratosphere. Therefore the distribution of long-range world-wide fall-out is largely controlled by the meteorology of the stratosphere and the mechanism of exchange with the troposphere.

*World-wide distribution of fall-out.* The characteristics of fall-out, delayed by storage in the stratospheric reservoir, may be studied by observing the concentrations in air and rain of the long-lived fission products strontium-90 (half-life 28 years) and caesium-137 (30 years). Strontium-90 has radiological importance because of its affinity for bone, caesium-137 because of its gamma-radiation. Two significant features have emerged from these studies.

The variation with latitude of strontium-90 in rainwater in 1962, a typical year, is shown in Fig. 1. The radioactivity is largely confined to the northern hemisphere, that is, the hemisphere of the major testing sites. Notable within the northern hemisphere are the equatorial minimum (despite the large test explosions in this zone), the maximum in middle latitudes and the increase above 60°N. This curve describes the *concentration* of strontium-90 in rainwater; the corresponding picture of the *deposition* of strontium-90, which of course is affected by the zonal distribution of rainfall and surface area, retains the minimum near the equator and the maximum in middle latitudes.

The second feature is the seasonal variation in
ong-lived fall-out concentrations. The curve in Fig. 2
·f the concentration of caesium-137 in surface air
neasured in England demonstrates an annual peak
usually in the month of May. The appearance of this
·eak in earlier years was attributed in part to the

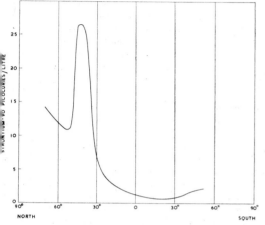

*Fig. 1. Strontium-90 content of rainwater with
latitude (1962).*

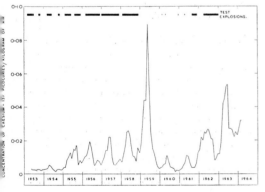

*Fig. 2. Caesium-137 in ground-level air at
Chilton, England.*

ffect of heavy weapon testing in the preceding au-
umn. However, evidence has accumulated to confirm
hat this phenomenon has a meteorological origin, in
·articular the seasonal peaks in 1960 and 1961 during
period free from large explosions.

*Stratospheric transport.* The seasonal and latitudinal
·ariations of fall-out have been explained by a meri-
·ional circulation that includes the subsidence of
·ld air over the winter pole into the lower stratosphere.
·rom here the fission products are available for trans-
·r into the troposphere in the spring, mainly at
·iddle latitudes where there is a major gap in the

tropopause. The stratospheric circulation was originally postulated by Brewer and Dobson to explain
observations of humidity and ozone. Alternatively
Feeley and Spar proposed that transfer in the stratosphere was best explained by turbulent mixing rather
than by an organized meridional circulation: this
follows from their interpretation of the stationary behaviour of radioactivity injected into the higher levels
of the stratosphere. In fact both transport mechanisms
are required to explain the distribution of fall-out
with latitude and time. The suggestion by Goldsmith
and Brown that the circulation becomes slower with
increasing altitude until above roughly 75,000 feet
mixing is by small-scale turbulent diffusion seems a
reasonable compromise in the absence of substantial
direct evidence from the stratosphere. This leads to
the concept of a stratosphere divided into two
regions. The lower region is capable of storage for up
to 9 months dependent upon season and latitude of
injection: the upper stagnant region is characterized
by a much longer storage time, possibly as long as
10 years.

*Bibliography*

BREWER A. W. (1949) *Quart. J. Roy. Met. Soc.* **75**, 351.
DOBSON G. M. B. (1956) *Proc. Roy. Soc.* A **236**, 187.
FEELEY H. W. and SPAR J. (1960) *Nature* **188**, 1062.
GOLDSMITH P. and BROWN F. (1961) *Nature* **191**, 1033.
Report of the United Nations Scientific Committee on
the Effect of Atomic Radiation (1962).

D. H. PEIRSON

**FIELD-EMISSION AND FIELD-ION MICROSCOPY;
RECENT DEVELOPMENTS IN.** Both of these techniques are now accepted as standard research tools and a
text-book on field-emission and field ionization has
been published (Gomer 1961).

Field-emission microscopy has been very successfully used for the study of chemical reactions on metal
surfaces. The effect of an adsorbed layer of gas on the
emission characteristics can be explained using classical electrostatic theory and treating the adsorbate
as a dipole layer. Many different gas-metal combinations have been investigated, especially refractory
metal systems, and the binding energies and migration
energies of various adsorbates have been measured.
The interest at present centres around problems associated with catalytic activity and oxidation processes.

The very high current density attainable with
field-emitters makes them attractive as point sources
of electrons. Multiple arrays of field emitters have
been operated together to provide extremely intense
sources of X rays. In pulsed operation such a source
can be used for high speed radiography, for example
in following the trajectory of a bullet through a solid
object. Point sources of electrons are also of interest
for X-ray microscopy and electron probe microanalysis. Conventional field emitters are rapidly blunted
by gas-sputtering in normal vacua ($\cong 10^{-5}$ torr) and

require ultra-high vacua for successful operation. The use of T-F-emission, is advocated for these applications. In T-F-emission the specimen tip is heated to the temperature required for thermionic emission but with simultaneous application of a high electrostatic field. Relatively large tip radii can then be used ($\sim 10\mu$), which are less susceptible to damage by gas-sputtering while at the same time providing electron sources an order of magnitude smaller than conventional thermionic sources.

While the electron microscope can only give contrast from groups of atoms, the field-ion microscope resolves the atoms themselves and is therefore finding increasing application in research. The hemispherical surface of the field-ion microscope specimen is prepared by electrolytic polishing followed by gradual field-evaporation in the microscope to a radius of from 100 Å up to 2000 Å. For *field-evaporation* the electrostatic field strength is raised by increasing the positive potential on the specimen until the surface atoms are evaporated as metallic ions without the need for appreciable thermal activation. Surface irregularities and protruding atoms are evaporated first, and the final tip surface is atomically smooth. The theory of field-evaporation and field-desorption (the evaporation of adsorbed atoms under a high electric field) has recently been given a complete treatment by Gomer and Swanson (1963). The presence of lattice defects or a second component reduces the regularity of the field-evaporated surface and leads to surfaces which are atomically rough. Such surfaces are characteristic of many solid solution alloys when imaged in the field-ion microscope.

The high ionization potential of helium ensures that contaminant gases in the microscope are auto-ionized in space before they can reach the metal surface and thus effectively eliminates contamination when the background pressure is below $10^{-6}$ torr. However, at the high electric fields needed to obtain a helium ion image many metals are themselves field-evaporated and in consequence field-ion microscopy was until recently limited to the study of refractory metals. The current-voltage characteristics of the helium field-ion microscope have now been measured for a number of metals and the specimen temperature has been shown to have an important influence on the threshold field for obtaining an image. By operating the helium ion microscope at specimen temperatures $\leqq 20°$K stable images can be obtained from most of the transition elements, including the iron group of metals.

The ionization potential of neon is considerably less than that of helium and the threshold field for ionization is some 40 per cent lower. Unfortunately, neon has a very low phosphor efficiency and some form of image intensifier is essential if satisfactory neon images are to be recorded. External image intensifiers have been used with great success, while post acceleration of the ions through a fine mesh grid and conversion of the ion image to an electron image have both proved feasible. The loss of resolution with neon as the image gas (due to its larger atomic radius) is not serious and neon ion microscopy offers considerable advantages for the non-refractory transition metals. Other image gases which have been employed are hydrogen and argon. The excellent potential resolution and image intensity attainable with hydrogen are somewhat marred by the chemical attack that often seems to occur in the presence of hydrogen when a high electrostatic field is applied to the metal. The large atomic radius and very low phosphor efficiency for argon militate against its successful use as an image gas.

Both field-ionization and field-evaporation are now reasonably well understood and the field-ion microscope is being applied to a wide range of research problems. In the metallurgical field, lattice defects in pure metals have been extensively studied and the technique has proved particularly successful in revealing the structure of grain-boundaries. Refractory alloys have also been investigated and attempts have been made to overcome the problems associated with the preferential evaporation of one component. In an investigation of the CoPt order-disorder transformation irregular field-evaporation has proved useful for following the development of order, since the fully ordered alloy evaporates regularly, like a pure metal. Radiation damage has also been studied and individual vacancies have been resolved, as well as the disordered zones and vacancy clusters resulting from complex damage processes. The field-ion microscope is also being used to study surface reactions and is proving particularly valuable for supplementing the information obtained with the field-emission microscope. The high resolution of the field-ion microscope makes it possible to characterize the structure of the surface and the nature of the adsorption sites. The field-emission microscope can then be used to follow the change in the electronic structure of the surface. Field-ionization is also being applied to mass-spectrometry. The mass spectrum obtained by field-ionization of large organic molecules is extremely simple when compared with those spectra obtained by thermal ionization or electron bombardment. The organic material is introduced into the vacuum chamber as a vapour and field-ionized over the surface of a chemically inert metal tip (usually gold or platinum). The mono-energetic ion beam is then focussed onto the slit of the mass-spectrometer. Recently, fine wires and sharp edges have been found to give good results in place of hemispherical points. These wire and "razor-blade" field-ion sources are much more robust than the conventional point sources.

*Bibliography*

Beckey H. (1961) *Symposium on Mass Spectrometry*, Oxford: Pergamon Press.
Brandon D. G. (1963) *Brit. J. App. Phys.* **14**, 474.
Gomer R. (1961) *Field Emission and Field Ionisation*, Harvard Memograph, Oxford: The University Press.

OMER R. and SWANSON L. (1963) *J. Chem. Phys.* **38**, 1613.

UELLER E. W. (1960) *Adv. Elect. Electron Phys.*, **13**, 83.

UELLER E. W. (1963) *J. Phys. Soc. Jap.* **18**, Suppl. II, 1.

D. G. BRANDON

## FLUORINE CHEMISTRY.

The element fluorine was rst isolated on June 26, 1886, by Prof. H. Moisson at he Ecole de Pharmacie in Paris. Following his discovery, little attention was devoted to fluorine chemistry until the 1930's and early 1940's when the mpetus of the wartime research on the preparation of adioactive uranium led to widespread interest not nly in the chemistry of fluorine itself but also in the reparation of a variety of fluorine-containing organic ompounds termed loosely "fluorocarbons". These atter materials have now assumed a prominent ole both in academic and industrial research laboraories and their preparation, properties, and uses will onstitute a major part of this discussion.

### Elemental Fluorine

*Occurrence.* Fluorine is widely distributed in the Earth's crust in rocks and is about thirteenth in order f abundance among the elements. Its abundance is reater than chlorine (excluding the various bodies of alt water), 5–10 times that of zinc and copper, and nany times that of lead (about 50).

Among the more than one hundred known fluorine-ontaining minerals, only a few contain a sufficient weight of the element to be of commercial interest. Among these are cryolite ($Na_3AlF_6$), fluorite ($CaF_2$), nd apatite [$CaF_2 \cdot 3 Ca_3(PO_4)_2$]. Fluorite or fluorpar is the prime source due to its wide abundance. t occurs on every continent, with large deposits in the Kentucky-Illinois area of the United States and also n Mexico and the U.S.S.R. The present estimated eserves are about seventy-five million tons.

In nature fluorspar occurs as a glassy mineral with hardness of 4, a specific gravity of about 3·18, and a melting point of 1418°C. It can have one of several olours, but is usually colourless.

*Preparation.* Elemental fluorine is prepared from he mineral ore by first forming hydrogen fluoride as:

$$CaF_2 + H_2SO_4 \longrightarrow CaSO_4 + 2 HF .$$

A mixture of hydrogen fluoride and potassium fluoride about 4 : 1) is then electrolysed in the melt (80–100°C) using a direct current to produce fluorine as

$$2 HF \xrightarrow[\text{KF}]{\text{Elect.}} F_2 + H_2 .$$

This is essentially the procedure used by Moisson with uitable modifications for industrial operation. A full discussion of this process may be found among the eferences cited in the bibliography.

*Properties.* Fluorine exists at room temperature as a greenish-yellow diatomic gas. The following physical properties are of interest:

| | |
|---|---|
| Boiling point | 84·93°K |
| Melting point | 53·54°K |
| Density at 80·0°K | 1·547 |
| Critical temperature | 144°K |
| Critical pressure | 55 atm |
| Heat of vaporization | 1558 cal/mole |
| Dissociation energy | 31–37 kcal/mole |
| Electronic configuration | $1s^2, 2s^2, 2p^5$ |
| Electronegativity | 4·0 |
| Atomic weight | 19·00 |

Chemically, fluorine is considered to be the most reactive known element forming stable bonds with many other elements including xenon and krypton. The carbon–fluorine bond is of particular interest. It is characterized by a high dissociation energy (110–120 kcal/mole), low polarizibility, short bond length, and relatively small atomic radii. Such properties impart to a highly fluorinated molecule both chemical and thermal stability. One other characteristic arising from these properties is weak intermolecular forces in liquid fluorocarbons (sometimes referred to as low internal pressures). This latter property is reflected in low boiling point, low surface tension, decreased solubility, and many other characteristics unique to the fluorocarbons.

Typical properties are listed below:

$$n - C_5F_{12}$$

| | |
|---|---|
| Boiling point | 29·32°C |
| Freezing point | − 125·6°C |
| Surface tension | 9·42 dynes/cm at 24·95°C |
| Molal volume | 177·8 cm³ at 20°C |
| Dielectric constant | 1·68 at 20°C |
| Density | 1·66 at 20°C |
| Heat of vaporization | 6510 cal/mole |
| Refractive index | 1·245 at 20°C |
| | (D line of sodium) |

### Methods of Fluorination

The two principal means of synthesizing a fluorocarbon are by replacement of hydrogen in an organic molecule by fluorine or by exchange of chlorine, bromine, or iodine with fluorine. The former technique necessitates the use of fluorine itself or an active metal fluoride, while the latter can be conducted using the fluorides of antimony, potassium, or sodium in most cases.

*Reaction with elemental fluorine.* Several chemical processes occur which include substitution of hydrogen, addition to an unsaturated carbon–carbon bond, fission of a carbon–carbon bond, and formation of polymers. Since direct fluorination is highly exothermic, carbon–carbon bond rupture takes place readily and the design of a reaction vessel should be primarily directed toward rapid and effective removal of heat from the reaction zone. A good design has been evolved which employs jets or other small openings in the reaction zone through which fluorine diluted with nitrogen is introduced. In addition, the reaction zone can be packed with a good heat conductor such as copper to aid in heat removal.

The reaction can best be explained on the basis of a radical process as follows:

$$F_2 \rightleftharpoons 2\,F\cdot$$

$$RH + F\cdot \rightarrow R\cdot + HF$$

$$R\cdot + F_2 \rightarrow RF + F\cdot$$

Initiation occurs readily by light or heat, usually proceeding rapidly at room temperature. The last step in the above sequence involves a chain transfer and thereby the whole process is a typical chain reaction.

Addition of fluorine to an unsaturated carbon–carbon bond takes place in a similar manner:

$$-\overset{|}{C}=\overset{|}{C}- + F\cdot \rightarrow -\overset{|}{C}F-\overset{|}{C}-$$

$$-\overset{|}{C}F-\overset{|}{C}- + F_2 \rightarrow -\overset{|}{C}F-\overset{|}{C}F + F\cdot$$

Polymerization may also occur, leading to higher molecular weight fluorocarbons.

*Reaction with active metal fluorides.* A second means of hydrogen replacement comprises the reaction of a higher fluoride of metals such as silver, cobalt, or lead with an organic molecule. The process is illustrated as follows:

$$2\,CoF_2 + F_2 \rightarrow 2\,CoF_3$$

$$RH + 2\,CoF_3 \rightarrow RF + 2\,CoF_2 + HF$$

The heat envolved (46 kcal/mole) is much less than that obtained in direct reaction with fluorine:

$$RH + F_2 \rightarrow RF + HF$$

$$\Delta H = -104\,kcal/mole$$

Thereby, the former reaction gives much less carbon chain fission and a higher yield of the desired product. It is the preferred procedure for synthesizing any sizable quantities of fluorocarbons.

Only a few metal fluorides act in the above manner. Their reactivity appears to be associated with the oxidation potential of the metal ion ($Ag^{+1}$ or $Co^{+2}$), the greatest reactivity with the highest potential. In order of decreasing effectiveness as fluorination agents they are: silver difluoride, cobaltic fluoride, ceric fluoride, manganese fluoride, plumbic fluoride, bismuth pentafluoride, chromic fluoride, mercuric fluoride, ferric fluoride, cupric fluoride, and stannic fluoride.

In conducting the reaction, fluorine is passed through a bed of the metal salt at elevated temperatures to form the active metal fluoride. The organic reagent is then passed through or over this salt bed at elevated temperatures to effect fluorination. In this manner a semi-continuous process can be achieved. The process

is used principally for the fluorination of hydrocarbons but, because of the high reactivity of the fluorination agent, only saturated fluorocarbons are obtained. It may also be used to fluorinate polychloro hydrocarbons, in which case replacement of chlorine as well as hydrogen by fluorine is achieved.

*Electrochemical fluorination.* A more recently developed method of substituting hydrogen by fluorine comprises the electrolysis of an organic reagent dissolved in anhydrous liquid hydrogen fluoride. In practice, the solution is placed in an electrolytic cell constructed of steel. Nickel plates serve as anodes while the cell body itself can serve as a cathode. A low voltage (5–6) direct current is applied and the cell contents cooled to minimize loss of hydrogen fluoride. Volatile fluorination products are removed as gases while the higher boiling products, which are insoluble in hydrogen fluoride, can be drained from the bottom of the cell.

The reaction mechanism is not readily apparent. Since the voltage is below that needed for the formation of elemental fluorine, it is possible that the fluorination involves a higher fluoride of nickel formed on the surface of the anode. The conductivity of the solution directly affects yield and efficiency. The best results are obtained using organic acids, amines, or ethers as starting materials since they generally are quite soluble in hydrogen fluoride. Fluorination of relatively insoluble compounds such as hydrocarbons requires the use of a soluble salt (lithium fluoride) to increase the solution conductivity.

The process works well with simple starting materials as acetic acid or tertiary alkylamines. As the carbon chain length increases, carbon–carbon bond rupture occurs as well as other side reactions, resulting in low yields of the desired fluorocarbon product.

*Halogen exchange.* Replacement of halogen, primarily chlorine, by fluorine is the principal method for preparing aliphatic fluorocarbons. In particular the fluorides of antimony (first described by F. Swarts) have found wide application and are the prime reagents used commercially. Antimony trifluoride can be used with compounds having the following structures:

$$RCX_2R \rightarrow RCF_2R$$

$$RCX_3 \rightarrow RCF_3$$

$$R\overset{|}{C}=\overset{|}{C}CX_3 \rightarrow R\overset{|}{C}=\overset{|}{C}CF_3$$

where R is hydrogen, alkyl, or aryl, and X represents chlorine, bromine, or iodine. Antimony trifluoride, however, is ineffective in replacing a single halogen atom on a carbon or a vinylic halogen.

A more effective and useful reagent is prepared from the pentavalent antimony chlorides. The fluorination of chloroform is illustrated as follows:

$$SbCl_5 + 3HF \rightarrow SbCl_2F_3 + 3\,HCl$$

$$SbCl_2F_3 + CHCl_3 \rightarrow SbCl_4F + CHClF_2$$

$$SbCl_4F + 2HF \rightarrow SbCl_2F_3 + 2\,HCl$$

In this manner continuous regeneration of the fluorination agent is possible.

Many other inorganic fluorides have been employed in halogen exchange, those of sodium, potassium, silver, mercury, and lead being the most useful. In particular, potassium fluoride in the presence of polar media such as amides, sulphones, and nitriles can be used to replace vinylic halogen as follows:

In many reactions, the fluoride ion acts as a strong nucleophilic agent. With fluorolefins, reversible addition may take place to produce a fluorocarbanion which may then add a proton from the solvent to yield a hydrogen fluoride addition product as:

$$F^- + CF_2{=}C{-} \rightleftharpoons \left[ CF_3\overset{\ominus}{C}{-} \right] \overset{H^+}{\longrightarrow} CF_3\overset{H}{C}{-}$$

A second possibility is a rearrangement reaction as:

$$F^- + CF_2{=}\overset{F}{CC}{-}\overset{|}{C}{-} \rightleftharpoons CF_3C{=}C{-}\overset{|}{C}{-} + F^-$$

The fluorides of lead and mercury are best prepared *in situ* from the oxide and hydrogen fluoride. Not only will these fluorides effect halogen exchange but can also add fluorine to unsaturated linkages in halogenated olefins.

*New fluorination agents.* In the past few years other fluorination agents have been discovered which are finding increasing use. While it is difficult to generalize on their chemistry, the following summarizes the present knowledge:

(1) Iodine pentafluoride ($IF_5$). Halogen exchange agent.
(2) Iodine heptafluoride ($IF_7$). Halogen exchange agent.
(3) Iodine monofluoride (IF) and bromine monofluoride (BrF). Add readily to unsaturated bonds in olefins.
(4) Bromine trifluoride ($BrF_3$) and chlorine trifluoride ($ClF_3$). Replace hydrogen or halogen with fluorine and add to unsaturated linkages.
(5) Perchloryl fluoride ($Cl_3OF$). Replaces active hydrogen with fluorine.
(6) Nitrosyl fluoride (NOF). When complexed with hydrogen fluoride it comprises a liquid halogen exchange agent.
(7) Sulphur tetrafluoride ($SF_4$). Replaces oxygen with fluorine as:

$$\begin{aligned} {>}C{=}O &\rightarrow {>}CF_2 \\ -CO_2H &\rightarrow -CF_3 \\ -CONH_2 &\rightarrow -CF_3 \\ -CHO &\rightarrow -CF_2H \\ MO \text{ and } RMO &\rightarrow MF_2 \text{ and } RMF_2 \\ (M = metal, &\quad R = alkyl \text{ or } aryl) \end{aligned}$$

## Uses of Fluorocarbons

Organic fluorine compounds have found wide application in two general areas, as refrigerants and aerosol propellants and as polymers for elastomeric or plastic applications. The former use, taking advantage of the chemical stability, low toxicity, and non-flammability of fluorocarbons, has developed into a major industry, producing several hundred million pounds per year of the various chlorofluoromethanes, ethanes, and ethylenes which constitute this general class of materials. A comprehensive discussion of this field can be found among the reference works cited.

More recently, fluorocarbon polymers have found increasing applications based on their properties of chemical and thermal stability and low solubility in organic solvents.

*Plastics.* The single most widely used fluorocarbon plastic is polytetrafluoroethylene ($-CF_2CF_2-)_n$. It is made as follows:

$$2 CHClF_2 \xrightarrow{Heat} CF_2{=}CF_2 + 2 HCl$$
$$\text{or } 2 CHF_3 \xrightarrow{Heat} CF_2{=}CF_2 + 2 HF$$
$$CF_2{=}CF_2 + Initiator \rightarrow (-CF_2CF_2Cl-)_n$$

A second polymer, polychlorotrifluoroethylene ($-CF_2 CFCl-)_n$, has also found wide application and is prepared as follows:

$$CCl_3CCl_3 + HF/Sb^{+5} \rightarrow CClF_2CCl_2F$$
$$CClF_2CCl_2F + Zn/EtOH \rightarrow CF_2{=}CFCl$$
$$CF_2{=}CFCl + Initiator \rightarrow (-CF_2CFCl-)_n$$

Two new fluoroplastics, now reaching the commercial stage, are polyvinylfluoride ($-CH_2CHF-)_n$ and polyvinylidenefluoride ($-CH_2CF_2-)_n$. The former is prepared as follows:

$$HC{\equiv}CH + HF \xrightarrow{Catalyst} CH_2{=}CHF$$

while the latter as follows:

$$CCl_3CH_3 + HF/Sb^{+5} \rightarrow CF_3CH_3$$
$$CF_3CH_3 \xrightarrow[Catalyst]{Heat} CH_2{=}CF_2$$

Both monomers are readily polymerized by free radical initiators.

*Elastomers.* The search for a rubber combining the stability and solvent resistance inherent in a fluorocarbon with the properties of a good elastomer has resulted in the commercial development of two different polymer systems, one with a carbon chain and the other a polysiloxane.

One polymer, polyheptafluorobutyl acrylate,

is prepared as follows:

$$C_3H_7CO_2H + HF \xrightarrow{\text{Electricity}} C_3F_7COF$$

$$C_3F_7COF + H_2O \rightarrow C_3F_7CO_2H$$

$$C_3F_7CO_2H + H_2/\text{cat.} \rightarrow C_3F_7CH_2OH$$

$$C_3F_7CH_2OH + CH_2{=}CHCOCl \rightarrow C_3F_7CH_2OCOCH{=}CH_2$$

Poylmerization using a redox system produces a rubbery gum. A second carbon chain polymer is made by copolymerization of vinylidenefluoride ($CH_2{=}CF_2$) and hexafluoropropene ($CF_3CF{=}CF_2$) using free radical initiators. On vulcanization, both polymers give rubbers of good tensile strength and a high degree of solvent resistance.

A third type, poly (trifluoropropyl) methylsiloxane ($CF_3CH_2CH_2Si(CH_3)O]_n$, not only offers solvent resistance but also low temperature flexibility not obtained in a carbon chain system. It is prepared as follows:

$$CF_3CH{=}CH_2 + MeSiHCl_2 \rightarrow CF_3CH_2CH_2Si(CH_3)Cl_2$$

$$CF_3CH_2CH_2(CH_3)Cl_2 + \text{hydrolysis and condensation} \rightarrow$$
$$\text{polymer}$$

The properties of the vulcanized fluorosilicone rubber are comparable to conventional silicone rubbers with the additional benefit of solvent resistance.

*Other uses.* Fluorochemicals are finding increasing use in a wide variety of applications. A brief listing includes blowing agents, solvents, fire extinguishing agents, anaesthetics, lubricants, and dielectrics.

*Bibliography*

HUDLICKY M. (1961) *Chemistry of Organic Fluorine Compounds*, Oxford: Pergamon Press.

LOVELACE A. M. *et al.* (1958) *Aliphatic Fluorine Compounds*, New York: Reinhold.

OSTEROTH D. (1964) *Chemie und Technologie Aliphatischer Fluoroorganischer Verbindungen*, Stuttgart: Ferdinand Enke.

SIMONS J. H. (Ed.) (1954, 1958, 1964) *Fluorine Chemistry*, Vols. I, II and V, New York: Academic Press.

STACEY M. *et al.* (Eds.) (1960, 1961, 1963) *Advances in Fluorine Chemistry*, London: Butterworths.

O. R. PIERCE

**FREE RADICAL, EXPERIMENTAL.** During the five years that have elapsed since the writing in 1959 of the articles in the main dictionary covering the above topic, free radical research has been the scene of intense activity. No revolutionary change has taken place in the field of enquiry but it has been greatly expanded.

The most important single factor in this expansion has been the application of gas chromatography to the problems of analysis. Gas chromatography was first widely used in 1954, but few investigations of free radicals were than planned to take advantage of the new technique. The first papers based on work in which the whole approach depended on gas chromatography appeared in 1955 and 1956. Today the full power of the method enables investigators to study an almost unlimited range of reactions instead of being limited to systems in which the key products are volatile. With the analytical restrictions removed kineticists can develop a fully articulated physical organic chemistry of the gas phase, most of which is concerned with free radicals.

Alongside this study of an increasing range of reactions there has been an increase in the interest taken in the details of molecular mechanics. Again this development is not confined to systems that contain free radicals but because free radicals are involved in the majority of gas phase reactions and because these details cannot normally be discerned in a condensed phase, radicals have played an important role. The aspect of molecular dynamics that has come under special scrutiny has been the role of energy in reactions. Spectroscopic methods have been developed along two lines. Electronically excited radicals and products of atom and radical reactions have been studied by ultra-violet spectra under conditions such that the vibrational states of the species can be discerned. Frequently the species have been produced by flash photolysis and followed by flash spectroscopy. Polanyi has achieved some success in observing directly by infra-red emission spectroscopy the energy distribution in simple products such as hydrogen chloride from the reaction of chlorine atoms with hydrogen. The interpretation of the results of these and other studies depends to a rapidly increasing extent on the use of digital computers. It is an unfortunate fact that the complexes of quantized oscillators which are real molecules are not tidy systems that lend themselves to elegant mathematical analysis.

The study of the reactions of ions generated within a mass-spectrometer has attracted considerable attention but although radicals are often involved it cannot be said that any behaviour peculiar to radicals can be discerned.

For many years hope has been expressed that molecular beam techniques would soon be applied to a wide range of kinetic systems. Those containing radicals were obvious candidates for study because of the generally high collision efficiency of radical processes. Progress has been made particularly by Herschbach but at the same time the very considerable difficulties in general application have become apparent. The techniques of study of reactions at very high temperatures have been advanced by improvement in shock tube methods. Several radical decompositions have been studied in this way, but the most suitable systems are still those in which the concentrations of reactants can be studied by optical spectroscopic techniques.

The chemistry of the atmosphere largely involves free radical processes and they have been intensively investigated. Reactions in the lower atmosphere have been studied because of their importance in smog formation. The space programme has provided the incentive and, to some extent, the new techniques for the investigation of reactions in the upper atmosphere where short wave-length solar radiation can dissociate most molecules. The chemistry involves not only the study of atoms and radicals in their ground electronic states, which has been the common concern in the past, but also electronically excited species.

Much work has been reported on free radicals frozen into an inert matrix. Both physical and chemical properties have been studied particularly by a very strong group working at the National Bureau of Standard, Washington, during the period under review.

*Bond dissociation energies.* The proper understanding of reactions of free radicals depends largely upon a knowledge of the heats of reaction and hence upon the heats of formation of the radicals. Recently, there has been a large accumulation of data on bond dissociation energies from which the heats of formation are derived. No new techniques of investigation have been developed but the range of the old methods has been extended and their reliability improved. The values obtained by kinetic studies have particularly been improved by better analysis. A list of the best values of some of the more important dissociation energies is given below (kcal mole$^{-1}$, 25°C).

| | | | |
|---|---|---|---|
| H—CH | 106 | H—CH$_2$ | 104 |
| H—CH$_3$ | 104 | H—C$_2$H$_3$ | 104 |
| H—C$_2$H$_5$ | 98 | H—C$_3$H$_5$ | 84 |
| H—i-C$_3$H$_7$ | 94·5 | H—i-C$_3$H$_7$ | 94·5 |
| H—t-C$_4$H$_9$ | 90·9 | H—C$_3$H$_5$ | 103 |
| H—CH$_2$C$_6$H$_5$ | 84 | H—CH$_2$CN | 79 |
| H—CHO | 92 | H—COCH$_3$ | 88 |
| H—COC$_6$H$_5$ | 79 | H—CF$_3$ | 104 |
| H—CCl$_3$ | 90 | H—NH$_2$ | 103 |
| H—NHCH$_3$ | 93 | H—N(CH$_3$)$_2$ | 88 |
| H—NHC$_6$H$_5$ | 80 | H—OH | 119 |
| H—OCH$_3$ | 102 | H—O$_2$H | 90 |

*Free radical reactions.* The range of known reaction types has not been much extended but recently many new examples have been adequately investigated.

*Radical elimination reactions and combinations.* Some ten to twenty studies of radical elimination reactions are reported each year. Most of these are undertaken in order to establish the strengths of bonds. Carrier techniques are particularly useful for this purpose and the familiar use of toluene has been extended to aniline, which is less liable to confusing side reactions, and to some substituted toluenes. At one time it was suggested that it was "normal" for simple unimolecular decompositions to have Arrhenius $A$ factors close to $10^{13}$ sec$^{-1}$. It has become apparent that this is an over-simplification and that it is likely that only those reactions in which an atom is eliminated should have such a low $A$ factor. Determinations of activation

energies partly based on this assumption are, therefore, suspect. The elimination of a free radical will normally involve a considerable positive entropy of activation. The problem of the high $A$ factors associated with the apparent decomposition of molecules into three fragments has not been satisfactorily resolved.

The measurement of rates of radical combination has proceeded slowly because of the difficulty of estimating the concentration of radicals in all but the most clean-cut systems. The rates of auto-combination of isopropyl and of tert.-butyl radicals have been found to be close to the rates of collision as with the smaller alkyl radicals. The corollary of these findings is that the symmetrical decompositions of the products have very high $A$ factors, say $10^{17}$ sec$^{-1}$.

*Disproportionations.* The last few years have seen the introduction of some order into the findings on the rates of disproportionations which are usually expressed relative to the parallel rates of combinations. This has been a direct result of improved analyses. It cannot be said that these reactions are yet understood. Only recently has it been established by measurements down to temperatures as low as $-50$°C that the reactions have a small temperature coefficient relative to combinations. They also appear to be subject to a small pressure effect. A rough linear relation exists between the rate of disproportionation and the entropies of the products of combination and disproportionation. This suggests a common precursor, either an activated complex or an activated molecule that moves towards one or other type of product according to some criterion of probability. The further understanding of disproportionation will probably depend upon the quantitative study of a wider range of reactants. It is not yet known for certain to what extent activated radicals disproportionate more rapidly.

*Transfer reactions.* The range of hydrogen transfer reactions that have been studied is steadily increasing and measurements of greater accuracy are being obtained with familiar systems. Measurements of reactions of wide ranges of compounds have been made with fluorine, chlorine and bromine atoms. A start has been made with the far less reactive iodine atoms. Methyl, ethyl, perfluoromethyl, ethyl and n-propyl have also been covered. Some sixty reactions of phenyl radicals in solution at one temperature have been reported. More restricted series are known for hydroxy and methoxyl radicals. As yet, less success has attended attempts to study the reactions of radicals containing nitrogen. The interpretation of the results has generally been based on the supposition that the reactions could be described in terms of a point moving over a potential energy surface. It has been demonstrated that the wave-length of the proton is comparable to the dimensions of the surface and that quantum-mechanical tunnelling must be allowed for. Because the transfer of hydrogen atoms is complicated by this consideration great interest attaches to the study of heavier atoms associated with shorter wave-length. Several iodine transfer rates have been measured.

*The addition and decomposition of radicals.* Even with the improved methods of analysis it has proved difficult to obtain accurate measurements of rates of additions in the gas phase. Much data exist for work in solution on vinyl polymerizations and this has not been significantly increased. Szwarc has however measured the rates of a very wide range of addition reactions of small radicals such as methyl and trifluoromethyl in solution. Although the rates of many radical decompositions have been found it has been difficult to make the measurements sufficiently precise to yield reliable energies of activation. Only very clean systems are satisfactory.

*Isomerizations.* As yet only a few isomerizations of radicals are well characterized. They fall into three classes: the conventional isomerization such as that of cyclopropyl to allyl; the internal transfer reaction in which a long alkyl radical bends back and abstracts a hydrogen atom from its tail; and vinyl shifts in which a vinyl group is transferred by a process that appears to involve a three-membered ring.

*Biradicals.* Understanding of the behaviour of biradicals has been greatly aided by the recognition of the different behaviour of species in singlet and triplet states. For methylene, in particular, it was difficult to decide which states were involved. It has now been shown that the ground state is the triplet but that the separation between the two states is small and that the singlet is the principal product of gas phase photolyses. The reactions of oxygen atoms in both singlet and triplet states have also been extensively studied. The investigation of activated molecules made by reactions of biradicals remains a primary concern of many investigators.

*Bibliography*

*Annual Review of Physical Chemistry*, Palo Alto: Annual Reviews Inc.
KONDRATIEV V. N. (1962) *Bond Dissociation Energies, Electron Affinities and Ionisation Potentials*, U.S.S.R.: Academy of Science.
PORTER G. (1961, 1963) *Progress in Reaction Kinetics*, Vols. 1 and 2, Oxford: Pergamon Press.

A. F. TROTMAN-DICKENSON

**FREQUENCY-CHANGER.** This is a circuit which translates signals lying in a given frequency band to another band or the same width, ideally without distortion of interaction between the various frequency components in the signal. It usually comprises a modulator with its local oscillator, and a filter to select the desired output frequency-band. The term is sometimes applied to a multi-electrode thermionic valve used as a modulator.

D. G. TUCKER

**FUEL CELLS.** What we loosely call the conversion of chemical into electrical energy is an old and well established process that is well exemplified by the lead-acid battery used to start automobiles. But this conversion does not generally extend to the chemical energy required for motive power. If petrol was as electrochemically reactive as lead, it might be possible to dispense with internal combustion engines and their attendant disadvantages. If the fuel-cell problem ever finds the most optimistic solution envisioned for it, fuel batteries will not only drive automobiles, but serve for many other applications as well. Today this solution remains a distant goal toward which only the first steps have been taken.

The middle of this century is witnessing unprecedented enthusiasm for the fuel cell. As a consequence, there has been an unfortunate tendency to call almost any electrochemical device by that name. For present purposes, a fuel cell is an electrochemical cell in which energy of reaction between a conventional fuel and oxygen (preferably from air) is converted directly and continuously into low-voltage direct-current electrical energy.

In this definition, "electrochemical cell" implies an anode (at which fuel is oxidized), a cathode (at which oxygen is reduced), an external circuit (in which an electron current can do work), and an electrolyte (through which the transfer of ions completes the circuit). A "conventional fuel" is a fossil fuel or a substance (such as hydrogen or ammonia) readily derived therefrom. One hopes "continuously" will mean "on demand for as long as fuel and oxygen are available via what will normally be continuous feed". Fuel cells make fuel batteries.

These definitions show the contrast between a fuel battery and a lead-acid battery, which may be taken to represent the ordinary types. Lead is not a conventional fuel. Lead dioxide, the cathode material is not oxygen. The lead-acid battery cannot deliver electrical energy continuously as the word is used above—this battery must be recharged, for it is an *accumulator* not a *converter* alone, as the fuel battery is.

Current nomenclature of fuel cells and batteries is confused. When the fuel is specified, it should begin the name as in "hydrogen fuel battery", "fuel" being retained to identify oxygen as the reactive material at the cathode. To differentiate between oxygen and air, one speaks of a "hydrogen-oxygen cell" or a "hydrogen-air battery".

No single, simple classification is satisfactory. That given below has the advantage of emphasizing the diversity and extent of the fuel-battery field.

If the fuel battery is to be a successful energy source, it should not only deliver electrical energy at low cost but it should have other advantages also. These other advantages will be weighted differently for the various applications, but they may all be grouped here under the general heading *Convenience.* Important factors making for convenience are:

High power from unit volume
High power from unit weight
Quietness
Cleanliness
Operation on air
Reliability

An arbitrary division of prospective fuel-cell advantages into Low Energy Cost and Convenience makes possible the simplified classification of fuel batteries shown in the table.

*Simplified Classification of Fuel Batteries*

| Type | Relative importance of | | Typical applications |
| | Low energy cost | Conven-ience | |
| --- | --- | --- | --- |
| Specialty | Minor | Major | Military and space |
| Industrial | Comparable | Comparable | Lift truck. Electric Locomotive |
| Central-station Large Small † | Major | Minor | D.C. for electrochemical industry |

† A methane fuel battery with a bank of lead-acid batteries might some day serve as a small "central station" in an "all-gas" home.

At this early stage in their development, fuel batteries are still so costly that capital investment required is currently an excellent criterion for the selection of the most promising applications. Specialty applications might tolerate as much as $ 10,000 per kilowatt installed: successful large central-station applications are unlikely if the cost per kilowatt installed greatly exceeds $ 100, which is an approximate value for present conventional central stations. The present capital cost of fuel cells will make it difficult for them to enter the consumer market.

The history of the fuel cell begins in 1839 with its discovery by Sir William Grove. After describing his first hydrogen fuel cell, he said: "I hope, by repeating this experiment in series, to effect decomposition of water by means of its composition"—or, in modern language, "to make a hydrogen fuel battery powerful enough to decompose water." Clearly, he meant not only to build a fuel battery, but to put it to work. He did what he set out to do in his fuel battery, part of which is shown in Fig. 1. The system shown there is of modern significance because it is in principle *regenerative*; that is to say, it could be rearranged to use electrical energy from an external source to generate hydrogen and oxygen, which could subsequently be recombined to yield electricity. In this way, energy—

for example, electrical energy from solar cells in space—could be stored until needed. Such a system may yet prove useful in space though unlikely to prove competitive under normal conditions on earth.

By 1889, Mond and Langer, also working in Great Britain, had greatly improved the performance of the hydrogen fuel cell, and shortly thereafter came the

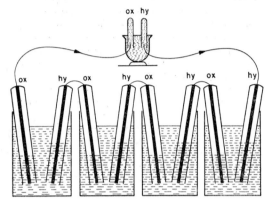

Fig. 1. *Four cells of Grove's hydrogen-oxygen battery. Twenty-six cells in series were needed to decompose water.*

growing realization of what "electricity direct from coal" might come to mean (Ostwald 1894). A period of great activity followed during which Jacques, an American, developed his "coal" fuel battery (Fig. 2) of

Fig. 2. *Jacques coal fuel battery of the last century.*

100 cells or more. In each cell, a carbon anode was centrally positioned in a cylindrical iron pot, the walls of which acted as cathode. The electrolyte was molten potassium hydroxide at 400 to 500°C. Air from a circular distributor plate at the bottom of the cell bubbled through the electrolyte to reach the cathode. The battery could deliver 16 amp at 90 volts, which is good performance for carbonaceous fuels even by present standards; and the battery was said to have operated for at least 6 months. The fatal flaw was this: during operation, the potassium hydroxide (costly!) was changed to potassium

carbonate (cheap!) *invariance* (see below) was not maintained, and operation consequently had to be intermittent.

Near the turn of the century, the character of fuel cell research began to change from exploratory to systematic, and this change is apparent in the many investigations conducted by E. Baur and his colleagues over most of Baur's fruitful and diversified scientific career. In part, the change was the natural consequence of failure to solve the fuel cell problem by purely inventive and wholly empirical methods; in part, it was the result of the growth of physical chemistry owing to the contributions of men such as Haber, Nernst, and Ostwald.

Among the notable advances that resulted were the "gas-diffusion" electrode, and the use of molten carbonate mixtures and of solid oxide-ion (e.g. $ZrO_2$–CaO) conductors as electrolytes. The "gas-diffusion" electrode, of highly porous structure, was particularly important because it came at about the time that gases were being generally recognized as preferred fuels for reasons of *reactivity* (see below). This (poorly named) electrode pointed the way toward modern electrode structures that keep gas and electrolyte *apart* to prevent objectionable loss yet *join* them (mainly in films of electrolyte present in electrode pores) well enough to make high current densities possible.

Before we turn to the modern fuel-cell era, which begins with the work of F. T. Bacon in 1932, we shall use a simple modern fuel cell for a brief discussion of principles. The cell in Fig. 3, the invention of W. T. Grubb and L. W. Niedrach, is unique in having a sulphonated solid polymer as electrolyte. Hydrogen, being a substance that readily relinquishes valence electrons, is oxidized at the anode; and oxygen is reduced to water at the cathode as it captures the valence electrons that flow to it from the anode through the external circuit. The circuit is completed internally by the ready migration through water in the polymer, of hydrogen ions from anode to cathode through the electrolyte. Under reversible conditions, the Gibbs free energy measures the tendency of the overall reaction (see Fig. 3) to occur, and hence the tendency of electrons to do work in the external circuit. When this tendency is exploited and work is done, the net electron flow is never random; direction is conserved throughout. Consequently, the fuel cell escapes the Carnot-cycle limitation, and this historically was its main attraction. This escape can be proved thermodynamically, and it is also made obvious by the fact that a fuel cell can do work at constant temperature—a heat engine cannot.

As thermodynamic calculations show, most conventional fuel cells have electromotive forces of about 1 volt under standard conditions. If the fuel cell is to develop reasonable power at reasonable efficiency, the product of cell voltage and current density (amp/sq ft or mA/cm²) must be optimized. Also, in a practical battery, a reasonable proportion of fuel must be consumed in a single pass through a cell. A reasonable goal for a good fuel cell is therefore the satisfactorily rapid oxidation of the fuel to carbon dioxide and/or water during operation on air at a terminal voltage near 0·6 V and a current density near 100 amp/ft² for a minimum of several months unattended.

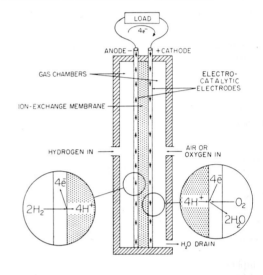

REACTIONS

| ANODE | OVERALL | CATHODE |
|---|---|---|
| $2H_2 = 4H^+ + 4e^-$ | $2H_2 + O_2 = 2H_2O$ | $4e^- + 4H^+ + O_2 = 2H_2O$ |

*Fig. 3. The ion-exchange membrane hydrogen fuel cell as a schematic representation of fuel-cell operation. Note that the overall reaction, the combustion of hydrogen, is the sum of anode and cathode reactions.*

The research requirements that must be met in the attaining of this goal can be succinctly summarized under the headings *Reactivity* and *Invariance*.

These headings do not include the obvious requirement that $I^2R$ losses should be minimized, as by close-spacing the electrodes and by providing an electrolyte of high ionic and negligible electronic conductivity.

*The reactivity requirements* are two:

*1. The stoichiometric requirement*

In the usual case, $H_2O$ and/or $CO_2$ should be the only oxidation products so that the quantity of electricity from one mole of fuel will be a maximum. For example, carbon oxidized to CO yields only 2 faradays per mole instead of the 4 faradays produced when $CO_2$ is the product. Water and $CO_2$ are the oxidation products of lowest energy. The fuel and oxidant must not react directly as a result, for example, of diffusion through the electrolyte, because no electricity is produced in such reaction. A "chemical short-circuit" of this kind is always undesirable and may be dangerous.

### 2. The kinetic requirement

Rapid electrode reactions mean high current density because the rates of these reactions determine the rate at which valence electrons are made available to the external circuit. Unimpeded electrode reactions mean (if $I^2R$ losses are low) high terminal voltage for the cell.

The reactivity requirements involve the rates and mechanisms of electrode reactions, and much more research into these matters will be needed if fuel cells are to succeed. The chief ways of meeting the requirements are: (1) by bringing the gas, electrode, and electrolyte properly into contact, as by using porous electrodes, the best known of which are the gas-diffusion electrodes (above); (2) by using electrocatalysts, either as electrodes or incorporated therein; (3) by selecting a more suitable electrolyte; (4) by raising pressure; (5) by raising temperature; (6) by proper choice of fuel; and (7) by chemical modification of the fuel.

Hydrogen is unique among conventional fuels. It presents no problem in stoichiometry, water being the only possible product if measures are taken to prevent the accumulation of hydrogen peroxide at the cathode. The anodic reactivity of hydrogen at the lower temperatures far exceeds that of the other conventional fuels.

The *invariance requirements* stem from the fact that the fuel cell is solely a converter of energy. The converter to start with should be as good as can be built, and it should remain invariant to ensure long life. Invariance (or long-life) requirements are:

1. No corrosion or side reactions.
2. Invariant electrolyte.
3. No change in electrodes. (Consumable carbon electrodes not included!)

The invariance requirements are more drastic and less obvious than might appear. A few examples follow. (1) Accumulation or production of partial oxidation products such as CO or $CH_3COOH$ should be avoided. (2) Electrocatalysts can be poisoned. (3) The pores in gas-diffusion electrodes can become filled with too much liquid or with gas, or with extraneous materials; in any such case, they are put out of action. (4) If the wrong ions carry current, the electrolyte will not remain invariant, and the anode and cathode reactions (ideally only fuel oxidized at anode; only oxygen reduced at cathode) may be thrown out of balance. (5) If the electrolyte is not an effective barrier, damaging (even catastrophic) mixing of gaseous fuel and oxygen may occur. In general, invariance requirements are met by a proper choice of materials, of structures, and of operating conditions, and the extent to which the requirements are met will determine how long a fuel battery can operate unattended.

Of the various (four, at least) efficiencies that are used in connexion with the fuel cell, the most significant is the *comparative thermal efficiency*, which is

$$\text{Eff}_{CT} = \frac{\text{Net electrical work out}}{\text{Heat of combustion of fuel consumed}}$$

In the denominator, conservative practice calls for the *higher* heat (water taken as liquid) of the fuel consumed in doing the work. The efficiency decreases markedly with current density, and this makes the current density-voltage curve one of the fuel cell's most important characteristics. Comparative thermal efficiencies from 70 per cent downward at practical current densities are not uncommon for fuel cells; such efficiencies easily exceed the ca. 40 per cent now attained in our best central stations. Efficiency is of course only part of the story; any energy source suited to its mission is a compromise involving much more than just efficiency alone.

The comparative thermal efficiency is a useful concept also for a fuel battery and for a system built around such a battery; in such cases, the net work out is obtained by subtracting the energy consumed by all auxiliaries (for example, pumps and regulatory devices). For regenerative or storage systems, the denominator will obviously be the total energy supplied to the system.

This is not the place for a description of engineering problems. As the fuel cell belongs to physical chemistry, so the fuel battery and the system belong to chemical engineering. The fuel battery is a complex in which transport of electricity, of heat, of mass, and of momentum must all occur at satisfactory rates if performance is to be acceptable and invariance is to be preserved. It seems fair to say that at least in the case of hydrogen as fuel, the main research is done, and the successful application of a fuel battery awaits the satisfactory solving of engineering problems. The more satisfactorily these are solved, the more closely will the efficiency, reliability, and life of the battery approach those of a representative cell.

The reason why the hydrogen devices are in this advanced position is the great electrochemical reactivity of that fuel. The fossil fuels themselves, though cheap and abundant, are so inert that they cannot be used in batteries today. Their eventual practical use is being brought nearer by the steady progress being made on this most difficult and most important aspect of the fuel-cell problem. The following are the main approaches under investigation: 1. To convert the fossil fuel by the action of water into carbon monoxide and/or hydrogen thus producing gases more reactive at the fuel-cell anode. 2. To operate at temperatures so high (above 500°C) that reactivity is satisfactory (though invariance can be a problem). 3. To find combinations of electrocatalyst and electrolyte that ensure adequate reactivity at moderate temperatures (100–500°C).

Intermediate between hydrogen and the fossil is a class of compromise fuels among which ammonia and methyl alcohol are to be found. These two substances currently appear the most promising for early commercial application because of the compromise (reactivity not too low, price not too high, energy not too low, handling not too inconvenient, availability reasonable) they represent.

The modern history of the fuel cell begins in 1932 with Bacon's investigation that led to the cell around which the battery system for Project Apollo is being built. Because current activity in the fuel-cell field is so great that even a naming of the many projects with their investigators is impossible here, it will be necessary to conclude with inadequate descriptions of two projects that show promise of early realization.

Late in the thirties, Bacon decided on an engineering approach to the fuel-cell problem. Confining himself to cheap electrocatalysts, he increased reactivity by increasing pressure and temperature. After an interruption owing to World War II, there resulted in 1960 a cell in which (at 200°C and 620 psig) the electromotive force (1·19 V) was closely approached by the open-circuit cell voltage (1·10 V), and in which a remarkable current density (near 1800 amp/ft$^2$ at just above 0·6 V) was attained. The cell had anode and cathode both of porous nickel, the cathode being protected from corrosion by a thin layer of lithiated nickel oxide; the electrolyte was strong potassium hydroxide. In these electrodes, gas and electrolyte were kept *apart* by a layer next the electrolyte of pores so fine as to remain filled with electrolyte, and gas and electrolyte were *joined* in thin films on the walls of the coarse pores. The Bacon battery in Fig. 4 is a landmark in fuel-cell history. In developing the Bacon cell for space applications, the Pratt and Whitney Company (U.S.) raised the potassium hydroxide from about 30 per cent by weight to above 75 per cent, which made possible a radical reduction in operating pressure (to 60 psig or less). To compensate for decreased reactivity, the temperature was raised to 250°C, an advantage as regards heat rejection in space. Electrode

*Fig. 4. Bacon hydrogen-oxygen battery. Forty cells in series, joined as in a filter press, constitute the battery. In the late fifties, such a battery on a stationary mount successfully operated a lift truck.*

preparation was improved, and the performance of experimental cells is noteworthy for the high efficiency attained at current densities near 400 amp/ft$^2$.

The cell of Fig. 3 has been developed for space applications by the General Electric Company (U.S.) as the heart of the system in Fig. 5. Two hydrogen-

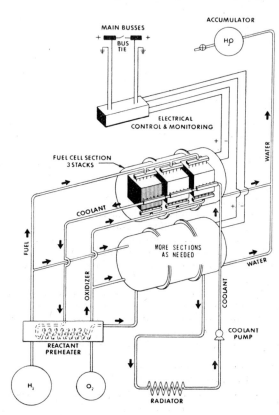

*Fig. 5. Schematic diagram of fuel-battery system planned for Project Gemini. The fuel cells are those of Fig. 3. The diagram indicates the engineering contributions necessary to make a fuel-cell application successful.*

oxygen batteries (sections) rated near 1 kilowatt each will supply on-board power and drinking water for two astronauts. Owing primarily to the nature of the electrolyte, a thin, solid ion-exchange membrane, the battery and system are relatively simple—for example, the product water can be transported to the reservoir irrespective of gravity and without the use of equipment containing moving parts.

At the time of writing, neither system has been proved out in space.

Space and military needs are today responsible for most of the greatly expanded fuel-cell activity, and it seems likely that successes in these areas will be followed by commercial applications. But the situation is complex enough to make prediction fallible.

Success in space seems near, yet even here prediction suffers owing to a lack of published information about invariance (especially life) and about successful gravity-free operation. Space requirements include reliability specifications absurdly rigorous by terrestial standards, but the long life and the low cost vital on earth are not vital in space at present. There is a greater tendency in Europe than in the United States to concentrate upon commercial applications, and it will be interesting to see whether European efforts of this kind lead to successful commercial devices sooner than do efforts in which the primary emphasis is not commercial. A case can be made for expecting fuel batteries to succeed in the following order: missions in space; submarine; lift truck; isolated energy sources (radar stations, buoys); locomotives—beyond which at this time prediction dare not go!

*Bibliography*

JUSTI E. and WINSEL A. (1962) *Cold Combustion Fuel Cells*, Wiesbaden: Franz Steiner.
LIEBHAFSKY H. A. and CAIRNS E. J., *Fuel Cells and Fuel Batteries*, New York: Wiley (In preparation).
MITCHELL W. Jr. (Ed.) (1963) *Fuel Cells*, New York: Academic Press.
YOUNG G. J. (Ed.) (1960) *Fuel Cells*, Vol. 1, New York: Reinhold.
YOUNG G. J. (Ed.) (1963) *Fuel Cells*, Vol. 2, New York: Reinhold.

H. A. LIEBHAFSKY

# G

## GALLIUM ARSENIDE, APPLICATIONS OF.

Gallium arsenide combines a number of exceptional properties which render it a most favourable material for high-frequency, high-temperature devices and for electroluminescence. These are

(1) An energy gap (1·35 eV) corresponding to low minority carrier density and, hence, low reverse currents in $p - n$ junctions at temperatures up to 300°C. This temperature may be compared with 85°C for germanium (0·78 eV) and 175°C for silicon (1·08 eV); theory suggests an even more favourable comparison.
(2) Low electron effective mass (0·072 m), with which is correlated
(3) high electron mobility (up to 8·500 cm²/V sec) and
(4) a central minimum in the conduction band, which favours optical transitions between conduction and valence bands and the shallow states associated with them, and gives a low lifetime in minority carriers.
(5) The availability of both $n$-type, $p$-type and semi-insulating ($> 10^6$ ohm cm) forms.

Nevertheless these advantages are only now beginning to make an impact on practical devices. It was found extremely difficult to prepare material containing less than $10^{16}$ impurity atoms/cm³ or with mobility much above 6000 cm²/V sec; attempts yielded highly-compensated often high-resistivity material associated with impurities such as carbon, silicon, oxygen and copper. Further degradation of the material may occur during device processing. No device technology has been found yielding the simplicity and surface passivation of the silicon planar transistor process, although some progress has been reported in the use of deposited oxides for masking the diffusion of impurities. Thus applications are limited to those where the special qualities of the material are required rather than to the replacement of silicon and germanium for ordinary uses.

*Diodes.* P-n junctions in GaAs show a 30 per cent higher breakdown voltage, $V_B$, than those in silicon of the same carrier concentration; as the mobility is higher in $n$-type material the breakdown voltage is some four or five times greater in $p^+-n$ junctions in GaAs than in silicon of the same resistivity, $\varrho$. This has shown advantages in varactor diodes, whose $Q$ is determined by the series resistance and junction capacitance. At frequency $f$ we have $Q = f_c/f$ where ideally for abrupt junctions $f_c = \{(V + \varphi)/V_B\}^{1/2}/(2\pi\varrho\varepsilon)$

and 　$V = $ applied bias voltage
　　　$\varphi = $ built-in voltage
　　　$\varepsilon = $ dielectric constant

In practice varactors intended for frequency multiplication are degraded by series resistance in the contacts and in the gallium arsenide outside the active region, so that values of $f_c$ three to five times worse than that predicted are presently found, viz. $f_c = 150$ Gc/s, $V_B = 10$; $f_c = 35$ Gc/s, $V_B = 100$. The higher operating temperature of GaAs varactors permits higher power to be stored in the varactor. The lower thermal conductivity (0·3 W/cm °C) than in silicon (0·84) is offset by increased strength which permits 0·002″ wafers to be handled. Further advances are to be expected from the use of accurately-controlled epitaxial layers deposited on highly-doped substrates.

Still better high-frequency performance has been found in small area varactors where values of $f_c$ determined from $Q$ measurements at 10 Gc/s exceed 250 Gc/s. One unit with $f_c$ of 348 Gc/s and a breakdown voltage of 11·8 V was reported: in another unit $f_c$ was as high as 730 Gc/s, but no value of $V_B$ was given. Microwave amplifiers at 9·375 Gc/s using 125 mW of pump power at 35·8 Gc/s have been constructed with a gain of 15 dB, noise figure of 2·5 dB and bandwidth of 16 Mc/s. This performance is already better than that obtained with silicon, despite the rudimentary state of the technology. The noise temperature was not correlated with $f_c$ but showed a strong negative correlation with the voltage exponent of capacity, as might be expected.

High frequency GaAs diodes have also been made using an evaporated gold Schottky barrier. Diodes using point contacts or bonded point contacts have shown advantages as detectors and as fast switching diodes where the negligible storage time is utilized.

*Tunnel diodes.* The low electron mass and "direct" gap led to early use of the material for tunnel diodes. Peak to valley current ratios of 70 : 1, current densities of $2 \times 10^5$ A/cm², and a peak current to capacity ratio of 15 mA/pF have been reported.

The higher energy gap leads to a wider negative resistance region: the voltage for peak current is typically 0·13 V and at the minimum valley current, 0·55 V. However, interest in such diodes for switching has declined, as units run above the peak current showed a gradual fall in the peak current which must be tightly controlled in this application. The deterioration does not appear to be noticeable in linear work where the dissipation is lower. A 1 Mc/s oscillator has been described showing a frequency change of $\pm$ 0·018 per cent over a temperature range of $-50°C$ to $+100°C$ and of 0·06 per cent with change in the bias point from 140 to 200 mV. Operation of diodes has been reported from $-196°C$ to $300°C$.

*Transistors.* Although the promise of GaAs for high frequency or high-temperature transistors has yet to be fulfilled, useful alloy-diffused and double-diffused devices have been described. A device with current gain of 160 up to 50 Mc/s falling to unity at 1370 Mc/s has been produced by alloying tin into a zinc diffused layer which is shaped into a 0·002″ mesa. Such devices have been limited in temperature by melting of the emitter alloy. Transistors with diffused tin emitters have now given satisfactory current gains and should prove capable of high temperature operation. Low storage times are obtained without the need for degradation of the material by lifetime killers.

*Electroluminescence.* In 1962 Nasledov *et al.* and Keyes and Quist reported efficient generation of radiation in forward-biassed p – n junctions in GaAs. The radiation was emitted at a slightly lower energy (0·04 eV) than the gap, typically at 0·83 $\mu$ at 80°K and 0·9 $\mu$ at 300°K. Other wave-lengths may be present but high efficiency is presently limited to emission of the recombination radiation between conduction and valence band and shallow levels associated with them. The efficiency of a light-emitting diode increases to a plateau at about 10 A/cm² when, at least at 80°K, it is limited largely by the high refractive index of the material which allows less than 2 per cent of light emitted isotropically in a semi-infinite slab of the material to emerge into vacuum. At room temperature practical diode lamps have efficiencies in the range 0·01–1 per cent, which may be somewhat increased by immersion optics. Such lamps form nearly point sources (say 0·010″ diameter) of nearly monochromatic light (typical half width 0·02 $\mu$) which may be modulated at frequencies up to 300 Mc/s, and should show the robustness and long life of a forward-biassed p – n junction rather than of a filament lamp. They have been used in tape readers, shaft encoders, tachometers and alarms and for simple secure communication systems of large bandwidth. As the current density is increased to about 2000 A/cm² at 80°K or 200 A/cm² at 4°K the stimulated emission of radiation becomes more important than spontaneous emission and, if optical feedback is permitted, stimulated, more or less coherent, radiation becomes dominant. Such a device is called the diode or injection laser. In typical GaAs diode lasers feedback is produced by reflection at a pair of accurately parallel and flat faces cleaved or polished at right angles to a flat p – n junction produced by diffu-

sion or liquid epitaxial regrowth. Only radiation in the plane of the junction at right angles to the faces is fed back by internal reflection. As the current density is increased a narrow peak in emission occurs on the long wave-length side of the spontaneous emission for directions within a degree or so of the perpendicular to the flat exposed faces. The half-width of the stimulated radiation may be as low as 1/60 Å. A coherent output has been obtained with an overall conversion efficiency of 30 per cent; peak powers up to 100 W and mean powers of 1 W may be achieved. Important applications of this very bright easily modulated source to ranging and communications are beginning to emerge.

Combinations of a GaAs lamp and a suitable detector, e.g. a silicon or germanium p – n junction diode, may be used to give a four-terminal device with complete electrical isolation between input and output. Integrated devices of this sort have been made in which lamp and detector are fabricated from the same block of semiconductor: some use semi-insulating GaAs to achieve electrical isolation. It has been variously claimed that such devices enable transistors of very low base resistance to be constructed, yield a controlled negative-resistance device and can give a fast simple solid-state relay. In general, these claims remain to be fully validated.

*Other applications.* GaAs solar cells have been produced with conversion efficiencies (10 per cent) comparable with silicon and with somewhat improved radiation resistance. Improved performance at elevated temperatures may allow the use of light concentrators giving an advantage in weight in satellite power systems. It is not yet clear whether these advantages will outweigh the low cost of silicon n-on-p type cells, even for satellite power supplies.

The steep absorption edge of GaAs associated with band-to-band transitions make it suitable for use as a long wave-length pass filter; the transmitted band extends far into the infra-red in appropriately processed material where free carrier absorption is absent. Evaporated GaAs filters have been prepared with properties not too dissimilar from the bulk.

Alloys of GaAs with InAs have been suggested for thermoelectric generation but suffer from a high vapour pressure at temperatures of interest. It has also been suggested that Hall effect devices may be made from GaAs, particularly where a large stable voltage output is required over a range of temperatures.

*See also*: Semiconductor devices.

*Bibliography*

DUMKE W. P. (1963) *Optical Masers*, New York: Wiley.
HILSUM C. and ROSE-INNES A. C. (1961) *Semiconducting III-V Compounds*, Oxford: Pergamon Press.
KRESSEL H. and GOLDSMITH N. (1963) *High Voltage*

*epitaxial GaAs microwave diodes, R.C.A. Review* **24**, 182.

Lax B. (1963) *Semiconductor Diode Lasers, Solid State Design* **4** (11), 26.

B. L. H. Wilson

**GAS LUBRICATION.** Like liquid lubrication theory, gas lubrication theory is well developed; however, only recently have we had adequate mathematical tools and adequate experimental facilities for verifying it. Gas bearings are distinguished by the manner in which the fluid pressure is developed. There are self-acting, externally pressurized, and squeeze-film types. Those which develop load capacity due to relative tangential surface motion are called self-acting. These have sometimes been called hydrodynamic. Those which develop load capacity primarily due to lubricant flow from an external supply through sources within the film are called externally pressurized; although the lubricating film in such a bearing is hydrodynamic, a load can be supported without relative motion of the bearing surfaces, and the bearing has been called hydrostatic. Those which develop load capacity due to relative normal surface motion are called squeeze-film bearings. The three classes of bearings may each be subdivided into configurations for supporting thrust and radial loads. The principal configurations are the same as those used with liquid lubricating films.

Though similar mechanisms operate, it is evident that self-acting gas films cannot support as heavy unit loads as liquid films. The viscosity of a gas, the factor that relates relative motion of the surfaces to load carrying pressure, may be several thousand times smaller than it is for a lubricating oil.

Not only are the viscosities of gases low, but they vary only slightly with temperature. The opposite is true of liquid lubricants. Thus, the temperature range over which a given gas film may operate is much wider than that for a liquid film. In addition, gases are stable and do not change phase or decompose like liquid lubricants at extreme temperatures or in the presence of radiation. Gas films have been successfully used as lubricants at cryogenic temperatures and at high temperatures ranging up to several thousand degrees Fahrenheit.

Gas lubrication also solves several contamination problems. First, the gas itself, usually air, is clean or easily kept clean and is plentiful as well. Secondly, a gas bearing can use a fluid within a part of a closed system, say gaseous oxygen in a liquid oxygen expansion turbine, thus eliminating contamination of the system from the outside. Since the environment itself can degrade liquid lubricants, gas bearings are being used in nuclear systems.

Finally, there is the question of rigidity. The gas bearing may not have large load capacity, but it can easily be made quite stiff so that it adequately resists bearing motion when the load changes. Indeed, the ability of gas films to resist changes in the position of the surfaces bounding them is as important practically as their high-speed, load-carrying characteristics. For example, this feature is widely used to separate magnetic head assemblies from memory disks, drums, and foils. The stiffness of a gas bearing can easily be made superior to that of a ball bearing for such applications as holding a shaft in position. (The unit film stiffness us much lower than that of a steel ball, but it acts on a much larger area.)

Although the use of gas as a lubricant was suggested over a hundred years ago by Hirn, it was not until just before the turn of the century that Kingsbury built the first experimental gas bearing. The first operational gyro bearings were used less than fifteen years ago. Nuclear submarines sailed under the polar ice cap navigating with such gyros. Air-driven and air-lubricated super centrifuges were used thirty years ago. However, the majority of applications have developed in the last ten years, and our detailed knowledge about gas bearings has grown within the last five years. The principal reason for this seeming tardiness is that heretofore the need just did not exist.

The self-acting bearing operates by the wedging of the fluid; the resistance to this wedging builds up the pressure. The load that can be supported by a given externally pressurized bearing is determined largely by the magnitude of the supply pressure. Increases in load capacity can be achieved by using multiple pressure sources, by increasing pressure, or by using recesses in which the pressure is nearly constant. The penalty, of course, is increased fluid flow.

True squeeze-film bearings cannot be created with incompressible lubricants. The only force available in such a film is a damping force which is proportional to velocity and does not have any steady-state, load-carrying, capacity. Squeeze-films do support loads in liquid-film bearings such as those in automobile connecting rods, but only because there is air separation during the pull-away part of the cycle and thus an air-vapour film on one side.

Current applications of gas bearings include the following. The slider bearing involves relative rectilinear tangential motion and usually has a rectangular surface. Such bearings are either fixed or pivoted (tilting pad). The sector thrust bearing is similar to the slider bearing except that relative tangential motion is circular. Journal bearings may have either partial or complete 360° films. Foil bearings involve one rigid and one flexible surface. The flexible surface can normally sustain only tension. Finally, conical and spherical bearings are used to sustain thrust in more than one direction.

These bearing configurations may be associated with plain or curved surfaces. Since journal or spherical radii are large compared to the film thicknesses, the surface curvature may usually be neglected for bearing film analysis. Journal bearings may be of several types. In the multiple lobe type, each of the bearing sections has a radius larger than the journal radius. The sections connect with discontinuous slopes. In the displaced section type, the complete bearing is cut across a diametral plane, and the two sections are moved in the plain a fraction of the radial clearance normal to the axis. The restricting land type has a radial clearance which is smaller near the bearing ends than in the middle.

As in the case of liquid film lubrication, gas film lubrication usually involves negligibly small fluid inertial forces. Therefore, the modified Reynolds number

$$Re = (\varrho UB/\mu)(h/B)^2$$

is small compared to one. The modified Reynolds number is the ratio of fluid inertia forces (which are proportional to density $\varrho$) to the fluid viscous forces (which are proportional to viscosity $\mu$); $h/B$ is the ratio of film thickness to length.

The normalized form of the Reynolds equation which relates pressures to the lubricating film thickness is

$$\frac{\partial}{\partial X}\left(H^3P\,\frac{\partial P}{\partial X}\right) + \frac{\partial}{\partial Y}\left(H^3P\,\frac{\partial P}{\partial Y}\right)$$
$$= \Lambda\,\frac{\partial}{\partial X}\,(HP) + \sigma\,\frac{\partial}{\partial T}\,(HP)$$

In this equation capital letters represent normalized quantities, density has been replaced by pressure because the film is essentially isothermal, and the viscosity has been assumed constant. The bearing number is

$$\Lambda = \frac{6\mu UB}{h^2 p_a}$$

and the squeeze number is

$$\sigma = \frac{12\mu\omega B^2}{h^2 p_a}.$$

In these relations, $U$ is a surface velocity, $B$ a bearing length dimension, $h$ a bearing film thickness, and $p_a$ a representative pressure which is usually taken to be ambient.

In discussing lubricating film compressibility, it is revealing to make a comparison with aerodynamics, where compressibility effects do not become noticeable until the fluid velocity (or that of the air foil through the fluid) reaches about half that of sound at the same conditions, $M = 0.5$. In fact, $M$ the Mach number, must often be about $0.7$ before it becomes a substantial factor. By contrast, a gas-lubricating film operates in almost a completely compressible region when the Mach number is as low as $0.3$.

The bearing number $\Lambda$ is a useful parameter for both liquid- and gas-lubricating films. In each case the load varies linearly with bearing number for small values of the bearing number. At the other limit, $\Lambda \to \infty$, the bearing load supported by a compressible film is finite. This can only happen if $(HP)$ is constant.

Thus, as $\Lambda \to \infty$, we have

$$P(X,\,Y) = \frac{H(O,\,Y)}{H(X,\,Y)}$$

where $H(O,\,Y)$ is the film thickness at the entrance of a film for motion in the $X$ direction. Note that a pressure discontinuity results at the end of the bearing when $H(B,\,Y) \neq H(O,\,Y)$. The load supported, as can be seen from Fig. 1, clearly shows the limiting behaviour at large and small bearing numbers.

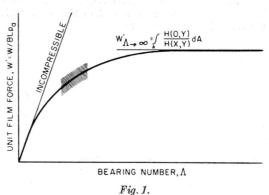

Fig. 1.

At large values of the squeeze number, the instantaneous film force is given by

$$P(X,\,Y,\,T) = \frac{H(X,\,Y,\,O)}{H(X,\,Y,\,T)}$$

and the load supported may be directly determined. This load capacity results because an equal change in film thickness above and below an average value results in a greater force when the film thickness is less than the average value. At small values of the squeeze number, the film is effectively incompressible, and no load can be supported. The maximum force in a squeeze-film bearing varies with $\sigma$ exactly as the steady load supported by a self-acting film does with $\Lambda$, and a figure like Fig. 1 results from plotting maximum force versus $\sigma$.

The behaviour of an externally pressurized gas bearing is most easily explained by reference to a flow diagram, as shown in Fig. 2. In this figure, the ambient pressure is available to the compressor and is that which exists at the output of the externally pressurized bearing, shown in Fig. 2. The supply pressure $P_s$, goes to the restricting orifice, where there is a pressure drop into the recess, which has a pressure $P_0$.

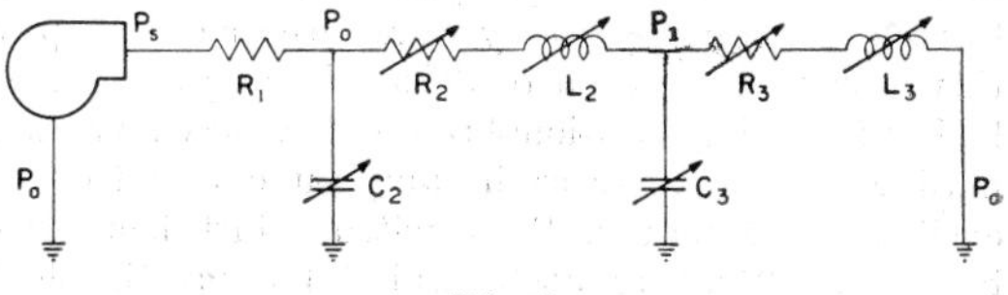

Fig. 2.

The pressure may also drop slightly from the recess at $P_1$ at the inlet of the lubricating film. If conditions are perfectly steady, $C_i = O$ and $L_i = O$. Then, with a constant pressure supply source, the ratio of $R_1 + R_2$ to $R_3$ determines the ability of the lubricating film to support a load at a particular film thickness and its resistance to changes in load. If the ratio is very large, $R_1 + R_2 \gg R_3$, the film thickness is large, and changes in the film thickness result in negligible changes in film force. Similarly, if the ratio is very small, $R_1 + R_2 \gg R_3$, the film thickness is small, and the film force is comparatively insensitive to film thickness. This behaviour is illustrated in Fig. 3.

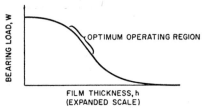

Load spacing curves for externally pressurized bearings

*Fig. 3.*

Under dynamic conditions the compressibility of the film ($C_3$) and of the gas in the recess ($C_2$), as well as the fluid squeeze-film effects ($L_3$) must be taken into consideration. Since these quantities, as well as $R_3$, are all non-linear with respect to the film thickness and are, in general, time-dependent, the complexity of the dynamic behaviour of an externally pressurized bearing is immediately evident. It is also immediately evident that resonance can be expected. In fact, resonance with $L_3$ and $C_2$ is the most likely. This phenomenon has been frequently observed in practice and is called pneumatic instability. By comparison, the externally pressurized incompressible film bearing does not have capacitance characteristics, and there is no corresponding resonance.

The dynamics of gas-lubricated journal bearings have recently received considerable attention. As the speed of a self-acting journal increases, several types of dynamic behaviour can occur. At low speeds, owing to imbalance, the shaft axis will orbit at shaft frequency in much the same manner as a passive vibrating system responds to an oscillating force. This "whirl" has a peak amplitude which corresponds to a "natural frequency". However, because of the non-linear character of the film forces, the amplitude and frequency of the resonance are a function of the amplitude of whirl. If the amplitude is not too high, further increase in velocity allows the shaft to pass this peak and move on to decreased amplitudes. At higher velocities, additional resonances may be passed until a sudden increase in amplitude due to self-excited whirl occurs. Self-excited whirl has often been called "half-frequency whirl" because the shaft orbits about the bearing axis, in the direction of

rotation, at just less than one-half the rotational speed. The fundamental cause of this kind of whirl is the phase relation between the radial displacement of the shaft and the resulting fluid force, i.e. the displacement and the force are not colinear. A consequence of a disturbance that pushes the shaft in a given direction is a component of the fluid film reaction at right angles to that direction. Under certain conditions, this force causes the shaft to orbit about some average position. Under other conditions, the whirl may grow in amplitude until failure occurs.

The tendency to self-excited whirl is unlike the ordinary resonant vibration of a passive dynamic system, which can be surmounted by increasing shaft speed. Self-excited whirl is avoided by making the film unsymmetrical, thus causing a rise in the frequency at which this whirl occurs. It may also be avoided by increasing the load on the bearing or by increasing the imbalance.

Shaft whip, a common resonance in liquid-lubricated bearings, is rarely a problem in gas bearings because the shaft is usually many times stiffer than the gas bearings.

The behaviour of unsteady gas lubricating films is complicated by the time-dependent density term in the Reynolds equation. This is particularly troublesome where repeated excitations give rise to re-enforced oscillations. Under these conditions the time constant,

$$t = \frac{12\mu B^2}{h(o)^2 p_a \pi^2 (1 + B^2/L^2)}$$

for plain parallel rectangular plates may be used to good advantage.

Several approximate solutions for self-acting gas bearing films have been obtained by making approximations of film geometry, by making mathematical simplifications during the analysis, or by using perturbation methods to reduce the non-linear Reynolds equation to a series of linear equations, the first few of which are solved. Another powerful approximation method involves replacing the Reynolds differential equation with a difference equation chosen to represent the film for any desired degree of accuracy.

Experimentation with gas bearings is difficult because it is necessary to use surfaces which are nearly perfect, to maintain desired alignment, and to measure spacing with high precision. Nevertheless, techniques for building and testing externally pressurized thrust bearings, self-acting sliders, and journal bearings have been developed.

*Definition of terms.* $Re = (\varrho\, UB/\mu)(h/B)^2$. The *modified Reynolds number* is a measure of the ratio of the fluid inertial to fluid viscous forces in a lubricating film. The film density is $\varrho$, effective tangential surface velocity $U$, breadth of film in the bearing $B$, viscosity $\mu$, and thickness $h$.

$\Lambda = 6\mu UB/h^2 p_a$. The *bearing number* is a parameter important both to liquid and gas film bearings where there is relative tangential surface motion. At low values of $\Lambda$, the lubricating film is effectively incompressible. At high values, the gas film is completely compressible, and the liquid film will separate where it diverges. The film viscosity is $\mu$, effective tangential surface velocity $U$, breadth of film in the bearing $B$, thickness $h$, and ambient pressure $p_a$.

$\sigma = 12\mu wB^2/h^2p_a$. The *squeeze number* is a parameter important both to liquid and gas film bearings where there is relative normal motion. At low values of $\sigma$, the lubricating film is effectively incompressible. At high values, the gas film is completely compressible, and the liquid film will separate when the surfaces pull apart. The film viscosity is $\mu$, frequency of relative normal oscillation of the bearing surfaces $w$, breadth of film in the bearing $B$, thickness $h$, and ambient pressure $p_a$.

*Bibliography*

CONSTANTINESCU V. N. (1963) *Lubrificaţia cu gaze*, Bucureşti: Editura Academiei Republicii Populare Romine.

FULLER D. D. (Ed.) (1959) *First International Symposium on Gas Lubricated Bearings*, Washington, D.C.: U.S. Government Printing Office, ACR–49.

GRASSAM N. S. and POWELL J. W. (Eds.) (1964) *Gas Lubricated Bearings*, London: Butterworths.

GROSS W. A. (1962) *Gas Film Lubrication*, New York: Wiley.

TIPEI N. (1962) *Theory of Lubrication*, Stanford: The University Press.

W. A. GROSS

**GEOMETRODYNAMICS.** Geometrodynamics is the study of the geometry of curved empty space and of the evolution of this geometry with time. It is therefore closely allied with the general theory of relativity. However, whereas the latter is a well-defined theory concerned chiefly with the interpretation of gravitational effects, geometrodynamics is a broad, often speculative, investigation of the extent to which all physical phenomena can be interpreted as aspects of the geometry of space-time.

Geometrodynamics has been brought to its present state of development largely through the efforts of John A. Wheeler; most of the work in the field has been done by him, his students, or his collaborators. He is also responsible for giving the subject its name.

To date there have been both successes and failures in this effort to geometrize physics. The investigation has brought to light many challenging mathematical and physical problems and has led to many insights into the richness of general relativity, classical electromagnetism, and other theories. It is hoped that the new ways of looking at things thus discovered will ultimately lead to significant advance in, for example, much the same way that Hamilton's formulation of mechanics laid the groundwork for quantum theory.

On the other hand, geometrodynamics has yet to suggest a new experiment or predict a new experimental result.

Geometrodynamics builds upon Einstein's general relativity as a foundation. The physical world is described in terms of a four-dimensional Riemannian geometry. The metric of this geometry is indefinite and not necessarily flat. Rather it is restricted by certain conditions (chiefly differential equations) on the metric tensor ($g_{ij}$); thus empty space itself, like the fields and particles of conventional theories, is endowed with a *dynamics*, a mechanism by which its present state determines its future evolution. Gravitational effects are interpreted, as they are in general relativity, as arising directly out of the curvature of space-time: The curved trajectories of particles in a gravitational field are actually the straightest possible (geodesic) trajectories, which are unavoidably deflected by the curvature of the regions through which they pass.

General relativity views the electromagnetic field (as it does all non-gravitational fields) as a non-geometric field which is related to the geometry of space-time only through the dynamical equation for the metric tensor, which equates a certain aspect of the curvature of space (described by a constant multiple of the Einstein tensor, $R_{ij} - \frac{1}{2}g_{ij}R$) to its energy-momentum content (described by the Maxwell stress-energy-momentum tensor, $T_{ij}$). Geometrodynamics reinterprets this equation as a definition of $T_{ij}$ in terms of the Einstein tensor. Furthermore, as has been mathematically demonstrated by Rainich and Misner, if $T_{ij}$ is known, then the electromagnetic field tensor $F_{ij}$ (that is, the electric and magnetic field vectors) is essentially determined. Thus geometrodynamics *defines* the electromagnetic field as a certain (very complicated) function of the curvature of space-time, thereby giving a geometric interpretation of electromagnetism. Maxwell's *free-field* equations for the electromagnetic field then become, after the electric and magnetic field vectors have been replaced by their geometric counterparts, the (geometro-)dynamical equations of an empty space possessing the characteristics of the gravitational and *charge-free* electromagnetic fields.

Electric charge is commonly thought of as a divergence in the electric field: The "lines" of electric flux in certain regions all seem to be diverging from (or converging upon) a singular point, the location of the charge; at the singularity, Maxwell's free-field equations must be generalized. It has been found, however, that geometrical interpretation of charge is most readily achieved, not by generalizing the geometrodynamical equations, but rather by generalizing the topology of space-time.

Consider, for example, a teacup having a hollow handle. Lines could be drawn on its interior surface converging toward one of the openings of the handle; the lines could continue unended and unbroken along the interior of the handle to emerge in a diverging pattern from its other end. To a person standing at some distance from the teacup and looking into it, the ends of the handle might appear as points, and he would think that the pattern of lines possessed actual points of divergence.

This is a two-dimensional analogy of the geometrodynamical view of our three-dimensional space: While electric charges macroscopically appear as points of divergence of the electric field lines, closer examination could reveal that the lines actually continue down a three-dimensional handle or "wormhole" (which is a concept mathematically sound though admittedly difficult to visualize) to emerge elsewhere in a region of opposite charge. The concept of a point charge is viewed as only a macroscopic approximation, since at some degree of fineness of observation a tunnel-like form would be apparent. The geometrodynamical equations can be shown to imply that the amount of flux passing through any wormhole is constant in time; this is the geometrical statement of charge conservation.

Current thinking in geometrodynamics is quite open in regard to topology. No reason at present is seen for supposing that wormholes constitute the only peculiar topological features of space-time, although possible physical interpretations of other possibilities have not yet been very fully investigated.

The effort to geometrically interpret other physical fields than the gravitational and electromagnetic fields has foundered upon a serious obstacle: All other fields are properly described only in terms of quantum mechanics. Therefore in order to proceed further one must know how to apply quantum concepts to the geometry of space-time. One is thus led to consider essentially the same problem as do those who seek to "quantize" general relativity. This problem has been attacked from various angles by several groups of competent investigators. The degree of present success is somewhat controversial. The more pessimistic claim that several issues of concept and principle remain unclarified, while even the most optimistic admit that actual application of a quantized theory will require considerable and tedious mathematical manipulation. In any event, at the present time geometrodynamical principles have been applied in any detail only to classical (non-quantum) fields, namely gravitation and electromagnetism.

Geometrical models have been found for classical particles as well as classical fields. These models should almost certainly be viewed as phenomenological, since, as with more conventional physical theory, actual particles are presumably related to the quantum aspects of fields. The particle models are of two general types. One is the wormhole opening or "mouth". Since there is no reason that a wormhole must be threaded by lines of flux, such a particle can either be charged or neutral. Also included in this type would be openings into other topological features of space-time.

The other type of particle model is the *geon*. (This name, coined by Dr. Wheeler, means gravitational-electromagnetic entity.) One could imagine constructing such a particle in the following way: A sufficient quantity of electromagnetic radiation is confined to a region of space so small that the gravitational attraction of the radiation for itself holds it together and prevents its rapid dissipation. Detailed mathematical analysis is required to demonstrate that this is really possible; such analysis has been made on the assumption that quantum effects are ignorable. However, the mass ($m$) to radius ($r$) ratio of a geon may be given a good classical estimate by equating the gravitational acceleration offered by such a mass at its surface ($Gm/r^2$, where $G$ is Newton's gravitational constant) to the centripetal acceleration required to hold particles moving at $c$, the speed of light ($c^2/r$). Thus the mass of a geon is $c^2/G$ times its radius. This means that a geon 0·001 metre in diameter would be about as massive as the Earth! In actual fact, however, quantum effects may not be neglected for such a geon; analysis shows that for any geon having a diameter less than that of the Earth's orbit, energy densities are so high that pair production is not negligible. It becomes even more clear why one should not hastily identify the geon with any object in the real world. Analysis has also been made of the so-called gravitational geon, which has in it no electromagnetic radiation, but consists only of gravitational energy holding itself together by its own gravitational attraction.

The geometry in the neighbourhood of an isolated spherical geon or an isolated spherical neutral wormhole mouth is, as would be expected, that of the famous Schwarzschild solution of the equations of general relativity. Thus, macroscopically viewed, these objects resemble particles.

Furthermore, they move like particles. That is, in a region of curved space, geons and neutral wormhole mouths pursue geodesic trajectories; moreover, charged wormhole mouths depart from geodesic motion according to the Lorentz force equation of electromagnetic theory. This is only true if the radius of curvature of space-time is large compared to the dimensions of the "particle"; otherwise effects occur which depend upon the particle's detailed structure. These facts follow from analysis of the geometrodynamical equations. They are of course closely related to the fact long known that the field equations of general relativity *require* that neutral particles follow geodesics and that charged particles obey the Lorentz equation: The equations of particle motion are not a separate postulate in general relativity.

Geometrodynamical "particles" have other interesting properties. Geons "decay", the electromagnetic energy trapped in them slowly leaking out as does light trapped in a glass sphere by total reflection. They also interact, with one another and with other objects, with the general result that the geon is disrupted. They thus constitute a phenomenological model of the transmutation physics characteristic of elementary particles. Wormholes might similarly transmute by sliding into and out of one another or perhaps even by breaking and joining.

Another concern of geometrodynamics is the explication of means by which physical measurements might be carried out in a wholly geometric manner. Deserving first consideration is the measurement of length and time, or, more correctly, of space-time interval, for, as is well-known from relativity theory, only the latter can have invariant geometric significance, its separation into length and time depending upon the state of motion of the observer.

The tools for measuring intervals are geodesic lines in the space-time continuum, which play the role that straightedge and compass do in plane geometry. Null geodesics may be physically realized by light rays of sufficiently weak intensity as not to disturb appreciably the metrical structure of space-time. Similarly, time-like geodesics are traced by geometrodynamical particles of sufficiently small mass. Two *events* near one another are selected and their separation is defined to be the standard interval to which all others are compared.

The measurement process, as originally described by Marzke, consists in connecting the standard interval and the interval to be measured with a certain pattern of null and time-like geodesics. (If the standard and unknown intervals enjoy a space-like separation, then they must both be compared with a third interval in their common future). A complete description of the pattern will not be given here. A fair idea of what is involved can be had by considering one important part of the pattern; the geodesic clock. This consists of two parallel time-like geodesics between which a null geodesic bounces back and forth. The interval between successive returns of the bouncing geodesic to either of the parallel geodesics is constant. It is by this device that the standard interval may in effect be "carried" from place to place. The actual measurement process must use many geodesic clocks to circumvent the effects of the distortions that the curvature of space-time causes in each single clock.

Thus any interval can be determined as a multiple of the standard interval. This means that, apart from a single overall scale factor (corresponding to the arbitrary choice of the standard of length), the metrical structure of space-time may be determined by geometric means.

The procedure most often suggested for measuring length and time, namely by the use of "ideal" measuring rods and clocks in the form of crystals and atomic clocks, is not acceptable in geometrodynamics for several reasons. First, since the efficacy of these devices depends upon quantum-mechanical considerations, their construction cannot at present be geometrically described. More importantly, there is no reason to necessarily assume that these devices do in fact measure the same values of length and time as a purely geometric method would. For example, it has been suggested by Dirac and others that certain atomic

"constants" might actually vary as the universe ages; this would cause atomic clocks to get out of step with geodesic clocks and even with one another. These remarks are not of course intended to deny that crystals and atomic clocks do in fact agree very well with geometric measurements and have great value from a practical standpoint.

It will be noted that time is geometrically measured in the same units as length (e.g., metres). The universal constant $c$ (300,000,000 m/sec) is the conversion factor between this geometric unit and conventional units (e.g. seconds). In fact, that is *all* it is; it no more needs a physical explanation than does the factor 36 used to convert the standard unit (yards) for measuring the length of fabric sold at retail to the standard unit (inches) for measuring its width.

In fact, all measurements in geometrodynamics are expressed in length units, and certain quantities usually conceived of as physical constants are shown to be only conversion factors. The constant $G$ is such a quantity, for $G^{\frac{1}{2}}/c^2$ converts between conventional and geometric electromagnetic field units and $G/c^2$ converts between conventional and geometric mass units. (Hence the mass of a geon in geometric units equals its radius.) That the electromagnetic field can be measured geometrically in terms of curvature has already been discussed. Mass similarly may be so measured. For example, a spherical mass exhibits a Schwarzschild metric in its vicinity; measurements of this metric suffice to determine the mass of the central body.

There of course remain constants that require explanation. These are all related to quantum-mechanical considerations. The square root of Planck's constant $h$, the masses of the elementary particles, and the elementary charge $e$ are each, geometrically, a length. Any one of them (say $h$) could be used to specify a fundamental length; then explanation is required for the magnitude of the others. This problem is of course unsolved in geometrodynamics as it is elsewhere. It is interesting to note that the lengths defined by $h$ and $e$ (around $10^{-34}$ m) are many orders of magnitude greater than those defined by elementary particle masses (around $10^{-52}$ m), although both are of course much smaller than any lengths encountered elsewhere in physics.

The distance $10^{-34}$ m will apparently have considerable significance in quantum geometrodynamics. Elementary considerations using the quantum principle can be made even without a detailed theory. These indicate that empty space (the vacuum) possesses significant probability amplitude of having a strongly curved and topologically complex structure on a scale of $10^{-34}$ m or smaller. That is, an isolated patch of vacuum, while macroscopically flat, will, if looked at on an extreme microscopic scale, be seen to have what is best described as a foamy structure. This "foam" possesses an enormous energy, alongside of which the energies in the macroscopic world can only be described as minute perturbations.

Research in geometrodynamics has not been concerned only with arriving at the world picture outlined above and with elucidating its mathematical details, most of which have been omitted here. It has been involved in many areas of theoretical and mathematical analysis, which are usually of common concern to both geometrodynamics and general relativity. Some of these areas are now briefly summarized. In each area one finds both completed studies and problems requiring continuing investigation.

There are the problems associated with solving the geometrodynamical equations. The search continues for exact solutions, for approximate solutions, for numerically integrated solutions, for algorithms for generating solutions, for methods of classifying solutions, and for physical interpretations of existing solutions.

Closely related are problems relating to the mathematical aspects of the equations. One seeks to define and solve the Cauchy and Dirichlet boundary-value problems, to find a variational principle from which to derive the equations, to express the equations in terms of canonical variables, to find mathematical expressions for "observables", the quantities of true physical significance, to discover geometric insights into the content of the equations, and to express the equations in more convenient forms.

There are problems associated with interpreting a theory using a curved space. One finds effort directed toward defining such intuitively familiar but strangely elusive concepts as mass, energy, momentum, wave, and radiation.

There are the problems, already mentioned, associated with introducing quantum principles into the theory. Aside from the major problem one also wonders: Where do the elementary particles fit in? What is the significance of such widely varying magnitudes as $10^{-52}$ m, $10^{-34}$ m, and $10^{-15}$ m (nuclear size)? What is the geometric interpretation of "spin"—the "non-classical two-valuedness" associated with fermions? (Or is there one?) How does one interpret a quantum theory that presumes to describe the whole universe and leaves no classical haven to which to retire?

And finally there are efforts to see to what extent geometrodynamics can shed light on other problems that confront physics. A very recent effort is an extensive analysis by Dr. Wheeler of the problem of gravitational collapse and the behaviour of large aggregates of matter.

The goal that geometrodynamics has set itself is that of giving a profound and aesthetically satisfying interpretation to physics. Only time will tell whether this end will be attained.

*Bibliography*

FLETCHER J. G. (1962) *Geometrodynamics* in *Gravitation: An Introduction to Current Research*. New York: Wiley.

RINDLER W. (1962) in *Encyclopaedic Dictionary of Physics* (J. Thewlis Ed.), **6**, 268, Oxford: Pergamon Press.

WHEELER J. A. (1962) *Geometrodynamics*, Vol. I of *Topics of Modern Physics*, New York: Academic Press.

J. G. FLETCHER

**GLASS, PHYSICS OF.** Glasses, as ordinarily understood, are made by fusing certain inorganic oxides at a high temperature and allowing the liquid to solidify without crystallizing. There is a certain range of temperature, roughly located by a transformation temperature $T_g$, above which the material displays liquid properties and below which it possesses properties more characteristic of the solid state. X-ray studies indicate that glasses, like liquids, possess only short range order. Only one or two diffuse rings are observed in the diffraction patterns of both glasses and liquids. Many physical properties of glasses, in both the solid and liquid states, can be qualitatively explained in terms of the random network theory. Certain oxides such as silica, $SiO_2$, boric oxide, $B_2O_3$, and phosphorus pentoxide, $P_2O_5$, are regarded as essential to glass formation and are called network formers. Other oxides such as soda, $Na_2O$, lime, $CaO$, lead oxide, $PbO$, and so on, are network modifiers and enter into the spaces or holes in the network. Commercially manufactured glasses are mainly fused silica, borosilicates like Pyrex, in which the main constituents are $SiO_2$ and $B_2O_3$, window glasses or soda-lime-silica glasses, and lead crystal glasses, containing $PbO$ as a principal constituent.

The specific heat of a glass below its transformation temperature is only slightly greater than that of the corresponding crystal. However, owing to the random structure the specific heat cannot be calculated by the methods of lattice dynamics. The best that can be done at the moment is to treat the glass as a collection of independent harmonic oscillators. According to quantum theory the specific heat at constant volume of a monatomic solid can be written $C_v = 3Rf(T)$ where $R$ is the gas constant and $f(T)$ is a function of the absolute temperature $T$. Above a certain temperature $\theta$, characteristic of each substance, $f(T)$ begins to approach unity and the specific heat becomes $3R$. For a molecular unit of $n$ atoms and molecular weight $M$ the classical specific heat is $3nR/M$. The specific heat of glass is usually still increasing near room temperature since $\theta$ may be in the range 400–600°C. Typical room temperature values of specific heat would be 0·18 cal $g^{-1}$ °$C^{-1}$ for a soda-lime-silica glass, 0·17 cal$^{-1}$ $g^{-1}$ °$C^{-1}$ for a borosilicate glass and 0·15 cal $g^{-1}$ °$C^{-1}$ for lead crystal glass. Heat capacity is often assumed to be approximately additive with composition. Thus Winkelmann suggested that the mean specific heat over the range 16–100°C should be written $C_m = \sum_i f_i b_i$ where $f_i$ is the weight fraction of the $i^{th}$ oxide and $b_i$ is its corresponding factor. The ratio of $b_i$ to the value which each oxide would contribute classically is shown in the table. It is seen that network formers are tightly bound whereas network modifiers contribute nearly their full classical value.

| Oxide | Mb/3nR |
|-------|--------|
| $SiO_2$ | 0·64 |
| $B_2O_3$ | 0·53 |
| $P_2O_5$ | 0·65 |
| $Na_2O$ | 0·93 |
| CaO | 0·90 |
| PbO | 0·96 |

The coefficient of thermal expansion of a glass is one of the most important properties from a technological point of view. It is often necessary to join two different glasses or a glass and a metal and successful sealing requires that the materials to be joined have nearly equal coefficients of expansion. Matching is possible to most materials since the coefficient of thermal expansion is greatly affected by changes in chemical composition. Fused silica has an extremely low coefficient of linear expansion ($5 \times 10^{-7}$ per °C) and is used as a reference standard. Pyrex has a coefficient of $32 \times 10^{-7}$ per °C but ordinary soda-lime-silica glasses have coefficients in the range $80–90 \times 10^{-7}$ per °C. Higher values (up to perhaps $150 \times 10^{-7}$ per °C) may be obtained by the addition of PbO and other oxides. Glasses expand isotropically and in the range from room temperature to 300°C the expansion curves are nearly linear (that is the coefficient of expansion is constant) and it is usual in the glass industry to quote mean coefficients from 0 to 300°C. The methods used for measuring the coefficient of expansion of glass are those commonly used for other materials. For example the change in length may be measured by means of an optical lever or alternatively two quartz plates separated by the specimen may be used as an interferometer. A differential method in which the expansion of a glass rod is compared with that of silica glass is used frequently. A number of sets of factors exist which enable the coefficient of expansion to be calculated if the composition of the glass is known. The relative effect of the various components is indicated by these constants. Thus the factor for $SiO_2$ is low because silica glass has a low expansion. $Na_2O$ and $K_2O$ both have large atomic volumes and so have high factors. The factors for other oxides are usually intermediate between these two values.

Glass is isotropic (because of its random structure). Hooke's law is found to be obeyed right up to its failure by brittle fracture. There is no permanent set in short experiments at room temperature and as seen in Fig. 1 the strain at fracture is still very small (20–30 per cent). Young's modulus $E$ can be measured either in uniaxial tension or in flexure by methods similar to those for metals. The shear modulus $S$ may be measured by using a torsional pendulum. Glass is also an ideal material to use in Cornu's experiment in which both $E$ and Poisson's ratio $v$ are measured at the same time. The shear modulus may then be calculated from $S = E/2(1 + v)$. More refined methods may take advantage of the fact that glass becomes birefringent under stress. Thus the wave-length of a high frequency sound wave can be determined by the diffraction of a plane polarized beam of light passing through a specimen which has been set into vibration by means of a quartz crystal. At room temperature the elastic

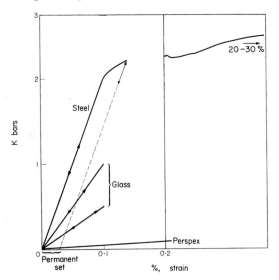

*Fig. 1. Stress-strain diagram,*

moduli range over a factor of 2 or 3 with composition. For fused silica $E$ is 730 kilobars and $S$ is 134 kilobars. 1 bar is equal to $10^6$ dyn cm$^{-2}$. It is equivalent to a pressure of 750·06 mm of mercury of density 13·595 g cm$^{-3}$ at 0°C, assuming standard gravity of 980·665 cm sec$^{-2}$. Most commercial glasses have values of $E$ in the range 500–800 kilobars. Glasses of high Young's moduli are needed in fibre reinforced plastics and special glasses can be prepared having moduli as high as 1400 kilobars. Boric oxide glass (which is not commercially usable) has a very low modulus of about 160 kilobars. Poisson's ratio of most glasses is about 0·25. It is rarely less than 0·2 (for fused silica it is 0·16) or more than 0·3. For most glasses Young's modulus decreases as the temperature increases. However, for silica and Pyrex the temperature coefficient is positive. There has been little systematic investigation of compressibility of glasses as a function of composition. Fused silica has a bulk modulus of 360 kilobars and for plate glass it is about 440 kilobars. The mechanical strength of glass under ideal conditions is very high but is generally limited by unfavourable surface conditions. The breaking strength of massive glass is about 0·7 kilobar, this value being increased by a factor of 100 for fine fibres. Strength is a complicated factor and is discussed by Stanworth and by Morey.

Viscosity gives a measure of the stiffness of a material. For glasses it is an extremely important quantity both from the point of view of understanding the glassy state and also from a technological point of view. Modern glassworks use automatic machinery and it is essential that the viscosity be exactly as intended at each stage of the process. Glass is a Newtonian liquid since it flows at a rate which is directly proportional to the stress applied. This prevents necking such as occurs in metals. The coefficient of shear viscosity $\eta$ can be measured over a much larger range in glasses than in normal liquids and thus measurements fall into two main types. At high temperatures and viscosities less than $10^7$ poise the concentric cylinder method is often used. In this method the liquid is sheared in the annular gap between two cylinders, one being fixed and the other rotated at constant speed. The viscous torque is balanced by an applied mechanical torque. At lower temperatures the viscosity is determined from the rate of elongation of a fibre of glass carrying a constant load. In this way a composite viscosity curve can be put together and it takes the form shown in Fig. 2. The curve is hyperbolic in appearance and there is no discontinuity in physical properties on vitrification. At the melting temperature the viscosity is of the order $10^2$ poise and glass is usually worked at temperatures corresponding to viscosities between $10^3$ poise and $10^7$ poise. There are several specific levels of viscosity which have been adopted for comparing different glasses on a temperature scale. The corresponding temperature are called viscosity reference points. These are marked in Fig. 2. Fused silica has an exceedingly high viscosity ($\sim 10^{10}$ poise) at 1400°C, the temperature at which most glasses are manufactured. Fortunately even a small amount of soda lowers the viscosity tremendously and a glass of the composition 25 per cent $Na_2O$ and 75 per cent $SiO_2$ has a viscosity of $10^2$ poise at 1400°C. The effect of compositional changes on viscosity has been investigated extensively. Oxides of the alkali metals are fluxes and lead oxide is also useful in lowering the viscosity of a given glass. Other oxides, such as lime, magnesia and alumina, stiffen the network and so increase the viscosity at high temperatures when replacing soda in soda-lime-silica glasses. The theory of viscosity is still being discussed but one approach is to regard it as a rate process and to write $\eta = A \exp B/RT$ where $A$ represents the frequency at which the flow unit can surmount the energy barrier $B$. However, $B$ is found to vary with temperature and generally an emprirical equation is used to describe the variation of viscosity with temperature. For example the equation $\log \eta = C + \dfrac{D}{T - T_0}$ fits the high temperature data quite well. $C$, $D$ and $T_0$ are constants.

Thermal conductivity is a property which is sensitive to the range over which order exists, particularly at low temperatures. The controlling process is the scattering of elastic waves and the thermal conductivity $K$ of a solid can be written as $K = \frac{1}{3} C_v u l$ where $C_v$ is the specific heat, $u$ is the velocity of sound and $l$ is the mean free path as limited by the disorder. The conductivity is very low because $l$ is small. Even quartz crystals which have been severely disordered by neutron irradiation show thermal conductivity at very low temperatures which is ten times as high as that of vitreous silica. At room temperatures glasses have thermal conductivities of the order 0·001–0·003 cal cm$^{-1}$ s$^{-1}$ °C$^{-1}$. For copper at the same temperature the conductivity is 1·0 cal cm$^{-1}$ s$^{-1}$ °C$^{-1}$. When the temperature is raised there is conduction due to radiation and its reabsorption. Radiation conductivity becomes important as low as 300°C and is largely responsible for discrepancies in high temperature measurements of conductivity. For example the apparent thermal conductivity of window glass at 1300°C is 0·20 cal cm$^{-1}$ s$^{-1}$ °C$^{-1}$. Conduction by radiation is important in heat transfer problems during the manufacture of glass. The effect of changes in chemical composition on thermal conductivity is not well known. Among common glasses, silica has the highest conductivity and high-lead compositions the lowest.

Like viscosity and thermal conductivity, electrical conductivity is a flow property. The flow of electric charge is limited by the volume resistivity $\varrho$. Commercial soda-lime-silica glasses have volume resistivities at room temperature of the order $10^{12}$ ohm cm$^3$. Thus

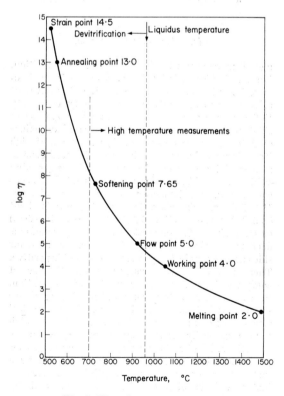

*Fig. 2. Viscosity-temperature curve.*

they may be regarded normally as insulators. The resistivity, however, varies exponentially with temperature according to $\varrho = \varrho_0 \exp H/\boldsymbol{R}T$ where $\varrho_0$ and $H$ vary only slowly with temperature; therefore glass becomes conducting as low as 200°C. According to the network theory the sodium ions are relatively loosely held in the glass network and it is to be expected that conduction will be ionic. This conclusion has been known since the early experiments of Warzburg and co-workers. These experiments also showed that Faraday's laws are obeyed. As the sodium content in the glass increases so the conductivity increases. For example the resistivity at room temperature decreases by a factor of $10^4$ when the soda content increases from 10 to 50 per cent. Glasses without alkali such as fused silica or glasses based on lime, alumina and boric oxide are insulators at even fairly high temperatures. The mechanism of conduction in such glasses is still being discussed.

The optical properties of glass are of vital importance in glass technology. Coloured glasses owe their colour to electronic transitions between the allowed energy levels of the metallic ions present in the structure. These energy levels depend on the charge on the central ion and on the electric field experienced by the ion due to its nearest neighbours. The most useful colouring agents are ions of the first transition series in the periodic table although rare earth ions and ions of the second transition series can also impart colour. Only minute amounts of some oxides are necessary to give appreciable colour. The classification of colours and colouring oxides is dealt with thoroughly by Weyl.

Surface tension is a property of the liquid state. It is a difficult property to measure in glasses because of the high temperatures and high viscosities involved. Few of the standard methods can be used but a common method is to heat a fibre of glass at temperatures such that the viscosity is $10^4$–$10^6$ poise. When such a fibre is heated its length at first increases and then decreases until the gravitational pull becomes equal to the surface tension. The surface tension of glasses varies only slowly with temperature and changes in composition can change it by a factor of about 2. For ordinary glasses surface tension is of the order of 300 dyn $cm^{-1}$.

Although glass is an ideal Hookeian solid for most practical purposes, deviations from Hooke's law were observed in the early part of the nineteenth century. With careful experimental techniques three regions of strain can be distinguished. In addition to the "instantaneous" strain there is a region of delayed strain, which takes a finite time for completion, and flow at a constant rate. A viscosity coefficient is sometimes associated with this third region and at room temperatures it is perhaps of the order of $10^{21}$ poise. Thus liquid-like flow is hardly detectable at room temperature and great permanence of shape can be expected from glass as a material. The delayed strain has a long relaxation time and its magnitude is very small

(probably much less than one per cent of the ordinary elastic strain). However, the relaxation time decreases rapidly as the temperature increases and the magnitude of the delayed strain also increases with temperature. At high temperatures it is difficult to separate delayed strains from pure viscous flow. On account of this time dependence of strain it is found that there is dissipation of mechanical energy under oscillating stress and internal friction peaks are often observed near room temperature at frequencies of 1 c/s. The damping depends on the composition of the glass. Lead glasses, for example, show less loss than soda-lime-silica glasses and indeed fused silica absorbs very little mechanical energy. Peaks in the damping curve are also observed at higher temperatures and attempts are now being made to form a mechanical spectrum in which the various absorption peaks are attributed to different atomic mechanisms.

There is a very close similarity between the electrical properties and mechanical properties of glass. If a fixed potential is applied across the plates of a condenser having glass as a dielectric the charge builds up in the same way as the development of strain under a fixed shearing stress. The part corresponding to the delayed elasticity is usually called the absorption current. This time dependence of polarization leads to power loss under alternating current operation. Losses are greatest for those glasses containing $Na^+$ ions. Low loss glasses are those containing heavy modifiers such as lead and barium. For all glasses, however, there is no very substantial change in power loss over the whole range of frequencies of technical interest, say from 1 kc/s to 1000 Mc/s. At the highest frequencies the loss begins to increase because the frequency is of the same order as the vibration frequencies of the ions within the structure.

Glasses were first studied thermodynamically because they apparently violate the Third Law of Thermodynamics. For example, the entropy of silica glass exceeds that of the crystal by 0·9 cal $mole^{-1}$ °$K^{-1}$ at the absolute zero. Below the transformation temperature $T_g$ differential properties such as specific heat, $C_p$, expansitivity, $\alpha$, and compressibility, $K$, are nearly equal to those of the crystal. Extensive properties such as entropy, enthalpy and volume are, however, continuous at $T_g$. Values of $C_p$, $\alpha$ and $K$ increase by a factor of 2 or more in passing through the transformation range. It is usual to separate the total contribution to physical properties above $T_g$ into a vibrational part and a configurational part, and the system is assumed to be in internal equilibrium. Below $T_g$ the molecular diffusion needed to allow configurational adjustments is exceedingly slow so that there remain only vibrational contributions to physical properties. Within the transformation range physical properties can be observed to change with time if the glass is held at a temperature at which the configuration is not in equilibrium. This is an important subject and is discussed by Jones, Morey and Stanworth.

*Bibliography*

JONES G. O. (1956) *Glass*, London: Methuen.

MOREY G. W. (1954) (2nd Edn) *The Properties of Glass*, New York: Reinhold.

STANWORTH J. E. (1950) *Physical Properties of Glass*, Oxford: Clarendon Press.

WEYL W. A. (1951) *Coloured Glasses*. Sheffield: Society of Glass Technology.

S. PARKE

**GOLAY CELL.** The Golay cell, in conjunction with a photoelectric amplifier, is used as a detector for electromagnetic waves from the ultra-violet to the microwave regions. The cell is pneumatic, consisting of a chamber and a ballast reservoir containing xenon-gas and separated by a mirror membrane (see figure). The chamber and reservoir are connected by a fine leak, which is adjusted so that pressures are the same on both sides of the mirror membrane and are not affected by changes in room temperature or constant radiation. The end of the chamber opposite to the mirror membrane is sealed by a window, the material of which depends on the transmission range required.

Since the cell is not affected by constant radiation a beam modulator (chopper) is placed in front of the window. The alternating radiation passing through the window warms the gas in the chamber, raises the pressure and hence causes a deformation of the mirror membrane, since the pressure in the reservoir will remain the same. A light beam reflected by the mirror membrane passes through a line grid on to a photocell. A deformation of the membrane mirror alters the intensity of the light reaching the photocell. The unit is calibrated to give the intensity of the radiation. (Reproduced by permission of Unicam.)

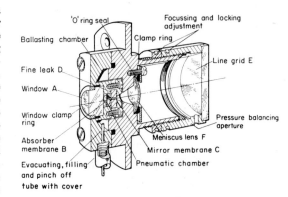

# H

**HELIUM-3, PROPAGATION OF SOUND IN.** When sound of fairly low frequency is propagated in liquid $^3$He, the propagation obeys the ordinary laws of hydrodynamics. The velocity of propagation agrees with the equation

$$c^2 = \left( \frac{\partial \varrho}{\partial p} \right) S \qquad (1)$$

at all temperatures so far investigated, and the absorption $\alpha$ is found to be in agreement with the result of hydrodynamics

$$\alpha = \frac{\omega^2}{2 \varrho c^3} \left( \frac{4}{3} \eta + \zeta + \frac{(\gamma - 1) \varkappa}{C_p} \right). \qquad (2)$$

(Here $\varrho$ = density; $p$ = pressure; $\eta$ = viscosity; $\zeta$ = second viscosity; $\gamma$ = ratio of specific heats; $\varkappa$ = thermal conductivity; $C_p$ = specific heat at constant pressure; $S$ = entropy.) In the two respects mentioned liquid $^3$He is a normal liquid, and demonstrates no particularly interesting properties.

Any experiment designed to measure the properties of sound in $^3$He involves disturbing the equilibrium of the liquid. When liquid $^3$He is disturbed in this way, it relaxes back to equilibrium with a characteristic time $\tau$. The usual laws of hydrodynamics can only hold if the frequency $\omega/2\pi$ of the sound wave satisfies the condition

$$\omega\tau \ll 1, \qquad (3)$$

which is equivalent to requiring that the liquid can come to local equilibrium within distances much shorter than the wave-length of the disturbance. Because the particles in liquid $^3$He obey a Fermi–Dirac distribution, $\tau$ increases as the temperature $T^{\circ}$K is reduced, and experiments give the value

$$\tau \sim 1{\cdot}7 \times 10^{-12} T^{-2} \text{ sec}$$

(which applies at the vapour pressure; $\tau$ falls with increase of pressure). It follows that if $\omega$ is kept constant and the temperature of the specimen is reduced, $\omega\tau$ increases until eventually it becomes very much greater than 1. There is considerable interest in the propagation of sound in the range $\omega\tau \gtrsim 1$, because it involves important problems in the theory of many-body systems. However, to attain a condition where $\omega\tau \sim 1$, one needs to use a frequency of 250 Mc/s and a temperature of $0{\cdot}05^{\circ}$K (or an equivalent combina-

tion), and these conditions are not easily met in practice. Most experiments performed so far have therefore satisfied condition (3).

In trying to understand the properties of liquid $^3$He, it is natural, since the atoms are widely spaced and have weak interactions, to try using a model of a perfect Fermi–Dirac gas. On this model it is predicted that if one tries to propagate sound at increasing values of $\omega\tau$, the velocity and attenuation should both rise, but when $\omega\tau$ becomes of order 1, the propagation of a wave motion should become impossible. As there is no mechanism for setting up local equilibrium within a wave-length when $\omega\tau \gg 1$, this result is not surprising. However, the perfect gas model has been found to be unsatisfactory as a description of liquid $^3$He (for example it cannot be reconciled with experimental measurements of the specific heat and magnetic susceptibility), and it has been replaced for practical purposes by the Fermi liquid theory, which is due to L. D. Landau.

Landau's theory describes liquid $^3$He at very low temperatures. It allows for very general interactions to occur between the atoms of $^3$He, and also takes account of the quantum statistics in the same way as does the gas model. For most of the properties of liquid $^3$He, good agreement is obtained between the Landau theory and experiment.

The predictions made by Landau's theory concerning the propagation of sound in liquid $^3$He are very different from those of the Fermi gas model. Landau predicts that a wave motion should propagate for all values of $\omega\tau$, even values very much greater than 1. This means that information can be passed on from one particle in the liquid to the next without the need for a collision between the particles. In fact the information is passed on by the interaction field. Landau calls this motion zero sound. If one performs an experiment by trying to propagate sound through $^3$He at increasing values of $\omega\tau$, one should see a smooth increase in the velocity of propagation (from 183 to 190 m/s at the vapour pressure), accompanied by a maximum in the absorption. Once $\omega\tau \gtrsim 1$, the absorption should fall, eventually becoming proportional to $T^2$.

The prediction of zero sound by Landau's theory is an important new result concerning systems of many particles, and several different measurements have been proposed to see whether zero sound can be found experimentally. The main experiments suggested are: acoustic measurements at frequencies of 1000 Mc/s or higher; scattering of light, X rays, or $\gamma$ rays (there is a frequency change on scattering related to the velocity of zero sound); and neutron scattering. Unfortunately all the experiments are rather difficult, either because they are at the limit of available techniques, or because they tend to heat the specimen, or both. At time of writing, the only measurement successfully carried out in the zero sound region is a measurement of the acoustic impedance. Here sound energy was radiated into a specimen of $^3$He, and the acoustic impedance was obtained from the amount of energy radiated for a given velocity of the source. While this experiment gives fairly close agreement with the theory, a definitive test for the existence of zero sound has yet to be made.

Transverse oscillations in liquid helium-3 are also discussed by Landau's theory. When $\omega\tau$ is small, these coincide with the damped viscous wave predicted by hydrodynamics. It is possible that when $\omega\tau$ is large there may still be a wave motion ("transverse zero sound"), but the parameters describing liquid $^3$He are not known accurately enough for a firm prediction to be made. It seems likely however that transverse zero sound should exist at pressures of a few atmospheres; no experimental verification has yet been attempted.

*See also:* zero sound.

*Bibliography*

ATKINS K. R. *Liquid Helium.*

G. A. BROOKER

**HERSCH CELL.** In 1952 Hersch reported an oxygen cell comprising a metallic silver cathode partially immersed in KOH solution which gave an almost linear relationship between current and oxygen concentration over the range 0·1–50 ppm $O_2$. Since then several developments in various fields have produced a most reliable instrument for measuring oxygen in solution or gases.

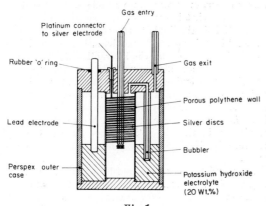

*Fig. 1.*

Figure 1 illustrates a commercially available Hersch cell incorporating advances over the original design.

When the silver and lead electrodes are connected through an external measuring meter (100 $\mu$amp), electrons pass from the lead electrode to the silver. When oxygen is present as an absorbed layer on the silver electrode, it combines with the electrons to produce hydroxyl ions according to the following process:

$$1/2 O_2 + H^+ - OH^- + 2e^- \rightarrow 2OH^-$$

At the anode, lead dissociates to form lead ions plus electrons

$$Pb \rightarrow Pb^{2+} + 2e^-$$

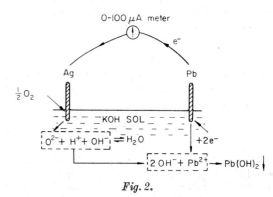

*Fig. 2.*

The $Pb^{2+}$ ion combines with $OH^-$ ions in solution to precipitate as lead hydroxide

$$Pb^{2+} + 2OH^- \rightarrow Pb(OH)_2\downarrow$$

The $OH^-$ formed at the silver electrode migrates to the lead electrode producing the overall process:

$$1/2 O_2 + Pb + H_2O \rightarrow Pb(OH)_2$$

and completing the cycle as shown in Fig. 2.

The rate of flow of current through the external circuit is a measure of the rate of cell reaction and therefore a measure of the rate of supply of oxygen to the electrode. This will depend on the rate of diffusion of oxygen through the porous membrane and the concentration gradient across it. Since the rate of removal of oxygen by the cell reaction is rapid, the oxygen concentration gradient is a function of the oxygen pressure in the test gas or solution. The cell output will also depend on the area of the silver electrode and the temperature. It is necessary therefore to calibrate the instrument before use.

Commercial instruments of the Hersch cell type have an output of $\sim 1$ $\mu$ amp/ppm $O_2$ and by the use of thermistors and transistor circuits, instability due to changes in ambient temperature, and long term output drift due to depletion of the electrolyte, have been reduced to a low value giving the cell an overall accuracy of $\pm 2$ per cent. By the use of a suitable external resistance ($\sim 150$ ohms) the oxygen concentration can be continuously recorded.

*Bibliography*

ALLSOPP P. J. (1957) Patent Spec. 24/453/1.

DEWEY D. L. (1961) *Research and Development*, No. 1 Sept.

DEWEY D. L. and GRAY L. H. (1961) *J. Polarographic Society* **7**, No. 1.

GRAY L. H. (1955) *Progr. Radiobiol.* 268.

HERSCH P. (1952) *Nature* **169**, 792.

HERSCH P. (1952) *The Chemical Age* **10**.

HERSCH P. (1954) Patent Spec. 707323, London: Patent Office.

HERSCH P. (1954) Baker Platinum Ltd. Publication, May.

HERSCH P. (1957) *Instrument Practice* **11**, 817, 937.

MACKERETH F. J. H. (1964) *J. Sci. Instrum.* **41**, 38.

B. RILEY

# I

**ICE FORMATION IN POLAR REGIONS.** Roughly one-half of the Earth's fresh water is in the form of ice, all but about 1 per cent of which is in the polar regions. By far the most abundant type is glacier ice, which covers 10 per cent of the world land surface and has a total volume of some $26 \times 10^6$ km$^3$. However, human activity is affected more significantly by relatively small quantities of snow, ground ice, and ice on rivers, lakes and the sea.

*Snow.* Snow originates when supercooled cloud droplets freeze. These water droplets, about $10\mu$ diameter, commonly supercool to $-20°C$ (even to $-35°C$ or lower) before freezing, usually on nuclei composed of clay minerals from terrestrial dust. A new crystal grows quickly at the expense of neighbouring droplets due to differences of saturation vapour pressure. Initial habit of the hexagonally symmetric crystal (columnar, platelike, dendritic) is governed by nucleation temperature, but subsequent growth depends on supersaturation of the environment. Atmospheric temperature and humidity variations modify a crystal during its fall to Earth.

The smallest particles ($\sim 20\mu$) fall through still air at about 10 cm/sec, following Stokes' law. Larger equi-dimensional particles, typical of blowing snow, have terminal velocities proportional to diameter, from 20 cm/sec at 100 $\mu$ diameter to 200 cm/sec at 1 mm diameter. More complex crystals, with greatest dimension 1–2 mm, fall at 30–100 cm/sec. Wind delays deposition and transports snow horizontally. Particles are suspended by turbulent diffusion at wind speeds above 10 m/sec; typical concentrations for winds of 15–30 m/sec are 0·4–20 g/m$^3$ at 2 m height, and 40–1000 g/m$^3$ at 3 cm height (where particles move also by *saltation*). Deposition occurs when the solid phase is over-concentrated in the surface layer. Falling and blowing snow is frequently charged ($10^{-16}$–$10^{-13}$ C/particle, +ve or −ve), highest charges occurring in strong winds and at low temperatures (below $-11°C$).

Initial bulk density of snow deposited in calm weather varies from about 0·05 g/cm$^3$ for *stellar* crystals (common at higher temperatures) up to 0·2 g/cm$^3$ for simple prismatic crystals (common at low temperatures). Wind-blown snow packs to densities exceeding 0·3 g/cm$^3$, and forms characteristic deposition-erosion patterns, primarily transverse in light winds (waves, ripples, barchans) and longitudinal in strong winds (*sastrugi*, dunes).

Mean snowfall rates vary from more than 200 g/cm$^2$·yr on coastal mountains to 10 g/cm$^2$/yr or less in the arid interiors of Antarctica and north Greenland, with 10–40 g/cm$^2$/yr representative of the range for Siberia and the Canadian barrens.

After snow is deposited, wind-borne impurities supplement those acquired during crystal growth. Snow in central Greenland and Antartica has the highest purity, e.g. Na$^+$, K$^+$, Cl$^-$ $\sim$ 0·01–0·1 mg/l, SO$_4^{--}$ $\sim$ 0·1–1·0 mg/l. In coastal regions purity is lower. Snow is generally acidic, with pH as low as 4; a typical pH for polar snow is 5·5. Wind-blown sand and dust form the principal mineral inclusions, although volcanic ash and cosmic particles may be detected. In snow, concentrations of $^{18}$O and D$_2$ vary directly with precipitation temperature, which itself varies seasonally and geographically. Some unstable isotopes, e.g. tritium, are also detectable. Certain algae, fungi, bacteria, moulds and insects can live and multiply on snow, some organically tinting the surface to various colours.

Deposited snow is changed by thermodynamic and mechanical processes, and most physical properties change in consequence. In dry snow, grains grow and bond together by vapour, surface and volume diffusion, and in the temperature-dependent sintering process angularities are subdued and the finest particles disappear. Thermal gradients stimulate vapour diffusion and convection in the pores, favouring grain growth in preferred layers. Surface meltwater may percolate and re-freeze in the snow as "pipes" or lenses of ice. Body forces compact snow by creep, a non-linear viscous process which is apparently Newtonian for small pressures ($<0·7$ kg/cm$^2$), but more strongly stress dependent at higher pressures. Creep rate changes exponentially with density (i.e. with porosity or void ratio): "compactive viscosity" for seasonal snow (warmer than $-10°C$) may increase from $10^6$ to $10^9$ g sec/cm$^2$ as density ranges from 0·1 to 0·45 g/cm$^3$, and for icecap snow ($-20°$ to $-30°C$) from $10^{10}$ to almost $10^{13}$ g sec/cm$^2$ as density increases from 0·35 to 0·60 g/cm$^3$. A Boltzmann, or Arrhenius, equation relates creep rate and temperature, with activation energy 7–20 k cal/mole.

Dry snow compacted to a density exceeding 0·8 g/cm$^3$ ceases to be permeable, the pores sealing off to form bubbles, and by convention it is then called glacier ice.

*Frost, rime and glaze.* Frost, rime and glaze are forms of ice deposited directly from the atmosphere onto cold, solid surfaces, including snow and ice. Frost (hoar), of which five distinct crystalline forms have been differentiated, condenses directly from vapour on a cold surface, to which it does not adhere strongly. Rime is milky-looking ice formed by instantaneous freezing of small supercooled droplets, with entrapment of air. It bonds quite firmly, building out in feathery shapes to windward of receiving surfaces. Glaze is a deposit of hard, clear ice formed by slow freezing of relatively large supercooled droplets, which stream over the receiving surface, excluding air and adhering strongly. Freezing rain is a type of glaze.

*Glaciers.* Glaciers are formed and sustained by accumulation and metamorphism of snow. They spread from the place of origin, where accumulation exceeds wastage, until overall ice loss balances overall accumulation. The term glacier embraces, in addition to perennial snow beds:

(a) Ice sheets which move outward in several directions, including continental ice sheets (Antarctica, $1.4 \times 10^7$ km$^2$; Greenland, $1.7 \times 10^6$ km$^2$) and ice caps (Arctic island, $10^3$–$10^4$ km$^2$).

(b) Streams with predominantly unidirectional flow, including independent valley glaciers and outflow glaciers from ice sheets ($10$–$10^3$ km$^2$), and also ice streams, which represent totally submerged outflow channels in ice sheets.

(c) Ice spreading out at the fringe of a glacier region—floating ice shelves and piedmont glaciers.

Mean depths range from about 200 m for minor valley glaciers to $1.5$ and $1.7$ km respectively for Greenland and Antarctica. Maximum depth in Antarctica exceeds 4 km.

Locally, net surface accumulation on a glacier is the algebraic sum of: precipitation ($+$), blown snow ($+$ or $-$), condensation ($+$), evaporation ($-$), and meltwater runoff ($-$). Negative accumulation is termed ablation. With steady-state conditions, temporal means of accumulation integrated over the total glacier surface sum to zero for a land-terminating glacier, and to the iceberg discharge ($+$ oceanic melt) for glaciers reaching the sea.

The local surface energy balance in a glacier is much influenced by accumulation and ablation, since transfer of sensible and latent heat is involved in the addition or removal of mass. The surface energy balance determines temperatures immediately below the glacier surface, so that they fluctuate diurnally and seasonally. Subsurface temperatures in dry snow vary sinusoidally through the year; below 10 m depth, wave amplitude is attenuated to less than 1°C, and the temperature at that depth closely approximates the mean annual air temperature at the surface above. Temperature distribution in deep ice is controlled by surface temperature, heat flow from the Earth, heat generation by shearing and sliding of the ice, and by movement of the boundaries occasioned by snow accumulation (or ablation) and ice flow.

A surplus snow layer buried by subsequent snowfalls compacts under the ever-increasing overburden, and constituent crystals slowly grow: on cold ice caps, crystals are less than 1 mm diameter near the surface, 1–2 mm at 50 m depth, and more than 4 mm at depths below 200 m. The transition from permeable snow to impermeable ice takes place usually between 50 and 100 m depth. Both crystal size and transition

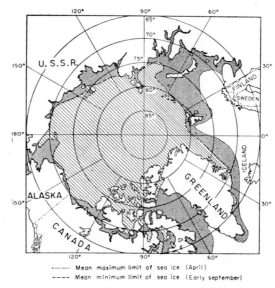

--------- Mean maximum limit of sea ice (April)
- - - - Mean minimum limit of sea ice (Early september)

*Fig. 1a. Distribution of Arctic Sea ice.*

*Fig. 1b. Currents controlling ice drift in the Arctic Ocean.*

depth depend on temperature and snow accumulation rate. (In the stagnant tongues of temperate glaciers, where stresses are low and temperature is at the melting point, crystals may be more than 10 cm across). Compression of impermeable ice is at the expense of air bubble volume, so that pressure is related to hydrostatic pressure. Bubble pressures of 300 b are probably common at the base of the Greenland and Antarctic ice sheets.

Ice flows laterally towards unrestrained edges by viscoplastic shearing and by slip on the rock bed,

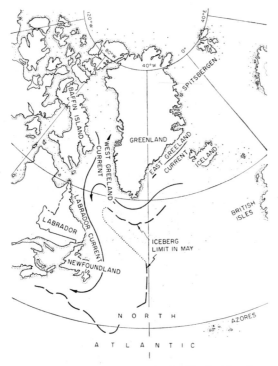

*Fig. 1c. Limits of icebergs in the North Atlantic.*

speeds being determined by ice properties and deviator stresses (hydrostatic pressure has no direct effect). In a wide uniform slab with steady-state laminar flow, shear stress at any depth is given by the product of overburden pressure and sine of surface slope. Maximum shear stresses, at the bed, are usually about 1 b. Edge restraints introduce transverse velocity gradients, while bed topography and accumulation cause longitudinal velocity gradients. Surface strains create characteristic features, notably crevasses, the cracks which typically form in patterns orthogonal to directions of principal tensile strain when strain rates approach $10^{-9}$ sec$^{-1}$. Crevasses, which are often concealed by snow bridges, are commonly 10–30 m deep and up to 10 m wide at the surface. Surface velocities on valley glaciers are mostly in the range 0·1–10 m/day, while unchannelled parts of ice sheet margins

move on the order of 0·1 m/day. Large ice shelves travel seaward at about 1 m/day.

The mechanical behaviour of ice under a given stress can be represented by a series combination of Maxwell and Kelvin–Voigt rheological models, but the steady-state (secondary) creep rate is not linearly related to stress. Secondary strain rate is commonly taken as proportional to the 3rd and 4th power of stress, although a hyperbolic sine relation better represents the transition from Newtonian toward St. Venant behaviour over a wide stress range. Flow laws relate stress and strain rate deviators, defined by the second invariants of their tensors. Strain rate varies with temperature according to the Boltzmann equation, with creep activation energy about 14 k cal/mole. The sole demonstrated glide plane of the ice lattice is (0001), although crystals oriented for hard glide can deform, perhaps by dislocation climb. Polycrystalline ice develops preferred crystal orientation under sustained shear, with the c-axes normal to the shear plane (a curious complication is the multiple maximum fabric often found in ice from glacier shear planes). In the absence of data, theoretical studies suggest that slip of ice on the glacier bed may be controlled by regelation if characteristic protuberances are small, or by stress concentration if they are large. Slip may be lubricated by a water film at the interface.

Climatic changes, or other perturbations of the glacier regime, create unsteady flow conditions. Disturbances apparently propagate through the glacier as kinematic waves, which may travel an order of magnitude faster than the ice itself is moving.

*Icebergs and "ice islands".* Icebergs are massive fragments broken (or "calved") from the seaward margins of land-based glaciers or floating ice shelves. They range in size from pieces as big as a house to enormous flat-topped slabs ("tabular bergs") tens of miles long and hundreds of feet thick. Dense glacier ice floats in sea water with about 88 per cent of its volume submerged, while less dense shelf ice floats about 83 per cent submerged.

The major sources of icebergs are the coasts of Antarctica and Greenland. Antarctica spawns icebergs at a rate in excess of $10^{12}$ tonnes/yr, chiefly from ice shelves, while Greenland's valley glaciers produce about $2 \times 10^{11}$ tonnes/yr. Moving with the ocean currents, Antarctic bergs travel in a general counterclockwise direction around the continent, the ones which avoid grounding gradually spiralling north into the open ocean to melt north of the Antarctic convergence. Greenland bergs drift generally southward to the North Atlantic and become a potential menace to shipping, especially near the Newfoundland Banks.

The so-called ice islands of the Arctic Ocean are actually tabular icebergs, up to 500 km$^2$ in area, broken from ice shelves on the north coast of Ellesmere Island and possibly other places. Those in the American sector, some of which have been occupied for long periods by scientific groups, drift in roughly elliptical clockwise paths on the Alaskan side of the Lomonosov Ridge at about 2 km/day.

*Ice on lakes and rivers.* As a lake surface cools below 4°C the thermal stratification of the water stabilizes due to water density decrease and, in the absence of mixing currents, the surface can freeze with warmer water below. Aggregations of tiny platelike or spicular crystals produce the initial ice cover. In this first skim of ice, crystal c-axes orient vertically if calm conditions prevail, but randomly if the surface is agitated during formation; subsequent growth tends to orient c-axes either vertically or horizontally. Soluble impurities concentrate along crystal boundaries, which may later be etched by preferential melting—strong sunshine in spring sometimes "candles" lake ice into a loose columnar structure.

Growth rate depends on net heat loss from the ice-water interface, i.e. on convection and conduction in the water, and conduction through the ice and snow, which itself is governed by thickness, conductivity, and surface temperature. Surface temperature depends largely on convective exchange with the air and radiation balance, so that snow cover is a major factor. Since air temperature is the dominating influence, ice thickness can be related to it within certain limits: thickness is proportional to the square root of "cumulative degree-days of frost", a quantity obtained by summation of mean daily negative air temperature since freeze-over. A proportionality factor accounts for local wind, radiation and snow conditions, size and situation of the lake, and physical constants of the ice.

In placid reaches, river ice forms in much the same way as lake ice, with some delay due to vertical mixing. In turbulent water, however, slight supercooling ($\sim 0.01°C$) is possible when cooling rates exceed $0.01°C/hr$ and tiny, slightly-buoyant crystals diffuse by turbulence into the stream to be transported to preferred deposition zones. These crystals, known as *frazil*, sometimes accumulate downstream under stable ice covers, restricting flow and causing flooding. They may also block the intakes of hydro-electric plants. Primary plate crystals are less than 1 mm across and about $20\mu$ thick. When the surface stream velocity is above some 3 m/sec there will usually be sufficient turbulence for complete diffusion of crystals; velocities less than 1 m/sec allow frazil to concentrate at the surface and so form a continuous ice sheet.

*Sea ice.* With salinities greater than 24·7 parts per thousand ($^0/_{00}$), water reaches its maximum density at temperatures below the normal freezing point, so that surface cooling on polar seas (salinity 30–35$^0/_{00}$) increases water density and promotes convection. Unless cooling is very rapid, a great depth of water reaches the freezing point ($-1.8$ to $-1.9°C$) before nucleation occurs. The first crystals to separate are usually free-floating disks of pure ice, which eventually mat and freeze together, trapping brine and air between them. Further growth, in what may be called

the columnar zone, produces preferred crystal orientation with c-axes horizontal. Gross salinity of the ice varies directly with freezing rate, mainly in the range 4–15$^0/_{00}$. Brine and air segregate in vertically elongated cells and, since phase equilibrium is preserved during temperature changes, brine volume fluctuates with temperature. At $-8°C$ and $-23°C$ respectively, $Na_2SO_4 . 10H_2O$ and $NaCl . 2H_2O$ precipitate. The temperature gradient promotes downward diffusion of brine, and in summer expansion and

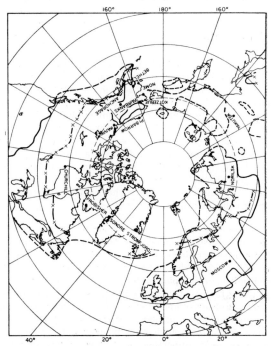

*Fig. 2a. Limits of permafrost and Pleistocene glaciation.* —— *approx. limit of Pleistocene glaciation;* ×–×–× *approx. limit of permafrost.*

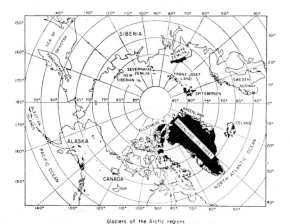

Glaciers of the Arctic regions

*Fig. 2b.*

10*

interconnection of brine cells permits gravity drainage. Thus sea ice freshens with time, the upper layers of perennial ice eventually becoming potable.

Heat flow through the ice, and hence growth rate, depends on basal temperature, true surface temperature (or flux), ice thickness, and conductivity. Effective thermal conductivity changes with temperature and salinity, since phase changes in the brine cells are involved. As with lake ice, thickness at any time can be estimated if the air temperature history is known: it is approximately proportional to the square root of "cumulative freezing degree-days", i.e. the time integration of air temperatures below sea water freezing point.

Ice which breaks away from its place of formation drifts with wind and current as *pack ice*, melting away if it reaches warmer waters in the open ocean, but refreezing in a matrix of new ice if constrained in bays or land-locked waters. Maximum ice thickness achieved by normal growth is about 3 m, but fresh surface meltwater infiltrating through cracks may lead to further accretion. A variety of cracks in sea ice result from tides, swells, or differential motion in pack ice. Wind shear or tidal effects acting on confined ice may buckle and thrust it into chaotic ridges, which can rise 3 m or more above the general surface and project 20–30 m below water level.

*Ground ice.* In cold regions, surface layers of the ground freeze and thaw seasonally. Depth of freeze-thaw is determined locally by the surface energy balance and by thermal properties of the soil, which depend mainly upon water content. Freezing and thawing indices, based chiefly on seasonal time-integrations of air temperatures below or above freezing respectively, are correlated with depth of freeze or thaw (1–2 m over much of the polar regions). Freeze or thaw of soils supporting man-made structures may have distressing consequences; thawing decreases bearing capacity in wet soil, while freezing, particularly of fine-grained ("frost-susceptible") soil, causes expansion, or "heaving".

Where mean annual surface temperatures are well below 0°C, ground temperatures beneath the freeze-thaw layer ("active layer") are also continuously below 0°C; rock and soil in this condition is termed *permafrost*. In general, ice persists indefinitely in permafrost, although supercooled water may be present. Apart from impurity and stress effects, freezing is inhibited by the properties of water films on fine particles—clay has been found with as much as 25 per cent unfrozen water at −10°C. Excluding glaciers, about 15 per cent of the Earth's land surface is underlain by permafrost, which is up to 600 m deep. Most is believed to have originated during the Pleistocene period.

*Ground ice* can be classed as (i) ice formed *in situ*, or (ii) buried ice. The first category can be split into four major sub-types:

(a) Ice lenses, produced by water migration to a freezing front, and responsible for most "frost heaving" (commonly 1 cm or so thick).

(b) Ice frozen in pores from the natural water content.

(c) Ice mounds (e.g. *pingos*, measuring tens of metres in width and height) formed by freezing ground water under hydrostatic pressure.

(d) Wedge ice, formed in contraction cracks from percolating water and condensation (typically 2–8 m deep and 0·5–3 m wide at the top).

*Buried ice*, also called "relict" or "fossil" ice, consists of massive bodies—remnants of glacier ice, snow beds, lake ice, etc. which have been buried by moraine, scree, earth flows, water-borne deposits, or wind-blown material.

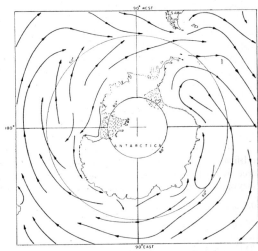

*Fig. 3a. Antarctic Ocean currents.*

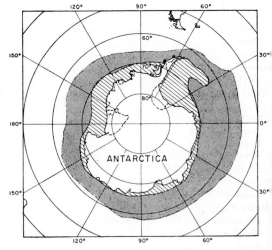

-------- Mean maximum limit of sea ice (September)
–––––– Mean minimum limit of sea ice (March)

*Fig. 3b. Distribution of Antarctic Sea ice.*

Ground ice develops characteristic surface patterns and landforms. Polygons, nets, circles, stripes, steps and mounds, with varying structures and size scales, denote particular types of ground ice. Earth flows (solifluction), *thermokarst* lakes, and certain drainage

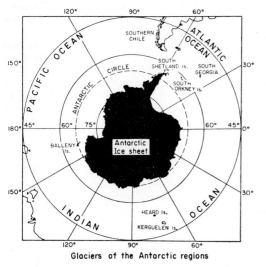

Fig. 3c.

modifications indicate ground ice effects. Rock glaciers (scree with interstitial ice, flowing downhill about 1 m/yr) are a kind of flowing permafrost.

*Bibliography*

*Bibliography of Snow, Ice and Permafrost* (with abstracts), compiled at Library of Congress, Washington, D.C. *for* Cold Regions Research and Engineering Laboratory, Hanover, N.H. (17 volumes, with about one new volume added each year).

*Glaciological results of the IGY*, issued variously by American Geographical Society, New York; Scott Polar Research Institute, Cambridge; Soviet Geophysical Committee, Academy of Sciences, Moscow.

*Journal of Glaciology*, published 3 times per year by the Glaciological Society, Cambridge, England.

1958) *Proceedings of the Easton Sea Ice Conference*, National Academy of Sciences, Washington, D.C.

1963) *Proceedings of Symposia* held by the *International Association of Scientific Hydrology*, International Union of Geodesy and Geophysics — Rome, 1954 (IASH pubs. no. 54 & 55); Berkeley, (IASH pub. no. 61).

*Reports of the Cold Regions Research and Engineering Laboratory*, Hanover, N.H., U.S.A.

1963) *Proceedings of the International Conference on Permafrost*, Purdue University, presented by the Building Research Advisory Board, National Academy of Sciences, National Research Council, Wasington, D.C.

M. MELLOR

**IMPERMEABLE CARBON.** Ordinary carbon and graphite materials are made by cementing a non-shrinking carbon filler such as petroleum coke of particle size 0·2–1000 μ with a binder such as pitch which has been derived from the distillation of coal tar. The proportion of filler to binder is very critical. The binder is chosen to give a high yield of carbon when heat treated but even under the best conditions this is unlikely to be greater than 60 per cent by weight. The material is shaped either by extrusion or moulding and then heat treated.

The pores in such a product range all the way from large cracks or voids down to defects not many times larger than the atomic spacing in the graphite crystal lattice. The wide spectrum of pores has several causes. The cracks (greater than 100 μ) are caused by major inhomogeneities in the filler-binder mix. The relatively macroscopic pores (approx. 100 μ–150 Å) arise from the method of manufacture, primarily from the voids formed during the escape of volatile matter from the binder. Additional macropores are already present within the grains of the coke filler and also result from the imperfections in the packing of the particles. Micropores (less than about 150 Å) appear to be formed where the local internal strain resulting from the extreme anisotropy of the graphite crystal exceeds the fracture strain. Because of the relative ease of cleaving the graphite crystal parallel to the "a" direction, such micropores are mainly oriented parallel to that direction.

It is evident therefore that all conventional carbon and graphite materials are porous. The bulk density obtained by weighing and measuring the external dimensions is low, 1·6–1·7 g/cm³, compared with the theoretical graphite crystal density of 2·26 g/cm³. Thus the total porosity is about 27 per cent.

The macropores are largely interconnected, thereby rendering such material permeable to fluids. The permeability is most easily measured by observing the flow rate of a gas under a given pressure gradient. The overall permeability coefficient is then defined as

$$K = pqL/\Delta pA$$

where
$K$ = overall permeability coefficient (cm²/s)
$p$ = gas pressure (dyn/cm²)
$q$ = flow rate measured at pressure $p$ (cm³/s)
$\Delta p$ = pressure drop across specimen (dyn/cm²)
$L$ = length of specimen (cm)
and
$A$ = cross-sectional area of specimen (cm²)

The permeability coefficient of ordinary carbon or graphite is generally about 1 cm²/s.

In a strict sense no completely impermeable carbon exists and the carbon or graphite materials which are marketed as impermeable are essentially materials of reduced permeability. Although the pore network in graphite is readily accessible to gases and aqueous solutions complete impermeability to gases is not always necessary. Thus for containing or transporting liquids such as liquid metals or fused salts at ordinary pressures it is sufficient that the graphite is not wetted. This characteristic leads to useful applications in metallurgy as a crucible material and for continuous casting. It is also the principle of the *mercury porosimeter* which is used to investigate pore spectra of carbons and graphites. The minimum pore entrance diameter, $d$, which is penetrated at pressure $p$ is given by

$$d = -4Y \cos \theta / p$$

where    $Y$ is the surface tension of mercury (480 dyn/cm)

and     $\theta$ is the contact angle (140°).

There are three methods by which a graphite of reduced permeability to gases may be made:

(i) by impregnation
(ii) by coating
(iii) by modification of the ordinary manufacturing process.

*Impregnation.* The self-lubricating properties of graphite make it ideal for many bearing applications. Where leakage of gases or liquids through a shaft or face seal has to be prevented, impregnated graphites are used. Various impregnants are used depending on the particular application and typical impregnants are tabulated below together with their maximum operating temperatures.

*Maximum Operating Temperature of Impregnated Graphites used in Mechanical Applications*

| Type of impregnant | Maximum temperature of operation (°C) | |
|---|---|---|
| | In air | In non-oxidising atmospheres or submerged |
| Paraffin wax | 50 | 50 |
| Chlorinated naphthalene wax | 125 | 125 |
| Thermosetting resins | 150–170 | 150–170 |
| Babbitt metal | 150 | 150 |
| Cadmium | 120 | 300 |
| Bronze | 315 | 315 |
| Silver | 370 | 815 |
| Copper | 370 | 925 |

Because of its chemical inertness and high thermal conductivity there is a considerable use for impervious graphite in chemical plant in applications involving heat exchange under corrosive conditions. The method of sealing the pores of the graphite, originally developed in the United States in 1936, is to impregnate it with a thermosetting resin which is then polymerized

*in situ.* Four types of resin are generally used, phenolic, modified phenolic, furan and cashew nut shell liquid. The corrosion resistance of such impregnated material is primarily dictated by properties of the particular resin impregnant used. The resins have not the thermal stability of the base graphite and breakdown of the resin limits the maximum operating temperature to 170–180°C. Although attempts have been made to increase this limiting temperature there is little commercial incentive.

Nevertheless there is a potential application of impregnated graphite as a canning material in nuclear reactors operating at high temperatures. For certain components graphite of permeability coefficient about $10^{-6}$ cm²/s has been required. This has been made by treating fine grain graphite with a thermosetting resin such as a furan resin or polyacrylonitrile followed by heat treatment up to 2000°C to convert the polymer to carbon. At such a low level of permeability great care has to be taken in processing the material otherwise the internal pressure build-up during decomposition of the impregnant can cause violent disintegration

*Coating.* An alternative method of reducing the pore structure in graphite is to coat the external surface. This is particularly useful when oxidation is a problem and although it is extremely difficult to achieve a completely impermeable coating, the large reduction in active surface by coating is often sufficient. A typical surface treatment involves reaction with silicon or decomposition of a silane to give a coating of silicon carbide. The graphite base material which normally has a low and anisotropic bulk thermal expansion behaviour has to be specially manufactured so that its thermal expansion characteristics are relatively close to those of the coating.

*Special manufacturing techniques.* Lower permeability graphite may be manufactured directly without recourse to impregnation or coating by careful modification of conventional processing techniques. Examples of this approach are the replacement of large proportions of the coke filler by carbon black and also the employment of infusible bonding agents such as organic thermosetting polymers. Exploitation of these techniques leads to graphites of permeability approximately $10^{-4}$ cm²/s but further reductions in permeability are rendered difficult because of the conflict between the need to allow binder decomposition products to escape yet at the same time retain an overall low permeability. This situation can paradoxically be overcome in a very special range of materials such as cellulose carbon where an impermeable end product can be obtained via a permeable intermediate. Provided the cellulose is formed in such a way that no large pores exist, then pores of the order of 500 Å which develop during carbonization close up by a process which may be considered analogous to sintering, when the cellulose carbon is heat treated to temperatures in excess of 1500°C. It appears that this is not a unique property of cellulose carbon but that provided a material of sufficiently fine porosity can be made, then pore closure will take place. Thermosetting organic polymers are an obvious alternative starting material for such a process. These latter materials are obviously exotic and can only be justified for use in novel components or arduous applications.

Although the permeability to gases can be reduced some twelve orders of magnitude, the mode of transport of those elements which are strongly adsorbed on graphite, such as for example caesium and strontium, appears to be by surface diffusion along internal pores. It is to be expected therefore that no obvious connexion has been found between the gaseous permeability constants and the diffusion parameters of such elements.

*Measurement of permeability.* The measurement of the permeability coefficient over such a wide range as has been mentioned above necessarily requires the use of several methods. Materials in the permeability range $1-10^{-4}$ cm$^2$/s can be studied by maintaining a vacuum on one side of the specimen and measuring the fall of pressure on the other side (or vice versa). With material of permeability coefficient between $10^{-4}$ and $10^{-6}$ cm$^2$/s a refinement of this technique is necessary such as by working with a higher ultimate vacuum on one side of the specimen and measuring the decay of vacuum with a Pirani gauge. Measurement of the permeability coefficient by observation of small pressure changes is unreliable when the permeability coefficient is less than $10^{-6}$ cm$^2$/s due to desorption of gases from the graphite and difficulties in sealing the specimen. Recourse has to be made to techniques which measure the transpiration of specific gaseous molecules such as $^{85}$Kr or He. The passage of $^{85}$Kr can be detected from its soft $\beta$ emission and He by a mass spectrometer. With such techniques permeability coefficients as low as $10^{-12}$ cm$^2$/s (i.e. of the same order of permeability as borosilicate glass) have been measured.

*Flow regimes.* In material of permeability coefficient ($K$) between 1 cm$^2$/s and $10^{-12}$ cm$^2$/s a variety of flow regimes can exist. In coarse pored material ($K$ approx. 1 cm$^2$/s) turbulent conditions can be set up, the rate of rise of permeability falling off at higher pressures. The results can be treated by assuming that the flow is the sum of two terms, one of which is due to viscous flow whilst the other is related to the kinetic energy of the gas.

At lower pressures and also with finer pored materials a plot of $K$ versus mean pressure is linear with a positive intercept on the $K$ axis. These results can be interpreted in terms of the Carman–Kozeny equation:

$$K = \frac{B_0}{\mu} \, p_m + \frac{4}{3} \, K_0 \sqrt{\frac{8 \, RT}{\pi M}}$$

where $p_m$ = mean pressure (dyne/cm$^2$)
    $B_0$ = viscous flow permeability constant (cm$^2$)
    $\mu$ = viscosity of gas (g/cm sec)
    $K_0$ = Knudsen flow permeability constant (cm)
    $R$ = gas constant (erg/°C)
    $T$ = absolute temperature (°K),
and   $M$ = molecular weight of gas

Again it is assumed that the overall flow is the sum of two terms, one viscous, the other in this case being molecular. The permeability constants $B_0$ and $K_0$ are properties of the graphite. Extensive measurements with many differing types of graphite have led to the following empirical relationship, which is at slight variance with simple theory, being proposed by Jenkins and Roberts:

$$B_0 = 0{\cdot}0032 K_0^{4/3}$$

In ultra-low permeability graphite ($10^{-8}$ cm$^2$/s > $K$ > $10^{-12}$ cm$^2$/s) the permeability coefficient $K$ is almost independent of mean pressure. This is because the mean free path of the gas is very much greater than the diameter of the pores and thus the viscosity term of the Carman-Kozeny equation disappears.

*Bibliography*

DAVIDSON H. W. and LOSTY H. H. W. (1963) *G.E.C. Journal* **30**, 22.

GIBB W. (1961) in *Encyclopaedic Dictionary of Physics*, (Ed. J. Thewlis) **3**, 489, Oxford: Pergamon Press.

HERING H. (1959–60), *Génie Atomique*, A7–XVI and A7–XVII, Paris, Bibliothèque des Sciences et Techniques Nucléaires.

MANTELL C. L. (1946) *Industrial Carbon*, New York: Van Nostrand.

Proceedings of the Fourth Conference on Carbon (1960) Oxford: Pergamon Press.

Proceedings of the Fifth Conference on Carbon (1962) 2 Vols., Oxford: Pergamon Press.

ROBERTS F. *et al.* (1961) *Progress in Nuclear Energy*, Series IV **4**, 105.

                          M. S. T. PRICE

# IONIZING RADIATION, MEASUREMENT OF.

## 1. Types of Ionizing Radiations; Principles of Detection Methods

This article deals with the techniques and methods used for the detection and measurement of ionizing radiations, viz. (a) energetic charged particles such as alpha and beta rays ($^4$He nuclei and electrons) emitted by radioactive materials or produced in particle accelerators (b) non-particulate gamma radiation emitted by radioactive materials, and X rays or "bremsstrahlung" radiation produced by bombardment of matter by energetic particles and (c) neutrons, i.e. uncharged nuclei of mass about equal to that of the hydrogen nucleus.

The coulomb-force interactions, between charged particles and the electrons of the material through which they are passing, account for the greater part of the energy loss of the particles. Energy is imparted to the electrons in the medium, producing both ionization and excitation of the constituent atoms, and these secondary electrons may have sufficient energy to produce further ionization or excitation effects. The particle can be detected by observation and measurement of these ionization or excitation effects in a suitable medium. The rate of energy loss (via these ionizing processes) varies over a wide range, depending on the mass, charge and energy of the ionizing particle. The rate of energy loss is larger for the more heavily charged and for the more slow moving particles. Heavy charged particles exhibit straight line paths of fixed finite range (depending on the particle type and its initial energy). Thus a 5 MeV alpha particle has a range in air of about 4 cm and the ionization per cm (specific ionization) varies from about $10^4$ to $10^5$ ion pairs/cm. Beta particles have a much lower specific ionization, of the order of $10^2$ ion pairs/cm in air and a 1 MeV $\beta$-particle has a path length of a few metres in air. Since the beta particle is of the same mass as the electrons of the medium there is a much greater probability of a large fractional energy transfer in a single collision process than for a heavy particle, and hence $\beta$-particles do not exhibit straight line tracks of well defined range.

X- and gamma-radiations interact with matter in a variety of ways (Compton collision, photoelectric absorption and positron-electron pair production processes) to produce energetic electrons. These secondary energetic electrons produce ionization and excitation effects which can be used as a means of detecting the primary gamma radiation. Neutrons can be detected by making use of suitable nuclear reactions which result in energetic charged particles which can produce ionization or excitation effects. For example, high energy neutrons (fast neutrons) may lose an appreciable fraction of their energy in an elastic collision with a nucleus of the medium; in a hydrogenous medium the average fractional energy transfer in a collision with a hydrogen nucleus is 0·5 and this energetic charged nucleus (proton) will produce an ionized track.

If the charged ions produced in a medium by an ionizing particle are free to move in the medium, e.g. ionized atoms and free electrons in a gas or electrons and electron-hole carriers in semiconductors and insulators, their presence can be detected by applying an electric field via electrodes in contact with the medium and measuring the resulting flow of charge in the external circuit connected to the electrodes. Radiation detectors using this effect can be categorized under the general heading of conduction counters or detectors; they include gas ionization devices which may or may not use gas multiplication effects and solid-state devices making use of electron and/or electron-hole conduction in semiconductors and insulators.

A second broad category of radiation detector, which also depends on ionization and excitation effects, is the scintillation detector. The light associated with de-excitation and recombination, following the passage of an ionizing particle, can be observed in gases; the light is emitted within about $10^{-9}$ sec and is primarily in the ultra-violet region. In certain organic solids some of the energy initially dissipated by the particle in producing ionization and excitation is transferred to neighbouring molecules and eventually re-radiated as light in the visible region. The overall time delay involved in the emission of this fluorescence radiation is usually of the order of $10^{-8}$ sec or less. In some inorganic crystals a similar transference of energy to neighbouring regions of the crystal lattice can also occur. In this case the light emitted is associated with imperfections due to impurity atoms in the lattice. The impurity centres have metastable states whose life-times determine the decay characteristics of the luminescence; the decay times are usually far greater than $10^{-8}$ sec and may be several hours. Phosphors with decay times greater than a few microsec are not usually of practical interest for scintillation counting. In a scintillation counter the luminescence radiation is observed by means of a photosensitive material, e.g. the photocathode of a photomultiplier tube. Electrons emitted from the photocathode are multiplied in further stages in the tube and overall gains of the order of $10^6$ are commonly achieved in multi-stage tubes.

Both conduction counters and scintillation counters can be used to register and measure individual ionizing events. The small pulse of current due to the collection of ionization produced by an ionizing particle in a conduction counter can be amplified in an a.c. coupled amplifier to the level necessary for operation of a pulse counting or scaling circuit or a mechanical register. A pulse-amplitude discriminator circuit is used between the pulse-amplifier and the counting circuit so that only pulses greater than some preset amplitude are registered. This serves to reject or minimize spurious counts due to pulses smaller than those produced by the ionizing events, e.g. due to electrical noise in the pulse amplifier or in the detector. In a scintillation counter the output signal from the output stage of the photomultiplier tube is usually much larger than that from a conduction counter and there is often no need for a separate high gain pulse amplifier. Both types of counters may also be used to measure the energy deposited by the particle in the counter, since the ionization or luminescence produced is related to the particle energy loss. If the detector is designed so as to absorb the whole of the particle energy, the pulse amplitude spectrum can be used to obtain the energy spectrum of the particles concerned. For spectrometry measurements of this kind multi-channel pulse amplitude analysers are used to record the pulse amplitude spectra.

## 2. Conduction Counters and Counting Systems

The ionization produced by charged particles, for a given total energy loss in a particular medium, varies

*Values of w (eV) = average particle energy dissipated in producing one carrier-pair.*

| Particle | Air | Argon | Hydrogen | Silicon | Germanium | CdS |
|---|---|---|---|---|---|---|
| $\alpha$ | 35·5 | 26·4 | 36·3 | 3·5 | 2·1 | 7·5 |
| $\beta$ | 34 | 26·4 | 36 | 3·5 | 2·1 | 7·5 |
| Proton | 36 | 26·4 | 32 | 3·5 | 2·1 | 7·5 |

only marginally for a wide range of particle types and energies. The average energy ($w$ in eV) dissipated in producing one carrier pair, in various media used for conduction counters, is given in the table.

In the case of gases, the value of $w$ is related to the ionization potential of the gas and is usually about three times the ionization potential. For semiconductors, $w$ is usually about three times the energy band gap between the valence and conduction energy bands.

*2.1. Mean-current gas ionization chambers.* The simplest form of conduction counter is the gas ionization chamber, consisting of a pair of electrodes (e.g. parallel plates, concentric cylinders or concentric spheres) with a suitable gas between them. An electric field is applied between the electrodes so as to separate the positive and negative ions formed in the gas and collect them at the electrodes. With a sufficiently high field no recombination of ions occurs and the mean current collected is equal to the rate of formation of the ions. Figure 1(a) shows a schematic diagram of an ionization chamber and d.c. amplifier used for this type of measurement and Fig. 1b shows a typical current/voltage characteristic for the chamber. The chamber is usually operated at a voltage in the "saturation" region, i.e. where all ionization produced in the gas is collected at the electrodes.

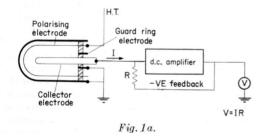

*Fig. 1a.*

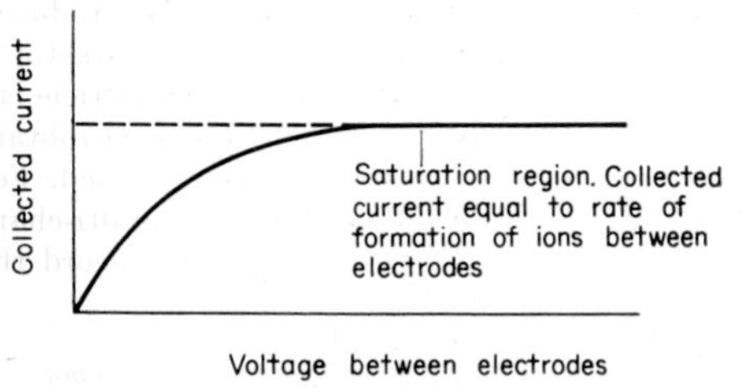

*Fig. 1b.*

The d.c. amplifier arrangement of Fig. 1(a) employs a high gain amplifier with negative feedback, so that the input terminal can be maintained at "virtual" earth potential. The output voltage $V$ is equal to $IR$, where $I$ is the ion chamber current and $R$ is the value of the resistor in the feedback line. The collector electrode is supported on an insulator between it and an earthed guard-ring electrode, i.e. there is a very small potential difference across the insulator. This technique minimizes the effect of leakage currents across the insulator. This method of measurement can be used with a suitable high-input impedance stage in the amplifier (electrometer valve or vibrating reed electrometer) for ionization current measurements as small as $10^{-15}$ amp. Mean ionization current measurements are widely used for the measurement of gamma-radiation dose-rate.

*2.2. Pulse-type ionization chambers.* The signal produced by the collection of ionization due to a single ionizing event depends on a number of factors, and especially on the speed of the ion collection. A charge "$q$" moving between parallel plate electrodes is equivalent to current flow $= qv$, where $v$ is the drift velocity of the charge, and the rate of rise of voltage across the electrodes (capacitance $C$) is given by $\dfrac{dV}{dt} = \dfrac{qv}{C}$. If a pair of ions ($q_+$ and $q_-$) are produced at a distance $x$ from the negative electrode the signal due to their collection is made up of two components, (a) a current $q_+v_+$ lasting for a time $\dfrac{x}{v_+}$ and (b) a current $q_-v_-$ lasting for a time $\dfrac{d-x}{v_-}$, where $d$ is the inter-electrode distance. The voltage signal across the electrodes will finally reach an amplitude $= q/C$, i.e. proportional to the energy dissipated in the counter, but only after both negative and positive carriers have been collected. In gases at atmospheric pressure the drift velocity of the positive ions ($v_+$) is of the order of $10^3$ cm/sec for an electric field strength of $10^3$ V/cm, i.e. the transit time of a positive ion and the duration of the associated signal is of the order of 1 millisec. The drift velocity of free electrons in gases is about a thousand times greater and the associated signal due to electron-collection in practical designs of ion chamber is often less than 1 microsec. There are very few applications of practical interest in which heavy-ion collection is measured in a pulse-type counter (because of the long-duration pulses) and it is usual to observe only the fast component of the signal due to electron collection, i.e. the pulse amplifier used has a low frequency cut-off so that slow pulses are rejected. In a counter of this type the peak pulse amplitude depends on the distance moved by the electrons before collection, i.e. it depends on the position of the primary ionization as well as its magnitude. Thus, simple two-electrode electron-collection counters can only be used for particle detection and counting purposes and not for particle energy measurements. Figure 2 shows a schematic diagram of a pulse-type counter connected to a high input impedance pulse amplifier. Pulse shaping networks are used to define and control the output pulse waveform. In Fig. 2, these take the form R–C "differentiating" and "integrating" networks. Electrical noise in the input stages of the pulse amplifier will impose a lower limit on the ionization (and particle energy) which can be detected in a simple gas ionization chamber. In practice this means that the method is limited to measurements on energetic short range particles such as alpha particles or fission fragments.

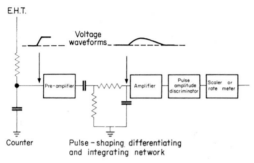

*Fig. 2 a.*

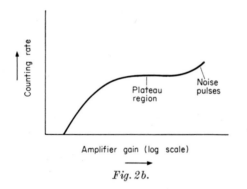

*Fig. 2 b.*

Fission ionization chambers are used for neutron counting purposes. A thin layer of $^{235}$U (or other fissile material) is deposited on an internal surface of the negative electrode, so that the energetic short-range fission fragments (resulting from neutron absorption in the $^{235}$U) produce ionized tracks close to the electrode. The electrons in the ionized tracks move towards the positive electrode and traverse the greater part of the inter-electrode spacing; thus, the greater part of the signal due to collection of the ionization appears as a fast "electron-collection" component. Figure 2(b) shows a typical counting-rate/amplifier-gain characteristic for a fission counter. At low amplifier gain levels only the larger pulses exceed the pulse-amplitude discriminator level and register as counts;

as the amplifier gain is increased a larger proportion of pulses are counted and in a good design of counter a plateau region is reached on the counting rate/gain curve which corresponds to 100 per cent detection of all fission pulses. At still higher gain levels the counting rate will increase rapidly due to spurious counts from electrical noise. This type of fission counter has important practical applications for neutron measurements in nuclear reactors, since it can be designed so as to be suitable for high temperature operation and can withstand high level radiation exposure.

*2.3. Gas multiplication. Proportional counters and Geiger–Müller counters.* The lower limit of particle energy detection can be reduced by making use of gas multiplication effects, thus increasing the charge collected. Counters with cylindrical electrodes, viz. a thin-wire anode placed along the axis of a cylindrical cathode, are commonly used to provide gas amplification factors up to $10^2$ or perhaps $10^3$, with an output signal proportional to the initial or primary ionization. The counter is operated with an anode voltage sufficiently high to produce secondary ionization processes near the anode wire. The electrons produced in the initial ionized track drift towards the centre wire under the influence of the applied electric field. As they approach the high field region near the anode the average energy gained in the interval between collision with the gas molecules increases to a level at which ionizing collisions can occur. A cascade of these processes can occur and stable operation can be achieved with gas gains of the order of $10^3$. The main component of the signal observed in the external circuit is due to the movement of the positive ions, formed in the secondary ionizing process, from the immediate region of the anode wire to a distance from it of a few wire diameters. The output signal amplitude is thus relatively independent of the original position of the ionized track produced in the counter, and is proportional to the quantity of initial ionization or to the particle energy dissipated. Some examples of gas proportional counters are described in a later section.

Gas amplification factors of $10^5$–$10^6$ can be obtained, but the output signal becomes independent of the initial ionization. The counter is then operating in the "Geiger-Müller discharge" region rather than in the "proportional" region. The discharge is no longer confined to a limited region of the anode wire, but spreads along the whole length of the wire as a result of further excitation and ionization processes. Special precautions are necessary in the Geiger-Müller type of counter to prevent the development of a continuous discharge. The discharge can be "quenched", i.e. prevented from developing into a continuous breakdown, by the use of an external circuit which reduces the applied voltage soon after the Geiger discharge has developed. Alternatively a self-quenching discharge can be obtained by the use of special additives to the counter gas. Some examples of Geiger-Müller counters are described in a later section.

*2.4. Examples of proportional counters; flow-through counters, X-ray counters and BF$_3$-filled neutron counters.* Figure 3(a) shows a cross-sectional drawing of a gas-proportional counter for counting alpha and or beta emissions from a radioactive source placed inside the counter. This type of counter is particularly useful for counting very low energy and short-range particles since there is no loss of particle energy between the source and the sensitive region of the counter. If the source is prepared as a very thin layer on a flat mounting plate the counting efficiency is 50 per cent ($2\pi$ acceptance angle). Two counters can be used side by side to provide a $4\pi$ acceptance angle and 100 per cent counting efficiency; in this case the source is mounted on a very thin supporting film.

After fitting the source in the counter the air is flushed out and replaced by a suitable counting gas (viz. non-electron-attaching gas such as methane or

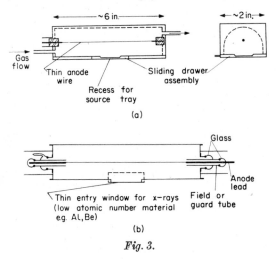

(a)

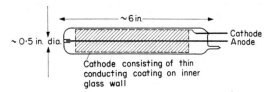

(b)

*Fig. 3.*

argon + methane). The counter is permanently connected via tubing to a compressed gas bottle and a small gas flow through the system ensures that the counter is maintained free from contamination with air.

The pulse amplifier, pulse-shaping and counting equipment used with a proportional counter is similar to that already described (Fig. 2). With an alpha-active source, the initial energy dissipated in the counter will usually be more than 1 MeV and the gas gain required is only 10–100 when a pulse amplifier with a gain of $\sim 10^5$ is used. When the counter is used for beta-activity measurements the particle energy dissipated in the counter (i.e. in a few cm of a gas at N.T.P.) is very much smaller and a correspondingly higher gas gain is required to produce pulses which will operate the amplitude discriminator circuit.

Proportional counters with a permanent gas filling can be used for the measurement of radiations which can penetrate the counter walls. One example is that of a proportional counter for the measurement of the energy of X rays and low-energy gamma-rays (see

Fig. 3(b)). The counter will usually be filled with a gas of high atomic number, so that there is a high probability of photoelectric absorption of a gamma photon in the gas with the subsequent emission of an electron whose energy is nearly equal to that of the absorbed photon. If this electron is completely absorbed or stopped within the sensitive region of the counter, the output pulse amplitude will be proportional to the gamma photon energy. This method is limited in practice to gamma energy measurements up to about 100 keV, because of practical difficulties in the design and operation of large high-pressure counters.

Boron trifluoride (BF$_3$) gas is used as a counting gas in proportional counters for neutron detection. Neutrons absorbed by $^{10}$B produce an unstable system which immediately disintegrates into a lithium nucleus and a helium nucleus ($\alpha$ particle) and these ionizing particles have a kinetic energy of 2·4 MeV. They have a short range ($\sim 1$ cm in BF$_3$ gas at N.T.P.) and only a small gas amplification factor is required to provide an adequate output signal from the counter. The cross-section for the ($n$, $\alpha$) reaction described is about $4000 \times 10^{-24}$ cm$^2$ for thermal neutrons; the total or macroscopic cross-section for a cylindrical counter 6 in. long and 1 in. diameter, and containing $^{10}$BF$_3$ gas at N.T.P., is about 8 cm$^2$. Thus the counting rate in a thermal neutron flux density of 1 neutron/cm$^2$ sec is about 8 counts/sec.

*2.5. Examples of Geiger–Müller counters; beta/gamma survey probes, end-window G.M. counters.* A unique advantage of the Geiger–Müller counter over other conduction counters is the high gas gain and large

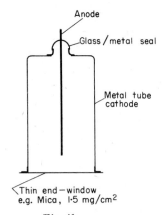

*Fig. 4a.*

*Fig. 4b.*

output signal available. The output charge can be in the region of $10^{-9}$ coulombs, sufficient for the operation of a pulse-amplitude discriminator circuit with little or no further amplification. A disadvantage in their use is an inability to discriminate between different types of ionizing particle since the output pulse is the same size for both large and small primary ionizing events. There is also a comparatively long "dead" time following a Geiger discharge before a further complete discharge can occur again and this limits the maximum counting rate which can be used. The "dead" period is governed by the time required for the positive ion sheath, produced near the anode, to move to the cathode and thus restore the initial quiescent state; the "dead" time is greatest in large diameter counters (e.g. 300–400 microsec), but may be only a few tens of microsec in small counters.

The large signal output of the Geiger counter makes it particularly attractive for use in portable $\beta$–$\gamma$ survey equipment. Figure 4(a) shows an outline drawing of a counter with a thin cylindrical outer wall suitable for detection of both gamma and beta-radiation. The wall may be made of glass (e.g. 30 mg/cm² thick) and the cathode may take the form of a thin conducting coating (graphite, SnO) on the inside surface. The gamma response of a Geiger counter is due mainly to interactions of gamma photons with the cathode wall, leading to secondary electrons which produce ionized tracks in the counter gas. Gamma counting efficiencies (i.e. counts per incident photon) are usually in the range 0·5–5 per cent, being highest for counters with a cathode wall of high atomic number. The count sensitivity per unit flux of gamma photons increases with increasing cathode area, and large counters, several feet in length and 1–2 in. diameter, may be used to provide a high counting-rate sensitivity for the measurement of low levels of gamma radiation.

Geiger counters with thin end-windows (see Fig. 4(b)) are widely used for the measurement of beta-activity in small sources. The source is prepared as a thin coating on a metal tray which is then mounted on a source holder in a defined position close to the thin end-window.

*2.6 Semiconductor counters.* There have been rapid developments in the design and use of silicon semi-conductor counters following the development of high-purity single-crystal silicon material. One form of silicon detector, a reverse-biased junction diode, is shown diagramatically in Fig. 5. These detectors can be regarded as the solid-state analogue of the parallel-plate gas-ionization chamber, with a sensitive volume extending over the depletion region in the neighbourhood of the reverse-biased p-n junction.

An important difference is that both positive and negative charge carriers have a high mobility and the collection of both carriers is observed even when a fast pulse amplifier is used. Thus the detector output is proportional to the particle energy dissipated in it, independently of the position of the particle track, and it can be used at high counting rates.

A thin p-type contact may be made by evaporating gold on to the surface of a suitably prepared n-type silicon base. Alternatively, impurity diffusion techniques may be used to prepare p-n or n-p junctions.

The depletion depth for an n-type silicon base material is given approximately by the relation $d = 0\cdot6 \sqrt{(\varrho V)}$ microns where $\varrho$ is the resistivity in ohm cm and $V$ is the reverse bias voltage. The sensitive depth need only be about 25 microns in order to absorb a 5 MeV alpha particle and this can be achieved with a reverse bias of about 10 volts on a p-n diode made from silicon of 250 ohm cm resistivity. Depletion depths up to 1 mm can be achieved using higher resistivity material and a higher reverse bias, e.g. with $V = 500$ and $\varrho = 8000$, the depletion depth is 1 mm, which corresponds to the range of a 14 MeV proton or a 0·6 MeV electron. Still greater sensitive thicknesses can be achieved using Li-doped intrinsic silicon.

The average particle energy dissipated in producing one ion pair (or carrier pair) in silicon is about 3·5 eV, i.e. about one tenth the energy required in gases. It is thus possible to use these detectors for particle energy measurements with a higher degree of resolution in energy than is possible with gas counters, because of the smaller random fluctuations in the amplitude of the counter response. The overall size of a solid-state detector required to absorb a particle of some given energy is considerably less than that of a gas ionization counter and this is an important advantage in many nuclear physics experiments. Silicon detectors of thicknesses up to a few mm's and with areas of up to a few cm² have been adapted to a variety of measurements of $\alpha$ and $\beta$ emissions from radioactive sources, to the measurement of high energy particles in particle accelerating machines and by making use of suitable neutron reactions for neutron counting and neutron spectrometry.

Figure 6 shows a pulse-amplitude spectrum obtained with a silicon detector in measurements of a ²⁰⁷Bi source. The emission from the ²⁰⁷Bi source consists of $\beta$-particles emitted from the nucleus with a continuum of energies, together with groups of monoenergetic electrons produced by internal conversion of gamma emissions from the nucleus in the K and L shells of the ²⁰⁷Bi atom. The spread in the observed line spectra

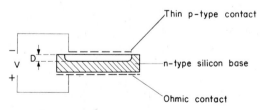

Depletion depth D=0·6 √ρV microns
(ρ =silicon resistivity in ohm cm )

*Fig. 5.*

(20 keV full width at half amplitude) is due mainly to electrical noise in the detector and in the input stage of the pulse amplifier used.

### 3. Scintillation Counters and Counting Systems

A wide variety of phosphors have been developed for use in scintillation counter applications. One of the earliest to be used was silver-activated zinc-sulphide (ZnS–Ag) in the form of a powder; this is particularly suitable for alpha counting, with the powder in the form of a thin film on a solid transparent base which can be optically connected to a photomultiplier tube. Large single crystals of inorganic phosphors are used to provide a very high efficiency for gamma detection and for gamma spectrometry. Organic phosphors in both solid and liquid form can similarly be used for high efficiency gamma detection; some liquid phosphors may also be used for "internal" $\beta$-counting, viz. a $\beta$-active material may be introduced as a solution into the liquid phosphor, thus enabling a 100 per cent

emission coefficient, $\delta$, of about 3 is obtained for incident electrons of 100 eV and this increases to a maximum value of 10 for 400 eV electrons. When a number ($n$) of secondary emission stages are cascaded, the overall gain is $\delta^n$, and gains of $10^6$ are readily achieved with 10 stages of secondar yemission. Figure 7 shows the arrangement of the photocathode and the secondary emitting dynodes in two forms of photomultiplier tube. The "focused" type of structure (Fig. 7(a)) usually has a much smaller spread in electron transit times and is used to provide full exploitation of the very short scintillation decay times available with some organic phosphors (i.e. a few nanosec). A wide range of sizes of photomultiplier tubes are available, with photocathode diameters from $\frac{1}{2}$ in. to 6 in. or so.

*3.1. Scintillation type gamma spectrometers.* For gamma spectrometry it is very desirable that a major constituent of the detection medium should be a material of high atomic number. This ensures that an appreciable fraction of the absorbed gamma photons

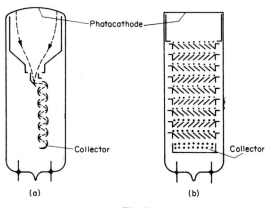

Fig. 6.            Fig. 7.

detection efficiency to be achieved. Organic phosphors which contain a high proportion of hydrogen have a high detection efficiency for fast neutrons because of the ionizing recoil protons produced by elastic collisions of fast neutrons with hydrogen nuclei.

Various types of photocathode have been developed for use in photomultiplier tubes for scintillation counting. The most generally useful is the oxidized caesium antimony ($Cs_3Sb$) cathode which is sensitive throughout the visible spectrum and has a peak sensitivity at about 4400 Å. The peak quantum efficiency (photoelectrons per photon) is about 12 per cent. Photoelectrons emitted from the photocathode are accelerated through a potential of a few hundred volts on to a secondary emitting dynode. A high secondary emission coefficient can be obtained from oxide layers produced on the surfaces of silver-magnesium, alumium-magnesium or beryllium-copper alloys, and the emission coefficient can be further enhanced by vacuum deposition of caesium on the surfaces. A secondary

produce secondary electrons of energy nearly equal to that of the photon, i.e. photoelectric absorption takes place. The pulse amplitude spectrum of the detector will then show a line structure which can be related to the incident gamma spectrum. These conditions are most readily achieved by the use of a scintillator as the detection medium, and the most suitable material available at present is thallium-activated sodium-iodide (NaI-Tl). This material can be grown as a large single crystal (up to 9 in. × 12 in.) and it has a high luminescence efficiency. (~10 per cent). The scintillation decay time is about 0·3 microsec. When used in conjunction with a photomultiplier with a $Cs_3Sb$ photocathode, the average number of photoelectrons produced, for 1 MeV particle energy absorbed in the phosphor, is about 5000. The width of the pulse-amplitude line spectrum due to photoelectric absorption is governed mainly by the statistical fluctuations in this average number of photoelectrons; the full width at half amplitude of a 1 MeV photoelectric peak is about 8 per cent for a 2 in. dia. × 2 in. long NaI(Tl) crystal. Figure 8(a) shows a typical pulse-amplitude spectrum obtained with a 2 in. diameter crystal and a source containing a mixture of gamma emitters. The spectrum is complicated by a continuum of pulse-amplitudes due to Compton collision processes but it is possible to subtract this component. In order to derive the source activity related to each gamma

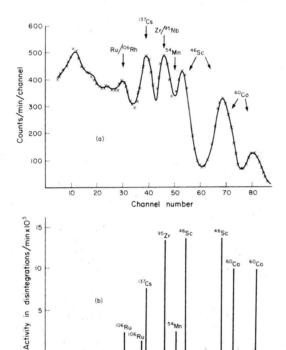

*Fig. 8.*

energy it is also necessary to make allowance for the variation of photo-peak sensitivity with energy. Figure 8(b) shows the source activity distribution with gamma energy derived from the experimental measurements of Fig. 8(a).

*3.2. Fast scintillation counters.* A major component of the scintillation light from crystalline stilbene phosphors has a decay time of about 6 nanosec, and some of the organic liquid and plastic phosphors have fast component decay times as short as 3 nanosec. These types of phosphors are used in scintillation detectors for time-of-flight particle spectrometry, i.e. where particle velocities are measured and/or selected in terms of their time-of-flight along evacuated flight tubes. Flight tubes several hundred yards in length are used and current forms of fast scintillation detectors enable time-of-flight measurements to be made with an accuracy of about 1 nanosec. The transit-time spread in the photomultiplier tube must be kept

as small as possible in order to achieve a fast rise time in the output pulse of the tube and tube full advantage of the speed of the phosphor. Tubes with "focused" dynode structures gives the best results, and it is possible to achieve an output pulse rise time of about 2 nanosec for a step-function input signal.

Combinations of fast phosphors and fast photomultiplier tubes are also used for high resolution fast-coincidence measurement for determination of the life-time of very short-lived isotopes produced by nuclear reactions. It is also possible to count random events at very high counting rates ($> 10^6$ per sec) with only a small error due to random coincidences, of two or more events.

### 4. Observation of Tracks of Nuclear Particles; Cloud Chambers, Bubble Chambers, Spark Chambers

Some experimental nuclear physics studies require the measurement of the relative positions of particle tracks. The track of an ionizing particle in a gas can be made visible by producing a condensation of a vapour and the formation of droplets of liquid along the track. This principle was used in one of the earliest forms of nuclear particle detector, the Wilson cloud chamber. The gas in the chamber is nearly saturated with a vapour and means are provided for a rapid adiabatic expansion of the gas. If an expansion follows closely on the formation of an ionized track, i.e. before recombination or diffusion of the ions can occur, a preferential condensation occurs along the line of the track. The cloud chamber is fitted with transparent windows so that the lines of liquid droplets can be illuminated and photographed.

Preferential vaporization of a liquid near its boiling point can similarly occur along the track of an ionizing particle and this effect is used in bubble-chambers. The liquids used are liquid gases and the lines of bubbles can be illuminated and photographed.

Extensive developments have been made in the last few years in the design and use of spark chambers for particle track imaging. One form of spark chamber contains a series of thin metal plates mounted parallel to one another with spacings of up to a few cm. Alternate plates are connected to the terminals of a high-voltage generator and the chamber is filled with a suitable gas mixture (e.g. argon plus some alcohol vapour). If the high voltage is applied as a short-duration pulse immediately after the passage of an ionizing particle through the electrode and gas assembly, spark discharges only occur between adjacent plates near the position where the particle has passed through the gas. Stereo-photographs are taken of these spark discharges, and the tracks of particles can be delineated from the positions of the sparks through the electrode assembly. Ancillary detectors are used to determine when an event of interest has occurred so that the chamber can be triggered into the discharge condition by application of the voltage pulse.

It is also possible to photograph the very faint tracks of light produced in a scintillator by an ionizing particle, by making use of high-sensitivity high-gain image intensifiers, but this method has had little application so far because of the technical difficulties involved and because the spark chamber technique is more flexible and can be more readily adapted for a wide variety of measurements.

*Bibliography*

ALLEN W. D. (1960) *Neutron Detection*, London: Newnes.

BIRKS J. B. (1963) *Scintillation Counters*, Oxford: Pergamon Press.

DEARNALEY G. and NORTHROP D. C. (1963) *Semiconductor Counters for Nuclear Radiations*, London: Spon.

PRICE W. J. (1958) *Nuclear Radiation Detection*, New York: McGraw-Hill.

SHARPE J. (1955) *Nuclear Radiation Detectors*, London: Methuen.

SNELL A. H. (1964) *Nuclear Instruments and their Uses*, New York: Wiley.

THEWLIS J. (Ed.) (1961–63) Various articles beginning *Bubble chamber, Cloud chambers, Radiation ionizing*, in *Encyclopaedic Dictionary of Physics*, Oxford: Pergamon Press.

W. ABSON

## IONIZING RADIATIONS, EXPOSURE OF POPULATIONS TO NATURAL SOURCES OF. *Introduction.*

It has been known since the beginning of the present century that most natural substances are slightly radioactive and in recent years it has been shown that living tissues, including those of man himself, also contain minute amounts of naturally occurring radioactive materials. The levels of this almost ubiquitous radioactivity are, however, very low compared with those in uranium- and thorium-bearing ores; the radium concentrations in most rocks, for example, are of the order of a million times less than that in a commercially workable ore such as pitchblende. Nevertheless, all living things on the Earth's surface continually receive small doses of ionizing radiation from this low-level environmental radioactivity. To this radiation dose, arising from terrestrial and human sources, must be added irradiation by the components of cosmic radiation which reach sea level and the lower altitudes. The general order of the dose, received by people living in areas of the Earth which are not specially radioactive, is about 1 rad in 10 years (a dose of 1 rad is defined as the absorption of 100 ergs per gramme of tissue); this is a very small dose, delivered at a very low-dose rate, but it provides a standard against which doses from the man-made radiations of twentieth century technology can be compared.

It is difficult, however, to assess the biological role of this natural radiation dose. Genetic effects are known to be produced in plants, insects and small mammals by radiation, but the effects have been studied in experiments in which the lowest dose is about 100 times greater, and the lowest dose-rate some 5000 times greater than the corresponding levels of radiation received by man from his natural environment. Nevertheless, it has been estimated that natural background radiation might be responsible for about 10 per cent of the known "spontaneous" mutation rate in man. Similarly, it is known that long-term somatic effects, such as the induction of tumours, can occur in man after rather large doses of radiation, but here again there is a big gap between the doses at which causal association has been found and the small doses received naturally. Such estimates as have been made by extrapolation on various models of carcinogenesis suggest that background radiation could be responsible at most, for example, for only a few per cent of the observed "spontaneous" incidence of leukaemia.

In considering the dose to populations, received from natural or other sources, it is necessary to choose the body tissues most likely to be relevant to possible radiation effects. It is reasonable for this purpose, to consider the dose to the gonads as relevant on genetical grounds and to take bone marrow and bone itself as tissues most likely to be appropriate to any consideration of somatic effects.

### Sources of Radiation External to the Body

*Cosmic radiation.* The dose-rate from the $\gamma$-ray and meson components of cosmic radiation has been shown to be 28 mrad/yr (Burch 1954; U.N. Report) at sea-level in geomagnetic latitudes greater than 50°N with a decrease to about 0·88 of this value below geomagnetic latitude 30°N. The dose-rate increases considerably with altitude and is approximately 45 mrad/yr at 5000 ft. and 88 mrad/yr at 15,000 ft. in northern latitudes. The dose-rate is lower inside buildings, because the "soft" component of the cosmic radiation is absorbed to an extent depending on the average mass per unit area of the overhead shielding. On the first floor of a typical two-storey house, the dose-rate is about 87 per cent of the unshielded value and on the ground floor, it is reduced by a further 5 per cent. The dose-rate from the neutron component of cosmic radiation is less certainly known. Its magnitude at sea-level is of the order of 2 mrad/yr and the tissue dose-rate has been calculated to be about 13 mrem/yr (the "rem" is a unit of dose which includes a factor representing the "Relative Biological Effectiveness" (RBE) of the radiation: Dose in rem = Dose in rad × RBE) but this value depends on the factor chosen to represent the greater biological effectiveness of irradiation by knock-on protons compared with electrons. It has been suggested that the total dose-rate from cosmic radiation could be of the order of 50 mrem/yr (U.N. Report).

### Table 1. Radioactivity in the environment

| Material | Radioelements | Radioactivity in $\mu\mu c/g$ ($\gamma$-ray equivalent of $^{226}$Ra) | Material | Radioelements | Radioactivity in $\mu\mu c/g$ ($\gamma$-ray equivalent of $^{226}$Ra) |
|---|---|---|---|---|---|
| *Rocks* Average granite | U: 4 $\mu g/g$ Th: 13 $\mu g/g$ K: 30,000 $\mu g/g$ | } 6·1 | *Water* Sea | U: 0.001 $\mu g/g$ K: 350 $\mu g/g$ | } $2 \cdot 8 \times 10^{-2}$ $\mu\mu c/cm^3$ |
| Alpine granite | | 12 | Thames | | $\sim 10^{-4}$ $\mu\mu c/cm^3$ |
| Clay (Yorks) | | 4·4 | | | |
| Sandstone (Yorks) | | 2 | *Air* | | |
| Limestone | | 1·4 | Country | Radon | $\sim 0 \cdot 3 \times 10^{-4}$ $\mu\mu c/cm^3$ |
| Brick | | 5 | City | Radon | $\sim 1$–$3 \times 10^{-4}$ $\mu\mu c/cm^3$ |
| Alum shales in Sweden | Ra: 60 $\mu\mu c/g$ Th: 1·5 $\mu\mu c/g$ K: 35,000 $\mu g/g$ | } 64 | | | |

### Table 2. Gamma-ray dose-rates in various localities

| Locality | Rock or soil | Site | Dose-rates in mrad/yr | | |
|---|---|---|---|---|---|
| | | | Outdoors | Indoors | |
| *England* London | London clay | 2 brick houses | 24,29 | 26,32 | Vennart (1957) |
| Sutton | Chalk | brick house | 31 | 47 | Vennart (1957) |
| | | brick hospital | 21–30 | 50–70 | Vennart (1957) |
| | | brick house | 48 | 77 | Spiers (1960) |
| Leeds | Sedimentary | stone house | 21 | 40 | Spiers (1960) |
| Broadway | Limestone | hotel | 37 | 60–80 | Spiers (1960) |
| Exeter | Sedimentary | granite house | 80–118 | 120 | Spiers (1960) |
| St. Ives | Granite | concrete house | 46–203 | 53 | Spiers (1960) |
| | | streets | 50–300 | | Spiers (1960), Willey (1958) |
| Carbis Bay | Granite | localized area | 515–600 | | Spiers (1960), Willey (1958) |
| *Scotland* Edinburgh | Sedimentary | Sandstone houses | 48·5 ± 0·6 | 48·5 ± 1·0 | Court-Brown et al. |
| Dundee | Sedimentary | Sandstone houses | 63 ± 2·3 | 63 ± 0·9 | Court-Brown et al. |
| Aberdeen | Granite | Granite 2-storey granite "bungalow" | 104 ± 1·2 | 89 ± 1·3 82 ± 1·9 | Court-Brown et al. Court-Brown et al. |
| Aberdeenshire | Mixed | Miscellaneous houses | 69·5 | 81·5 | |
| *"Active Areas"* France | Granite | Various | 180–350 | | U.N. Report |
| Nile Delta | Monazite | 6 villages | 110–400 | | U.N. Report |
| India | Monazite | Kerala | population weighted mean: 1300 | | U.N. Report |
| Brazil | Monazite | Rio de Janeiro and Esperito Santa* | 500 (Average) | | U.N. Report |
| | Volcanic | Minas Gerais and Goias* | 1600 (Average) | | U.N. Report |

\* Localized areas in the States named.

*Local gamma radiation.* The most variable part of the environmental radiation is $\gamma$ radiation from the Earth's surface and from the materials of which houses and buildings are made. Out-of-doors the $\gamma$-ray dose-rate depends on the nature of the surface rocks and soils because they incorporate radioactive materials to very variable extents. Indoors, the $\gamma$-ray dose-rate depends on the building materials and in part on the house construction. The principal radioelements responsible for the local $\gamma$ radiation are the elements of the uranium and thorium series and the naturally-occurring radioisotope of potassium $^{40}$K. Natural radioelements such as those of the actinium series, $^{87}$Rb, $^{14}$C, $^{3}$H and a number of others, contribute insignificantly to the local $\gamma$ radiation, either because the abundance is too low, or because of the nature and energy of the radiation emitted. Generally, igneous rocks, for example granites, are relatively high in radioactivity and sedimentary rocks much lower. In this country, limestone and chalk give the lowest background dose-rate and the granites the highest. In some regions of Cornwall, the presence of uranium-bearing ores give localized areas of high $\gamma$ radiation and in some regions of the world, the occurrence of fairly extensive areas of monazite sands produce dose-rates more than ten times greater than those found in this country and in most areas which are not specially radioactive. Some examples of the radioactive content of rocks and other materials are given in Table 1.

The levels of radioactivity in surface waters are much below those in the lithosphere; in sea water the radioactivity is of the order of one hundred times less than in rocks and is very largely due to potassium; in many surface fresh waters, the levels are lower again by two orders of magnitude. On the other hand, waters drawn from deep wells are frequently very radioactive (U.N. Report; Hurst 1957; Muth *et al.* 1957); the radium contents of some spa waters can be as high as 0·1 $\mu\mu$c/g, sufficient sometimes to induce a belief that herein lie the medicinal properties the waters are claimed to possess. Typical levels of radioactivity in waters are also included in Table 1, together with a few values for the natural radioactivity of air which is contributed mainly by radon and its decay products.

Some typical dose-rates from local $\gamma$ radiation in this country are given in the first part of Table 2 (Spiers 1960); in the second part of Table 2 mean dose-rates are given, with the standard errors of the means, for four localities in Scotland, where extensive surveys were made to determine the average population dose (Court Brown *et al.*). The dose-rates in the houses in these surveys were averages determined by measuring at three representative points, in the living room, kitchen and a bedroom. In the granite houses a difference is evident between two or more storey houses and bungalow type houses with the bedrooms partly in the roof.

Dose-rates of comparable magnitude have been reported from a number of countries (U.N. Report), viz. Austria, France, Japan, Sweden and U.S.A. Dose-rates in the range 180–350 mrad/yr have been reported for rather extensive granitic, schistous and sandstone areas in France and rather similar values in a few places in the monazite areas of the Northern Nile Delta (U.N. Report). The highest dose-rates affecting any considerable numbers of people (U.N. Report) are those found in the monazite area in Kerala, India, where for some 100,000 inhabitants the population weighted mean dose-rate is 1300 mrad/yr, and in Brazil where the average dose-rate for a population of 30,000 is reported to be 500 mrad/yr. In the areas of high dose-rate so far investigated, the zones of high radioactivity are very localized, in that they extend as thin strips, many kilometres in length, but only fractions of a kilometre in width. The problems of determining a representative population dose are therefore considerable.

### Sources of Internal Irradiation

*Potassium-40.* The largest contribution to the internal irradiation of the soft tissues of the body arises from $^{40}$K. Potassium is present in most body tissues, except fat, and although the potassium content of body organs varies from about 0·05 to 0·31 per cent, the average for body tissues as a whole can be taken as about 0·2 per cent by weight. Because about 89 per cent of the $^{40}$K disintegrations are by $\beta$-emission of maximum energy 1·32 MeV and 11 per cent by electron capture with a $\gamma$-ray emission of energy 1·46 MeV, most of the internal dose from $^{40}$K is provided by the $\beta$ radiation. The average soft tissues dose-rate is approximately 18 mrad/yr from $\beta$ radiation and 2 mrad/yr from $\gamma$ radiation, making a total of 20 mrad/yr. In hard bone the potassium content is probably no more than a quarter of that in soft tissues and the consequent dose-rate about 7 mrad/yr, whereas in trabecular bone the dose-rate is more nearly controlled by the marrow content of potassium which is about 0·2 per cent and the average marrow dose-rate is about 15 mrad/yr (U.N. Report).

*Carbon-14.* A small dose to all body tissues is contributed by $^{14}$C which is a $\beta$ emitter with a maximum energy of 0·155 MeV. Taking the $\beta$ emission as 14 $\beta$'s per min per g carbon and the average carbon content of the whole body as 18 per cent by weight leads to a dose-rate of approximately 1 mrad/yr. The dose-rate to soft tissues is a little less than this and that to bone about 50 per cent higher (U.N. Report; Libby 1955).

*Elements of the radium and thorium series.* After initial uncertainties as to the order of magnitude of the $^{226}$Ra content of the body (Muth *et al.* 1957; Hursh 1957 it was established that a level of 100 $\mu\mu$c could be regarded as typical of many populations, although there is some indication that the world average may be somewhat lower and that 75 $\mu\mu$c of $^{226}$Ra is a more representative value, with about 80 per cent of this amount in the skeleton and 20 per cent in soft tissues (U.N. Report). There is also evidence that $^{228}$Ra (mesothorium-I) is present in cremation ashes (Mayneord *et al.* 1958) and that a total body content of 50 μμc may be fairly representative, again with about 80 per cent in the skeleton. A more recent finding is that $^{210}$Pb (and hence $^{210}$Bi and $^{210}$Po) is present in the skeleton in excess of that which could be attributed to the $^{226}$Ra content (Holtzman 1960; Hursh 1960; Hill 1961) and that a $^{210}$Pb content of 200 μμc may be assumed as typical (U.N. Report). It appears that all three elements $^{226}$Ra, $^{228}$Ra and $^{210}$Pb are acquired mainly and independently through their presence in food. In calculating dose-rates to bone or other tissues from these elements, allowance must be made where necessary for the partial escape of the daughter products in the disintegration series. In the case of $^{226}$Ra there is ample evidence that a considerable fraction of the radon escapes from the mineral bone and is eliminated in the breath, the fraction retained in bone being about 35 per cent. The $^{220}$Rn (thoron) from $^{228}$Ra has a sufficiently short half-life (51·2 s) to make it doubtful if any appreciable amount escapes and, in calculating the dose, 100 per cent retention of daughter products is assumed. Because most of the energy dissipated in the $^{228}$Ra, $^{226}$Ra and $^{210}$Pb chains is contributed by α particles it is necessary to assume a value for the RBE in order to express the dose in rem and add it to the dose from other natural sources. An RBE of 10 is usually taken although evidence is lacking for the actual value of the RBE for α particles at the low dose-rates associated with background radiation.

*Table 3. Dose-rates to body tissues from natural sources*

| Source of radiation | Dose-rate in mrem per year | | | |
|---|---|---|---|---|
| | Gonads | Bone marrow | Cortical bone | |
| | | | Oste-ocyte | Hav: Canal |
| *External* | | | | |
| Cosmic radiation | | | | |
| (a) Charged particles | 24 | 88 | 88 | 88 |
| (b) Neutrons* | 13 | | | |
| Local γ radiation | 50 | | | |
| Radon in air* | 1 | | | |
| *Internal* | | | | |
| $^{40}$K | 20 | 15 | 7 | 7 |
| $^{14}$C | 1 | 1 | 1 | 1 |
| $^{226}$Ra* (60 μμc in bone) | 1 | 1 | 20 | 12 |
| $^{228}$Ra* (40 μμc in bone) | 1 | 2 | 28 | 17 |
| $^{210}$Pb* (200 μμc in bone) | 1 | 1 | 15 | 9 |
| Total all sources | 112 | 108 | 159 | 134 |

\* RBE taken as 10. In the case of $^{210}$Pb, 50 per cent equilibrium of the daughter products is assumed, following the U.N. Report.

*Summary of Dose-rates from Natural Sources*

A summary of dose-rates to various body tissues is given in Table 3 for people living in areas of the world where neither the local rock and soil nor the local drinking waters are specially radioactive. The results are calculated for the gonads, the marrow in trabecular or spongy bone and two sites in hard or cortical bone; in the latter, it is the cells situated in the bone itself, the osteocytes, and cells lining bone surfaces such as the walls of the Haversion canals, which are the living tissues principally at risk from radiation arising from the radioactivity of the mineral bone

In estimating the contributions from external sources of radiation some allowance must be made for the average times spent indoors and out-of-doors; this can only be a very approximate allowance and a period of 6 hours out-of-doors has been arbitrarily assumed Shielding of the gonads, bone and bone marrow by overlying tissues reduces the dose-rate to all these tissues by a factor of about 0·63 (Spiers and Overton 1962). Allowing for these factors, typical dose-rates for the charged-particle cosmic radiation and for local γ radiation have been taken respectively as 24 and 50 mrem/yr. A very small external contribution arises from the radon content of the air and a similarly small internal contribution from $^{14}$C in tissues. The dose-rates to the gonads and bone marrow from the suggested extra-skeletal contents of $^{226}$Ra, $^{228}$Ra and $^{210}$Pb are also very small.

In those parts of the world where the levels of the local γ radiation are as high as indicated in Table 2 the γ-ray contribution may well dominate the dose-rate to all body tissues and values will be upwards of ten times those in Table 3.

*Bibliography*

Burch P. R. J. (1954) *Proc. Phys. Soc.*, London **67A**, 421
Court Brown W. M. *et al. Brit. Med. J.* **1**, 1753.
Hill C. R. and Jaworowski Z. S. (1961) *Nature* **190** 353.
Holtzman R. B. (1960) *Argonne National Laboratory Report*, ANL 6199, 94.
Hursh J. B. (1957) *Brit. J. Radiol. Suppl.*, No. 7, 45
Hursh J. B. (1960) *Science* **132**, 1666.
Libby W. F. (1955) *Science* **122**, 57.
Mayneord W. V. *et al.* (1958) *Proc. 2nd Int. Conf. Peaceful uses Atomic Energy*, Geneva **23**, 150.
Muth H. *et al.* (1957) *Brit. J. Radiol.*, Suppl. No. 7, 54
*Report of the United Nations Scientific Committee on the Effects of Atomic Radiation*, (1962) General Assembly Official Records 17th Session, Suppl. No. 16 (A/5216)
Spiers F. W. (1960) in *The Hazards to Man of Nuclear and Allied Radiations*, 2nd Report to The Medical Research Council, H.M.S.O., London, Appendix D
Spiers F. W. and Overton T. R. (1962) *Phys. Med Biol.* **7 (1)** , 35.
Vennart J. (1957) *Brit. J. Radiol.* **30**, 55.
Willey E. J. B. (1958) *Brit. J. Radiol.* **31**, 31.

F. W. Spiers

# K, L

**KAISER EFFECT.** The irreversible characteristic of the acoustic emission phenomena in metals named after its discoverer J. Kaiser in 1950. The extremely low energy acoustic pulses produced during initial deformation of a metal are not observable in subsequent restressing up to the original limit of stress. The effect corresponds to the elimination of the plastic region of the stress–strain curve due to work hardening of the metal. Exceeding the previous stress limit or reversal of the stress as in the Bauchinger Effect reintroduces the acoustice mission response. The Kaiser effect shows promise as a sensitive research tool in the study of physical deformation and work hardening mechanisms.

B. H. SCHOFIELD

**LIAPUNOV FUNCTIONS.** In many problems of *dynamic analysis*, and in *automatic control theory*, it is necessary to investigate the stability of isolated equilibrium points of the system of real first order ordinary differential equations

$$\frac{dx_i}{dt} = f_i(x_1, x_2, ..., x_n, t) \quad (i = 1, 2, 3, ..., n). \quad (1)$$

It may be necessary first to reduce the original system equation or equations to the form (1) by the introduction of new "state variables", $x_i$, and equations: for example

$$\frac{d^2x}{dt^2} + a_1 \frac{dx}{dt} + a_2 x = 0$$

may be expressed in the form

$$\left.\begin{array}{l} \dfrac{dx_1}{dt} = x_2 \\[2mm] \dfrac{dx_2}{dt} = -a_2 x_1 - a_1 x_2 \end{array}\right\} \quad (2)$$

by taking $x_1 = x$ and $x_2 = \dfrac{dx}{dt}$.

Equilibrium points in the $n$-dimensional "state space" of $x_1, x_2, ..., x_n$ are those points at which $f_i(x_1, x_2, ..., x_n, t) = 0$ for each $i$ and all $t$, and, without loss of generality, the origin $x_i = 0$ ($i = 1, 2, ..., n$) may be taken to be the point of equilibrium under investigation.

Under mild conditions on the $f_i$ ("Cauchy-Lipschitz" conditions) there is a unique solution of the system (1) from given initial conditions giving rise to a "trajectory" in the state space. The origin is said to be "locally asymptotically stable" if all trajectories starting in some finite neighbourhood of the origin tend to the origin as $t \to + \infty$.

In 1892 the Russian mathematician A. M. Liapunov published an ingenious method of investigating such stability by setting up a suitable function $V(x_1, x_2, ..., x_n)$ giving rise to a nest of closed contours surrounding the origin, each contour being given by $V(x_1, x_2, x_3, ..., x_n) = c$, and the nest by increasing values of $c(> 0)$ with $V(0, 0, 0, ..., 0) = 0$. The sign of the rate of change of $V$, $\dfrac{dV}{dt}$, following the trajectories of the system (1) is then examined. $\dfrac{dV}{dt}$, is easily found by partial differentiation of $V$ and substitution from the system equations (1):

$$\frac{dV}{dt} = \sum_{i=1}^{n} \frac{\partial V}{\partial x_i} \frac{dx_i}{dt} = \sum_{i=1}^{n} \frac{\partial V}{\partial x_i} f_i(x_1, x_2, ..., x_n, t)$$

If $\dfrac{dV}{dt}$ is always negative then all the trajectories starting in the nest must pass inwards through the contour surfaces $V(x_1, x_2, ..., x_n) = C$ and thus tend to the origin as $t \to + \infty$.

For example, the function $V = \left(\dfrac{a_1^2 + a_2 + a_2^2}{a_1 a_2}\right) x_1^2 + \dfrac{2}{a_2} x_1 x_2 + \dfrac{(a_2 + 1)}{a_1 a_2} x_2^2$ has the derivative $\dfrac{dV}{dt} = -2x_1^2 - 2x_2^2$ following the trajectories of the system (2). $V$ gives rise to a nest of closed surfaces if and only if it is a positive-definite quadratic form in $x_1$ and $x_2$, for which $a_1$ and $a_2$ must both be positive. These conditions on $V$ are necessary and sufficient stability conditions for the origin.

Suitable functions $V$ are known as "Liapunov functions" and some skill is required in finding suitable functions for a particular problem. The advantage of the method is that useful information may be gained about stability without actually finding any solutions of (1). This is particularly valuable when the system (1) is non-linear. In many problems the method leads to a sufficient but not necessary stability criterion found by taking a positive-definite $V$ and imposing conditions for $\dfrac{dV}{dt}$ to be negative-definite, or negative semi-definite but not identically zero.

New applications of the method and techniques for constructing Liapunov functions are topics of current research.

*Bibliography*

CHETAYEV N. G. (1961) *Stability of Motion*, Oxford: Pergamon Press.

HAHN W. (1963) *Theory and Application of Liapunov's Direct Method*, New York: Prentice-Hall.

LA SALLE J. and LEFSCHETZ S. (1961) *Stability by Liapunov's Direct Method with Applications*, New York: Academic Press.

LIAPUNOV A. M. (1947) *Ann. Math. Studies No. 17*, Princeton: The University Press.

P. C. PARKS

**LIGHT GAS GUNS.** Guns using conventional propellants are limited to muzzle velocities less than 3 km/sec. While this may be adequate for their normal uses, guns are also required which can fire shots suitable for the study of aerodynamic ballistics, aerophysics and impact. For aerodynamic ballistics, velocities of up to 8 km/sec (the re-entry velocity for an orbiting vehicle) are required, and for these studies it is frequently necessary to test finned or winged models. These are launched from a gun barrel by using sabots which enclose the model and fit the bore, but these assemblies are naturally more fragile than shells and need to be launched with the least possible shock loading. For aerophysics, velocities up to 72 km/sec (the highest velocity for meteorites) are desirable, and for impact, relative velocities of particles which might be encountered by space vehicles can be as high as 85 km/sec.

The basic limitation in the use of conventional propellants is that, however high the pressure generated, the propellant gas is heavy, and must be accelerated with the model. If a light gas, such as helium or hydrogen, is used instead of conventional propellant, the mass of propellant gas is reduced. For guns where "breech" pressure and temperature are predetermined, this effect is often considered in terms of a limiting velocity, a few times the speed of sound in the gas, at which the pressure available for driving the shot falls to zero. In most light gas, guns however, the "breech" pressure and temperature vary during the launching cycle, so this concept is of limited value.

The simplest light gas gun has a single stage (Fig. 1). The driving chamber (breech) is filled with light gas, and a shear disk retains the shot. To obtain high pressure and temperature in the driving chamber, the gas is heated either by electrical discharge or by the combustion of hydrogen and oxygen in a mixture diluted with helium (the "steam gun"). Using this principle, the muzzle velocity has been increased to 6 km/sec.

The introduction of electrical energy usually means that electrode material contaminates the light gas, and the combustion of hydrogen and oxygen introduces steam which is relatively heavy, so that other means of compressing and heating the gas would be preferable. In the two-stage gun (Fig. 2) there is a pump tube which is a single-stroke compressor containing a light gas initially at a pressure of a few atmospheres. The driving chamber is the same as in the single-stage gun, and contains propellant, hydrogen/oxygen/helium or helium. When released, the driving gas pushes the piston along the pump tube, raising the pressure to a peak of about $10^4$ atm. Near the peak of the stroke, the shot is released, projecting it along the barrel. Muzzle velocities of 10 km/sec with a shot weighing 0·1 g and 6 km/sec with 1000 g have been reached.

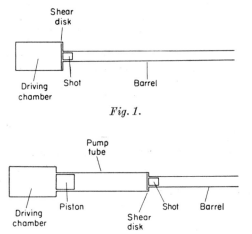

*Fig. 1.*

*Fig. 2.*

Beyond this, attempts have been made to introduce a third stage, with propellant driving a shock wave into helium. The heated helium is then used to drive the piston, as in the two-stage gun. Another possible improvement results from extra heating of the gas in the pump tube, either by heating before compression, or discharging an arc during or after compression. In specific instances these methods have improved performance, but not beyond the figures quoted for the two-stage gun.

It is possible to increase the relative velocity between the shot and the air through which it flies after leaving the gun, by blowing the air in the opposite direction. In such a counterflow tunnel, a relative velocity of 13·5 km/sec has been achieved for aerophysics studies. A similar idea has been tested in which a target to be impacted has been projected against an oncoming shot and subsequently recovered.

There have been several theories for predicting the performance of light gas guns. The motion of a piston driven by propellant may be determined by the usual interior ballistic equations or measured empirically d the compression and launching cycles may be scribed analytically by differential equations. The rm of the equations depends on whether the piston accelerates rapidly enough to generate a shock wave the pump tube, but in any case the equations must e integrated either graphically or with the aid of an ectronic computer. Attempts have been made to clude departures from perfect gas laws and other fects, but many simplifications are necessary to make e equations tractable, and the calculations are most seful in indicating trends of performance.

The most common measurement used to check erformance is muzzle velocity. A useful measurements is the pressure at the shear disk, but this ressure rises from a few atmospheres to about 4 atm, and a transducer with this range and a sponse time of 10 microsec would be required. ressures may be deduced, however, from the motion piston and shot. The piston position may be determined by using wire contact switches, which close an ectrical circuit as the piston passes them. A similar stem may be used in the barrel, provided that ionization from the shock wave ahead of the shot, or in e ambient low pressure in the barrel, does not terfere with the electrical insulation. An alternative, hich may also be affected by strong shock wave mization, is to use a microwave antenna which enerates a standing wave pattern in the barrel and sponds to the position of the shot as it passes rough nodes and antinodes.

With these techniques for measuring performance, nd improved understanding of the theory of light as guns, it is expected that their performance will be ill further improved.

*ibliography*

LASS I. I. (1963) *Hypervelocity Launchers, Review No. 22* University of Toronto Institute of Aerophysics.

National Symposium on Hypervelocity Techniques, Denver (1960) New York: Institute of Aerospace Sciences.
Second National Symposium on Hypervelocity Techniques (1962) New York: Plenum Press.
Third National Symposium on Hypervelocity Techniques (1964) University of Denver.

D. F. T. WINTER

**LINEAR ENERGY TRANSFER.** A pseudonym for "stopping power" as applied to ionizing radiation; used mainly by radiation chemists and radiation biologists. Values are usually quoted as amount of energy lost per unit distance travelled.

A. R. ANDERSON

**LUBRICATION, WEEPING.** This occurs with bearings made of soft, permeable material whose pores are soaked with liquid. Application of load to such a bearing pressurizes the liquid, and the pressurized liquid supports the major part of the load hydrostatically. Because hydrostatic support is frictionless the friction coefficient of a weeping bearing is at first very low, but it rises as the liquid leaks away and more and more of the load is carried by the solid skeleton of the material. The friction will return to its original low value if the load is removed and the bearing material allowed to expand and recharge itself with liquid.

It is thought that the cartilage which forms the bearing material of animal joints is a weeping bearing. In cartilage the pore structure is very fine grained, wringing out of liquid is very slow, and the friction remains remarkably low throughout applications of load lasting many minutes. Most other soft, porous materials are too coarse to retain the pressurized liquid long enough for its effect to be noticed.

*See also*: Animal joints, lubrication of.

C. W. McCUTCHEN

# M

## MAGNETIC FIELDS, HIGH, GENERATION OF.

High magnetic fields can be generated in a variety of ways. Water-cooled magnets, iron magnets, pulse magnets, cryogenic magnets (cooled to liquid $N_2$ temperature or below) and superconducting magnets may be employed, the final choice being dictated by consideration of power, height, space, duty-cycle, required field magnitude, duration and volume.

The relative importance of these considerations usually determines the method with little overlapping of choice. For example if an experiment requires a relatively small experimental volume at helium temperature and a field between 30 and 60 kgauss, a superconducting solenoid, perhaps using iron cores, is a logical choice. For fields below 30 kgauss in modest volumes, a conventional iron magnet is usually a flexible and economically sound choice. A very large diameter high field magnet, especially if pulsed, would probably be cryogenic. D.C. fields over large volumes in the 30–60 kgauss range will dictate the use of superconducting coils and for higher fields, at least at the present, cryogenic coils. The highest d.c. fields will undoubtedly be generated by water-cooled copper magnets for many years to come and the very highest fields by short pulse time magnets.

## MAGNETIC FIELDS IN GALAXIES.

The galaxies of the spiral and irregular types definitely appear to contain magnetic fields, as far as can be inferred from observation of light and radio waves from galaxies. Magnetic fields probably play a major role in giving these galaxies their present form. Since the presence of a magnetic field at cosmic distances can only be deduced from its interaction with gas it is not known whether the elliptical galaxies which contain very little gas (less than 1 per cent by mass) also have associated magnetic fields. These fields are neither connected to nor exert any influence on the stars which form the major part of the mass of any galaxy but they are believed to pervade the tenuous dust and gas. The gas itself can be neutral at a temperature of $100°K$ or ionized at a temperature of $10^{4°}K$ and above or a cosmic-ray gas moving at relativistic velocities.

In general all components of the gas are "frozen" into the magnetic field in the following way. The charged particles, ions or electrons, are bound to the field and spiral around the magnetic field lines. Any increase in velocity perpendicular to the field direction only serves to increase the radius of gyration of the atom and as the average radius of gyration of the thermal gas is $\sim 10^{-13}$ parsec the "freezing in" of the ions is virtually complete. The neutral gas is in turn bound to the heavy ions through the high collision cross-section for momentum transfer of $\sim 10^{-14}$ cm². The corresponding mean free path is $\sim 10^{-4}$ parsec. This length may be compared with 10 parsecs for the diameter of neutral hydrogen clouds and 200 parsecs for the thickness of the neutral gas layer in our own galaxy. Alternatively the time for a magnetic field to diffuse through a typical predominantly neutral (1 in $10^4$ are ions) interstellar gas cloud may be estimated and is found to be $10^{10}$ years for the magnetic field strengths found in the disk of the galaxy. Since this is of the order of the life-time of a galaxy it can be seen that the coupling of the magnetic field and the gas is complete. On the other hand the more energetic cosmic rays are not so readily held within the disk of the galaxy. For example, an interstellar cosmic-ray particle with an energy of $10^{15}$ eV has a radius of gyration of $0·1$ parsec in a field of $10^{-5}$ gauss while a particle of energy $10^{18}$ eV would have a radius of 100 parsec. The life-time of the former in the disk of the galaxy would be $\sim 10^8$ years while that of the latter would be only $\sim 10^4$ years. Thus low energy cosmic rays ($\sim 10^{15}$ eV) are effectively held in the disk of the galaxy by the magnetic field.

The various pieces of evidence for the existence, the structure and the strength of the magnetic field in our own galaxy and external systems will now be reviewed. The discovery of the polarization of starlight was the final clue to the existence of an ordered magnetic field at least in the nearby spiral arms. The measurements of various phenomena which have given us a clearer picture of the arrangement, strength and effects of the magnetic field in our own galaxy will now be considered along with the evidence for magnetic fields in external galaxies.

*The polarization of starlight.* The observed linear polarization in the light of some stars which also undergoes absorption by interstellar dust grains is generally interpreted as being due to the reflection of some of the light by aligned dust grains. The theory of Davis and Greenstein proposes that the alignment arises from the paramagnetic absorption of the spinning grains which in the most stable configuration

e up with their (short) axes of rotation parallel to
magnetic field. The observed position angle of the
arization then indicates the direction of the magne-
field. Fields of $10^{-5}$–$10^{-4}$ gauss were originally
ught necessary for this process to work but smaller
ds of $\sim 10^{-6}$ gauss may be sufficient if the grains
tain a proportion of ferromagnetic material.

The direction of the magnetic field in space can be
luced from the change of the scatter of position
gle and percentage polarization with galactic
gitude. For example the alignment is close and the
rcentage polarization is relatively high when looking
pendicular to the field, as in Fig. 1, while the align-
nt is random and the percentage polarization is
all when looking along the magnetic field lines. The
gnetic field traced from the polarization of starlight
rthogonal to the line of sight at galactic longitude
$= 140°$ and is apparently inclined at $40°$ to a
pothetical circular spiral structure. These measure-
nts probably refer to a region of interstellar space
the galactic plane within 1 kiloparsec of the Sun in
directions because the bulk of the stars studied for

interstellar polarization were at distances of no more
than 2 kiloparsecs.

Nebulosities in the solar vicinity frequently exhibit
marked elongations often but not always parallel to
the galactic plane. Shajn has noted that the direction
of extension is generally parallel to the optical polarization and has suggested that these nebulosities may be
good tracers of the magnetic field in their vicinity.
This structure arises because of the tendency for
nebulosities to extend *along* the magnetic field lines
where the field offers no resistance to motion rather
than in the perpendicular direction where the magnetic
field prevents diffusion.

*Synchroton emission from the galaxy.* The detection
of radio emission from the interstellar regions of the
galactic disk and halo provides further information
about the galactic magnetic field. The non-thermal
component of this emission is believed to arise from
the synchrotron process whereby electrons moving at
relativistic velocities generate emissions in the electromagnetic spectrum as they spiral in a magnetic field.
In the case of the interstellar medium the electrons are
the cosmic-ray electrons and the expected emission
$J(\nu)$ per unit volume at a frequency $\nu$ is

$$J(\nu) \propto N(>E)\, B^{1+\frac{s}{2}}\, \nu^{-\frac{s}{2}}$$

where $N(>E)$ is the number of cosmic-ray electrons
with energy greater than $E$, the value which gives
radiation characteristically at frequency $\nu$, $s$ is the
energy spectral index of the cosmic-ray electron
energy distribution and $B$ is the magnetic field. The
observed value of the radio frequency spectral index
$(s/2)$ is 0·6–0·8 at frequencies above 400 Mc/s and the
cosmic-ray electron energy spectral index $(s)$ is found
to have a compatible value of 1·6. The magnetic field
required to give the observed radio emission can be
computed from the above equation to be $\sim 3 \times 10^{-5}$
gauss on the assumption that the emission arises uniformly throughout the interstellar medium. However,
there is some evidence that the emission is irregularly
distributed both in the disk and the halo and that some
of the emission is associated with regions of high
localized fields as found for example in the remnants
of supernovae. Indeed the Crab nebula, the remnant
of the supernova of 1054 A.D., was the first astronomical object in which the synchrotron process was
clearly recognized. The magnetic fields required to
explain the radio emission here are $\sim 10^{-3}$ gauss.

Another characteristic of synchrotron emission is
that it is linearly polarized perpendicular to the
magnetic field. This polarization at radio and optical
wave-lengths has been observed in the Crab Nebula.
More recently very careful observations of the background radio emission from the galaxy have demonstrated the existence of linear polarization at deci-
metre wave-lengths. Figure 2 shows a plot of the
408 Mc/s polarization measurements at various points
in the galactic anticentre along with optical polarization results for the area which indicate that the radio

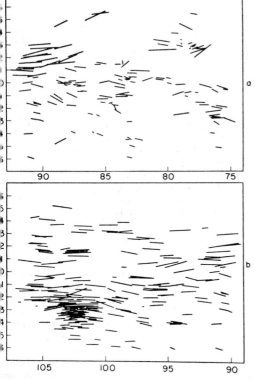

*Fig. 1. The optical polarization of starlight. The
length and position angle of each line indicate the
relative amount of polarization and the plane of
vibration. The coordinates are galactic latitude
and longitude. (By courtesy of Dr. W. A. Hiltner
and the Editor of the Astrophysical Journal.)*

polarization near $l^{II} = 140°$ $b^{II} = +5°$ is perpendicular to the magnetic field direction inferred from the optical data. In this small region of space the plane of polarization appears not to be rotated by the Faraday effect in the intervening medium (see below). Further observations at a nearby frequency confirm this result which suggests that the nearby galactic magnetic field is tangential at $l^{II} = 140°$.

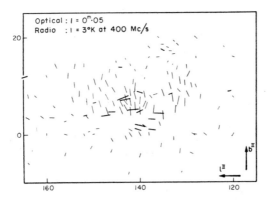

*Fig. 2. Optical and radio polarization in the vicinity of $l^{II} = 140°$ and $b^{II} = +5°$. The radio polarization is orthogonal to the magnetic field and the optical polarization is parallel to it.*
*(By courtesy of Professor C. A. Muller.)*

### Direct Measurements of Galactic Magnetic Fields

*1. The Zeeman experiment.* The Zeeman effect at radio frequencies provides a direct means of measuring the very weak interstellar magnetic field. In a longitudinal magnetic field the left and right hand circular polarization vectors of a spectral line are displaced 2·8 Mc/s per gauss. The neutral hydrogen spectral line at 1420 Mc/s is suitable for this experiment since narrow spectral features due to individual neutral hydrogen clouds having a width of 10–20 kc/s are available. Sensitivities corresponding to displacements of ~5 c/s have been achieved for several spectral features. Results so far indicate an upper limit to the general magnetic field in the solar vicinity of ~ $7 \times 10^{-6}$ gauss. Observations taken at several obsertories show that any general ordered magnetic field passing through interstellar clouds is significantly less than $3 \times 10^{-5}$ gauss—the value put forward in a number of theories based largely on the magnetic field required to give the observed non-thermal radio emission from the disk. Attempts to resolve this situation have included suggestions that the field may have a complex arrangement within interstellar clouds giving the low mean value measured by the Zeeman experiment or that the field may not penetrate the clouds which are supposed diamagnetic. This situation has not yet been adequately resolved.

*2. Faraday rotation.* The measurement of the Faraday rotation in the magneto-ionic medium of the nearby spiral arms provides a means of obtaining the local magnetic field although it is less direct than the Zeeman experiment. The rotation of the plane of polarization in a medium of depth $L$ parsecs with electron density $N$ cm$^{-3}$ and a line of sight component of the magnetic field $H_{11}$ oersted is

$$= 1·4 \times 10^5 N H_{11} L \lambda^2 \text{ radians}$$

where $\lambda$, the wave-length of observation, is in metres. Measurements at several frequencies enable $N H_{11} L$ to be evaluated. This technique can now be applied to the extragalactic radio sources which have been found to be linearly polarized. The rotation observed shows a tendency towards a decrease for sources away from the galactic plane indicating the possible existence of a magnetic field in the plane. A value for $H$ can be determined if $N$ and $L$ can be evaluated. A relationship giving $N^2 L$ is directly available from the measurement of the thermal radio emission from ionized hydrogen in the galactic plane. Since the distribution of ionized hydrogen perpendicular to the plane is fairly well known, $L$ can be determined and hence $H$ can be calculated. A preliminary estimate suggests $H$ is $2$–$8 \times 10^{-6}$ gauss depending upon the clumpiness in the distribution of ionized hydrogen.

### Problems related to the Existence of a Galactic Magnetic Field

*1. Cosmic rays.* Galactic magnetic fields are thought to have a major influence upon the properties of cosmic rays ranging from their storage to their acceleration. A characteristic of cosmic rays is the isotropy in their direction of arrival at the solar neighbourhood regardless of energy up to $10^{17}$ eV. This effect can be attributed to the presence of the galactic magnetic field which causes cosmic rays to move in spiral paths about the magnetic lines of force. A relativistic cosmic ray of energy $E$ eV executes a spiral with radius $R$ parsecs in a field of $H$ oersted given by

$$R = 10^{-21} \frac{E}{H}.$$

Thus any cosmic ray with an energy less than $10^{17}$ eV will have a radius of gyration less than 10 parsecs for a field of $10^{-5}$ oersted; this distance is to be compared with 200 parsecs, the thickness of the disk which traps the cosmic rays. A further influence of the magnetic field is to contain the cosmic rays in the galaxy. The relevant consideration here is that the magnetic energy density should be comparable with or greater than the cosmic-ray energy density of $10^{-12}$ erg cm$^{-3}$ if the cosmic rays are to be held in the galaxy. The corresponding magnetic field is $\geq 5 \times 10^{-6}$ gauss, so that the values of the field determined experimentally are adequate to retain all the cosmic rays except those which are so energetic that their gyroradii are comparable with the thickness of the disk, namely ~ $10^{18}$ eV.

Cosmic rays of this energy and greater have been observed and although they are rare it must be concluded that they eventually escape from the disk or alternatively are extragalactic in origin and have penetrated the halo and disk fields by virtue of their high energy.

Supernova explosions are generally believed to be significant sources of cosmic rays in the galaxy. However, if they are the only source the integrated energy spectrum would be expected to deviate significantly from the straight line spectrum observed. The basic mechanism proposed by Fermi for accelerating cosmic-ray particles may well account for this discrepancy and for the apparent insufficiency of supernovae to produce the total observed flux of cosmic rays. The Fermi process envisages the acceleration of cosmic rays on interaction with moving magnetic fields. The theory predicts a linear power law for the energy spectrum of the cosmic rays which requires magnetic field velocities of $\sim 100$ km/s at some time during the history of the galaxy to give the observed spectral index of 1·6. An alternative origin of cosmic rays is proposed in the Steady State theory of cosmology developed by Hoyle and Gold. Here the cosmic-ray particles are formed as decay products of newly created neutrons and are accelerated by a Fermi process to an energy of $\sim 10^{-12}$ erg cm$^{-3}$ with an energy spectrum similar to that observed in the galaxy.

*2. Star formation.* A major requirement in theories of star formation from the general interstellar medium is to remove or greatly reduce the effect of the field in the condensing protostar. The difficulty arises because the gravitational pressure promoting the contraction increases inversely as the radius while the magnetic pressure increases inversely as (radius)$^4$. Thus a stage is reached in which the contraction will cease unless the magnetic field can be released from the gas. In the early stages of contraction fields any greater than $\sim 2 \times 10^{-6}$ are an embarrassment and at later stages it is suggested that the ionization and thus the means of coupling the gas to the field is reduced to a sufficiently low value by the masking of the ionizing radiation from the gas by the dust included in the contracting cloud complex.

*3. Spiral structure of galaxies.* A striking property of spiral galaxies is the persistence of the spiral structure in the presence of differential galactic rotation which by itself would produce a new arm every $5 \times 10^8$ years. Such a high rate of arm production would give $\sim 20$ arms in the lifetime of a galaxy and so obliterate any existing structure unless some other forces were at work. The galactic magnetic field is a likely means of providing the forces required to maintain a simple spiral structure in galaxies. Magnetic forces arising from fields of the size discussed above are important on the scale of spiral arms and are greater than gravitational effects on this scale. Magnetic fields have no influence upon the motion of stars once they have formed and it is only necessary to calculate the forces on the gaseous component of the disk when considering spiral structure. Two main lines of approach have been made to this problem. One suggests that the magnetic field is of galactic dimensions embracing the disk and halo and possibly involving a general circulation of gas through the disk and halo; the spiral arms are then perturbations on this general field. The other approach considers that the spiral structure is locally stabilized by its magnetic field which may have a cylindrical or helical configuration in the arm. A helical field provides additional stability for a given field strength due to magnetic tension along the field lines. At present the observational data is inadequate to provide a basis for any detailed theory although theoretical investigations have shown the fundamental importance of magnetic fields in the spiral structure of galaxies.

### Magnetic Fields in External Galaxies

*1. Optically normal galaxies.* External galactic systems provide a means of studying some aspects of the large scale distributions of magnetic fields not readily detectable in our own galaxy. For example an investigation of the polarization of the light from globular clusters surrounding the spiral galaxy M 31 has provided a clue to the structure of its magnetic field in a way similar to the studies of the polarization of starlight in the Milky Way. The field, again assumed parallel to the observed polarization, was found parallel to the spiral structure of M 31.

The non-thermal radio emission from a number of the nearer optically normal galaxies has been investigated in detail. Since this emission arises from the synchrotron mechanism the distribution of emission will reflect the combined distribution of cosmic-ray electrons and magnetic fields. Magnetic fields will extend for at least as far as the radio emission and may be co-extensive with it if an appreciable fraction of the cosmic rays escape from the disks of galaxies and extend well beyond the limits of the magnetic field. The spiral galaxies M 31 and IC 342 have extended coronae and it is concluded that they have appreciable magnetic fields up to distances equal to the galactic disk radius above the plane of the galaxy. Other spiral galaxies typified by NGC 253 have their radio emission concentrated near the centre of the galaxy suggesting a more concentrated magnetic field distribution. The radio emission from the Large and Small Magellanic Clouds which are irregular galaxies are more or less coextensive with their stellar populations and suggest similar distributions of magnetic fields. On the other hand no radio emission has so far been observed from elliptical nebulae thus indicating a deficiency of magnetic fields or cosmic rays or both in these systems.

*2. Intense extragalactic radio sources.* The intense extragalactic radio sources which often have diameters 10 to 20 times that of their parent galaxies require much higher magnetic fields and cosmic-ray densities than the normal galaxies to account for their emission. These sources are also believed to emit radio

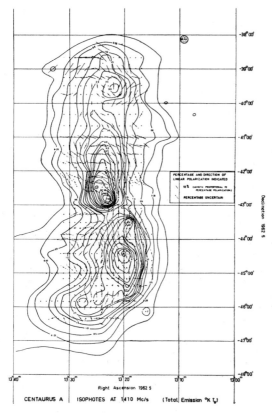

Fig. 3. Linear polarization in the 1410 Mc/s polarization of the radio source Centaurus A.
*(By courtesy of Mr. B. F. C. Cooper.)*

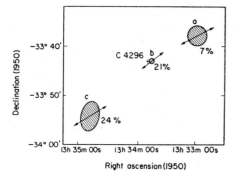

Fig. 4. The structure and polarization of the radio source 13-33. The direction and relative amount of linear polarization of each component is shown.
*(By courtesy of the Editor of Nature.)*

waves through the synchrotron mechanism; this was confirmed when linear polarization was discovered in many of them. Estimates can be made of the strength of the magnetic field if it is assumed that there is equipartition between the magnetic and the cosmic-ray particle energy. The total energy of the Cygnus A radio source is computed to be $3 \times 10^{60}$ ergs and its magnetic field to be $2 \times 10^{-4}$ gauss. Such high magnetic fields could originate in the catastrophic event which produced the radio source or in the compression of the galactic or intergalactic magnetic field. The energy in the double central core and double outer components of Centaurus A is $7 \times 10^{56}$ and $2 \times 10^{59}$ ergs respectively and the corresponding magnetic fields are $8 \times 10^{-5}$ and $6 \times 10^{-6}$ oersted. The Centaurus A source whose contours are plotted in Fig. 3 is associated with the optical system NGC 5128.

The distribution of linear polarization across several extragalactic radio sources has been measured at a number of frequencies and consequently the distribution of magnetic field directions in the source can be determined. In Centaurus A, for which some of the 1400 Mc/s position angles are shown in Fig. 3, the polarization and the magnetic field direction in the weaker outer double components vary significantly from place to place so that the net polarization is about 5 per cent. The percentage polarization of one component of the inner double (not resolved in Fig. 3) is 15 per cent suggesting a more aligned field. The polarization measurements of the source 13-33 are of particular interest because they show a high percentage polarization of greater than 20 per cent over a large volume of space. The overall extent of the source whose brightness distribution is shown schematically in Fig. 4 is 200 kiloparsecs. The intrinsic position angle of the polarization in each component is parallel to the axis of the source and thus the field is perpendicular to the axis. This orientation of the magnetic field is similar to that expected in the shock front associated with an explosion from the galaxy closely associated with the central component. Further study of the distribution of magnetic fields in intense extragalactic radio sources will help in the understanding of these complex objects.

*Bibliography*

(1963) *Astrophysical J.* **137**, 153.
(1963) *Mon. Not. Roy. Astron. Soc.* **126**, 343, 353, 369.
(1963) *Mon. Not. Roy. Astron. Soc.* **126**, 489.
Woltjer L. (Ed.) (1962) *Interstellar Material in Galaxies*, New York: Benjamin.

R. D. Davies

**MAGNETIC FILMS FOR RANDOM-ACCESS STORES.** One essential facility in a computer is to be able to store information until required. This information includes data to be processed, arithmetical processes to be carried out and where to find the ext data and instructions. The information is re-presented in the form of binary digits (zeros and ones) using bi-stable elements.

There are two forms of storage in use at present; one form provides sequential access and the other random access to the storage elements. An example of the first form is a magnetic drum which carries a layer of magnetic material (similar to a recording tape) round its curved surface. The magnetic material rotates under writing and reading heads similar to those used in tape-recorders, giving sequential access to the storage elements. The advantage is low cost because the electronic circuits are shared by a large number of storage elements. The disadvantage is the low speed arising from the time taken for the drum to rotate. Access times are of the order of one milli-second.

Random-access stores at present use ferrite cores. These cores are rings of ferrite material which can be magnetized in either sense round the ring. An array (matrix) of cores is wired in rows and columns so that the operating circuits can be shared. Because the cores have a square hysteresis loop, a selected core can be switched by applying a pulse of current to the appro-priate column wire and a similar pulse to the appro-priate row wire, neither pulse alone being sufficient to produce switching. It is essential that the cores in the same row and in the same column as the selected element should not be appreciably disturbed. This arrangement provides random access to the storage elements and is therefore much faster than sequential-access stores. However, both the storage elements and the electronic circuits are more costly for a given number of elements. In spite of the greater speed of the ferrite-core stores they are still limiting the speed of operation of the computer.

The unique properties of magnetic films offer the possibility of providing a new storage element. Magnetic film stores have two possible applications. The first is in large stores where the speed of operation is comparable with that of ferrite-core stores but advantage is taken of the comparatively simple construction, which can lead to lower costs. The second application is in small stores operating at speeds considerably greater than those possible with simple ferrite-core stores.

There are two basic constructions of magnetic film stores. In the construction most widely studied and developed, the films are deposited on flat glass or metal plates. The conductors usually comprise copper strips on polyester foil and are laid on top of the films. In the other construction, the films are deposited on tubes or rods and usually one of the operating currents is applied along the axis of the tube or rod and the other current is applied round the circumference. This arrangement has the advantage compared with flat films that larger output signals can be obtained by using thicker films (because of the closed circum-ferential flux path). However, larger drive powers are required in order to overcome the larger back e.m.f.

of the film elements and because of the high impedance of the wiring compared with that of strip conductors. Assembly is also likely to be more difficult.

The following description applies particularly to planar films.

*Preparation of magnetic films.* For storage applica-tions, the magnetic film must have an easy axis in its plane, so that the magnetization is stable in either direction along this axis in the absence of an external field. As discussed later, the magnetic properties must be such as to give a reasonable tolerance on the drive currents: these currents must be large enough to operate any selected storage cell satisfactorily without disturbing the information in any unselected cell. Finally, the chemical composition of the film should be chosen so that magnetostrictive effects are small: otherwise the magnetic properties are very sensitive to strain in the film, whether produced during prepara-tion, by subsequent handling or by temperature changes.

In practice, the required properties can be obtained by deposition in vacuum, either by evaporation or sputtering, of films based on 80 : 20 Ni–Fe and having a thickness of the order of 1000 Å ($10^{-4}$ mm). Films have also been prepared by electroplating and by chemical deposition. A magnetic field is applied during deposition in order to align the easy axis of the film in the required direction. The films are deposited either on glass plates, which provide a readily available flat smooth surface, or on smooth metal plates. The metal plates have the advantage that short current pulses along the conductors on top of the film are reflected, giving close coupling to the film elements and low impedance couductors. The films may be either continuous, individual elements being areas located by the conductor pattern, or broken into isolated elements, usually circular or rectangular.

*Hysteresis loops of magnetic films.* The ideal be-haviour of the magnetic film storage elements can be illustrated by their hysteresis loops. The direction of magnetization can be reversed by a sufficiently large magnetic field applied along the easy axis. Alternating magnetic fields along this axis produce the square hysteresis loop shown in Fig. 1(a), where the magneti-zation along the field direction is plotted against the applied field. This loop shows that as the magnetic field is increased, switching can occur only when the field reaches the coercivity, $H_c$. As the film is rotated in its plane, the loop becomes narrower and the sides are no longer vertical. In the extreme case, with the easy axis perpendicular to the applied field, the line loop of Fig. 1(b) is obtained. As the field is increased from zero, the magnetization rotates from the easy direction until the field reaches the value, $H_K$, which is large enough to align the magnetization and pro-duce saturation. Dispersion in the orientation of the easy axis produces hysteresis in the line loop unless the drive field is small, typically less than $0.5\,H_K$. The "anisotropy field", $H_K$, can then be measured experimentally by extrapolating the low-field line loop.

The square hysteresis loop of Fig. 1(a) becomes narrower also when a transverse bias field $H_T$ is applied perpendicular to the easy axis. When $H_T \gtrsim H_K$ it becomes an S-shaped loop without hysteresis. When $H_T = H_K$, the ideal loop (Fig. 1(c)) has an infinite slope at the origin, representing infinite permeability. However, the maximum value of permeability obtained when the bias is increased through $H_K$ is reduced rapidly by dispersion in the orientation and magnitude of the anisotropy field.

*General properties of magnetic films.* Magnetic fields applied along the easy axis of the film have no effect until they exceed a threshold value and reverse the magnetization. In principle it is possible to distort the

difficulties can be avoided by taking advantage of the unique properties of the films when operated by orthogonal fields.

The behaviour of magnetic film elements in the presence of two orthogonal fields $H_L$, $H_T$, respectively along and perpendicular to the easy axis has been largely explained using a simple rotational model. The magnetization $M$ of the element, when inclined at angle $\theta$ to the easy axis, is subject to the torques shown in Fig. 2. If $H_T$ is kept constant and $H_L$ increased, switching (irreversible rotation) will commence when the magnetization is inclined at an angle $\theta_0$ to the easy axis where

$$H_T M / 2K = \cos^3 \theta_0$$

and

$$H_L M / 2K = \sin^3 \theta_0.$$

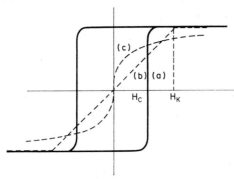

*Fig. 1. Magnetic film hysteresis loops (simple model):*

(a) *along the easy axis*
(b) *perpendicular to the easy axis*
(c) *along the easy with transverse bias field equal to $H_K$.*

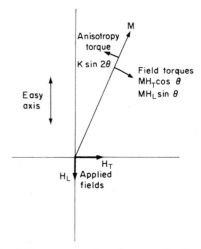

*Fig. 2. Torques exerted on the magnetization vector M.*

conductor pattern so that both row and column conductors apply fields along the easy axis to switch any selected element, neither field alone being sufficient to produce switching. In practice, however, the coercivity of the deposited film varies too widely. The maximum field available for switching an element is limited to double the field which will not quite switch an element having the lowest coercivity. If the coercivity varies, this maximum field may not be sufficient to switch an element having the highest coercivity. Even if it is able to do so, the tolerances on the operating currents will be small. Furthermore, the maximum field is usually sufficient to switch the elements only by the comparatively slow process of nucleation and growth of domains, as in ferrite cores. Fields applied along the easy axis must be comparatively large to give switching by the fast process of coherent rotation of the magnetization. If the field is inclined to the easy axis, rotational switching can occur but the tolerances on the pulses and on the properties of the films are considerably reduced. These

As $H_T$ is increased from zero, $\theta_0$ increases and the field $H_L$ necessary to produce switching decreases, approaching zero when $H_T = 2K/M$. This behaviour is utilized in the operation of magnetic film elements by orthogonal fields. The field $2K/M$ is the anisotropy field, $H_K$, of the film.

In practice, the switching behaviour is more complicated than that of the simple model. The solid curve of Fig. 3 indicates the combination of fields necessary to produce switching in the ideal film. There is often a small region below the curve where switching may occur by nucleation and movement of domain walls when successive pulses of transverse field are applied in the presence of a longitudinal field. As the fields are increased towards those represented by the ideal switching curve, the creep effects become more pronounced. For example, with a 50 c/s hard-direction field, wall velocities increasing from 0·3 to 100 μ/s have been observed as the fields are increased.

A further complication, revealed by electron microscopy is that the magnetization shows a ripple attern. The ripple lines are oriented perpendicular to he mean magnetization direction and the local magnetization within successive bands is oriented alternately to either side of the mean direction. As one or both fields are increased in the region of incoherent rotation of Fig. 3, the more favourably oriented magnetization bands are switched first: the magnetization of the other bands is then prevented from rotating by demagnetizing fields and is subsequently switched by wall movement. It has been suggested that ripple effects are associated with the micro-scale dispersion of the easy axis arising from a combination of uniaxial and randomly-oriented anisotropies. As

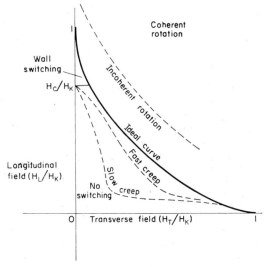

*Fig. 3. Switching mechanisms produced by various applied fields (not to scale).*

dispersion and ripple are increased, for example by successive heat treatment, the properties of the film become more complicated and less like those of the ideal model.

*Basic mode of operation of magnetic film matrices.* In order to obtain adequate tolerances on the current pulses and the film properties it has been found essential to operate the film elements with orthogonal fields and to use a "word-selection" arrangement. The word conductors comprise a set of parallel strips aligned parallel to the easy axis. The digit conductors comprise a similar set of strips but aligned perpendicular to the easy axis. They may also serve as sense conductors, detecting the output signal from the films, or separate sense conductors similarly aligned may be used. When a continuous film is used the cross-over points of the conductors locate the storage elements in the film.

The sequence of operations is as follows:

(i) A current pulse is applied along the selected word conductor to rotate the magnetizations of

the elements in the selected word perpendicular to the easy axis. This gives positive or negative output signals on the sense conductors, depending on the previous setting.

(ii) Using the same or a subsequent word pulse to ensure that the magnetization is in the hard direction and therefore in a sensitive state, writing pulses are applied along the digit conductors to tilt the magnetization slightly in one or other direction.

(iii) The word current is removed and the magnetization of each element of the selected word falls back along the easy axis in the required direction under the influence of the digit currents.

(iv) The digit currents are removed.

The advantage of the word-selection arrangement is that, provided that the easy axis is correctly aligned with respect to the conductors, the tolerances on the operating currents are wide. Excessive word current can do no harm, as the magnetization cannot rotate more than 90° and this current is applied only to selected film elements. The digit current required to determine which way the magnetization falls back is much less than the minimum current necessary to switch the unselected elements along the same row which are not subject to the word current. A further advantage is that the outputs are positive or negative, which are easier to distinguish than the signal or noise outputs of the same polarity obtained in "square-loop" systems. The only critical feature is the orientation of the easy axis, which tends to be variable from plate to plate, particularly at the corners of the plate.

*Testing magnetic films.* The ultimate test of storage elements is whether they will operate satisfactorily in a computer, but there are several other tests which give a useful guide to the properties of the elements. For example, it is important to know the range of currents over which the elements will operate. A convenient form of testing is to subject the film element to a write-disturb-read cycle. Writing is carried out by using both word and digit current pulses, disturbing by using a series of digit pulses alone (of similar amplitude but reversed polarity compared with the writing digit pulse) and reading by applying a word pulse alone and measuring the output voltage. This voltage is then plotted as a function of the digit current used for writing and disturbing to give the "tolerance curve", as in Fig. 4. In the absence of a digit current, the film element may have a tendency to fall back to one easy direction when the word current is removed, because of use of the bias or tilted modes or because of misorientation of the easy axis (skew). There will then be a minimum current to write in the opposite direction. This will set the lower limit, represented by OA in Fig. 4 for the digit current. The upper limit will be set by the lowest current OB or OC at which polarity of the output voltage is reversed by the disturb pulses. For a given skew, α, the minimum current needed for writing in the direction opposite to that in which the magnetization would fall back in the absence of a digit current is proportional to $H_K\alpha$. The maximum permissible current for writing is proportional to $H_c$. For wide current tolerances $H_c/H_K\alpha$ should be as large as possible. Increasing $H_c/H_K$ will increase the tolerance to skew for a given current tolerance.

The tolerance curve may also be used for studying the interaction effects between neighbouring elements, which limit the packing density of the elements. The curves obtained with adverse settings of surrounding elements can be compared with those obtained when the tested element is operated in isolation.

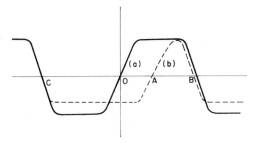

*Fig. 4. Typical tolerance curves:*

(a) *aligned film*

(b) *misoriented film.*

Measurement of the hysteresis loops enables $H_c$, $H_K$ and dispersion to be determined. Measurement of the average local easy-axis orientation can be made by means of a probe-head carrying orthogonal drive and sense conductors. When the drive conductor is accurately aligned along the easy axis, there will be no output in the sense conductor produced by pulses applied to the drive conductor. In addition, a more detailed knowledge of the behaviour of magnetic films can be obtained by supplementing the electrical measurements by observations of the domain configurations produced in the film by any particular arrangement of magnetic fields. Reflection of plane polarized light from the surface of the film produces a rotation in the plane of polarization which depends on the direction of magnetization of the film at the point of reflection. It is thus possible to observe domains in the film as light patches on a dark background or vice versa.

*Other modes of operation.* Two modes have been devised which avoid the necessity of providing digit currents to write "zeros". In the "tilted" mode, the conductors are slightly inclined to the easy axis of the film. In the "biased" mode, a steady magnetic field is applied in the zero direction. Both of these modes, however, result in reducing the tolerances on the operating currents and the film properties.

Double-element storage cells using a double-word conductor give wide tolerances to misorientation of the easy axis. The two parts of the word conductor are close together and connected in series. This ensures that the easy-axis orientation is similar for both elements and that the effects of its misorientation cancel. Double-element cells are also useful in reducing noise due to coupling between the drive and sense circuits. This noise is particularly pronounced near the corners of the film plate and in very fast operation of the film elements. Furthermore, these cells result in reduced interaction between adjacent words.

There is also the possibility of operating magnetic film elements in non-destructive readout modes.

*Future developments.* Magnetic film should provide a low-cost, high-speed storage medium for use in computers. There is still much work to be carried out on improving methods of preparation, relating the many process variables to the film properties and achieving greater reproducibility. There is also still much to be learned about the mechanism by which the film acquires an easy axis and about the processes involved in the nucleation and movement of domains.

*Bibliography*

BRADLEY E. M. (1962) *J. Appl. Phys.* **33**, 1051.

COHEN M. S. (1963) *J. Appl. Phys.* **34**, 1841.

JAMES J. B., STEPTOE B. J. and KAPOSI A. A. (1963) *The Radio and Electronic Engineer* **25**, 509.

MACLACHLAN D. F. A. (1962) *Brit. Comm. and Electronics* **9**, 602.

MIDDELHOEK S. (1962) *Z. angew. Phys.* **14**, 191; (1962) *I.B.M. J.R.* and *D.* **6**, 140.

MOSSMAN P. (1964) *Proc. I.E.E.* **111**, 1411

MOSSMAN P. and WILLIAMS M. (1963) *J. Appl. Phys.* **34**, 1175.

PRUTTON M. (1960) *Brit. J. Appl. Phys.* **11**, 335.

RAFFEL J. I. *et al.* (1961) *Proc. I.R.E.* **49**, 155.

RUBENS S. M. (Sept. 1963) *Electro Technology* **72**, No. 3, 114.

SMITH D. O. (1958) *J. Appl. Phys.* **29**, 264.

WILLIAMS M. (1962) *Proc. I.E.E.* **109** B, Suppl. 21, 186.

WILLIAMS M. *et al.* (1965) *Proc. I.E.E.* **112**, 280.

R. C. KELL

**MAGNETOHYDRODYNAMIC GENERATION OF ELECTRICITY.** *1. Introduction.* When any conductor moves across a magnetic field an e.m.f. is generated and this is the case whether the conductor is solid (such as a metal or semiconductor), liquid (such as a liquid metal or an electrolyte) or gaseous. In this context all gaseous conductors are referred to as plasmas, although that term is sometimes reserved for good gaseous conductors, i.e. highly ionized gases.

A basic type of generator, which is quite simple and direct in concept, can consist of a pair of electrodes in a fast moving ionized gas such as a flame, (see Fig. 1) with a transverse magnetic field. The electric power produced comes from the energy of the moving gas stream. This type of generation has been termed magnetohydrodynamic generation and represents one of the several direct generation processes in which interest has revived within the last few years.

*2. Electrodynamics.* Even when considerable efforts are made to enhance the electrical conductivity $\sigma$ of the gas, it remains six or more orders of magnitude less than that of copper and is typically only 30 mho/m.

This means that the power density is much less than in a conventional alternator (although power densities in excess of 1 MW/m³ have been measured) so that a much greater volume of magnetic field and interaction space is required. Also the magnetic Reynolds number $R_m = \sigma\mu vl$, which is a measure of the induced relative to the inducing field, is very small (only about 0·03 per metre of characteristic length $l$, taking $v$, the gas velocity, to be of the order of 1000 m/s; $\mu$ is the permeability of the medium) so that the magnetic field distortion due to the gas flow is not noticeable except over very large distances.

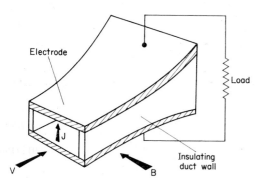

*Fig. 1. Basic magnetohydrodynamic generation configuration.*

These considerations demonstrate the differences between the plasmas used for magnetohydrodynamic generation and those required for thermonuclear fusion, which are completely ionized and may have conductivities even higher than that of copper. In controlled thermonuclear research the magnetic field is usually distorted drastically, for example it may be "swept up" by the plasma, and it may (hopefully) be used for containment purposes. That is not possible with the partially (less than one part in $10^3$) ionized plasmas considered here.

Provided there is a complete circuit, current will flow, and there is then a retarding force $JB$ on the gas (where $J$ is current density and $B$ magnetic induction) which does work at a total rate of $JBv$. Of this, internal ohmic losses account for $J^2/\sigma$ leaving an output of $JE$ (where $E$ is electric field). Thus $JB = J^2/\sigma + JE$ and $J = \sigma(vB - E)$. This is just the generalized form of Ohm's law for a conductor moving through a magnetic field; the signs have been chosen to emphasize the fact that the current flows through the magnetohydrodynamic region against the direction of the electric field $E$ since power is being generated.

In an alternator, since $\sigma$ (the electrical conductivity) is very large (for copper) it is possible to have $E \cong vB$ (to within say 0·3%) and still obtain large current densities. In a magnetohydrodynamic generator with a very much lower conductivity, $E$ will have to differ appreciably from $vB$ to achieve any reasonable current density. The output power density may be written as $JE = \sigma v^2 B^2 K(1 - K)$, where $K = E/vB$ represents the ratio of the external to the total circuit resistance, or the fraction of total generated power actually extracted.

For maximum power density at each point, and hence minimum volume of generator and magnetic field (and presumably cost), a matched load should be used, i.e. $K = \frac{1}{2}$, and it can be shown that for maximum average power density $K$ should be even lower.

*3. Thermodynamics.* The energy for producing the electric power $JE$ can be removed from the gas in two forms: kinetic, in which case the gas is slowed down, and potential, in which case the gas velocity is not changed but its temperature falls, and this is the situation which is found in any conventional turbine. The terminology of the turbine field may therefore be carried over and the two types may be referred to as impulse and reaction magnetohydrodynamic generators respectively. In general, as in the gas turbine field a practical device will be a combination of these two types.

The small stage efficiency of a turbine is defined as the fraction of the total isentropic power actually extracted for the same pressure drop, and in a constant velocity magnetohydrodynamic generator is simply $K$. Thus the internal ohmic losses play the same role in a magnetohydrodynamic generator as do frictional losses in a turbine, that is to say they do not represent true energy losses since the heat produced remains in the gas but they do represent increases in entropy and pressure drop which necessitate larger compressors and compressor powers. The compressors themselves may add considerably to the total cost or weight of a generator and they represent a decrease in efficiency. In a gas turbine the compressor power may be twice as much as the total heat power input so that a high turbine efficiency is particularly important. Although the ratio is likely to be less with magnetohydrodynamic systems, it is still true that as far as compressor requirements are concerned the value of $K$ should be as close to unity as possible. A compromise will therefore have to be reached between these requirements ($K \sim 1$) and those for minimum volume of magnetohydrodynamic generator and magnetic field ($K < \frac{1}{2}$); at present values of $K$ in the region of $\frac{3}{4}$ are envisaged.

*4. Conductivity.* The gas or plasma consists of three sorts of particles, electrons, ions and neutrals, and provided the electrons may move freely around the circuit their effective mobilities will be orders of magnitude higher than those of the ions so that the conduction is virtually entirely electronic. The resistance of the plasma is then governed by collisions between the electrons and the other particles. Now the total electron collision frequency is the sum of the electron-ion and the electron-neutral collision frequencies and conductivity is inversely proportional to collision frequency so that the total conductivity is given by

$$\frac{1}{\sigma} = \frac{1}{\sigma_{en}} + \frac{1}{\sigma_{ei}}$$

where $\sigma_{en}$ and $\sigma_{ei}$ are the conductivities due to electron/neutral and electron/ion interactions respectively.

For electron-ion collisions (as in a fully ionized gas) Spitzer has shown that

$$\sigma_{ei} = aT_e^{3/2}/\ln \Lambda$$

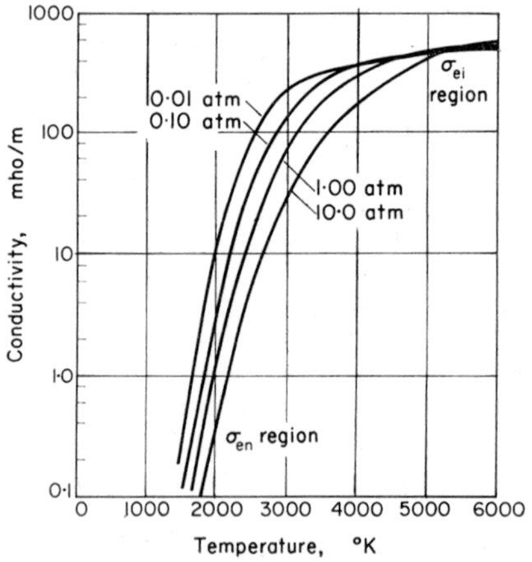

Fig. 2. Variation of plasma electrical conductivity with electron temperature, showing regions where electron-ion and electron-neutral collisions predominate.

where $\ln \Lambda$ is a very slowly varying function of pressure, $p$, and electron temperature, $T_e$, and $a$ is a constant, and this process will predominate when the gas is more than about one part in a thousand ionized. As full ionization is approached the number of electrons cannot change very greatly and so $\sigma_{ei}$ has a relatively slow variation with electron temperature (see Fig. 2).

However, although the ion collision cross-section is much larger than the neutral cross-section, if there are sufficiently few ions the electron-neutral collisions must predominate and the conductivity will be controlled by the number of electrons. The degree of ionization $f$ may be obtained from Saha's equation and for small degrees of ionization is of the form

$$f = bp^{-1/2}T_e^{5/4}\exp\left(-eV_i/2kT_e\right)$$

where $V_i$ is the ionization potential, $k$ is Boltzmann's constant, $e$ electronic charge, and $b$ a constant. Thus the number of electrons, and hence the conductivity $\sigma_{en}$, varies extremely rapidly with temperature (roughly as $T_e^{13}$ over ranges of interest) and slowly (as $p^{-1/2}$) with pressure as shown in Fig. 2.

For high conductivities a low ionization potential material must be used, (potassium is 4·3 V, caesium 3·9 V) but this is not necessary for all the gas molecules. In practice up to 1 per cent or so of alkali metal is used to "seed" the gas.

Experimental measurements agree well with conductivities computed as above.

Because of the difference in mass between the electrons and the other particles the ohmic heating due to the current through the plasma, which initially goes entirely to the electrons, may not be immediately distributed to the other particles. The random electron energy will therefore be increased above that of the gas molecules and the kinetic electron temperature will be higher than for thermal equilibrium conditions.

This effect may give significant enhancement of electrical conductivity in magnetohydrodynamic devices.

Unfortunately the effect is likely to be important only in monatomic gases such as helium or argon, which might be used in a closed-cycle nuclearly-heated device. With molecular gases the electrons can lose energy to gas molecules by inelastic collisions (exciting vibrational and rotational states) so that it is rapidly distributed throughout the gas and little enhancement is obtained.

Various other means may also be envisaged which will produce elevated electron temperatures and enhanced ionization including direct current, radio frequencies, microwaves, ultra-violet, X rays, and even higher energy nuclear radiations, as well as bombardment by high velocity electron, ion or neutron streams and contact surface ionization. None of these methods has yet been shown to be practicable, but there is considerable interest in this subject.

One effect produced by high magnetic fields is that the conductivity becomes a tensor quantity so that the current $\mathbf{J}$ does not flow in the direction of the field $(\mathbf{E} + \mathbf{v} \times \mathbf{B})$. This phenomenon is known as the Hall effect and is similar to that occurring in solids. Microscopically the electrons perform circles (or cycloids) in the fields at the cyclotron frequency thereby losing forward motion, until they suffer collisions.

If collisions are sufficiently frequent compared with the cyclotron frequency, $\omega_c$, the forward slowing will be negligible and normal scalar conductivity will be found. The angle through which the electrons rotate in a mean collision time $\tau$ is $\omega_c\tau$, and if this is at all large Hall effects will be pronounced; $\omega_c\tau$ is also the tangent of the angle between the direction of current flow and the field, $(\mathbf{E} + \mathbf{v} \times \mathbf{B})$.

If the electric field is forced to be perpendicular to the flow direction by having long equipotential electrodes along the top and bottom of the duct (see Fig. 3(a)) then, with large $\omega_c\tau$, the current which is at an angle must have an axial Hall component which represents an ohmic loss. Also the component of interest, the transverse component, is reduced by a factor $1/(1 + \omega_c^2\tau^2)$ so that the effective value of conductivity is also reduced by this factor. This geometry is therefore only practicable for very small $\omega_c \tau$.

If the current is all forced to be transverse to the flow direction by breaking the electrodes up into segments with each opposite pair connected through its own separate load, as in Fig. 3(b) there are no unnecessary ohmic losses and the full effective conductivity is restored. There is, of course, an axial Hall electric field set up so that each separate circuit is at a different potential, and by connecting the separate circuits

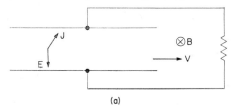

(a)

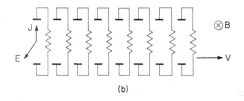

(b)

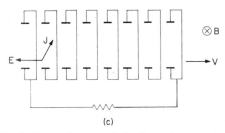

(c)

*Fig. 3. Types of magnetohydrodynamic generator:*
*a) Continuous electrodes, tranverse E;*
*b) Segmented electrodes, transverse J;*
*c) Hall generator, axial E.*

in series it may be possible to obtain quite high voltages. This geometry is useful for the medium values of $\omega_c \tau$ (say from one to five) often found in practice, and is the one that generally has been preferred experimentally.

Finally, it is possible by short circuiting opposite pairs of electrodes as in Fig. 3(c) to force the electric field to be axial and to use the axial voltage and current. This type of generator is called a Hall-type generator and is appropriate to large values of $\omega_c \tau$.

**5. Plasma flow.** The velocity of the gas comes from expanding it, after heating or combustion, through a nozzle.

Now for optimum power density, $\sigma v^2$ should be a maximum and this gives two conflicting require-

ments, firstly that the temperature should be as close to the stagnation temperature as possible for high conductivity and secondly that the gas be expanded to as low a temperature as possible for maximum velocity. A compromise must therefore be reached, and an optimum temperature ratio found which will in turn give an optimum Mach number. This Mach number varies with $\gamma$ (the ratio of the specific heats) and hence with the type of gas and its temperature, but is generally not too far below or above sonic. There may, of course, be difficulties with supersonic flows if shock waves are formed.

Various cases such as constant temperature, constant area, constant velocity etc., have been considered but since the optimum Mach number does not vary very rapidly, the constant Mach number generator may well prove to be close to the optimum case.

The usual quasi-one-dimensional approximation is used and the equation of motion (force per unit volume) giving the acceleration $v\,dv/dx$ produced by the pressure drop along the duct in direction $x$ and the magnetic force on the current is $\varrho v\,dv/dx + dp/dx + JB = 0$ ignoring frictional drag, while the energy equation (power per unit volume) $\varrho v\,d(c_p T + \frac{1}{2} v^2)/dx + JE = 0$ states that the total change in flow energy (potential $c_p T$ and kinetic $\frac{1}{2} v^2$) is due to the electrical output $JE$, ignoring heat transfer. Here $\varrho$ is gas density, $c_p$ specific heat at constant pressure and $T$ gas temperature.

Now the main difficulty in solving these equations arises from the complicated dependence of conductivity, and hence current density, upon pressure and temperature. However, if $J$ is eliminated between them the variation of various parameters can be found

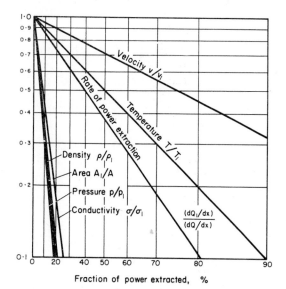

Fraction of power extracted, %

*Fig. 4. Variation of duct parameters with amount of power extracted for the constant Mach number case.*

without a detailed knowledge of the conductivity variation (see Fig. 4).

To find how these parameters vary with distance, however, one of the equations must be integrated and a detailed knowledge of the variation of $\sigma$ is required as given in section 3.

*6. Open cycles.* Most of the practical and experimental difficulties stem from the high temperatures that are required to produce sufficient thermal ionization and conductivity in the gas.

The duct side-walls present considerable problems since they must retain their electrical insulating properties while containing the very high temperature, high velocity gas flow.

With flames from fossil fuels all the known refractories are rapidly oxidized at the temperatures envisaged, about 2500°K, except for the ceramic oxides such as MgO, $ZrO_2$, etc. Unfortunately, these oxides are all attacked by the alkali metals which are necessary for seeding. Much experimental work has been done with such materials which last for short periods but for large scale applications water-cooled surfaces are usually envisaged. These will have very large surface heat losses (up to 4 MW/m²) which are typically of the order of 0·5 per cent of the total enthalpy per duct diameter length. However, provided the scale is sufficiently large to give a small surface/volume ratio (the duct diameter should be at least 1/20 of its length), the losses will not be prohibitive. It is therefore possible, to give a minimum size for which a magnetohydrodynamic generator can be economic. Estimates vary from a hundred or so megawatts upwards, but the sizes are comparable with those of present-day generating units, and so are not prohibitive. For this

reason magnetohydrodynamic generators are considered primarily for large scale power generation.

The problem of avoiding short circuiting effects if metal walls are used may be tackled by segmenting and/or some other technique such as coating with a thin layer of ceramic. A possible insulating duct wall is shown in Fig. 5.

The electrodes present even more severe difficulties, since they must not only withstand the very high temperature environment, but must also emit electrons if large voltage drops are to be avoided.

To emit electrons thermionically the surface must run hot, but it will then corrode rapidly and would need to be fed in continuously as in a carbon arc. If the electrodes are cold and do not emit, then voltage drops which may be 100 V or more must be tolerated just as in an electric arc.

End effects may be important and can be minimized by suitably tailoring the magnetic field. Provided a sufficiently large scale device is contemplated, they will probably not be prohibitive.

The problem of producing a sufficiently high flame temperature economically is a severe one. Oxygen can be used but is costly to produce. Air, on the other hand, will need to be heated to very high temperatures, greater than 1500°C, if comparable temperatures are to be achieved, and such preheaters present economic as well as practical problems. Pebble-bed type preheaters have been demonstrated at these temperatures but are by no means well established.

The first successful experiments were carried out by Rosa in 1960 using hot argon from a plasma jet seeded with potassium, and 10 kW were generated for short periods.

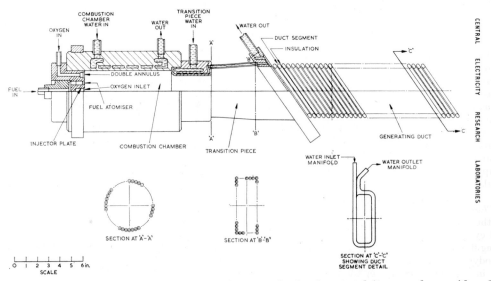

*Fig. 5. A tubular water-cooled duct. This type of construction has been tested for many hours with seeded flames (3000°K) near sonic velocity. Segmented electrodes and insulating duct walls are provided by water-cooled tubes with insulating material between them. The tube slope is governed by the Hall effect.*

These were closely followed by similar experiments
y other workers and a few months later 10 kW was
tained for periods up to 50 minutes from a device
urning fuel oil in oxygen, with potassium seeding
ssolved in the form of an octoate in the fuel; the
uct and electrodes were made from materials such as
agnesia and zirconia.

In 1962 experiments were reported on an apparatus
ving a thermal input of about 20 MW which burnt
cohol, with dissolved KOH for seeding, in oxygen; a
agnetohydrodynamic output of 1300 kW has since
en recorded. The initial experiments used ablating
aple wood walls, but subsequently more permanent

*Fig. 6. Small-scale generation experiment. Graphite electrodes in
a seeded flame seen through the pole pecie of the magnet.*

lls have been tested. More recently a larger self-
cited system has generated tens of MW.

These, and other experiments on a smaller scale,
ch as the one shown in Fig. 6, have demonstrated
at many of the basic principles of magnetohydro-
namic generation are well understood, and larger
le experiments are being undertaken to investigate
e possibility of instabilities and other problems
sociated with extracting large fractions of the total
thalpy electrically. Various pulsed experiments
ve also been reported.

*7. Closed cycles for nuclear applications.* In many
ys the experimental difficulties are less severe if a
sed cycle is used, since there is then a choice of
rking fluid, which may be selected for its electrical,
rmodynamic and chemical properties. The noble
ses, in particular helium and argon, have found
our here as well as mercury.

Caesium can be used for seeding despite its high
st since there is total recovery and this means that

gas temperatures may be several hundred degrees lower
to achieve the same thermal ionization as with potassium; elevated electron temperatures may also be possible since monoatomic gases can be chosen. In the presence of inert gases various materials such as graphite
and tantalum are relatively stable at high temperatures and so may be used for the duct construction.

Various experiments with closed cycles are already
being carried out.

The biggest drawback with closed cycle applications is that a very high temperature heat exchanger
must be provided with exit gas temperatures in excess
of 2000°C for thermal ionization. With fossil fuels, all
the material difficulties already discussed
would arise in such a heat exchanger so
that it is more natural to associate closed
cycles with nuclear applications. Even
so, the prospect of producing in pile heat
exchangers with exit gas temperatures
in excess of 2000°C is formidable and
well beyond the scope of present nuclear technology, even taking account
of current work on high temperature
reactors, and most workers ensage the
use of non-thermal ionization with rather
lower gas temperature. The research on
nuclear applications therefore appears
to be longer term than that on open
cycles.

*8. A.C. Generation.* The possibility of
generating alternating current in a magnetohydrodynamic device is attractive
from various points of view. The most
obvious advantage would be that the
generated power could be coupled out
inductively, thereby avoiding the necessity of having electrodes with their attendant practical difficulties which have
already been discussed.

Also, if the conductivity of the gas flow were varied
it would be possible to have a high temperature in the
interaction regions while the mean temperature which
the materials of the duct would have to withstand
could be much lower.

Furthermore, most of the magnetohydrodynamic
devices already discussed produce direct current, while
large scale applications may require alternating current, so that the need for expensive d.c. to a.c. conversion equipment would be avoided.

One system that has been studied in detail has a
travelling magnetic wave coupled to a uniform moving plasma. The plasma is slowed down by the wave
and gives up energy to it. Unfortunately the amounts
of reactive (imaginary) power stored in the varying
magnetic field are very large compared with the useful
(real) power produced unless enhanced conductivities
can be obtained, and the capacitances required to
handle such reactive power at conventional frequencies
(50 or 60 c/s) would be impractically large and costly.

The alternative is to have a (basically) static magnetic field and then some other physical quantity such as conductivity must be varied. Thring has suggested that this might be achieved by periodically injecting oxygen and seeding materials to give high combustion temperatures and hence regions of high conductivity. The available power which can be coupled out inductively is then $R_m$ smaller than if electrodes are used, and the system would appear to be subject to Rayleigh-Taylor instabilities which would prevent the non-conducting gas from being slowed down by the conducting portions.

It may be that the use of liquid metals, which have much higher conductivities, blown along by their own vapour, thereby avoiding the usual frothing and separation problems, and/or higher frequencies will overcome some of the difficulties in this line of research.

*9. Magnetic field.* The cost of supplying large magnetic fields over large volumes is likely to prove one of the major items in magnetohydrodynamic systems. Since the length of the generator decreases as $B^2$, it is obvious that it is worthwhile to consider very high fields. Above about 20,000 gauss there is little to be gained by using iron cores, and air-cored systems are to be preferred. These may consume up to 10 per cent or even more of the output power if conventional techniques are used but cryogenic or superconducting magnets may enable very high fields to be attained with negligible power consumption.

*10. Economics.* The magnetohydrodynamic generator is a direct generation device and a basic system might be quite cheap to build. It is likely to have a relatively low efficiency, particularly when any compressor energy requirements are deducted, so it is only likely to be used by itself in cases where efficiency and fuel cost or weight are unimportant.

Where long periods of time are involved, as in central power station generation where 20 or 30 years' life is required, the cost of fuel, and hence the efficiency, becomes important so that the magnetohydrodynamic generator must be used in conjunction with some other cycle. The exhaust gases are still at around 2000°K and may therefore be passed directly into, for example, the boiler of a steam system, so that the magnetohydrodynamic generator may be regarded as a thermodynamic "topping" device.

Thus the overall efficiency of the MHD generator used in conjunction with a steam plant that may approach 40 per cent efficiency, might be in the region of 50 per cent, a very worthwhile advance. Furthermore, since waste heat is passed on to the steam plant the capital cost of the MHD plant can be much more than for conventional plant.

There are a number of large scale engineering problems associated with integrating MHD plant with subsequent steam plant. These include the recovery of seed material which will be necessary both economically and to avoid subsequent excessive corrosion and the construction of high temperature air preheaters for open cycle systems, and the attainment of sufficiently high reactor temperatures for closed cycles

*11. Acknowledgements.* The work was carried out at the Central Electricity Research Laboratories, Leatherhead, and is published by permission of the Central Electricity Generating Board.

*See also:* Elevated electron temperature. Rayleigh-Taylor instabilities.

*Bibliography*

COOMBE R. A. (Ed.) (1964) *Magnetohydrodynamic Generation of Electrical Power*, London: Chapman and Hall
FORREST J. S. (Ed.) (1962) *Gas Discharges and the Electricity Supply Industry*, London: Butterworths.
LINDLEY B. C. (Ed.) (1963) *Magnetoplasmadynamic Electrical Power Generation*, London: I.E.E.
McGRATH I. A. *et al.* (Eds.) (1963) *Advances in Magnetohydrodynamics*, Oxford: Pergamon Press.
MANNAL I. C. and MATHER N. W. (Eds.) (1962) *Engineering Aspects of Magnetohydrodynamics*, Columbia: The University Press.
MHD (1965) *Proceedings of the International Symposium on MHD Electrical Power Generation.* Paris: O.E.C.D.
ROSA R. J. (1961) *Phys. of Fluids* **4**, 182.

D. T. SWIFT-HOOK

**MASERS.** *1. Introduction.* Since the previous (Dictionary) article was written, the most spectacular development in the MASER field has been the successful extension of stimulated emission amplifier and oscillator techniques into the infra-red and optical regions, so that MASER has now come to stand for "Molecular" amplification rather than "Microwave" amplification. The present article is confined to developments in the microwave region. Here, the correctness of current maser theory as regards noise temperature has been confirmed with great accuracy, and many solid-state maser amplifiers have been installed in centres of radio astronomical research. The low noise input temperature of the maser has been used to advantage in satellite communications (Project Echo), and microwave spectroscopy. The maser oscillator is still under investigation as a frequency standard, emphasis has moved now to the atomic hydrogen maser. The microwave maser has also been used in a new experimental test of special relativity. Techniques have improved to the extent that travelling-wave masers have been constructed in the millimetre region using dielectric slowing only, and in the 1 Gc/s region, using a Karp slow-wave structure. Broadbanding of cavity masers has also been studied. New maser materials are still being discovered, and the theory of "cross-relaxation", a spin-spin interaction phenomenon which has a large bearing on maser performance, has been well established.

*2. Application of masers.* The microwave multilevel solid state maser has been used for the following projects in radioastronomy; reception of radar echoes from Venus; radiometric observations of Venus, Mars, Jupiter, Saturn and other known, or suspected radio sources (e.g., comet Burnham, Tycho Brahés supernova); hydrogen line surveys; and measurements of sky temperature. Both the ammonia maser and the solid-state maser have been used as pre-amplifiers in electron paramagnetic resonance spectrometers, and the expected improvements in sensitivity have been achieved.

In the experimental test of special relativity, two identical ammonia masers were mounted in a horizontal plane with their molecular beams parallel, but travelling in opposite directions. The masers could be rotated together about an axis perpendicular to this plane. On a simple "ether" theory, provided that the molecular vibrations remain unaffectected by passage through ether, a change in relative frequency should be found if the maser axes are set parallel to the direction of "drift" through the ether, and then turned through 180°. Special relativity predicts no such change in relative frequency. The experiment demonstrated that any change in relative frequency on rotation of the masers was less than 1 part in $10^{12}$, the limiting accuracy.

*3. Atomic hydrogen maser.* The atomic hydrogen maser is an atomic beam device, which operates as follows. Atomic hydrogen is generated in a r.f. discharge tube and a collimated beam of atoms is allowed to pass down the axis of a six-pole permanent magnet system which gives an inhomogeneous magnetic field. This focuses atoms in the $[F = 1,\ m = 0]$ and $[F = 1,\ m = 1]$ states into an aperture in a Teflon-coated quartz bulb, which is situated in the centre of a cylindrical resonant cavity operating in the $TE_{011}$ mode. The cavity is tuned to the $[F = 1,\ m = 0] \rightarrow [F = 0,\ m = 0]$ hyperfine transition frequency, of approximately $1420 \cdot 405$ Mc/sec. The atoms make random collisions with the Teflon coated bulb wall, and eventually leave through the entrance. The hydrogen atoms have a very small interaction with the Teflon wall, and are not seriously perturbed even though they remain in the bulb for several seconds and undergo $\sim 10^{15}$ collisions with the wall. The resonance line is so sharp under these conditions that self-excited maser oscillations at the appropriate frequency can occur. The first order Doppler shift is greatly reduced by the fact that the velocity of atoms in the bulb, suitable averaged, is close to zero. Since there will be some dependence of transition frequency on any magnetic field to which the atoms are exposed in the bulb, the Earth's magnetic field is cancelled with concentric Mu-metal shielding cylinders. An accuracy of better than $10^{-13}$ is thought to be achievable with the atomic hydrogen maser. Independent resetability to $10^{-12}$ has been achieved so far, and r.m.s. frequency deviations of 5 M $10^{-14}$ over 5 days and 1 M $10^{-13}$ over 15 minutes have been recorded.

*4. Solid-state maser materials and techniques.* Most masers reported continue to use ruby ($Cr^{3+}$ in $Al_2O_3$)

as the active medium. Push-pull pumping concentrated ruby ($\sim 0 \cdot 2$ per cent $Cr^{3+}$) allows operation at liquid nitrogen temperature at 3 cm wave-lengths. $Fe^{3+}$ and $Cr^{3+}$ in $TiO_2$ (rutile) have also proved very useful at liquid helium temperatures. The high dielectric constant of rutile ($\sim 100$) has been utilized in the design of mm wave-length travelling wave masers using dielectric slowing only, and to reduce to practicable proportions the dimensions of travelling-wave masers at 21 cm wave-length. The problem of broadbanding a cavity maser is exactly similar to that of "matching" a passive resonant circuit with the same *modulus* of effective Q-factor. The phenomenon of "spin-harmonic coupling", or "cross relaxation" (to be discussed below) has yielded the "high temperature" maser operation alluded to above, the enhancement of maser action under certain conditions, and the use of pump frequencies lower than the signal frequency. Finally, an interesting new maser material (because of the high separation of $\sim 54$ Gc/s between the $1 \pm \frac{1}{2} >$ and $1 \pm \frac{3}{2}$ doublets in zero magnetic field) is emerald, $Cr^{3+}$ in beryl ($Be_3Al_2(SiO_2)_6$).

*5. Cross-relaxation.* Cross-relaxation or 'spin-harmonic coupling' is an energy interchange process between neighbouring paramagnetic moments (spins). It can be illustrated by considering an energy-level scheme involving 3 levels, of energies $W_1 < W_2 < W_3$. Firstly, suppose levels 2 and 3 are very close in energy, and that neighbouring ions, A and B, are in states 1 and 3 respectively. It can be shown quantum-mechanically that there is a definite probability that A will make the transition $1 \rightarrow 2$ while B makes the transition $3 \rightarrow 1$; the small energy discrepancy is taken up by the spin-spin (dipolar) interaction of the whole assembly of spins. The process is a resonant one, having maximum probability when the energy differences $W_2 - W_1$ and $W_3 - W_1$ are exactly equal. There is a "line width" for the cross-relaxation process, of the same order of magnitude as that for the levels themselves. The above process is one that can defeat the 3-level pumping scheme, equalizing the populations of all 3 levels if the cross-relaxation probability is large enough. Cross-relaxation is dependent on concentration (distance between neighbouring ions) while spin-lattice relaxation, which normally depletes level 2, in a 3-level pumping scheme, is temperature dependent. Hence it is possible to obtain maser action with some concentrated materials at higher temperatures ($\sim 77°K$) where spin-lattice processes are dominant, but not at lower temperatures ($\sim 4 \cdot 2°K$) where the cross-relaxation process becomes dominant.

The name "spin harmonic coupling" arose because of the following possible cross-relaxation mechanism. Suppose $(W_2 - W_1) = n(W_3 - W_2)$, where $n$ is a small integer, greater than unity. Then it is possible for one spin to make the transition $1 \rightarrow 2$, while $n$ make the transition $3 \rightarrow 2$. Such processes may be helpful or detrimental to maser action, depending on the ratios of the energy gaps, the spin-concentration, and the number of levels involved. The outcome for any particular system can be predicted by thermodynamical considerations, and should be considered in maser design.

*Bibliography*

BROTHERTON M. (1964) *Masers and Lasers*, New York: McGraw-Hill.

GRIVET P. and BLOEMBERGEN N. (Eds.) (1964) *Quantum Electronics III*, Paris: Dunod.

SIEGMAN A. E. (1964) *Microwave Solid State Masers*, New York: McGraw-Hill.

SINGER J. R. (Ed.) (1961) *Advances in Quantum Electronics*, New York: Columbia University Press.

TROUP G. (1962) in *Encyclopædic Dictionary of Physics* (J. Thewlis Ed.), **4**, 498, Oxford: Pergamon Press.

TROUP G. (1963) (2nd Edn) *Masers and Lasers*, London: Methuen.

                         G. TROUP

**MEKOMETER.** *1. Introduction.* The mekometer is an electronic device for the measurement of distances by means of a light beam modulated at microwave frequency. It is based on the so-called linear electro-optic effect in certain crystals, which was discovered at the turn of the century by Pockels and until recently lay dormant in the optics text books.

Pockels discovered that some crystals change their optical properties when a strong electric field is applied along one axis, the optic axis of the crystal. In these circumstances, the velocity of light passing through the crystal parallel to the optic axis depends on the direction of the vibrations in the light wave—its plane of polarization. The result is that components of the light in two planes at right angles will be out of step as they pass through the crystal. The light is then said to be elliptically polarized and the crystal to be birefringent, that is having two refractive indices. The degree of ellipticity depends upon the potential difference applied; the direction of rotation of the emergent light vectors depends upon the direction of the applied field. Thus if the applied field is alternating the emergent light will be elliptically polarized in opposite senses during successive half-cycles of the alternating field.

The first commercial use of the Pockels effects was in the United States (Billings 1960; Mason 1946) where ammonium dihydrogen phosphate (ADP) and potassium dihydrogen phosphate (KDP) were used to amplitude-modulate light at low frequencies for putting sound on film.

Towards the end of 1960 the NPL became interested in the possibility of modulating light with ADP in the microwave region. Measurements of the dielectric constants at 35 Gc/s (35,000 Mc/s) yielded the d.c. values thus indicating that microwave light modulation was practicable.

The velocity of light, under given atmospheric conditions, can be used as a length standard. Its value in vacuum is known to better than one part in a million.

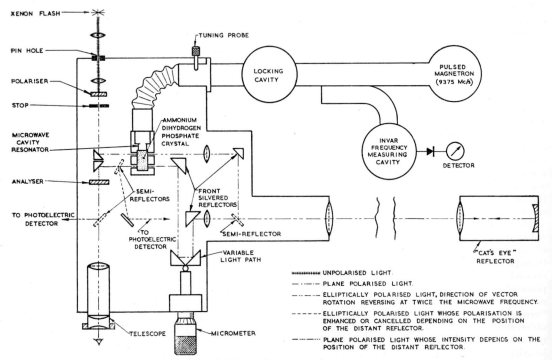

*Fig. 1. NPL microwave mekometer.*

Two devices exist in the survey field which employ this principle, the Geodimeter and Tellurometer. The former employs a light beam modulated in intensity by a nitrobenzene Kerr-cell at 30 Mc/s and the latter a microwave beam modulated at a similar frequency. The accuracy of these instruments, which have a limiting sensitivity of a few millimetres, is such as to make them very useful for the measurement of distances of a mile or more.

It was realized that the Pockels effect offered light modulation frequencies far in excess of 30 Mc/s and hence the highly accurate measurement of short distances; furthermore since the Pockels effect was linear with voltage the ADP crystal could act, in effect, as its own detector. A photocell would not be required that responded to the modulation frequency. In fact it is only in the last few months (1964) that a travelling wave phototube has become available which could operate in the microwave region and this is a cumbersome and expensive device.

By September 1960 patents had been filed and in early 1961 the mekometer had become a reality using the Pockels effect at 9·375 Gc/s in a device for the accurate measurement of distance (Froome and Bradsell 1960, 1961).

The Americans were almost at the same stages of development in this field and in 1961 produced a travelling-wave light modulator using KDP at 9·25 Gc/s (Kaminow 1961). The present article is concerned with the instrument developed at the NPL.

*2. The mekometer.* In designing the mekometer it was realized that there existed efficient electronic measuring devices to cover distances above a kilometre, but a device for the accurate measurement of shorter distances would find many uses in large scale metrology.

Figure 1 shows the layout of the mekometer. An ADP crystal (10 mm long, 8 mm wide and 5 mm thick), is situated in a cavity resonator coupled to an X-band magnetron which delivers 1·5 microsecond pulses of 7 kW peak power at a frequency of 9·375 Gc/s.

Parallel plane polarized light from a xenon flash tube, synchronized with the magnetron, passes through a hole in the cavity resonator at a position of maximum electric field. It emerges elliptically polarized, the direction of rotation of the vectors reversing for successive half-cycles of the microwave field.

After traversing the distance to the "cat's eye" reflector and back via the variable light path the elliptical polarization-modulated light is returned through the crystal, where the ellipticity is either increased or decreased depending on its instantaneous phase relative to that of the electric field in the cavity resonator. The light pencil then passes through a "crossed" polarizer, hence its intensity depends on the difference of the distance between the crystal and the reflector from an integral number of half wavelengths. The half wave-length at 9·375 Gc/s is approximately 16 mm.

An actual recording of the output light intensity variation when the distant reflector was slowly moved is shown in Fig. 2, the distance between modulator and reflector being 50 metres.

A balanced photoelectric detector system is employed comprising two photomultipliers, one viewing the light after it has passed the analyser and the other monitoring the intensity of the light pencil after it has retraversed the ADP crystal. This arrangement takes care of intensity fluctuations due to vagaries of the xenon flash. Each photomultiplier has a tuned circuit in its anode which is made to "ring" by the pulses provided by the 1·5 microsecond light flashes; the resulting oscillations are amplified and rectified, the difference between the two currents is displayed on a centre-zero metre.

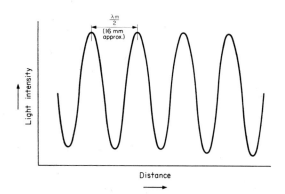

*Fig. 2. Recording of the photoelectric detector output as the distance between the mekometer and the distant reflector is increased.*

The micrometer operated variable light path incorporated in the instrument allows the measured distance to be adjusted to an exact number of half wave-lengths, at which point the photoelectric detector reads zero; thus an accurate measurement of the "excess fraction" of a half-wave is made. If in addition to this excess fraction the whole number of half wave-lengths is known, the exact distance from the ADP crystal to the reflector can then be computed from the velocity of light and the modulating frequency. The velocity in air is derived from measurements of temperature, pressure and humidity using the usual refractive index formula.

If the distance to be measured is not known sufficiently well to predict the whole number of half wave-lengths involved, more than one modulation frequency can be used and the "method of exact fractions", well known in the field of optical interferometry, can be employed to derive the distance.

The method of exact fractions requires the distance to be estimated and the fractional parts, produced by dividing this distance by each wave-length, to be measured. The fractional parts are then used to calculate the distance. With the mekometer the distance is increased by a variable light path so that it is exactly a whole number of wave-lengths, for each wave-length used, the increase in distance being used in place of the fractions.

A simple formula of the form

$$\text{Distance} \quad A + B + C - D$$

where $A$, $B$, $C$ and $D$ are derived from the micrometer readings and the respective modulation wave-length serves to calculate the distance being measured. The greater the number of modulating frequencies used, differing from each other by only a few per cent, the less chance of ambiguity there is. In the case of the mekometer three or four frequencies will suffice.

The mekometer can measure ranges of approximately 50 metres (160 feet) to an accuracy of about 0·05 mm (0·0002 ft).

*3. Recent developments.* The instrument described is intended for use under laboratory conditions for the highly accurate measurement of distances up to 100 metres.

At the present time the possibilities of a commercial instrument are being investigated. This device would be of small physical size, fully portable and measure distances up to 5000 feet. An accuracy of 1 part in 30,000 and an inherent error of 0·002 ft, important only at short range, is envisaged. At the longer ranges this accuracy could be greatly improved if the refractive index of the air were measured.

The instrument, which is still to be called the mekometer, will operate at a frequency of between 600 and 1000 Mc/s. Four frequencies will probably be used including three differing from the primary frequency by 5 per cent, 0·25 per cent and 0·025 per cent respectively. All frequencies will be generated from stable quartz crystal oscillators.

The fractional parts of the wave-lengths will be measured by micrometer, or similar means, and a suitable choice of primary frequency will allow the range to be calculated, in a few seconds, from a simple formula.

The most recent NPL instrument, Mekometer II (1965), operating in the u.h.f. region, offers automatic refractive index compensation and direct read-out of distance. The quartz crystal frequency standards are replaced by coaxial cavity resonator wave-length standards which make the new instrument independent of the velocity of light.

*Bibliography*

BILLINGS B. H. (1960) *Colloquia of the International Commission for Optics*, Optics in Metrology, May 1958, London: Pergamon Press.

FROOME K. D. and BRADSELL R. H. (1960) Patent Application No. 32675.

FROOME K. D. and BRADSELL R. H. (1961) *J. Sci. Instrum.* **38**, 458.

KAMINOW I. P. (1961) *Phys. Rev. Letters* **6**, 528.

MASON W. P. J. (1946) *Phys. Rev.* **69**, 173.

RUSHTON E. (1961) *Brit. J. Appl. Phys.* **12**, 417.

ZHELUDEV I. S. and VLOKH O. G. (1958) *Kristallografiya* **3**, 39. (English translation; (1959) *Soviet Physics-Crystallography* **3**, 647.)

R. H. BRADSELL

# MICROELECTRONICS SYSTEMS AND TECHNIQUES.

Microelectronics systems and techniques have been made possible by the changeover from valves to transistors, and although transistorization has greatly reduced the weight and size of electronic equipment using valves, microelectronic techniques offer an enormous gain over systems making the best possible use of conventional transistors and subminiature components.

There are two main microelectronics systems—thin film circuits and semiconductor integrated circuits. Thin film circuits using subminiature valves had been possible before the invention of the transistor in 1948, but the small size and low operating voltage of the latter has made it more compatible with thin film component techniques. Semiconductor integrated circuits were not possible before the invention of the transistor.

*Thin film integrated circuits.* In any typical tubular component most of the available volume is taken up by material which plays no part in its electrical performance. In a cracked carbon resistor the active film element has a volume of only one five-hundredth of the total component volume. Similarly, in a ceramic dielectric fixed capacitor, about one two-hundred and fiftieth of the volume is effective, the remainder consisting of the ceramic tube, case and connexions. The heat dissipation of the cylindrically shaped component is inefficient because of the low surface area/volume ratio and an improvement can be made by opening out the cylinder into flat strips. Flat components can be fabricated in many ways but the almost universal current methods employ evaporating and sputtering techniques, fixed resistors being evaporated from nickel-chromium wires and tantalum resistors sputtered from tantalum foil. Masks are made through which the evaporation or sputtering takes place, thus forming the shape of the resistors required, control of resistance value being carried out by accurate regulation of deposition time.

Fixed capacitors are similarly fabricated by evaporation of dielectrics such as silicon monoxide from powders, and tantalum pentoxide from tantalum metal sputtered in an oxidizing atmosphere. The top and bottom electrodes of the capacitors are usually of evaporated aluminium. Wiring line patterns (normally in gold) are evaporated through masks; finally, small transistors are attached to the circuit. Evaporated or deposited thin film transistors are being made experimentally of cadmium sulphide, silicon, and other semiconductor materials, and these may eventually replace individually attached transistors, thus forming a two-dimensional deposited thin film integrated circuit.

*Semiconductor integrated circuits.* In this type of circuit, a single crystal of silicon is used not only to provide various transistor functions, but also to include resistive and capacitive elements. A single bar of silicon might be considered as being a resistor, with the resistance determined by the resistivity of the material, its cross-section, and its length. Isolation from other components can be provided by depletion layers or areas produced by diffusion techniques, in which a diffused layer is selectively shaped by photolithographic methods to isolate the resistance element. Thin-film nickel-chromium resistors may be evaporated or tantalum resistors sputtered through masks on to an insulating layer of thermally grown silicon dioxide. Capacitors may be formed by using the capacitive effect of a reverse biased diode. By varying this bias the capacitance may be made to vary, thus giving the effect of a variable capacitor. The thermally grown silicon dioxide itself may be used as a dielectric with an aluminium counter-electrode.

The transistors and diodes are formed in the correct positions in the crystal structure by diffusing the required impurities through masks made very accurately by photolithography. Isolated areas are formed by multiple diffusions.

A problem with this form of circuit is that cross-coupling occurs with closely spaced components. This makes the high frequency operation difficult, and to avoid this, multi-chip construction is often used. In this method the block is cut into two or more pieces, providing more effective isolation and therefore making possible higher frequency circuits. Connexions are made between chips and also to the outer leads. Microwelding techniques are used for the external connexions (between packages, etc.), while thermo-compression bonding is commonly used for internal connexions, i.e. on the surface of the single multi-circuit silicon slice, or between individual chips.

Recent isolation techniques include the use of silicon monoxide layers and etching away of areas under the contact strips, which improves the high frequency performance considerably.

Semiconductor integrated circuits are sometimes packaged into 8 or 12 pin TO5 transistor cases, approximately $\frac{1}{4}$ in. diameter $\times$ $\frac{1}{4}$ in. high, or sometimes into flat cases with tape leads, which are even smaller ($\frac{1}{4} \times \frac{1}{8} \times 1/32$ in.). An illustration of the latter is shown in the figure, in which the transistor and component areas can be clearly seen.

An indication of the packing densities possible in microelectronics is given in the table below.

*Terminology.* Classifications of microelectronics have been provisionally agreed by the International Electrotechnical Commission (I.E.C.), as shown in the next column.

### Component Packing Densities for Low Power Electronic Circuits

| | Approx. No. of components per cu. in. |
|---|---|
| *Conventional components* | |
| Pre-war valves with standard components | 1 |
| War-time miniature valves with standard components | 4 |
| Subminiature valves with miniature components | 7 |
| Transistors with subminiature components | 20 |
| *Microelectronics* | |
| Microassemblies (Micromodules, Ministac, Cordwood, Pellet, etc.) | 50–150 |
| Thin Film Circuits (achieved in equipment) | 250 |
| Semiconductor Integrated Circuits (achieved in equipment) | 1200 |
| (experimentally achieved) | 10,000 |
| *Future* | |
| Circuit fabrication using electron beam methods (theoretical) | > 10,000,000 |
| *Note* | |
| Neuron density in a human brain | > 1,000,000,000 |

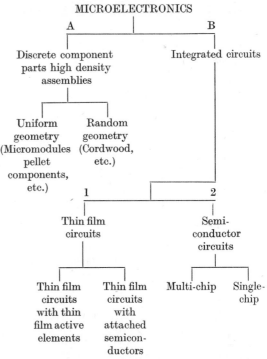

MICROELECTRONICS

A hybrid circuit is any combination of 1 and 2 or A and B.

Some definitions of microelectronics systems have been provisionally agreed by I.E.C., as follows:

*Microelectronics.* That entire body of technology which is associated with, or applied to, the realization of electronic circuits with a degree of miniaturization greater than that usually obtained with conventional methods and/or parts.

*Microstructure.* A structure of high component density the parts of which may be assembled or may be integrated and which for the purpose of commerce and specification is considered indivisible.

*High density assembly.* A microstructure in which the various components and devices are realized and tested separately before assembled and packaged.

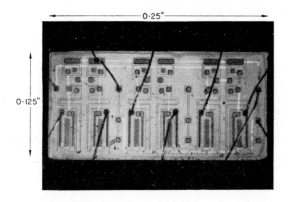

*Integrated circuit.* A microstructure in which a number of circuit elements are inseparably associated on or within a continuous body.

*Thin film integrated circuit.* An integrated circuit composed entirely of thin film circuit elements and interconnexions deposited on supporting material.

*Semiconductor integrated circuit.* An integrated circuit composed of circuit elements and interconnexions, the circuit elements being realized entirely within one or more blocks of semiconductor material.

*Hybrid integrated circuit.* An integrated circuit using a combination of thin film and semiconductor techniques.

*Hybrid microstructure.* A microstructure consisting of one or more integrated circuits in combination with one or more discrete devices or components.

*Bibliography*

DUMMER G. W. A. and GRANVILLE J. W. (1961) *Miniature and Microminiature Electronics*, London: Pitman.
KEONJIAN E. (1963) *Microelectronics: Theory, Design and Fabrication*, New York: McGraw-Hill.

               G. W. A. DUMMER

**MICROWAVE ULTRASONICS.** *Introduction.* The term microwave ultrasonics is now used to describe very high frequency vibrations in solids which are artificially generated. Since their range of frequencies lie between 1 and 24 Gc/s, they are of a similar nature to the thermal vibrations in a solid where the dominant lattice vibrational frequency is approximately given by $hv = kT$, so that frequencies of 1 and 24 Gc/s would correspond to the dominant phonons at 0·05°K and 1·0°K. This explains why such waves are alternatively described by the term "microwave phonons" but it is important to stress that microwave phonons, unlike thermal phonons, are coherent, polarized in a given direction, and have a single frequency.

Normal quartz transducers can be used to produce ultrasonic waves up to frequencies near 1 Gc/s but the extremely thin and fragile wafers of quartz preclude any great increase in the useful production of ultrasonic power above 1 Gc/s by this method.

*Experimental Methods of Generating Microwave Phonons*

It has now been shown experimentally that one can generate microwave ultrasonics in two ways. Firstly, using rods or thick wafers of piezoelectric crystals or secondly, using the ferromagnetic resonance of thin films.

*A. The piezoelectric method.* In this method, a suitably cut and oriented piezoelectric crystal is placed in a region of high electric field within a resonant microwave cavity and Fig. 1 illustrates the simple example for the case of a quartz rod with the $x$ crystallographic axis of the crystal orientated along the axis of the rod. The electric field in the re-entrant cavity is then perpendicular to the end face of the rod. When electromagnetic power is passed, in pulses, into the cavity an acoustic wave is stimulated at the free face of the quartz rod which then progresses down the rod with the characteristic velocity of sound.

The type of wavefront produced in the quartz is determined by the flatness and orientation of the end

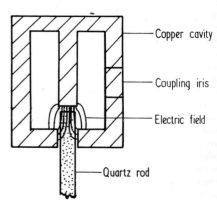

*Fig. 1. Re-entrant microwave cavity showing approximate electric field configuration.*

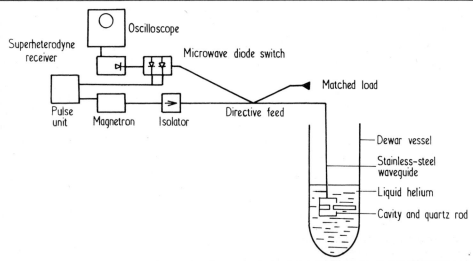

Fig. 2. Experimental arrangement for the generation and detection of microwave ultrasonics.

of the rod. Since the wave-length of 10 Gc/s ultrasonics in quartz is about 5000 Å it is clearly necessary to work with an optically flat face. A second requirement is that this face be perpendicular to a pure mode axis of the piezoelectric material in order that a plane wave should propagate straight down the specimen.

To observe the pulses of acoustic power generated by this method the inverse piezoelectric effect is used either by placing an identical resonant cavity at the other end of the rod or by using a single cavity as a transmitter and receiver cavity. In either case it is essential to arrange the second end face of the rod to be optically flat and also parallel to the first end to within very fine limits or the reflected or detected sound wave will have the wrong phase.

A typical experimental arrangement working at 10 Gc/s is shown in Fig. 2 where a single microwave cavity is used as a transmitter and receiver. Power from a magnetron is fed in 1 μsec pulses, 1 millisecond apart, down a suitable waveguide system through a directive feed into the cavity. The acoustic signal generated in the rod is reflected at the end of the rod and returns with the normal acoustic delay to produce a signal in the cavity. The overall conversion of power into and out of the rod is inefficient and a reduction factor of about $10^6$ is common.

The reflected signal is passed back through the directive feed where a fraction of it proceeds to the receiver which is normally a very sensitive super heterodyne system. This receiver, being so sensitive, requires protection when the magnetron driving pulse is operating and this can be obtained using crystal diode switches in the receiver arm of the system. The detected signal is then displayed on an oscilloscope giving a series of acoustic echoes of decaying amplitude.

A typical form for an echo decay pattern is shown in Fig. 3 which indicates that the decay is far from

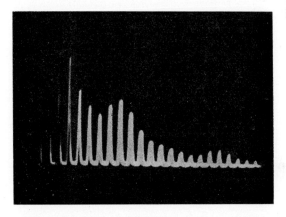

Fig. 3. Typical microwave acoustic echo pattern.

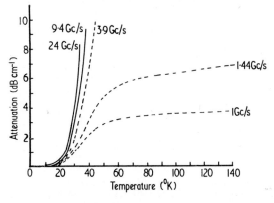

Fig. 4. Variation of attenuation with temperature for compression waves in quartz. (After Bömmel and Dransfeld, and Jacobsen (1960).)

exponential as might have been expected but this can be due to the lack of parallelism in the end faces of the quartz rod, defects and strains in the rod, and misorientation of the crystallographic axis of the quartz.

The experiments which are reported in the literature range in frequency from 1 to 24 Gc/s and all are conducted at relatively low temperatures. The curves of attenuation against temperature shown in Fig. 4 clearly indicate that for an experiment at, say, 10 Gc/s, the maximum temperature allowed is around 20°K, above which measurements will be increasingly difficult in quartz specimens of any thickness. In fact, most experiments are now conducted in the liquid helium range from 4·2 to 1·4°K unless effects are specially investigated as a function of temperature.

. *B. The ferromagnetic resonance method.* When a ferromagnetic film, in a steady magnetic field normal to the film, is subjected to an alternating magnetic

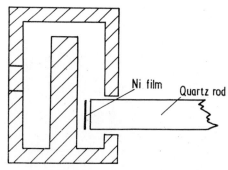

*Fig. 5. Cavity for generation of microwave phonons by ferromagnetic resonance in a nickel film.*

field perpendicular to the steady field, the magnetization of the film precesses about the steady field direction. By magnetostrictive coupling, this produces a rotating shear strain over the surface of the film.

Using a nickel film on the end of a quartz rod it has been possible to produce acoustic waves of frequencies up to 9·5 Gc/s. The experimental arrangement is shown in Fig. 5 where the nickel film is situated in the intense magnetic field of a microwave cavity similar to that used in the section A above.

This method is thought to be more efficient than the piezoelectric method for generating microwave ultrasonics but is limited in its usefulness because it already requires a static magnetic field acting on the specimen thus precluding any experiment involving a variation of magnetic field.

Of the two methods of generation outlined above, the first is by far the most widely used. It is not, however, without its difficulties since a surprising variation in the behaviour of specimens is observed from a group of quartz rods all prepared in the same way from the same quartz crystal. In work at frequencies of 10 Gc/s and above it is necessary to select, by trial, a good transducer.

### The Properties and Uses of Microwave Ultrasonics

The problems of generating microwave ultrasonics are now well-known and can, to a great extent, be overcome. This has produced a considerable amount of experimental effort aimed at elucidating the behaviour of these waves and their interactions in solids.

*1. Interaction with thermal vibrations in quartz.* Figure 4 is a summary of results obtained for the attenuation of microwave ultrasonics in quartz. This attenuation arises because the forces binding atoms together in a solid are not purely harmonic so that when a microwave phonon passes into a region already disturbed by a thermal phonon the forces between atoms are slightly different from the equilibrium forces it would encounter in an undisturbed crystal. It is possible, in this way, for energy to be converted from the coherent microwave acoustic vibration into the incoherent thermal distribution of lattice vibrations. Further work is in progress to extend our knowledge of these interactions and to investigate the non-linear effects in such crystals as MgO, where mixing can occur between sound waves of differing frequency.

*2. Interaction with lattice defects.* The striking difference in behaviour between rods, similarly prepared, has led to an investigation of the effect of strain and crystal defects on the propagation of microwave ultrasonics.

A very small transverse strain applied to a quartz rod such as is described above (A) is sufficient to cause a large change in the decay pattern observed. It has also been shown that irradiating quartz with $\gamma$ or X rays will produce a large increase in the attenuation of such waves.

*3. Interaction with paramagnetic impurities in crystals.* The possibility of observing interactions between microwave phonons and resonant electron spin systems was successfully demonstrated soon after microwave phonons had first been generated and it is in this direction that the most progress has been made in the application of this new technique.

Experimentally, it is required to generate phonons at one end of a specimen and observe an electron spin resonance signal at the other end. Figure 6 gives an outline of the experimental arrangement where A and B may be a single quartz rod with end B containing a paramagnetic impurity or B may be a separate material containing the paramagnetic centres with A acting merely as a transducer. When using two separate materials it is necessary to have the rod B with its end faces cut and aligned with the same precision as for the quartz rods. The transducer and specimen must then be bonded together so that the end faces of the combined specimen are still accurately parallel and the bond must be mechanically strong. Such bonds have been achieved using a thin film of indium, frozen vacuum grease and various epoxy resin compounds but it is not easy to produce a good bond, the best giving a power loss of as much as 20 dB.

Generation of the microwave phonons is by the piezoelectric method (A) and the observation of the electron spin resonance signal requires the use of any standard form of electron spin resonance spectrometer. A variety of different experiments have been performed in this field, which fall into two groups.

*A. The effect of microwave ultrasonics on resonant paramagnetic systems.* In these experiments the main interest lies in observing a normal electron spin resonance signal from the paramagnetic sample and then allowing ultrasonic power to pass into the system.

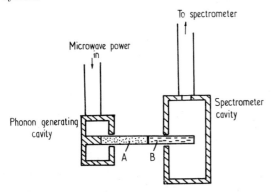

*Fig. 6. Cavity system for investigating electron-spin–phonon interactions.*

It has been shown in this way that the absorption signal produced by defect centres in quartz (induced by $\gamma$ irradiation) can be reduced in size when ultrasonic power is incident on the system. Similar results were obtained using the manganese impurity centre in quartz.

The chromium impurity in ruby is a system which has been widely investigated in connexion with its use as an active element in masers and lasers. Using ultrasonics it is possible to show that the transition probabilities for this ion under the influence of ultrasonic power follow the expected form.

In each case considered above the experimental requirements are that a specimen has energy levels whose separation may be changed with a magnetic field. The levels are adjusted to be in resonance with incident electromagnetic energy (to observe the e.s.r. signal) and also with the incident ultrasonics. Power from the sound wave is then used to stimulate spins from the lower to the upper level hence reducing the number available for observation in the e.s.r. spectrometer.

A final experiment in this section was devised in an attempt to produce a method of detecting the presence of microwave ultrasonics which does not require a phase coherent wavefront as does the piezoelectric method.

In essence, a paramagnetic system is chosen, having three energy levels ($Fe^{++}$ in MgO has been used) with a separation which can be varied by a magnetic field. Figure 7 illustrates the situation. Transitions are induced between the levels 1 and 3 using an electromagnetic quantum $h\nu_m$ and a quantum of ultrasonic energy $h\nu_u$. By tuning the spectrometer cavity to $\nu_m$ and producing pulses of phonons at frequency $\nu_m$ the spectrometer will show pulses of absorption when excited with microwaves at a frequency $\nu_m$. In ideal circumstances this could be a useful and sensitive detection device for microwave ultrasonics.

*B. The effect of resonant spin system on the propagation of microwave ultrasonics.* In this section the interest

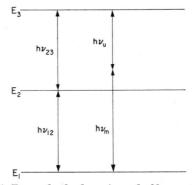

*Fig. 7. Energy level scheme for a double quantum transition using a phonon at frequency $\nu_u$ and a photon at $\nu_m$.*

shifts to the way in which the attenuation and velocity of microwave ultrasonics can be modified by resonant magnetic systems in the material through which the waves propagate.

Early theoretical work showed two main effects. The first was an absorption of ultrasonic energy when the magnetic system was brought into resonance with the ultrasonic frequency. This absorption varies in intensity depending on the strength of the coupling between the spin system concerned and the crystalline electric field which is being modulated by the ultrasonic wave. Some paramagnetic systems such as $Fe^{++}$ and $Ni^{++}$ in MgO crystals show the theoretically-predicted strong effects and in both cases it is possible to arrange the experimental conditions to give complete absorption of ultrasonic pulses. The absorption arises because power from the sound wave is used to raise the effective spin temperature of the paramagnetic system.

The second theoretical suggestion was that a resonant spin system would cause a medium to become dispersive and the velocity of sound waves should decrease. This effect has also been experimentally demonstrated using the systems mentioned above. Both the dispersion and absorption should show the characteristic line shapes such as are observed in ordinary electron spin resonance.

More detailed work has shown (very recently) that the early theoretical predictions and experiments are not sufficient to explain all the processes involved in this interaction. It is now clear that the mechanisms of pure absorption and dispersion are only observed in specially selected cases or in systems where the coupling between spins and lattice is comparatively weak.

*4. The interaction of microwave ultrasonics with charge carriers in metals and semiconductors.* It has been possible to show experimentally that the charge carriers in cadmium sulphide can, under the influence of an electric field, have a drift velocity equal to the velocity of microwave ultrasonics in the material. Under such conditions electrical power can be transferred into ultrasonic power with a consequent amplification of the ultrasonic wave. These early experiments have given rise to a considerable interest in the wide aspects of the interaction of such waves with charge carriers and it now seems likely that such phenomena as cyclotron resonance and oscillatory absorption of microwave ultrasonics will be observed at increasingly high frequencies. One point of particular interest is that, for microwave frequencies, the electrical skin depth of a metal is very small so that cyclotron resonance is only observed in surface layers. However, the use of microwave ultrasonics eliminates this difficulty since there is no problem with skin depth and waves may be propagated through the bulk specimens.

*Bibliography*

AL'TSHULER S. A., KOCHELAEV B. I. and LEUSHIN A. M. (1962) *Sov. Phys. Uspéckhi* **4**, 880.
BÖMMEL H. E. and DRANSFELD K. (1960) *Phys. Rev.* **117**, 1245.
JACOBSEN E. H. (1960) *Quantum Electronics*, Columbia: The University Press.
JACOBSEN E. H. and STEVENS K. W. H. (1963) *Phys. Rev.* **129**, 2036.
ROWELL P. M. (1960) *Brit. J. Appl. Phys.* **14**, 60.
TRUELL R. and ELBAUM C. (1962) *Handbuch der Physik* XI, II.

P. M. ROWELL

**MILLIMETRE WAVE TECHNIQUES.** Considerable attention has been given to special methods of generating, transmitting and detecting radiations of millimetre and submillimetre wave-lengths and to the development of associated measuring equipment.

Millimetre and submillimetre wave instrumentation is used in the field of microwave gas spectroscopy to measure absorption frequencies due to molecular rotational transitions, for research on solid-state materials and for analysing energy gaps in superconductors. A major reason for interest in millimetre waves has been the requirement for additional communication channels because at these wave-lengths enormous bandwidths can be made available.

Space communications, radar and telemetry are fields particularly applicable to millimetre waves where lower atmosphere absorption of these waves can be ignored and advantage can be taken of narrow aerial beams and high accuracy readily attainable with reasonable sized components. Highly-directive beams, which confine intelligence to a small volume and make jamming difficult, are attractive in the design of secure communication systems. The propagation characteristics at millimetre wave-lengths of the highly-ionized plasma produced in controlled-fusion research can be related to such plasma properties as temperature, electron density and confinement time.

Many other uses would undoubtedly be found for millimetre waves if power sources and techniques were cheap and readily available.

*Transmission Methods*

At wave-lengths below a few millimetres use of conventional rectangular waveguides for energy transmission and measurements becomes difficult due to very small dimensions and high attenuations. Difficulties of making components in the small waveguide involved also become severe and the performances of such components deteriorate with decreasing wave-length. Manufacturing tolerances of $\pm$ 0·0001 in. are necessary to achieve good performance. Other methods of transmission and component design are thus needed for submillimetre wave-lengths. There is considerable advantage, however, in extending normal microwave techniques as far as practicable but it seems probable that not far from 1 mm wave-length these will be abandoned and replaced by optical and quasi-optical techniques.

The internal dimensions and theoretical attenuations of standard rectangular waveguides for the millimetre wave-length range are listed in Table 1. Losses in these waveguides can be calculated from formulae derived on the assumption that surfaces are perfectly smooth. Measured losses may be considerably greater, however, than those calculated from these formulae as shown by the figures in Tables 2 and 3.

*Table 1. Characteristics of standard rectangular waveguides*

| Waveguide reference | Frequency range (Gc/s) | Internal dimensions (in.) | Attenuation* (dB/100 ft) |
|---|---|---|---|
| WG 22 | 26·5– 40 | 0·280 × 0·140 | 17·3 |
| WG 26 | 60– 90 | 0·122 × 0·061 | 59·7 |
| WG 29 | 110–170 | 0·065 × 0·0325 | 154 |
| WG 32 | 220–325 | 0·034 × 0·017 | 405 |

\* For pure copper and calculated at a frequency of 1·5 times the cut-off frequency.

*Table 2. Characteristics of rectangular waveguides at 35 Gc/s*

| Material | D.C. resistivity (microhm-cm) | Measured attenuations at 35 Gc/s (dB/m) | Ratio of measured to calculated attenuations † |
|---|---|---|---|
| Tellurium-copper* | 1·88 | 0·95 | 1·64 |
| Standard-silver | 2·11–2·22 | 0·78–0·95 | 1·20–1·55 |
| Copper | 1·78 | 0·85–0·88 | 1·55–1·57 |
| Silver-lined copper* | 1·63‡ | 0·65 | 1·24 |
| Brass | 6·75–6·89 | 1·38–1·99 | 1·29–1·85 †† |
| Aluminium | 2·90 | 1·24–1·37 | 1·77–1·93 |

* Only one sample available for examination.
† Calculated values obtained by neglecting surface roughness.
‡ Assumed value.
†† The value of 1·85 was recorded for a sample which had numerous cracks up to 0·002 in. deep along its length.

(By Courtesy of The Institution of Electrical Engineers.)

*Table 3. Characteristics of rectangular waveguides at 70 Gc/s.*

| Material | D.C. resistivity (microhm-cm) | Measured attenuations (dB/m) | Ratio of measured to calculated attenuations* |
|---|---|---|---|
| Copper | 1·72–1·75 | 3·5–5·3 | 1·7–2·5 |
| Brass (60/40) | 6·9 | 5·9 | 1·4 |
| Brass (70/30) | 6·49–6·5 | 5·5–5·9 | 1·3–1·5 |
| Brass (90/10) | 4·29 | 5·3 | 1·6 |
| Standard-silver | 2·15 | 3·1 | 1·4 |

* Calculated values obtained by neglecting surface roughness.
(By Courtesy of The Institution of Electrical Engineers.)

The discrepancies between theoretical and experimental values may be partly attributed to surface roughness but there may be an increase in the resistivity of a waveguide surface caused by work hardening, and tarnish films also produce losses.

The conventional rectangular waveguide will continue to be the principal one used at frequencies up to 40 Gc/s and may also be employed for very restricted transmission distances up to about 100 Gc/s. Various other types of waveguide that have been considered for transmission of short millimetre and submillimetre wavelength energy and the general problems associated with their design are given in Fig. 1. The $H_{01}$-mode circular metal guide and the H-guide have special interest. Assuming the dielectric has negligible loss, the attenuations of these guides decrease as frequency increases. To obtain the low attenuation of the $H_{01}$ circular mode, waveguide of sufficiently large diameter must be employed and the pipe is then above cut-off for many other modes. Mode conversion, particularly to the $H_{11}$ mode, occurs at bends or discontinuities. This may be suppressed with a helically-wound wire lining on the waveguide interior which tends to suppress wall currents of modes other than the $H_{01}$ mode. Accurately-wound helices, to tolerances of about 0·001 in. in 2 in. guide are required, however, for satisfactory operation. A dielectric coating on the interior of circular waveguide can suppress conversion to the $H_{11}$ mode by making its propagation constant different from that of the $H_{01}$ mode. A combination of helical lining and dielectric coating can improve the stability of the $H_{01}$ mode in the presence of guide bends.

In the H-guide no currents flow in the longitudinal direction in the conducting strips so that propagation is unaffected by gaps in the walls in the transverse direction. Therefore, no connectors are needed to join H-guide sections. Another advantage of H-guide in the millimetre-wave region is that cross-sectional dimensions are greater than those of rectangular waveguides for the same frequency. Modified structures with improved properties have been derived from the original H-guide form (e.g. double-slab guide, groove guide).

The dielectric image line is simply a strip of low-loss dielectric material on a metal plane. Conduction loss in the image plane is reduced by the use of a high-conductivity metal. To obtain low dielectric loss and single-mode operation at millimetre wave-lengths the dielectric material must be small in cross-section and must have low loss tangent and dielectric constant. Satisfactory materials are polystyrene foam (which has been processed to have small cell size and uniform density) or Teflon which is available in thin narrow tape form.

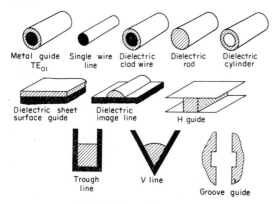

Fig. 1. Transmission systems.
(*By courtesy of the Institute of Electrical and Electronic Engineers Inc.*)

About the smallest practicable diameter for a polystyrene foam rod with a semicircular cross-section is 0·125 in. (3·17 mm). The calculated attenuation of such a rod mounted on a copper image plane is shown in Fig. 2. A second curve is shown for Teflon and a rod diameter of 0·025 in. (0·64 mm). This was chosen

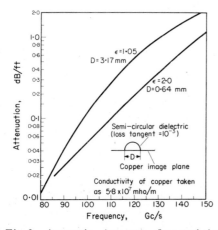

Fig. 2. Attenuation–frequency characteristics of dielectric image lines. (*By courtesy of the Institute of Electrical and Electronic Engineers Inc.*)

because a semicircular cross-section with this diameter has the same area as a tape 0·002 in. × 0·125 in. Obviously further reduction of attenuation by reduction of dielectric cross-section has only limited possibility. Lower attenuation may be obtained, however, by using a flattened cross-section (rectangular or elliptical) with the longer transverse direction parallel to the image plane.

Results of measurements on thin dielectric tape lines (Teflon 0·002 in. thick) using polished copper image planes 4 in. wide in lengths of 5, 12 and 20 ft are shown in Table 4. Theoretical losses for semi-circular dielectric cross-section lines of the same area are also given. The measured attenuation values are considerably lower than for rectangular waveguids at these frequencies.

Table 4. Attenuations of dielectric image lines

| Frequency Gc/s | Tape width (in.) | Average measured loss (dB/ft) | Theoretical loss for semi-circular dielectric cross-section dB/ft |
|---|---|---|---|
| 105 | 0·25 | 0·14 | 0·63 |
| 140 | 0·125 | 0·43 | 0·70 |
| 140 | 0·25 | 0·98 | 3·2 |

(By courtesy of the Institute of Electrical and Electronic Engineers.)

Surface-wave propagation on dielectric-coated or uncoated metal wires at millimetre wave-lengths has been studied. Table 5 gives the theoretical attenuations for a number of wires. Some measured attenuations at 105 and 140 Gc/s were found to be two to two and a half times higher than calculated values for small diameter wires (0·080 and 0·128 in. diameter) and approximately three to seven times higher than

Table 5. Attenuations of single-wire transmission lines

| Wire diameter (in.) | Frequency (Gc/s) | Calculated attenuation (dB/ft) | Measured attenuation (dB/ft) |
|---|---|---|---|
| 0·080 | 105 | 0·07 | 0·14 |
| 0·128 | 105 | 0·05 | 0·10 |
| 0·375 | 105 | 0·02 | 0·12 |
| 0·500 | 105 | 0·015 | 0·10 |
| 0·080 | 140 | 0·08 | 0·20 |
| 0·128 | 140 | 0·06 | 0·11 |
| 0·375 | 140 | 0·03 | 0·10 |
| 0·500 | 140 | 0·02 | 0·08 |

(By courtesy of the Institute of Electrical and Electronic Engineers.)

calculated values for larger wires (0·375 and 0·5 in.). The excess loss has been attributed to copper and radiation losses at surface imperfections, irregularities, bends and nylon supports along the wires. The measured attenuations for the single-wire lines in this frequency range are, however, much lower than for conventional rectangular waveguides. To obtain the lowest losses the wire line must be quite straight and uniform.

Even a very light coating of low-loss dielectric has a relatively drastic effect on the attenuation (see Fig. 3). For the example chosen the attenuation is greater by more than an order of magnitude for the coated wire.

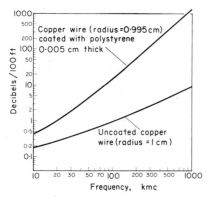

*Fig. 3. Attenuation–frequency characteristics of dielectric-coated and uncoated copper wires. (By courtesy of the Institute of Electrical and Electronic Engineers Inc.)*

Some waveguides illustrated in Fig. 1 may be standard ones for transmitting power in the frequency range 100–300 Gc/s. At higher frequencies practical problems will be such as to make it necessary to turn to optical or quasi-optical types of transmission system.

A need for low-loss transmission over moderate distances (say 30 ft) has been met by using simple overmoded rectangular $H_{01}$ waveguides with linear launching tapers. Preliminary experiments with such a system at 280 Gc/s, using a linear taper 19 in. long from waveguide WG32 to waveguide WG16 (0·9 × 0·4 in. internal) indicated an insertion loss of 2 dB for the taper and a loss less than 0·1 dB/ft in the straight overmoded waveguide compared to the theoretical loss of 0·04 dB/ft.

### Generation of Millimetre Waves

Many investigations have been concerned with extending conventional microwave valve design to the shortest possible wave-lengths. Reflex klystrons have been produced with power outputs greater than 100 and 40 mW at wave-lengths of 4 and 2·5 mm respectively. Other developments have occurred in the design of magnetrons and suitable slow-wave struc-

tures for backward-wave oscillators for millimetre wave-lengths. For example, in 1957 a power of a few tenths of a milliwatt at a wave-length of 1·5 mm was generated by devising suitable slow-wave structures of the ridge-loaded ladder type formed by winding a flat tape over a frame and brazing. It was concluded at the time that inevitable circuit losses of the construction were likely to prevent further reduction in wave-length. More recently a valve has been fabricated, using a circuit of the slotted-cavity type, with a bandwidth of 10 per cent and an output power up to 15 mW in the 0·7 mm wave region.

Four fundamental problems plague conventional valves operating in the low millimetre range:

(1) physical size and tolerance, (2) heat dissipation (3) circuit losses and (4) cathode and starting current densities.

Figure 4 shows a cross-section of the central portion of a klystron for 4 mm waves the resonant

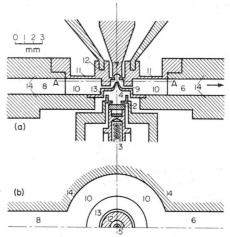

*Fig. 4. Central portion of type DX 151 reflex klystron for 4 mm waves. a) axial section, b) transverse section along horizontal plane through A-A in (a). 2 L cathode; 3 heater; 4 focusing electrode; 5 Resonant-cavity; 6 output waveguide; 7 repeller; 8 waveguide with shorting plunger (not shown); 9 quarter-wave transformer; 10 annular space; (9, 10 and 8 together form the matching transformer between the resonant cavity and the output waveguide); 11 thin wall allowing copper section 12 to be moved axially in relation to the copper section 13 by means of an external turning mechanism (mechanical tuning); 14 copper block in which central portion is mounted.*

Dimensions:
*Diameter of opening in electrode 4    1·0 mm*
*Diameter of holes in 12 and 13    0·25 mm*
*Diameter of resonant cavity 5    1·6 mm*
*Width of waveguides 6 and 8    3·6 mm*
*Height of resonant cavity 5    0·7 mm*
*Height of waveguides 6 and 8    1·8 mm*
*(By courtesy of Philips Electrical Ltd.)*

frequency of which can be varied continuously over a range of 10–15 per cent. One difficult problem in such tubes is accurate alignment of electrodes; in the present case to within 10 μ. This was solved satisfactorily by mounting the electron gun and the reflector so that after the tube is sealed off they can be moved fractionally in various directions perpendicular to the tube axis. For this purpose a flexible diaphragm is used as part of the tube wall. The L cathode employed gives the required current density of 2·5 A/cm² for an average life of several thousand hours. The emitting surface must be smooth to within about 1 μ for the electron-beam to have the correct shape. The electron-beam diameter at the narrowest point is about 0·12 mm.

In another class of investigations into the generation of short millimetre waves, in which the undulator and the rebatron are outstanding examples, high-energy, tightly-bunched electron streams are used. The undulator requires a linear accelerator of some few MeV to produce bunched electron beams with velocities about 0·98 that of light. These pass through a static periodic magnetic field which gives them transverse accelerations. The electrons thus radiate and due to the high beam velocity the observed frequency of such radiation is shifted by the Doppler effect to the millimetre-wave region. Similar requirements on the electron beams are necessary in the rebatron but here a suitable resonant structure is required capable of sustaining or abstracting harmonics from the highly bunched electron streams. Such devices have yielded significant data but do not seem capable of providing easily available, economical and convenient sources of millimetre-wave radiation. Further, they are pulsed and for many applications a C.W. source is preferable.

Much research work has been done above 100 Gc/s using energy obtained as higher harmonics of lower-frequency sources. One approach has been to apply filtration to the non-sinusoidal output of conventional or modified klystrons and magnetrons to select energy components at harmonics of the fundamental frequency; a second approach is to use the primary oscillator to drive a non-linear device such as a crystal.

A typical "crossed-guide" crystal harmonic generator for generating 2 mm waves by frequency doubling is shown in Fig. 5. It consists of a point-contact diode situated at the junction between the 4 mm waveguide 1 and the 2 mm waveguide 2. Projecting through these two waveguides is a tungsten cat's whisker (approximately 0·002 in. diameter) which, together with a silicon crystal 4, forms the diode. The lower end of the cat's whisker is welded to an insulating pin 5. The other end is etched to a fine point. The silicon wafer, which can be finely adjusted in the vertical direction by a differential screw is brought into contact with the whisker. The part of the whisker running through the two waveguides acts as a coupling probe and because of the crystal non-linearity the current induced in the coupling probe by the incident 4 mm waves contains a component with twice the

frequency. This produces a 2 mm wave in the upper waveguide.

A large range of results can be obtained and the 2 mm output is always about 20 dB lower than the 4 mm input. Silicon which has been bombarded with positive ions is found to be superior to silicon dices from commercial microwave crystals. Techniques have been established for the formation of germanium-titanium and gallium-arsenide/phosphor-bronze crystal/whisker junctions.

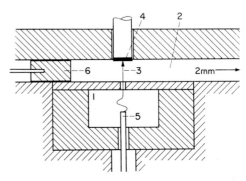

*Fig. 5. Cross-section of frequency multiplier.*
*1 is 4 mm waveguide; 2 is 2 mm waveguide; 3 cat's whisker; 4 silicon crystal; 5 insulated pin carrying the cat's whisker; 6 shorting plunger in the 2 mm waveguide. 3 and 4 together form a point-contact diode.*
*(By courtesy of Philips Electrical Ltd.)*

Theoretical treatments on the possible efficiencies of harmonic generators are available. The conversion capabilities of positive non-linear resistances have been considered; the results are represented by an ideal limit curve in which the efficiency varies as $1/n^2$ where $n$ is the harmonic number. General results for the power relations in non-linear capacitors and inductors have also been developed. The non-linear (voltage-dependent) capacitance of back-biased semiconductor junctions is used extensively for low-noise amplification and harmonic generation at lower microwave frequencies. For millimetre-wave applications the main problems are those of obtaining suitable materials and the required higher performance or $Q$ value of the junction. A loss of some 9 dB has been quoted for conversion of power at 24 Gc/s to 48 Gc/s using a point-contact gallium-arsenide non-linear capacitance diode.

Investigations of other non-linear media such as ferrites and plasmas, which can handle large power inputs, have also given significant results. With an arc harmonic generator, using a fundamental 5 W source of C.W. power at 35 Gc/s harmonics as high as the 21st (735 Gc/s) have been detected with a *Golay cell*.

Many other generation schemes have been suggested including the Cerenkov radiator, the use of multiple-quantum transitions and the possibility of mixing two laser outputs to provide a suitable difference frequency.

### Millimetre Waveguide Components

Most standard microwave components have been extended to cover frequency ranges up to 300 Gc/s. These techniques can possibly be applied to the longer submillimetre wave-lengths but at some point it becomes necessary to concede that waveguide components cannot be indefinitely scaled down in size. Fabrication techniques include use of drawn waveguide, copper-gold eutectic bonding, precision milling, spark-erosion and especially electroforming. Dimensional tolerances are tight—on stainless-steel usually used in the electroforming processes they are normally 0·0001 or 0·0002 in. with surface finishes of 5 micro-inches or better.

Slotted-line standing-wave indicators have been produced for frequencies up to 300 Gc/s. In certain models in waveguides WG26, WG29 and WG32 sizes the construction is quite conventional; the main body is split down the centre of the broad walls of the main guide. The two halves are milled and burnished, then small parts of the bottom walls are removed so that after the parts have been gold-plated and assembled a slot 0·01 in. wide and 0·015 in. deep is left. The probe was originally turned to 0·003 in. diameter but more recently a Wollaston wire has been used, which after soldering to the top of an adjustable depth pin, is etched to leave a 0·001 in. platinum-wire probe. This has improved the performance considerably and would allow the slot width to be decreased. The probe couples out into an inserted electroformed cross-guide and so into a separate diode detector mount. The probe carriage moves on two precision-ground steel bars which fit into jig-bored holes in the main body.

Reflection coefficient can be measured, instead of the standing-wave ratio, with a form of bridge circuit using a variable impedance. The bridge (a hybrid Tee) is balanced by making the reflection coefficient of the variable impedance equal to that of the unknown impedance. The principle of the variable impedance is illustrated in Fig. 6. At flange 1 the waveguide has a rectangular cross-section; via a transition 2 this becomes a circular guide 3 which contains two absorbing vanes 4 and 5 of thin metallized slivers of mica. Vane 4 is fixed perpendicular to the incident electric field E; vane 5 is mounted on a plunger 6 which can be both rotated and displaced axially. If the planes of 4 and 5 subtend an angle $\theta$, the component $E \sin\theta$ parallel to vane 5 is absorbed by this vane; the component $E \cos\theta$, perpendicular to 5, is reflected by plunger 6. Now $E \cos\theta$ has a component $E \cos\theta \sin\theta$, which is absorbed in vane 4, and a component $E \sin^2\theta$ which emerges from 1 as a reflected wave. Thus, the modulus of the reflection coefficient is equal to $\cos^2\theta$ and is varied by turning vane 5. The reflection-coefficient argument is varied by moving the plunger axially.

In a 2 mm-wave tuner of novel design the stub does not slide along the waveguide, as in the normal sliding-screw type, but turns about an axis 3 (Fig. 7) so that it describes a pivotting motion. In the millimetre-wave region this design has the advantages of no vertical movement due to surface irregularities, less friction,

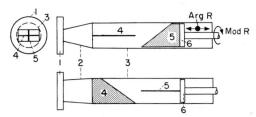

Fig. 6. *Axial sections of the variable impedance along two mutually perpendicular planes, and end view. 1 flange with rectangular opening; 2 transition from rectangular to circular cross-section; 3 circular waveguide; 4 fixed vane; 5 vane capable of being rotated and axially displaced, in 3 and fixed to plunger; 6. When 5 is rotated, the modulus of the complex reflection coefficient is changed; when 5 is displaced axially, the argument is changed.*
(*By courtesy of Philips Electrical Ltd.*)

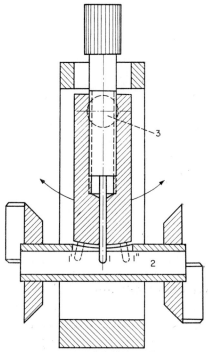

Fig. 7. *Schematic representation of pivotting screw tuner. Stub 1 can be screwed in and out, but instead of sliding lengthwise it pivots about spindle 3. The extreme positions are 1' and 1''.*
(*By courtesy of Philips Electrical Ltd.*)

13*

smoother adjustment and short axial length (therefore small loss—less than 0·1 dB).

One rotary attenuator in waveguide WG29 uses 0·002 in. thick metallized mica vanes, the central movable vane having a maximum attenuation appreciably greater than 40 dB. The insertion loss is 1·4 dB. For less accurate measurements guillotine attenuators are used in which a differential screw motion moves a 0·002 in. thick metallized mica vane vertically through a 0·005 in. slot in the waveguide. Variable short circuits are frequently of the non-contacting type with a cylindrical, choked piston moving in precision electroformed waveguide. At 140 Gc/s voltage reflection coefficients of 0·92–0·94 are normally obtained. The designs of rotary directional couplers, ferrite isolators, hybrid Tees, power measuring equipment using dry calorimeters or Wollaston-wire bolometers and many other devices have proceeded well into the millimetre-wave region. A range of components for the 3 mm band has been developed in low-loss $H_{01}$ circular waveguide.

A number of optical-type components for millimetre wave-lengths have been constructed. These include the Michelson and Fabry-Perot interferometers and the double-prism attenuator and directional coupler. In addition proposals for an absorption wavemeter, standing-wave indicator and matching unit have been made based on the double prism.

Crystal diodes, extremely thin (10–20 microinches) Wollaston-wire bolometers and, in some cases, Golay cells are generally used for the detection of millimetre waves. Photoconductive detectors have enjoyed considerable success at infra-red wave-lengths and have recently been demonstrated, using indium antimonide, at wave-lengths as long as a few millimetres.

*See also: Golay cell.*

*Bibliography*

COLEMAN P. D. (1963) *State of the Art: Background and Recent Developments—Millimetre and Submillimetre Waves, I.E.E.E. Transactions on Microwave Theory and Techniques,* **MTT-11** 271.
CULSHAW W. (1961) *Millimetre Wave Techniques, Advances in Electronics and Electron Physics.*
HARVEY A. F. (1959) *Optical Techniques at Microwave Frequencies, Proc. I.E.E.* **106,** Part B 141.
F. A. BENSON

## MONTE CARLO METHOD.

### 1. General Description of the Method

The Monte Carlo method is a method of solving numerically problems arising in mathematics, physics and other sciences, by constructing for each problem a random process whose parameters are equal to the required quantities, making observations on the random process, and estimating from these observations the values of the parameters of the process. More strictly, the Monte Carlo method is usually understood to involve the construction of an artificial random process on which observations can be made by ordinary computational means (for example, pencil, paper, numerical tables, and calculating machines), rather than mere observation of a physical system which possesses a random element.

There are two main types of problem which can be handled by the Monte Carlo method, the probabilistic and the deterministic. In the former, the actual random processes of the problem are simulated by means of suitably chosen random numbers, and from a sufficiently large number of observations using these random numbers conclusions are drawn regarding the the behaviour and properties of the original processes. Problems of this type to which the method has been successfully applied include investigations in neutron physics, in questions of queueing and congestion, and in the behaviour of epidemics. For such problems the method has two main advantages: (i) it may give at least the main properties of the solution in cases, which often arise in practice, which are too complicated to allow of analytical treatment, (ii) the experimental investigation of the artificial random process is often much more economical in time and in money than would be a similar examination of the actual physical process, and it often permits investigation at values of parameters which in practice would be wasteful or dangerous. Large-scale electronic computers have been very useful for this kind of work: realistic simulation of actual systems often involves such a degree of complication that many operations are required to obtain a single observation, and only these high-speed machines can accumulate a sufficiently large number of observations in a reasonable time.

The second type of problem is usually concerned with a more purely mathematical process: examples are the solution of algebraic or differential equations, the evaluation of definite integrals, and the inversion of matrices. Here we have a deterministic problem which we can pose analytically but, perhaps, cannot solve: we seek a probabilistic problem which has the same analytic expression as the original one, and from a Monte Carlo examination of the probabilistic problem we obtain the solution of the original one. The most obvious example is the boundary value problem for Laplace's equation: since this equation occurs in the study of the random diffusion of a particle in a region bounded by absorbing barriers, we can simulate this diffusion process by means of a random walk in the appropriate region, examine the properties of the random walk process by the Monte Carlo method, and apply the results to the original problem. It is to be observed that this is the reverse of the situation which obtained after the relation between boundary value problems and random processes was first discovered. At that time solutions of boundary value problems given by the theory of differential equations were used to shed light on the corresponding random processes.

On the whole, Monte Carlo methods have been more successful with the first than with the second type of problem. Perhaps this was to be expected, since for many problems of the first type no alternative to the Monte Carlo solution is available, whereas for many of the mathematical processes of the second type a number of satisfactory numerical techniques exists. With some cases of the second type, however (for example, the evaluation of certain multiple integrals) the Monte Carlo method is better than the available alternatives. As with other numerical techniques, the method should be applied to a given problem in such a way as to minimize the error, and an estimate of the error in the solution should be made. Within the twenty years or so of its existence the method has at times been compared unfavourably with the alternatives. This is perhaps because it was applied to unsuitable problems and because insufficient attention was given to methods of minimizing the errors. It will be seen below that in some cases it is possible by rearranging the problem to reduce considerably the errors attaching to Monte Carlo estimates. In recent years much more attention has been given to these techniques, and it is now recognized that the Monte Carlo method has considerable actual and potential value, particularly with complicated problems which defy analytical treatment.

In what follows a brief account is given of the way in which the method can be applied to some problems of the types mentioned above. Since random numbers are an essential ingredient of the method, the reader is referred to the article "Random and Pseudo-random Numbers" for a description of methods of generating and testing them and of arranging that they reproduce the properties of various probability distributions. Techniques for reducing errors are more easily illustrated with reference to the deterministic than the probabilistic type of problem, so that they are treated in that order, in sections 2 and 3 respectively.

## 2. Application of the Monte Carlo Method to Deterministic Problems

### 2.1. Evaluation of definite integrals.

Suppose that it is required to evaluate the integral $I = \int_a^b f(x)\,dx$, $f(x)$ being an integrand which is bounded in $(a, b)$. Without loss of generality we can take it that $0 \leq f(x) \leq 1$. A simple method of estimating $I$ using random sampling numbers is as follows.

Let $\xi$ be a random number which is uniformly distributed on $(a, b)$: then the mean or expected value of $f(\xi)$ is $\int_a^b f(\xi)\,d\xi/(b-a) = I/(b-a)$. If now we draw $N$ values $\xi_i$ of $\xi$ $(i = 1, 2, ..., N)$, the average of the resulting values $f(\xi_i)$ will approximate to the expected value of $f(\xi)$ if $N$ is large: that is, $\sum_{i=1}^{N} f(\xi_i)/N \doteq I/(b-a)$. It follows that the quantity $\theta_1 = (b-a) \sum_{i=1}^{N} f(\xi_i)/N$ is an estimate of $I$.

As an example of a modification of this method which may give greater accuracy for the same number of observations, consider how $I$ might be estimated using *stratified sampling*. Here we divide the interval $(a, b)$ into $m$ sub-intervals $(a_k, b_k)$ with lengths $l_k$ $(k = 1, 2, ..., m)$, so that $a = a_1 < b_1 = a_2 < b_2 = a_3 < \ldots < b_{m-1} = a_m < b_m = b; l_1 + l_2 + l_3 + \cdots + l_m = b - a$. We then estimate the value of the integral $I_k = \int_{a_k}^{b_k} f(x)\,dx$ by the above method, using $N_k$ values of a quantity $\xi^{(k)}$ which is uniformly distributed in $(a_k, b_k)$, and since $I$ is the sum of these $m$ integrals $I_k$, we get as a second estimate of $I$ the quantity $\theta_2 = \sum_{k=1}^{m} (l_k/N_k) \sum_{i=1}^{N_k} f(\xi_i^{(k)})$. It is clear that the expected value of $\theta_2$, as of $\theta_1$, is $I$.

To compare the accuracies of $\theta_1$ and $\theta_2$ we may evaluate their variances. Now $\theta_1$ is the average of $N$ quantities $(b-a) f(\xi_i)$, each of which has variance $(b-a) \int_a^b f^2(x)\,dx - I^2$, so that the variance of $\theta_1$ is this quantity divided by $N$. Similarly $\theta_2$ is the sum of $m$ numbers, the $k$th one being the average of $N_k$ quantities $l_k f(\xi_i^{(k)})$, so that the variance of $\theta_2$ is

$$\theta_2 = \sum_{k=1}^{m} \left( l_k \int_{a_k}^{b_k} f^2(x)\,dx - I_k^2 \right) \Big/ N_k.$$

We can arrange that the total number of observations is the same in the two methods, that is that $\sum_{k=1}^{m} N_k = N$, but the variance of $\theta_2$ will depend on the way in which the $N$ observations are divided among the $m$ sub-intervals. One possible procedure is to take the $N_k$ to be proportional to the $l_k$, that is $N_k = N l_k/(b-a)$. In that case var $\theta_2 = (b-a)$ $\left( \int_a^b f^2(x)\,dx - \sum_{k=1}^{m} (I_k^2/l_k) \right) \Big/ N$. From the inequality

$$0 \leq \sum_{k=1}^{m} l_k((I_k/l_k) - I/(b-a))^2 = \sum_{k=1}^{m} (I_k^2/l_k) - I^2/(b-a)$$

it follows that var $\theta_2 \leq$ var $\theta_1$.

These expressions illustrate the fact that a Monte Carlo estimate based on $N$ observations will have variance $\sigma^2/N$, $\sigma^2$ being the variance of a single observation and being characteristic of the particular method employed. The error will thus be of order $N^{-1/2}$. Any modification which reduces $\sigma^2$ will increase the accuracy.

The variance $\sigma^2$ will usually be estimated from the observations themselves by means of the standard statistical formula $\left[ \sum_{i=1}^{N} x_i^2 - \left( \sum_{i=1}^{N} x_i \right)^2 \Big/ N \right] \Big/ (N-1)$, $x_i$ being the $i$th observation.

If we have two equally accurate Monte Carlo estimates of the same quantity, the variance of a single observation being $\sigma_1^2$ in method 1 and the number of observations $N_1$, with corresponding meanings for $\sigma_2^2$ and $N_2$, we have $\sigma_1^2/N_1 = \sigma_2^2/N_2$. To assess the efficiency of method 2 relative to method 1 it is reasonable to use the ratio of the times the methods require to achieve equal accuracies. If single observations require $n_1$ and $n_2$ units of time respectively in the two methods, this ratio is $n_1 N_1/n_2 N_2 = (n_1/n_2)$ $(\sigma_1^2/\sigma_2^2)$, that is, the product of a *labour ratio* and a *variance ratio*. The first of these depends on the details of the methods and on the computing facilities available; the second is usually fairly easy to assess by the methods given above.

In the example just given it is reasonable to suppose that $n_1/n_2 \doteqdot 0\cdot7$. From actual arithmetical examples, however, even with only a few strata, it will be found that the reduction in variance achieved with method 2 far outweighs this, so that the stratification results in a considerable overall gain in efficiency.

This example illustrates how a change in the method of approach to the problem can increase the efficiency. Many methods for reducing the variance have been devised and are discussed in the books named below. One such method, for example, writes $f(x)$ in the form $\varphi(x) + (f(x) - \varphi(x))$, where $\varphi(x)$ is a function whose integral over $(a, b)$ can be evaluated analytically and the difference $f(x) - \varphi(x)$ is numerically small over $(a, b)$; the integral of this difference over $(a, b)$ is then evaluated by a Monte Carlo method, with high accuracy because of the small variability of the integrand.

Similar methods can be used in the evaluation of multiple integrals. The order of convergence of the estimates, $N^{-1/2}$, is not high, but has the advantage that it is independent of the order of the integral. On the whole, the Monte Carlo method has higher accuracy than the use of quadrature formulae when the order exceeds about four. It is to be noted that if the order of the integral can be reduced by analytic integration with respect to one or more of the variables, this will result in an increase in the accuracy of the Monte Carlo estimate of the integral.

### 2.2. Solution of systems of linear algebraic equations.

A number of methods have been proposed for solving linear equations by simulating some random process. The following is an outline of one such method.

Let the system of equations to be solved be written in matrix form $\mathbf{Ax} = \mathbf{b}$, $\mathbf{A}$ being an $n \times n$ matrix and $\mathbf{x}$ and $\mathbf{b}$ $n$-dimensional column vectors. If $\mathbf{A}$ can be written in the form $\mathbf{I} - \mathbf{B}$, $\mathbf{I}$ being the unit matrix and $\mathbf{B}$ a matrix whose latent roots have moduli less than unity, we can write the solution of the given equations in the form $\mathbf{x} = \mathbf{A}^{-1}\mathbf{b} = \mathbf{b} + \mathbf{Bb} + \mathbf{B}^2\mathbf{b} + \ldots$ . The $m$th component of the vector $\mathbf{x}$ is then given by

$$x_m = b_m + \sum_{i=1}^{n} B_{mi} b_i + \sum_{i=1}^{n} \sum_{j=1}^{n} B_{mi} B_{ij} b_j + \ldots, \; B_{ij}$$

being the element in the $i$th row and $j$th column of $\mathbf{B}$ and $b_i$ being the $i$th component of $\mathbf{b}$.

The Monte Carlo method uses a sequence of uniformly distributed random numbers to construct a random number $y_m$ whose mathematical expectation is equal to the expression for $x_m$. Details of this

process may be found in the books referred to in the bibliography. The arithmetic mean $\bar{y}_m$ of $N$ such numbers $y_m$ is taken as an approximation to the required $x_m$, and the accuracy of $\bar{y}_m$ is estimated as indicated above.

The advantage of this particular method is that a single component $x_m$ of the unknown vector $\mathbf{x}$ can be determined without any of the other components being known. The number of operations required is thus proportional merely to the number of equations, not to a higher power of this number as with standard numerical methods. This again illustrates the value of Monte Carlo methods in multidimensional problems.

### 3. Application of the Monte Carlo Method to Probabilistic Problems

*3.1. Neutron diffusion.* Historically, the first use of the Monte Carlo method as a genuine research tool was in connexion with problems in neutron diffusion encountered in work on the atomic bomb. When a neutron enters a fissile medium, the distance it travels before it collides with an atom is a random variable, and when a collision takes place there are several possibilities, absorption, scattering and fission, each of which has a characteristic probability which is determined from the physics of the situation and may be taken to be known. The Monte Carlo method consists of a direct simulation of the physical process. If the atoms of the medium are randomly distributed, the free path of a neutron before collision will have a negative exponential probability distribution. Assuming that a neutron enters the medium, we generate a random number to give the length of its path before collision, and we use other random numbers to determine completely the result of the collision, for example whether it is fission, and if so what are the energies and directions of the emergent neutrons. In this way it is possible, by following the histories of large numbers of neutrons, to find the distributions of the numbers, energies and directions of the neutrons in specified regions of the medium. Similar considerations will apply if the incident particle is a photon. The method has been successfully applied to investigations on the effectiveness of shields and on the criticality of nuclear reactors.

To increase the efficiency of the process many refinements of the technique have been devised. Thus "weighting" methods are available to avoid losing paths through absorption, or to increase or decrease, without introducing bias, the number of particles being followed if it becomes inconveniently small or large. Similarly, methods are available for examining the distribution of particles penetrating shields of given thickness which simultaneously give the same information for shields of smaller thickness.

*3.2. Congestion and queueing problems.* A type of probabilistic problem which has attracted a great deal of attention in recent years concerns *congestion*. Here a system which is working within its capacity may, owing to the operation of chance, appear temporarily to be overloaded, and the extent to which it is overloaded, and the length of the period of overloading, may be important practical questions. The chance fluctuations here are not in the nature of small perturbations on a deterministic pattern, but are essential to the problem and may have large effects. The most obvious example is the queue or waiting line. Here some kind of service is provided and customers or demands arrive to receive this service. The service facility in general has limited capacity, and customers arriving while it is engaged have to wait, usually in a queue in order of arrival. Either the times taken to provide service for individual customers, or the intervals between arrivals of successive customers, or both, are random variables. Questions of interest in queueing problems include the probability distributions, or mean values, of length of queue, of waiting time of customers from arrival to entry into service, and of the length of the "busy period" or time-interval between successive periods at which the service facility is free.

The number and variety of fields in which these problems occur is considerable. They appear in such things as telephone service, road traffic, and industrial practice, and solution of the problems can have important practical consequences. Within the last twenty years or so there has grown a large body of mathematical theory on the subject, and although this necessarily starts with simplifying assumptions it has achieved results of some generality. But it is obvious that all sorts of complications can arise in these questions, and that the Monte Carlo method is very suitable for dealing with them. Here the method is used as a direct simulation of the process. The histories of individual customers are followed, random numbers being used to represent their arrival and service times and any other features of the system which are probabilistic. Monte Carlo methods in fact form an extremely valuable tool in investigating these and similar problems.

In this article it has been possible to give only the barest outline of the nature, techniques and applications of the Monte Carlo method. For fuller accounts the reader should consult the following books. The first and third give a comprehensive description of the subject as it stands at present.

*Bibliography*

HAMMERSLEY J. M. and HANDSCOMB D. C. (1964) *Monte Carlo Methods*, London: Methuen; New York: Wiley.

MEYER H. A. (Ed.) (1956) *Symposium on Monte Carlo Methods*, New York: Wiley.

SHREIDER Y. A. (Ed.) (1964) *Method of Statistical Testing (Monte Carlo Method)*, Amsterdam: Elsevier.

A. J. HOWIE

**MUSICAL INSTRUMENTS, PHYSICS OF.** The essential difference between a musical sound and a noise lies, as far as physics is concerned, in the regularity of its wave form. Pure white noise—whose wave form is a random series of pulses and whose sound resembles sustained applause—can be imagined to continue indefinitely at the same level and yet it has no musical quality at all. On the other hand, a pure tone—whose wave form is sinusoidal and whose sound is like a very poor flute—has a specific musical pitch. All forms of musical sound lie somewhere between these two extremes, and have some degree of regularity. The subjective quality we call pitch—the highness or lowness of a note—depends chiefly on the frequency, though for pure tones a large change in intensity sometimes gives the sensation of a change of pitch. If we wish to build a musical instrument therefore it must produce waves in the air with some degree of regularity, and the frequency of their repetition must be variable if tunes are to be played.

The necessary regularity can be achieved in various ways. Rotation is one possible source; the siren, for example, produces musical tones by the rotation of a perforated disk in front of an aperture which chops the air stream into "puffs" of a definite repetition frequency. Recently photoelectric sirens have been produced in which opaque patterns on a transparent rotating disk interrupt a light beam and produce periodic variations in the output of a photocell on which the light subsequently falls. Rotation is also used in some electronic organs; in some the rotating element is a soft iron disk with a number of teeth which move near to a magnet surrounded by a coil, in others a plate carrying engraved wave forms spinning close to another plate causes periodic variations in capacity between the two elements. From both these systems periodically-varying currents may be derived and can be amplified and fed to a loudspeaker to produce musical sounds. In order to play tunes the speed of roation can be varied though inertia makes this method unwieldy; a more usual method is to provide a separate source for each sound required and to select by switches or mixers of some kind.

Vibration is a more obvious source of musical sound and vibrating strings, plates, and air columns form the main sources in conventional musical instruments. Again the pitch can be changed to play tunes by providing one vibrator per note; this method is used for example in the piano and in the organ, and strictly speaking they are collections of instruments, each with its own playing mechanism brought into action by depressing a key. Alternatively, the shape or state of strain of the vibrator can be altered to change the pitch; this makes the number of vibrators necessary to cover the required range much smaller. In the string family, for example, the vibrating length of the string is changed by altering the position of the fingers of the left hand, and in the woodwind the shape of the air cavity is changed by opening and closing holes in the pipe. In the strings the tension is altered for tuning purposes before performance. A third method of pitch-changing involves changing the mode of vibration. Most vibrating systems are capable of vibrating in different modes according to the way in which they are clamped and stimulated. The classical demonstration of this phenomenon is the Chladni plate. The modes for linear vibrators such as strings and pipes tend to have simple numerical ratios; thus for a string clamped at each end, or for a cylindrical pipe open at both ends, the sequence has the frequency ratios $1 : 2 : 3 : 4 : 5$, etc. For a conical pipe open at the wide end and either closed or open at the narrow end the same sequence arises, but for a cylindrical pipe closed at one end the sequence is $1 : 3 : 5 : 7$, etc. When modes have such simple numerical ratios they are said to be harmonic modes. The bugle uses such a sequence, and in the other brass instruments the same sequence forms the basis of pitch-changing but keys are also supplied to alter the effective length of the pipe, and hence to enable gaps to be filled. The sequence of harmonics for a string clamped at each end would correspond to the following notes if the fundamental were "C":

| 1 | 2 | 3 | 4 | 5 | 6 | 7 | 8 | 9 | 10 |
|---|---|---|---|---|---|---|---|---|----|
| C | C' | G' | C'' | E'' | G'' | ? | C''' | ? | E''' |

The seventh and ninth harmonics are notes which are not found in normal Western musical scales. The harmonic sequence is not exact for most practical vibrators because of end conditions; for example, the amount of air beyond the open end of a pipe which takes part differs for different modes, and hence the pitches are not in strict harmonic ratios. For strings and pipes which are very long compared with their diameter, however, the approximation is surprisingly good.

The sounds produced by simple vibrators or rotating devices are very thin and dull, and comparison with conventional instruments reveals large differences in tone. Comparison of oscillograph traces also shows big differences. It can be seen, for example, that whereas a pure tone of frequency 440 cycles has a pure sinusoidal wave form, that for an oboe, flute or violin playing a note of the same pitch is much more complex. It does, however, repeat at the same basic frequency and a close examination suggests that higher frequency components are probably present as well. It has been pointed out already that vibrating systems can operate in different modes, and the possibility of vibration in several modes at once is worth investigating. If mixtures of modes are heard together does the result sound more like a conventional instrument and does its wave trace acquire the necessary extra complexity? Investigation shows that it does, and in fact it is relatively easy to produce a single steady note by electronic means which has the same wave shape and which sounds very like a real instrument. Unfortunately, however, when tunes are played on such electronic devices the result is easily

distinguishable from that of a real instrument. Where does the difference lie?

Most natural vibrating systems produce very weak sounds on their own (consider, for example, a sound produced by a tuning-fork held in the hand). An amplifier of some kind forms an integral part of most instruments. The most usual form is some kind of resonator which may be either sharply tuned or may cover a broad band. A tuning-fork placed on a table or blackboard is an example of a system using a broad-band amplifier. In instruments which have a separate generator for each note, each may have its own amplifier (for example, the reed pipes in an organ) and then the proportions of harmonics for each can be made the same; each amplifier can be sharply tuned to suit the individual note. Alternatively, many primary vibrators may be attached to one amplifier, as for example the many strings of a harp or piano, and then the characteristics of the amplifier may have different effects at different frequencies; for this arrangement, and for instruments which involve only one or two vibrators on a common amplifier, the characteristics of the amplifier may predominate in the sound produced. The result is the so-called "*formant*" characteristic. Let us suppose, for example, that an instrument has a very simple formant characteristic with a single broad peak, say in the region of 1000 c.p.s. When a note of fundamental 300 c.p.s. is played the third harmonic will tend to predominate; when a note of fundamental 500 c.p.s. is played it will be the second harmonic which predominates and for a 1000 c.p.s. fundamental the fundamental itself will predominate, thus the harmonic content, and hence the quality of sound, varies at different pitches. The clarinet provides a good example of this phenomenon and has three distinct ranges, the low or chalumeau register, the middle or clarion register, and the upper or high register. Formants also play a part in differentiating between other instruments of the same type, e.g. between a number of different violins. Many other factors also contribute to the formant; for example in violins the wood resonance as well as the air resonance plays a part, and in the woodwind the radiation characteristics of the open holes, from which most of the sound emerges, are important.

A further complication arises from the need for amplification; the two components—primary vibrator and amplifier—form a coupled system and assume the characteristics of a forced vibrating system. The solution of the appropriate differential equation includes a steady-state term and a transient term; this transient term—the so-called *starting transient*—is of great importance in distinguishing between instruments and between different examples of the same instrument. It is even involved in distinguishing between the tone produced by different players on the same instrument. It has been described as an argument between the primary vibrator and the amplifier which is ultimately settled by compromise and gives way to the steady state. Its importance can be demonstrated in many ways; for example, if a tape-recording is made, the transient part of a series of notes can be artificially wiped out; the quality becomes remarkably changed. An alternative is to reverse the tape so that the steady state is heard before the transient. A particularly effective demonstration of this sort can be performed with the piano. All audiences agree that the reversed piano-recording sounds like an organ. The organ, of course, has a relatively slow build-up and a relatively slow decay. A normal piano has a violent percussive transient followed by an exponential decay; when reversed, therefore, the piano has an exponential build-up not unlike that of the organ, but terminates very suddenly. The interesting fact that all audiences associate a reversed piano-recording with organ tone demonstrates the extreme importance of the initial part of each note in recognition compared with the middle and end.

Many other factors are involved in determining the complex tone of musical instruments, such as vibrato, noise, combination tones, etc., but space does not permit a full discussion of them. Perhaps it could be mentioned in passing that in the analysis of the tone of an instrument such as the bassoon very little of the energy is in the fundamental component; the formant characteristic leads to peaks in the region of the third, fourth and fifth harmonics. It seems likely that the "missing" fundamental is supplied at least partially by difference tones generated by the strong higher harmonics; it may well be, however, that psycho-physiological phenomena are also involved. We shall conclude with a brief description of the physical characteristics of some actual instruments.

In stringed instruments the primary vibrator is a string clamped at both ends and therefore giving a full series of harmonics. The body of the instrument provides broad-band amplification. Its response curve is far from flat and this accounts for the special characteristics of individual instruments of the same type. The vibration is initiated by bowing which involves "stick-slip" phenomena; the friction between the bow and the string depends on their relative velocities and is greatest when the two are not in relative motion, thus a continuous movement of the bow across the string provides an energy feed with a periodicity which depends on the natural period of the string, and hence can lead to large-scale vibrations.

In the woodwinds the air column is the primary vibrator and the initiation is either by edge tones (for the flute and recorder families) or by a reed (in the oboe and clarinet families). Transients arise partly as a result of the initiating mechanism itself and partly as a result of interactions between the initiator and the pipe. There are three principal influences on the tone. The first is the harmonic content; the flute is effectively a cylindrical pipe open at each end and has a full series of harmonics, the clarinet is cylindrical and effectively closed at one end and has only odd harmonics, and the oboe is conical and hence has a full series of harmonics. The second is the formant characteristic which depends very largely on the radiation characteristics of the finger and key holes, except for the very lowest notes in which it depends on the shape of the bell. The third is the starting transient which is particularly characteristic in the case of the reed instruments.

In the brass instruments the initiating mechanism is provided by the lips of the player, and the pipe is partly cylindrical and partly conical with a large flare or bell at the end. Change of pitch is largely by change of mode and, since there are no holes in the pipe, the bell plays a very large part in determining the tone. The small bell of the trumpet, for example, radiates best at high frequencies and gives the characteristic brilliance to the tone. The much larger bell of the French horn radiates effectively at lower frequencies and gives rise to the mellowness.

In the organ both reed and edge-tone initiation find a place; each pipe, however, can have its formant and transient characteristics adjusted by the organ-builder and, since each has only to produce one note, no compromises are necessary. The transients vary from stop to stop, and have considerable influence on the final tonal quality. In electronic organs a reson-ably acceptable organ quality can only be achieved if attention is paid to the simulation of the transients as well as to the correct harmonic mixture.

Recently a new group of instruments—Les Structures Sonores, Lasry-Baschet—has been introduced and is perhaps the first new application of purely mechanical vibrators and amplifiers for some consider-able time. These instruments involve transverse vibrations of metal bars in which vibrations can be initiated through the longitudinal vibrations of glass rods attached to them. Vibrations in the glass rods are in turn induced by stroking with wet fingers. In other examples in this group the metal bars are made to vibrate by striking. The amplification is provided by various unconventional but effective devices such as air-cushions and horns of many curious shapes. Finally, mention should be made of the use of electronic computers to produce musical sounds to order by calculating and then producing the precise required wave form.

*Bibliography*

BARTHOLOMEW W. T. (1942) *Acoustics of Music*, New York: Prentice Hall.
BENADE A. H. (1960) *Horns, Strings and Harmony*, New York: Doubleday.
BOUASSE H. (1929) *Instruments a Vent*, Paris: Librarie delagravie.
SEASHORE C. E. (1938) *Psychology of Music*, New York: McGraw-Hill.
TAYLOR C. A. (1965) *Physics of Musical Sounds*, London: E.U.P.
WINCKEL F. (1960) *Phänomene des Musikalischen Hörens*, Berlin: Max Hesses.
WOOD A. (1962) (6th Edn) *The Physics of Music*, London: Methuen.
                C. A. TAYLOR

# N

**NEUTRINO, RECENT WORK ON.** Recently, the emphasis in experimental neutrino physics has been on the use of high energy proton synchrotrons as sources of neutrinos and antineutrinos. The accelerated protons are deflected on to a target to produce π-meson beams, and sufficient shielding is used so that only neutrinos arising from the decay of the π-mesons pass through to detectors. Using such an arrangement with the Brookhaven 30 GeV synchrotron, Danby, Gaillard, Goulianos, Lederman, Mistry, Schwartz and Steinberger were able in 1962 to demonstrate the existence of two different types of neutrino. In a 10-ton aluminium spark chamber, the neutrino interactions produced only μ-mesons, indicating a fundamental difference between neutrinos emitted in company with μ-mesons and those emitted with electrons. A subsequent experiment with the CERN 25 GeV synchrotron has confirmed this finding. In addition, the latter experiment showed that leptons were conserved in neutrino reactions, and that the neutrino accompanying the μ-meson decay of the K-meson was identical with that in the decay of the π-meson. These neutrino beams will make possible the study of weak interactions at high energies, and in this context the Geneva experiment has already indicated the existence of the so-called "*intermediate boson*".

There continues to be no experimental evidence of electromagnetic interaction for either type of neutrino. The current best estimate for the *mass of the electronic neutrino*, obtained from the end-point of the ³H ß-spectrum, sets upper limits of 700 eV or 200 eV depending on the form of Fermi couplings assumed in the interaction, and the experiments are consistent with a value of zero. From energy-momentum balance in the π-meson decay, the upper limit to the *mass of the μ-mesonic neutrino* is 3 MeV, but the uncertainty here may be due to the form in which the neutrino mass emerges from the kinematic equations.

The high interaction mean free path of the neutrino has given interest to the measurement of neutrino fluxes in cosmic radiation, and the field of experimental neutrino astrophysics appears likely to yield much information of astronomical and cosmological value.

*Bibliography*

BERNSTEIN J. *et al.* (1963) *Phys. Rev.* **132**, 1227.
DANBY G. *et al.* (1962) *Phys. Rev. Letters* **9**, No. 1, 36.
PONTECORVO B. (1963) *Soviet Physics—Uspekhi* **6**, No. 1, 1.

J. E. GORE

**NEUTRON GENERATOR.** Neutrons can be produced by various neutron-emitting nuclear reactions. These reactions occur when high energy projectiles strike certain target materials. Various requirements such as desired neutron flux, neutron energy, field uniformity, flux stability, desired mode of neutron production (continuous or pulsed) will determine the most suitable reaction and consequently the most suitable method and apparatus for neutron production. In a neutron source a target material is bombarded by projectiles or γ-rays emitted from a radioisotope. In a neutron generator, however, the target material is bombarded by either projectiles which are produced in

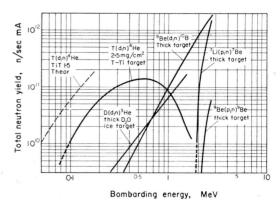

*Fig. 1. Neutron yield from various neutron emitting reactions.*

an ion source and accelerated to high energy in a section of the generator called the accelerator, or by X rays generated by such projectiles. The table gives the most useful nuclear reactions used in accelerating machines for the production of neutrons, and the diagram in Fig. 1 shows the neutron yield as a function of the bombarding energy of the positive ion particles for the first four reactions listed in the table.

As can be seen from Fig. 1 the reaction T³ (d, n) ⁴He is at present the best neutron producing reaction at low ion bombarding energies. This is often refered to as d-t reaction because deuterium and tritium are the interacting particles, one of them usually present in the accelerated ion beam and the other in the target. However, even a mixture of both can be used in the

| Reaction | Projectiles | Bombarding energy | Neutron energy |
|---|---|---|---|
| $^2$H (d, n) $^3$He | deuterons | <100 keV | 3·3 MeV |
| $^3$T (d, n) $^4$He | deuterons | <100 keV | 14·3 MeV |
| $^9$Be (d, n) $^{10}$B | deuterons | <1·5 MeV | 4·4 MeV |
| $^7$Li (p, n) $^7$B | protons | <1·88 MeV | 30–500 KeV |
| $^9$Be ($\gamma$, n) $^8$Be | electrons (producing X rays) | <1·66 MeV | 1–2 MeV |
| $^2$H ($\gamma$, n) $^1$H | electrons (producing X rays) | <2·2 MeV | 1–2 MeV |
| U ($\gamma$, n) U | electrons (producing X rays) | <9 MeV | 1–2 MeV |

ion beam and at the target, in which case the resulting neutron yield will be somewhat lower than in non-mixed systems. A tritium target consists of tritium gas absorbed in a thin layer of titanium or zirconium evaporated *in vacuo* on a copper or stainless steel foil. Such targets with accelerators of comparatively simple and compact design can give a neutron output of about $10^{11}$ neutron/s. During the last ten years a number of papers have appeared in the literature describing the construction and performance of simple low voltage positive ion accelerators intended for the production of neutrons. With rising interest in high intensity neutron sources for industrial or medical application a number of firms have produced portable, low voltage neutron generators which are commercially available at a moderate cost. Two main types of neutron generator are available: the sealed neutron generator tube and the demountable portable accelerating machine. The first type was developed with the intention of reducing the size of the accelerator tube to the minimum and avoiding the conventional vacuum pumps which require much space and power. To indicate its compactness and simplicity the sealed tube is often referred to as neutron source rather than a neutron generator but in fact they are identical in principle and most of their component parts are similar. However, their construction differs considerably. In the first type the construction is as used in the design of various types of sealed-off devices, e.g. X-ray tubes, while an accelerating machine consists of several metal and glass components assembled together in a vacuum tight manner so that the entire assembly can be continuously pumped by one or several vacuum pumps.

Sealed neutron generator tubes have been described in the literature by several authors. A cross-section of a typical tube is shown Fig. 2. In such a tube deuterium gas is usually stored in a replenisher which may consist fo an indirectly heated nickel sponge impregnated with titanium deuteride or a simple tungsten filament coated with zirconium deuteride. If heated, the deuteride will release deuterium and when cooled off it will absorb it rapidly. Pressure in the tube may be measured and controlled by a pressure sensing element such as the Pirani gauge. The operating pressure range is usually between 1 and 20 mTorr. Deuterium is ionized in the ion source. In sealed tubes a Penning type ion source is most frequently used with an axial magnetic field obtained by an electromagnet or a permanent magnet. This source produces mostly molecular deuterium ions which give considerably lower neutron yield than atomic deuterium ions.

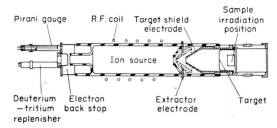

*Fig. 3. Sealed neutron generator tube with an r.f. ion source.*

Other ion sources too, such as r.f. type ion sources, may be used with sealed tubes as shown in Fig. 3. The ions produced in the ion source are accelerated by the potential difference across the gap between the ion source and the target plate. A potential of about 70–100 kV obtained from an H.T. unit is used for this purpose. As mentioned above neutrons will be emitted at the target under ion bombardment. The very short range of ion penetration confines considerable impact heat to a thin surface layer of the target. This is very detrimental to the tritiated titanium targets because absorbed tritium will be desorbed if the temperature rises much over 200°C. Efficient cooling of the target is therefore very important and is usually effected by water or oil. Also, due to various other reasons, such as dilution and displacement of tritium in the target by deuterium from the beam as well as sputtering of the titanium layer, the life of the target is limited. Since it is a complicated procedure to replace the target in a sealed tube, the method of

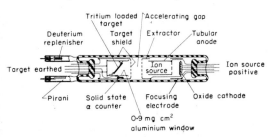

*Fig. 2. Sealed neutron generator tube with Penning-type ion source.*

using a mixed beam, i.e. a beam composed of deuterium and tritium gas, has been found to be of considerable advantage in extending the life of the target and thus the life of the tube. In such a tube the gas replenisher is originally filled with the mixture of the two gases and during the operation of the tube the target is formed by driving into the absorption layer ions of both gases. The target is thus self-replenishing for the entire life of the tube. Sealed tubes are often used in pulsed operation and due to their small size they are particularly suitable for oil-well logging or any application for which isotopic neutron sources are normally used. The life-time of sealed tubes is about 100 hours, which in pulsed operation, with pulses as short as 1 μsec and neutron output as high as $10^{10}$ n/sec offer many advantages over isotopic sources.

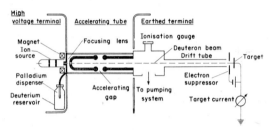

*Fig. 4. Schematic diagram of an ion accelerator.*

Low-voltage positive-ion accelerating machines using the d–t reaction are an alternative which is often used as a high flux neutron generator. Many machines, varying in design, have been built at universities and research laboratories, and others developed by firms are commercially available. The operating voltage is usually in the region of 100–400 kV which does not require a large accelerating tube or much space for the housing of the machine. A schematic diagram of a typical accelerating machine is shown in Fig. 4. The deuterium gas is stored in a reservoir bottle at a pressure of several atmospheres and is mostly dispensed in a controlled manner via a directly heated palladium tube into the ion source. Many types of ion source have been used with demountable neutron generators: Penning-type sources, r.f. ion sources of different design and duoplasmatron ion sources. The simplest seems to be the r.f. ion source described by C. D. Moak, which is used in many modern accelerating machines. The ion current delivered by this source consists of up to 90 per cent of atomic deuterium ions and the ratio of the gas and power consumption to the output ion current is very favourable. This is particularly useful from the point of view of the power required for the supply of equipment contained within the high voltage terminal. It is always preferable to have the ion source end of the generator connected to the positive output of the H.T. unit so that the target end is at ground potential and thus easily accessible for the insertion of samples for the neutron irradiation or change of target. The high voltage terminal may contain, in addition to the ion source, power supplies, the focusing voltage supply and the beam pulsing unit if pulsed operation is required. The ions are extracted from the ion source plasma through a channel in the negative probe electrode. The channel, which is usually a few millimetres in diameter and several millimetres long, serves also as a diaphragm to separate the source region where the pressure is a few mTorr from the rest of the accelerator where the pressure is several orders lower. A divergent ion beam emerges from the source and is focused in the focusing stage. This is accomplished by use of either aperture-type or cylindrical lenses as part of the accelerating system or by the use of an Einzel lens. The accelerating tube may be designed in various ways. For lower voltage a single accelerating gap may be used but for higher voltages several gaps or a constant gradient accelerator tube employing conventional "dished" electrodes may be preferred. The ion optical system must be capable of focusing the beam at the distance of the target so that the spot diameter is similar to that of the target. Even illumination of the target is essential since uniform heat dissipation will ensure the best target utilization. The design of the accelerating tube and the vacuum system should be such that a vacuum as free as possible from hydrocarbon or other contaminating vapours is obtained. Ion-type vacuum pumps are often used in place of oil-diffusion pumps. Low vapour pressure materials must be used throughout the tube and vapour traps (liquid air traps or Chevron baffles) are used to ensure absence of contaminants. If contaminant vapours are present in the system during the course of operation a carbonaceous layer will form on the target surface. In penetrating this film the deuterons will suffer a loss of energy which will in turn cause a drop in neutron yield. An additional cold trap is often used adjacent to the target to keep the region clean. At the present state of developments, tritium targets have a half-life of 60 min or more at 1 mA of ion current, depending on the concentration of tritium in the titanium. Targets with tritium concentration from 1 Curie/in² to 15 Curies/in² are available. Such targets give total neutron outputs of about $10^{11}$ neutrons/sec which corresponds to a fast neutron flux of about $10^{10}$ n/cm² sec and a thermal flux of approximately $10^9$ n/cm² sec. It is expected that this might be improved by at least one order of magnitude in the future. The limitation lies at present in the target cooling problems. If the permissible operating temperature of tritium targets could be increased so that they could be used at greater ion current densities, a higher neutron flux could readily be obtained. High current ion sources capable of producing many milliamperes of current are available and could be used in this case. The problems of short target life and rapid decreasing of neutron yield can be considerably overcome by using a rotating target. In such an assembly a strip of target material is clamped onto a rotating drum which is water- or mercury-cooled. The drum can be rotated very slowly, e.g. 1 rev/24 hrs which will give an essentially constant neutron flux through the whole period, or it can be rotated fast to ensure the best target cooling and therefore a long life at high current densities. The strip in the rotating target assembly has a many times greater target area than the simple disk target. This means also less time spent in changing targets and a lower cost per unit area.

The applications for low-voltage positive-ion accelerating machines are very varied. One of the most important industrial applications is considered to be activation analysis. Samples to be analysed are irradiated in a fast or thermal neutron flux and the induced radioactivity analysed by methods usual in gamma-ray spectrometry. The presence of many elements can

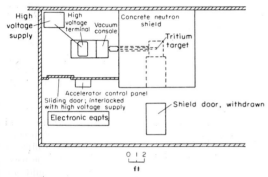

*Fig. 5. Plan of a neutron generator laboratory.*

be determined to a fraction of one per cent. Not less important are the applications of neutron generators in biology and medicine. Many interesting effects produced by neutron irradiation can be studied. A further important use of the accelerating machine is in connexion with sub-critical reactor assemblies. It enables the study of parameters used in reactor engineering. In this case the target which is at the end of a long drift tube is inserted into the reactor assembly. For most other applications a shield will be required around the target. This is usually built of concrete bricks but wax, polyethylene or water may also be used as moderators to provide the necessary shielding. A typical plan of an accelerator neutron generator laboratory is shown in Fig. 5.

*Bibliography*

A Table of Accelerators—Neutron Sources (Dec. 1960) *Nucleonics* **18**, 66.

GIBBONS D. and OLIVE G. (1964) *Applications of a Neutron Generator in Radioactivation Analysis,* AERE–R4576.

MEINKE W. W. and SHIDELER R. W. (Mar. 1962) *Activation Analysis, New Generators and Techniques Make it Routine, Nucleonics* **20**, 60.

OLIVE G. *et al.* (1962) *A Review of High Intensity Neutron Sources and Their Application in Industry,* AERE–R3920.

D. COSSUTTA

**NOBLE GAS COMPOUNDS.** The six gases, helium, neon, argon, krypton, xenon and radon were discovered at the end of the nineteenth century. Their lack of chemical properties resulted in their being placed in a unique group in the periodic table, and led to the name Inert Gases. The development of the electronic theory of valence around 1920 gave adequate theoretical reasons to expect the inert gases to be devoid of any ability to enter into chemical combination with other elements. As the valency of an element and its chemical properties are dependent upon its ability to gain, lose, or share electrons in order to attain the relatively stable filled outer shell of electrons, the inert gases, which already have such an electronic structure, would seem to have nothing to gain by combining with other elements.

A number of attempts to form chemical compounds were made at various times between the discovery of the gases and 1962. The best that could be said is that some of these attempts were inconclusive, and early in 1962 no stable species were known in which an inert gas was chemically bound to another element.

Early in 1962 Neil Bartlett in the course of some experiments with platinum hexafluoride ($PtF_6$) found that xenon gas reacted with $PtF_6$ vapour to form a compound which he postulated as $Xe^+PtF^-_6$. A group of scientists at the Argonne National Laboratory quickly confirmed this reaction of xenon, and extended the work to include other hexafluorides. They eventually concluded that xenon should react with fluorine and in the late summer of 1962 Claassen, Selig, and Malm prepared and characterized xenon tetrafluoride, $XeF_4$. This was the first simple compound of an inert gas, and opened up a fruitful field of research.

*Compounds of xenon.* Four fluorides of xenon have been reported: xenon difluoride, $XeF_2$; xenon tetrafluoride, $XeF_4$; xenon hexafluoride, $XeF_6$; and xenon octafluoride, $XeF_8$. The evidence for the octafluoride is somewhat tenuous, but the other three are well-established and well-characterized compounds. The fluorides may be formed by heating mixtures of xenon and fluorine at varying temperatures, pressures, and mole ratios. The conditions used for each fluoride are indicated in the table. The conditions given in this table are not necessarily the optimum ones for each reaction, they merely represent those which have been used successfully to prepare the respective fluorides.

*Conditions for preparation of xenon fluorides*

| Fluoride | Xe/F$_2$ Ratio | Pressure (atm) | Temperature (°C) | Time (hr) |
|---|---|---|---|---|
| XeF$_2$ | 20:1 | 60 | 400 | 2 |
| XeF$_4$ | 1:5 | 6 | 400 | 1 |
| XeF$_6$ | 1:20 | 60 | 300 | 16 |
| XeF$_8$ | 1:16 | 200 | 620 | not given |

Generally, there is some degree of cross-contamination; but careful purification allows $XeF_2$, $XeF_4$, and $XeF_6$ to be prepared in the pure state.

Several other methods have also been devised for the preparation of the difluoride and tetrafluoride. These include irradiation of xenon-fluorine mixtures with ultra-violet light, $\gamma$ rays and electron beams, and passing an electric discharge through mixtures of the gases.

*Xenon difluoride.* Xenon difluoride is a colourless, crystalline solid which melts at about 140°C, and has a vapour pressure of about 3 mm at 25°C and a heat of vaporization of 12·3 kcal/mole. It dissolves in water to give a solution containing undissociated xenon difluoride molecules, which have a half-life of about 7 hours at 0°C. Eventually, the difluoride is hydrolysed to xenon and hydrogen fluoride. The difluoride is soluble in anhydrous hydrogen fluoride (10 moles/100 g HF at 25°C) and can be recovered, unchanged, from such solutions. The xenon difluoride molecule is linear in both the gas phase and the crystalline state. The crystal structure has been determined by both X-ray and neutron diffraction. It is tetragonal with $a = 4·315 \pm 0·003$ Å and $c = 6·990 \pm 0·004$ Å. The space group is $I\,4/mm$ with xenon atoms at 0, 0, 0 and $\frac{1}{2}, \frac{1}{2}, \frac{1}{2}$ and fluorine atoms at $0, 0, z; 0, 0, \bar{z}, + b \cdot c$. This leads to a density of 4·32 g/cm³ based on two molecules in the cell. The structure exhibits linear F—Xe—F molecules aligned along the tetrad axis, with Xe—F bond distance of $2·00 \pm 0·01$ Å. Each fluorine atom has one fluorine neighbour at 3·02 Å and four at 3·09 Å, and each xenon atom has eight non-bonded fluorine neighbours at 3·41 Å.

Xenon difluoride is reduced by hydrogen at 300°C to give xenon and hydrogen fluoride. With antimony pentafluoride and tantalum pentafluoride it forms the complexes $XeF_2 \cdot 2SbF_5$ and $XeF_2 \cdot 2TaF_5$.

*Xenon tetrafluoride.* Xenon tetrafluoride is a colourless, crystalline solid which melts at about 114°C, has a vapour pressure of about 2·5 mm at 25°C and a heat of vaporization of 15·3 kcal/mole. It reacts with water according to the following equation:

$$3XeF_4 + 6H_2O \rightarrow XeO_3 + 2Xe + 1·5\,O_2 + 12HF$$

Xenon tetrafluoride is sparingly soluble in anhydrous hydrogen fluoride (0·2 moles/1000 g at 25°C). The molecule is square-planar with a central xenon atom surrounded by four equivalent fluorine atoms. The crystal is monoclinic with $a = 5·050 \pm 0·003$ Å, $b = 5·922 \pm 0·003$ Å, $c = 5·771 \pm 0·003$ Å, and $\beta = 99·6 \pm 0·1°$. The space group is $P\,2_1/n$ with two molecules in the unit cell, leading to a density of 4·04 g/cm³. The xenon atoms are in positions 0, 0, 0; and $\frac{1}{2}, \frac{1}{2}, \frac{1}{2}$ and the fluorine atoms at $\pm (x, y, z), \pm (\frac{1}{2} - x, \frac{1}{2} + y, \frac{1}{2} - z)$. The Xe—F bond length is $1·95 \pm 0·01$ Å and the F(1)—Xe—F(2) angle is $90·0 \pm 0·1°$. Each fluorine makes eight contacts with fluorines of other molecules at distances of 2·99–3·26 Å and two intramolecular

contacts of 2·74 Å. Xenon tetrafluoride is diamagnetic with a magnetic susceptibility of $-50·6 \times 10^{-6}$ e.m.u./mole.

*Xenon hexafluoride.* Xenon hexafluoride is a colourless, crystalline solid at room temperature. Around 42°C it changes colour, becomes yellow and melts to a yellow liquid at about 46°C; this colour change is reversible. The vapour pressure at 25°C is about 25 mm and the vapour is yellow. The heat of vaporization is about 13 kcal/mole. Xenon hexafluoride reacts violently with water at room temperature, but carefully controlled reactions at low temperatures result in hydrolysis according to the following equation:

$$XeF_6 + 3H_2O = XeO_3 + 6HF.$$

Xenon hexafluoride can also be hydrolysed to give xenon oxide tetrafluoride ($XeOF_4$) by reaction with the stoichiometric amount of water:

$$XeF_6 + H_2O = XeOF_4 + 2HF.$$

The hydrolysis of xenon hexafluoride in basic solution yields a somewhat different result. A white crystalline material is precipitated which turns out to be an octavalent salt derived from perxenic acid ($H_4XeO_6$).

$$2XeF_6 + 4Na^+ + 16OH^- \rightarrow Na_4XeO_6 + Xe + O_2$$
$$+ 12F^- + 8H_2O.$$

A similar result is obtained by passing ozone through a basic solution of $XeO_3$. Solutions of perxenate salts are powerful oxidizing agents.

Xenon hexafluoride is very soluble in anhydrous hydrogen fluoride (10 moles/1000 g at 25°C). The resulting solution is conducting and the formation of $HF_2^-$ and $XeF_5^+$ has been postulated. Xenon hexafluoride is reduced by hydrogen to xenon and hydrogen fluoride at 35°C. It reacts with a number of other fluorides to form complexes such as $XeF_6 \cdot CsF$, $XeF_6 \cdot RbF$, $XeF_6 \cdot BF_3$, and $XeF_6 \cdot SbF_5$.

The structure of xenon hexafluoride has not been established. The high reactivity of this material makes it difficult to handle in the conventional containers used for X-ray and spectroscopic measurements. By analogy with the other known hexafluorides one would expect the molecule to exhibit $O_h$ symmetry in the vapour state. The present experimental evidence suggests a structure which is either octahedrally symmetrical or slightly distorted from such symmetry.

The xenon fluorides are all thermodynamically stable and can be stored in pre-fluorinated nickel or monel containers. The difluoride and tetrafluoride can also be stored in well-dried glass equipment. The hexafluoride reacts with glass and forms xenon oxide tetrafluoride; if this is allowed to remain in the glass, it eventually forms xenon trioxide. The average xenon-fluorine bond energy is approximately 30 kcal/mole in each of the fluorides.

*Xenon oxide tetrafluoride.* As previously mentioned, this material is made by hydrolysis of xenon hexafluoride. It is a colourless liquid at room temperature with a vapour pressure of about 30 mm at 25°C, and a heat of vapourization of 9·0 kcal/mole.

The crystal structure has not been determined, but examinations of the vibrational and microwave spectra have shown the molecule to have $C_{4v}$ symmetry with the xenon and oxygen on a line through the centre of a square plane of four fluorine atoms. The xenon is in the same plane as the fluorines and the O—Xe—F angle is close to 90°C. The interatomic distances are Xe—O, 1·71 Å and Xe—F, 1·90 Å.

*Xenon trioxide.* Xenon trioxide is a product of aqueous hydrolysis of either xenon tetrafluoride or xenon hexafluoride. The aqueous solution is quite stable, but on evaporation it yields a colourless, crystalline, non-volatile solid which is dangerously explosive. The conditions under which the dry trioxide may be safely handled have not been determined and any sample must be treated as a potential hazard, with explosive power about equivalent to that of T.N.T.

The xenon trioxide crystal is orthorhombic with $a = 6·163 \pm 0·008$ Å, $b = 8·115 \pm 0·010$ Å and $c = 5·234 \pm 0·008$ Å. The space group is $P2_12_12_1$ with four molecules in the unit cell and a density of 4·55 g/cm³. The atoms are in the general positions $x, y, z; \frac{1}{2} - x, \bar{y}, \frac{1}{2} + z; \frac{1}{2} + x, \frac{1}{2} - y, \bar{z}; \bar{x}, \frac{1}{2} + y, \frac{1}{2} - z$ with the following parameters:

|     | $x$ | $y$ | $z$ |
| --- | --- | --- | --- |
| Xe | 0·9438 ± 0·0003 | 0·1496 ± 0·0003 | 0·2192 ± 0·0004 |
| $O_1$ | 0·537 ± 0·004 | 0·267 ± 0·004 | 0·066 ± 0·006 |
| $O_2$ | 0·171 ± 0·005 | 0·096 ± 0·004 | 0·406 ± 0·006 |
| $O_3$ | 0·142 ± 0·004 | 0·454 ± 0·003 | 0·142 ± 0·006 |

The average Xe—O bond length is 1·76 Å and the average O—Xe—O angle is 103°C. Each xenon has three close oxygen neighbours from other molecules at an average distance of 2·86 Å.

*Xenon tetroxide.* The perxenates, mentioned previously, are the only stable, octavalent compounds of xenon. The action of concentrated sulphuric acid on a perxenate results in the evolution of xenon tetroxide which is an unstable compound. Condensed samples of this material have been known to explode at temperatures as low as −40°C. However, it is possible to handle the vapour somewhat more easily and a study has been made of the infra-red spectrum. The gaseous tetroxide appears to be a tetrahedral molecule of symmetry $T_d$.

## Compounds of Other Noble Gases

*Radon.* The high radioactivity associated with this element makes studies on even the microscale extremely difficult. Evidence has been obtained for the formation of a radon fluoride of unknown chemical formula.

*Krypton.* Two fluorides of krypton have been reported; the difluoride and the tetrafluoride. The difluoride is a colourless, crystalline material which has been formed by irradiation of solid krypton and fluorine at 4°K, and identified by studying the infra-red spectrum of the solid. It has also been prepared by passing an electric discharge through a krypton fluorine mixture at −195°C and by subjecting a similar mixture to irradiation in an electron beam at −150°C. Krypton difluoride is unstable at room temperature and dissociates into krypton and fluorine. The infra-red spectrum of the solid and gas indicate a linear molecule.

Krypton tetrafluoride has been reported as the product produced when an electric discharge is passed through a krypton–fluorine mixture at −195°C. These are the same conditions mentioned above for the preparation of the difluoride. Some controversy appears to exist with regard to which fluoride is formed, or whether a mixture is formed. The reactive nature of these compounds makes the resolution of this problem difficult, but present evidence suggests that the difluoride is the reaction product.

*Bonding in noble gas compounds.* The discovery of compounds of the noble gases does not mean that new theories of chemical bonding are needed. The types of bonds in these compounds are analogous to those in the interhalogen compounds, which have been known for many years, for example, $ICl_2^-$ and $ICl_4^-$ are comparable to $XeF_2$ and $XeF_4$. Despite the fact that such bonding is well-known there does not yet seem to be any satisfactory theoretical explanation of the bonding in noble gas compounds. A suggestion has been made that the binding involves localized, electron-pair bonds with eight electrons contributed by the xenon, and one by each fluorine atom. Thus, in xenon tetrafluoride there would be a total of twelve electrons in six electron-pairs. Four of these pairs form the coplanar xenon–fluorine bonds and the other two are atomic lone pairs. The theoretical models derived by this assumption lead to molecular shapes more or less in agreement with those experimentally determined. The drawback is that this type of bonding would involve hybrid orbitals involving promotion of xenon 5p electrons to the 5d level. The energy required for such a promotion is high, and may be unreasonably large.

A second theory makes use of a molecular orbital model. If in xenon difluoride all the valence shell orbitals are considered filled except $2p_z$ of each fluorine atom and $5p_z$ of each xenon atom then there remain four valence electrons to be placed in molecular orbitals made up from linear combinations of the three atomic orbitals. The xenon fluorides would then contain linear F—Xe—F bonds of this "three centre-four electron" type. The bonding has also been discussed on the basis of a valence-bond resonance model in which one considers structures of the type F⁻Xe⁺—F and F—Xe⁺F⁻ between which resonance takes place.

Each of the models has been used to try to explain the observed experimental facts obtained from spectroscopic, NMR and Mössbauer studies. The success has been quite limited and much more detailed calculations and more experimental work is needed before adequate explanations can be developed.

*Bibliography*

CHERNICK C. L. (1963) *Chemical Compounds of the Noble Gases, Record of Chemical Progress* **24**, No. 3.

HYMAN H. H. (Ed.) (1963) *Noble Gas Compounds*, Chicago: The University Press.

MALM J. G. *et al.* (1965) *The Chemistry of Xenon, Chem. Rev.* **65**, No. 2.

MOODY G. J. and THOMAS J. D. R. (1964) *Noble Gases and their Compounds*, Oxford: Pergamon Press.

SELIG H. *et al.* (1964) *The Chemistry of the Noble Gases, Scientific American* **210**, No. 5.

C. L. CHERNICK

**NOISE FROM PLASTICALLY DEFORMED METALS.** Acoustic emission is a generic term denoting the occurrence of deformation induced sound pulses or elastic waves in metals. This term includes those sounds that are usually audible as well as those that are not. The audible sounds are more commonly known as the phenomenon *"tin cry"* which is produced by the twinning deformation so common in the tetragonal and hexagonal crystallographic structures. Two other processes which may also produce audible sounds are (1) sudden reorientation of large grains in a polycrystalline material, and (2) the martensite transformation associated with the heat treatment of steel. The inaudible emission is of greater import since it is ubiquitous in metals; being related to the more efficacious deformation process, slip.

The latter emission is of extremely low energy requiring amplification of the order of $10^6$–$10^7$ for satisfactory observation and analysis. Crude estimates place the source energy in the range of one to ten electron volts. In the typical detection procedure, a piezoelectric transducer is affixed to one end of a tensile specimen with the transducer output being transmitted through electronic amplifiers to suitable monitors, e.g. oscilloscope, loudspeaker, chart recorder, magnetic tape. Rochelle salt and ADP (ammonium dihydrogen phosphate) piezoelectric transducers have been utilized with considerable success. Subsequent deformation of the tensile specimen produces acoustic emission which is readily observed on the monitors.

Historically, knowledge of the audible acoustic emission is quite ancient, most likely observed in man's first attempts to form tin. The first scientific studies of record were not undertaken until the twentieth century, circa 1928, by a number of investigators in Europe.

Partly as a consequence of instrumentation limitations of the time, only a superficial view of the more salient characteristics of the phenomenon were tenable;

the prolific emission at much lower energy escaping detection. Little scientific significance was attached to the audible acoustic effect, at that time. In 1953 however, J. Kaiser undertook a study in Germany of a variety of metals including those that normally do not exhibit twinning as well as those that do. Employing the more advanced instrumentation of high sensitivity he detected the low energy emission in all of the metals investigated: cast iron, copper, lead, tin, aluminium, brass, and steel. This initial effort by Kaiser was expanded in studies in the United States and Europe, some of which are still in progress. The present state of these experimental findings is reviewed in the following discussion. For the most part the

*Fig. 1. Burst-type emission typical of zinc, carbon steel, 2024 alimunium—amplifier gain $10^6$.*

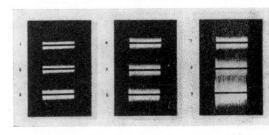

*Fig. 2. Amplitude rise of high frequency emission from aluminium. Frame 1: start of stressing. Frame 9: onset of yielding. Amplifier gain $10^6$.*

experimental results concern observations on single crystal specimens, however, the emission behaviour is essentially identical for polycrystalline metals of usual grain size found in commercial materials. It should be noted that conclusions based on the early experiment attributed the emission to the friction of crystallite aggregations, i.e. shear between the grains of polycrystalline metals. The single crystal findings suggest a mechanism on a much smaller scale than grain boundary.

Evidence of acoustic emission becomes apparent almost immediately upon stressing many metal particularly those that do not exhibit a "linear response of the stress-strain curve. In other metals appears at stress levels $\frac{1}{3}$ to $\frac{1}{2}$ the nominal engineering yield point. Characteristically all metals exhibit two distinct types of emission referred to as burst type and high frequency type. See Figs. 1 and 2. The relative activity of the two types varies with different metal and alloys with pure zinc and pure aluminium re presenting the extremes; zinc being predominantly burst type and the aluminium almost entirely high frequency type.

As observed on an oscilloscope screen the burst type appears as an exponentially decaying ring down pulse with a periodicity of occurrence long relative to the pulse time constant. The high frequency type is somewhat more subtle and appears as an amplitude rise of the general background noise level with occurrence periodicity of the pulses comparable to their time constant. In contrast to the burst type which does not change amplitude significantly as deformation proceeds, the high frequency rises to appreciable amplitude levels usually becoming maximal just prior to gross yielding and again prior to final failure. Of the two types the high frequency is most interesting from the deformation mechanics view.

Significant characteristics of the acoustic emission are: it is irreversible (*Kaiser Effect*), it is a volume as opposed to a surface phenomenon, the amplitude level (high frequency type) is proportional to strain rate, the emission signal repetition rate ranges from a few hundred counts per second to 75,000 per second at normal loading rates, evidence indicates a direct correlation between the number of fine slip lines and the number of acoustic pulses. These acoustic pulses evidence the discontinuous nature of deformation on a micro-scale suggestive of the macro Portevin-Le Chatelier Effect and stepped stress-strain curve. In conjunction with these characteristics the extremely low energy of the emission is suggestive of a micro-mechanism source such as dislocation blocking and unblocking.

The irreversibility characteristic is most unique and aside from its physical implications offers the possibility of application as a tool with non-destructive potentialities. Once the material is subjected to a given state of stress and deformation, during which considerable emission is generated, subsequent re-stressing up to the previous stress level will not induce emission. Exceeding the previous stress level produces an abrupt appearance of the emission at usual tensile loading rates. This property is the most positive evidence of the intimate relationship between emission and permanent deformation. The appearance of the emission in aluminium, for example, at $\frac{1}{3}$ the nominal yield strength demonstrates the existence of localized plastic deformation. Ageing, heat treating, or otherwise restoring the material to the unstressed state will of course restore the emission; the degree of restoration depending on the completeness of the treatment.

The fundamental mechanism giving rise to the inaudible emission has not been conclusively identified to date. Present results indicate a source intimately associated with the dynamic response of dislocation blocking and unblocking. Experiments on gold single crystals served to demonstrate that the emission is primarily a volume effect, as well as to show that surface oxide layers, such as exist on aluminium, are not generators of the acoustic pulses. Cold working of

the specimen surfaces and etchant removal of the surface material had negligible effect on the emission characteristics and the irreversibility as reflected in stress levels. Variation in strain rate did, however, show a strong influence on the emission energy manifested by the amplitude changes. An auspice of the latent possibilities of the emission phenomena was demonstrated in these studies on gold. A gold specimen, annealed following deformation then subsequently deformed again, exhibited mechanical twinning type emission pulses; a deformation mode hitherto not associated with this material at room temperature. Electron microscopy later revealed a number of minute areas presenting mechanical twin-like markings. The emission behaviour also provides some insight into the enigma of the mechanism by which elastic energy is dissipated in materials. Further study should aid in a better knowledge of the molecular processes which occur when solids are deformed, and hence of the relation between molecular structure and macroscopic physical properties.

Other acoustic emission phenomena noteworthy are: the pulses emitted during etchant removal of surface material (the converse polarity being quiescent), the existence of the high frequency emission during the propagation of a fatigue crack and, emission during the solidification of volume contracting metals.

*See also:* Kaiser effect. Tin cry.

*Bibliography*

BECKER R. and OROWAN E. (1932) *Z. Physik* **79**.

BORCHERS H. and TENSI H. M. (1962) *Piezoelektrische Impulsmessungen während der mechanischen Beanspruchung von AlMg 3 und A199. Z. Metallkde.* **53**.

KAISER J. (1953) *Archiv für das Eisenhüttenwesen* **50**.

LEAN J. B. *et al.* (May 1958) *Sur la formation d'ondes sonores, au cours d'essais de traction, dans des eprouvettes metalliques.* Comptes rendus des seances de l'Academie des Sciences, Volume 246.

SCHOFIELD B. H. (1963) *Acoustic Emission from Metals—Its Detection, Characteristics and Source.* Proc. Symp. on Physics and Non-destructive Testing, Southwest Research Institute Publ.

SCHOFIELD B. H. *et al.* (April 1958) *Acoustic Emission Under Applied Stress*, WADC Technical Report 58-194, ASTIA Document No. AD155674.

B. H. SCHOFIELD

**NON-DESTRUCTIVE TESTING, RECENT DEVELOPMENTS IN.** The scientific understanding of the behaviour of materials is a subject currently attracting widespread interest and support. Materials development in general has tended to suffer from the handicap of practice pre-dating theory by too large a margin. The explanation of plastic deformation processes, which are intimately linked to mechanisms of failure has, for example, only had a chance to be applied in retrospect to the rapid empirical development of high strength alloys. The introduction of ceramic, plastic and composite materials into engineering practice has also largely been without comprehensive theoretical support. It is perhaps the semiconductor field where materials science theory and practice are advancing in closest harmony at the present time.

One particular aspect of materials development where technology has very noticeably outstripped theory is that of quality inspection. Developments and refinements in non-destructive testing are primarily being channelled into developing more sensitive techniques for detecting and locating defects and other structural variables. Defect significance is still very ill-defined and tolerances continue to be largely specified with inadequate theoretical backing. In fact the available effort in non-destructive testing development would seem to be disproportionately distributed at the present time in view of the urgent need to determine more quantitatively the effects of material variability on service behaviour. Rationalization will only be achieved by a wider application of physics to materials inspection. Indeed one of the most encouraging recent developments has been the wider recognition of this need, accompanied, as it has been, by improved liaison between physicists, metallurgists, ceramists and engineers. This is a prelude to what might well develop into a concerted attack on the fundamental problems of materials evaluation.

McGonnagle has indicated how the rather special needs imposed on materials by nuclear engineering during the last two decades have caused considerable emphasis to be placed on the development and use of non-destructive tests. Paradoxically the sophistication and ingenuity introduced into the technology to meet these and similar needs, whilst acting as a spur to other precision industries, has become an embarrassment to some of the more conventional industries. This is because of the ability to reveal quality variations that are well below the level of significance when related to the performance requirements of their products. This is leading to the realization that two types of inspection requirement exist. For assessing new developments and pre-production routes in any industry there is everything to be gained from introducing and utilizing the most sensitive and revealing methods of inspection and evaluation based on as many significant physical variables as can be monitored non-destructively. The information obtained in this way can be used to locate, and thence to control, the sensitive fabrication variables and also to define, very precisely, acceptable tolerances. Reliance can then justifiably be placed on more economic and more rapid methods of quality control, tailored to the particular requirements of the ultimate product.

Stanford has listed the many variables that influence the service behaviour of a product and has emphasized that the scope of non-destructive testing should be broadly interpreted to embrace all aspects of quality. It should not be restricted to the location of defects, which was the narrow front on which the technology was first based. By adopting this enlightened approach, non-destructive testing is automatically transferred from an empirical technology growing up in isolation into one of the fringe subjects of physics. It is in this context and with this widened background of knowledge and experience that present day developments are largely being fostered.

Most of the recent technical improvements in industrial inspection techniques have been centred on the range of physical properties and associated techniques already tabulated by Homès and Thewlis. However, apart from the pure defect detecting techniques of visual, penetrant and magnetic particle inspection the most widely practised non-destructive testing methods in industry continue to be based on the passage of electromagnetic radiation (X and $\gamma$ radiography) and elastic waves (ultrasonics) through the material under test. Consequently, it is in these techniques where most progress and development is apparent.

Radiographic development projects are currently very diverse. The basic requirement of producing image contrast by differential absorption remains unchanged but the means both of generating the radiation and of recording the image have been very considerably improved and extended.

The range of radiation energy in commercial equipments now extends from a few keV to tens of MeV. Any further extension of this range is probably only of academic interest since the minimum absorption values for engineering materials can now be achieved, so that increased energy will not produce any corresponding increase in the penetration ability of the radiation. The increased availability of high-output high-energy transportable betatron and linac X-ray sources tailored to the needs of industrial radiography has tended to limit the growth of radioactive gamma ray source techniques. Although the radioisotope sources have obvious advantages for certain site inspection problems, they have relatively low radiation outputs and hence often require unacceptably long exposure times when applied to thick section radiography.

Apart from improvements in high energy techniques, high speed X-ray sources are now being more widely introduced. *Field emission X-ray tubes* are available that can operate over long periods producing pulses of less than 100 nanosecond duration at up to 300 kV with sufficient radiation output to give single shot radiographs of rapidly moving objects. These are invaluable for studying ballistic and explosive phenomena, but are also being usefully applied to such industrial applications as observing the characteristics of two-phase liquid-gas systems and studying dynamic processes such as precision metal casting.

Very high definition microradiography is now possible using *X-ray microscopes* in which the X-ray source is reduced to a focal spot less than 1 micron diameter by means of electron focusing coils. Useful magnifications up to 1000 are possible and this enables very fine microstructural detail to be revealed in thin slices of materials. Radiations from 5 to 60 kV are now possible from such minute focal spots and this provides a very valuable semi-non-destructive technique to complement optical microscopy.

There is growing interest in the possibilities of using thermal neutron beams for radiography. The majority of the techniques so far developed have ultimately used a nuclear reactor as the neutron source but there are obvious advantages in exploring the possibility of using accelerator sources. A number of current programmes are aimed in this direction. Since the variation of neutron capture and scatter cross-sections with atomic number is very different from that of X rays many inspection problems that cannot be tackled by conventional X-ray or gamma-ray techniques lend themselves to solution by *neutron radiography*. For example hydrogenous materials have a very high scatter cross-section for thermal neutrons and so are suitable subjects for neutron radiography. On the other hand uranium and steel behave similarly and the attenuation is such that large sections of either can be radiographed. Another interesting application of this technique is that highly radioactive materials can be radiographed since the method of detecting and recording the neutron beam can be made quite insensitive to the gamma ray background.

The problem of radiographing radioactive components has also led to increased interest in recording radiographs on colour film. The total range of optical density in colour film emulsions is less than with conventional radiographic films so that a fairly high background level of radiation can be superimposed on a *colour radiograph* without masking the essential detail. This reduced density range, coupled with the subjective advantage of introducing a hue variation in addition to a density variation, has led to more widespread interest in this form of recording for general radiography. Polaroid rapid process films have also been especially developed with radiographic applications in mind. In fact, this method of instant recording may go some way to bridge the gap between conventional film radiography and direct-view fluoroscopic techniques.

There have been some considerable advances in direct viewing recently, extending the fluoroscopic technique into the megavoltage region. The number of technique variants is now confusingly large but in all of them a fluorescent screen image is viewed through an image intensifier either directly or indirectly by means of an optically linked closed circuit television system. Sensitivities approaching those of film radiography are being claimed for some of these sophisticated arrangements. They have, of course, the potential advantage of lowering radiographic inspection costs and providing immediate information for in-line inspection problems. A somewhat different method or direct-view radiography has been developed around the *X-ray* sensitive *Vidicon* which has now reached commercial production. The charge produced by the X-ray image on the front face of the converter tube is read off by a scanning electron beam. The reso-

lution of such a system is limited solely by the size and scan spacing of the electron beam. Sensitivities equal to those of film radiography are readily obtainable and the image enlargement inherent in the system allows very small samples such as semiconductor components to be inspected very readily and very effectively. At present the Vidicon tubes have a limited sensitivity so that only thin and relatively non-absorbent sections can be examined, whilst the area of the Vidicon surface is also rather limited.

Quantitative data on X-ray absorption variations in a sample can normally only be obtained from a radiograph from a subsequent densitometric scan. It has recently been demonstrated that these operations can conveniently be combined so that quantitative information can be obtained directly. In this *scintillographic* technique the sample is scanned with a narrow beam of X rays or $\gamma$ rays and the transmitted beam is directed into a scintillation counter. The counter output is "quantized" into discrete levels and fed to a facsimile recorder. In this way the variation in transmitted intensity is recorded pictorially, each predetermined intensity range being displayed in a particular shade of grey on the record. This shows promise of being a particularly useful technique for recording the homogeneity of multiphase or composite materials. Radiation scanning techniques, in general, are a growing branch of non-destructive testing and are finding application for studying density variations in powder and sinter compacts. Since very sensitive and very small semiconductor radiation detectors are becoming available, very rapid methods of inspecting samples for small scale non-uniformity can now be devised. Autoradiation intensity variations from radioactive specimens can also be monitored in this way and the distribution of fission products determined selectively by using energy discrimination techniques to isolate particular isotopic radiations.

Radioactive isotopes now form the basis of many well established and well proven inspection techniques. They are used in metrology to measure sheet and wall thickness and to inspect coating layers and have the obvious advantage of being completely contactless. These and the complementary X-ray gauges are very appropriate developments for use in automated manufacturing processes.

Neutron activation analysis is now providing another very sensitive analytical technique for detecting and measuring the levels of trace constituents. Again on the analytical side, a small portable *X-ray fluorescence spectrometer* has been assembled around a radioactive exciting source. This is proving a useful instrument for surface composition analysis.

In ultrasonics much of the emphasis in recent developments has been directed towards more quantitative defect detection. More sensitive piezoelectric ceramic materials have been introduced and high intensity and more localized beams have been produced by using focusing transducers or acoustic lens and mirror systems.

There has been a considerable improvement in recording ultrasonic signals. *Facsimile recording* is now well established and by "quantizing" the signal before recording, quantitative intensity contours can be mapped out. Isometric recording with a single probe or a probe array can provide a very elegant three dimensional presentation of individual defects although it has not been generally introduced because of the associated reduction in inspection speed.

*Ultrasonic image converter* tubes are now commercially available although there is still room for further development both of the electronic tubes themselves and of the ultrasonic systems designed to use them. There have been parallel developments in Russia, Britain and America and two basic systems have evolved. One uses a high velocity electron scanning beam to excite secondary emission from the quartz plate as a means of reading off the piezoelectric charge pattern. The other uses a low velocity scanning beam to read off the charge pattern by neutralizing the piezoelectric charge point-by-point and returning the quartz plate to cathode potential. The great advantage of these systems is the high potential speed of inspection. It has been shown that this advantage is coupled with a sensitivity and resolution as high as can be achieved by the more conventional probe scanning methods.

*Ultrasonic resonance techniques* are in widespread use for measuring tube and plate wall thickness when only one side is exposed. A more recent instrument in this field uses a pulsed beam of variable frequency and the resonance is detected as an interference minimum in the pulse reflected from the component. This ultrasonic interference micrometer gives an accuracy of better than 0·1 per cent on tubes as small as 0·070″ diameter and can cope with wall thicknesses in the range 0·005″ to 0·25″.

Ultrasonic wave propagation in engineering materials has not yet been fully analysed theoretically and some of the material variables that affect propagation make exact computation rather intractable. One particular aspect that is receiving attention both theoretically and experimentally at the present time is propagation through anisotropic materials. One interesting application is that a purely isotropic medium under stress exhibits anisotropic elastic constants and this has the effect of providing a velocity differential between the shear waves polarized in two of the directions of principal stress. This *acousto-elastic effect* is being developed as a means of measuring the level of elastic stresses applied to specimens and of determining third-order elastic constants. For non-isotropic media, such as most metals, similar effects, generally of a larger magnitude, are produced when anisotropy is introduced by preferred orientations of the grains. This provides a sensitive method of detecting this form of microstructural variation.

The velocity of ultrasonic waves travelling in any particular mode depends on the value of one of the elastic constants of the material. By measuring veloci-

ties on test samples values of these elastic constants can be determined. Dynamic measurements of this kind hold out some hope of getting at grips with the problems of developing non-destructive tests to monitor the mechanical strength characteristics of materials.

Ultrasound is both absorbed and scattered in its passage through materials and attenuation measurements of this type constitute a branch of non-destructive testing for monitoring microstructural variables which is still in its infancy. Absorption can occur over different frequency ranges spanning natural resonance frequencies through ultrasonics to hypersonics by hysteresis and relaxation processes. These are produced by interactions between the vibrations and grains, dislocations, conduction electrons, magnetic domains and at the highest frequencies (around $10^9$ cycles/sec) lattice phonons. By analysing effects such as these the non-destructive evaluation of some of the basic lattice properties comes within sight.

Scatter is a very pronounced attenuating factor for ultrasonic beams passing through polycrystalline metals and a number of different mechanisms have been suggested to account for the loss, the operative one depending on the ratio of the wave-length to the size of the grains which act as the scattering sites. This effect has been used to provide a rapid method of monitoring grain size and hence the effectiveness of heat treatments in producing a uniform grain structure. By using facsimile recording a pictorial presentation of grain boundary networks in uranium has been made by such a technique.

It is because ultrasonic wave propagation is influenced by so many structural features that it is the obious basis for a considerable programme of further development. It should enable non-destructive tests to be developed to provide closer control of a large number of factors that affect the quality and behaviour of a component in service. The introduction of semiconductor materials into the ultrasonic field shows promise of initiating a further wave of active development. Already semiconductor transducers have been produced for generating hypersonic frequencies, semiconductor amplification of ultrasound has been demonstrated and semiconductor surfaces are being developed for the face plate of an image converter tube.

When many variables can affect a test result the problem of selectivity becomes important. In the case of ultrasonics the problem is not so acute, as attenuation and velocity are two distinct measurable variables, and many of the effects, in any case, are a manifestation of relaxation processes and so are frequency dependent. In the case of *eddy current testing* selectivity is more of a problem since the impedance of the test coil be the only variable, although this can generally be resolved into amplitude and phase components. The problem of selectivity is currently receiving close study and techniques are being developed using modulation analysis, pulsed eddy current systems, dual frequency operations and magnetic saturation. All of these variants help to improve the discrimination of defects signals and minimize effects due to dimensional tolerances, coil spacing and residual magnetic constituents in the sample being examined. Wall thickness measurements are also possible with eddy current techniques and the development of the Eddyfax system using fascimile recording has enabled tube wall thickness contours varying in steps of around 0·0001″ to be plotted directly.

Discrimination is part of the rapidly growing subject of signal analysis which has evolved around statistical communications theory. It is likely that non-destructive testing technology will materially benefit by cross fertilization of ideas and techniques with those active in this branch of the communications field. Problems of extracting signals from noisy backgrounds and studying the total information content of a signal have immediate counterparts in all branches of non-destructive testing. Current work on frequency analysis of ultrasonic pulses to supplement the limited information obtainable from consideration of signal amplitudes alone is already pointing the way to more informative inspection techniques. The wider adoption of sophisticated data processing techniques and computer analysis of results and the use of microminiaturized circuity are again directions in which non-destructive testing may well extend to improve its effectiveness and usefulness.

In a discussion on current developments in non-destructive testing it is appropriate to look closely at developments in physics in general, for academic research in quite unrelated fields may in fact provide the embryo of a future non-destructive testing technique. Microwave techniques are one such field where the inspection possibilities are already being studied both for inspecting non-metallic materials and for precision metrology. Infra-red scanning techniques for measuring surface temperature contours are also beginning to find an application in industrial inspection and the introduction of pyroelectric sensing elements may well raise the sensitivity of inspection techniques for those variables that can be monitored by the precise measurement of thermal transients. The thermal comparator is another instrument in this field that now enables thermal conductivities of a whole range of materials to be measured non-destructively. Fibroptic principles are probably capable of much further development both as light guides and sensing elements, whilst coherent light sources are also beginning to enter the non-destructive testing field with their ability to observe the Brownian motion and hence measure the mass of submicroscopic colloidal particles. The Mössbauer effect of recoilless emission and absorption also has possibilities in non-destructive testing and a number of applications for environment sensing and materials analysis are currently being studied.

Finally, a deeper appreciation of the mechanisms of material failure is required and again slow but noticeable progress is being made in this direction. Photoelastic stress coatings are being used to study local deformations around natural defects. In addition, exoelectron and ultrasonic emission are two new experimental techniques currently being applied in an attempt to elucidate aspects of this problem.

This brings the subject full circle back to the introductory remarks. The continued sophistication of non-destructive testing techniques is fundamentally pointless unless the raison d'être for introducing non-destructive testing into a manufacturing process becomes more scientifically established and it is on this aspect of the subject that development at the present time is weakest.

*Bibliography*

BALIKOV O. I. (1960) *Ultrasonics and its Industrial Applications* (Trans. by Consultants Bureau).

BANKS B. *et al.* (1962) *Ultrasonic Flaw Detection in Metals*, London: Illiffe.

BLITZ J. (1963) *Fundamentals of Ultrasonics*, London: Butterworths.

BLOKHIN N. A. (1957) *The Physics of X-Rays*, Translation AEC-tr4502.
*Proceedings of the 4th International Conference on Non-Destructive Testing*, London: Butterworths.

GOLDMAN R. (1962) *Ultrasonic Technology*, New York: Reinhold.

LAMBLE J. H. (Ed.) (1962) *Principles and Practice of Non-Destructive Testing*, London: Heywood.

McGONNAGLE W. J. *Proceedings of the Symposium on Physics and Non-Destructive Testing*, (1) ANL 6346 (2) ANL 6515 (3) and (4) S. W. Research Institute Publications.

McGONNAGLE W. J. (1961) *Non-Destructive Testing*, New York: McGraw-Hill.

McMASTER R. C. (Ed.) (1959) *Handbook of Non-Destructive Testing*, Amer. Society for Materials Evaluation, New York: Ronald Press.

STANFORD E. G. (1962) in *Encyclopaedic Dictionary of Physics*, 5, 35; Oxford: Pergamon Press.

STANFORD E. G. (Ed.) *Progress in Non-Destructive Testing* 1, 2 and 3, London: Heywood.

STANFORD E. G. (Ed.) *Progress in Applied Materials Research* 4 and 5, London: Heywood.

R. S. SHARPE

## NUCLEAR REACTOR, ADVANCED GAS-COOLED TYPE (A.G.R.).

This is the name given to graphite moderated reactors cooled by carbon dioxide designed to operate at temperatures materially higher than the magnox reactors based on the design of Calder Hall. The principal change necessary to permit this higher operating temperature is in the fuel element. The metallic uranium rod, roughly 1 inch in diameter, canned in magnox (magnesium alloy) of the Calder Hall type stations is replaced by a cluster of fuel pins, typically 0·4–0·6 in. in diameter each, made up of pellets of uranium oxide canned in stainless steel. This compensates for the lower thermal conductivity of uranium oxide, and provides a more refractory can, so permitting the higher operating temperature and a long irradiation life.

A prototype Advanced Gas-cooled Reactor with a designed thermal power of 100 MW and net electrical output of 28 MW has been built by the Atomic Energy Authority at Windscale and started operation in August 1962. Full power was first reached in January 1963 and since February 1963 it has been supplying electricity to the national grid.

The reactor core is 15 feet in diameter and 14 feet high, with 253 vertical channels, 5 in. in diameter, set on a triangular lattice with a pitch of 10·75 in. It is contained in a cylindrical mild steel pressure vessel with hemispherical ends. The core is surmounted by a neutron shield to protect the upper part of the reactor vessel from irradiation, thereby facilitating inspection when the reactor is shut down. The shield is made up of tubes extending the core channels, surrounded by graphite and boron steel to thermalize and capture neutrons escaping upwards from the core. The hot coolant flows through these tubes and is collected in a manifold—the "hot box"—situated above the neutron shield. This manifold is connected by short, straight horizontal ducts to the bases of the four heat exchangers. These ducts are insulated and contained within larger concentric ducts, the annular spaces between them providing return passages for the coolant. The neutron shield is also effective in minimizing neutron streaming down these ducts.

The tubes leading from each channel through the neutron shield are extended above the hot box to the top dome of the reactor vessel where they are connected to the charge tubes leading from the reactor vessel to the charge floor above. Thus each core channel has its individual charge tube, simplifying refuelling and permitting control of coolant flow in each individual channel during operation. It also provides for extensive temperature measurement of fuel elements and coolant and for direct detection of a burst fuel can.

The usual problem of accommodating the thermal expansion of the vessels enclosing the reactor and heat exchangers and their linking ducts was solved in an elegant manner by supporting all the vessels at duct level, thereby eliminating problems of vertical expansion, and by mounting the heat exchangers on ball bearing supports horizontal expansion is compensated by radial movement of the heat exchangers away from the reactor vessel. A sliding joint and bellows allows for the differential expansion between the inner and outer connecting ducts.

To permit the use of mild steel for the pressure retaining vessels of the reactor and heat exchangers it was necessary to protect them from contact with coolant at the reactor outlet temperature. This is achieved in the heat exchangers by enclosing the water and steam tubes in an insulated box within the pressure

shell. The hot coolant from the reactor flows up inside this box, over the tubes, and gives up much of its heat before coming in contact with the shell. It then flows down the annular space between box and shell to the centrifugal impeller which is mounted in the base. The impeller drives it back to the reactor through the annulus between inner and outer ducts and on arrival at the reactor the coolant is deflected by baffles over the whole inner surface of the reactor pressure vessel before returning to the base of the core. In addition, a fraction of the cool gas is diverted downwards through channels in the moderator formed by recesses in the graphite bricks, thereby ensuring that the temperature of the main mass of the moderator is kept as even as possible, and close to the coolant inlet temperature. This simplifies the design of the graphite structure and provides enhanced safety under fault conditions.

The original fuel clusters, each composed of 21 elements, were assembled in pairs inside a graphite sleeve to form an assembly. Four of these assemblies, linked together by a central tie-rod, form a fuel stringer 14 feet long. The fuel stringer is attached to a plug stringer which reaches from the top of the reactor core to the top of the charge tube, giving an overall length of 50 feet and providing shielding inside the charge tube. The combined stringer can be loaded into or withdrawn from the reactor in one operation by a single refuelling machine with the reactor operating at full power. The discharged fuel is stored in a rotary drum for its cooling period, the fission product heating being removed by circulating carbon dioxide. The reactor normally carries about 200 channels of fuel containing some 33,000 individual fuel elements, many of them being experimental elements of various types. The remaining channels contain control or safety rods, absorber, or non-fuel samples for irradiation. Four of these channels form test loops for irradiations under special conditions, or where there is a risk that the experiment might cause contamination of the channel.

The use of uranium oxide fuel and stainless steel cans makes some enrichment of the uranium necessary. In order to provide experimental flexibility the Windscale AGR fuel contains 2·5 per cent U 235, but for a large commercial AGR designed purely for power production the initial enrichment would be reduced to about 1·6 per cent. The enrichment of replacement fuel would naturally depend on the burn-up required.

The carbon dioxide pressure in the Windscale AGR is 270 psi and it is designed for a temperature of 575°C, while a commercial reactor of this type could operate at much higher temperatures, some proposals having already reached the range 650–700°C.

The operation of the prototype reactor has two main purposes, to prove the reliability of the system and to provide a test bed for improved designs of fuel elements and other reactor components.

When the reactor was shut down for a month for routine maintenance in September 1964 after 18 months of operation at full power, it had exported

292 million units of electricity, representing a load factor of 78 per cent. Losses due to being shut down or running at reduced power for reasons connected with the experimental programme were equivalent to 9·3 per cent of time at full power, giving an operational availability of 87·3 per cent. During this period no fuel element failures were detected, and at the end of it the mean channel burn-up was 4370 MWD/tonne, with peak figures of 6750 MDW/tonne for a channel and 9450 MDW/tonne for the most highly rated element. These figures compare with a design peak figure of 10,000 MWD/tonne and give every confidence that this figure can be materially improved upon in practice. The reactor has also operated steadily at a thermal power above the design figure of 100 MW, and plans are in hand to re-blade the turbine and increase the electrical output by 25 per cent.

One aspect of fuel performance which it was important to investigate under operating conditions was the consequence of a leak developing in a fuel element. The continued absence of naturally occurring defects ultimately led to the loading of a fuel assembly containing a can with a hole 0·01 in. in diameter drilled in it. In the first instance, this was put into one of the experimental loop channels, but after several weeks no increased activity could be detected and it was transferred to a standard channel. It remained there without any measurable change in activity arising until it had been irradiated for a total of 5 months, indicating that such a leak does not deteriorate and that there is no urgent need to detect or remove faulty fuel elements. Further experiments will be carried out with cans having more and larger holes, and with holes of varying shapes; if these confirm the stability of the fuel in contact with the coolant it could lead to great simplification of the detection equipment for defective fuel elements.

It was also essential to dispel by operating experience the possibility that the reactor life might be curtailed due to the reaction which takes place in the presence of radiation between carbon dioxide and graphite, resulting in the formation of carbon monoxide, and a consequent gradual erosion of the moderator structure. This reaction has long been known, but under the operating conditions of magnox reactors the rate of graphite removal is not great enough to threaten the integrity of the moderator throughout the life of the reactor. However, the greater intensity of radiation energy in the core of an AGR, associated with the higher fuel ratings, leads to a greater reaction rate, which must be reduced to achieve a similar integrity. Experiments in Materials Testing Reactors showed that the addition to the coolant of small proportions of hydrocarbon gases such as methane was very effective in achieving this reduction, and that careful control of the gas composition could also prevent the formation of carbonaceous materials through the accumulation of carbon monoxide in the coolant, which had been considered a possible disadvantage of AGRs. A coolant within a range of composition based on these tests has been in use in the Windscale AGR since February 1964 to provide a full-scale long term confirmatory test. Progress is followed in a variety of ways. Samples of graphite labelled with carbon-14 were loaded into the core and a continuous monitor measuring the activity released gives an indication of the rate of attack. Results over the first six months indicated that the coolant mixture in use had reduced the reaction rate well below the minimum acceptable for a reactor life of thirty years; indeed graphite removal was only at about half the rate in the Calder Hall reactors. Long term confirmation will be obtained by direct physical measurements on samples of graphite distributed throughout the core. Work continues both to establish the precise mechanism of the reaction and to achieve even greater reduction in its rate, and improved graphites are being developed, but it is already clear that the graphite/$CO_2$ reaction is not a cause for concern in AGRs.

The second main objective in operating the prototype was to investigate improved types of fuel element, principally in terms of advances in performance and reductions in fabrication costs. The main lines of development were higher temperatures, longer and thicker fuel elements, and cans with thinner walls and three important groups of experiments are in progress; involving the irradiation of:

(1) Fuel elements of greater diameter with the same fuel rating, to study the effect of higher centre temperature.
(2) Fuel elements of standard diameter with a higher rating, sufficient to produce the same centre temperature as the thicker elements.
(3) Fuel elements with a higher can surface temperature.

A longer term development is the use of coated fuels, with uranium oxide or carbide dispersed in an inert ceramic such as beryllia or silicon carbide, dispensing with a metallic can and permitting still higher surface temperatures.

The present situation (Oct. 1964) on the commercial development of Advanced Gas-cooled Reactors is that the Atomic Energy Authority have carried out design studies for a twin reactor generating station with a range of electrical outputs extending beyond 1000 MW, using a pre-stressed concrete pressure vessel to house each reactor and its associated heat exchangers. The operating temperature will be such that the steam produced in the heat exchangers will be of suitable quality for the most modern 500 MW turboalternator sets; the first time that this has been achieved in a nuclear power station design. The information gained from these studies has been supplied to the industrial consortia who have been invited by the Central Electricity Generating Board to submit tenders incorporating Advanced Gas-cooled Reactors for the next nuclear power station, which is to be sited at Dungeness, Kent.

J. A. DIXON

## NUCLEAR REACTOR, BOILING WATER TYPE.

The principal feature of boiling water reactors is that bulk boiling of the water used as moderator and coolant is allowed to take place in the reactor core. The reactor is contained in a pressure vessel and the steam generated is at the vapour pressure of the water at the operating temperature. This is opposed to the situation in a pressurized reactor system where there is an excess pressure. Almost all boiling water reactors employ ordinary water as the moderator and coolant through heavy water has also been used notably in the Halden Boiling Heavy Water Reactor.

Because of the general use of ordinary water as moderator these reactors must employ enriched uranium as fuel. In most of the existing boiling water reactors the fuel is in the form of cylindrical pellets of enriched uranium dioxide contained in tubes to form fuel element rods. Plate type sandwich fuel elements have also been used. Other materials besides $UO_2$ may also be present in the fuel element, notably thorium dioxide $ThO_2$ to allow some production of $^{233}U$. Zircaloy is used as the cladding in most of the reactors but stainless steel has also been used. In the low power SL–1 reactor in Idaho which suffered the accident in 1961 aluminium containing 1 per cent nickel was used as cladding for plate type elements.

Boiling water reactors have been developed because of their potential as power producing reactors. Typical of these is the Dresden Nuclear Power Station at Morris, Illinois, which has been running since 1959. It operates at a pressure of about 1000 psi (coolant outlet temperature, 547°F) and has an electrical output of 180 MW. A more recent plant, the Bodega Bay Atomic Park reactor, in Idaho, has, however, a projected electrical output of 325 MW with the same steam conditions.

Whilst basically the reactor part of all boiling water reactors is in principle similar, there are three different ways in which the reactor may be linked to the steam turbine. In the first type, "*direct-cycle*", steam generated in the reactor pressure vessel passes directly to a turbine. The first boiling water reactor (as opposed to an experimental assembly), the Experimental Boiling Water Reactor (EBWR) at Le Mont, Illinois, is of this type. The second type, "*indirect-cycle*", utilises an intermediate heat exchanger so that the secondary steam operates the turbine. An example of this type is the Elk River Plant in Minnesota which also incorporates a coal-fired superheater. In the third type, "*dual-cycle*", steam is fed to the turbine both directly from the reactor vessel and also from a secondary heat exchanger. The Dresden Nuclear Power Station is an example of this type.

The advantages of boiling water reactors stem particularly from their relative simplicity from the engineering view-point and use of water which is both a good moderator and heat transfer medium. In addition they have been demonstrated to be stable in operation over a wide range of conditions. Particular disadvantages of these reactors is the relatively low operating temperature and power density. In addition, as has been mentioned, enriched uranium must be used as fuel and so breeding is not possible. Furthermore, the power output, which is determined by the core size, is limited by the size of pressure vessel which can be made.

The feasibility of boiling water reactors was demonstrated first by a well known series of reactor experiments in the United States—the *Borax reactor experiments* in Idaho. There have been five of these commencing in 1953 and they have, in particular, determined the effect of steam voids in the reactors and the stability of the reactors. After being used to demonstrate that boiling water reactors were possible technically and also stable, the first experimental reactor Borax-1 was deliberately destroyed. A control rod was ejected very quickly and the resulting power excursion observed. It was established that the action of steam void formation shut the reactor down so quickly that the peak power was reached even before the control rod was completely out of the reactor.

*Bibliography*

HOGERTON J. F. (1963) *The Atomic Energy Deskbook*, London: Chapman and Hall; New York: Reinhold.
International Atomic Energy Agency (1959 and 1962) *Directory of Nuclear Reactors*, Vols. I and III.
KRAMER A. W. (1958) *Boiling Water Reactors*, New York: Addison–Wesley.

J. F. HILL

## NUCLEAR REACTOR, PACKAGED.

A packaged nuclear reactor is a power system utilizing a nuclear reactor for the production of electricity and heat for terrestrial applications. Packaged systems are designed and constructed to be lightweight and highly portable or mobile. Factory assembly and testing prior to shipment result in a minimum amount of time for field assembly, erection, and activation. Such nuclear plants are required in areas where power requirements are not large enough for central station plants, i.e. in the range from 100 to 50,000 kW (electrical), and in which the generation of power by conventional fossil fuels is costly because of logistic and fuel delivery problems or where mobility of power sources is required.

A number of packaged nuclear plants have been developed and built. These systems, whose characteristics are enumerated in Table 1, represent nuclear power units which can fulfil a wide variety of missions over a broad power range. This power range is categorized as follows:

(L)    Low power—300–1000 kW (electrical)
(M)   Medium power—1000–10,000 kW (electrical)
(H)   High power—10,000 kW (electrical) and above

The mission requirements determine whether a system is stationary (S), portable (P), or mobile (M). Packaged reactors which have been designed and built

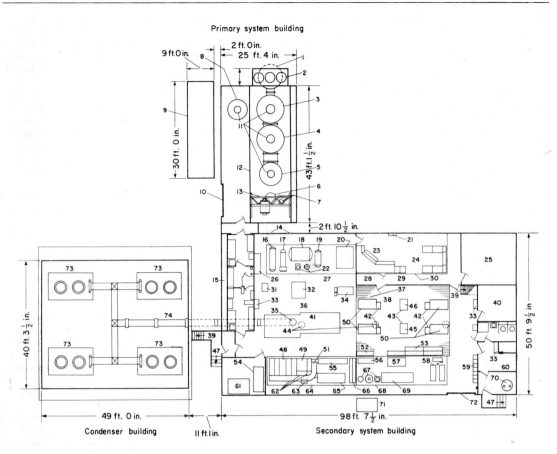

*Fig. 1. PM-3A General Plant Arrangement.*

*1 Void containment vessel;*    *2 Shield water air blast coolers;*    *3 Steam generator containment vessel;* *4 Reactor containment vessel;*    *5 Spent fuel storage containment vessel;*    *6 New core storage;*    *7 New core segment-storage;*    *8 Waste disposal containment vessel;*    *9 Miscellaneous for contaminated waste storage;*    *10 Wooden rollup door;*    *11 Manhole;*    *12 Crane track;*    *13 Gantry crane;*    *14 Openings for cables;*    *15 Decontamination package;*    *16 Hotwell;*    *17 Heater;*    *18 De-aerator;*    *19 Evaporator;* *20 Condensate storage tank;*    *21 Fire protection system control panel;*    *22 Pumps;*    *23 Control console;* *24 Instrument repair shop;*    *25 Office;*    *26 Sampling equipment;*    *27 Heat Transfer apparatus packages;* *28 Console door;*    *29 Passage;*    *30 Control package;*    *31 Oil conditioner;*    *32 Oil cooler;*    *33 First aid cabinet;*    *34 Chemical feed tanks;*    *35 Turbine exhaust;*    *36 Work room;*    *37 $\frac{1}{6}$ in. steel plate floor cover non-skid surface;*    *38 Work room;*    *39 Up;*    *40 Ward room;*    *41 Turbine generator;*    *42 Tool stand;* *43 Welding benches;*    *44 Roof ventilators;*    *45 Gas;*    *46 Arc;*    *47 Down;*    *48 High-voltage switchgear;* *49 Amplifier;*    *50 Steel and wood top work benches;*    *51 Low-voltage switchgear;*    *52 Bins;*    *53 Metal bins;*    *54 Water storage;*    *55 Motor control centre;*    *56 Lathe;*    *57 Bench;*    *58 Air compressor;* *59 Lockers;*    *60 Control office;*    *61 Snow melter;*    *62 Cable trays;*    *63 CB test equipment;*    *64 Switch-gear package;*    *65 D.C. and a.c. vital bus quipment;*    *66 Battery;*    *67 Auxiliary boiler;*    *68 Maintenance package;*    *69 Diesel generator;*    *70 Sump and pump;*    *71 Fuel oil storage tank pad;*    *72 Wooden rollup door;*    *73 Air-cooled condensers;*    *74 Steam line.*

*Table 1. Packaged nuclear reactors*

| Reactor | Location | Core life yr | Power heat | | | | Fuel | | |
|---|---|---|---|---|---|---|---|---|---|
| | | | MW TH | Net kW (electrical) | Steam BTU/hr × 10⁻⁶ | Electrical eff. % | wt. kg U | Enrichment % | Material |
| SM–1 (APPR–1) | Ft. Belvoir Va. | 2·0 | 10·7 | 1925 | — | 19·2 | 24·2 | 93 | 26, $UO_2$-SS |
| SM–1A (APPR–1A) | Ft. Greely Alaska | 1·0 | 20·2 | 1640 | 42 | 21 | 24·2 | 93 | 26, $UO_2$–SS |
| PM–1 | Sundance, Wyoming | 2·0 | 9·37 | 1000 | 7·0 | 13·6 | 31·3 | 93 | 28, $UO_2$–SS |
| PM–2A | Camp Century Greenland | 1·3 | 10 | 1570 | 1·0 | 16·2 | 19·9 | 93 | 26, $UO_2$–SS |
| PM–3A | McMurdo Sound Antarctica | 2·0 | 9·36 | 1500 | — | 16·0 | 31·3 | 93 | 28, $UO_2$–SS |
| MH–1A | Ft. Belvoir Va. | 1·0 | 45 | 10000 | — | 25·5 | Inner[2] 1447 Outer[2] 1459 | 4·07 4·67 | $UO_2$ $UO_2$ |
| ML–1 | NRTS Idaho | 0·35 | 3·3 | 330 | — | 10 | 52·6 | 93 | $UO_2$ and $UO_2 \cdot BeO$ |

*Table 1. (Continued)*

| Burnable poison | | | Coolant | | | | | |
|---|---|---|---|---|---|---|---|---|
| Material | Location | wt. % | Flow rate (GPM) | Material | Temp. °F in/out | Outlet PSIA | Reactor Δ PSI |
| $B_4C$ | Fuel | 0·08 | 3862 | $H_2O$ | 428/448 | 1200 | 4·5 |
| $B_4C$ | Fuel | 0·08 | 7350 | $H_2O$ | 423/433 | 1200 | 7·7 |
| B–SS | Lump (rods) | 0·27 | 2200 | $H_2O$ | 447/479 | 1300 | 8·5 |
| $B_4C$ | Fuel Cermet | 0·08 | 4890 | $H_2O$ | 500/518 | 1750 | 3·9 |
| B–SS | Lump | 0·27 | 2200 | $H_2O$ | 447/479 | 1300 | 3·9 |
| — | — | — | 8000 | $H_2O$ | 470/510 | 1600 | 11·6 |
| Cd Alloy Ag Shim | Lump (tubes) | | 24·9 lb/sec | Air or $N_2$ | 791/1200 | 289 | 24 |

*Notes:* (1) Max. heat flux includes hot spot factors.   (2) Equilibrium core.   (3) 100° ambient.

| Fuel | | | | Fuel clad | | | Control rods | |
|---|---|---|---|---|---|---|---|---|
| Burnup % U max/avg. | Shape | Fuel thick. mils | Centre temp. max/avg. °F | Thick. mils | Material | Surface temp. max/avg. °F | Number shape | Poison material |
| 57/36 | Plate | 20 | 595/470 | 5 | 304L SS | 578/465 | 7 Box | $B_4C$–Fe |
| 57/36 | Plate | 20 | 610/487 | 5 | 304L SS | 578/476 | 7 Box | $B_4C$–Fe |
| 53/27 | Tube | 30 | 660/530 | 7.5 | 347 SS | 588/480 | 6 Wye | $Eu_2O_3 \cdot 2TiO_2$–SS |
| 36/57 | Plate | 20 | 632/539 | 5 | 304L SS | 617/534 | 5 Box | $Eu_2O_3$–SS |
| 53/27 | Tube | 30 | 660/530 | 7.5 | 347 SS | 588/480 | 6 Wye | $Eu_2O_3 \cdot 2TiO_2$–SS |
| MWD/MT 14,880 | Pellet | 456 | 3180 | 23 | 347 SS | 615/600 | 12 Cross | B-10, SS |
| 25,250 | Pellet | 456 | 2340 | 23 | 347 SS | | | |
| | Pellet | 177 | 2650/1960 | 30 | Hastelloy X | 1750/1430 | 6 Blade | Shim-scram Cd–In–Ag Regulating 202 SS |

| Coolant | Conversion | | | | Heat | |
|---|---|---|---|---|---|---|
| Heat flux (1) (BTU/hr/ft (2)) $\times 10^{-5}$ max/avg. | Type | Temp. °F | Press. PSIA | Flow $10^{-3}$ (lb/hr) | Turbine exhaust in. hg. abs. | Condenser system |
| 1.89/0.65 | Steam | 387 | 213 | 37 | 2.4 | ST/$H_2O$ |
| 3.20/1.14 | Steam | 382 | 200 | 70 | 2.5 | ST/$H_2O$ |
| 3.66/0.71 | Steam | 417 | 300 | 34 | 9 | ST/Air |
| 1.78/0.72 | Steam | 459 | 465 | 37 | 8 | ST/Glycol to Air |
| 3.66/0.71 | Steam | 417 | 300 | 35 | 6 | ST/Air |
| 4.07/1.18 | Steam | 426 | 330 | 171 | 2 | ST/$H_2O$ |
| 1.30/0.78 | Gas (3) Brayton | HP 368 in 791 out LP 907 in 489 out | HP 322 in 315 out LP 121 in 120 out | 87.8 | 486°F in 133°F out 190°F in 180°F out | Pre-cooler Gas/Air Moderator Cooler |

are of three types: pressurized water reactors such as the SM–1A, PM–3A and MH–1A (the numeric indicates the plant model in a series, the letter designates the plant as an operational field plant as opposed to a prototype); boiling water reactors such as the SL–1; and advanced systems such as the ML–1, utilizing a gas-cooled, water-moderated reactor and the MCR (Mobile Compact Reactor), which employs liquid metal as coolant and working fluid. Some of the outstanding design features of the plants shown in Table 1 are enumerated.

*SM–1 and SM–1A.* The SM–1, originally called the Army Package Power Reactor, was designed and constructed to demonstrate and test reactor components and equipment to be used in later packaged plants and is used as a training reactor for the technicians and operators assigned to field plants. The SM–1A is essentially a duplicate of the SM–1 and represents a field application of the technology developed as a result of the design and construction of the SM–1.

*PM–1 and PM–3A.* The PM–1 and PM–3A packaged nuclear plants utilize pressurized water reactors and produce 3-phase, 60 cycle electricity generated at a line voltage of 4160 V. Both plants are similar in general arrangement (see Fig. 1) with the reactor immersed in shielding water from which air blast coolers remove heat generated by gamma radiation or transferred to it through primary system thermal insulation. A plant decontamination package is used to control personnel access to the primary system area and contains primary system coolant charging pumps, chemical addition and water analysis equipment, and a personnel and equipment decontamination area.

In the power conversion system (Fig. 2) steam flows to the turbine from the steam generator; a part is extracted from the turbine at a single point after which it passes through a feedwater heater and into the deaerator before reaching the condensers. Dual condensate and feedwater pumps are used; one of each is driven by an electric motor, the others by small steam turbines. In the case of the PM–1 an evaporator-reboiler off the main steam line is the source of steam for space heat and condensate make-up water. The air cooled condenser system automatically maintains turbine back pressure. The power plant control console provides for one-man operation. Four interconnected tanks, which enclose all primary (nuclear) system components, make up the containment system for the PM–3A. These tanks are shipped separately and interconnected at the site. The void tank which houses some components was added to lower system design pressure by providing additional expansion volume. The predicted peak pressure after a maximum credible incident is 126 psi. The containment system is periodically tested to assure that the daily leak rate is less than 0·5 per cent of the contained volume. The tanks are fabricated from T–1 steel, selected because

of its high strength and low nil ductility transition temperature of −195°F. Vessel wall thickness is $\frac{1}{4}$ in. with stainless steel cladding added to the inside of the reactor and spent fuel storage tanks. The PM–1 plant is not contained, although the plant arrangement is identical.

Primary system instrumentation and controls are electric with full use made of magnetic amplifiers and transistors. The maintenance package contains maintenance and auxiliary equipment. Switchgear, motor control centre, and the vital a.c./d.c. bus are located in the switchgear package. Batteries, main-

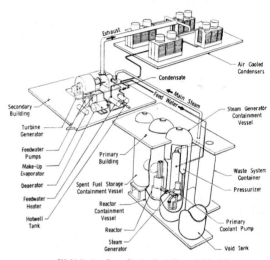

PM-3A Nuclear Power Plant - - Basic Flow and Orientation

*Fig. 2. PM–3A Nuclear power plant.—Basic flow and operation.*

tained on trickle charge, drive an a.c. generator to provide emergency instrument power. Liquid waste processing equipment is housed in the waste disposal tank with gaseous waste bottled in cylinders or retained in an expansion tank in the waste disposal tank.

The type 347 stainless steel reactor pressure vessel (Fig. 3) is $9\frac{1}{4}$ ft high by 3 ft 8 in. outside diameter and consists of a 2 : 1 ellipsoidal bottom, a cylindrical shell $2\frac{1}{16}$ in. thick, a conical transition section joining the shell to the lower flange and a spherical head with reinforced area for attaching six magnetic jack control rod actuators. Type 347 stainless steel is used to eliminate the effect of rising nil ductility transition temperature with increasing exposure of the pressure vessel to neutron radiation. Coolant enters through the inlet nozzle, passes through an orifice plate to flow between two thermal shields into the core shroud area, through the core into the upper plenum and out the outlet nozzle. The steam generator is of the vertical type with integral steam drum and separators. The primary coolant pump is a vertically mounted, centrifugal canned motor pump. The pressurizer is used to hold pressure and limit pressure swings; pressure decrease is accomplished by spraying primary coolant into the steam dome. No valves are used in the main loop, all field joints being flanged.

The reactor core (Fig. 4) is composed of six peripheral bundles and one centre bundle. Each bundle contains fuel tubes, lumped poison rods and for the peripheral bundles, one control rod each. Core active length is 30 in. and its diameter 23 in. The core contains 741 fuel tubes of 0·5 in. outside diameter with a 44 mil wall thickness. Lumped burnable poisons are used to control long term reactivity and to minimize the number of control rods required.

5500 ft above sea level. Severe climatic conditions with surface temperatures ranging from −75°F to +35°F and winds from 5 to 100 mph velocity dictated the location of the entire power plant in tunnels below the snow surface.

Plant packages are placed in tunnels approximately 24 ft wide and 24 ft deep which are covered with arched plates and snow back-fill. Enclosures for the packages consist of insulated panels and roofs. Enclosure and tunnel ventilation removes heat lost from equipment and piping. Differential settlement between plant packages is taken up using jacks and special piping connexions. The turbine, condenser, and heat

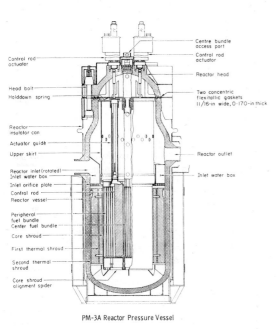

Fig. 3. *PM–3A Reactor pressure vessel.*

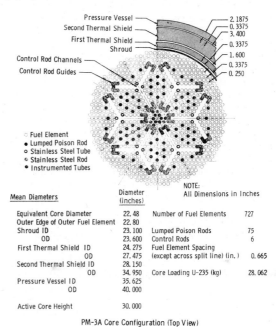

| Mean Diameters | Diameter (inches) | NOTE: All Dimensions in Inches | |
|---|---|---|---|
| Equivalent Core Diameter | 22. 48 | Number of Fuel Elements | 727 |
| Outer Edge of Outer Fuel Element | 22. 80 | | |
| Shroud ID | 23. 100 | Lumped Poison Rods | 75 |
| OD | 23. 600 | Control Rods | 6 |
| First Thermal Shield  ID | 24. 275 | Fuel Element Spacing | |
| OD | 27. 475 | (except across split line) (in. ) | 0. 665 |
| Second Thermal Shield ID | 28. 150 | | |
| OD | 34. 950 | Core Loading U-235 (kg) | 28. 062 |
| Pressure Vessel ID | 35. 625 | | |
| OD | 40. 000 | | |
| Active Core Height | 30. 000 | | |

Fig. 4. *PM–3A Core configuration (top view).*

The PM–1 site is 6650 ft above sea level in a location selected for its attributes as a remote operational site. Mean annual snowfall is 110 in. with a mean January temperature at 9°F with a minimum of below −40°F. Wind gusts up to 100 mph are fairly common. The PM–3A reactor site is on Observation Hill at McMurdo Sound at an elevation of 300 ft above sea level. The base is completely isolated from April 1 to October 1 each year with temperature extremes from +40 to −60°F.

*PM–2A.* The PM–2A packaged power plant (Fig. 5) is a pressurized water system whose power producing cycle is basically the same as that of the PM–3A except that the method of rejecting heat utilizes a steam-to ethylene glycol-to air system. The plant was designed for the conditions imposed by a remote ice cap site, which required the plant to be located in an ice cap at an elevation of approximately

exchanger skids are arranged with their long axis on a common centreline for the most direct connexion of interconnecting piping.

*MH–1A.* The design of the MH–1A is based on four criteria for a mobile plant: (a) proven technology, (b) ease of operation, (c) maximum reliability, and (d) maximum safety. The power plant is contained within a floating mount, a converted Z–EC2 class "Liberty" ship. The reactor and equipment which contains high-pressure radioactive fluid is enclosed within a vapour containment vessel which is a 31 ft diameter right cylinder with hemispherical heads, with an overall length of 41 ft. It is installed horizontally amidships. Grounding and collision protection is provided to assure the integrity of the containment system. The collision barrier consists of reinforced hull structure in the form of increased deck plating and heavy horizontal stringers. The collision energy absorption capability of the barrier is 100,000 ft tons. The containment vessel is also designed to withstand external hydrostatic forces resulting from vessel submergence. The steam electric conversion system is set forward of the containment vessel while the refuelling and decontamination areas are located aft and above the vessel. The space immediately surrounding the containment vessel is enclosed by concrete shielding.

The reactor core utilizes low enrichment fuel, in two zones. Fuel will be partially replaced after the equivalent of 1 year at full power. The main turbine generator produces 11,500 kW (electrical) (gross). Electrical energy is distributed at several voltages and two frequencies. Waste heat from the turbine is rejected to a surface condenser which is cooled by sea water.

*ML–1.* The ML-1 plant is a high temperature, gas-cooled, water moderated reactor with compact power conversion equipment. The plant operates as a conventional Brayton closed-cycle gas turbine plant using a nuclear reactor as a heat source. Oxygenated nitrogen ($0.5$ vol% $O_2$ + $99.5$ vol.% $N_2$) is the reactor coolant and conversion system working fluid. The plant consists of two major packages, the reactor "skid" and the power conversion "skid". Equipment

for control of the plant is provided in a control cab and auxiliary subsystems used in startup and shutdown are located on an auxiliary skid. The control cab is located at a distance of 500 ft from the power plant.

The reactor is a heterogeneous, water moderated type, fuelled with enriched uranium dioxide. The core consists of 61 fuel elements located in pressure tubes connecting upper and lower plenum chambers. Each fuel element contains 18 pins (rods) which are filled with ceramic fuel pellets to make up an active core length of 21 in. Reactor core diameter is approximately 22 in., reflected top (8 in.) and bottom (6 in.) with water–tungsten–stainless steel combinations. Reactor heat is removed by the coolant gas flowing through the pin type fuel elements. A demineralized water moderator surrounds the pressure tubes and an integral radiation shield surrounds the reactor. The reactor core, shielding, and pressure vessel assembly are enclosed in a tank containing borated water during operation to provide neutron shielding. The reactor is controlled by six pairs of semaphore type control blades placed near the circumference of the core which operate in the moderator spaces between the fuel elements.

Major components of the power conversion skid are a turbine compressor set and reduction gear, precooler with fans, recuperator, switchgear, generator and

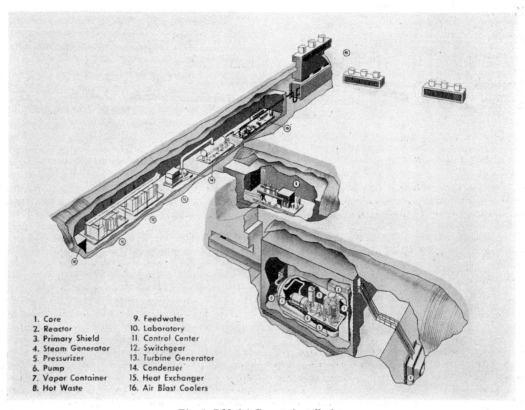

| | |
|---|---|
| 1. Core | 9. Feedwater |
| 2. Reactor | 10. Laboratory |
| 3. Primary Shield | 11. Control Center |
| 4. Steam Generator | 12. Switchgear |
| 5. Pressurizer | 13. Turbine Generator |
| 6. Pump | 14. Condenser |
| 7. Vapor Container | 15. Heat Exchanger |
| 8. Hot Waste | 16. Air Blast Coolers |

*Fig. 5. PM–2A Remote installation.*

starting motor, piping and valves. Gas leaves the reactor, is expanded in the turbine, passes through the low pressure side of a regenerative heat exchanger (recuperator), through a precooler and to the compressor. The compressed gas passes through the high pressure side of the recuperator where it is heated and goes into the reactor to complete the cycle. The output from the turbine is used to drive a direct coupled compressor. The output from the turbine shaft is used to drive the alternator after shaft speed is reduced to 3600 rpm through a gear box.

The power conversion skid also contains a vacuum pump for system purging and a lubricating system for the rotating machinery. Auxiliary systems include make-up water treatment equipment, gas make-up system, waste gas storage equipment, chemicals, and tools.

The factors which influence the compactness of a packaged plant are reactor system, conversion system and auxiliary system size, method of transportation, and ultimate use.

Mobile packaged plants, such as the MH–1A and ML–1 differ in the method by which mobility is obtained. The ML–1 modules are transported on standard wheeled vehicles or semi-trailers (primary transportation mode) which do not form an integral part of the nuclear plant. High mobility is achieved through the use of a minimum number of plant modules, which make up a very compact plant and which can be connected or disconnected in a minimum time. The ML–1 design is such that the plant is required to deliver rated power within 12 hours after arrival at an operating site and be disassembled and loaded within 36 hr from reactor shutdown following operation for extended periods at full power. Secondary modes of transportation are railway, ship or barge, snow sled, and C–133, C–130 and C–124C aircraft. Secondary shielding is not included in the plant but so-called expedient shielding, i.e. local aggregate, wood, distance, is used as required at the operational site. The very short activation and deactivation times dictate the use of highly compact reactor and conversion systems. Therefore, gas is chosen as the working fluid coupled with fully enriched nuclear fuel and a high speed gas turbine which directly drives the compressor and generator gear box. Cooling requirements are minimized by rejecting heat at a high temperature through the use of a precooler and regenerative heat exchanger. A tabulation of plant packages with auxiliary systems is given in Table 2.

*Table 2. ML–1 Package breakdown*

| Package | Weight (tons) |
|---|---|
| Power conversion | 15 |
| Reactor | 15 |
| Control cab | 2·5 |
| Auxiliary system | 6 |

Mobility in the case of the MH–1A is obtained by the use of a seagoing barge as a mount for a high power output nuclear plant. The physical characteristics of the mount are given in Table 3. Compactness in the nuclear plant is not as demanding as in the case of the ML–1 and, consequently, low enrichment fuel in a more conventional pressurized water power system can be employed. The activation and deactivation time for the MH–1A is less than that for the ML–1 since the mount carries its own reactor shielding and refuelling equipment and can be moved very rapidly.

*Table 3. MH–1A Mount characteristics*

| | |
|---|---|
| Length, overall | 441 ft, 6 in. |
| Beam, molded | 65 ft, 0 in. |
| Depth, to upper deck | 37 ft, 4 in. |
| Draft, max. operating | 17 ft, 4 in. |
| Displacement at max. draft | 8750 tons |

The portability of packaged nuclear plants is a requirement dictated by method of transportation and ability to be relocated. In general, portable plants, when located, become semi-permanent installations although installation and activation time must be as short possible. Portability requirements for plants such as the PM–1 have been dictated by the mission capability of C–130 type aircraft. Modular or skid type construction is utilized in which total package weight including the skid cannot exceed 30,000 lbs in weight or 8 ft 8 in. square by 30 ft long in size with a load distribution under 1080 lbs per running foot. G-loadings to be taken during transportation are 8 G in the forward, aft and vertical directions and 5 G in the lateral direction. Some typical module sizes and weights are given in Table 4.

*Table 4. Portable plant 30,000 lb module tabulation*

| Component | No. Modules | |
|---|---|---|
| | PM–3A | PM–2A |
| Powerplant | 18 | 31 |
| Buildings and foundations | 30 | 36 |
| Tools and operating supplies | 4 | 4 |

Portable plants constructed to date have been pre-assembled at the factory, the modules interconnected, and all systems, except for the reactor, test operated at full operating capacity. These provisions limit the amount of field construction to interconnexion of the plant modules and check out of the plant subsystems prior to approaching full power operation. Although plant designs exist in which the module panels are insulated and can form the basic plant housing at site, this type of housing has not been employed. Sheet metal, insulated panel, rapidly constructed buildings have been employed, placed on rough timber foundations with both wood and metal floor covering. In the case of both PM–1 and PM–3A, secondary reactor shielding is provided by emplacement of primary system tanks in the earth, and backfilling around the tanks with careful attention given to earth shield cooling especially in areas of permafrost. Typical site assembly and activation time spans are given in Table 5.

*Table 5. Portable plant assembly and activation times*

|  | Span time (days) | Man-hours |
|---|---|---|
| PM–1 | 161 | 26,000 |
| PM–3A | 77 | 56,000 |
| PM–2A | 80 | 60,000 |

Maintenance is accomplished in portable plants with equipment provided as part of a small shop equipped with drill press, lathe, welding equipment, and a wide assortment of tools which are used in all but the heaviest of maintenance. Refuelling equipment is provided as an integral part of the plants.

J. F. O'Brien

**NUCLEAR REACTORS, NEW FUELS FOR.** The development of new fuels is primarily directed towards improvements in the economics and safety of power reactors. Improvements are sought in the direction of cheaper fabrication processes, higher burn-up, higher thermal efficiency, minimum distortion during life, minimum parasitic capture of neutrons and maximum safety of operation. It is convenient to consider developments in terms of the various types of fuel material.

*Metal fuels.* Metal fuels have been used extensively in power reactors, notably in the British Magnox stations. On the basis of early irradiation data it has often been stated that the burn-up limit of unalloyed uranium at a maximum fuel temperature of 600°C was about 3000 MWd/t (megawatt-days per ton) this limit being imposed by the fission gases causing excessive swelling. While the addition of molybdenum up to 20 at.% has allowed this limit to be extended it has also demanded considerable enrichment of the uranium. Recently, however, it has been demonstrated that the addition of about 1000 ppm iron and 1000 ppm aluminium allows the uranium to attain a burn-up of 10,000 MWd/t without excessive swelling. These additions were originally made to effect a grain refinement to eliminate wrinkling or anisotropic growth of the individual crystals. Electron microscope studies have shown that the additions form U-Fe and U-Al intermetallic compounds which are present as very small particles. Fission gas bubbles nucleate at the surfaces of these particles where they are an-

chored. In the absence of the particles the bubbles are seen to move under the influence of stress fields and thermal gradients and when bubbles meet they coalesce to form larger bubbles. Once the bubble diameters exceed a certain size (around 0·1 $\mu$) the gas pressure is no longer balanced by surface tension forces and the metal swells. Thus, not only has a technological advance been achieved but also an explanation of the underlying physical processes.

*Oxides.* $UO_2$ is used extensively in thermal reactors, mainly in the form of sintered pellets which are often ground to close diametral tolerances in order to reduce or eliminate sheath distortion and reduce the fuel/sheath interface temperature difference. Pellets may also have dimpled faces to minimize thermal expansion effects. In an attempt to lower fabrication costs studies have been made of vibratory compaction processes in which carefully sized particles of dense $UO_2$ are fed into a metal tube, which is sealed at the bottom end, and are caused to pack down to a high density by sonic or ultrasonic vibration of the tube. Using three sizes of particle, densities as high as 90 per cent of the theoretical (X-ray) density have been achieved. The process is suitable for active recycling of fuels and for the controlled addition of plutonium enrichment. However, the resulting fuel has an even lower thermal conductivity than sintered pellets and under irradiation the powder usually rapidly redistributes itself to form a central cavity which is surrounded by large, dense crystals. For many applications, particularly where a high fission gas release can be tolerated, this type of fuel is likely to be adopted. Among its attractions, it permits the fabrication of fuel elements of unusual shapes.

An alternative approach to improving the performance of $UO_2$ is to attempt to increase its thermal conductivity. Workers at the A.E.C.L., Chalk River, Canada, laboratories have found that $UO_2$ with an oxygen: metal ratio of less that 2·0 (typically 1·98) has a higher thermal conductivity than stoichiometric material in the temperature range 800–1400°C due, it is thought, to the presence of a fine dispersion of uranium metal. Other workers, notably at the Martin Co., U.S.A., have shown that the incorporation of radially oriented molybdenum fibres improves the thermal conductivity by useful amounts.

Uranium-plutonium oxide solid solutions are of interest where plutonium enrichment is sought. Fabrication and irradiation studies of mixed oxides at the 1, 10 and 15 wt.% Pu levels have shown few differences in behaviour from $UO_2$. Plutonium readily forms oxygen-deficient structures and care must be exercised in choice of sintering conditions if control of stoichiometry is required. Under irradiation (U, Pu) $O_2$ shows less tendency to thermal stress cracking than $UO_2$. Fission gas release characteristics seem to be very little altered by plutonium additions.

*Interstitial compounds.* Carbide fuels are receiving considerable attention for potential fast reactor and sodium-cooled thermal reactor use due to their having higher thermal conductivities than oxide combined with comparable fission product migration kinetics. It has been demonstrated that UC and (U, Pu)C can be fabricated in high density sintered and cast forms and by vibratory compaction. These materials have been irradiated to burn-ups as high as 25,000 MWd/t with very low swelling and release of fission products. However, for certain applications the chemical reactivity of carbides would appear to restrict or eliminate their use. By studying the alloying behaviour of carbides these difficulties may be overcome. UC forms alloys of four different types:

(1) An insoluble system with another carbide, e.g. UC–SiC; thus SiC may be used to form a coating on UC with which it will not react and which is more oxidation-resistant.

(2) Partial solubility with another carbide–VC, WC, TiC.

(3) Complete solubility with another carbide–ZrC, NbC. These solid solutions are stronger than UC and have higher melting points. They have been adopted, therefore, as cathode materials for nuclear diodes since these must operate with surface temperatures approaching 2000°C. HfC is also completely soluble and may prove to be a suitable burnable poison.

(4) Eutectics are formed with certain metals, notably Fe, Cr and Cu having melting points of about 1100–1200°C. These may find application where the metal imparts improved corrosion resistance and as a means of forming a metallurgical bond with the can.

With UC showing favourable properties for a reactor fuel attention has been turned to other uranium compounds having a similar structure i.e. UN, US and UP. The table below compares some of their properties.

The nitride, phosphide and sulphide have all been fabricated to high densities and preliminary irradiation studies have shown promising behaviour. UN has been irradiated to 3·5 per cent burn-up of U atoms and exhibited low gas release and swelling. UN is also reported to be more resistant than UC to corrosion by boiling water at 100°C.

An interesting development of a safe thermal reactor fuel is uranium-zirconium-hydride which was originally used by General Atomics in their TRIGA research reactor and was later adopted for the SNAP compact, low-power space reactors. The fuel comprises 8 per cent U, 91 per cent Zr and 1 per cent H by *weight* or 49 per cent hydrogen by atoms, representing *nearly* as high a concentration of H atoms per unit volume as in water. The intimate association of fuel and moderator atoms ensures a prompt negative temperature coefficient of reactivity of approximately $-1·3 \times 10^{-4}$ per degree centigrade in the SNAP reactors where it is used at coolant temperatures up to 650°C. The hydride dissociates at high temperatures and the cladding must be fairly impervious to hydrogen.

*Cermets.* Dispersion fuels have certain advantages over bulk fuels in terms of safety and dimensional stability. Improvements have been made in dispersion fuels both with metal and ceramic matrices. In the former case improvements have been in the directions of higher strength matrix materials and more ideal dispersions of the ceramic particles. Nimonic alloys and the refractory metals W, Mo and Nb have been used as matrix materials. In the latter case vapour deposited coatings on $UO_2$ granules have permitted cermets with as low as 20 per cent metal phase to be fabricated by isostatic pressing techniques.

It has long been argued that a cermet has the best chance of achieving high burn-up when the ceramic phase consists of large spheres of low density evenly dispersed in a dense metal matrix. Techniques have been developed for making $UO_2$ spheres up to 700 μ diameter of varying density, coating these with steel powder and compacting to a high density by swaging, rolling or isostatic pressing to produce cermets of near-ideal structures. Under irradiation these have shown marked improvements in performance over less perfect structures; for example, a 50 vol.% $UO_2$–stainless steel cermet plate-type element irradiated at a surface temperature between 600°C and 650°C has been shown to fail at burn-ups in excess of 5 per cent U atoms ($5·6 \times 10^{20}$ fissions/cm³) when fabricated by the original methods and to be capable of exceeding 10 per cent burn-up ($11·2 \times 10^{20}$ fissions/cm³) when made by the improved technique.

*Ceramic dispersion fuels.* The development of dispersion fuels with non-metallic matrices has been mainly directed towards the use of uncanned fuels for high temperature gas cooled reactors. The elimination of the metallic can gives improved neutron economy which, together with higher coolant temperatures giving higher thermal efficiencies, leads to cheaper power costs. The main requirements of such fuels are a very high degree of fission product retention, good dimensional stability under irradiation together with adequate

| Material | Melting point °C | Uranium density g/cm³ | Thermal conductivity cal/cm² °C | Thermal neutron cross-section of 2nd element (barns) |
|---|---|---|---|---|
| UC | 2350 | 12·9 | 0·050 | 0·004 |
| UN | 2650 | 13·5 | 0·044 | 1·88 |
| UP | 2640 | — | — | 0·20 |
| US | 2450 | ~ 10·0 | 0·026 | 0·52 |
| UO₂ | 2800 | 9·6 | 0·008 | 0·0002 |

mechanical strength and high thermal conductivity in order to reduce thermal stresses.

In the oxide systems dispersion fuels based on beryllia matrices appear the most promising due to its superior nuclear and thermal properties compared with other oxides. Use has been made of similar fabrication techniques to those adopted for cermets, that is prior coating of $UO_2$ or $(U, Th) O_2$ spheres with BeO in order to achieve a uniform dispersion of fuel in the matrix. To achieve a high degree of fission product retention matrix densities of $> 98$ per cent theoretical are required to eliminate open porosity and fine grain size is desirable in the beryllia to reduce the effects of fast neutron bombardment. Fuels of this type containing 30 vol.% fissile/fertile phase can achieve burn-ups of $10^{21}$ fissions/cm$^3$ at temperatures of 1000–1200°C with very small changes in dimension and low release of fission gases ($< 1$ per cent). The main contributor to fission gas release is recoil from fuel particles at the surface and hence the incorporation of a dense fuel free surface layer is the major fabrication problem now being tackled.

Fuels based on graphite matrices such as used in the O.E.C.D. Dragon reactor at Winfrith, England have achieved a major break through since the coated particle fuel concept has been demonstrated, changing the whole philosophy of these reactors from what was essentially fission product emitting fuel, at least as far as the gaseous fission products, to fission product retaining with cleaner primary coolant circuit. Typically the fuel kernel consists of a sintered or fused spherical particle of $(Th, U)C_2$ of 200–500 $\mu$ in diameter coated with a carbon or carbide (e.g. SiC) layer of the order of 100 $\mu$ in thickness. The preferred method of coating is from the gaseous phase at high temperatures in a fluidized bed. Pyrocarbon so deposited is highly orientated with the "c" direction predominantly radial and exhibits very low diffusion coefficients for the rare gases particularly in this direction. Typically $R/B$ (rate of release/rate of birth) values for short lived fission gases can be less than $10^{-6}$ at temperatures of 1400°C. The release rates of Sr, Ba and Cs are, however, some 2–3 orders of magnitude higher, and have led to the development within the O.E.C.D. Dragon Project of a modified type of coating which consists of a SiC layer sandwiched between pyrocarbon. Release values from this type of coating shows values for the Group II and III metals comparable to the fission gases.

High burn-up tests have shown that up to 30 per cent burn-up of all heavy metal atoms in the fuel kernel can be achieved at an irradiation temperature of 800°C with particle coating failure rate around 1 per cent and $R/B$ values in the range $10^{-4}$–$10^{-5}$. It has also been demonstrated that burn-ups up to about 10 per cent can be obtained in the temperature range up to 1600°C with low $R/B$ values and failure rates. A reasonable correlation between release rate and particle failure exists and hence the main metallurgical problem is in determining the causes for coating failure. Fission fragment damage to the inner layers

of the coating and swelling of the fuel kernel are probably the two major effects. Oxide coated (BeO or $Al_2O_3$) particles of $UO_2$ have also been investigated and have shown good high temperature performance but are prone to failure on low temperature irradiation. It is confidently expected therefore that coated particles can be made that will meet the immediate requirements for high temperature reactors; the next stage in the development is to incorporate these in a matrix which is impermeable enough to act as a second barrier to fission product diffusion and in addition to have adequate chemical and physical properties which will permit direct cooling of the fuel bodies in the coolant stream.

In the development of fuels with improved performances it is unlikely that dramatic discoveries will be made; too much exploratory work has been done during the past ten years for this to occur. More probably, the pattern of development will follow that set by metal fuels, i.e. a better scientific understanding of the processes occurring in the fuel during irradiation will suggest the most profitable directions for improvements.

*Bibliography*

(1964) *International Conference on Carbide Fuels*, New York: Macmillan.
(1963) I.A.E.A. Conference on "New Reactor Fuel Materials", Prague.
(1962) *Plutonium as a Power Reactor Fuel*, American Nuclear Society Topical Meeting, USAEC, Report HW-75007.
(1962) *Ceramic Matrix Fuels Containing Coated Particles*, USAEC, TID-7654.
BELLE J. (Ed.) (1961) *Uranium Dioxide*, U.S. Govt. Printing Office.

B. R. T. FROST and J. B. SAYERS

**NUCLEIC ACIDS, STRUCTURE OF.** *Introduction.* In all species there is a characteristic resemblance between parents and progency. For such similarities to be possible, information must be transmitted from generation to generation. It is now clear that this information is stored in the structure of molecules. As the new organism grows this information is used to control its development. Work during the last two decades has shown that molecules called nucleic acids have fundamental roles in the transmission of the genetic information from one generation to the next and in its translation into structural features in the new organism. Further it has been realized that if these processes are to be understood at the molecular level it is necessary to determine the structure of the molecules involved.

Nucleic acids are polymer molecules built in a linear array from monomers called nucleotides. Each nucleotide consists of a sugar ring which is covalently bonded to both a phosphate group and a planar

oup called a base. There are two kinds of nucleic
id: deoxyribonucleic acid (DNA) and ribonucleic
id (RNA). They can be distinguished by the chemi-
al structure of the nucleotides from which they are
omposed (Fig. 1). The sugar of the RNA nucleotide
iffers from that of DNA in having one of the hydro-
ens attached to $C_2$ replaced by a hydroxyl group. For
oth RNA and DNA, polymerization can be consider-
d as taking place by the formation of a covalent bond
etween the phosphate 01 of each nucleotide and the
ugar C3 of another nucleotide. The various nucleo-
des along a polynucleotide chain differ only in the
ature of the base. For DNA there are four commonly
ccurring bases: adenine, thymine, guanine and cyto-
ne (Fig. 2). RNA differs from DNA in having uracil
istead of thymine as one of the four most commonly

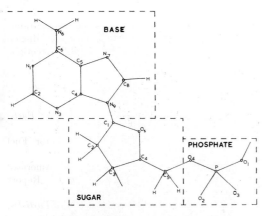

*Fig. 1. Chemical structure of a deoxyribonucleic
acid nucleotide in which the base is adenine. Polymerization can be considered as taking place by the
formation of a covalent bond between the phosphate
01 of each nucleotide and the sugar C3 of another
nucleotide.*

ccurring bases. Uracil has a hydrogen attached to
5 instead of the methyl group of thymine. It has
ot yet been possible to determine the complete nuc-
otide sequence of any naturally occurring nucleic
cid.

*The molecular structure of DNA.* A great deal of
vidence has been accumulated which identifies DNA
s the molecule in whose structure the genetic infor-
ation is coded. All organisms contain DNA except
few viruses which contain RNA. In those organisms
hich are built up from cells it is located in the cell
ucleus in structures called chromosomes. It also
ccurs in sperm which can be thought of as packages
f DNA. DNA has been extracted from a large variety
f cells and characterized by physical and chemical
echniques. It is a polymer with a molecular weight of
any million; in solution it behaves as a semi-rigid

rod. Chargaff determined the base composition of
DNA from a large variety of sources and found that
for DNA from a particular organism the number of
adenines always equalled the number of thymines and
similarly the number of guanines equalled the number
of cytosines. However, the ratio of adenines to guanines varied considerably with the source of the
DNA.

First Astbury and then Wilkins and his collaborators
found that it was possible to prepare fibres of DNA in

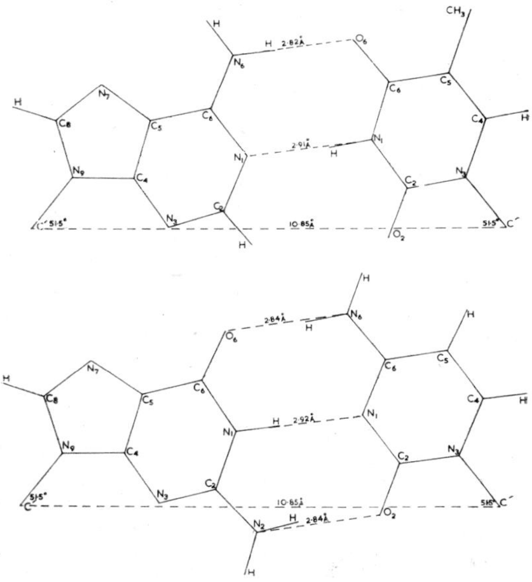

*Fig. 2. Watson-Crick base-pairing scheme for
DNA (Refined by Arnott).*

which the long polymer molecules were approximately
parallel to the length of the fibre. X-ray diffraction
patterns from these fibres indicated that there were
regions in the fibre where the DNA molecules were
arranged in a regular three-dimensional array. This
could only be possible if the molecules had a very
regular structure since any irregularities would disturb
the crystallinity of the sepcimen. The overall intensity
distribution in the X-ray diffraction patterns indicated that the molecule had a helical shape and measurements on the patterns gave the pitch of the helix,
the approximate molecular diameter and the number
of repeating units per helix pitch. Clearly a repeating
unit must consist of one or more nucleotides. Watson
and Crick used physical and chemical data as the
basis of a molecular model for DNA. In their model
two polynucleotide strands were held together by
hydrogen bonds between bases in different strands.
The pairing was such that adenine was always hydro-
gen-bonded to thymine and guanine to cytosine, and
although the shapes of the individual bases vary considerably, the dimensions of the two base pairs are

very similar (Fig. 2). Therefore it was possible for the sugar-phosphate chain to have a regular conformation which was independent of the sequence of the base pairs along the molecule. In their model the two-stranded structure was coiled up into a right-handed helix such that there were ten nucleotide pairs per pitch of the helix. The plane of each base pair was perpendicular to the helix axis and the sequence of atoms in one sugar-phosphate chain was the reverse of that in the other. Since an adenine in one chain was always paired with a thymine in the other it can be seen that the sequence of bases in one chain determines that in the other, and therefore all the information required to specify the base-pair sequence in a DNA molecule is contained in one chain. This relation has important implications since it is known that the amount of DNA in a cell doubles before the cell divides and the composition of the DNA in the two daughter cells is the same as that in the parent cell. In fact all the cells of an organism contain DNA with the same base composition. The complementary base sequence in the two chains of the Watson and Crick model for DNA suggest a mechanism whereby DNA molecules can copy themselves. This process of duplication or replication, as it is sometimes called, can be imagined to take place by the ends of the two poly-nucleotide strands of the DNA becoming unravelled in the same way that a two-stranded rope might be taken apart. Once they were unravelled the hydrogen bonding groups on the bases would be exposed and could serve as a template for the synthesis of a new DNA chain from nucleotides in the vicinity of the DNA. It is likely that the presence of two partly synthezised chains on the unravelled chains of the original DNA molecule would encourage further unravelling and so the process would be repeated until two new complete DNA molecules had been formed. This theory of DNA duplication is necessarily naive and it should be stressed that as yet it is unproven. Clearly errors in such a copying process are more likely to occur if there are other molecules in the neighbourhood of the replicating DNA which have similar chemical structures to DNA bases and base-pairs. Many molecules of this type fall into the class of chemicals called mutagens which are characterized by their ability to produce changes in the genetic constitution of an organism. Such changes are called mutations.

From a detailed comparison of the observed and calculated X-ray diffraction patterns, Wilkins and his collaborators have refined the model proposed by Watson and Crick. Although some of their modifications have been considerable their most recent model (Fig. 3) retains the essentials of the original hypothesis, i.e. a right-handed helical conformation with specific base-pairing (Fig. 2) and polynicleotide chains running in opposite directions. Since at normal pH both DNA and RNA are completely ionized, the alkali metal salts of the acids are usually studied. It has been found that the conformations assumed by DNA n the fibre depends on the relative humidity of the

fibre environment and the counter-ion present, e.g when the relative humidity of the environment of sodium DNA fibre is decreased from 92 to 75 per cen there is a very marked conformational change. Thi change is reversible and does not involve the breakin of any hydrogen or covalent bonds. It is produce simply by rotation about single covalent bonds. Sinc there are eleven such bonds in each DNA nucleotid there is clearly ample scope for conformational change

The low humidity form of sodium DNA is called ⱯΑ and the high humidity form B. X-ray diffractio patterns from chromosomal material and sperm head were sufficiently well defined to show that the DN in these structures, even though bound to protein was also in the B-form. It appears therefore that th B form is not an artefact of the extraction procedur and is the biologically important conformation o

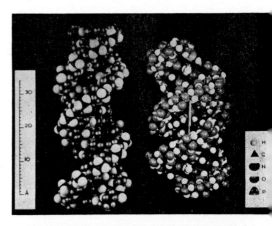

*Fig. 3. Molecular models of (left) deoxyribonucleic acid in the B conformation and (right) the helical regions of ribonucleic acid. The ribonucleic acid conformation is very similar to the A conformation of DNA (see table).*

DNA. The other conformations of DNA are artefact of drying but their determination has given adde confidence in the Crick-Watson hypothesis since the account for quite different diffraction patterns wit models which are clearly related to the B-form. Th various conformations differ in the number of residue per helix turn, the helix pitch and the nucleotide con formation. They are compared in the table.

*The stability of the molecular structure of DNA.* Th stabilization of the DNA secondary structure i generally regarded as being due principally to hydro phobic interactions between successive base-pairs an hydrogen bonds between the two bases in each base pair. Hydrophobic and van der Waal's interaction involving other groups in the DNA also make a con tribution to the molecular stability. The effect o relative humidity and the nature of the counter-io

*Various structures of DNA in fibres*

| Configuration | Number of nucleotide pairs per turn of helix | Inclination of base-pairs to helix axis | Salt | Relative Humidity | Lattice | Unit cell dimensions | | | | Crystallinity |
|---|---|---|---|---|---|---|---|---|---|---|
| | | | | | | a (Å) | b (Å) | c (Å) | β (Å) | |
| A | 11·0 | 70° | Na, K, Rb | 75% | Monoclinic | 22·24 | 40·62 | 28·15 | 97·0 | Crystalline |
| B | 10·0 | 88·2° | Li | 66% | Orthorhombic | 22·7 | 31·2 | 33·7 | | Crystalline |
| | | | Li | 75–90% | Orthorhombic | 24·4 | 38·5 | 33·6 | | Semicrystalline |
| | | | Li, Na, K, Rb | 92% | Hexagonal | 46 | | 34·6 | | Semicrystalline |
| C | 9·3 | 85° | Li | 44% | Orthorhombic | 20·1 | 31·9 | 30·9 | | Semicrystalline |

on the conformation assumed by DNA has already been described. The conformation of sodium DNA in the fibre can sometimes be changed by sudden stretching of the fibre. The detailed molecular structure of the stretched conformation has not yet been elucidated. However, optical birefringence measurements and a preliminary analysis of the X-ray diffraction pattern indicate that stretching produces a molecule with a smaller molecular diameter in which the base-pairs are inclined at about 45° to the helix axis rather than perpendicular to it as in the B-form. This change in molecular conformation is reversible and the normal DNA conformation is regained when the tension on the fibre is reduced or the relative humidity of its environment is increased.

X-ray diffraction studies have indicated that in water solution at room temperature, neutral pH and moderate ionic strenghts, DNA over a wide range of concentration is in the B conformation. Conformational changes in solution can be produced in a number of ways, e.g. by highly alkaline or highly acidic pH, by very low ionic strength, by raising the temperature of the solution. When a solution of DNA is heated the breakdown of the secondary structure occurs suddenly at a temperature which is determined principally by the ionic strength of the solution and the base composition of the DNA. This structural change involves breaking of the hydrogen bonds linking the two DNA strands and because it is a co-operative phenomenon i.e. all the hydrogen bonds in the molecule are broken during a very small increase in temperature) and by analogy with the solid–liquid transition it is called the the melting temperature of the DNA. The melting temperature of DNA increases with its cytosine content. This has been attributed to the extra hydrogen bond of the guanine-cytosine base-pair but this explanation may be too naive. If after melting, the DNA solution is cooled quickly the original conformation is not regained. This is not surprising since it is

extremely unlikely that the two strands of each molecule will line up opposite each other so that the original base-pairs can be reformed. There is still, however, considerable base-stacking, base-pairing and helical, structure in this so-called denatured DNA.

It is expected that the base-pairs formed will include others besides those proposed by Watson and Crick (e.g. adenine–guanine, thymine–cytosine) and because these pairs are of unequal size the sugar-phosphate chains will no longer be able to assume regular conformations. Some of these base-pairs might be expected to occur between nucleotides in the same polynucleotide chain by the chain folding back on itself and others between nucleotides in different chains as in undenatured DNA.

DNA preparations from higher organisms usually contain a number of molecular species differing in molecular weight and base composition. One of the reasons for this is that in general DNA from different chromosomes in the same cell will differ in this way. However, many DNA preparations from bacteria and bacterial viruses are very homogeneous and if one of these preparations is heat denatured and then cooled very slowly a large fraction of the original DNA structure may be regained. The resultant DNA is called renatured and has been shown to be biologically active.

Changes in the DNA structure can also be produced by adding other molecules to a DNA solution. Molecules of particular interest are those which are important biologically, e.g. mutagens, carcinogens and antibiotics. Proflavine and ethidium are both molecules which contain a planar group which is about the same size and shape as a DNA base-pair. Proflavine has been used to produce mutations in viruses which have been attributed to the insertion or deletion of a base-pair during DNA replication. X-ray diffraction studies of complexes of DNA and proflavine and DNA and ethidium indicate that the planar group of these molecules can slip in between successive base-pairs in the DNA. This requires some uncoiling of the DNA helix at the point of insertion. It may be that an interaction of this nature is responsible for the insertion and deletion mutations described above.

Changes in the conformation of DNA in solution can be followed in a number of ways, e.g. by optical rotary dispersion, ultra-violet absorption spectra and buoyant density in CsCl density gradient measurements. There is a 20 per cent increase in the optical density at 2600 Å of a DNA solution following heat denaturation.

*The molecular structure of RNA.* The information defining the characteristics of an organism is stored in the base sequence of its DNA but is expressed to a large extent in terms of the proteins which it synthesizes. Proteins are also polymer molecules consisting of a linear arrangement of monomers called amino acids. There are twenty different commonly occurring amino acids and their sequence along the protein chain determines its characteristics. The relation between the information stored in the four-letter language of DNA base sequence and its expression in the twenty-letter language of protein amino acid sequence is called the genetic code. Protein synthesis takes place at sites in the cytoplasm of the cell called ribosomes which are complexes of protein and an RNA called ribosomal RNA.

Communication between the ribosomes and the cell nucleus where the genetic information is stored is effected by an RNA appropriately enough called messenger RNA. Messenger RNA is synthesized in the nucleus under the control of the DNA and it has been shown that genetic information is stored in its base sequence. In the cytoplasm the messenger RNA interacts with ribosomes in such a way that the arrangement of amino acids in the growing protein chain is determined by this base sequence. Another RNA called transfer RNA is involved in this process. Its function is to collect free amino acids in the cytoplasm and take them to the ribosomes for incorporation into protein. There is at least one kind of transfer RNA molecule for each of the twenty different amino acids. Clearly precise mechanisms of molecular recognition are involved in these processes suggesting that these molecules have very well defined structures and it is hoped that the determination of these structures will indicate how they function.

It is only during the last few years that physical and chemical studies have been made on well characterized fractions of the total cellular RNA. The molecular weight of ribosomal RNA may be as high as one million, messenger RNA typically a few hundred thousand and transfer RNA about 25,000. It should be noted the name transfer RNA describes a biological function whereas what is generally studied is a chemical fraction called soluble RNA which is rich in transfer activity. Information on the nucleotide sequence

of messenger RNA has come from a technique similar to that used for renaturing DNA (previous section). The DNA which controls the synthesis of a particular messenger RNA is heat denatured and held in single strand form with the hydrogen bonding groups on the bases exposed. Under certain conditions this DNA will interact with the messenger RNA to form a hybrid molecule in which part of the DNA polynucleotide chain has a messenger RNA molecule hydrogen bonded to it in much the same way that the two DNA strands are held together in undenatured DNA. Studies of hybrid molecules of this kind suggest that in some virus at least messenger RNA is synthesized using one strand of the DNA as a template.

Early physical and chemical studies on RNA were made on preparations which consisted predominantly of ribosomal RNA. RNA's from a number of plant viruses (e.g. Tobacco Mosaic Virus), which contain no DNA, were also studied. It appeared that all these RNAs had a rather similar molecular structure which whilst it had considerable base-pairing and helical character was much less regular than that discovered for DNA. X-ray diffraction patterns from RNA fibres indicated that the helical regions were probably two stranded and that the structure changed little with the relative humidity of the fibre environment. However, the X-ray diffraction data was inadequate for the molecular structure to be determined in any detail. This only became possible when yeast RNA preparations were obtained from yeast which contained molecular fragments with a molecular weight of about 10,000. If concentrated gels of these fragments were allowed to dry slowly regions were formed in which the fragments were arranged in a regular array. These regions could be oriented by stretching the gel just before it dried. X-ray diffraction patterns from fibres produced in this way had a similar overall intensity to those from other RNA preparations but were much better defined. They indicated that these fragments had a very regular molecular conformation rather similar to the A form of DNA (Fig. 3). It also became clear that the helical regions of the microsomal and viral RNA studied previously must also have the same conformation rather similar to that of A DNA. The lack of definition in the patterns from these molecules was attributed to either irregularities in the helical regions or to regular helical regions being separated by irregular regions which prevented them from packing regularly. Since in the cell ribosomal RNA is bound to protein it is possible that the conformation of the RNA is changed during extraction. However, parallel studies of ribosomes and extracted RNA indicated that the conformation of the RNA is little changed by extraction.

More recently X-ray diffraction patterns have been obtained from RNA extracted from both Reo virus and Wound Tumour virus. They indicate that the molecule is two stranded with a conformation as regular as that of DNA. Although these patterns have not yet been interpreted in detail it appears that the conformation assumed by these molecules is very similar to that of the yeast RNA fragments and therefore all RNA's appear to have similar molecular structures which are independent of the relative humidity.

*The structure of viral nucleic acids.* One of the most striking results of studies on a wide variety of cells from many different organisms is the similarity of the mechanisms of transmission and translation of genetic information. For example DNA's from a wide variety of sources have very similar molecular structures despite the wide variation in their base-sequence. In this respect a number of viruses are the exception which proves the rule. Some plant viruses (e.g. Tobacco Mosaic and Turnip Yellow Mosaic) and Polio virus have no DNA and contain only RNA and protein. In these cases the RNA is presumed to correspond to the messenger RNA of higher organisms. The conformation of RNA in Tobacco Mosaic Virus is quite different from that of the extracted RNA. In the virus the RNA appears to interact in a specific way with the viral protein so that as an extended single strand it assumes a helical conformation with a radius of about 40 Å and a pitch of about 23 Å. Reo virus and Wound Tumour virus also contain no DNA and as was noted in the previous section their RNA has a two stranded conformation with the same degree of regularity as DNA. The virus MS2 also has no DNA and like Tobacco Mosaic Virus has a single stranded RNA. The virus multiplies in a bacterial host where it produces a two-stranded replicating form which has one chain like that of the original virus and the other with a complementary base sequence so that the molecule has a highly regular conformation. A similar situation is observed with the virus $\varphi$X-174 which consists of protein and a single-stranded DNA. This virus also multiplies in a bacterial host via a two-stranded replicating form. A further interesting feature of this single stranded viral DNA is that its two ends seem to be joined together so that it is "circular". The virus $\varphi$X-174 has an approximately spherical shape, *fd* is another virus which has a single-stranded DNA but in contrast to $\varphi$X-174 it is filamentous.

*Synthetic polynucleotides.* Certain enzyme systems have been isolated which are capable of synthesizing polynucleotides from single nucleotides. In this way polynucleotides of known composition can be synthesized and used as model compounds in experiments designed to determine the structure and function of naturally occuring nucleic acids. By substituting such synthetic RNA's for messenger RNA in protein synthesizing systems much information has been obtained on the nature of the genetic code. For example the polypeptide produced under the control of a synthetic messenger containing uracil as the only base contains phenylalanine as the only amino acid. This indicates that a sequence of uracils in the messenger RNA codes for phenylalanine in the synthesized protein.

The molecular conformation assumed by synthetic polynucleotides in solution is in some cases very sensitive to pH. For example at neutral pH polyadenylic acid (a synthetic RNA containing adenine as the only base) has a rather irregular single strand structure but if the pH is reduced to 5·5 a two-stranded complex is formed in which the polyadenylic strands are held together by hydrogen bonds between adenines in different chains. The base pairs are stacked on top of each other as in DNA but in contrast to DNA the atomic sequence in the two chains is in the same direction. Fibres containing the molecule in this form have been prepared and used in X-ray diffraction studies which were adequate for a detailed molecular model to be built. The stability of this structure is attributed to the protonation of adenine at N1 (Fig. 2). A similar structure has been proposed for polycytidylic acid (a synthetic RNA in which cytosine is the only base) at acid pH. If solutions of polyadenylic (poly A) and polyuridylic (poly U) acid are mixed together at neutral pH there is a decrease in the optical density of solution. For naturally occurring polynucleotides an increase in the optical density of the solution is associated with a change to a less ordered structure, e.g. following thermal denaturation of DNA. Therefore the decrease in the optical density following mixing of poly A and poly U has been taken to indicate the formation of ordered structure. The decrease in optical density as a function of the relative amounts of poly A and poly U mixed together has been found to depend on the ionic content of the solution. The details are not yet well defined but it appears that maximum decreases are obtained for mixtures containing poly A and poly U in molar ratios of 1 : 1 and 1 : 2. These observations have been taken to indicate the formation of a two-stranded poly A + poly U complex in the first case and a three-stranded poly A + poly U complex in the second. X-ray diffraction patterns have been obtained from fibres of both two- and three-stranded material. However, a detailed interpretation of these patterns in terms of molecular models has not yet been published. Other synthetic polynucleotides also appear to form complexes of the poly A + poly U type. Of particular interest is that formed by polycytidylic acid and polyinosinic acid (inosine has a very similar chemical structure to guanine). Fibres of this complex give X-ray diffraction patterns which resemble those obtained from Reo virus RNA but again no detailed molecular model has been described.

*Acknowledgements.* I should like to thank Sir John Randall, F.R.S. for provision of facilities and encouragement and Professors M. H. F. Wilkins, F.R.S. and F. Hutchinson and Drs. S. Arnott and W. Muller for discussion.

*Bibliography*

CHARGAFF E. (1963) *Essays on Nucleic Acids*, Amsterdam: Elsevier.

CHARGAFF E. and DAVIDSON J. N. (Eds.) (1955) *The Nucleic Acids I*, New York: Academic Press.

CHARGAFF E. and DAVIDSON J. N. (Eds.) (1955) *The Nucleic Acids II*, New York: Academic Press.

CHARGAFF E. and DAVIDSON J. N. (Eds.) (1960) *The Nucleic Acids III*, New York: Academic Press.

CRICK F. H. C. (1957) *Nucleic Acids, Scientific American* **197** (3), 188.

HURWITZ J. and FURTH J. J. (1962) *Messenger RNA, Scientific American* **206** (2), 41.

JORDON D. O. (1960) *The Chemistry of the Nucleic Acids*, London: Butterworths.

SINSHEINER R. L. (1962) *Single-Stranded DNA, Scientific American* **207** (1), 109.

SPIEGELMAN S. (1964) *Hybrid Nucleic Acids, Scientific American*, **210** (5), 48.

STEINER R. F. and BEERS R. F. (1961) *Synthetic Polynucleotides*, Amsterdam: Elsevier.

WILKINS M. H. F. (1961) *Molecular Configuration of Nucleic Acids, Science* **140**, 941.

W. FULLER

**OLBERS' PARADOX.** Olbers' paradox is, roughly speaking, that the sky is dark at night. H. W. M. Olbers, writing in 1826, was the first to consider the cosmological significance of this fact.

Imagine a universe uniformly populated with stars which have no systematic motions and which have always been shining. Then it is easy to show that the brightness of light at an arbitrary point in empty space is equal to that at the surface of a star. Thus, the night sky should be as bright as day, and much brighter than the day sky on Earth.

This is the paradox. Its resolution by modern cosmological theory rests mainly on two points. First, that the stars have not been shining for infinite time; secondly, that owing to the red shift of the light from distant galaxies, arising from the expansion of the Universe, this light is less energetic than it would be on the purely static hypothesis. These two factors are sufficient to reduce the light intensity of the night sky to its observed value (about $10^{-13}$ erg cm$^{-3}$).

W. B. Bonnor

**OPTICAL DIFFRACTOMETER.** The use of optical diffraction patterns—or optical transforms as they are often called—in studying X-ray diffraction problems is described under the heading "Crystal-Structure Determination, Optical Methods of". Diffraction patterns can be prepared with very simple apparatus if the diffracting objects—or masks—are small enough. For example, if a handkerchief is stretched taut and held close to the eye, and the eye is focused on to a distant point source of light, the diffraction pattern of the mesh of the handkerchief can be seen without any additional apparatus. If masks representing portions of crystal structures are to be prepared rapidly, however, it is convenient to make them fairly large, and it then becomes difficult to observe their diffraction patterns without special apparatus; the angles of diffraction involved may be less than one degree. A device designed for this purpose is usually described as an optical diffractometer; there have been many different designs, but there are certain basic conditions which must be satisfied if successful results are to be obtained, and for convenience these requirements will be discussed in terms of the optical diffractometer described by Hughes and Taylor (1953) and by Taylor and Thompson (1957, 1958).

Figure 1 shows the optical arrangement. Light from a primary source A—usually a compact-source mercury-vapour lamp—is focused by means of a lens and prism on to a secondary source B, a pinhole which may be from 5 to 100 microns in diameter. A filter to isolate one spectral line is also usually included in this part of the system. The secondary source B is in the back focal plane and on the axis of a lens C whose focal length may be about 1·5 metres. A second lens D is usually—though not necessarily—an identical lens and forms an image of B in its focal plane F after reflection by a mirror E. The Fraunhofer diffraction pattern, or *optical transform*, of a mask placed at 0 between the lenses will appear at F and may be photographed directly on a film or observed by means of a microscope. Figure 2 shows two typical masks, together with their optical diffraction patterns. The masks are reproduced full size, and the pattern enlarged 25 times.

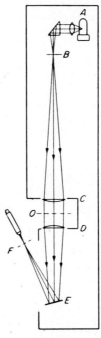

*Fig. 1. General arrangement of optical diffracto-meter.*

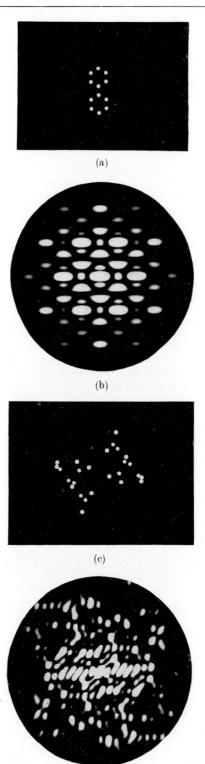

(a)

(b)

(c)

(d)

In an ideal diffractometer a perfectly parallel beam of constant amplitude and phase across the whole aperture should fall on the mask, and the phase changes introduced between the mask and the focal plane F should be identical for all beams parallel to a given direction. Such conditions are very difficult to secure; there are many possible causes of imperfect operation. The lenses C and D must be of very high quality, fully corrected for spherical aberration and free from bubbles, striae, defects in blooming, dust, grease, etc. In a telescope or other image-forming system each point in the image receives light from all parts of the lens and a small local defect is of little consequence; in a diffractometer such a local defect might occur opposite one hole in the mask and could change the whole pattern. As in all optical systems careful alignment is necessary; all the components must be coaxial, and in order to ensure this certain mechanical design conditions are imposed. Each optical component must be fitted in an adjustable mount permitting both lateral translation and tilt; it is not adequate to assume that the optical axis of a lens, for example, passes through the centre of the mount and is normal to it. Details of a recommended alignment procedure are given by Taylor and Thompson (1957).

The illuminating system is of great importance; in particular it is necessary to choose the size of the

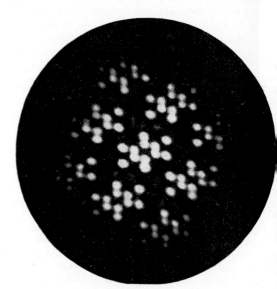

*Fig. 3. Optical Fourier synthesis of hexamethyl-benzene (after A. W. Hanson).*

*Fig. 2. (a) mask representing simple planar molecule, (b) optical diffraction pattern of (a), (c) mask representing a pair of centrosymmetrically-related non-planar molecules, (d) optical diffraction pattern of (c).*

pinhole used as the secondary source B in relation to the mask whose transform is being prepared. If the pinhole is too large the illumination may not be coherent enough over the whole mask area; if it is too small the photographic exposure may become too long. The system must be extremely rigid to minimize vibration effects; this is usually ensured by clamping the optical components to a steel girder ($8'' \times 4''$ I-section is suitable) and supporting the girder itself on resilient mounts. Precise focusing is also important and special techniques have to be adopted because the normal criteria (e.g. sharpness of line) do not apply.

Masks for use in the diffractometer can be prepared either by creating holes in opaque material, for example by drilling in metal, or by punching paper or old X-ray film (the latter has been found to be particularly convenient). Some kind of pantograph punch is often used for the rapid preparation of masks from a large-scale drawing. An alternative is to prepare the mask photographically either by direct photography or by photo-etching on copper foil; masks on a much smaller scale may be made this way. Photographic plates alone are not very satisfactory because of phase changes introduced by non-uniformities in the emulsion or in the glass. If such a plate is sandwiched between optical flats with cedar-wood oil the phase effects can be satisfactorily minimised (Bragg and Stokes 1945). The technique used for etching is similar to that used for printed circuits on copper-coated plastic sheets; the plastic backing is later removed.

The diffractometer may also be used for *Fourier synthesis*. For this purpose the mask represents the amplitudes of the orders of diffraction which occur on

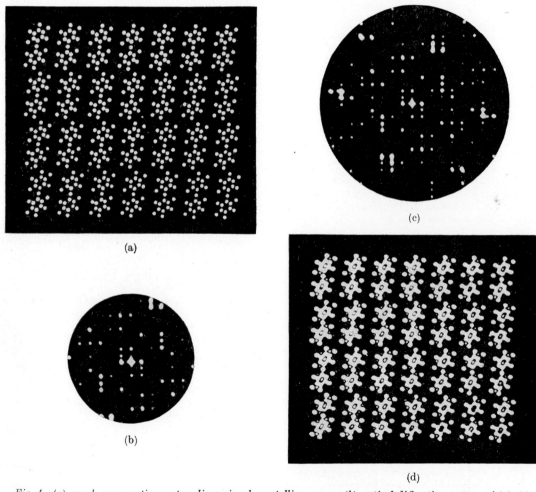

*Fig. 4.* (a) *mask representing a two-dimensional crystalline array,* (b) *optical diffraction pattern of* (a), (c) *portion of* (b) *restricted by a circular aperture,* (d) *recombined image formed from* (c) (*compare series-termination effects in Fourier synthesis*) (*after B. J. Thompson*).

an X-ray photograph, and it is necessary to introduce the correct relative phases; several techniques for achieving this have been described using either mica and polarized light or pieces of mica tilted at different angles. Figure 3 shows a Fourier synthesis of hexamethylbenzene prepared optically by Hanson's (1952) method.

The diffractometer can easily be modified to permit a double process of *Fourier transformation*, sometimes referred to as *image recombination*. The modification consists of substituting for the microscope a further lens which will form a real image of the mask 0 on a photographic plate G. Thus the object is at 0, its transform appears at F, and the transform of the transform, i.e. an image, appears at G. If various modifications of the transform are made at F the recombined image at G will also be modified; for example, the Abbe theory of limit of resolution can be demonstrated. Figure 4 shows an instance in which (a) is the mask, (b) its complete transform, (c) is a portion selected by an aperture placed at F, (d) is the recombined image produced at G.

*Bibliography*

BRAGG W. L. and STOKES A. R. (1945) *Nature* **156**, 332.
BUERGER M. J. (1950) *J. Appl. Phys.* **21**, 909.
HANSON A. W. (1952) *Nature* **170**, 580.
HOSEMANN R. and BONART R. (1957) *Kolloid-Zeitschrift* **152**, 53.
HUGHES W. and TAYLOR C. A. (1953) *J. Sci. Instrum.* **30**, 105.
LIPSON H. (1961) in *Encyclopædic Dictionary of Physics*, (J. Thewlis Ed.) **2**, 231; Oxford: Pergamon Press.
TAYLOR C. A. and LIPSON H. (1964) *Optical Transforms: Their Preparation and Application to X-ray Diffraction Problems*, London: Bell.
TAYLOR C. A. and THOMPSON B. J. (1957) *J. Sci. Instrum.* **34**, 439.
TAYLOR C. A. and THOMPSON B. J. (1958) *J. Sci. Instrum.* **35**, 294.
C. A. TAYLOR

**OPTICAL MICROMETER.** The purpose of an optical micrometer is to sub-divide a scale, either linear or circular, that may be viewed through an optical instrument such as telescope, microscope or projector.

The most widely used type of optical micrometer is a flat parallel plate of glass (refractor block) which is

| Tilt of parallel plate | Approximate error when $d$ is evaluated with respect to $i$ | Approximate error when $d$ is evaluated with respect to tan $i$ |
|---|---|---|
| $\pm\ 5°$ | 1 : 1000 | 1 : 6000 |
| $\pm\ 7^{1}/_{2}°$ | 1 : 400 | 1 : 3000 |
| $\pm 10°$ | 1 : 200 | 1 : 1500 |

interposed in the optical path of the instrument. A tilt of the plate will cause an image displacement which is determined linearly in terms of the amount of the angular tilt (Fig. 1).

The *tilting plate micrometer (or parallel plate micrometer)* follows a tangent law reasonably closely. This can be seen from the table of theoretical linear and

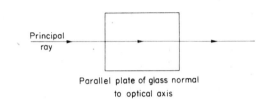

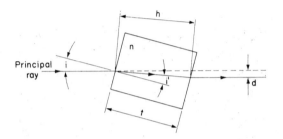

Parallel plate of glass normal to optical axis

*Fig. 1.*

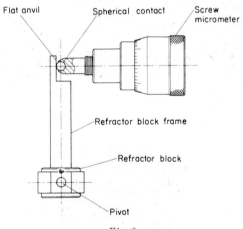

*Fig. 2.*

tangential errors for given tilts of the refractor block (see table).

The best arrangement for translating the angular movement of the refractor block into a convenient and accurate linear scale is to use a mechanical linkage which obeys the tangential law. For example, the refractor block can be mounted in a pivoted metal frame having a lever arm terminating in a flat anvil. This flat anvil must be parallel to a radial centre line through the pivot of the refractor block. The anvil can then be actuated by means of a conventional screw micrometer fitted with a spherical contact. The centre of the spherical contact being in line with the refractor block pivot when the block is normal to the optical axis of the instrument (Fig. 2).

When the tilting plate micrometer is applied to a microscope or projector, the refractor block is usually situated between the objective glass and its image plane. When applied to a telescope, it is more convenient to position the refractor block in front of the objective glass, thus avoiding the effect of change of magnification when using the instrument at varying distances from the object being viewed.

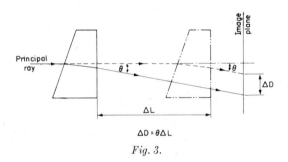

*Fig. 3.*

In some microscopes and projectors, the sub-division of the viewed scale is achieved in two steps, firstly from a scale on the eyepiece or screen graticule in the image plane to give the intermediate breakdown, and then a parallel plate micrometer for the final breakdown. This system requires the optical micrometer to operate asymmetrically with respect to the main optical axis of the instrument, which gives rise to secondary errors that could be compensated by the introduction of a weak lens situated between the refractor block and the image plane.

For precise sub-division it is best to use the optical micrometer perpendicular to the optical axis and to restrict the tilt angle to $\pm 5°$.

Another less used form of optical micrometer is a glass wedge interposed in the optical path of a microscope or telescope. When the wedge is moved along the optical axis this will also cause lateral displacement of the image seen through the eyepiece (Fig. 3).

The *travelling wedge type of micrometer* is little used in modern instruments as it demands more space than the tilting block version.

S. C. Bottomley

**ORGANIC SEMICONDUCTORS.** *An organic semiconductor is an organic solid the electrical conductivity of which is not ionic but electronic and in which the conductivity increases with a rise in temperature.* Hall effects and thermoelectric effects have been obtained with some organic materials, thus confirming the electronic nature of the conductivity. Further evidence on the same point comes from the constancy of the current when this is passed for a long time.

The *resistivity* of organic solids ranges from about $10^{20}\ \Omega$ cm for the paraffins, a value too high to be easily measured, to about $10^{-2}\ \Omega$ cm for $Q^+(T_2)^-$ where $Q^+$ denotes a quinolinium ion and T a tetracyanoquinodimethane molecule, and $6 \times 10^{-6}\Omega$ cm for a graphite-bromine compound.

In general the resistivity is lowered by an increase in the extent of unsaturation or aromaticity in the molecule of the solid; a decrease in its ionization energy; an increase in its electron affinity; or an increase in its electronic polarizability. Whilst there are good grounds for holding these broad conclusions experimental problems of electrode contacts, space-charges, the lack of electrical neutrality and perfect purity have made quantitative work difficult.

The *temperature dependence of resistivity* usually obeys a relation of the type $\varrho = \varrho_0 \exp(E/kT)$. Values of $E$ vary from less than $0.1$ eV for perylene-$Br_2$ ($\varrho = 8\ \Omega$ cm at room temperature) or cyananthrone ($\varrho = 10^5\ \Omega$ cm) to $2$ eV for diphenylbutadiene or 2-amino-4,6-dimethylpyrimidine. On simple theory $1/\varrho_0$ is proportional to the mobility of the carriers and to the number of species capable of yielding carriers.

The resistivity drops with an increase in *pressure*. For isoviolanthrone, increasing the pressure from 300 to 8000 kg cm$^{-2}$ lowered $\varrho$ by a factor of 500. This is of the same order as that of the similar effect in selenium or phosphors. It is explained qualitatively by the increased bandwidth arising from an increased overlap at higher pressures of orbitals on neighbouring molecules, and by a greater polarization energy.

*Lattice imperfections* in various samples of flavanthrone vary in number with the method of preparation and X-ray diagrams have been used to detect the extent of the imperfections. The resistivity was greater the more imperfections were present. If this result is general it is important since it suggests that such physical imperfections are not the primary source of current carriers. Chemical impurities are another matter.

*Simple molecular crystals* are mostly p-type conductors and the electrons are therefore usually trapped deeply enough not to affect the currents. The value of $E$ (in the expression for $\varrho$) falls usually between $0.3$ and $1.3$ eV (e.g. $\beta$-carotene; phenothiazine; anthracene; pyrene; etc). With a number of exceptions the values of $E$ and $\varrho$, for mobilities of the order observed, do not indicate intrinsic conductivity.

In most cases the conductivity is therefore extrinsic and impurity donors or acceptors, electrode effects and grain boundary effects must be considered. For anthracene, Inokuchi's value of $E$ ($1.4$ eV) is compatible with intrinsic conductivity and is not far from that expected (see below). Sometimes the slope of the curve log $\varrho$ vs. $1/T$ changes at a certain temperature, indicating a change of mechanism. Amongst aromatic substances, although the relative arrangement of the hexagonal rings affects the properties, increasing the number of rings lowers both the activation energy and the value of $E$. The conductivity has therefore been associated with the $\pi$-electrons.

Introducing quinone linkages into molecules of large aromatic hydrocarbons results in a lower resistivity. Flavanthrone conducts $10^4$ times as well as well as indanthrone which differs from it by having $2\,CH$ groups instead of $2\,N$ atoms in each molecule. Acridine however is not very different from anthracene. Sulphur compounds made from aromatic hydrocarbons show a conductivity much greater than that of the parent compounds as well as $E$ values of $0.1$–$0.3$ eV. They contain also a large number of unpaired spins.

*Bimolecular solids* in which one molecular species is easily ionized to give electrons and the other has a high electron affinity, frequently have low resistivities. For perylene-iodine, for example, $\varrho = 8\Omega$ cm. The greatest conductivity is found when there are two acceptor molecules to each donor, a result which is similar to that found with tetracyanoquinodimethane compounds, but probably not for the same reason since the majority carriers are reported to be holes (in violanthrene-iodine) or holes plus electrons ($2$ perylene-$3I_2$). Mobilities are lower in these compounds being less than $10^{-3}$ cm$^2$ V$^{-1}$ sec$^{-1}$. Values of $E$ are lower than in simple molecular crystals. Some of these compounds (e.g. phthalocyanine: chloranil) obey the equation for intrinsic semiconductors.

*Dyes.* Seventy-seven organic dyes are known to be semiconductors of which about half are n-type, the others p-type. None is clearly an intrinsic semiconductor. When the anion is coloured, oxygen increases the conductivity of p-type dyes, hydrogen behaves oppositely and nitrogen is without effect. The oxygen traps electrons and the hydrogen traps holes. The sign of the carrier generally agrees with the sign of the charge on the coloured ion. Dyes are frequently also good photoconductors.

A few *free radicals* occur as solids. On the simplest band picture of a crystal the band of highest energy would be only half-filled and the conductivity expected naively to be large, and $E$ to be small. Diphenyl picryl hydrazyl (DPPH) is one such solid, but its conductivity is as low as that of many non-radical solids. This is true also of other solid free radicals. DPPH has a low value of $E$ in the expression for $\varrho$, but even this can be taken as partial evidence in favour of the expectations of the simplest band model, as it may well be caused by the presence of an impurity.

*Polymers* have been much studied especially by Pohl with a view to developing compounds of high conductivity. Most are p-type. If unsaturation extends throughout the polymer then large conductivities should occur at least within the polymeric molecule. In pyrolytic graphites the conductivity is high ($10^4\,\Omega^{-1}$ cm$^{-1}$) but in polyphenylene it is low ($10^{-15}\,\Omega^{-1}$ cm$^{-1}$). Polycopperphthalocyanine has a high conductivity ($0.05\,\Omega^{-1}$ cm$^{-1}$ at room temperature) and gives a measurable Hall effect, showing that the conduction is p-type. Polysulphurnitride has a high conductivity ($300\,\Omega^{-1}$ cm$^{-1}$). The values vary with the method of preparation.

Pyrolysis of polymers can introduce unsaturation and markedly increase the conductivity. Pyrolysis also introduces unpaired electrons, the number of which rises with an increase in pyrolysis temperature. In polymers which are almost entirely carbon the peripheral carbon atoms with unattached valencies can act as electron acceptors and make the material p-type. At higher pyrolysis temperatures the electrons so bound are liberated and make the conduction n-type. In pyropolymers the number of carriers is often as high as $10^{19}$ cm$^{-3}$ and impurities therefore relatively insignificant.

Nickel doping of pyropolymers can increase the mobility and so the conductivity by 100 times, especially if the pyrolysis temperature is relatively low.

Polymers of xanthene, tetrachlorophenyl-thioether, ketones and metal phthalocyanines resemble ordinary non-polymeric materials. Nylon, polytetrafluoroethylene and polystyrene are very good insulators.

*Mobilities* $\mu$ of both electrons and positive holes in simple molecular crystals are often of the order of $1$ cm$^2$ sec$^{-1}$ V$^{-1}$ and generally anisotropic. In charge-transfer complexes the mobility is lower, ranging commonly from $10^{-5}$ to $10^{-3}$ cm$^2$ V$^{-1}$ sec$^{-1}$. Accurate measurements were made first by Kepler and Le Blanc using a pulse technique. Typical results are anthracene $\mu_+ = 0.98$; $\mu = 0.54$ cm$^2$ V$^{-1}$ sec$^{-1}$. In the compounds of tetracyanoquinodimethane the acceptor molecules T and the ions T$^-$ lie with their planes parallel, and the direction of greatest conductivity is perpendicular to these planes.

In general the mobility in organic solids decreases as the temperature is raised, a result consistent with the application of the band theory of solids and with the extension of the orbital occupied by the carrier over many lattice spacings. Calculations of $\mu$ by the band-model agree with the observed values to rather better than an order of magnitude, even when phonon interaction is neglected. Such interactions, however, are to be expected, as in the interaction of the moving carrier with the electrons in polarizable orbitals of neighbouring molecules. Glarum and Siebrand have discussed the phonon interaction and Lyons and Mackie, on a classical basis, the polarization energies. In neither case is a complete quantum mechanical treatment yet available, involving for electronic polarization the introduction of excited molecular states.

Conductivity depends upon both the number of carriers and their mobilities and is thus a twofold problem.

In the case of the *formation of carriers* intrinsically and thermally, one of the bulk molecules acts as a donor of an electron, another as an acceptor. Considerations of energy and entropy lead to the energy $E$ in the above expression for $\varrho$ being evaluated as $E = (I_c - A_c)/2$ when coulombic interaction between the charges formed is negligible, where $I_c$ is the energy necessary to remove an electron from a molecule of the crystal to infinity, and $A_c$ is the energy liberated on replacing this electron in the crystal at a site distant from the original site. $I_c$ and $A_c$ differ from the analogous molecular quantities $I_G$ and $A_G$ by a polarization energy $P$: $I_c = I_G - P$, and $A_c = A_G + P$, where $P$ is a positive quantity. Such classical arguments as used here are valid in the case of narrow energy bands, since they yield the position of the centre of the energy band. Analogous expressions can be derived for the case where guest molecules act as donors or acceptors. In this case, however, $I_c$ and $A_c$ refer to the guest in the host lattice. The expression for the intrinsic case is replaced by one which includes a term dependent on the concentrations of the guests. Another case is the formation of carriers at a surface of the crystal, either that in contact with an adsorbed gas or that at the electrode. The various cases have not yet been clearly sorted out.

In strongly conducting substances like $Q^+(T_2)^-$ (*see above*) the free radicals Q are the (hypothetical) donors all of which donate electrons to T. The number of electrons equals the number of $Q^+$ ions and is spread over twice this number of T molecules. The movement of the electron from $T^-$ to a neighbouring T in a similar environment to the original $T^-$ requires zero energy. Consequently electron motion is expected to be easy and the observed temperature coefficient of resistivity is almost zero, as is observed. The molecule T has one of the highest electron affinities of all organic molecules and the radical Q seems likely to have a low ionization energy: the carriers are formed at low temperatures and consequently no formation energy appears in the temperature coefficient. These compounds, unlike many organics, are n-type.

If the mole-ratio of donor to acceptor (T) is raised from 1 : 2 to 1 : 1 the conductivity drops by four orders of magnitude, explicable by the relative lack of neutral acceptors onto which electrons from negative ions can move.

*Electron-spin resonance* spectra quantitatively detect molecules with an unpaired electron if present in a concentration greater than $10^{12}$ cm$^{-3}$. They have therefore been used in favourable cases to detect such species in organic semiconductors. In some cases such as perylene-iodine and the polyacenequinone radical polymers the number of unpaired spins varies with the conductivity (but this is not always so). Here the number of unpaired electrons found differs from that calculated from the number of molecules present and the observed value of $E$.

*Ion-pair formation* in organics has often been disregarded, but since the formation of ion-pairs in an organic crystal (at equilibrium and containing equal numbers of opposite charges) is expected at room temperature to remove almost all the carriers, such formation is important in the understanding of conductivity. Almost all ion-pairs will be nearest-neighbour pairs. The energy necessary to dissociate such a pair in a crystal of low dielectric constant is a fraction of an electron volt, an amount rather lower than the energy usually found by the temperature coefficient of resistivity. The thermal dissociation of ion-pairs formed in irradiated samples provides a possible explanation of the increase of photoconductivity with temperature, which is frequently observed. (An alternative theory of this phenomenon is that the temperature dependence is associated with the trapping of excitation energy and its thermal release.)

An ion-pair in a crystal represents an excited state and subject to restrictions imposed by symmetry such an excited state must mix with other excited states characterized by the excitation of individual molecules (a process often requiring too much energy for it to occur thermally). The theory of excited states therefore requires consideration both of excitons in which the charges are separated and also of those in which they are both centred on the one molecular site.

*Ultra-pure organic solids* are hard to obtain and many of the observed results are likely to depend on impurities. Anthracene has been purified using zone-refining in the final stage and impurities detected by taking the fluorescence spectrum of the solid at $4°$K. The related hydrocarbon, naphthacene, decomposes in a zone-refiner. Such problems with large hydrocarbons and with dyes, many of which are semiconductors, have not been overcome. However, a compensating factor is that for an impurity to affect the electrical results it must have better donor or acceptor properties than the bulk molecules. For substances of large $A_c$ values, such as iodine crystals ($A_c > 5$ eV), impurities in the iodine are unlikely to affect the observed properties of say the perylene-iodine solid.

*Melting* an organic semiconductor lowers the mobility to about $10^{-3}$ cm$^2$ V$^{-1}$ sec$^{-1}$, but the conductivity rises by several powers of ten. Carriers are thus formed about $10^5$ times as much in the melt as in the solid. The mechanism of conductivity might also change to an ionic type since the mobility is of the right order for an ionic mechanism and increases with an increase in temperature. The temperature coefficient of the conductivity in the melt is 0·4 eV which is distinctly in excess of the 0·15 eV expected from viscosity measurements. The increase in $P$ resulting from a greater orientation of neighbouring molecules around a carrier in the melt as compared to the solid explains this tentatively, since the increase in $P$ lowers the energy of formation of carriers from neutral molecules.

*The photoejection of electrons* from organic solids has been observed with yields often about $10^{-4}$–$10^{-3}$ electrons/quantum. The threshold measures $I_c$, which can also be calculated from $I_G$ by estimating $P$, which is often about 1·6 eV.

*Photoconduction* has been observed in many organic semiconductors. Additives which quench the fluorescence of the crystal quench the photoconduction also. Generally the absorption of light by the crystal makes photoconduction observable, and the threshold of photoconduction has often been identified with the edge of the (lowest-energy) well-defined optical absorption band system of the crystal, although some experiments have shown photoconduction at lower energies still. The absorption spectra of molecular crystals are more complex than those of valence crystals and are described not in terms of valence and conduction bands but in terms of excited states of the individual molecules and crystal interactions resulting in exciton motion.

Excited states of the crystal in which the electron and hole separated by a large distance are optically inaccessible and the primary action of the light to form a state involving an ion-pair, which later dissociates thermally or in a suitable collision, seems indicated. There is some evidence also that collision between two excitons can cause photoconduction and photoemission.

In photoconduction studied with the electrodes on the crystal surface carriers are formed by the migration of excitons to the surface where the nature of the adsorbed gas affects the results. For example adsorbed electronegative gases such as oxygen increase the conductivity of p-type crystals. The movement of carriers through the bulk is of general occurrence, but again the mode of formation of carriers in the bulk is not established.

*Trapped charges* occur frequently in organic crystals. Persistent internal polarization has been reported to follow light absorption. The liberation of carriers from traps has been observed by forming the carriers at low temperatures by exposure to light and measuring the currents produced as the crystal is warmed and the carriers liberated from the traps. In anthracene a trap depth of ca. 0·8 eV was obtained in this way. Carriers in the same material were also liberated by near infra-red radiation and the effects seen as a more rapid rise and/or decay of the photocurrent following the switching on or off of the light. Since trapped charges give rise to space-charge effects, infra-red radiation in suitable cases can reduce the latter. In the presence of trapped charges the photocurrent falls to zero over a long time period. Anthracene is in fact a photoelectret. Carriers can also be liberated from traps by light absorbed by the bulk crystal. In this case light falling on a central area in a crystal subject to an applied field produces trapped charges which give rise to a field in the crystal which is opposed to the applied field. Switching off both the light and the field and re-illumination of the same area yields a photocurrent in the reverse direction to that found in the original experiment.

The presence of *radical-ions* in both semiconduction and photoconduction has been detected in solids of high carrier concentration by electron spin resonance. In some cases other methods have detected the presence of radical-ions. Thus in lamellar complexes in which layers of donor and acceptor crystals are in contact the absorption spectrum of one or other of the ions has been measured. (In these experiments the dark conductivity of the donor layer was increased by a factor of $10$–$10^7$ and the photoconductivity increased also.) Some spectral evidence indicates the formation of doubly charged ions as well as singly charged radical ions.

An increase in the *light intensity L* increases the photoconductivity $I_{\mathrm{ph}}$. $I_{\mathrm{ph}} \propto L^n$. Frequently $n = 1$ but $0·5$ and other values have been found. Although space-charge effects have not always been eliminated in the work done, a value of $0·5$ for $n$ has been interpreted as indicating a current limited by recombination of oppositely charged carriers. One theory of carrier generation by light suggests their formation in a collision of two excitons: singlet-triplet collisions being expected to yield carriers at a higher rate than either singlet-singlet or triplet-triplet collisions. The two-exciton theory is consistent with $n = 1$ and does not necessarily require $n = 2$.

The voltage dependence of photocurrents is frequently but not always linear. Space-charge effects certainly account for some of the non-linearity but at sufficiently high field strengths (or high temperatures) space-charges become less important and ultimately insignificant. The results seem to exclude the applied field as an agent in the formation of charge carriers. If the applied field were an agent a non-linearity is expected at high fields following linearity at low fields. The opposite has been observed.

When the light shines through one of the electrodes in a "sandwich" cell arrangement a marked rectification effect is observed. The sign of the illuminated electrode which yields the greater current has been taken to agree with the sign of the majority carrier.

*Applications.* In substances of biological significance such as fibrous and globular proteins, globin, haemoglobin, glycine and polyglycine values of $E$ in the expression for $\varrho$, taken on dried material, range from $1·1$ to $1·6$ eV. Since such values are mostly thermally inaccessible, Szent-Gyorgyi's suggestion that semiconduction might occur widely in proteins has not been substantiated by these measurements. The conductivity of dry proteins can be increased by bombardment with hydrogen atoms or by adding acceptor molecules such as chloranil, which can inject carriers into the protein.

Conductivity through an organic system has been postulated also in the cytochrome system. Electron spin resonance indicating unpaired electrons has been found also in irradiated chloroplasts at 90°K and charge movement has been suggested in the chloroplast membrane at an early stage of photosynthesis. A theory of charge transfer as the primary mode of action of certain tranquillizing drugs such as chlorpromazine has been given. Certainly molecules of these drugs do have low ionization energies. Charge transport has been envisaged too as occurring after the absorption of light in the retina. In all these biological systems the mechanism is still unclear but in some or all of them biological semi and photoconductors may yet prove to be present.

Few tests of organic semiconductors as non-biological catalysts have been made but unpaired electrons on the surface of dimethylaniline-chloranil type solids have been detected by their catalysis of the para-hydrogen conversion at low temperatures. Catalysis of the $H_2 + D_2 \rightarrow 2HD$ reaction was not effected by the same solids.

One theory of sensitization of photographic emulsions has invoked photo-induced charge transfer between the dye and the inorganic crystal. Such effects would have features in common with the observed sensitization by cyanine dyes of the photoconductance of cadmium sulphide crystals and the thermoelectric effect and photo-e.m.f. in pinacyanol—CdS, malachite green—CdS, and perylene—stannic oxide systems. Photovoltages have also been observed with alkali metal-aromatic hydrocarbon systems but energy conversion by organics is very inefficient as yet.

Organic semiconductors have been used in electro-photography, but not yet commercially.

*Conclusion.* After Pocchetino, Byk and Borck, and Volmer made measurements in the early years of this century, organic semiconductors were relatively neglected until after 1950. Much work since then has made the outlines of the problems clearer. This class of compounds is distinctly different in electrical behaviour from both ionic and valence crystals, many properties needing for their interpretation not only determinations of mobility but also ionization, affinity and polarization energies.

*Bibliography*

ELEY D. D. (1959) *Organic Semiconductors, Research* **12**, 293.

GARRETT C. G. B. (1959) *Organic Semiconductors* in *Semiconductors* (Ed. N. Hannay) New York: Reinhold.

GUTMANN F. (1962) *Organic Semiconductors, Rev. Pure Appl. Chem.* **12**, 2.

GUTMANN F. and LYONS L. E. *Organic Semiconductors* (in press), New York: Wiley.

HARRIS D. J. (1962) in *Encyclopaedic Dictionary of Physics* (Ed. J. THEWLIS) articles beginning *Semiconductor*, Vol. 6, Oxford: Pergamon Press.

INOKUCHI H. and AKAMATU H. (1961) *Electrical Conductivity of Organic Semiconductors, Solid State Physics* **12**, 93.

KALLMANN H. and SILVER M. (Eds.) (1961) *Electrical Conductivity in Organic Solids*, New York: Interscience.

KEARNS D. *Organic Semiconductors, Progress in Solid State Chemistry* (in press).

LYONS L. E. (1963) *Electron Transfer across the Boundaries of Organic Solids* in *Physics and Chemistry of the Organic Solid State*, Vol. 1 (Eds. D. Fox, M. M. LABES and A. WEISSBERGER) New York: Wiley.

POHL H. A. (1962) in *Modern Aspects of the Vitreous State* (Ed. J. D. MACKENZIE) New York: Butterworths.

TERENIN A. N. (1961) *Proc. Chem. Soc.* (London) 321.

L. E. LYONS

# P

**PERFECT NUMBERS.** Prime numbers have fascinated man from classical times, and the simple dichotomy of the natural numbers into prime and non-prime, or *composite* numbers has been the usual prolegomenon to further research. The Greeks trichotomized the composite numbers according to whether they are greater than, equal to, or less than, the sum of their factors (including unity but excluding the particular number itself). An example of the first or *deficient* class is 27 (factor sum $= 1 + 3 + 9 = 13$); one of the second or *perfect* class is $28 = 1 + 2 + 4 + 7 + 14$; and the third or *abundant* class has 30 as a typical member. This sort of classification is by no means obvious and may even seem trivial; however, it has won a permanent place in number theory because of the greater rarity and difficulty of identification of the perfect numbers.

The first three perfect numbers, 6, 28, and 496, are easily checked (although to check that there are no other examples below 500 is not so easy). The subsequent increase in size is prodigious; the sixth is already in the billions, and the sixteenth is a giant of 1327 digits. All known perfect numbers are even (it is likely, but unproved, that odd examples do not exist), and fit the Euclid–Euler formula; this is, in effect, an instruction to seek a pair of prime numbers, say $p$ (the smaller) and $M_p$ (the larger), that will fit the equation

$$2^p - 1 = M_p;$$

then for each and every such pair there corresponds a perfect number

$$M_p(M_p + 1)/2 = (2^p - 1) \, 2^{p-1}.$$

The symbol $M_p$ is used in tribute to Descartes's friend Marin Mersenne who published some remarkable conjectures about such primes in 1644. Nowadays all numbers of the form $2^p - 1$ (where $p$ is *any* prime) are called Mersenne Numbers, and among these the rare Mersenne Primes are of course key to the perfect numbers.

The Euclid–Euler formula is deceptively simple, for no-one has been able to locate Mersenne Primes except by trial and error. The advent of electronic computer help in the 1950's revived a flagging interest and since then the number of authenticated Mersenne Primes has risen sharply from 12 to 23. Computer programming is based on the Lucas–Lehmer theorem: *The number* $M_p = 2^p - 1$ $(p > 1)$ *is prime if and only if* $S_{p-1} = 0$ *(mod* $M_p$*), where* $S$ *is a member of the sequence* $S_1 = 4$; $S_n = S_{n-1}^2 - 2$, and it will be seen that this involves the handling of numbers far in excess of $M_p$. It is therefore noteworthy that the recent checking of $M_p = 2^{11213} - 1$ for primality took only 135 minutes of machine time (ILLIAC II, University of Illinois), and this fact may be interestingly juxtaposed with a statement concerning a very much smaller $M_p$ made in 1878 by Edouard Lucas: "To verify by known methods ... that $2^{257} - 1$ is a prime, the whole population of the globe, calculating simultaneously, would require

*The Known Perfect Numbers*

| Order | $p$ | Perfect number : $(2^p - 1) \, 2^{p-1}$ | Date of discovery |
|---|---|---|---|
| 1 | 2 | 6 | B.C. |
| 2 | 3 | 28 | B.C. |
| 3 | 5 | 496 | B.C. |
| 4 | 7 | 8 128 | $< 100$ A.D. |
| 5 | 13 | 33 550 336 | 1456 |
| 6 | 17 | 8 589 869 056 | 1603 |
| 7 | 19 | 137 438 691 328 | 1603 |
| 8 | 31 | 2 305 843 008 139 952 128 | 1750 |
| 9 | 61 | contains 37 decimal digits | 1883 |
| 10 | 89 | contains 54 decimal digits | 1911 |
| 11 | 107 | contains 65 decimal digits | 1914 |
| 12 | 127 | contains 77 decimal digits | 1876 |
| 13 | 521 | contains 314 decimal digits | 1952 |
| 14 | 607 | contains 366 decimal digits | 1952 |
| 15 | 1279 | contains 770 decimal digits | 1952 |
| 16 | 2203 | contains 1327 decimal digits | 1952 |
| 17 | 2281 | contains 1373 decimal digits | 1952 |
| 18 | 3217 | contains 1937 decimal digits | 1957 |
| 19 | 4253 | contains 2561 decimal digits | 1961 |
| 20 | 4423 | contains 2663 decimal digits | 1961 |
| 21 | 9689 | contains 5834 decimal digits | 1964 |
| 22 | 9941 | contains 5985 decimal digits | 1964 |
| 23 | 11213 | contains 6751 decimal digits | 1964 |

more than a million of millions of millions of centuries." (Incidentally, the number

$$2^{11213} - 1 \sim 28 \times 10^{3374}$$

is currently the largest known prime).

Present knowledge is summarized in the accompanying table. It is not known whether there is an infinitude of perfect numbers, but the limits of feasible exploration may soon be reached. The prime-exponent $p$ characterizing the 100th perfect number is unlikely to be less than $10^{15}$, which means that the number itself is physically unrecordable in decimal notation.

Some progress has been made towards an asymptotic density function for Mersenne Primes, and the latest conjecture, by D. B. Gillies, is that the number of $M_p$ smaller than any given natural number $G$ approximates to $2(\ln \ln G)/(\ln 2)$.

*Bibliography*

GILLIES D. B. (1964) *Mathematics of Computation*, **18**, 93.
SHANKS D. (1962) *Solved and Unsolved Problems in Number Theory*, Washington, D. C.: Spartan Books.
UHLER H. S. (1952) *Scripta Mathematica*, **18**, 122.

N. T. GRIDGEMAN

## PERMEATION OF GASES THROUGH SOLIDS.
A solid wall, free of holes and separating a gas from a vacuum space, may be permeated by the gas. The process consists of three main steps: (1) Adsorption and solution of the gas in the surface layer next to the gas; (2) Diffusion of the dissolved gas down its concentration gradient in the solid to the vacuum side: (3) Going out of solution and desorption to the vacuum side.

The process differs from flow through a hole by being dependent on gas *partial* pressure difference, by being an activated diffusion process, exponentially dependent on temperature, and by being highly selective for the particular gas–solid couple involved.

Both specific solubility and mobility are involved in permeation. If, at constant temperature, Henry's law is followed, so solubility $S$ in the solid is proportional to the gas pressure, and if the diffusion is Fickian, with the diffusion coefficient $D$ independent of concentration then $P = D \times S$ where $P$ is permeation rate.

The usual units are: (1) Permeation rate, $P$, in cubic centimetres of gas (STP) per second per square centimetre area per millimetre thickness per 1 cm Hg gas partial pressure difference. The amount permeating at constant temperature is directly proportional to time, area, pressure difference and inversely proportional to thickness. (2) The diffusion coefficient $D$ is in square centimetres per second. (3) $S$ is solubility in cubic centimetres of gas (STP) per cubic centimetre of solid for 1 atm (76 cm) applied.

For these particular dimensions of the units, $7 \cdot 6 \, P = D \times S$.

The diffusion constant can be measured, as Daynes showed, by the time lag between application of gas pressure to a wall and the attainment of steady state flow through that wall to a vacuum. For a wall thickness $d$ in cm and a lag $L$ in seconds, $D = d^2/6L$.

Another method which is well adapted to gases dissolved in solids is the progress of degassing with time. The fraction $E$ of gas remaining is determined at various times. A. B. Newman gives tables and solutions of the diffusion equations, related to $E$ for various shapes of solids, and to the very useful Fourier number, $Dt/a^2$. The latter involves the characteristic dimension "$a$" in cm (as $\frac{1}{2}$ thickness for a slab, radius for a sphere), the time $t$ in sec and $D$ the diffusion coefficient in cm²/sec. The rate of take-up of gas by the solid can also be followed. In this case $F$ = fraction of equilibrium solubility attained = $1 - E$, for a particular time, temperature and solid shape, and from the take-up, the diffusion constant $D$ can be calculated. The Fourier number also occurs in heat transport problems.

There are many valid analogies between mass transport, occurring in permeation, and heat flow in solids. Fourier's law of linear diffusion applying to heat flow in one dimension in solids is:

$$\frac{\partial T}{\partial t} = h \frac{\partial^2 T}{\partial x^2}$$

with $T$ = temperature, $t$ = time, $x$ = distance, and $h$ = thermal diffusivity. The latter is related to thermal conductivity $K$ by the equation $h = K/Cd$ where $C$ = specific heat and $d$ = density and $Cd$ is the heat capacity per unit volume. The units for $h$ are cm²/sec in the c.g.s. system.

Fick's law for matter transport in solids is a special case of the Fourier law and is expressed:

$$\frac{\partial c}{\partial t} = D \frac{\partial^2 c}{\partial x^2} .$$

Here the concentration $c$ replaces the temperature $T$ involved in the driving force for the process, and $D$ has the same dimensions as $h$, cm²/sec. Note also the similarity, as $D = P/S$, for permeation rate and solubility per unit volume. Very many of the equations for heat flow as given by Crank, and Carslaw and Jaeger can be applied to mass transport in solids, when suitable boundary conditions are used. The activation energy for the overall permeation process follows an Arrhenius type law: $P = P_0 \exp. (Q/\boldsymbol{R}T)$, giving the temperature dependence of permeation rate. In the case of helium permeating through various glasses, $Q$ varies from about 5000 cal/mole for vitreous silica to 12,000 for an alumina glass of low permeability. Oxygen through vitreous silica has a value of 27,000 cal for the heat of permeation.

This permeation energy $Q$ includes the activation energy for diffusion plus the heat of solution. For many materials each may show exponential dependence on temperature.

16*

However, care must be taken in interpreting results in the neighbourhood of transition regions. Some polymers, near the glass temperature, show a change of sign in heat of solution. In many other systems, heat of permeation and of diffusion are quite similar in value indicating small heat of solution.

Generalizations for the passage of gases through solids are few, but the following remarks may be made:

*1. Inorganic glasses.* He, $H_2$, $D_2$, $T_2$, Ne, Ar and $O_2$ have been measured through vitreous silica and other glasses. The rate is fastest through the open structure of vitreous silica, particularly for helium, and for all the gases the rate varies as the first power of the pressure. This indicates that the undissociated molecule is moving as such. At 700°C the hydrogen rate is one tenth the helium rate, and oxygen is lower by a factor of a million. The rate of helium through crystalline quartz is exceedingly low, contrasting with the high rate through silica glass.

*2. Metals.* No rare gas permeates any metal under purely thermal activation. Ion impact or nuclear disintegration gives some penetration and the gas can subsequently come out. Most metals are permeated by hydrogen, which moves rapidly through Pd, Fe and Ni and very slowly through tungsten. Its rate varies as the square root of the gas pressure applied, indicating passage through the metal as atoms, not molecules. Hydrogen may also be made to permeate iron from solution, as by electrolysis, acid attack, enamelling procedures or aqueous corrosion.

Oxygen gas permeates silver selectively.

*3. Polymers.* All gases permeate all polymers. The rate for water is apt to be high, and there are many specific effects for other polymer–gas couples. All gas rates vary as the first power of the pressure. Increased crystallinity in polymer structure leads to decrease in gas permeability. In many cases of high permeability, the diffusion coefficient is apt to be dependent on concentration.

*Bibliography*

BARRER R. M. (1951) *Diffusion In and Through Solids*, Cambridge: The University Press.

CARSLAW H. S. and JAEGER J. C. (1959) *Conduction of Heat in Solids*, Oxford: Clarendon Press.

CRANK J. (1956) *The Mathematics of Diffusion*, Oxford: Clarendon Press.

DAYNES H. (1920) *Proc. Roy. Soc.* **97** A, 286.

DUSHMAN S. (1962) (2nd Edn) *Scientific Foundations of Vacuum Technique*, New York: Wiley.

JOST W. (1960) *Diffusion in Solids, Liquids and Gases*, New York: Academic Press.

NEWMAN A. B. (1931) *Chem. and Met. Eng.* **38**, 710.

NORTON F. J. (1961) *Trans. 8th Vac. Symp. and 2nd, Intl. Congress*, Oxford: Pergamon Press.

ROGERS C. E. (1964) in *Engineering Design for Plastics*, (Baer E. Ed.) London: Chapman & Hall.

TOMLIN D. H. (1961) in *Encyclopaedic Dictionary of Physics* (Ed. J. Thewlies) **2**, 387, 398; Oxford: Pergamon Press.

F. J. NORTON

**PHASE-INTEGRAL METHODS.** Many problems concerning the propagation of waves in inhomogeneous media may be reduced to an ordinary differential equation of the form

$$\frac{d^2w}{dz^2} + h^2 q(z, h) w = 0. \tag{1}$$

where $h$ is a parameter (usually real, positive and large) and $q(z, h)$ a function of position. Physically $z$ is real, but mathematically $z = x + iy$ is allowed to be complex. Only for special choices of the profile $q(z, h)$ can this equation be solved in terms of the various standard transcendental functions; otherwise, approximate analytical techniques must be used. Phase-integral methods form one such approximate technique; moreover the method enables a physical interpretation of the approximate solutions to be kept to the fore throughout the theory. If a time factor $e^{i\omega t}$ (usually suppressed) exists in $w$, and if propagation takes place along $Oz$, then

$$w = \exp(\pm ihq^{\frac{1}{2}}z)\, e^{i\omega t}$$

if $q$ is a constant. The speed of propagation is given by

$$v = \omega/(hq^{\frac{1}{2}}).$$

If the free-space value of $q$ is unity, the free-space velocity equals $\omega/h \equiv c$. Then generally $v = c/q^{1/2}$, so $q^{1/2}$ behaves like the refractive index of the medium. Even when the medium is inhomogeneous, $q^{1/2}$ is still regarded as the refractive index locally, and the object of phase-integral methods is to demonstrate the form and behaviour of the approximate solutions in terms of the refractive index of the medium.

*Historical.* The basis of the method was laid by Stokes (1857, 1871), who investigated the phenomenon of the discontinuous changes that take place in the arbitrary constants associated with the asymptotic expansions of certain second order differential equations. Physically, approximate solutions of (1) were considered by Lord Rayleigh (1912) and by Gans (1915), but their systematic treatment was first exploited by Jeffreys (1923). From that time onwards, the method was particularly applied to problems in quantum mechanics, with specific applications to potential barriers and potential wells. The origin of the name *W.K.B. solutions* (after Wentzel, Kramers and Brillouin) dates from these early applications. Later, the method was applied to the propagation of radio waves in the ionosphere, and the basic ideas were generalized to embrace higher order differential equations, the characteristic waves propagated in the medium representing the physical interpretation of the approximate solutions. Mathematically, improvements in the technique are associated with the name of Langer, and the improved Langer approximation has often been used in problems demanding the solution for $w$ in the neighbourhood of the transition points, namely points where $q$ vanishes.

Problems in wave mechanics demand solutions for real $z$ only, but problems in radio propagation, where dissipative effects occur owing to the collision frequency, necessitate the use of the complex $z$-plane. Under these circumstances, transition points are complex points. The approximate W.K.B. solutions are valid only in restricted domains of the complex $z$-plane (or for certain ranges along the $x$-axis). The object of the method is to trace solutions from one domain of validity into a neighbouring domain of validity in the following sense: Given a linear combination of the two W.K.B. solutions in one domain, find the corresponding linear combination in a neighbouring domain. In the theory of the asymptotic solutions of transcendental linear differential equations, it is found that the linear combination of the fundamental asymptotic solutions must be changed from domain to domain (usually as $\arg z$ varies); this change constitutes the *Stokes phenomenon*. In the present problem, the domains are associated with the transition points, and the Stokes phenomenon demands that the linear combination of the W.K.B. solutions be changed from domain to domain around the transition points.

Mathematically, solutions are sought that are *uniformly asymptotic* with respect to the large parameter $h$, where this usually means that the error involved in the approximation must be of the form $f(z, h)/h$ times the approximation, where $|f(z, h)| < M$ for all points $z$ in the domain and $h > h_0$. Few physical applications really consider the nature of the errors involved, and under these circumstances, Jeffreys has called these solutions *crude approximations*. It must be recognized that any discussion of errors in specific problems is often a difficult matter, and hence is usually neglected.

*Approximate solutions.* If $w = \exp\left(\int \varphi \, dz\right)$, equation 1 becomes

$$\varphi^2 + \varphi' + h^2 q = 0,$$

a Riccati equation. Expansion of $\varphi$ in descending powers of $h$ yields the W.K.B. solutions

$$w_{1,2} = q^{-1/4} \exp\left(\pm ih \int q^{1/2} dz\right). \qquad (2)$$

On the other hand, if we introduce the new independent and dependent variables $w = (\sqrt{z'}) X$, $z = z(\xi)$, we obtain

$$X'' + h^2 q z'^2 X = g X, \qquad (3)$$

where $g = \frac{3}{4} z''^2/z'^2 - \frac{1}{2} z'''/z'$. A prime denotes differentiation with respect to $\xi$.

The W.K.B. approximation is obtained by placing $-h^2 q z'^2 = 1$ and neglecting the term $gX$. Then

$X = \exp(\xi)$, where

$$\xi = \pm ih \int_0^z q^{1/2} dz,$$

with a suitable branch chosen for $q^{1/2}$; it is convenient to choose $z = 0$ as the lower limit, where $z = 0$ is a zero of $q$. We thus arrive again at solution (2). Equation 3 can now be transformed into an integral equation, from which the errors involved and the domains of validity of these solutions can be deduced. Certainly they are not valid at the transition point owing to the factor $q^{-1/4}$. The Stokes phenomenon implies that these solutions are not valid all round a transition point; various sectors require different combinations of the two solutions. A necessary condition for the validity of the W.K.B. solutions is

$$\left| \frac{3}{4} z''^2/z'^2 - \frac{1}{2} z'''/z' \right| \ll 1. \qquad (4)$$

On the other hand, we may choose $\xi = -h^2 q z'^2$ in equation 3, yielding

$$\xi = \left( \pm \frac{3ih}{2} \int_0^z q^{1/2} dz \right)^{2/3},$$

while if the right-hand side is neglected, we have

$$X'' = \xi X,$$

the Airy equation with standard solutions denoted by $Ai(\xi)$ and $Bi(\xi)$. Then

$$w_1 = q^{-1/4} \left( \int_0^z q^{1/2} dz \right)^{1/6} Ai\left[ \left( \frac{3ih}{2} \int_0^z q^{1/2} dz \right)^{2/3} \right], \qquad (5)$$

equivalent to the improved Langer approximation usually expressed in terms of Bessel functions of order one-third. The Airy integral $Ai(\xi)$ and its companion function $Bi(\xi)$ possess known series solutions, asymptotic series, contour-integral representations and are extensively tabulated. The solutions are valid around and at the transition point $z = 0$. Moreover, such a solution is not confined only to a particular sector around $z = 0$; it embraces the whole complex $z$-plane around $z = 0$. It ceases to be valid in domains near and around other transition points given by $q = 0$. But such a solution spans various gaps that exist in the domains of validity of the ordinary W.K.B. solutions.

The asymptotic forms of $Ai(\xi)$ are required; these are deduced from a Laplace-type contour integral representation, using the method of steepest descents. If $\xi$ is real, large and positive, we have

$$Ai(\xi) \sim \tfrac{1}{2} \pi^{-1/2} \xi^{-1/4} \exp\left(-\tfrac{2}{3} \xi^{3/2}\right) \left[1 + 0\left(\xi^{-3/2}\right)\right], \qquad (6)$$

an exponentially small expression. If $\xi$ is real, large and negative, we have

$$Ai(\xi) \sim \pi^{-1/2}(-\xi)^{-1/4} \left\{\cos\left[\tfrac{2}{3}(-\xi)^{3/2} - \tfrac{1}{4}\pi\right] + \right.$$
$$\left. + 0\left[(-\xi)^{3/2}\right]\right\}. \qquad (7)$$

representing an attenuated oscillatory graph. Here, $\arg(-\xi) = 0$. For general values of $\arg \xi$, the domain around $\xi = 0$ must be divided up into sectors, bounded by the lines

$$\arg \xi = 0, \pm \tfrac{2}{3}\pi, \quad \text{known as Stokes lines,}$$

$$\arg \xi = \pm \tfrac{1}{3}\pi, \pi, \quad \text{known as anti-Stokes lines;}$$

a branch cut must be introduced somewhere, say in the range

$$\pi < \arg \xi < \tfrac{4}{3}\pi.$$

Form (6) persists for $|\arg \xi| < \pi$, while form (7) is valid for $\tfrac{2}{3}\pi < \arg \xi < \tfrac{4}{3}\pi$ where $-\tfrac{1}{3}\pi < \arg(-\xi) < \tfrac{1}{3}\pi$. $\mathrm{Ai}(\xi)$ is exponentially small (subdominant) for $|\arg \xi| < \tfrac{1}{3}\pi$, and exponentially large (dominant) in other sectors bounded by anti-Stokes lines. Zeros of $\mathrm{Ai}(\xi)$ exist along the anti-Stokes lines if trigonometrical functions are used in the asymptotic form, namely for $\arg \xi = \pi$.

*One transition point.* Consider, now, solution (5), the standard cases arising when $q$ is real along the real $z$-axis. To introduce a transition point, assume $q(x) < 0$ for $x > 0$, and $q(x) > 0$ for $x < 0$. The asymptotic forms of (5), valid only when $|\xi|$ is large, are identical with the W.K.B. solutions (2). When $h$ is large, $z$ may be quite near the transition point, subject to condition (4). For $x > 0$, let

$$h \int_0^x q^{1/2}\, dx = -iM, \quad M > 0,$$

and for $x < 0$,

$$h \int_0^x q^{1/2}\, dx = -L, \qquad L > 0;$$

then solution (5) yields its asymptotic forms:

$$2q^{-1/4}\sin(L + \tfrac{1}{4}\pi) \leftarrow w \rightarrow (-q)^{-1/4}\mathrm{e}^{-M}, \qquad (8)$$

the overall approximation (5) for $w$ giving rise to these two W.K.B. solutions in the ranges $x < 0$ and $x > 0$ respectively. Strictly, the asymptotic forms both branch out from the one solution $w$; however, in this case, the overall solution $w$ may be discarded. The connexion formula is valid in both directions, the oscillating function connecting on to the exponentially small function and conversely. This connexion formula across a transition point is due to Jeffreys. His second formula is based on $\mathrm{Bi}(\xi)$ appearing in solution (5). Its asymptotic form reads

$$q^{-1/4}\cos(L + \tfrac{1}{4}\pi) \rightarrow (-q)^{-1/4}\mathrm{e}^M; \qquad (9)$$

the oscillatory left-hand side implies the dominant right-hand side, but not conversely.

If $q$ is not real along the real axis, the transition point may occur at an arbitrary point $z = a$ in the complex plane. These two connexion formulae are not then directly applicable, but a simpler approach is possible without using ready-made formulae. Solution (5) is still used, with $z = a$ substituted for $z = 0$ as the lower limit of integration. This solution can be decomposed into its various asymptotic forms depending on the value of $\arg \xi$. The boundaries of the various domains depend on $\arg \xi$; we find that anti-Stokes lines occur when

$$\mathrm{Im} \int_a^z q^{1/2}\, dz = 0,$$

and Stokes lines when

$$\mathrm{Re} \int_a^z q^{1/2}\, dz = 0,$$

where $h$ is assumed to be real and positive. A series of rules is then formulated for tracing a linear combination of the two W.K.B. solutions over the various lines from sector to sector; a branch cut leading from the transition point must also be included. In a sector bounded by anti-Stokes lines, one solution $w_1$, say, is subdominant (exponentially small) and the other solution $w_2$ is dominant. If in a sector just prior (in the clockwise sense) to an anti-Stokes line we have a linear combination $Aw_1 + Bw_2$, then in the following sector (anticlockwise) the dominancy of the two terms is changed. If in a sector just prior to a Stokes line we have the solution

$$w = Aw_1 + Bw_2$$
$$= A \times \text{subdominant W.K.B. solution}$$
$$\quad + B \times \text{dominant W.K.B. solution,}$$

then in the sector following the Stokes line we have

$$w = (A + iB)\, w_1 + Bw_2$$
$$= (A + iB) \times \text{subdominant W.K.B. solution}$$
$$\quad + B \times \text{dominant W.K.B. solution.}$$

The coefficient of the dominant term remains constant in order to preserve continuity, but the coefficient of the subdominant term changes, the discontinuity thereby introduced being smaller than the intrinsic error in the dominant term. By this means (care being taken with a branch cut) the Stokes phenomenon is wholly embraced, and solutions can be traced over large domains of the complex $z$-plane.

*Reflection coefficients.* These may formally be calculated. If the time factor $\mathrm{e}^{i\omega t}$ is assumed to be suppressed in the solution $w$, and if $q \rightarrow 1$ as $\mathrm{Re}\ z = x \rightarrow -\infty$, such that the W.K.B. solutions are proportional to $\mathrm{e}^{\pm ihx}$, then $\mathrm{e}^{-ihx}$ represents an incident wave and $\mathrm{e}^{ihx}$ a reflected wave. The reflection coefficient is defined as the ratio of their coefficients. More generally, if a phase-reference level $x = b$ is chosen, then the incident wave from $x = -\infty$ will be

$$w_1 = q^{-1/4}\exp\left(-ih \int_b^x q^{1/2}\, dx\right),$$

while the reflected wave will have the form

$$w_2 = q^{-1/4}\exp\left(ih \int_b^x q^{1/2}\, dx\right).$$

It follows that $w = w_1 + R w_2$ along the negative $x$-axis. Along the positive $x$-axis, we require the solution to satisfy the physical condition that $w$ should be exponentially small—that is, if only one transition point is in question. If the transition point is $z = a$ (generally complex), then a sector bounded by anti-Stokes lines radiating outwards from $a$ must contain this portion of the positive $x$-axis. An exponentially small (subdominant) solution for $w$ is chosen in this sector; this is traced round to the negative $x$-axis by the rules just stated, yielding

$$R = \mathrm{i} \exp\left(-2\mathrm{i}h \int_b^a q^{1/2}\,\mathrm{d}z\right). \qquad (10)$$

The whole of the above discussion can be generalized to embrace transition points of order $n$, namely when $q$ possesses the factor $(z - a)^n$.

*Two transition points.* Many problems involve two complex transition points $z = a$ and $z = b$ in the domain of interest. The differential equation would be of the form

$$\frac{\mathrm{d}^2 w}{\mathrm{d}z^2} + h^2 (z - a)(z - b)\, p(z, h)\, w = 0.$$

This equation can be reduced approximately to and compared with the Weber equation of the parabolic cylinder functions,

$$X'' + (n + \tfrac{1}{2} - \tfrac{1}{4}\xi^2)\, X = 0, \qquad (9)$$

possessing exactly two transition points. The solutions of this equation are known analytically as readily as those of the Airy equation, but the asymptotic forms are more complicated in that they involve the gamma function in their coefficients. Ordinary W.K.B. solutions in the domain well away from the transition points, namely where

$$|z - \tfrac{1}{2}(a + b)| \gg \tfrac{1}{2}|b - a|,$$

may be traced around using the properties of these asymptotic forms and the associated Stokes phenomenon. Nevertheless, applications have usually involved the patching together of the ordinary W.K.B. solutions associated with the two transition points separately, assuming conditions under which the W.K.B. solutions are valid between the transition points, namely near $z = \tfrac{1}{2}(a + b)$.

In particular, if $q$ is real along the $x$-axis, two important possibilities arise involving two transition points:

   i. The *potential well.* $q < 0$ for $|x| > a$; $q > 0$ for $|x| < a$.

   ii. The *potential barrier.* $q > 0$ for $|x| > a$; $q < 0$ for $|x| < a$.

A second real type of potential barrier may be formed by allowing transition points at $z = \pm \mathrm{i}a$ ($a$ real) and with $q > 0$ for all $x$.

The potential well is such that oscillatory W.K.B. solutions exist on the real axis $|x| < a$, while for $x > a$ and $x < -a$ evanescent solutions are required. For $x > a$, we choose

$$w = A(-q)^{-1/4} \exp\left(-h \int_a^x (-q)^{1/2}\,\mathrm{d}x\right),$$

and we may then use Jeffreys's connexion formula (8) to trace this solution into the range $|x| < a$. Similarly, for $x < -a$, we choose

$$w = B(-q)^{-1/4} \exp\left(+h \int_{-a}^x (-q)^{1/2}\,\mathrm{d}x\right),$$

and use the connexion formula to trace this into the range $|x| < a$. The equivalence of these two solutions in the range $|x| < a$ yields the condition

$$h \int_{-a}^a q^{1/2}\,\mathrm{d}x = \pi(\tfrac{1}{2} + n), \qquad (11)$$

$n$ being a large positive integer. This provides a series of eigenvalues for the parameter $h$. For the harmonic oscillator, $q = a^2 - x^2$, and we obtain $h = (1 + 2n)/a^2$; in this case the exact solution can be expressed in terms of Hermite polynomials. Equation 11 is valid even if the W.K.B. solutions are not valid approximations in the range $|x| < a$, but the present method does not prove this fact.

The potential barrier with real transition points is defined by $q < 0$ for $|x| < a$. For $x > a$, an upgoing W.K.B. solution is chosen, namely

$$w = T q^{-1/4} \exp\left(-\mathrm{i}h \int_a^x q^{1/2}\,\mathrm{d}x\right).$$

The use of Jeffreys's connexion formulae now provides just to the left of $x = a$ a linear combination of the subdominant W.K.B. solution and the dominant W.K.B. solution on the Stokes line joining the two transition points. The difficulty of the process is that the combination changes its character just to the right of the second transition point $z = -a$; the solution that had been dominant now becomes subdominant, and conversely. Yet the subdominant solutions must be retained in the presence of the dominant solutions, even though their magnitude is exponentially smaller than the error in the dominant terms. On the Stokes line, the respective subdominant coefficients are vague, being in the process of changing in keeping with the Stokes phenomenon. But Jeffreys's second connexion formula is devised to give the best possible answer under these circumstances, even if it is strictly not permissible to use it. This difficulty is only overcome by the use of the approximate solutions in terms of the parabolic cylinder functions. The W.K.B. method now yields for the reflection coefficient $R$

$$R = \mathrm{i}\left[1 - \tfrac{1}{2}\exp\left(-2h \int_{-a}^a (-q)^{1/2}\,\mathrm{d}x\right)\right],$$

the phase-reference level being the lower transition point $x = -a$.

In more complicated problems where the two transition points are complex (this includes absorption owing to energy losses), Jeffreys's formulae are no longer applicable. The overall representation in terms of the parabolic cylinder functions and the deduction of the approximate values of the reflection and transmission coefficients do not appear to be available in the literature.

*Quantum mechanics.* The theory has been extensively applied in quantum mechanics, where problems involving one or two transition points often arise. The traditional quantum condition

$$\oint p \, dx = n\boldsymbol{h},$$

where $\boldsymbol{h}$ is Planck's constant, is refined by phase-integral methods for the potential well to

$$\oint p \, dx = \left(\tfrac{1}{2} + n\right) \boldsymbol{h}.$$

The theory of the hydrogen atom provides a constant source of confusion. The separated form of the wave function contains the factor $R$ dependent on the radius $r$ and satisfying the equation

$$R'' + \frac{2}{r} R' + \left[\frac{8\pi^2\boldsymbol{m}}{\boldsymbol{h}^2}\left(H + \frac{\boldsymbol{e}^2}{r}\right) - \frac{l(l+1)}{r^2}\right] R = 0,$$

or if $R = S/r$,

$$S'' + \left[\frac{8\pi^2\boldsymbol{m}}{\boldsymbol{h}^2}\left(H + \frac{\boldsymbol{e}^2}{r}\right) - \frac{l(l+1)}{r^2}\right] S = 0,$$

possessing a well between the two transition points $r_1$ and $r_2$. Condition (11) is often used to produce the energy levels. But this method is entirely wrong, even though useful results are produced. Physically, $R$ must remain finite at $r = 0$, but the W.K.B. solution (subdominant for $r$ just less than $r_1$) nevertheless tends to infinity as $r \to 0$. Langer has overcome this difficulty by means of a preliminary transformation of the equation, using

$$r = e^x, \quad R = e^{-1/2 x} U,$$

yielding

$$U'' + \left[\frac{8\pi^2\boldsymbol{m}}{\boldsymbol{h}^2}(He^{2x} + \boldsymbol{e}^2 e^x) - \left(l + \frac{1}{2}\right)^2\right] U = 0.$$

Equation 11 then gives

$$H = -2\pi^2\boldsymbol{m}\boldsymbol{e}^4/\boldsymbol{h}^2 n^2,$$

where $n$ is an integer.

*Ionospheric reflection.* The reflection of radio waves provides another fruitful field of application of phase-integral methods. Propagation of electromagnetic waves in an anisotropic horizontally stratified medium yields second order differential equations for the electric field components. In particular, if the medium is isotropic (when no external magnetic field is imposed), the second order equations can be reduced to type (1), where the function $q$ is complex owing to dissipative forces caused by the presence of a collision frequency of the free electrons with the heavier positive ions. The transition points are therefore complex, and the phase-integral formula (10) yields results $|R| < 1$. When the magnetic field is present, the equations can best be arranged in first order form, using two electric and two magnetic field components as the four dependent variables. The four simultaneous equations may be treated by the phase-integral method by means of a preliminary transformation. If the four dependent variables are represented by a column matrix $\mathbf{e}$, then the first order simultaneous differential equations take the form

$$\frac{d\mathbf{e}}{dz} = \mathbf{T}\mathbf{e},$$

where $\mathbf{T}$ is a complicated $4 \times 4$ matrix whose elements are functions of $z$. The transformation $\mathbf{e} = \mathbf{R}\mathbf{f}$ diagonalizes the matrix $\mathbf{T}$ if $\mathbf{R}$ is chosen to consist of the characteristic vectors of $T$, namely $\mathbf{R}^{-1}\mathbf{T}\mathbf{R} = \mathbf{D}$, where $\mathbf{D}$ is the diagonal matrix consisting of the characteristic roots of $\mathbf{T}$. This transformation reduces the simultaneous equations to the form

$$\frac{d\mathbf{f}}{dz} = \mathbf{D}\mathbf{f} - \mathbf{R}^{-1}\frac{d\mathbf{R}}{dz}\mathbf{f}. \tag{12}$$

The transition points occur at the complex roots of the determinantal equation $\det \mathbf{R} = 0$, and are known as the *reflection* of *coupling points* depending on their nature. At such points, two characteristic roots are equal in value. For high enough frequencies, the approximate W.K.B. solutions of the four equations (12) (in which the non-diagonal elements of $\mathbf{R}^{-1} \, d\mathbf{R}/dz$ are neglected) represent the four characteristic waves that can be propagated in the medium. They are propagated independently, except they are coupled together around the transition points, and phase-integral methods may be utilized for dealing quantitatively with these coupling processes. It is usual to reinterpret the method in terms of phase-integrals integrated along certain contours in the complex $z$-plane between and around the transition points. Details of this process are still rather speculative, though greatly used. The actual positions of the transition points (calculated by means of electronic computers) furnish valuable physical insight into the propagation processes.

*Bibliography*

BREKHOVSKIKH L. M. (1960) *Waves in Layered Media*, New York: Academic Press.

BUDDEN K. G. (1961) *Radio Waves in the Ionosphere*, Cambridge: The University Press.

BUDDEN K. G. (1961) *The Waveguide Mode Theory of Wave Propagation*, New York: Prentice-Hall.

FRÖMAN N. and FRÖMAN P. O. (1965) *J.W.K.B. Approximation*, Amsterdam: North-Holland.

HEADING J. (1962) *An Introduction to Phase-integral Methods*, London: Methuen.

JEFFREYS, H. and JEFFREYS B. (1956) *Methods of Mathematical Physics*, Cambridge: The University Press.

JORDAN D. W. (1961) in *Encylcopaedic Dictionary of Physics* (J. THEWLIS Ed.) **1**, 269, Oxford: Pergamon Press.

KEMBLE E. C. (1937) *The Fundamental Principles of Quantum Mechanics*, New York: McGraw-Hill.

WAIT J. R. (1962) *Electromagnetic Waves in Stratified Media*, Oxford: Pergamon Press.

                J. HEADING

## PHOTOELASTIC METHOD APPLIED TO MATERIALS RESEARCH.

Photoelasticity has, until recently, played little part in research into the physical behaviour of modern materials. Obviously, a technique which assumes elastic, linear, isotropic and homogeneous properties for both model and structure in order to transfer data from one to the other is of little direct use in studying such subjects as plasticity, heterogeneity, anisotropy, etc., in materials. There are two particular fields, however, in which photoelasticity can be applied directly to the physical properties of materials. These are:

1. Research into the properties of materials which themselves exhibit the photoelastic effect, such as glass and some crystalline materials,
2. Studies of surface deformation (both elastic and plastic) in any material to which recently developed photoelastic coatings can be applied.

*Research into the physical properties of birefringent materials.* Perhaps the most common materials, apart from modern transparent plastics, to exhibit stress-birefringence are glass and some ionic crystals.

A considerable amount of research has been carried out into the mechanical properties and failure mechanisms of glass, and the bibliography gives details of some of the work that has been carried out in this field.

Many ionic crystals also exhibit stress-birefringence, and this phenomenon has been used by some investigators to study the deformation mechanisms of such crystals. This stress birefringence, however, is not the same as for structurally isotropic materials, and Nye has pointed out that the stress distribution in the crystal can be deduced from the distribution of birefringence only provided that the relative orientation of principal stress axes and crystallographic axes is taken into account.

Photoelastic data are particularly valuable as a means of determining dislocation movements in ionic crystals such as MgO under plastic deformation. If we observe a slip band in such a crystal in a circular polariscope system, the birefringence observed may be used to determine the position and density of the extra half-planes.

If our microscope system is set up to for observations under dark field conditions (crossed polarizer and analyser, crossed quarter-wave plates), then when polarizer and analyser axes are aligned with the slip line, a rotation of the analyser either clockwise or counter-clockwise will reduce the birefringence to zero on one side of the slip line. The number of degrees rotation will give the magnitude of the birefringence (180 degrees = 1 fringe order) and hence the stress magnitude on either side of the slip line, knowing the stress-optic coefficient and crystal thickness. The direction of rotation (clockwise or counter-clockwise) will give the sign of the stress (tensile or compressive). If we know the sign and magnitude of the stresses, the position and density of dislocations along the slip line can be calculated, assuming that the side containing the extra half planes is in compression. For this type of measurement it is essential to pre-calibrate the rotation of the analyser against a stress of known sign (such as a small specimen under compression), so that there can be no confusion as to sign; i.e. for a given instrument, such as a polarizing microscope, we would be able to say that, with polarizer and analyser crossed at 0 and 90 degrees and both quarter-wave plates inserted with their axes crossed and at 45 degrees to polarizer and analyser, for a slip band oriented along the 0–180 degrees axis of the microscope stage, a clockwise rotation of the analyser would reduce the birefringence to zero for tensile stresses, counter-clockwise rotations being required for compressive stresses.

An inherent assumption that is made in such determinations is that stresses normal to the slip band are insignificant in magnitude in comparison with stresses parallel to it. In conditions where such an assumption cannot be made, stress separation procedures would have to be applied in order to determine the nature and magnitude of the individual stresses involved.

*Materials research with photoelastic coatings.* The photoelastic coating technique is a development of the photoelastic method whereby the actual structure to be analysed is directly coated with a photoelastic plastic. Thus strains produced in the structure are directly transmitted to the coating and readings of birefringence can be taken directly from the structure using a reflection polariscope.

Although such a technique was suggested as long ago as 1930, it was not until 1953 that instrumentation and plastics were developed that made such a method feasible as a practical stress analysis tool. Subsequent improvement and development have produced a stress analysis tool which is ideally suited to some of the problems inherent in materials research.

The coating may be applied in several ways; for materials with a naturally reflective surface, in which it is desired to observe fine details of surface strain at fairly high strain levels (e.g. observations of grain boundary stresses; slip systems at the surface of metallic single crystals, etc.), the most suitable method is to brush a thin coating (0·005–0·030 in.) directly on to the surface.

For materials with non-reflective surfaces such as wood, concrete, rubber, etc., the method found most suitable in practice is to bond the photoelastic coating to the material with a reflective cement. Even for materials such as steel and aluminium which do have reflective surfaces, reflective cement is preferred, as it does not necessitate prior polishing of the surface to be bonded, and the reflection characteristics obtained (approximately cosine distribution) are preferable to the predominantly specular reflection otherwise obtained.

For flat surfaces, standard sheets of plastic may be cast or purchased commercially of relatively constant thickness and strain-optic coefficient. Alternatively, for curved surfaces, a sheet of the required thickness may be cast, which when partially polymerized may be removed from the casting frame and shaped over the surface. After complete polymerization the shaped coating may be removed, cleaned, measured from point to point to determine if there are any significant variations in thickness, and bonded to the structure with reflective cement.

For normal stress analysis purposes on areas greater than a few square inches, a large field reflection polariscope is used, fractional orders of relative retardation being determined by means of the Tardy compensation technique.

For observation of smaller areas, a small field microscope polariscope of the reflection type is used, fractional fringe orders being measured with a Babinet compensator.

Sensitivities that may be obtained with such commercially available equipment are: $\pm 2$ degrees for directions of principal strains, $\pm 10$ micro-in./in. shear strain magnitude for a normal (3 per cent maximum elongation) coating $\frac{1}{8}$ in. thick. Both types of instrument have oblique incidence attachments for the separation of principal strains, together with adaptors for the attachment of photographic equipment and stroboscopic light sources for the investigation of dynamic strains.

Maximum strains measured can vary from 3 to 200 per cent depending on the type of coating used.

For most materials a high modulus (450,000 lbs/sq-in.) coating is used having high optical sensitivity and a maximum elongation of between 3 and 10 per cent. For the study of materials such as rubber, coatings are available with a Young's Modulus as low as 350 lbs/sq-in. and a maximum elongation of up to 200 per cent.

The advantage of photoelastic coatings in the field of materials research is that surface strains are quantitatively determined in a continuous manner at every point of a coated surface.

The technique enables studies of material properties to be carried out not only in the elastic range, but under conditions of plastic deformation.

Residual stresses in materials may be determined by various stress-relief methods, and the technique has also been used to study thermal stress problems.

*See also*: Tardy and Senarmont compensation methods.

*Bibliography*

BACON G. E. (1961) *Babinet compensator*, in *Encyclopaedic Dictionary of Physics* Vol. 1, Oxford: Pergamon Press.

BOND W. L. and ANDRUS J. (1956). *Phys. Rev.* **101**, 1211.

BRILLIANTOW N. A. and OBREIMOW I. W. (1934) *Phys. Z. Sowjet.* **6**, 587; **12**, 7 (1937).

BULLOUGH R. (1958) *Phys. Rev.* **110**, 620.

DERESIEWICZ H. (1963) in *Encyclopaedic Dictionary of Physics* (Ed. J. Thewlis) **7**, 56; Oxford: Pergamon Press.

EDGERTON H. E. and BARSTOW F. E. (1941) *Further studies of glass fracture with high speed photography*, *J. Amer. Cer. Soc.* **24**, 131.

GATES J. W. (1962) in *Encyclopaedic Dictionary of Physics* (Ed. J. Thewlis) **5**, 546; Oxford: Pergamon Press.

HOLISTER G. S. (1961) Recent Developments in Photoelastic Coating Techniques. *J. R. Aero. Soc.* **65**, 610.

HOLISTER, G. S. (Oct. 1962) *The Photoelastic Method Applied to Materials Research, Applied Materials Research* **1**, No. 3.

HOLISTER G. S. (1965) *Experimental Stress Analysis, Principles and Methods*, Cambridge: The University Press.

KEAR B. K. and PRATT P. L. (1958) *Acta Metall.* **6**, 457.

KOLSKY H. (1952) *A photoelastic investigation of the hardness of plastics and glass, J. Soc. Glass Tech.* **36**, 56.

NAISH J. M. and WEBB E. R. (June 1952) *Strain Patterns in Toughened Glass, Proc. Phys. Soc.* (London) B **65**, 457.

NYE J. F. (1949) *Proc. Roy. Soc.* A **198**, 190; (1949) A **200**, 47.

OBREIMOW I. W. and SCHUBNIKOV L. W. (1927) *Z.Phys.* **41**, 907.

PUGH E. M. *et al.* (Jan. 1952) *Glass cracking caused by high explosives, J. Appl. Phys.* **23**, 48.

SCHWEIGER H. (1955) *Photoelastic shock investigation on thin glass bars, Ann. Phys.* (Leipzig) **16**, No. 3–4, 119. In German.

VAN ZEE A. and NORITAKE H. M. (May 1958) *Measurement of Stress-Optical Coefficient and Rate of Stress Release in Commercial Soda-Lime Glasses, J.Amer. Cer. Soc.* **41**, No. 5.

ZANDMAN F. (1960) Maximum Shear Strain Measurements and Determination of Initial Yielding by the Use of the Photoelastic Coating Technique. *A.S.T.M. Spec. Publ.* No. 289.

ZANDMAN F. *et al.* (Sept. 1963) Photoelastic Coating Analysis in Thermal Fields. Experimental Mechanics.

G. S. HOLISTER

**PHOTOELASTIC STRAIN-GAUGES AND TRANSDUCERS.** The photoelastic strain gauge is constructed from a thin, small plate of birefringent plastic, having a reflective backing and a covering of a sandwich sheet of quarter-wave plate and polarizer, producing circularly polarized light when viewed in normal incidence. It is bonded to a structure with an epoxy cement and observed in white light.

The gauges are constructed to various shapes and sizes and have residual stress patterns built into them by conventional stress freezing techniques so that stress magnitude can be determined from the movement of the frozen stress pattern.

What information can be extracted from the gauge depends on:

1. the dimensions and shape of the gauge,
2. the way in which the gauge is bonded to the structure,
3. the nature of the residual fringe pattern in the gauge,
4. the direction of viewing,
5. the orientation of the gauge relative to the principal strain directions in the structure to which it is bonded.

As with more conventional strain gauges, photoelastic gauges may be divided into two basic types:

a) *Uniaxial gauges*, designed to measure the strain in the direction of the gauge axis, and used in cases where the directions of principal strain are known.

b) '*Rosette*' *gauges*, used for the determination of the direction and magnitude of the principal surface strains where no information is available as to the principal strain directions.

Figure 1 illustrates a typical uniaxial gauge. The gauge contains a residual fringe pattern of linear gradient (equally spaced fringes), incorporates a graduated scale and is designed so that a principal strain differences in the gauge of approximately 1000 micro-in./in. causes one fringe to move a whole fringe

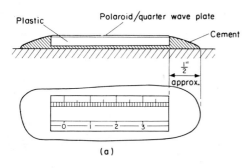

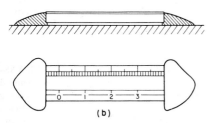

*Fig. 1.*

order, i.e. a given fringe will move to the position previously occupied by the fringe next to it.

When a gauge is bonded over its entire surface and is oriented along a direction of principal strain the shift of pattern $\Delta$ is proportional to $\varepsilon_1 - \varepsilon_2$. Therefore $\varepsilon_1 - \varepsilon_2 = \Delta_{\text{divisions}} \times A$ where $\Delta$ is the shift of fringe pattern (divisions) and $A$ is the constant of the gauge in micro-inches per inch per division.

When the gauge is bonded at the ends only the structural strain in a direction perpendicular to the gauge is not transmitted to the gauge. The gauge is strained along its length and the lateral contraction of the gauge is related to the longitudinal strain by

$$\varepsilon_2 = -\mu_1 \varepsilon_1 = -0{\cdot}35\,\varepsilon_1.$$

The strain along the gauge length is therefore given by

$$\varepsilon_L = \Delta \cdot \frac{A}{1{\cdot}35}$$

where $\Delta$ is the shift of pattern between no load and load conditions ($\Delta = n_L - n_0$) and $A$ is the gauge constant.

A slight modification of this type of gauge is the "*torque gauge*", in which the frozen stresses and resultant fringes are inclined at 45° to the gauge axis, the fringe movement along the gauge axis being proportional to pure shear conditions only. Such gauges may be bonded along the axis of a rotating shaft subjected to torque and the shaft torque measured by viewing the fringe movement with a stroboscopic light source.

The gauge must be bonded all over its lower surface, fringe shift being proportional to $\varepsilon_1 - \varepsilon_2 (= 2|\varepsilon_{1,2}|$ for pure torque since $\varepsilon_1 = -\varepsilon_2$). It is customary to calibrate the gauge in units of torque directly on the shaft.

There are two types of photoelastic "rosette" gauge available commercially.

The first type, called the "*strain compass*" by its developer, Dr. G. U. Oppel, makes no attempt to determine quantitatively the strain magnitudes, but gives a rapid means of determining the nature (tensile or compressive) and orientation of the principal strains provided a sufficiently high strain level exists.

The strain compass consists of a small disk of photoelastic material, about one inch in diameter, with a small hole in the centre and a reflective lower surface.

It may be bonded to a structure by a quick setting cement such as Eastman 910 applied around its periphery only. Any strains in the structure are transmitted to the disk, producing stress concentration patterns around the hole when viewed in circularly polarized light. With a little practice, the nature of the applied strains may be determined from the type of fringe pattern produced. The directions of principal strain are given by the axes of symmetry of the pattern.

The rosette gauge developed by Holister, Zandman and Redner is designed to determine the directions and separate magnitudes of the principal strains at the point of application.

The gauge (Fig. 2) consists of (1) a photoelastic disk with a central hole; (2) a circular polarizer held on top of the disk by a centring button that fits in the hole; (3) a circular scale engraved on the button for direction measurement, and (4) a reflective lower surface.

The photoelastic disk has a built-in frozen stress pattern of concentric circular fringes which are used as references for determination of stress magnitudes. Under load the fringes distort into a symmetrical pattern, the axes of synnetry giving the directions of principal strain (the "minor" axis gives direction of $\varepsilon_1$, the greater principal strain; the "major" axis gives the direction of $\varepsilon_2$, the smaller principal strain—see Fig. 2).

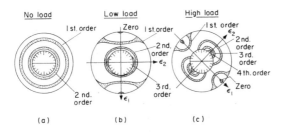

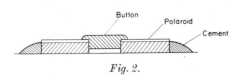

*Fig. 2.*

The fringe movement along the minor axis is a function of the magnitude of $\varepsilon_2$, the fringe movement along the major axis is a function of the magnitude of $\varepsilon_1$. The gauge is bonded to a structure with an epoxy cement on the outer edge only as shown, a generous fillet being built up to transmit strains effectively from the structure to the gauge. Readings of fringe movement along the axes of symmetry (measured with an illuminated comparator for maximum accuracy or estimated for an order of magnitude reading) may be converted to readings of strain magnitude in micro-in./in. by referring to a series of reference curves supplied with the gauge. Accuracies obtained are approximately $\pm 3°$ for principal stress directions and $\pm 50$ micro-in./in. for strain magnitudes.

A similar concept has been employed by Hiramatsu in Japan and I. Hawkes *et al.* at the Postgraduate School of Mines, Sheffield University, to develop photoelastic transducers constructed from glass and designed for the measurement of both residual stress and stress build-up in underground excavations and large civil engineering structures.

To determine residual stresses, a photoelastic transducer has been designed by I. Hawkes which is in the form of a glass disk diametrally loaded by two

adjustable plattens contained in a cylindrical steel cylinder enclosing the glass disk.

The transducer is inserted into a hole bored into the rock surface and cemented into place. The whole assembly is then overcored with a diamond coring tool. The rock stresses in the vicinity are relieved by overcoring and the elastic rebound of the overcored specimen produces a change in the fringe pattern in the glass disk when viewed in circularly polarized light, the change in fringe order at the centre of the disk being proportional to the component of residual stress along the axis of diametral loading. Three such measurements must be taken to determine the direction and separate magnitudes of the principal strains in the surface of the rock.

Stress build-up in structures and underground excavations is measured by means of a "bore-hole plug stress meter". This consists of a cylindrical glass plug with a central axial hole, the length being between 2 and 4 in. depending on the sensitivity required and the outer diameter about $1\frac{1}{4}$ in. It is fitted with polarizing filters at each end and an illuminating bulb at the rear.

When bonded into a bore-hole in a structure or rockface, any stress build-up produces stress concentration patterns around the central hole similar to those observed in the "strain compass". The directions and magnitudes of the principal stresses may be determined from the type of fringe pattern produced. If this plug is used in materials for which the ratio of plug modulus to material modulus is greater than 2 it can be shown that it behaves as a rigid inclusion and consequently the readings are proportional to stress rather than strain.

All of these photoelastic gauges and transducers have the advantage of extreme simplicity and robustness with the additional advantage that direct readout may be obtained without the use of additional instrumentation. They are limited in their sensitivity compared with more conventional gauges and their use is restricted to areas where they may be observed by eye or camera.

*Bibliography*

HAWKES I. *et al.* (1961–62) *Trans. Inst. Mining and Metall.* **71**, 581.
HIRAMATSU Y., OKA Y. and OGINO S. (1961) *Mem. Fac. Engng. Kyoto* **23**, 90–110.
HOLISTER G. S. (Jan. 1963) *Applied Materials Research.*
OPPEL G. U. (1961) *Proc. Soc. Exp. Stress Anal.* **18**, 1.
ZIENKIEWICZ O. C. and HOLISTER G. S. (Eds.) (1965) *Stress Analysis—Numerical and Experimental Methods*, New York: Wiley.

G. S. HOLISTER

**PHYSICAL TECHNIQUES IN MEDICINE.** The use of physical techniques in medical investigation is steadily increasing, particularly with the advances in electronics technology. Because of the specialized nature of the requirements, much of the apparatus has been developed in medical physics and electronics departments, and the advice of medical physicists is being sort at an early stage by commercial equipment manufacturers. As a result, the field of activity of medical physics is rapidly expanding beyond the traditional sphere of radiation therapy.

*Patient monitoring equipment.* With the general application of transistors to electronic circuits, it has been possible to reduce the size of individual recording channels. This has led to a tendency to make more use of multi-channel recording systems. With these, it is possible to monitor signals arising from a patient or experimental subject, such as the *electro-cardiogram* (E.C.G.), *electro-encephalogram* (E.E.G.), blood pressure, respiratory pattern, and body temperatures. These signals must be displayed, and progress has been evident in the design of suitable display devices. For example, compact transistorized oscilloscopes are available for routine use in the operating room. They are sufficiently sensitive to be able to record either the E.C.G. or the E.E.G., and are hermetically sealed against the ingress of explosive anaesthetic vapours. A direct writing recorder is needed when a permanent record is necessary. Ink pens writing on ordinary paper, and hot stylus pens writing on heat sensitive paper are usable up to about 100 c/s. This is adequate for many biological signals, but a higher frequency response is needed for recording action potentials and phono-cardiographic signals. Ultra-violet recorders offer the advantages of a frequency response which can extend to more than 1000 c/s, and a trace which is made visible a few seconds after exposure to the ultra-violet beam. An alternative system is the Swedish ink-jet recorder. Here, a fine glass capillary supports a small cylindrical permanent magnet between the poles of an electromagnet. The output current of the recording channel amplifier passes through the electromagnet coils. As a result, the permanent magnet is deflected in sympathy with the signal. A jet of filtered ink under high pressure, is sprayed out of the capillary on to a moving roll of absorbent paper. The frequency response of this arrangement is linear up to 5000 c/s and is usable up to 900 c/s. The ink-jet recorder is replacing the combination of a cathode-ray tube plus camera normally used for the recording of action potentials.

Functions such as the blood pressure, respiration, and blood flow have to be changed by means of a suitable transducer, into a corresponding electrical signal which can be fed into a recording system. The pressure transducers encountered in present-day work are nearly always of the differential transformer or unbonded strain gauge types. The development of the more sensitive silicon strain gauges, has made possible the multiply implantation of these small elements in experimental animals. Increasing attention is being given to the development of electrode systems which can be implanted on a long term basis. It is important to be able to study the bodily functions of humans and animals during their norma lfree activity. This has led to the production of miniature telemetering radio transmitters, usually working on a frequency modulation basis. The signal is received on a V.H.F., F.M., receiver and fed to a recorder. Examples of this method include the recording of the *electromyogram* of exercising subjects, and the heart rate of athletes. The use of endo-radiosondes (radio pills) is well established for telemetering details of the pH of stomach contents, intestinal temperatures and presures. The necessity for producing very small circuits has aided the design of compact electrometer circuits based around diodes having a capacity which is voltage sensitive. These electrometers are finding application in gas chromatography with flame ionization detectors.

Whilst on the subject of the chronic implantation of electrodes, mention should be made of the fact that workers at the Burden Neurological Institute in Bristol have implanted up to sixty recording electrodes in the brain of a human patient, and kept them there for several weeks.

Pressure transducers for the direct measurement of blood pressure using a needle or catheter have been available for some time. Recent work has been directed towards the design of miniature transducers which can be mounted at the tip of a cardiac catheter, one of the advantages of which is to eliminate the effect of fluid damping. The last few years has seen a concentration of activity on methods of measuring blood flow. Currently, electromagnetic flowmeters are most used. The flow meter detection heads can be implanted in animals for some months. They can also be used routinely in acute studies in the operating room on surgical patients.

*Data processing.* With the increasing availability of physical methods of measurement, it is possible to rapidly accumulate a large amount of information in the course of a relatively short time. Hence medical physicists are applying themselves to methods of data storage and processing. The tendency is to use analogue recording for visual inspection of the biological signals, at the same time recording these signals on a frequency modulation type recording system. This is capable of handling signals down to d.c. An audio channel is included to take a commentary. The great advantage of magnetic tape recording is that the signals are subsequently available in an electrical form. At a convenient time, they can be fed into devices such as a servo-multiplier, adding or subtracting circuit, or a digitizer. Commercially available systems have been produced mainly for industrial data logging applications, and are only marginally fast enough for medical work. Under these conditions, it is advantageous to be able to record at one tape speed, and then to play back at a slower speed. In this way the frequency of the signals is reduced to a value which can be handled by the processing equipment. The availability of central digital computers is making possible the collection of data from many hospitals, and the issue of a statistical breakdown of the results. A striking example of this system in operation is concerned with a nationwide study in America of perinatal statistics. This is concerned with the factors affecting the birth of children. Ninety items of information on each of 300,000 "births" was recorded in 1962. Once a month, the computer produces a complete breakdown of each hospital's statistics. These include types of birth, abnormality and mortality, weights, days of stay, sex, multiple births, pre-natal care and other items.

More hospitals are being built with intensive care suites, where patients can be looked after, perhaps immediately after an operation, when they need extra nursing attention. As a result, systems are emerging which will receive information on body temperature, respiration and pulse frequency, from sensors connected to the patients. Each patient is "interrogated" at regular intervals by a central scanner, and the information recorded on a chart.

*The application of physical methods to the treatment of patients.* The use of transistors has made it possible to implant electric stimulating circuits within the body for long periods. Perhaps the most striking case arises in the use of cardiac pacemakers. When, due to disease, the heart is beating too slowly, it is sometimes possible for the application to the heart of a regular train of electrical pulses to restore the normal heart rate. One approach is to encapsulate the transistorized pulse circuit in epoxy resin, and then to embed it, in a surgically prepared cavity, below the skin. Flexible leads are run to the heart muscle from the pacemaker. A former disadvantage of this arrangement was that the pacing rate was set by the circuit constants, and could not be altered. In the latest equipment this has now been overcome. Implantable pacemakers are now being developed which can be triggered from the heart's natural pacemaker, if this is working at a normal frequency, but is not able to initiate normal cardiac rhythm. During exercise, when the heart rate increases, this type of electronic pacemaker will increase its pacing rate in sympathy. An alternative scheme is to use a pulse generator external to the body, feeding into the primary coil of a pulse transformer. The secondary coil is implanted inside the body, with leads connected to the heart. Now the pacing rate can be altered by the patient.

In the condition known as "fibrillation", the muscles of the heart act randomly, and not in synchronism. As a result, the heart does not work efficiently, the cardiac output fails, and death can rapidly ensure if the normal rhythm of the heart is not restored. An electric shock applied to the heart is frequently effective in restarting the normal rhythm. Instruments for applying a suitable controlled shock are known as defibrillators. It has been shown recently that the timing of the shock, in regard to an abnormal cardiac cycle, is important in correcting the rhythm. This has given rise to a greater sophistication of the defibrillation apparatus.

Success is being achieved in the use of implanted transistor circuits for the stimulation of the bladder and rectum. These units find application in most cases where patients are unable to exercise their normal control.

From the above, it can be seen that the use of physical methods is becoming widespread in medicine. Physicists are working on the estimation of ultrasound, the evaluation of respirators, the design of analytical techniques such as infra-red gas analysis, the design of miniature blood flow meters, the application of analogue computers, the actuation of artificial limbs from the electromyogram, and many other topics in addition to their work with radioactive isotopes and radiotherapy units.

D. W. HILL

**PLASMA, GENERATION AND APPLICATIONS OF.** The word "*plasma*" was coined by Irving Langmuir in 1928 to denote a luminous, ionized, electrically conducting gas. This definition covers such disparate examples as the glow discharges found in neon and fluorescent tubes, electric arcs and thermal plasmas used for heating, lightning phenomena, the atmospheres of the Sun and stars, and the wakes of meteorites and space vehicles. Since ionization of a neutral gas requires considerable energy, in order to form a plasma it is necessary to heat or otherwise excite a cold neutral gas.

Plasmas can be generated by adiabatic compression of gases. The plasmas formed at high altitudes around re-entering space vehicles are generated in this manner. Plasmas can also be generated in the laboratory using the shock tube in which a *driver gas* under high pressure is allowed to expand rapidly, compressing a second low pressure gas. There is some evidence that plasmas are also generated during explosions. These methods have the disadvantage that they only produce the plasma for a few milliseconds.

The most convenient method for continuously generating plasmas is the use of electrical energy. It is necessary to distinguish two main types of plasmas generated electrically. The term *cold plasma* (or low pressure plasma or glow discharge) denotes a plasma characterized by high electric fields in which the electrons, because of their high mobility, reach temperatures many times higher than that of the atoms and ions which remain near room temperature. The term *thermal plasma* (or high pressure plasma) is used to denote a plasma which is at or near thermal equilibrium in the sense that a single temperature describes most of its properties and the Saha equation then gives the relationship between temperature and degree of ionization.

These two types of plasma can be understood in terms of an imagined experiment in which one measures the voltage necessary to pass a current through a gas yielding a graph such as shown in

Fig. 1. (Such an experiment is difficult in practice because power supply and electrode requirements are different in the different ranges.) At low currents (say below 0·1 amp), high voltages are required. This is the glow discharge or cold plasma region, and the resulting plasma is luminous but not hot. In this region the field is high enough so that the relatively few electrons present can gain enough energy between collisions to excite or ionize the atoms on impact. Because the electrons are so light compared to the atoms, very little kinetic energy, or heat, is imparted to the atoms and this energy is radiated as light. Under these conditions the "temperature" of the electrons is hundreds of times higher than the atoms.

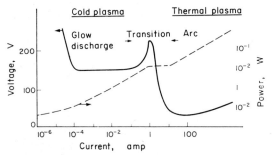

*Fig. 1. Relationship of voltage to current, showing glow discharge and arc region, and showing increase of power with current. (The curve is hypothetical, since the electrodes suitable for one region are not suitable in the other, but the indicated magnitudes are correct.)*

Since cold plasmas can emit light without becoming hot, they find application in neon and fluorescent tubes and more recently gas lasers where thermal equilibrium must be avoided. They can be formed by using a direct or alternating current discharge between electrodes, or by using radio frequency or microwave energy, in which case no electrodes are necessary. These plasmas are generally, but not necessarily, formed at low pressure, and a large body of knowledge has been accumulated concerning their atomic properties—collision cross-section, mobilities, etc.

If the current in the discharge of Fig. 1 is sufficiently increased, there will be a sudden transition to a thermal plasma when the kinetic energy of the atoms can no longer be conducted or radiated away. (This is one method of initiating thermal plasmas, but they can also be started by a high voltage spark or simply touching the electrodes together.) At equilibrium the properties of the plasma depend only on the temperature and pressure. Figure 2 shows the degree of ionization as a function of temperature for several gases, at one atmosphere. The amount of energy necessary to reach these temperatures for representative gases is shown in Fig. 3. The electrical and thermal conductivities for two representative gases are shown in Fig. 4. The conductivity of a well ionized gas is

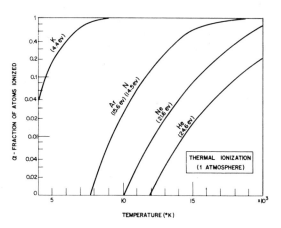

*Fig. 2.*

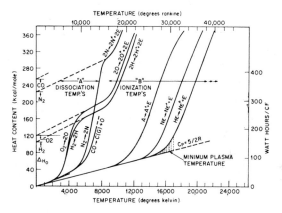

*Fig. 3.*

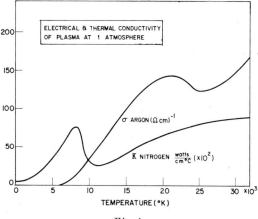

*Fig. 4.*

approximately 100 $(ohm\text{-}cm)^{-1}$. Because of this low resistance, the plasma can be maintained electrically by passage of currents from a few to a few million amperes (either through electrodes or inductive coupling).

Temperatures have been measured in the arc, and typical results are shown in Fig. 5 for a 200 amp argon plasma. This also shows the calculated magnetohydrodynamic pressures and a probe for measuring the streaming of plasma resulting from these pressures. Peak plasma velocities of 17,000 cm/sec were measured in this plasma. The heat transfer to the electrodes is due to this streaming, to ion recombination and to electron condensation. Heat transfer rates are typically 50 times greater with arcs than with flames of comparable velocities.

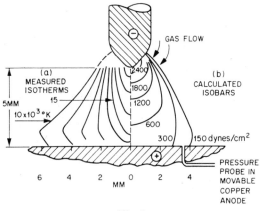

*Fig. 5.*

In some applications the prime objective is to generate a very high temperature gas, while in the electric arc the plasma is largely incidental and the effects produced at the electrodes are the chief purpose of the process. In the arc a plasma is formed between the electrodes and maintained by passage of current. The electrodes are intensely heated by this current and the high voltage drops occurring in the cathode and anode regions, and this electrode heating can be used for lighting (the carbon or zirconium arc light), metal heating (in welding or electric furnace work), or producing chemical reaction (in high intensity arcs).

*Bibliography*

FINKELNBURG W. and MAECKER H. (1956) *Elektrische Bogen und thermische Plasma*, Handbuch der Physik **22**, 254, Berlin: Springer-Verlag. (Available in English translation as Document ARL 62-302.)

SOMERVILLE J. M. *The Electric Arc*, London: Methuen.

T. B. REED

**PLASMA TORCHES.** A plasma torch is a device used to heat gases electrically to very high temperatures. The hot gases can be used for such purposes as chemical synthesis, metal fabrication, and re-entry

simulation in the same manner that ordinary combustion flames are used. However, the temperatures produced with plasmas are usually two to three times higher than those with chemical flames, and any gas can form a plasma while chemical flames must be composed of combustion products such as carbon dioxide.

Either cold plasmas or thermal plasmas can be used but generally thermal plasmas produce higher temperatures and are easier to control at higher power levels. Thermal plasmas are generated electrically, either using electrodes to carry the current to the plasma or using induction heating to induce currents in the plasma. Electrode plasma generation can be traced from the familiar electric arc in Fig. 1. If an electric arc operated from a hot cathode (a) has a hole

EVOLUTION OF THE PLASMA TORCH FROM THE OPEN ARC

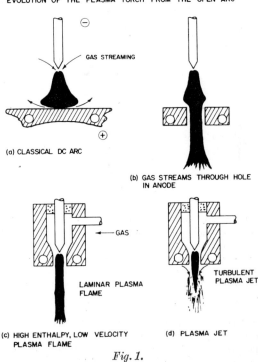

*Fig. 1.*

in the anode (b), a plasma will be forced by magnetohydrodynamic pressure through the anode as a plasma flame. If now this anode is formed into a nozzle-chamber (c) and gas is forced through this nozzle, a laminar plasma with temperatures at the core in excess of 10,000° K is produced. As gas flow is increased, the voltage and power of the plasma torch increases correspondingly, and there is a transition to a turbulent plasma jet (d) characterized by somewhat lower temperatures, but much higher velocities and heat transfer rates. Probably the first plasma torch was the Gerdian arc developed after World War I in Germany in which the arc column is surrounded by tangentially flowing water to cool and constrict the column. A modification of the cold cathode, 7000 V, 1000 amp torch, for use in chemical synthesis used at Hüls in Germany during World War II is shown in Fig. 2(b). A number of variations of electrode design have been used. Fig. 2(a) shows a tangential gas flow around a cold cathode for use with oxidizing reactive gases.

Electrode plasma torches were first developed for commercial use in the mid-1950's. The first successful commercial application was metal cutting in which a hot gas jet melts a well defined kerf through heavy metal plate. Such a process has long been familiar using oxy-acetylene, but plasma jets will cut any

Because of the absence of electrodes, the *induction plasma* can be operated in any gas and has typically larger cross sections and lower velocities than electrode torches with correspondingly lower heat transfer and larger dwell times. Although it is too early to predict all the possible uses for induction plasma torches, a number have been already found. The induction plasma is used for growth of crystals melting above 1800°C using the Verneuil geometry, and here the plasma has the advantages of low velocity, controllable atmosphere, and higher temperatures. These same properties have made the induction plasma useful in spheroidization of powders, especially nuclear fuels, such as uranium oxide, with their very high melting

COLD CATHODE PLASMA TORCHES

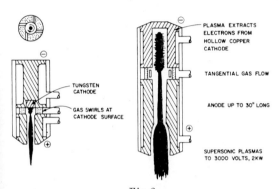

*Fig. 2.*

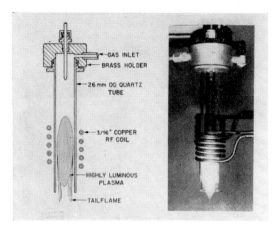

*Fig. 3.*

metal whereas oxy-acetylene cutting is only useful with steel. High velocity plasma jets are also useful for spraying metals and ceramics as protective coatings, and they have been used for chemical synthesis. More recently electrode plasma torches have been used to generate the high temperature gases required for re-entry simulation in rocket research, and torches of 1–10 megawatts have been built for this purpose.

Such electrode generated plasmas have characteristics associated with the electrode-plasma interfaces. Due to the high current density at these points, the plasma is "pumped" magnetohydrodynamically, giving rather high velocities and a characteristically low cross-section to the plasma. Some gases attack the electrodes chemically or electrically. Considerable power loss is often associated with the electrode-plasma interface. Instead of using electrodes to transmit power to the plasma, the plasma can be heated inductively just as metals and other conductors are. This method of plasma generation was first used in 1960 and Fig. 3 shows a torch embodying this principle. Figure 4 shows temperatures measured spectroscopically in this torch. Because of skin depth considerations, a typical 1 in. diameter induction plasma will operate in the range from 1 to 20 megacycles, and larger plasmas allow lower frequencies.

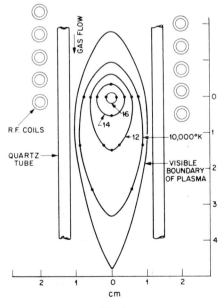

*Fig. 4.*

points. It would also seem to be especially applicable for high temperature chemistry and spectroscopy, where the absence of electrodes makes injection of material into the plasma simpler than it is with electrode torches.

Although cold plasmas may not seem at first to be useful in high temperature technology, diatomic gases such as hydrogen, nitrogen, and oxygen when passed through a sufficiently high electric field will dissociate and give back this energy of dissociation to the gases or when reaching some surface to be heated. Gas temperatures of 5000°K have been measured in cold plasma torches. However, monatomic gases in such devices "will not burn paper". Cold plasma torches may find special applications for some spectroscopic and chemical studies where it is desirable to have high excitation with low gas temperatures.

*Bibliography*

FINKELNBURG W. and MAECKER H. (1956) *Elektrische Bogen und thermische Plasma, Handbuch der Physik* **22**, 254, Berlin: Springer-Verlag. (Available in English translation as Document ARL 62-302).

JOHN R. R. and BADE W. L. (1961) *J. Am. Rocket Soc.* **31**, 4.

REED T. B. (1962) *Plasma Torches, International Science and Technology*, June.

SOMERVILLE J. M. *The Electric Arc*, London: Methuen.

T. B. REED

**PROCESS CONTROL.** A control mechanism has been defined as a system or device which exerts a restraining, governing or directing influence. The subject of process control is wider than this as it embraces the study of control mechanisms in themselves and also the actual defining in physical terms of the control means to be employed. In process control work it is often the latter part which occupies the major effort: the actual control mechanism employed is generally much simpler than those used, for instance, in missile systems. It is not required to perform with the same precision (though a high order of accuracy of measurement may be required) and consequently does not demand the same degree of study. It is sometimes thought that process control is synonymous with "automatic control" or "automation". It is true that automatic systems are often used but it should not be assumed that this is essential or that the study of process control is confined solely to automatic systems. For instance, the subjects of Cybernetics (Ashby) and Ergonomics (Singleton) are also involved.

In order to define the control means it is obviously necessary to study the process. This involves determining the variables to be measured in order that the process shall function to a specification, how they shall be measured and the relationship between the input and output variables. In this way a mathematical model of the process may be built up in which the measured functions are stated in terms of their dependent variables. At this stage it is possible to construct the control loops.

There is a basic similarity between a process control loop and a servomechanism. The chief difference is that a process control mechanism is usually controlling to a fixed value and is subject to a variety of external disturbances for which it has to correct. Though this is true only for the simple loop—with complex loops and optimising control this distinction disappears—the process control loop obeys the same physical laws as a servomechanism and it is necessary to study the stability of the system in the same manner, using basically similar techniques.

There is an increasing awareness for the need to study the control of a process as a discipline in its own right and, as modern plants grow more complex, the need to do this will undoubtedly grow.

The major industries concerned today are the aeronautics, chemical, electrical, machine tool, oil and steel industries. References to three books on the subject are given in the Bibliography (Campbell, Considine, Grabbe *et al.*).

Physics has been defined in an earlier volume of this work as embracing the study of all those parts of natural philosophy which can be explored by observations and experiments. It has long been associated with the development of measurement techniques, which are also an essential part of process control. Process control is thus essentially a branch of applied physics, which has its own philosophy based on an understanding of feedback and feedforward networks and measurement techniques.

*Future trends.* A process is generally run on the basis of maintaining all input conditions fixed and operating on the raw materials in a fixed manner throughout, so as to maintain a consistent quality of output. To do this a number of independent process control loops are set up to control the various parameters at fixed values; the values are then only changed if it is desired to change the quality of the product, or to cater for changes in process conditions such as a change in raw material or catalyst efficiency.

In a well organized plant the aim is to adjust the control valves so that the plant is run at maximum efficiency and the criterion of the efficiency is normally profitability. The technique for doing this is called process optimization and various statistical routines are available for this purpose. However, it is not always practicable to collect the necessary data and perform all the calculations required by manual means in the time available; the modern concept is to link all individual control loops together through a central computer, programme the computer to carry out the necessary optimization calculations and to make the necessary adjustments to the plant. An example of this can be seen in the processing of steel billets where a computer can be used to work out the lengths into which each billet should be cut so as to fulfill production demands with the minimum wastage.

A further example is in electricity distribution where supply from each power station may be adjusted to suit the overall demand at the minimum cost.

This technique is relatively new and is still developing fairly slowly in this country but is likely to become increasingly important in the future.

*Bibliography*

ASHBY R. W. (1956) *An Introduction to Cybernetics*, London: Chapman and Hall.

CAMPBELL D. P. (1958) *Process Dynamics*, London: Chapman and Hall.

CONSIDINE D. M. (1957) *Process Instruments and Controls Handbook*, London: McGraw-Hill.

GRABBE E. M. *et al.* (1961) *Handbook of Automation, Computation and Control* (3 Vols.), New York: Wiley.

NEWMAN E. A. (1961) in *Enxyclopaedic Dictionary of Physics* (Ed. J. Thewlis) **2**, 74; Oxford: Pergamon Press.

PORTER A. (1962) in *Encyclopaedic Dictionary of Physics* (Ed. J. Thewlis) **6**, 465; Oxford: Pergamon Press.

QUASTLER H. (1961) in *Encyclopaedic Dictionary of Physics* (Ed. J. Thewlis) **2**, 248; Oxford: Pergamon Press.

SINGLETON (1962) *The Industrial Use of Ergonomics*, London: H. M. Stationery Office.

L. J. POSTLE

**PROTON MAGNETOMETER.** This instrument, based on the phenomenon of nuclear free-precession, is used principally for measuring the total intensity of the geomagnetic field. It is absolute, highly precise (1 part in 100,000), portable and rapid. Consequently, it is displacing suspended-magnet types of magnetometer both in magnetic observatories and in geological and geophysical surveying. It is not suitable for making measurements within a laboratory because it can only function when remote from the gradients associated with iron and steel objects. It has been used in rockets for measuring the geomagnetic field at high altitudes but it is not suitable for measuring the very weak fields encountered in space. Its performance as a field surveying instrument has made the magnetic detection of buried archeological remains a practical proposition.

Although the proton magnetometer measures the total magnetic field intensity it can be adapted (when it is known as the *vector proton magnetometer*) to measure components by the use of Helmholtz annulling coils. It is not then so easily portable.

The free-precession of protons was first detected by M. Packard and R. Varian in 1953, having been predicted theoretically by Bloch in 1946. A nucleus having a magnetic moment, $\mu$, and an angular momentum, $p$, will have a precession frequency in a magnetic field of intensity, $F$, given by

$$f = \frac{\mu F}{2\pi p}.$$

For the proton the *magnetogyric ratio* $(\mu/p)$ (the inverse of the gyromagnetic ratio) is 26751·3 sec$^{-1}$ oersted$^{-1}$ so that

$$f = 4257.6 \times F \text{ cycles per second.}$$

$F$ varies between 0·24 and 0·7 oersted over the Earth's surface, giving precession frequencies in the range 1000–3000 cps. The instrument consists of two parts, firstly a detector in which a sufficient number of protons precess in phase to induce a significant voltage in a coil surrounding them, and, secondly, electronic circuits (usually transistorized) which measure the frequency $f$ with high precision. For outdoor use the instrument can be powered from miniature accumulators having a capacity of several amp-hours.

*The detector.* A sketch of a typical "detector-bottle" is shown in Fig. 1. It is filled with a liquid such as

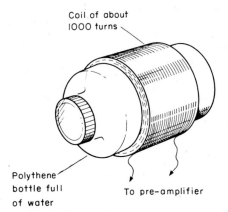

*Fig. 1. Detector bottle for proton magnetometer.*

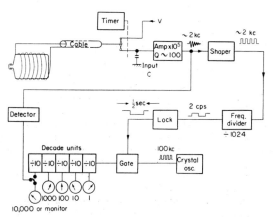

*Fig. 2. Proton Magnetometer: block diagram. The decade meters record the number of 100 kc/s oscillator pulses that pass through the gate while it is held open for the duration of 1024 proton pulses. A locking unit ("lock") ensures the gate subsequently remains closed until about 0·1 sec after another polarization has been completed.*

water or methyl alcohol which has a large number of protons per $cm^3$ and a suitable relaxation time (see later). The coil serves both for detection and *polarization*, by means of suitable switching relays in the instrument (see Fig. 2). During polarization a steady current of the order of an amp is passed through the coil, creating a field of several hundred oersted along the axis. This tends to align the protons (but because of thermal agitation the number aligned parallel only exceeds those aligned anti-parallel by about 1 part in $10^7$) and a magnetic moment per $cm^3$ of about $10^{-7}$ gauss is induced.

When the polarizing current is shut off, the lined-up protons precess in phase about the geomagnetic field direction. The alternating voltage induced in the coil due to this rotating magnetic moment is a maximum if the axis of the bottle is positioned roughly perpendicular to the geomagnetic field (i.e. the axis horizontal and East-West) and is of the order of a few microvolts.

It is essential that the final shut-off of the polarizing field is clean and occurs in a time short compared to the period of free precession. This is so that the protons do not "follow" the resultant field as it swings round but remain perpendicular to the geomagnetic field.

It should be noted that only the *amplitude* of the signal is dependent on the orientation of the detector; the frequency is determined solely by the magneto-gyric ratio of the proton and the total magnetic field strength, $F$, acting at the detector. The precession signal does not last indefinitely; its amplitude decays exponentially, the time to decay by a factor $e$ being termed the relaxation time. The relaxation time is finite because perturbing inter- and intramolecular fields cause the protons to lose phase coherence and to align in the geomagnetic field. Low viscosity lengthens the relaxation time because rapid molecular diffusion smooths out the perturbing magnetic fields. As the temperature of a liquid is lowered towards freezing point the relaxation time decreases, and on solidification it becomes very short indeed because the smoothing-out effect is entirely lost.

Methyl alcohol and distilled water have convenient relaxation times of about 3 sec. To achieve this, the liquid must be free from all paramagnetic impurities and dissolved oxygen (*free* oxygen is paramagnetic) should be removed by bubbling nitrogen through just before sealing. Benzene has a relaxation time of about 20 sec, and while this is valuable for special purposes it is not suitable for routine use. Firstly, the number of protons per $cm^3$ is low and in consequence the signal strength from a given volume is weak. Secondly, the signal strength is exponentially dependent on the polarizing time reaching $(1 - e^{-1})$ of its saturation value (for a given current) when the polarizing time is equal to the relaxation time. The rapidity of measurement is an important feature in many applications and consequently a longer than necessary relaxation time is undesirable. For airborne measurements even 3 sec is too long, and the relaxation time is reduced by addition of a ferric salt.

If the field being measured is non-uniform the signal decays more rapidly than is appropriate to the relaxation time because protons in different parts of the detector-bottle experience different fields; consequently, they precess with different frequencies and rapidly lose phase coherence. If the overall difference between different parts of the detector is $10^{-4}$ oersted the signal reaches zero in just over a second. Consequently, the proton magnetometer cannot be used anywhere near steel-framed buildings etc.

*Frequency measurements.* The most commonly employed technique is shown schematically in Fig. 2. Essentially it consists in counting the number of cycles of a 100 kc/s crystal-controlled oscillator that occur during the time taken for the proton precession signal to complete a fixed number of cycles. If the counting period is arranged to be about 0·5 sec the precision obtainable is 1 part in 50,000. The proton precession signal is amplified selectively and then, after shaping, it is frequency divided by a binary chain. For a ten-stage chain, the factor of division is 1024 ($= 2^{10}$), so that for a field strength of 0·48 oersted, giving a precession frequency of about 2000 c/s, the output of the dividing chain is about 2 c/s (see Fig. 3). This is applied via a lock unit to a gate such

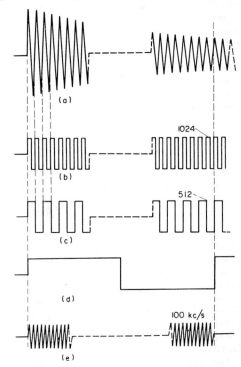

Fig. 3. *Proton Magnetometer: waveforms.* (a) *Proton signal from amplifier, $f = 2000$ c/s.* (b) *Proton signal after squaring, $f = 2000$ c/s.* (c) *After frequency division by one binary stage, $f = 1000$ c/s.* (d) *After frequency division by ten binary stages, $f = 2$ c/s.* (e) *Gated output of 100 kc oscillator.*

that the first positive-going edge entering the lock unit opens the gate and the next closes it, the gate then remaining closed until the unit is reset. While the gate is open the output of the 100 kc/s oscillator is fed to a decade chain. The operation of each decade unit is such that one pulse appears at the output for every ten that enter at the input, and if the number entering is not a multiple of ten the remainder after division by ten is indicated on the corresponding digital meter. Thus the 5-digit number indicated by these five meters equals the number of 100 kc/s oscillator pulses that get through the gate. Since the gate remains open for exactly 1024 precession periods the precession frequency is given by

$$f = \frac{100,000 \times 1024}{(\text{decade count})}$$

so that the field strength is given by

$$F = \frac{24,051 \cdot 1}{(\text{decade count})} \text{ oersted}.$$

The sensitivity can be increased both by counting for more proton precession periods (4096 gives, for example, an accuracy of 1 part in 200,000) or by increasing the frequency of the crystal oscillator (say to 1 Mc/s). The *absolute* accuracy depends on the calibration of the crystal-controlled oscillator and on the precision with which the proton magnetogyric ratio is known. A good oscillator will remain constant to 1 part in 100,000 over the temperature range 10–40°C and better performance can be obtained if the ambient temperature is held constant. At present the internationally agreed value for the magnetogyric ratio is 26751·3 sec$^{-1}$ oersted$^{-1}$; this is the value obtained by Bender and Driscoll in 1958 with an uncertainty of $\pm 7$ parts per million.

*Bibliography*

AITKEN M. J. (1961) *Physics and Archaeology*, New York: Interscience.
BENDER P. L. and DRISCOLL R. L. (1958) *A Free Precession Determination of the Proton Gyromagnetic Ratio*, I.R.E. Trans. I–7, 176.
BLOCH F. (1946) *Nuclear Induction, Phys. Rev.* **70**, 461.
NELSON J. H. (1960) *The Gyromagnetic Ratio of the Proton, J. Geophys. Res.* **65**, 3826.
PACKARD M. and VARIAN R. (1954) *Free Nuclear Induction in Earth's Magnetic Field, Phys. Rev.* **93**, 941.
STANDLEY K. J. (1961) *in Encyclopaedic Dictionary of Physics*, (Ed. J. Thewlis) **4**, 411; Oxford: Pergamon Press.
WATERS G. S. and FRANCIS P. D. (1958) *A Nuclear Magnetometer, J. Sci. Instr.* **35**, 88.
WHITHAM K. (1960) *Methods and Techniques in Geophysics* (Ed. Runcorn), 156.

M. J. AITKEN

**PULMONARY MECHANICS.** The subject encompasses the application of principles of static and dynamic mechanics to questions related to the lungs and associated structures.

To promote brevity and lucidity, the discussion is limited mainly to the respiratory system of one mammal: man. The basic pulmonary structure is considered at this time to be essentially the same in all mammals and to have evolved from the fish's gill. The mammalian lungs consist of two conical organs each one being highly convoluted and occupying one side of the thoracic cavity. They are individually enclosed and separated by the pleura, the central partition of which contains the heart and associated large blood vessels. Extensions of the pleura attach the lungs to the oesophagus or gullet. That part of the pleura in contact with the surrounding thoracic cage provides contiguity between lung tissue and the latter. Each lung communicates through a bronchus with the trachea and thence, through the nose or mouth to the outside air. Within the lungs, the bronchi branch into continually finer extensible tubular airways until they reach the acini, where are found most of the more or less hemispherical lung *alveoli* which serve as an area for gaseous exchange and where by far the major part of volume change of lung tissue is thought to occur. The average alveolar diameter is about 200 microns, with a wide distribution of diameters among individual alveoli. Estimates of their number range from $1 \cdot 5 \times 10^8$ to $1 \times 10^9$ with a total area estimated to be between 5 and 200 m². The tubular airways of the lungs have a cartilagenous and muscular structure. The alveoli are enclosed in a network of capillaries, collagenous, and elastic fibres, but no muscle. The lungs also contain many lymphatic channels which are more or less muscular, and are sometimes considered, but not known certainly, to reach the alveolar surface. Tissue elements are defined largly in terms of their histochemical properties. A list of those elements which make up all of the respiratory structures would be long, and assignment of mechanical function to them would be largely impossible at the present state of knowledge. The adult lungs weigh about 1 kg and their combined tissues are denser than water.

The major attempts to treat this system mechanically, consider it as an engine describable in terms of volume as the single independent variable, with pressure the dependent variable. Thus, the equation: $P = \alpha(V) + \beta(\dot{V}) + \gamma(\ddot{V})$, is used to describe the pressure-volume relation of the entire lung-cage system, or its parts, where $P$, $V$, $\dot{V}$, and $\ddot{V}$ indicate, respectively, pressure, volume, steady rate of volume change, and volume acceleration, and the coefficients $\alpha$, $\beta$, and $\gamma$ are functions of the volume as indicated. The equation separates the total driving pressure into elastic, flow resistive, and inertial terms, respectively, and can be valid only if all segments of the respiratory system move in a fixed relationship to the total motion. It implies that the effective pressures at all points of the individual types of participating surfaces are everywhere the same at each individual type of surface. These surfaces are: the interface between alveoli and air, the pleural surface between lungs and thoracic cage, and the external surface of the thoracic cage, or body surface. Pulmonary physiologists recognize generally that the basic equation probably cannot apply rigorously, but use it as a guide in approaching this complicated subject.

The separate terms of the basic equation will be considered in turn, with emphasis on the coefficients which usually are of greatest interest. Contraction of the muscular *diaphragm* and muscles of the thorax produce enlargement of the thoracic cage and an equal enlargement of the lungs which act spatially in series with it. The concomitant reduction of pressure in the airways, greatest in the alveoli, permits air to pass therein. Relaxation of respiratory muscles permits contraction of extended elastic respiratory tissues to produce normal *expiration*, but expiration can be forced. The driving pressure in each case is equal to the sum of the three terms given in the equation, and is influenced by the abdominal muscles and contents and the heart with its attendant structures.

The coefficient, $\alpha$, of the elastic term, $\alpha(V)$, has been termed elastance. The reciprocal of elastance, or compliance, is used perhaps more frequently and is recorded as 1/cm water. It is dependent upon extent, but not rate of change of volume. It is measured only under what is thought to be static conditions over a given range of volumes with volume held constant for few seconds at the end of each volume change. The pressure differential sought in measurement of compliance of the lungs exists between the pleura and the lumina of the alveoli, where the latter pressure in these circumstances is the same as pressure in the closed mouth with air passages open and all muscles relaxed which are external to the lungs and involved grossly in respiration. Pressure in the oesophagus is considered, generally, to be equal to pressure in the pleura; an assumption not finally proved since pleural pressures are not known to be uniform over all the pleural surfaces. Compliance of the lung-thoracic cage system is measured by plotting volumes of the system against the corresponding pressure differentials between the lumina of the alveoli and the outside air. Although the lungs and thoracic cage are in series spatially, their elastic properties add as springs in parallel, hence the reciprocal of the compliance of the entire system is considered equal to the sum of the reciprocals of the compliances of the lung and the thoracic cage. Compliance of the thoracic cage alone can, therefore, be estimated from the separate compliances of the lungs and total compliance. Over ordinary volume changes, the volume-pressure relationship of the lung-thoracic cage system may be linear. It is sometimes concave toward the pressure axis under these conditions, and especially so when the volumes of air exchanged become large. Pressures are then higher at a given static volume during inflation than during deflation, resulting in hysteresis loops which also are observed using excised lungs of animals, or anaesthetized animals with the thorax intact. Some of the hysteresis is relatively independent of time and,

on excised lungs, is sometimes not changed greatly when pressures are held constant at the corresponding volumes for periods of up to 40 minutes. The specific respiratory tissues and their properties which affect the production of compliance curves are unknown. Compliance of the thoracic cage is thought to be due to stretching of muscles and ligaments. Compliance of the lung has sometimes been partitioned between the lung tissue and surface forces at the alveolar-air interface. The effect of properties of lung tissue is known to be complicated and is obscure because of the many types of tissue elements involved. Attempts to assess the effects of surface forces have been made, and here are placed arbitrarily into one of three categories: (1) The alveolar surface acts like a film with one surface and with constant surface tension, over the end of a tube. As air is forced into the free end of the tube, the radius of curvature of the surface passes from infinity to an ever decreasing value until it becomes equal to the radius of the tube and the surface is hemispherical. If further expansion of the surface is not permitted, its radius of curvature never exceeds this value and the collapsing pressure exerted at the surface passes from zero to a maximum according to the fundamental capillarity equation. If radius of curvature of the surface were permitted to exceed that of the tube, the retractive force arising from capillarity would decrease with increasing volume, and elastic properties of the lung would be determined mostly by lung tissues. Attempts to determine a precise relationship between changes of alveolar radius of curvature and movements of the lung associated with breathing have not been decisive. (2) Surface tension at the alveolar-air interface varies with the area of the alveolar surface. It increases during inspiration as the alveolar area increases. The reverse occurs during expiration, but with the production of a hysteresis loop perhaps similar in shape to the volume-pressure hysteresis obtainable during measurement of compliance. This relationship between alveolar area and surface tension is sometimes considered to be an important agency for stabilization of the alveolar structure of the lung against capillary forces. (3) The alveolar structure of the lung is stabilized by virtual absence of contractile pressure due to capillarity which, although a function of area, remains always so small as to be negligible. Change of contractile pressure with respect to radius of curvature also is not important for the same reason. Moreover, area of the alveoli does not change necessarily upon expansion or contraction of lung tissue. An example of such behaviour is the small, if actual, change of area of a paper bag when emptied and filled with air. Hysteresis loops might be explained under this category by closure of alveolar spaces at the end of expiration which would require higher pressures to open on inspiration. They might also be explained on the basis of elastic and rigid components of structure without reference to surface tension.

Surface active material has been obtained from mammalian lung by a process which is thought to displace it from the alveolar-air interface. It contains protein and lipid. The protein fraction has not been analysed. The lipid fraction is mostly phospholipid, but unesterified cholesterol, fatty acids, and triglycerides have been identified as constituents. When the area of this material, spread as a film on a surface balance, is diminished slowly, surface tension falls in a non-linear manner. As the area is restored slowly to the original, the surface tension rises, again non-linearly, to form a closed hysteresis loop with the direction of generation indicating an endergonic process. Such observations have been interpreted to indicate that surface active material in the lung alveoli confers upon lung tissue a significant part of its static hysteresis behaviour. Hysteresis loops demonstrated on the surface balance have not been studied extensively under static conditions so there can be no rigorous comparison at present between them and static hysteresis loops obtained from excised lungs or the intact respiratory system.

Alveolar area or volume need not change for adequate ventilation of the lungs which may be accomplished by suitable rhythmic compression and rarefaction of air within a rigid chamber enclosing a subject. Movements of respiratory tissues cease because pressure changes in the chamber are transmitted equally to the inside of the lung and to the outside of the body. If the equilibrium static surface tension of alveolar surfactant on a surface balance can be used as a model for its magnitude in the alveoli under these conditions, it would seem to be sufficiently great to produce instability of alveolar spaces if La Place's equation of capillarity and generally held values for the physical forces controlling movements of body fluids apply. The very low static surface tension said to be produced by bubbles formed by removal of surface active material from alveolar surfaces has not been reproduced on the surface balance. Such low surface tensions thereon have been obtained only during continuous diminution of the surface area of the film even when replenishment of surface active material was continuous and rapid.

A considerable field which may be amenable to experimental methods appears to be open on these points.

The coefficient, $\beta$, of the flow resistive term, $\beta(\dot{V})$, is measured as (cm water) sec/1 under dynamic conditions and denotes resistance to steady flow of gases moving in and out of the lungs and concomitant resistance to respiratory movements of the lungs and thoracic tissuess. There is no proven theory relating pressure and flow in highly branched, distensible systems. Such a system in the lungs does not necessarily follow volume changes passively, but may take an active part due to innervation of its musculature, and is influenced by inhaled gaseous chemicals, small particles, and recent volume history of the lungs. That part of the flow resistive term attributable to flow of gas is, therefore, often considered to be the sum of pressures required to accommodate Poiseuille and turbulent flow of air through the tubular air passages

and is measured as pressure in the lumina of the alveoli minus atmospheric pressure on expiration: the reverse on inspiration. Measurement of pressure in the lumina of the alveoli involves evaluation of small pressure changes produced by breathing movements inside a rigid chamber surrounding a subject. Simultaneous measurements of atmospheric pressure and instantaneous air flow permit estimation of airway flow resistance alone and separate from other resistances. Airway flow resistance decreases with increased lung volume and increases with increased breathing rate. It is normally less on inspiration than on expiration during gentle breathing. At least half of it appears to exist in the upper part of the respiratory system (near the port of entry of air).

The sum of the resistances of lung tissues which accompanies steady change in their enclosed volume and termed pulmonary tissue resistance, plus airway flow resistance, is termed pulmonary resistance, and is measured by plotting air flow against the pressure across the lung tissue (pressure in the alveoli minus pressure in the pleura, or esophageal pressure) after subtracting the corresponding pressure exerted against elastic recoil of the lung. The final curves are concave toward the pressure axis. Pulmonary tissue resistance is obtained by subtracting airway flow resistance from pulmonary resistance. It is about $\frac{1}{5}$ of the pulmonary resistance, or $\frac{1}{4}$ of the airway flow resistance. That (pressure) sec/1 required to overcome resistance to motion of all tissues involved in breathing has been termed total resistance. It has not been considered extensively because necessary knowledge is small concerning resistance of the thoracic cage and its attendant structures which accompanies steady change of its enclosed volume.

The final term, $\gamma(\ddot{V})$, expresses the pressure required to produce volume acceleration and is considered to be due primarily to mass inertia of gases and tissues and to constitute only about 0·5 per cent of the total pressure at rest and not more than 5 per cent of the total during heavy exercise. It is often included in the flow resistive term. Units of the coefficient, $\gamma$, are (cm water) sec²/1.

The work of breathing is defined as the integral of pressure times volume increment over the total change in volume. The work of moving lungs, thoracic cage, and gas has been measured by plotting the pressure required to produce respiratory movements of an anaesthetized subject against volume of air moved. The values obtained are only 0·5 kg m/min. at rest, but rise to 250 kg m/min. at a maximum breathing rate of 200 1/min. Based on attempts to measure added oxygen consumption of respiratory muscles due to breathing, their mechanical efficiency has been estimated at 5–10 per cent using about 3 per cent of the total oxygen consumption of the adult when resting.

*Bibliography*

BARACH A. L. (1944) *Principles and Practice of Inhalation Therapy*, Philadelphia: Lippincott.

BUTLER J. (1957) *Clin. Sci.* **16**, 421.

COMROE J. N., JR. *et al.* (1962) *The Lung, Clinical Physiology and Pulmonary Function Tests*, Chicago: Yearbook Publ.

ENGEL S. (1962) *Lung Structure*, Springfield, Illinois: C. C. Thomas.

MEAD J. (1961) *Physiol. Revs.* **41**, 281.

MENDENHALL R. M. (1963) *Arch. Environ. Health* **6**, 74.

MILLER W. S. (1950) *The Lung*, Springfield, Illinois: C. C. Thomas.

PATTLE R. A. (1965) *Physiol. Rev.* **45**, 48.

ROSSIER P. H. *et al.* (1960) (Luchsinger, P. C. and Moser K. M., Eds. and translators from German) *Respiration, Physiologic Principles and Their Clinical Application*, St. Louis: C. V. Mosby.

SUTNICK A. I. and SOLOFF L. A. (1963) *Am. J. Med.* **35**, 31.

VON HAYEK H. (1960) (Krahl V. E., translator from German) *The Human Lung*, New York: Hafner.

WYSS O. A. M. (1963) *Ann. Rev. Physiol.* **25**, 143.

R. M. MENDENHALL

**PULSE RADIOLYSIS.** A technique used in radiation chemistry whereby high instantaneous concentrations of chemically-reactive species are produced in liquids by a short (2–10 μsec) intense pulse of electrons from an electron linear accelerator. Absorption spectroscopy is commonly used to identify the transient intermediates and to follow their subsequent reactions.

A. R. ANDERSON

# Q, R

**QUANTUM ELECTRONICS.** The term "quantum electronics" has evolved from the idea that some electronic systems such as microwave and optical masers which are best described using quantum mechanics should be called quantum electronics. Following general practice, the field of quantum electronics encompasses masers, optical masers (lasers), phonon phenomena, and electric and magnetic resonance phenomena. While it is true that these phenomena may be described using classical electromagnetic theory, a complete and more exact description involves quantum mechanical concepts.

The invention of the practicable maser by Gordon, Zeiger, & Townes in 1955 initiated the field. Later the term quantum electronics was applied to technical meetings involving masers, and in January 1963, the first issue of the Institute of Electronic and Electrical Engineers was termed, "The Quantum Electronics Issue."

Quantum electronics generally involves atomic and molecular transitions and the interaction of the atoms and molecules with electromagnetic or vibrational waves. In a sense, all electrical phenomena are encompassed by quantum mechanics; however, quantum electronic phenomena are arbitrarily restricted to cases where the energy levels of the atoms or molecules are sharp and of importance to the form of the electromagnetic interaction.

*See also*: Masers.　　　　　　J. R. SINGER

**RADIATION CHEMISTRY.** Radiation chemistry is the study of the chemical effects of high energy radiation, i.e. ionizing radiation, and includes many classes of chemical reactions induced by radiation, viz. oxidation and reduction, decomposition, synthesis, hydrogenation, polymerization, and molecular rearrangement, as well as kinetic studies. It should be clearly distinguished from the subject of Radiochemistry which concerns the study of radioactive materials and their chemical behaviour.

High energy radiation includes both penetrating electromagnetic radiation, e.g. X and $\gamma$ radiation, and swiftly moving atomic particles such as electrons, protons and neutrons. In all cases energy is transferred along tracks of electrically charged particles—$\gamma$ radiation produces electrons, and neutrons produce recoil ions and atoms—so that radiation chemistry is essentially the study of the chemical effects brought about by electrically charged atomic particles. Ionization or excitation by high energy radiation is an extremely rapid process ($< 10^{-16}$ s), producing activated molecules which are extremely unstable and promptly undergo secondary reactions either spontaneously or in collision. Important secondary processes include unimolecular dissociation, energy transfer, ion-molecule reactions, and charge transfer. The latter two are not generally encountered in other branches of chemistry, while the two former processes are also found in photochemistry. The system rapidly attains thermal equilibrium, the primary products being converted to stable molecules or to chemically-reactive free radicals and ions, which react with each other or with solute and solvent molecules.

The initial energy of the photon or charged particle is usually much greater than typical excitation, ionization or bond energies of molecules, e.g. the energy of incident radiation is usually expressed in keV or MeV, and bond energies in eV; 1 eV is equal to 23·05 kcal per mole. It follows that the initial act of energy absorption is markedly non-specific, in contrast to photochemistry where absorption of quanta is highly specific and depends on allowed optical transitions in individual molecules. Thus while absorption of ultra-violet light produces well defined excited states of atoms and molecules, the non-specificity of high energy radiation produces complex spectra of ions and electronically-excited states. Another basic difference from photochemistry, particularly important in condensed phases, is that the absorption of light quanta produces excitations distributed at random throughout the medium, while the absorption of high energy radiation results in a drastically non-uniform distribution of excitations and ionizations.

The contemporary impetus in radiation chemistry coincides with the development of nuclear reactors and the intense radioactive sources made available by the production of artificial isotopes such as $^{60}$Co, $^{137}$Cs, $^{210}$Po, $^{239}$Pu, as well as the irradiated fuel rods and the reactors themselves. At the same time the rapid development of X-ray generators, and charged particle accelerators such as cyclotrons and Van de Graaff generators, has provided a great variety of sources for the radiation chemist. Of all these developments, however, the present widespread study of radiation chemistry is due principally to the production of $^{60}Co$ $\gamma$ radiation sources.

Quantitative aspects of radiation chemistry are expressed in the radiation chemical yield which is defined as either,

(a) Ion Pair Yield =

$$\frac{\text{No. of molecules (or atoms) produced (or consumed)}}{\text{No. of primary ionizations}} = \frac{M}{N}$$

or

(b) $G$ value =

$$\frac{\text{No. of molecules (or atoms) produced (or consumed)} \times 100}{\text{Energy absorbed in eV}}$$

The $G$ value is most commonly used, even for gas phase radiolysis, but it has no fundamental significance, and is simply chosen so that most yields cover a convenient numerical range. The relationship between the two definitions of yield is,

$$M/N = GW/100$$

where $W$ is the mean energy per ion pair.

Primary dosimetry to measure the energy input is carried out by calorimetry (condensed phase), ionization measurements (gases), and charge input measurements (for well-defined ion beams). As these methods are often difficult and time consuming, the common practice is to use secondary chemical dosimeters for which the radiation chemical yield has been unequivocally established for given conditions of radiation type and dose rate. The principal chemical dosimeters are based on the oxidation of aqueous ferrous sulphate (Fricke dosimeter) for condensed systems, and the decomposition of nitrous oxide for gas phase radiolysis.

In radiation chemistry, two general types of radiation are distinguished,

(a) lightly ionizing radiation, e.g. electrons, X and $\gamma$ radiation, which produces primary species in condensed systems in small discrete volumes ("spurs") place between species from adjacent so well separated that no interaction can take spurs;

(b) densely ionizing radiation, e.g. protons and helium ions, which produces primary species so close together that the individual spurs overlap, resulting in a continuous volume of excitations and ionizations with cylindrical symmetry.

These two cases are simply a manifestation of the rate of energy transfer to the system, which results in important quantitative differences in chemical behaviour, and are referred to as LET effects, i.e. effects due to changes in linear energy transfer from the incident radiation. There is a wide range of conditions between the two extremes, and the effect of changing LET can best be illustrated by reference to the radiation chemistry of water.

The radiolysis of water has been studied more extensively than any other topic in radiation chemistry, principally because of its relevance to biological systems, and of the widespread importance of aqueous solutions in other branches of chemistry. Pure water exposed to lightly-ionizing radiation is stable, but decomposes to give $H_2$ and $H_2O_2$ when exposed to densely-ionizing radiation.

This behaviour can be understood on the simple model that radiolysis of water leads ultimately to free H atoms and OH radicals, which in the case of $^{60}Co$ $\gamma$ radiation are produced throughout the liquid in individual spurs separated on the average by $\sim 5000$Å (for a 0·5 MeV electron). The average number of radicals per spur is estimated at between 2 and 12 and the instantaneous concentration of radicals in the spur is $\sim 1$ mole/l. These radicals react with each other, producing $H_2$ and $H_2O_2$ and re-forming $H_2O$, in competition with concentration-controlled diffusion of the radicals into the bulk of the solution. As a result of these competitive processes, a large fraction of the radicals escapes from the spurs without reacting, and then reacts with the molecular products, $H_2$ and $H_2O_2$, which would otherwise accumulate in the bulk of the liquid. These processes, represented by equations 1–6, result in a chain mechanism leading to the re-formation of water and explaining its apparently stability to lightly-ionizing radiation.

spur reactions
$$\begin{cases} H_2O & \rightarrow H + OH & (1) \\ H + OH & = H_2O & (2) \\ H + H & = H_2 & (3) \\ OH + OH & = H_2O_2 & (4) \end{cases}$$

reactions in bulk of liquid
$$\begin{cases} H + H_2O_2 = H_2O + OH & (5) \\ OH + H_2 & = H_2O + H & (6) \end{cases}$$

In the case of a 5 MeV alpha-particle, as an example of densely-ionizing radiation, the separation of the individual spurs is only 8 Å, so that very few radicals escape from the track without reacting, and $H_2$ and $H_2O_2$ build up in solution. In the special case of very high instantaneous dose rates from electron linear accelerators ($> 10^7 \times$ dose rate from $^{60}Co$ $\gamma$ radiation sources), lightly ionizing radiations can lead to the continuous production of molecular products. This effect at high dose rates is due to the overlapping of individual tracks and should be distinguished from the intra-track effects produced by radiation of differing LET. It should also be distinguished from dose-rate effects at lower dose rates, which are generally due to chemical chain reactions.

With $\gamma$ radiation the continuous production of $H_2$ and $H_2O_2$ can be effected by boiling or by adding low concentrations of chemically-reactive solutes, generally called "scavengers". In the former case the steady-state concentration of molecular products in solution is reduced, thus inhibiting the re-formation chain reaction, while in the second case the back reaction is inhibited by reactions of the solute with radicals escaping from the spurs. By measuring the resultant chemical change in various solutes, and by measuring the yields of $H_2$ and $H_2O_2$, it is possible to determine both the yield of radicals escaping from the spurs and the yield of molecular products formed by reactions within the spurs. These yields of radicals and molecular products surviving track reactions are generally regarded as "primary chemical yields" and are designated in one of three common ways, viz. $g(X)$, $G_X X$, $G_X$. In this article the convention $g(X)$ is adopted for the radical and molecular product yields, while the yields obtained directly by experiment are designated by $G(X)$.

Determination of the radical and molecular product yields ($g(X)$), from experimentally measured yields ($G(X)$) invariably requires knowledge or interpretation of the mechanism of the postulated chemical reactions. This is well illustrated in the radiation-induced oxidation of oxygenated ferrous sulphate in sulphuric acid solution (Fricke dosimeter).

$$H_2O \rightarrow H, OH, H_2O_2, H_2 \tag{7}$$
$$Fe^{2+} + OH = Fe^{3+} + OH^- \tag{8}$$
$$H + O_2 = HO_2 \tag{9}$$
$$Fe^{2+} + HO_2 = Fe^{3+} + HO_2^- \tag{10}$$
$$HO_2^- + H^+ = H_2O_2 \tag{11}$$
$$Fe^{2+} + H_2O_2 = Fe^{3+} + OH + OH^- \tag{12}$$

From the above series of reactions it follows that the stoichiometry can be represented as,

$$G(Fe^{3+}) = 2g(H_2O_2) + 3g(H) + g(OH) \tag{13}$$
$$G(H_2) = g(H_2) \tag{14}$$

where $G(Fe^{3+})$ and $G(H_2)$ are measured yields. In the absence of $O_2$, reaction 9 cannot occur and is replaced by,

$$Fe^{2+} + H + H^+ = Fe^{3+} + H_2 \tag{15}$$

In the absence of $O_2$, therefore,

$$G(Fe^{3+}) = 2g(H_2O_2) + g(OH) + g(H) \tag{16}$$
$$G(H_2) = g(H_2) + g(H) \tag{17}$$

Equations 13, 14, 16, and 17, can be solved for $g(H_2)$ and $g(H)$, and a separate determination of $g(OH)$ or $g(H_2O_2)$ allows all the primary chemical yields to be determined. The radiolytic reduction of aqueous $Ce^{4+}$ gives $G(O_2) = g(H_2O_2)$; radiolysis of aqueous $HCOOH/O_2$ gives $G(CO_2) = g(OH)$.

Measurement of the primary chemical yields with radiation of differing LET shows that the radical yields decrease and the molecular product yields increase with increasing LET as illustrated in the figure; the smoothed curves represent generally-accepted data for irradiations in acid solution. At high values of LET it is seen that a significant yield of the hydroperoxy radical is observed due to reaction 18 occurring in the track.

$$OH + H_2O_2 = HO_2 + H_2O \tag{18}$$

It has been shown in many chemical systems that the primary chemical yields are essentially independent of scavenger concentrations up to $10^{-3}$ M, but that solutes in higher concentrations begin to interfere with reactions occurring in the spurs or tracks. These reactions have been the subject of many theoretical and experimental studies in terms of a "radical diffusion kinetic" model, i.e. a model which considers reactions of radicals within the spurs or tracks in competition with the time-dependent expansion of the primary volume of excitation.

While the general concepts of water radiolysis can be clearly understood on the basis of the simple model of water dissociating to give H and OH radicals, it

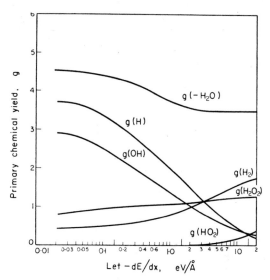

*Effect of LET on primary chemical yields in the radiolysis of water ($0.8NH_2SO_4$ solutions).*

is known that the reducing and oxidizing species can exist in various acidic and basic forms, corresponding to the loss or gain of a proton. Thus the reducing radical can react as the neutral H atom, the solvated electron $e_{Aq}^-$, or as $H_2^+$, while the corresponding forms of the oxidizing radical are OH, $O^-$, and $H_2O^+$. These various forms are inter-related as follows,

$$e_{Aq}^- \overset{H^+}{\rightleftharpoons} H \overset{H^+}{\rightleftharpoons} H_2^+$$
$$O^- \overset{H^+}{\rightleftharpoons} OH \overset{H^+}{\rightleftharpoons} H_2O^+$$

and the quantitative contribution from any form depends critically on pH.

In contrast to the detailed understanding of aqueous radiation chemistry, the interpretation of organic radiation chemistry is still to a large extent semi-empirical. Much current effort, however, is devoted to (a) obtaining more accurate data on yields; (b) clarifying the effects of radiation intensity, total dose, changes in LET, and the reactions of added radical scavengers; and (c) comparisons between radiolysis in the gas and liquid phases which provide some evidence on the relative importance of bimolecular and unimolecular reactions. The results of these studies should lead to a continuity of interpretation similar to that for water. One of the principal difficulties in developing quantitative kinetic understanding of organic radiation chemistry lies in the fact that the track is not so well defined as in water. Since chemically-reactive free radicals and other species derived from the decomposition of organic molecules can react with the solvent molecules, it is more difficult to define a sharply-delineated track volume, in which radical–radical reactions predominate. Present work, however, demonstrates that in aliphatic hydrocarbons good quantitative correlations can be derived on the basis of homogeneous kinetics, while for aromatic molecules application of modified radical diffusion models is furthering understanding of the basic chemical processes. Processes which have to be considered include bimolecular reactions between radicals or excited molecules, energy transfer, ion-molecule reactions, and unimolecular decompositions.

In general, the effect of changing LET is less pronounced in organic systems than in water but at high values of LET, marked changes are produced in the decomposition of aromatic molecules. Current studies on LET effects in both aliphatic and aromatic molecules are providing valuable information on competition between various possible processes.

Another general problem in organic radiation chemistry concerns the distinction between molecular detachment processes and radical abstraction reactions, particularly with regard to $H_2$ formation. It is a common feature of organic systems that addition of a radical scavenger, e.g. $I_2$, $Fe^{3+}$, reduces the yield of $H_2$ to a small residual yield, the "unscavengable $H_2$", which is assumed to arise from unimolecular dissociation processes. Confirmation of molecular detachment processes has been obtained using tracer techniques, particularly by irradiating mixtures of a substance and its deutero analogue. Radiolysis of mixtures of $CH_4$ and $CD_4$ produces $H_2$, $D_2$ and HD; the two former by unimolecular dissociation and the HD by abstraction reactions involving H and D atoms. In many cases however, e.g. alcohols, acetone, interpretation of yields with mixtures of H and D isotopes is somewhat equivocal, and it appears that the "unscavengable $H_2$" originates in processes which are more complex than unimolecular dissociation.

The most striking single generalization concerning the radiolysis of organic molecules is the smaller radiation sensitivity of aromatic hydrocarbons and their simple derivatives compared with simple aliphatic and alicyclic hydrocarbons. The greater radiation stability of aromatic molecules springs from the fact that many of their valency electrons are shared, and electronic excitation leads quite often to internal conversion and light emission rather than to dissociation. Gas evolution is generally very small in aromatic hydrocarbons, but this gives a false impression of their radiation stability, as high molecular weight condensed products ("polymers") are inevitably formed in high yield, e.g. with $^{60}Co$ $\gamma$ radiation:

benzene gives $G(H_2) = 0.04$, $G(\rightarrow \text{polymer}) = 0.9$

cyclohexane gives $G(H_2) = 5.6$, $G(\rightarrow \text{polymer}) \sim 4$

Mixtures of aromatic and unsaturated cyclic hydrocarbons, e.g. bezene and cyclohexane, give yields of certain products which are lower than those calculated on the basis of direct additivity of radiation effects in the separate components. This "protection effect" is believed to arise from interference of the aromatic molecules at a very early stage in the complex processes of energy distribution, following the initial act of energy absorption from the incident radiation. However, identification of the specific processes has not been unambiguously achieved.

Hydrogen is the most prominent product in the radiolysis of aliphatic hydrocarbons, with methane being the second most important product. Yields of methane increase with increasing number of methyl groups so that it is produced more effectively from branched-chain hydrocarbons than from straight-chain hydrocarbons.

Studies of the radiation chemistry of gases have developed from the discovery of chemical changes produced by electrical discharges during the 19th century. Major studies in gas radiolysis include:

(a) radiation decomposition of single gases; $O_2$, $CO_2$, CO, $CH_4$, $NH_3$, $N_2O$,

(b) radiation-induced oxidation of $H_2$, CO, $N_2$, $NH_3$, $C_2H_2$,

(c) hydrogenation of $C_2H_4$, $C_2N_2$, $D_2$, $Cl_2$,

(d) polymerization of unsaturates, e.g. $C_2H_4$, $C_2H_2$, and

(e) gas-solid interactions, e.g. $CO_2$/graphite reaction.

Recent advances in analytical techniques, particularly gas chromatography, have shown that radiation-induced processes in gases are in many cases extremely sensitive to traces of impurities, and that secondary reactions involving products can participate at extremely low percentage conversion of the parent molecules ($< 0.1$ per cent). Thus the interpretation of primary chemical processes is by no means unambiguous, nor indeed are many of the initial chemical yields well established.

Interpretation of gas phase reactions has varied from purely ionic mechanisms, involving clustering of molecules around primary ions, to completely free radical mechanisms, but nowadays the importance of both ionic and radical processes is clearly recognized. Data from mass spectrometry studies of cracking patterns and ion-molecule reactions are widely invoked in the interpretation of gas phase radiolysis, but the low operating pressures of mass spectrometers ($< 1 \text{ mm (Hg)}$) introduce difficulties in extrapolating to the higher pressures ($\sim 1$ atm) generally employed in radiation chemical studies. Much current work is devoted to the careful establishment of initial chemical yields at low percentage conversion of the parent molecule, studies of the effects of low concentration of additives, the use of rare gases as diluents to study the chemical manifestations of charge and energy transfer processes, and the use of exchange techniques to clarify radiolytic mechanisms.

The use of ionizing radiation in high polymer chemistry is another well established branch of radiation chemistry. The effect of radiation on polymers is controlled by their tendency to degradation (chain scission) or to formation of covalent bonds between macromolecules (cross-linking). In some cases, e.g. polymethyl methacrylate, polyisobutylene, scission is the dominant effect, while in others, e.g. polyethylene, polystyrene, cross-linking predominates. These cross-linked materials exhibit new physical properties, better resistance to chemicals and to heat, and improved mechanical properties.

In addition to the use of radiation to produce changes in properties, the application of high energy radiation to polymer chemistry can be generally classified in two ways;

(a) studies of graft polymerization, and

(b) general studies of radiation induced polymerization.

The production of graft polymers is an extremely important field, as potentially it offers the prospect of producing polymers with pre-selected properties. High-energy radiation is a particularly useful tool, as many radiation-induced polymerization reactions can be initiated at modest temperatures. Radiation-induced polymerizations are generally radical-initiated processes although there are several well-established cases of ionic polymerizations, which are distinguished by proceeding at lower temperatures, e.g. polymerization of isobutylene and of styrene. Polymerizations in solution, in suspension, and in the solid state are also studied. Polymerization of acrylamide in the state shows that the polymer forms in very localized volumes, which appear to be parallel to certain preferred lattice orientations, indicating that polymerization occurs only in regions of trapped radicals in the monomer.

The general features in the radiolysis of solids are controlled by the greatly decreased mobility of radicals and ions, in comparison with that in liquids or gases. Coloration of ionic crystals, e.g. alkali halides, by ionizing radiation is produced by electrons being trapped at various defects in the crystal, resulting in the absorption of light. The accompanying chemical changes can be investigated by subsequent dissolution or thermal degradation of the crystal, so that when irradiated alkali halides are dissolved in water, $H_2$ is evolved, and halogens and hydroxyl ions are formed in solution. Decomposition of polyatomic ions has been studied, e.g. $KClO_4$ gives chlorate, chloride and oxygen; $KClO_3$ gives chloride, chlorite, hypochlorite, and oxygen; nitrate gives nitrite and oxygen. Detailed studies of nitrates have shown marked differences in nitrite yields ranging from 2 for $Ba(NO_3)_2$ to $\sim 0.02$ for $LiNO_3$. If it is assumed that the initial yields of O atoms from the various nitrates are unlikely to differ markedly, the observations on the yields of nitrite can be correlated quite well with the packing of the crystal lattice, the mechanism involving competition between O atoms escaping from the point of origin and recombining with trapped nitrite ions.

Due to the lower mobility of radicals, studies in solid phase radiolysis provide valuable sources of information on the identification of chemically-reactive intermediates responsible for the observed radiation chemistry, e.g. trapped electrons, radicals and ions have been identified in ice and in glassy solids. Many deductions about the nature of the intermediates have been largely inferential, based on kinetic interpretations of the resultant chemical changes, but there is currently much emphasis on their positive identification by physical techniques. Such techniques include electron spin resonance spectrometry, ultra-violet and infra-red absorption spectroscopy, emission spectroscopy and electrical conductivity methods.

In addition, the development of "pulse radiolysis" using very high instantaneous dose rates of electrons from electron linear accelerators, in conjunction with absorption spectroscopy provides much valuable information on the nature, and absolute rate constants for reactions of intermediates in liquids. In this way the solvated electron in water radiolysis, $e_{Aq}^-$, has been positively identified and its properties extensively investigated. It has also been possible to identify radicals by e.s.r. in irradiated liquid hydrocarbons, using high steady dose rates from an electron Van de Graaff generator.

While radiation chemistry is a very important and active branch of chemistry, there are very few examples of its application to industrial processes, although much work is devoted to this aspect, including the interesting concept of "chemo-nuclear" reactors, i.e. the use of nuclear reactors to produce chemicals on a large scale. Present industrial applications include modifying the properties of plastics, a small-scale process for the production of ethyl bromide, and the increasing use of radiation in medical sterilization. Perhaps the widest application is in the field of nuclear technology, where detailed information on certain radiation chemical processes is vital to the reactor design. These problems include the radiation-induced decomposition of water moderators and of organic moderators, the radiolytic oxidation of graphite by $CO_2$, and a host of other problems connected with the radiation stability of numerous materials used in the construction of nuclear reactors.

*See also:* Linear energy transfer. Pulse radiolysis.

*Bibliography*

ALLEN A. O. (1961) *The Radiation Chemistry of Water and Aqueous Solutions*, New York: Van Nostrand.

CHAPIRO A. (1962) *Radiation Chemistry of Polymeric Systems*, New York: Interscience.

HART E. J. and PLATZMAN R. L. (1963) *Mechanisms in Radiobiology*, in *Radiation Chemistry*, New York: Academic Press.

LIND S. C. (1961) *Radiation Chemistry of Gases*, New York: Reinhold.

Radiation Chemistry (1963) *Nucleonics* **19** (10) October.

SWALLOW A. J. (1960) *Radiation Chemistry of Organic Compounds*, Oxford: Pergamon Press.

A. R. ANDERSON

## RADIOACTIVE MATERIAL, MIGRATION OF.

The alpha-emitting isotope, $^{210}$Po, which has a half-life of 138·40 days, is known to be very difficult to confine or to keep localized within some sort of confinement. It is well appreciated that, because of their radioactivity, most radioisotopes are generally quite easy to detect in far lesser amounts than could be detected if these isotopes were stable, and, because of this fact, it has long been recognized that droplets of spray and dust particles so small as to be invisible to the eye can nevertheless carry with them relatively large and objectionable amounts of radioactivity; thus the spread of radioactive contamination is often a serious problem and always a factor to be guarded against in the handling of radioactive material. However, $^{210}$Po apparently spreads so readily and so extensively as to constitute a special case. This isotope has been observed to migrate upstream against a current of air and to translocate under conditions where it would appear to be doing so of its own accord. Workers disagree as to whether solutions exhibit this effect to a greater or lesser extent than solid elemental polonium or solid compounds of polonium, but all agree that extra special precautions must be taken in handling this isotope.

No experiments have been published which have been designed to elucidate the mechanism of this migration, although there are several schools of thought. Three factors which have a bearing on the case are (1) the relatively large quantities of polonium which can now be readily made (by neutron irradiation of bismuth, followed by beta decay of the $^{210}$Bi), (2) the relatively high specific activity (or activity per unit weight) of the material so produced, by virtue of its moderately short half-life, and (3) the high energy (5·30 MeV) of the emitted alpha particle. The first two factors suggest that possibly the migration of polonium may not be greatly different from that of other radioisotopes but rather that conditions are likely to be found where the migration of polonium is more easily and readily observed than is the migration of other isotopes.

It has been postulated that when one of these polonium alpha particles is emitted, sufficient recoil energy is imparted to other atoms with which the decayed atom had been associated so as to move the small agglomerate a short distance, and that as other atoms in this agglomerate decay by alpha emission the recoil energy from each decay moves the agglomerate a little farther in random directions until finally a gross movement of radioactivity is noticed. However, this migration has not been observed (or at least not to the same extent) with other isotopes such as $^{242}$Cm, which has a similar half-life (163 days) and an even more energetic alpha (6·110 and 6·066 MeV). It may be argued that the chemical and physical forces holding together the elemental material or its compounds or solutions would be different for different elements and that this might contribute to differences in apparent migration from one element to the next. In the case of solids, another contributing factor may be the disruptive effect caused by the displacement of other atoms by the entrapped helium formed when alpha particles which have not escaped from the surface of the solid are stopped and capture two electrons.

Another theory attributes the migration of polonium to its relatively high vapour pressure.

Still another school of thought rejects both of these theories and holds that polonium is transported only by outside agents such as flies, mice, eddy currents known to be present at the faces of some hoods, human carelessness, etc. While these factors may apply in some cases, other cases have been observed in which it would appear to be possible to rule out such external agents.

L. STANG, JR.

## RANDOM AND PSEUDO-RANDOM NUMBERS.

For many investigations using statistical sampling, particularly those described as Monte Carlo methods, it is desirable to have, or be able to produce, sequences of numbers purporting to be random. The following is a brief account of methods of constructing such sequences, and of some of their properties.

*1. Generation and testing of random and pseudo-random numbers.* It will be seen in section 2 that if we have random numbers which are uniformly distributed, that is, are as likely to take any one value as any other within their range of variation (for example, single-digit uniform decimal numbers are numbers capable of assuming any one of the values 0, 1, ..., 9, each with probability 0·1), it is possible in principle to construct from them random numbers whose distribution has any specified form. The important practical question therefore is to be able to construct uniformly-distributed random numbers.

There are physical processes which appear to behave in a random way and which can be made to produce sequences of uniformly-distributed random numbers. Examples are the emission of noise from electronic tubes or of radiation from a radioactive source, or the position in which a carefully-balanced roulette wheel comes to rest. All of these have been used to produce sequences of numbers purporting to be uniformly distributed. With a radioactive source, for example, if the average number of particles radiated within a fixed interval of time is large, the probability that the actual number radiated within the interval will be even is very close to 0·5; if therefore we record a 0 when the number is even and a 1 when it is odd we shall get a sequence of uniformly distributed binary random numbers.

The question of whether the numbers produced are actually random is almost a philosophical one, but it does not concern us here. What we want to know is: do these numbers behave *for practical purposes* as would a sequence of random numbers? We interpret "for practical purposes" as meaning "as regards the satisfaction of certain tests of randomness". One could go on indefinitely proposing properties of truly random sequences and testing whether a given sequence possessed these properties with a reasonable degree of accuracy. In fact, one subjects the sequence to a few standard tests and if it passes these accepts it as random. One such system of tests which is sometimes used is that proposed by Kendall and Babington Smith (1938 and 1939a). It consists of four tests of a fairly simple type, for example that each digit should appear in the correct proportion of cases. For each of these tests the frequencies of the various cases actually observed in the sequence under test are compared with those theoretically expected, the basis of the comparison being some statistical criterion such as the $\chi^2$, which indicates how great must be differences between the two sets of frequencies before they can be regarded as disagreeing. A sequence of numbers satisfying all these tests is accepted as random, however the numbers may have been constructed.

Some electronic computers contain circuits designed to generate random numbers by one or other of these physical processes, but it is now more common to produce sequences by specially constructed arithmetical algorithms. Each of the numbers in such a sequence is obtained by carrying out definite operations on previous numbers of the sequence, so that the sequence is deterministic. On the other hand, with suitably chosen algorithms the numbers produced satisfy tests of randomness, so that they may be regarded as random. Such numbers are therefore called *pseudo-random*. Pseudo-random numbers have considerable advantages: a sequence of them can be exactly reproduced if necessary, they do not require special circuits in the computer, and the algorithms by which they are constructed are often very suitable for programming. Of course the number of numbers, with a specified number of digits, which can be constructed is finite, so that a sequence of pseudo-random numbers necessarily has a period after which it repeats itself. In any given case it is necessary to make sure that this period is large compared with the number of numbers likely to be used in the investigation. Many algorithms have been proposed which satisfactorily meet these requirements. A series of them is based on the use of residues: for example, Lehmer has used the recurrence relation $a_{n+1} = k\,a_n$ (mod $M$), where $k = 23$ and $M = 10^8 + 1$, which generates a sequence $a_n$ of eight-digit decimal numbers with period 5882352. For fuller information on the subject, the reader should consult the books on Monte Carlo methods referred to in the bibliography.

For comparatively small investigations, or for those who have not access to electronic computers, tables of numbers purporting to be random have been published: references to some of these are given in the bibliography.

*2. Generation of random numbers whose distribution is not uniform.* If $X$ is a random variable, the properties of its probability distribution may be described by its distribution function $F(x)$, which is the probability that $X$ has a value not exceeding $x$: $F(x) = \Pr(X \leq x)$. For example, if $X$ is uniformly distributed on the interval $(a, b)$, $F(x) = 0$, $x < a$, $= (x - a)/(b - a)$, $a \leq x \leq b$, and $= 1$, $b < x$.

The derivative $f(x)$ of $F(x)$, i.e. the probability density function of $X$, if it exists, describes the distribution of $X$ as satisfactorily as does $F(x)$. For the above uniformly distributed random variable, $f(x) = 1/(b - a)$ for $a \leq x \leq b$ and $= 0$ for other values of $x$.

A simple theorem enables us to construct random numbers whose distribution function has an assigned form $F(x)$ from numbers whose distribution is uniform on the interval $(0, 1)$: it is that if $X$ has distribution function $F(x)$, the random variable $F(X)$, say $Y$, is uniformly distributed on $(0, 1)$. Thus if $y_n$, $n = 1, 2, \ldots$, are a sequence of numbers uniformly distributed on $(0, 1)$, the numbers $F^{-1}(y_n)$, where $F^{-1}(y)$ is the solution for $x$ of the equation $y = F(x)$, have distribution function $F(x)$.

This method depends on the ease with which the equation $y = F(x)$ can be solved for $x$. In many important cases this does not present serious difficulty. For particular distributions, however, for example the normal, special methods are available—see the books on Monte Carlo methods referred to in the bibliography.

*Bibliography*

Hammersley J. M. and Handscomb D. C. (1964) *Monte Carlo Methods*, London: Methuen; New York: Wiley.

Kendall M. G. and Babington Smith B. (1938) *Randomness and random sampling numbers, J.R.S.S.* **101**, 147.

Kendall M. G. and Babington Smith B. (1939 a) *Second paper on random sampling numbers, J.R.S.S.* Suppl. **6**, 51.

Kendall M. G. and Babington Smith B. (1939 b) *Tables of random sampling numbers, Tracts for Computers*, 24; Cambridge: The University Press.

Rand Corporation (1955) *A million random digits with* 100,000 *normal deviates*, Glencoe, Illinois: Free Press.

Shreider Y. A. (Ed.) (1964) *Method of Statistical Testing (Monte Carlo Method)*, Amsterdam: Elsevier.

A. J. Howie

**RAYLEIGH-TAYLOR INSTABILITIES.** These instabilities can occur at the interface between two accelerated liquids when one fluid is accelerated by another less dense one.

The problem was discussed in general terms by Lord Rayleigh and more specifically by Sir Geoffrey Taylor (1950) who showed that a disturbance corrugating the surface with wave number $k$ would grow exponentially with time constant

$$[-(\varrho_2 + \varrho_1)/(\varrho_2 - \varrho_1)(\boldsymbol{g} + g_1)\,k]^{1/2}$$

where $\varrho_1$ and $\varrho_2$ are the densities of the upper and lower liquids respectively, $\boldsymbol{g}$ is the acceleration due to gravity and $g_1$ is the forced acceleration. The initial disturbance of the interface will increase exponentially with time until it has attained a magnitude which is no longer small in comparison with the wave-length. If the horizontal surface of a liquid at rest under gravity is displaced into the form of regular small corrugations and then released, standing oscillatory waves are produced. Theoretically, a liquid could exist in a state of unstable equilibrium with a flat lower horizontal surface supported by air pressure. If small wave-like corrugations were formed on its lower surface and then released, they might be expected to increase exponentially so long as their height was small compared with the wave-length. This instability of the lower surface of a liquid would disappear if the liquid were allowed to fall freely and would pass over into stability if the liquid were forced downwards with an acceleration greater than that of gravity. Similarly, the initial stability of the upper surface of a liquid might be expected to pass over into instability if the liquid were given a downward acceleration greater than that of gravity. In simple terms, instability can be expected when one fluid is accelerated by another less dense one.

In this context, the fluids may be liquid, gaseous or even hypothetical as for example when a conducting gas or plasma is acted upon by a magnetic field which in vacuo behaves like a fluid with transverse pressure $B^2/2$.

D. T. Swift–Hook

## REDUNDANCY TECHNIQUES IN COMPUTING SYSTEMS.

*1. Introduction.* Modern computing systems contain many devices or *modules* and many *connexions*. We call such modules and their connexions, *elements*. Networks of these elements, known as *modular nets* are required to execute at high speeds, long sequences of precise computations at high levels of reliability. In analogue computers (wherein physical quantities are generally represented by such quantities as voltage, current, impedance, and so forth) precision is limited by intrinsic fluctuations or *noise* present in the nets comprising the computer. In digital computers (wherein physical quantities are represented by numbers) arbitrarily high precision may be obtained by increasing the number of digits used in the representation of the given objects of computation. However, such an increase in precision is obtained at the cost of decreased *reliability*. That is, the longer the digital expansion, the greater is the expected number of errors

in the computation due to possible malfunctions or failures of the many elements comprising modular nets. For example, the overloading of electrical networks may cause short-circuiting and breakdown of modules such as valves or transistors. The contacts of a relay switching network may stick from time to time resulting in occasional errors in the output of the network; similarly, noise in modules comprising gating circuits for example, may result in occasional output errors. A power failure may result in complete loss of function of modular nets. In addition the modules used in the system may be old or may have suffered radiation damage, and so forth. Finally the connexions may be faulty; dryjoints, mistakes in wiring, short or open-circuiting etc. all may contribute to the overall unreliability of the system. We call these various sources of error *faults*.

In general, *faults* may be split into two classes: *malfunctions* whose effect is to produce *transient*

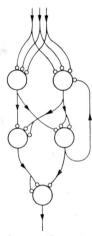

Fig. 1. A computing automaton A.

*errors*, and *failures* whose effect is to produce *stationary errors* in computing systems. Associated with this classification are different definitions of what is meant by reliability. In the case of transient errors we use, as a reliability measure, the probability of system malfunction when given the probability of modular and connexion malfunction. In the case of stationary errors we use as our measure the expected lifetime to failure of the system when given the lifetime to failure of modules. The problem of reliable system design is that of finding ways of designing systems which operate at requisite levels of reliability given fixed levels of elemental reliability. We may express this formally as follows: Consider the computing system or *automaton* A shown in Fig. 1.

Let $N$ be the number of modules comprising A, and let $t_i$ be the number of times that the output of the $i$-th module $(i = 1, \ldots, N)$ influences a single output of A, and let $\sum_{i=1}^{N} t_i = Q < \infty$. That is, there can be loops within A but their effect is of finite duration. Let $p$ be the probability of any single modular malfunction and let $P$ be the probability that $A$ malfunctions. It is easily seen that

$$(1 - p)^Q \leq 1 - P \qquad (1)$$

This equation implies, for fixed values of $p$, a positive lower bound on $P$, i.e. arbitrarily high reliability cannot be directly attained from A alone. Similar considerations apply to the cases of connexion errors and of failures.

Evidently an automaton A' must be designed which will compute the same function as A, and which will contain mechanisms for limiting errors due to faults. In general, there exist several possible strategies. A' may incorporate a *fault-detection and location* mechanism coupled to an element removal and replacement mechanism, or else the latter mechanism may be automatic. Alternatively A' may be designed so that faults have little or no effect on the system, i.e. a *fault-masking* mechanism may be incorporated. Yet another strategy is to make A' *adaptive*. Faulty elements are given progressively less "weight" in the system than correctly functioning elements. In what follows we give a brief outline of some of these strategies, and the resultant automata.

*2. Fault-masking of transient errors.* We consider first the case of malfunctions resulting in transient errors. We distinguish between *contact networks* (c-nets) and *gating networks* (g-nets). The basic difference between these types of network is that in the former, the output of each element, given faultless operation, uniquely specifies the input, whereas in the latter such is not the case. Thus relay switching circuits are c-nets, whereas networks of modules such as NOR-gates are g-nets. In masking the effects of transient errors in such nets a basic principle employed is that of *replication by redundancy*. That is, more elements are used in the design of A' than in A, and the extra elements are used to provide repetitions of the basic operations executed in A. For example, consider an idealized relay whose malfunctions comprise the failure of its contacts to open or close when signalled to do so by the current flowing in the coil. If the relay is energized the contact is closed with probability $a$, open with probability $1-a$. If the relay is not energized, the contact is closed with probability $c$, open with probability $1-c$. If $a>c$, we have what is called a *make-contact*; of $a<c$ we have a *break-contact*. In general to preserve closure of a contact network (i.e. to guarantee that at least one essential contact is closed) we need only construct a parallel contact network, and conversely to preserve an open circuit we need only construct a series network. Replacement of the single relay with associated make contact of Fig. 2 by the parallel relay network with associated series parallel contact network of Fig. 3 will produce a large improvement in reliability.

If each of the contacts $x_1, \ldots, x_4$ has the probability $p$ of being closed, then the probability of the network being closed is

$$h(p) = 1 - (1 - p^2)^2 = 2p^2 - p^4 \qquad (2)$$

A plot of this function is shown in Fig. 4.

Clearly if $a < 0.618 < c$ we will now have a better relay in respect of closure than the single relay system. For example if $1 - a = c = 0.01$, the multiterminal circuit makes errors when the coils are energized, with probability $3.96 \times 10^{-4}$, and when the coils are not energized with probability $2 \times 10^{-4}$.

*Fig. 2. Single relay X with make-contact x.*

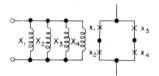

*Fig. 3. Series–parallel relay network.*

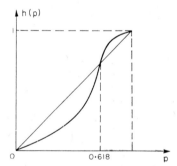

*Fig. 4. Plot of $h(p)$, the probability of closure of the 4-contact network, against $p$ the closure probability of a single relay.*

In general many multicontact relay networks have this property characterized by a polynomial of the form

$$h(p) = \sum_{n=0}^{m} A_n p^n (1 - p)^{m-n} \qquad (3)$$

where $m$ is the number of contacts in the network, and $A_n$ is the number of ways in which a subset of $n$ contacts can be selected so that if those contacts are closed and the remainder are open, then the total network is closed. Such networks are sometimes called *Hammock nets* after Moore and Shannon (1956).

In g-nets fault-masking of transient errors, via replication by redundancy is used both at the network level, and at the elemental level, in a somewhat similar manner to that of c-nets. The simplest example (attributed to von Neumann, 1952) is the *triplication* of an entire g-net (see Fig. 5) followed by a so-called majority vote-taker or *majority organ*. This module computes a function such that if at least two of the g-nets, give similar outputs, then so does the majority organ. It follows that only if at least two g-nets are simultaneously malfunctioning does an error propagate through the majority organ. More succinctly, we say that the majority organ is single-error insensitive. Assuming independence of malfunctions in g-nets and

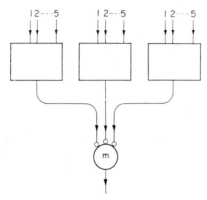

Fig. 5. A triplicated gating network plus majority organ m.

in the majority organ, we may calculate the probability of network malfunction. Let $p$ be the probability of malfunction of each g-net, and $\delta$ that of the majority organ. Then the probability of at least two of the g-nets being in error is

$$\theta = 3p^2(1 - p) + p^3$$
$$= 3p^2 - 2p^3 \tag{4}$$

and the probability of system malfunction (assuming of course no interconnexion errors) is

$$P(\delta, p) = (1 - \delta)\theta + \delta(1 - \theta)$$
$$= \delta + (1 - 2\delta)(3p^2 - 2p^3) \tag{5}$$

The graph of this function of $p$ is shown in Fig. 6.

It can be shown that by suitable modification of the redundant network (i.e. by using two layers of majority organs, one layer triplicated) and by iteration of such networks, that the steady-state value $P(\delta, p) \sim P_0$ will be attained, provided $\delta < 0.0073$. For example, an error level of $P = 2 \times 10^{-2}$ can be maintained in such a system provided $\delta \leqq 0.0041$ and $p < 0.5$. Unfortunately, such a design requires the use of a very large number of redundant modules. Thus, if $\mu$ is the longest chain of logical operations to be executed in such a network, the approximately $3^{\mu}$ modules are required in the redundant network. Consequently the

procedure is impracticable for all but a small range of values of $\mu$.

Similarly it may be shown that malfunctions in g-nets may be controlled by elemental replication, rather than by replication of entire g-nets. It is known that networks executing arbitrary logical functions may be composed from so-called *universal logical elements*, and that *mutatis mutandis*, the majority organ is universal in this sense. Thus any Boolian function may be computed by g-nets comprising only majority organs. The elemental replication technique (introduced by von Neumann, op. cit.) known as

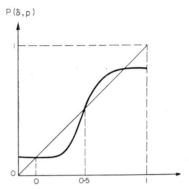

Fig. 6. Plot of $P(\delta, p)$, the probability of error of the triplicated network of Fig. 5.

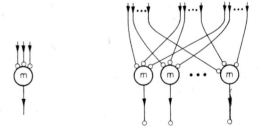

Fig. 7. The technique of multiplexing.

*multiplexing* entails the replacement of all single connexions in the given g-nets by groups of connexions or *bundles*, and all modules of the g-nets by aggregates of similar modules. See Fig. 7.

It can be shown, with suitable randomization of the input and output bundles to and from aggregates, and in the absence of malfunctions, that these redundant g-nets approximate the function computed by a single module, i.e. the function computed by the majority organ, sometimes called the *quorum function* (Moore and Shannon, op. cit.). In the presence of malfunctions, however, an error-controlling network is required to maintain the approximation. It was shown by von Neumann that the network of Fig. 8 will control transient errors. The redundancy created by the multiplexing is used in a certain *encoding and decoding* of the signal patterns within the network. The details of this code are as follows:

Each bundle comprises $n$ connexions, each the carrier of a binary impulse (signal or no-signal). There are thus $2n$ distinct signal patterns in the bundle ranging from $(111 \dots 1)$ to $(000 \dots 0)$. If no malfunctions were to occur, these extreme levels of excitation would be the only ones existent in the bundle. Let the number of ones in any pattern be $x$, and set a fiduciary level $\Delta$ so that $n \geq x \geq (1 - \Delta) n$ represents signal (1), $\Delta n \geq x \geq 0$ represents no-signal (0), and any intermediate level of excitation $(1 - \Delta) n > x > \Delta n$

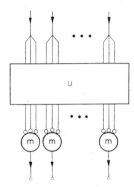

Fig. 8. *An error-controlling network.*

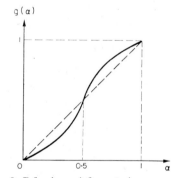

Fig. 9. *Behaviour of the restoring organ.*

represents the occurrence of a malfunction. The problem of obtaining reliable computation is now reduced to the problem of maintaining levels of excitation throughout the multiplexed network sufficiently close to $x = n$ or else 0. Von Neumann was able to show that the g-net of Fig. 8 had only two stable states of activity corresponding to the above extrema. This network has therefore the property that with probability fairly close to one, any input bundle with excitation level fairly close to one of the extrema gives rise to an output bundle with excitation level even closer to the extremum in question.

The basic function of the box labelled $U$ in Fig. 8 is to maintain the statistical independence of the inputs to the next layer of majority organs, and in fact it executes a suitable permutation of these inputs sufficient to maintain such independence. Under such

conditions if $\alpha n$ of the $n$ inputs are active (transmitting signals) then the probability that any majority organ is active is

$$\alpha^* = 3\alpha^2 - 2\alpha^3 = g(\alpha) \qquad (6)$$

For sufficiently large $n$, it is therefore highly probable that approximately $\alpha n$ outputs will be active. Figure 9 shows the graph of the function $g(\alpha)$. It will be seen that only the levels $g(\alpha) = 0$ or else 1 are stable, and they represent the asymptotic behaviour of the network for any initial activity level $\alpha \neq 0.5$. This is clearly an error-controlling process in that with respect to the chosen code, any initial state corresponding to malfunctions of some majority organs will be restored, by suitable iterations of the process to a final state corresponding to the signal states 1 or 0. This process is clearly analogous to the use of certain circuits in actual automata for *amplification* rather than for gating or detection, etc.

Equation 6 is obtained on the assumption that the error-controlling network is not itself subject to internal malfunctions. It can be shown if malfunctions are assumed to occur with some non-zero probability $p$ in all elements of the redundant g-net, that there exists an optimum fiduciary level $\Delta$, so that for sufficiently large redundancy $n$, the probability of network malfunction $P(n, p)$ can be made lower than any given fidelity level $\delta$. In particular for $p < 0.0107$ and $\Delta = 0.07$, and for $n$ sufficiently large,

$$P(n, p) \sim (1/\sqrt{2\pi k}) \exp(-k^2/2) \qquad (7)$$

where $k = 0.062 \sqrt{n}$. The table gives some figures for this result for $p = 0.005$.

| $n$ | $P(n, p)$ |
|---|---|
| 1000 | $2.7 \times 10^{-2}$ |
| 2000 | $2.6 \times 10^{-3}$ |
| 3000 | $2.5 \times 10^{-4}$ |
| 5000 | $4.0 \times 10^{-6}$ |
| 10,000 | $1.6 \times 10^{-10}$ |

$P(n, p)$ *as a function of n.*

It will be seen that a redundancy greater than 1000 is required before $P$ drops below $p$. Such a design is therefore extremely inefficient. These results concerning system and component replication can be viewed as no more than existence proofs of the theorem that reliable automata may be contructed from networks of unreliable elements.

We note, however, that the basic elements comprising such networks are all majority organs, and as such are "simple" in the sense that they compute functions of only three variables. (Certain other universal elements can be used, e.g. the Sheffer-stroke organ—neither $x$ nor $y$, and this is even simpler.) If more "complex" elements are used which compute functions of an indefinitely large number of variables, a rather different existence theorem can be proved, wherein the levels of redundancy required for the synthesis of reliable automata can be minimized (Winograd and Cowan 1963). The basis for this theorem is to be found in the theory of *error-correcting codes*. Such codes are presently used in the design of reliable long range communication and telemetry systems, and in the design of reliable computer storage systems. These codes are essentially rules of transformation to and from one set of symbols to another. Thus if $x_1 x_2 \ldots x_k$ is an initial message to be transmitted through a noisy channel, what is in fact transmitted is a sequence of signals $x'_1 x'_2 \ldots x'_n$ where $x'_\beta = e'_\beta(x_1, x_2, \ldots, x_k)$, $(\beta = 1, 2, \ldots, n)$. The function $e_\beta$ is called an *encoding function*. In general this means that any $x'_\beta$ is a function of many $x_\alpha$, $(\alpha = 1, \ldots, k)$, and a fortiori, any $x_\alpha$ is represented in many of the $x'_\beta$. It is this multiple representation which permits the message to be recovered from a signal sequence distorted by transmission through some noisy channel. That is, at the receiver a *decoding function* operates according to the transformation $x_\alpha = d_\alpha (x''_1, \ldots, x''_n)$ where $x''_\beta$ is a (possibly) distorted version of $x'_\beta$. Thus the message $x_1 x_2 \ldots x_k$ is reconstructed. It will be seen that there are essentially two parameters associated with such a code. The ratio $(n/k)$ which is a measure of the *redundancy* in the code, and the *complexity* inherent in the coding functions, which we relate to the number of variables over which the functions $e_\beta$ and $d_\alpha$ range. The associated probability of decoding error $P_c$ averaged over a certain collection of codes which map long sequences of messages into long sequences of signals, and conversely, has been related to the redundancy $(n/k)$ and to the *noise* in the channel is a famous theorem (Shannon 1948). It was proved that noisy communication channels could be described by a certain number (the *channel capacity*) which specified a certain minimum redundancy required for any level of reliable communication through the channel. Thus if $C$ is the channel capacity and $R$ is the so-called *rate* of transmission of information through the channel (the reciprocal of redundancy) then on the average, for the specified collection of codes

$$P_c \sim 2^{-n(C-R)} \tag{8}$$

Evidently arbitrarily small error-rates ($P_c \to 0$) can be obtained in principle, provided $R < C$, i.e. provided $k < n\,C$, by using sufficiently long message and signal sequences, and preserving the ratio $(k/n)$. However, this implies that the complexity of the requisite coding equipment increases fairly rapidly with $n$.

Consider now the modular net shown in Fig. 1. As we have noted, equation 1 implies for fixed values of $p$, a positive lower bound on $P$. To circumvent this, and to obtain arbitrarily small values of $P$ with fixed $p$, we design a redundant automaton $A'$ in which error-correcting codes are imbedded. We first consider $k$ *copies of the design for $A$*, rather than one copy. How-

ever, the copies may operate on different sets of inputs or on different programmes. We replace each set of $k$ corresponding modules and associated connexions by an aggregate $n$ modules with associated connexions, via the following transformation. Let each set of $k$ modules compute the functions $f_{i\alpha}(\mathbf{x}_i)$ ($i = 1, \ldots, N$; $\alpha = 1, \ldots, k$) where $\mathbf{x}_i = (x_1, x_2, \ldots, x_s)$ represents the set of input connexions to the $i$-th module of A. Then the functions computed by the $n$ modules of A' are given by:

$$f'_{i\beta} = e_\beta \begin{pmatrix} f_{i1}(d'_1(\mathbf{x}''_{in}), d'_1(\mathbf{x}''_{2n}), \ldots, d'_1(\mathbf{x}''_{sn})), \\ f_{i2}(d'_2(\mathbf{x}''_{in}), d'_2(\mathbf{x}''_{2n}), \ldots, d'_2(\mathbf{x}''_{sn})), \\ \cdots\cdots\cdots\cdots\cdots\cdots\cdots \\ f_{ik}(d'_k(\mathbf{x}''_{in}), d'_k(\mathbf{x}''_{2n}), \ldots, d'_k(\mathbf{x}''_{sn})) \end{pmatrix} \tag{9}$$

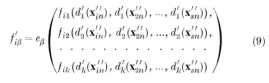
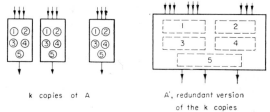

k copies of A          A', redundant version
                       of the k copies

*Fig. 10.*

where the decoding function $d'_\alpha$ is such that it equals $d_\alpha$ if $x''$ is not an external input, and equals the identity function otherwise. Thus the $k$ modules which compute the functions $f_{i1}, f_{i2}, \cdots, f_{ik}$ are replaced by $n$ modules which compute the functions $f'_{i1}, f'_{i2}, \cdots, f'_{in}$ respectively.

It will be seen that each of these functions operates on at most $ns$ input variables, whereas each of the $f_{i\alpha}$ operates on $s$ input variables. Thus each of the modules computing $f'_{i\beta}$ is at most $n$ times more complicated than the modules computing $f_{i\alpha}$. What is gained by this increased complexity is that each module of A' *decodes all its inputs and so corrects errors in modules feeding directly into it, executes its requisite function, and then encodes this computation for transmission to the next aggregate of modules*. Since an $(n, k)$ error-correcting code is obviously imbedded in A' so that the structure of A' is isomorphic to that of A, we may combine equations 1 and 8 to obtain a formula relating $P$ the probability that A' malfunctions, to $P_c$, the average error probability for sufficiently large $n$ and $k$, of $(n, k)$ codes i.e.

$$P \leq 1 - (1 - 2^{-n(C-R)})^Q \tag{10}$$

Equation 10 implies that $P$ may be made arbitrarily small, for fixed $Q$ and $p$, by increasing $n$ in such a way that the ratio $(n/k)$ is maintained, provided that the modules used have a capacity greater than $(k/n)$.

In somewhat similar fashion, it can be shown provided errors of interconnexion are not too frequent, that modular redundancy can control such errors.

It is clear that this solution to the problem of synthesizing reliable automata from unreliable elements, differs radically from von Neumann's solution. In his construction redundancy was introduced locally, each module being replaced by $n$ copies. This corresponds to the use of an $(n, 1)$ error-correcting code, with associated coding equipment constructed from the simplest of modules. In the Winograd–Cowan theorem, redundancy is introduced non-locally over an aggregate of $k$ modules via an $(n, k)$ error-correcting code, with the associated (complex) coding equipment imbedded into the $n$ modules comprising the redundant aggregate. In a sense, this construction represents the other extreme to von Neumann's solution, wherein complexity is minimized at the cost of redundancy, to attain arbitrarily small error-rates. In this case redundancy is minimized at the cost of complexity. However, it is a crucial requirement of the theorem that modular malfunctions do not become more probable with increased modular complexity. This is not a requirement that can be met by present technological methods of module fabrication, and so neither of the designs represents a practical method for obtaining very high levels of reliable computation. For fixed levels of reliable computation however, some application seems feasible. See for example, Tryon (1926) wherein a $(4, 1)$ code is used for single-error correction.

It is clear that suitable combinations of the two techniques of transient error control, multiplexing and coding, may lead to practical solutions to the problem. Given modules of fixed complexity a certain amount of non-local coding may be possible resulting in a lowered error-rate for certain aggregate of modules. If still lower error-rates are required there will be a level at which further application of the coding technique will require the use of more complex modules than are available. Such modules would have to be constructed from the (less complex) given ones, in which case (since the number required of these less complex modules, increases rapidly with $ns$), there will exist a level where the outputs of the complex modules are statistically independent of the inputs ($C = 0$). This suggests a scheme in which coding is applied at the lowest (elemental) level of organization until either complexity or channel capacity is used up, whence multiplexing is applied *to the coded aggregates*. It seems reasonable to expect that this method will result in efficient designs for controlling transient errors. That is, for given requirements of reliability, and for given modules, the method will result in some kind of minimum redundancy design. A special case of this technique is to be found in Pierce (1964).

*3. Control of stationary errors.* In case the errors in A are stationary, resulting from permanent failures of its elements, the techniques discussed in § 2 do not provide completely satisfying answers to the reliability problem, which now becomes the problem of designing an automation A' whose expected lifetime to failure is arbitrarily longer than the expected lifetime to failure of any of its elements. (In certain cases the problem is that of designing integrated circuit configurations so as to increase the *yield* of usable circuits resulting from a given manufacturing process.)

An important technique recently discovered uses *adaptive* majority organs (Pierce 1962). A basic defect in the multiplexing technique in respect of stationary errors, is that since some inputs to majority organs are permanently in error, a consistently reliable minority may be "outvoted" by a consistently unreliable majority. Such a limitation may be overcome by the generalized majority organ of Fig. 11. This module

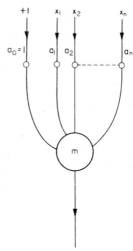

*Fig. 11. Generalized majority organ.*

computes the quorum function

$$x' = \text{signum}\left[a_0 + \sum_{i=1}^{s} a_i x_i\right] \tag{11}$$

all inputs $x_i$ taking the values $\pm 1$. If input errors are statistically independent, the vote-weights $a_i$ can be chosen so that the output $x'$ is the digit most likely to be correct, under the assumption that what is required of the module is a reliable computation of the quorum function on the $x_i$. If the error probability of the $i$-th input is $p_i$ then the weights $a_i$ giving such an output are:

$$a_0 = \ln \frac{(a\ priori\ \text{probability of} + 1)}{(a\ priori\ \text{probability of} - 1)}$$

$$a_i = \ln (1 - p_i/p_i) \quad i = 1, ..., s \tag{12}$$

If an input is completely random, $p_i = 0.5$, then $a_i = 0$, and increases thereafter monotonically as $p_i$ approaches zero or one. These requisite settings may be obtained by comparing the inputs $x_i$ with the output $x'$ of the module, and counting the number of coincidences between the values. Thus in a cycle of $M$ operations, if $P_i$ is the number of coincidences and $Q_i$ the number of disagreements, the settings

$$a_i = \ln (P_i/Q_i) \tag{13}$$

will give equation 12. It can be shown, provided suitable limits are placed on the possible values of the $a_i$, that automatic setting of these weights controlled by feedback from the output is feasible, and that it can be used practically, to optimize the weights.

If errors have the further property that they are catastrophic, i.e. they tend to influence neighbouring elements, a modified technique can be used in which a threshold of unreliability $\theta$ is set, so that if $(P_i/Q_i)$ ever exceeds $\theta$, $a_i$ is set to zero. A practical implementation of this is to couple $\theta$ to a fuse-blowing mechanism.

Several other techniques exist for dealing with stationary errors which generally comprise a combination of fault-detection and location (which is a special case of fault-masking), with either a component replacement mechanism, or with a switching mechanism, which switches redundant aggregates of unused components (or newly serviced components) into the system, upon receipt of a signal from the fault-location circuits (Löfgren 1960, Griesmer et al. 1962). It seems likely that the newly discovered error-locating codes (Wolf and Elspas 1963) will find application here.

*Bibliography*

GRIESMER J. H. *et al.* (1962) in *Redundancy Techniques in Computing Systems* (Wilcox R. H. and Mann W. C. Eds.), Spartan Books: Washington, D.C.

KAUTZ W. H. (1962) in *Redundancy Techniques in Computing Systems* (Wilcox R. H. and Mann W. C. Eds.), Spartan Books: Washington, D.C.

LÖFGREN L. (1958) Information and Control **1**, 2, 127.

LÖFGREN L. (1962) in *Biological Prototypes and Synthetic Systems* (Bernard E. and Kare M. Eds.), New York: Plenum Press.

MOORE E. F. and SHANNON C. E. (1956) *J. Franklin Institute*, **262**, 191, 281.

PIERCE W. H. (1962) in *Redundancy Techniques in Computing Systems* (Wilcox R. H. and Mann W. C. Eds.), Spartan Books: Washington, D. C.

PIERCE W. H. (1964) *J. Franklin Institute* **277**, 55.

SHANNON C. E. and WEAVER W. (1949) *Mathematical Theory of Communication*, Illinois: The University Press.

TRYON J. G. (1962) in *Redundancy Techniques in Computing Systems* (Wilcox R. H. and Mann W. C. Eds.), Spartan Books: Washington, D.C.

VON NEUMANN J. (1956) *Automata Studies* (Shannon C. E. and McCarthy J. Eds.), Princeton: The University Press.

WINOGRAD S. and COWAN J. D. (1963) *Reliable Computation in the Presence of Noise*, Cambridge, Mass: M.I.T. Press.

WOLF J. K. and ELSPAS B. (1963) *I.E.E.E. Trans. on Information Theory*, April, 1963, 113.

J. D. COWAN

**RENEWAL THEORY.** Renewal theory (or the theory of recurrent events or of self-renewing aggregates) is a mathematical model for a variety of physical situations. It is used, for example, in studying the theory of counters, storage problems and the reliability of complex equipment.

The theory postulates a sequence of events which occur at times $\tau_0, \tau_1, \tau_2, \ldots$ in such a way that the interval $\tau_i - \tau_{i-1} = X_i$ is a *random variable*. It is assumed that $\tau_0 = 0$. These events may, for example, represent the time at which a faulty component is replaced and the interval $X_i$ is the length of life of the $i$th component. Alternatively the events may be the absorption of a particle by a counter and $X_i$ is the interval between the $i$th and $(i-1)$th particles.

It is assumed that the random variables $\{X_1, X_2, \ldots\}$ are statistically independent and positive and have identical distribution functions given by $P_r(X_j \leq x) = F(x)$. Then renewal theory is, inter alia, concerned with the properties of the following derived random variables:

i) $S_r = X_1 + X_2 + \cdots + X_r$, the time up to the $r$th renewal (or event);

ii) $N_t$, the number of renewals (or events) in the interval $(0, t)$, i.e. the interval not including 0 but including $t$;

iii) $U_t$ and $V_t$, the backward and forward recurrence times from an arbitrary time $t$, being the time from $t$ to the last event and the next event, respectively. $V_t$ is sometimes referred to as the residual life time or the excess variable of $t$.

It is convenient to distinguish two cases namely that in which the random variables $X_i$ take only integer values $r = 1, 2, 3, \ldots$ and their distribution function is of the form $F(x) = \sum_{r=1}^{x} p_r$, where $p_r = P_r(X_i = r)$, and that in which $F(x)$ has a continuous element. The former will be referred to as the discrete case and the latter as the continuous case.

*Discrete case.* This case also covers, by suitable choice of the time scale, the so-called periodic case where $X_i$ takes values $k, 2k, 3k, \ldots$, where $k$ is an integer. An extended treatment is given by Feller (1957).

If $\pi(z) = \sum_{r=1}^{\infty} z^r p_r$ is the probability generating function of $X_i$ (i.e. the coefficient of $z^r$ in $\pi(z)$ is $p_r$) then the coefficient of $z^k$ in $\{\pi(z)\}^r$ is $P_r(S_r = k)$.

Interest in the random variable $N_t$ usually concentrates on its *mean* or expected value $E(N_t) = \sum_{n=0}^{\infty} n P_r(N_t = n)$ and its *variance* $\text{var}(N_t) = E(N_t^2) - E^2(N_t)$ where $E(N_t^2) = \sum_{n=0}^{\infty} n^2 P_r(N_t = n)$.

If $u_k$ is defined as the probability that some event occurs at time $k$ then

$$u_k = \sum_{r=1}^{\infty} Pr(S_r = k)$$

and is the coefficient of $z^k$ in

$$1 + \sum_{r=1}^{\infty} \{\pi(z)\}^r,$$

assuming that $u_0 = 1$. Then

$$U(z) = \sum_{k=0}^{\infty} u_k z^k = 1/[1 - \pi(z)].$$

If $t$ is an integer, $E(N_t) = \sum_{k=1}^{t} u_k$, giving

$$\sum_{l=1}^{\infty} E(N_t) z^l = \pi(z)/(1 - z)[1 - \pi(z)].$$

It may be shown that $u_k \to 1/\mu$ as $k \to \infty$, where $\mu = \sum_{r=1}^{\infty} r P_r(X_i = r)$ is the mean length of interval between successive events provided $\pi(1) = 1$, (i.e. that an event is certain to recur sooner or later). It follows that for large $t$, $E(N_t) \sim t/\mu$.

Similar results may be obtained for $\text{var}(N_t) = E(N_t^2) - E^2(N_t)$. The value of $\text{var}(N_t)$ for finite $t$ may be obtained from the fact that

$$\sum_{l=1}^{\infty} E(N_t^2) z^l = \pi(z)[1 + \pi(z)]/(1 - z)[1 - \pi(z)]^2$$

with the corresponding limiting result that for large $t$

$\text{var}(N_t) \sim t\sigma^2/\mu^2$, where $\sigma^2 = \sum_{r=1}^{\infty} r^2 P_r(X_i = r) - \mu^2$ is the variance of the intervals between successive events.

A special case of some interest is when the probability distribution of the intervals between successive events is given by

$$Pr(X_i = r) = p_r = (1 - p) p^{r-r}, \quad r = 1, 2, ...,$$

$$0 < p < 1.$$

For this distribution $\mu = 1/(1 - p)$ and $\sigma^2 = p/(1 - p)^2$.

In this case $\pi(z) = (1 - p)z/(1 - pz)$ and $P_r(S_r = k)$ is the coefficient of $z^k$ in $(1 - p)^r z^r/(1 - pz)^r$, that is

$$Pr(S_r = k) = \binom{k-1}{r-1} (1-p)^r p^{k-r}, \quad k = r, \quad r+1, ...$$

Also the probability that some event occurs at time $k$ is given by $u_k = (1 - p) = 1/\mu$, for all $k \neq 0$. The events resulting from this (geometric) distribution of intervals between events may therefore be regarded as random, that is, equally likely to occur at any time. In this case $E(N_t) = t/\mu$ and $\text{var}(N_t) = tp(1 - p) = t\sigma^2/\mu^3$, for all $t$.

It may be deduced from this special case that whatever the distribution of the intervals between events, a long time after the start of the process the events behave as if they were random, at least locally.

Although renewal theory is here taken to refer to a single sequence of events of the above form, the name is sometimes used in a more restricted sense to refer to a number of independent sequences of such events to describe, for example, the behaviour of a complex piece of apparatus consisting of a number of compo-

nents each subject to failure independently. This situation in the discrete case is discussed by Feller (loc. cit.)

*Continuous case.* It is here supposed that the intervals between successive events are random variables $X_i$, $i = 1, 2, ...$, where the distribution function of $X_i$ is given by $Pr(X_i \leq x) = F(x)$. (It is sometimes convenient to consider $X_1$ as having a different distribution from the remaining $X_i$ but this case will not be considered here. For a more detailed discussion of this special case and of renewal theory in general, reference should be made to Cox (1962).) Another survey of the field, which includes a bibliography of the theory of counters was given by Smith (1958).

As before $S_r = X_1 + X_2 + \cdots + X_r$ and it is supposed that the distribution function of $S_r$ is given by $Pr(S_r \leq x) = F_r(x)$.

The distribution function $F_r(x)$ is defined by the recursive relation

$$F_r(x) = \int_0^z F(z - y) \, dF_{r-1}(y), \quad \text{where} \quad F_1(x) \equiv F(x).$$

Thus $F_r(x)$ is the $r$-fold convolution of $F(x)$.

Alternatively if the Laplace-Stieltjes transform of $dF_r(x)$ is given by

$$f^*_r(s) = \int_0^{\infty} e^{-sx} \, dF_r(x),$$

the convolution integral above may be written

$$f^*_r(s) = f^*(s) f^*_{r-1}(s) \quad \text{or} \quad f^*_r(s) = \{f^*(z)\}^r.$$

The expression for $F_r(x)$ may be obtained from this last equation by a suitable contour integral or more conveniently by using, for example, tables given by Erdelyi *et al.* (1954).

The probability distribution of $N_t$, the number of events occurring in $(0, t)$ is given by

$$P_r(N_t = r) = P_r(S_r \leq t < S_{r+1})$$

$$= F_r(t) - F_{r+1}(t), \quad \text{with} \quad F_0(t) = 1.$$

Thus the generating function $F(t, z) = \sum_{r=0}^{\infty} z^r P_r(N_t = r)$ may be written

$$G(t, z) = 1 + \sum_{r=1}^{\infty} z^{r-1}(z - 1) F_r(z).$$

In the terms of Laplace transforms

$$G^*(s, z) = \int_0^{\infty} e^{-st} G(t, z) \, dt = [1 - f^*(s)]/s[1 - zf^*(s)].$$

Interest again focuses on the renewal function $H(t) = E(N_t)$, which is given either by $H(t) = \sum_{t=0}^{\infty} r P_r(N_t = r)$ or equivalently by $\partial G(t, z)/\partial z \big|_{z=1}$.

Thus $H(t) = \sum_{r=1}^{\infty} F_r(t)$. If $F(t)$ is absolutely continuous $H(t)$ may be differentiated to give $dH(t)/dt = h(t)$

the renewal density (or intensity). From the generating function $G^*(s, z)$, the result

$$H^*(s) = \int_0^\infty e^{-st} H(t) \, dt = f^*(s)/s[1 - f^*(s)].$$

This is equivalent to the integral equation of renewal theory

$$H(t) = F(t) + \int_0^t H(t - u) \, dF(u)$$

or the similar equation in terms of the renewal density, where this exists,

$$h(t) = f(t) + \int_0^t h(t - u) f(u) \, du, \quad \text{where} \quad f(t) = dF(t)/dt.$$

These equations may readily be obtained by direct arguments.

In general an explicit expression for $H(t)$ (or equivalently $h(t)$) is not obtainable from $H^*(s)$. If however we define $\mu$ and $\sigma^2$ to be the mean and variance of the intervals between events, i.e.

$$\mu = \int_0^\infty x \, dF(x) \quad \text{and} \quad \sigma^2 = \int_0^\infty x^2 \, dF(x) - \mu^2,$$

then provided, $\sigma^2 < \infty$, for large $t$ we have

$$H(t) = t/\mu + (\sigma^2 - \mu^2)/2\mu^2 + o(1).$$

Higher moments of $N_t$ may also be obtained from the generating function $G(t, z)$ (or equivalently from $G^*(s, z)$). If the $k$th factorial moment of $N_t$ is

$$H_{(k)}(t) = E\{N_t^{(k)}\} = \sum_{r=0}^\infty r^{(k)} P_r(N_t = r),$$

where $r^{(k)} = r(r - 1) \dots (r - k + 1)$, then

$$H^*_{(k)}(s) = \int_0^\infty e^{-st} H_{(k)}(t) \, dt = k! [f^*(s)]^k/s[(1 - f^*(s))]^k.$$

Expressions for the Laplace transforms of the moments of $N_t$ about zero (i.e. $E\{N_t^k\} = \sum_{r=0}^\infty r^k P_r(N_t = k)$) may be obtained in terms of Stirling numbers (see, for example, Takacs (1960)).

Of particular interest is the second moment of $N_t$ about its mean (i.e. the variance) which is related to the factorial moments by the relation

$$\text{var}(N_t) = H_{(2)}(t) + H_{(1)}(t) - [H_{(1)}(t)]^2.$$

Explicit expressions for this variance are not normally available but for large $t$, provided $\sigma^2$ and $\beta$ are finite,

$$\text{var}(N_t) = \sigma^2 t/\mu^3 + (15\sigma^4 + \mu^4)/12\mu^4 - 2\beta\sigma^3/3\mu^3 + o(1),$$

where $\mu$ and $\sigma^2$ are the mean and variance of the intervals between events and $\beta$ is the coefficient of

skewness of the distribution of these intervals, i.e.

$$\beta = \int_0^\infty (x - \mu)^3 \, dF(x)/\sigma^3.$$

Asymptotic results for higher moments of $N_t$ have been given by Smith (loc. cit.)

A useful distributional approximation for $N_t$ is that, provided $t$ is large, then $N_t$ is normally distributed with mean $t/\mu$ and variance $\sigma^2 t/\mu^3$, that is

$$P_r(N_t < r) \sim \int_{-\infty}^k \exp(-x^2/2) \, dx/(2\pi)^{\frac{1}{2}},$$

where $k = (\mu r - t)/\sigma(t/\mu)^{\frac{1}{2}}.$

A special case of some interest is the homogeneous Poisson process. This is the case where the density function of the time between successive events is given by

$$f(t) = \lambda e^{-\lambda t},$$

with Laplace transform $f^*(s) = \lambda/(\lambda + s)$.

Then $G^*(s, z) = 1/(\lambda + s - \lambda z)$. The coefficient of $z^r$ in $G^*(s, z)$ is $\lambda^r/(\lambda + z)^{r+1}$, yielding, on inversion, the result

$$P_r(N_t = r) = (\lambda t)^r e^{-\lambda t}/r!, \quad r = 0, 1, \dots,$$

the Poisson distribution.

The renewal function $H(t)$ is obtained from the Laplace transform $H^*(s) = \lambda/s^2$ to be $H(t) = \lambda t$, or it may be obtained from the Poisson distribution above. The corresponding renewal density is given by $h(t) = \lambda$. Thus if the independent intervals between successive events have an exponential distribution, the events are said to be random, i.e. are equally likely to occur at any time and the number of events in an interval of length $t$ have a Poisson distribution.

The asymptotic results for a general interval distribution $F(t)$ show that a long time after the start any renewal process settles down to random behaviour.

*Recurrence times.* The backward recurrence time of the renewal process, that is the time to the previous event from an arbitrary point of time is $U_t$ having distribution function $K(t, x) = P_r(U_t \leq x)$.

It may be shown that

$$K(t, x) = \sum_{n=1}^\infty \int_{t-x}^t [1 - F(t - u)] \, dF_n(u)$$

$$= \int_{t-x}^t [1 - F(t - u)] \, dH(u).$$

Then $\lim_{t \to \infty} K(t, z) = \int_0^x [1 - F(u)] \, du/\mu.$

A similar result is obtainable for the forward recurrence time $V_t$ (the residual life time or excess variable of $t$), namely

$$P_r(V_t \leq x) = L(t, x) = \int\limits_{t}^{t+x} [1 - F(t + x - u)] \, \mathrm{d}H(u),$$

and

$$\lim_{t \to \infty} L(t, x) = \int\limits_{0}^{z} [1 - F(u)] \, \mathrm{d}u/\mu,$$

as before.

In the special case when $F(t) = 1 - e^{-\lambda t}$ both $K(t, x)$ and $L(t, x)$ take the value $1 - e^{-\lambda x}$ (since $H(u) = \lambda u$ and $\mu = 1/\lambda$). Since $t$ is an arbitrary moment of time the fact that both the backward and the forward recurrence times have the same distribution as the interval between successive events illustrates again the random nature of the events arising from the exponential distribution of intervals.

The limiting properties of these recurrence times (for large $t$) also explain the meaning of the asymptotic results for the moments of the number of events in an interval of length $t$. If a renewal process is such that the distribution function of the time before the first event is $\int\limits_{0}^{z}[1 - F(u)] \, \mathrm{d}u/\mu$, and the interval between successive events after the first is $F(t)$ then $H(t) = t/\mu$ whatever the functional form of $F(t)$. Thus the density of events is constant and this process (the equilibrium renewal process) appears to exhibit the property of randomness. Because the forward recurrence time at an arbitrary time $t$, has asymptotically the distribution $\int\limits_{0}^{x}[1 - F(u)] \, \mathrm{d}u/\mu$, the equilibrium renewal process may be regarded as an ordinary renewal process which started a long time before. This randomness is apparent only, for an examination of higher moments shown, for example, that in the random case $\mathrm{var}(N_t) = \lambda t$, whereas in the equilibrium renewal process $\mathrm{var}(N_t)$ depends upon the functional form of $F(x)$.

*Extended renewal process.* In ordinary renewal processes it is supposed that the intervals between successive events, $X_i$, are positive random variables. The behaviour of the process when the assumption of positivity has been dropped has been discussed by Karlin (1955). The most important result is concerned with the asymptotic behaviour of the mean number of events, namely

$$E(N_{t+T}) - E(N_t) \sim T/\mu,$$

where, as before, $\mu(>0)$ is the mean interval between events. Both the ordinary and the extended process may for some purposes be regarded as *random walks*.

*Application to the theory of the Type I counter.* As an example of the use of renewal theory we consider the

type I counter. In such a counter the arrival of the $i$th particle is followed by a "dead" interval of length $Y_i$ where $Y_i$ is an independent random variable having distribution function $G(y)$. If the particles arrive at the counter in a homogeneous Poisson process the length of time between the recording of the $i$th and the recording of the $(i + 1)$th particle is a random variable $X_i = Y_i + Z$, where $Z$ has an exponential distribution with parameter $\lambda$, i.e. $p(Z \leq z) = 1 - e^{-\lambda z}$. The distribution function of $X_i$ is therefore

$$F(x) = \lambda \int\limits_{0}^{x} G(y) \, e^{-\lambda(x-y)} \, \mathrm{d}y.$$

The equivalent relation after Laplace transformation is

$$\int\limits_{0}^{\infty} e^{-sx} \, \mathrm{d}F(x) = f^*(s) = \lambda g^*(s)/(\lambda + s),$$

where

$$g^*(s) = \int\limits_{0}^{\infty} e^{-sy} \, \mathrm{d}G(y),$$

and it is assumed that $G(0+) = 0$.

Renewal theory may now be used to obtain the Laplace transform of the expected value of $N_t$, the number of particles recorded in time $t$ (assuming that a particle was recorded, but not counted, at time $t = 0$). This is

$$H^*(s) = \lambda g^*(s)/[\lambda + s - \lambda g^*(s)].$$

Similarly the Laplace transform of higher moments could be obtained. In general explicit inversions of these expressions may not be possible. If $\mu$ and $\sigma^2$ are the mean and variance of the "dead" time and $\beta$ is the coefficient of skewness of the distribution of this time, we may use the asymptotic results to obtain

$$E(N_t) = \lambda t/(1 + \lambda\mu) + \tfrac{1}{2}[(1 + \lambda^2\sigma^2)/(1 + \lambda\mu)^2 - 1] + o(1),$$

and

$$\mathrm{var}(N_t) = (1 + \lambda^2\sigma^2) \, \lambda t/(1 + \lambda\mu)^2 +$$
$$+ \, 15[(1 + \lambda^2\sigma^2)^2 + (1 + \lambda\mu)^4]/12(1 + \lambda\mu)^4 -$$
$$- \, 2(2 + \lambda^3\sigma^3\beta)/3(1 + \lambda\mu)^3 + o(1).$$

*Bibliography*

Cox D. R. (1962) *Renewal Theory*, London: Methuen.
Erdelyi A. *et al.* (1954) *Tables of Integral Transform*, Vol. 1. New York: McGraw-Hill.
Feller W. (1957) (2nd Edn) *Probability Theory and its Applications*, New York: Wiley.
Karlin S. (1955) *Pacific J. Math.* **5**, 229.
Smith W. L. (1958) *J. Roy. Statist. Soc.* **B 20**, 243.
Takacs L. (1960) *Stochastic Processes*, London: Methuen.

F. Downton

## SATELLITE ORBITS, CHANGE OF, IN SPACE.

The orbit of a satellite about a planet can most conveniently be described by regarding the orbit as an ellipse which slowly changes its size, shape and orientation under the action of three perturbing forces. These are: (1) the departure of the planet's gravitational field from spherical symmetry; (2) the drag of the planet's atmosphere; and (3) the influence of external forces, which, for Earth satellites, are due primarily to the Sun and Moon. In this article the basic elliptic orbit is defined, and the effects of the three perturbations are described in turn. The discussion centres on Earth satellites, but at the end the application to other planets is briefly outlined.

The orbit of a satellite about the Earth may be defined by five parameters called *orbital elements*, with a sixth to specify the angular position of the satellite in its orbit.

Two angles define the orientation of the orbital plane in space, namely the inclination $i$ to the equator (see Fig. 1) and the right ascension of the ascending node, $\Omega$. The ascending node $N$ is the point where the satellite crosses the equator going north, and $\Omega$, the right ascension of $N$, is the angular distance of $N$ from the first point of Aries, $\gamma$. Thus in Fig. 1, $\Omega$ is given by the angle $\gamma CN$, where $C$ is the Earth's centre. $\Omega$ is measured positive eastwards.

Three further orbital elements define the size and shape of an elliptic orbit and its orientation in the orbital plane. The size is specified by the semi-major axis $a$ (see Fig. 2) and the shape by the eccentricity $e$ (less than 1). From Fig. 2 the minimum distance from the Earth's centre, at perigee $P$, is $r_p = a(1 - e)$, and the maximum (apogee) distance is $a(1 + e)$. The fifth orbital element, specifying the direction of perigee, $CP$, is the argument of perigee, $\omega$, the angle between the ascending node and perigee, measured round the orbit in the direction of motion (see Fig. 1).

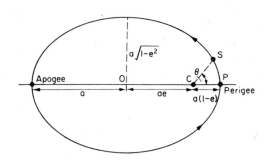

*Fig. 2. Diagram showing semi-major axis $a$ and eccentricity $e$ of orbit, with satellite at $S$.*

The sixth orbital element, specifying the angular position of the satellite at a given time, can be defined in several ways: for example, it can be taken as the true anomaly at epoch, the value of the angle $\theta$ in Fig. 2 at a specified time. This sixth element is not relevant to the present article.

If the Earth were spherically symmetrical, had no atmosphere and were isolated from extraterrestrial influences, the orbit would remain an ellipse of fixed size and shape, in a fixed plane: the orbital elements $a$, $e$, $i$, $\Omega$ and $\omega$ would remain constant. The perturbing forces cause changes in the elements; but these changes are slow, i.e. are very small in the course of a single revolution, and that is why the representation of the orbit in terms of $a$, $e$, $i$, $\Omega$ and $\omega$ is so useful.

The effects of the Earth's departure from spherical symmetry will be described first. The Earth's gravitational potential $U$ at an external point distant $r$ from the centre may, ignoring the very small variations with longitude, be written as a series of spherical

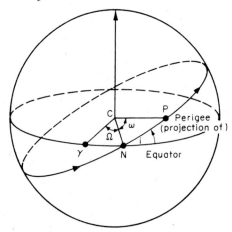

*Fig. 1. Projection of satellite orbit on unit sphere with centre at the Earth's centre, showing inclination $i$ and right ascension $\Omega$.*

harmonics,

$$U = \frac{GM}{r}\left\{1 - \sum_{n=2}^{\infty} J_n\left(\frac{R}{r}\right)^n P_n(\sin\varphi)\right\} \qquad (1)$$

where $G$ is the gravitational constant, $M$ the Earth's mass, $R$ the Earth's equatorial radius, $\varphi$ the geocentric latitude and $P_n(\sin\varphi)$ is the Legendre polynomial of degree $n$ and argument $\sin\varphi$. The $J_n$ are constants, and $J_1$ is zero if the Earth's centre of mass lies on the equator. The first term on the right-hand side of (1) represents the potential due to the basic spherical Earth, $GM/r$. The term in $J_2$, the second harmonic, represents the main effect of the Earth's oblateness, and $J_2$ is over 400 times larger than any of the higher $J_n$. The third harmonic, the term in $J_3$, represents an asymmetry about the equator, the so-called pearshape effect; and so on. Values for $J_2 - J_7$ as currently determined are as follows: $10^6 J_2 = 1082\cdot64$, $10^6 J_3 = -2\cdot6$, $10^6 J_4 = -1\cdot5$, $10^6 J_5 = -0\cdot1$, $10^6 J_6 = 0\cdot6$, $10^6 J_7 = -0\cdot4$, with likely errors of about $0\cdot1$ (see *Satellite orbits, discoveries from*).

The departure of the gravitational field from spherical symmetry (i.e. the existence of non-zero $J_n$) has two major effects on satellite orbits. First, the orbital plane rotates about the Earth's axis in the direction opposite to the satellite's motion, so that if $i < 90$ there is a steady, or *secular*, decrease in the angle $\Omega$ in Fig. 1, while $i$ remains constant. The rate of change, $\dot\Omega$, is given by

$$\dot\Omega = -\left(\frac{GM}{a^3}\right)^{\frac{1}{2}}\left(\frac{R}{p}\right)^2\cos i\left[\frac{3}{2}J_2 - \frac{15}{4}J_4\left(\frac{R}{p}\right)^2 \times\right.$$
$$\times \left\{\left(1 - \frac{7}{4}f\right)\left(1 + \frac{3}{2}e^2\right) + \left(\frac{7}{4}f - \frac{3}{4}\right)e^2\cos 2\omega\right\} +$$
$$\left. + \text{terms in } J_6, J_8, \ldots + \text{terms in } J_3\, e \sin\omega, J_5 e \sin\omega\ldots\right]$$
$$\qquad (2)$$

where $p = a(1 - e^2)$ and $f = \sin^2 i$. If numerical values are inserted, and only the $J_2$ term is retained, equation 2 reduces to

$$\dot\Omega = -9\cdot97\,(R/a)^{3\cdot5}\,(1 - e^2)^{-2}\cos i \quad \text{degrees per day.}$$
$$\qquad (3)$$

The second major effect of the gravitational field is to cause a steady change in the direction of perigee in the orbital plane. The argument of perigee $\omega$ changes at a rate given by

$$\dot\omega = \left(\frac{GM}{a^3}\right)^{\frac{1}{2}}\left(\frac{R}{p}\right)^2\left[3J_2\left(1 - \frac{5}{4}f\right) - \frac{15}{2}J_4\left(\frac{R}{p}\right)^2 \times\right.$$
$$\times \left\{1 - \frac{31}{8}f + \frac{49}{16}f^2 + f\cos 2\omega\left(\frac{3}{8} - \frac{7}{16}f\right)\right\} +$$
$$\left. + \text{terms in } J_6, J_8, \ldots + \text{terms in } \frac{J_3\sin\omega}{e}, \ldots\right] \quad (4)$$

If numerical values are inserted and only the $J_2$ term is retained, equation 4 reduces to

$$\dot\omega = 4\cdot98\,(R/a)^{3\cdot5}(1 - e^2)^{-2}(5\cos^2 i - 1) \quad \text{degrees per day.}$$
$$\qquad (5)$$

Equations 3 and 5 indicate that the changes in $\Omega$ and $\omega$ are quite rapid. For a near-Earth orbit, with $R/a = 0\cdot95$ and small $e$, equation 3 shows that the orbital plane rotates at a rate between 8 deg/day for a near-equatorial orbit and zero for a polar orbit. From equation 5, $\dot\omega$ is zero when $\cos i = 1/\sqrt5$, or $i = 63\cdot4°$, the *critical inclination* as it is called, so that the latitude of perigee then remains constant. $\dot\omega$ is positive for $i < 63\cdot4°$, and for near-equatorial orbits can be as large as 16 deg/day; $\dot\omega$ is negative for $i > 63\cdot4°$ and for polar orbits $-\dot\omega$ can be as large as 4 deg/day.

The main effect of the odd harmonics is to produce oscillations in the elements $i$ and $e$: the main oscillation due to $J_3$ has the same period as $\omega$—at least 22 days and usually several months—and is known as a *long-period* oscillation. The variations in $i$ and $e$ may be expressed as

$$i = i_E + \frac{eJ_3R}{2J_2p}\cos i\sin\omega + \text{terms in } \frac{eJ_5}{J_2},$$
$$\ldots + \text{terms in } \frac{e^2J_4}{J_2}, \ldots \qquad (6)$$

$$e = e_E - \frac{J_3R}{2J_2p}\sin i\sin\omega + \text{terms in } \frac{J_5}{J_2},$$
$$\ldots + \text{terms in } \frac{eJ_4}{J_2}, \ldots \qquad (7)$$

where suffix $E$ denotes values when $\omega = 0$ (perigee at the equator). The oscillation in $e$ is the more important, since it corresponds to a variation of several kilometres in the perigee distance $r_p$. Inserting numerical values for $J_3$ and $J_2$ in (7), we obtain

$$r_p - r_{pE} = -7\cdot6\sin i\sin\omega \text{ km}. \qquad (8)$$

Thus the perigee point is $7\cdot6\sin i$ km nearer the Earth's centre when it is at northern apex (most northerly point) than when it is at the equator. Equation 8 is not accurate when $i > 45°$: the term in $J_7$ must be added.

The variation of the gravitational potential with longitude, which is not allowed for in equation 1, gives rise to small short-period oscillations in the orbital elements with periods of 1 day, $\frac{1}{2}$ day, etc. One of the most important of these oscillations is an along-track perturbation with an amplitude of about $0\cdot5\sin^2 i$ km and a period of $\frac{1}{2}$ day.

Next the effect of the Earth's atmosphere on the orbit will be described. This effect is only important if the perigee height is less than 1000 km. A satellite moving with velocity $V$ relative to ambient air of density $\varrho$ is subjected to a drag $D$ acting in the direction opposite to $V$. The drag may be written

$$D = \frac{1}{2}\varrho V^2 S C_D, \qquad (9)$$

where $C_D$ is the drag coefficient and $S$ is a reference area. If $S$ is taken as the cross-sectional area perpendicular to the direction of motion, $C_D$ usually has a value near 2·2. Equation 9 shows that $D$ is proportional to the air density $\varrho$, and since air density decreases very rapidly as height increases (by a factor of about 100 as height increases from 200 to 500 km) the drag encountered at heights near that of perigee is much greater than at other points in the orbit (provided the orbit is appreciably elliptic). To a first approximation, therefore, the effect of air drag is to retard the satellite as it passes perigee, thereby causing the height at the subsequent apogee to decrease, while the perigee height remains almost constant; $a$ and $e$ decrease while $r_p' = a(1 - e)$ decreases only very slightly.

In developing the theory for the effect of the atmosphere, it is useful to assume that, at heights near that of perigee, $\varrho$ varies exponentially with distance $r$ from the Earth's centre,

$$\varrho = \varrho_p \exp\left\{-\left(\frac{r - r_p}{H}\right)\right\}, \tag{10}$$

where $\varrho_p$ is the density at perigee height and $H$, known as the "density scale height", is constant. In a spherically symmetrical atmosphere with constant $H$ and (at a given height) no variation of density with time, eccentricity can be expressed in terms of time $t$ as

$$\frac{e}{e_0} \simeq \left(1 - \frac{t}{t_L}\right)^{\frac{1}{2}}, \tag{11}$$

with an error which does not normally exceed 0·003, though it can be larger in special circumstances. In (11) the initial eccentricity $e_0$ is assumed less than about 0·2, and $t_L$ is the total lifetime of the satellite. In practice the Earth's upper atmosphere is not quite spherically symmetrical, $H$ is not quite constant, and density varies with sunspot activity and from day to night. Nevertheless equation 11 remains useful as a first approximation.

While the eccentricity $e$ is decreasing, the perigee distance $r_p$ can be expressed in terms of $e$ as

$$r_{p0} - r_p = \frac{H}{2}\left[\ln\frac{e_0}{e} - (e_0 - e)\left(1 - \frac{e + e_0}{2}\right)\right.$$
$$\left. - \frac{H}{a_0}\left\{\frac{5}{2}\ln\frac{e_0}{e} - \frac{3(e_0 - e)}{4e_0 e}\right\}\right], \tag{12}$$

where suffix $o$ denotes initial values and $0·02 < e < 0·2$. Since $H$ is of order 50 km for perigee heights of 200–600 km, $H/a_0$ is small, of order 0·006, and the first term within square brackets in (12) is the main one. Equation 12 thus shows that the maximum change in $r_p$ as $e$ decreases from $e_0$ to $0·1e_0$ (that is, by (11), during the first 99 per cent of the satellite's life) is only about 1·2$H$, i.e. of order 60 km. Equation 12 is little affected by atmospheric oblateness; can be used when $H$ varies with height if $H$ is evaluated at a height $\frac{3}{2}H$

above mean perigee height; and is unaffected by variations of density with time.

During the final near-circular phase of a satellite's life, the decrease in perigee distance can be much greater. If we write $z = ae/H$, the change in perigee distance may be expressed as

$$r_{p1} - r_p = H\left\{\ln\frac{z_1 I_1(z_1)}{z I_1(z)} - (z_1 - z) + 0\left(\frac{H}{a}\right)\right\}, \tag{13}$$

where $r_{p1}$ is the initial value of $r_p$, $z_1$ is the initial value of $z$, and $z_1 < 3$, corresponding to $e_1$ less than about 0·02. $I_1(z)$ is the Bessel function of the first kind and imaginary argument, of order 1. As $e$, and hence $z$, decreases towards zero, $I_1(z) \to 0$ and the drop in perigee height can, from (13), become very large. In practice the satellite's life in orbit ends a little before $e$ reaches zero, usually at the time when the perigee height has decreased to about 120 km: the satellite then makes its final fiery plunge through the lower atmosphere.

The third effect to be described here is that of the luni-solar perturbations: these are due to the gravitational attractions of the Sun and Moon, and also to solar radiation pressure. The Earth's oblateness causes steady secular changes in $\Omega$ and $\omega$; the atmosphere produces secular changes in $a$ and $e$; in contrast, luni-solar perturbations generally produce only *periodic* variations in the elements. The analytical expressions for these perturbations are too lengthy to quote here, and a general description must suffice.

The gravitational attractions of the Sun and Moon cause virtually no change in $a$ (since to the first order they do not change the energy of the satellite), but give rise to oscillatory changes in $e$, $i$, $\Omega$ and $\omega$. For near-Earth satellites the amplitude of these oscillations is usually small: for $a = 900$ km and $e = 0·2$ the perigee distance usually oscillates with an amplitude of about 1 km. For more distant or more eccentric orbits the effects are larger, with amplitudes increasing as $ea^{5/2}$. If the apogee distance is more than about $\frac{3}{4}$ of the Moon's distance of 380,000 km, the orbit may be not merely perturbed but completely transformed as it passes near the Moon. The normal, small luni-solar oscillations are extremely complex, containing components with fifteen different periods: the main period can be anything from 18 days upwards. When the size and orientation of the orbit are such that the period is several years, i.e. near-resonance occurs, the luni-solar oscillations can be important even for near-Earth satellites. If the oscillation in perigee height is large, because of either resonance or a large orbital eccentricity, luni-solar perturbations can be responsible for ending the satellite's life by bringing perigee into the lower atmosphere.

Solar radiation pressure, like the luni-solar gravitational perturbations, causes oscillations in $e$, $i$, $\omega$ and $\Omega$, if the whole orbit is in sunlight. If, as is more usual, part of the orbit is in the Earth's shadow, the semimajor axis $a$ also usually suffers a small change, while the amplitudes of the oscillations in the other elements are slightly reduced. For satellites of normal construction the effect of solar radiation pressure is small, the amplitude of the oscillation in perigee distance usually being less than 1 km; but if the area-to-mass ratio is very large, as with balloon-satellites, the amplitude of the oscillation can be very great. For example, with the Echo 1 balloon, the perigee height suffered an oscillation having an amplitude of 600 km and a period of about 11 months.

Close satellites of planets other than the Earth will in general suffer perturbations similar to those of Earth satellites. Most planets in the solar system are symmetrical about their spin axis and flattened at the poles like the Earth, so that their gravitational fields cause secular changes in $\Omega$ and $\omega$. (The Moon, however, does not conform to this pattern and orbits about the Moon need special treatment.) Most of the planets have atmospheres, some of them probably much denser than the Earth's, so that atmospheric perturbations may be even more important. Solar perturbations are similar for satellites of other planets, but "lunar" perturbations are almost absent unless the orbit passes near one of the planet's natural satellites.

*Bibliography*

EHRICKE K. A. (1962) *Space Flight, Volume II: Dynamics.*, Princeton: Van Nostrand.

KING-HELE D. (1964) *Theory of Satellite Orbits in an Atmosphere*, London: Butterworths.

D. G. KING-HELE

**SATELLITE ORBITS, DISCOVERIES FROM.** The orbit of a satellite close to the Earth departs from a fixed ellipse in a fixed plane as a result of two major forces: the first is caused by the departure of the Earth's gravitational field from spherical symmetry; the second is the drag of the atmosphere. By observing actual satellites, determining their orbits, noting the orbital perturbations which occur and comparing these perturbations with those expected on the basis of theory, we can deduce the forces acting on the satellite and hence determine the gravitational field and the properties of the upper atmosphere. This rather indirect procedure has proved to be a most powerful (and inexpensive) method of geophysical research.

The determination of the gravitational field can best be described in two stages. First we assume that the gravitational potential $U$ at an external point distant $r$ from the Earth's centre is independent of longitude and may be written as a series of spherical harmonics,

$$U = \frac{GM}{r} \left\{ 1 - \sum_{n=2}^{\infty} J_n \left( \frac{R}{r} \right)^n P_n(\sin\varphi) \right\} \quad (1)$$

where $G$ is the gravitational constant, $M$ the Earth's mass, $R$ the Earth's equatorial radius, $\varphi$ the geocentric latitude and $P_n(\sin\varphi)$ the Legendre polynomial of degree $n$ and argument $\sin\varphi$. The $J_n$ are constants whose values are to be determined, and the term in $J_n$ is known as the $n$th harmonic.

As explained in the preceding article (*Satellite orbits, change of, in space*), the main effect of the $J_n$ terms is to change the right ascension of the ascending node $\Omega$, and $\dot{\Omega}$ may be written in the form

$$\dot{\Omega} = AJ_2 + BJ_4 + CJ_6 + \cdots KJ_2^2$$
$$+ \text{ small periodic terms in } J_3, J_5, \ldots \quad (2)$$

where the coefficients $A, B, C, \ldots$ are functions of the orbital elements $a, e, i$. Every satellite for which an accurate observed value of $\dot{\Omega}$ can be obtained provides one equation of the form (2), which is essentially a linear equation in $J_2, J_4, J_6, \ldots$. Thus if results are available for $m$ satellites with appreciably different orbital elements, we have $m$ simultaneous equations between $J_2, J_4, J_6, \ldots$ which can be solved for as many of the even $J_n$ as seems appropriate, assuming the higher-order $J_n$ are zero. Similar methods can be used with the argument of perigee $\omega$, instead of $\Omega$, though $\omega$ is usually less accurately known than $\Omega$.

The odd harmonics, the terms in $J_3, J_5$, etc., give rise to oscillations in eccentricity and inclination, and the observed amplitude of these oscillations provides a linear equation between the odd-numbered $J_n$, and almost independent of the even $J_n$. With results from a number of satellites, several of the odd $J_n$ can be determined.

Before these methods became available, only $J_2$ had been determined and was generally believed to have a value of $1091 \times 10^{-6}$. The results from satellites have shown that the value is too high, and have allowed the evaluation of a large number of higher harmonics. Values up to about the ninth harmonic have been reliably established, and the following is a consistent set: $10^6 J_2 = 1082 \cdot 64$, $10^6 J_3 = -2 \cdot 56$, $10^6 J_4 = -1 \cdot 52$, $10^6 J_5 = -0 \cdot 15$, $10^6 J_6 = 0 \cdot 57$, $10^6 J_7 = -0 \cdot 44$, $10^6 J_8 = 0 \cdot 44$, $10^6 J_9 = 0 \cdot 12$. The errors (standard deviations) in these values are likely to be about $0 \cdot 1$.

These results can be interpreted in terms of the Earth's shape, i.e. the geoid or sea-level surface. The values of the even harmonics imply that the spheroid which best fits the observed values has a polar diameter $D_p$ $42 \cdot 77$ km less than the equatorial diameter $D_E$. ($D_E = 12756 \cdot 33$ km, $D_p = 12713 \cdot 56$ km, so that the flattening $f = (D_E - D_p)/D_E = 1/298 \cdot 25$). The values of the odd harmonics show that the north pole is about 40 metres further from the equator than the south pole: the north pole is elevated by about 10 metres above the reference spheroid, while the south pole is depressed by about 30 metres below the spheroid. This is the "pear-shape" effect.

The coefficients $J_n$ describe how the shape and gravitational field of the Earth vary with latitude, but they do not give any indication of variations with longitude because they represent an average over all longitudes. When the variations with longitude are taken into account, it is usual to write the gravitational potential $U$ in the form

$$U = \frac{GM}{r}\left[1 + \sum_{n=2}^{\infty}\sum_{s=2}^{n}\left(\frac{R}{r}\right)^{n} P_{ns}(\sin\varphi)\, \times$$

$$\times\,\{C_{ns}\cos s\lambda + S_{ns}\sin s\lambda\}\right] \qquad (3)$$

where $\lambda$ is the longitude (positive eastwards), the $P_{ns}$ are the associated Legendre functions and the $C_{ns}$, $S_{ns}$ are numerical coefficients. The coefficients $J_n$ in equation 1 reappear in this more generalized form as the coefficients $-C_{n0}$.

Since satellites tend to sample all longitudes impartially as the Earth spins beneath them, they are little affected when the gravitational field shows slight variations with longitude. However, the terms $C_{ns}$ and $S_{ns}$ do cause small periodic oscillations in the orbital elements, usually with periods of $1/s$ days, and the values of sets of $C_{ns}$ and $S_{ns}$ can be found if these small oscillations are accurately determined.

At present the sets of values of $C_{ns}$ and $S_{ns}$ obtained by different investigators are rather widely varied, and it is probably best not to quote them individually here. The shape of the geoid which emerges from these studies, however, seems to be fairly well established. Relative to a spheroid with the dimensions quoted earlier, the most important features of the geoid are a depression, about 60 m deep, just south of India and an elevation 60 m high near New Guinea. There are also elevations of order 30 m, centred in France and the South Atlantic, and a depression of 30 m near the south pole.

Various properties of the upper atmosphere at heights of 200–1000 km can be determined from changes in satellite orbits. First, and most important, the air density at a height a little above the perigee height of the satellite can be determined from the rate of decrease of the orbital period $T$. If the eccentricity $e$ is between 0·02 and 0·2 and $\dot{T}$ denotes the rate of change of $T$, the air density $\varrho_A$ at a height $\frac{1}{2}H_p$ above perigee is given most simply by

$$\varrho_A = -\frac{0.157\dot{T}}{\delta}\left(\frac{e}{aH_p}\right)^{\frac{1}{2}}\left\{1 - 2e - \frac{H_p}{8ae}(1-10e)\,+\right.$$

$$\left. +\, O\left(e^2,\frac{H^2}{a^2e^2}\right)\right\}. \qquad (4)$$

Here $a$ is the semi-major axis and $H_p$ is the value of the "density scale height" $H$ at the perigee height of the satellite. $H$ is a measure of the gradient of density, $H = -\varrho\, dr/d\varrho$, and usually takes values between 25 and 80 km for heights between 200 and 600 km. If $m$ is the mass of the satellite and $S$ its cross-sectional area perpendicular to the direction of motion, $\delta$ is given by $\delta = FSC_D/m$, where $F$ is a factor which allows for atmospheric rotation and usually has a value between 0·9 and 1·1, and $C_D$ is the drag coeffi-

cient, which usually takes a value near 2·2. Equation 4 has the advantage that it is insensitive to errors in the value assumed for $H_p$, an error of 25 per cent in $H_p$ giving an error of only 1 per cent in $\varrho_A$. Probably the largest error in using (4) comes from $\delta$: the cross-sectional area may be slightly in error unless the satellite is spherical, and the drag coefficient is also uncertain to the extent of about 5 per cent. Consequently the numerical values of density obtained from (4) are liable to errors of about 10 per cent. Fortunately, this error is not of much significance in view of the very large variations in density which occur.

The analysis of satellite orbits has shown that at heights above 200 km the density of the upper atmosphere is strongly under solar control, in two

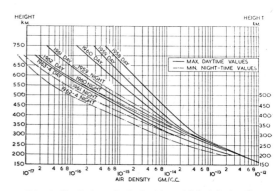

*Fig. 1. Variation of air density with height for the years 1958–64.*

quite distinct ways. First, it responds to the presence or absence of the Sun, the density (and temperature) being higher by day than by night, the "day-to-night effect". Second, the upper atmosphere responds to solar activity, air density (and temperature) being higher when the Sun is more active. Figure 1 shows day and night-time density at heights of 160–800 km, from 1958 (sunspot maximum) to 1964 (sunspot minimum), as determined from the orbits of 45 satellites which had low enough perigees and for which values of $m$ and $S$ were known. The variations with solar activity and from day to night are small at 200 km height but very large at greater altitudes. At a height of 600 km, where the effects are usually greatest, the maximum daytime density exceeds the minimum night-time density by a factor of between 4 and 10, while the daytime density in 1958 exceeded that in 1964 by a factor of about 30.

The variations in $\varrho_A$ over shorter time-intervals can also be determined from equation 4. Thus for the day-to-night variation, it is found that in equatorial and middle latitudes the density increases during the morning to a maximum at about 14 hr local time, declines during the late afternoon and evening to a minimum which is almost attained by midnight, and then remains nearly constant between midnight and dawn before beginning its morning increase again.

The dependence of air density on solar activity is shown by direct comparison with indices of solar activity. The indices most frequently used are: (i) the radiation energy on a wave-length of 10·7 cm, which probably gives an indication of the extreme ultra-violet radiation energy; and (ii) the geomagnetic planetary index $a_p$, which provides perhaps the best indication of the effect of streams of particles from the Sun impinging on the Earth. Figure 2 shows the rate of change of orbital period due to air drag $\dot{T}_A$ for Explorer 9, which is proportional to the air density at

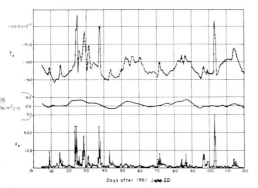

*Fig. 2. Rate of change of orbital period $\dot{T}_A$ for Explorer 9, with 10·7 cm, solar radiation energy S and geomagnetic index $a_p$. (After Jacchia and Slowey, Smithsonian Astrophys. Obs. SP. Rpt. 125.)*

a height of about 650 km, compared with the geomagnetic index $a_p$ and the 10·7 cm radiation energy S. It is seen that for almost all the sharp peaks in $\dot{T}_A$, there are corresponding peaks in $a_p$. If these peaks are ignored, the variation of $\dot{T}_A$, that is of air density, corresponds quite well with that of S: both tend to exhibit the 27-day recurrence which is characteristic of solar activity (because the Sun rotates once about every 27 days, relative to the Earth). Figure 2 also shows transient increases of density by a factor of up to 4, in response to solar activity.

It can be concluded that solar radiation and streams of particles from the Sun heat the upper atmosphere, thus controlling temperature and hence density at heights above 200 km, causing day-to-day fluctuations, a 27-day recurrence tendency and a massive change during the course of the 11-year sunspot cycle. Also, because solar radiation is absent, the temperature and density on the dark side of the Earth are lower than on the sunlit side for heights above 250 km.

Another and a much smaller variation of density, determined from analysis of satellite orbits, is a semi-annual oscillation, with density reaching minima in January and July, and maxima in April and October.

The July minimum is usually lower than the January one and the October maximum is usually higher than the April one, thus implying the existence of an annual variation superposed upon the semi-annual one. These effects are probably also controlled by the Sun, though the exact mechanism is uncertain.

The second orbital change which reveals properties of the atmosphere is the decrease in the perigee height, which to a first approximation is given by

$$r_{p0} - r_p \simeq \frac{H}{2} \ln \frac{e_0}{e} \qquad (5)$$

if the eccentricity $e$ is between 0·2 and 0·02. Thus if the initial values $r_{p0}$ and $e_0$, and a pair of later values $r_p$ and $e$, are accurately known, the density scale height $H$ can be obtained. This method shows that at heights of 200–500 km $H$ tends to decrease as solar activity declines; at 300 km height $H$ decreased from 60 km in 1958 to about 40 km in 1964. The parameter $H$ is useful because it indicates the temperature, being approximately proportional to the air temperature divided by its mean molecular weight. At heights above about 300 km, temperature is almost independent of height, but varies with solar activity, from about 700°K (by night) or 900°K (by day) at sunspot minimum, to about 1500°K (by night) or 2000°K (by day) at a strong sunspot maximum.

The third relevant orbital change is in the inclination $i$ of the orbit to the equator. The rotation of the atmosphere subjects the satellite to a small lateral force, which causes the inclination to decrease slightly (if $i < 90°$). This decrease $\Delta i$ is proportional to the rotational speed of the atmosphere, which may be expressed as $\Lambda$ times that of the Earth, say. As a first approximation, for $0·05 < e < 0·2$, $\Delta i$ is given by

$$\Delta i = \frac{1}{6} \Lambda \, \Delta T_d \sin i \left\{ 1 - 4e + \left( 1 - 4e - \frac{2H}{ae} \right) \cos 2\omega \right. +$$
$$\left. + O(e^2, H^2/a^2e^2) \right\} \qquad (6)$$

where $\Delta T_d$ is the change in orbital period expressed as a fraction of a day. The method can only be used for satellites which suffer a large change in period, preferably at least 10 minutes. $\Delta i$ is then usually of order 0·1°. This technique has not yet been very widely used because of the difficulty of determining $\Delta i$ accurately enough. The results obtained suggest that at heights of 200–300 km the upper atmosphere is rotating faster than the Earth, with $\Lambda = 1·4 \pm 0·2$. The reason for this is not known.

*See also*: Satellite orbits, change of, in space.

*Bibliography*

Cook A. H. (1963) *Space Science Reviews* **2**, 355.
Jacchia L. G. (1963) *Rev. Mod. Phys.* **35**, 973.

KING-HELE D. G. (1962) (2nd Edn) *Satellites and Scientific Research*, London: Routledge.

KING-HELE D. G. (1962) *Progress in the Astronautical Sciences* Vol. 1, Amsterdam: North-Holland.

D. G. KING-HELE

**SATELLITE TRIANGULATION.** *Introduction.* This term is used to describe the technique by which artificial satellites are used as triangulation targets to perform a triangulation in space, and determine the positions of observing stations. Strictly speaking a satellite triangulation should make use only of measured angular quantities, but measured linear quantities can also be included in the solution.

Compared with the methods of classical geodesy, we have the advantage of a simultaneous solution in three dimensions (instead of separately for planimetry and elevation), of being completely independent of the deflexion of the vertical, and of a much simpler mathematical formulation. Satellite triangulation can reach large distances whereas the methods of classical geodesy cannot.

Although three-dimensional triangulation techniques had been proposed much earlier, it is only with the advent of artificial satellites that a simple and accurate space triangulation scheme could be made possible, by using satellites as elevated targets. Artificial satellites are at such high altitudes that the uncertainty in the atmospheric refraction correction—one of the most serious drawbacks in three-dimensional triangulation—is considerably reduced particularly if the direction of a satellite is determined with respect to the star background. The use of the star field to determine directions eliminates also completely the use of the vertical: the directions are expressed in terms of topocentric right ascensions and declinations, an excellent universal reference system for directions.

Satellite triangulation schemes can be divided into two general groups, depending on whether the orbit of the satellite is accurately known, and thus the satellite's position can be used as a known parameter, or the orbit is not reliable enough so that the satellite is used only as a triangulation target of unknown position.

In the first method, a three-dimensional resection is made, enabling the direct determination of geocentric coordinates of the observing station (Fig. 1). In the second method, we materialize and solve a complete three-dimensional net, using as vertices the ground stations and the satellite (Fig. 2). Since satellites move with a speed of about 8 km/sec, all observations referring to the same satellite position should be made simultaneously—or reduced to fictitious simultaneous observations. To perform a space triangulation with only angular measurements, we need as in the case of a surface triangulation, a base line to provide the scale, and the coordinates of at least one known point to be used as origin. In the case of a surface triangulation we also need the orientation of at least one line, but in a satellite triangulation in which absolute directions are directly observed, the space orientation of the entire net is obtained automatically.

*The reference systems.* Cartesian coordinates have definite advantages, due primarily to their very simple metric properties, and are exclusively used in satellite triangulation, rather than the system of ellipsoidal coordinates $(\varphi, \lambda, H)$ more commonly used in geodesy.

An ideal Terrestrial System will be centred at the centre of gravity of the Earth, the 3-axis directed towards the mean axis of rotation of the Earth and the 1–3 plane parallel to the mean Meridian of Greenwich. But since the position of the centre of gravity of the Earth with respect to the crust is not known, this terrestrial system is replaced by a number of geodetic (cartesian) systems which however, as can be easily proved, are parallel to the terrestrial system to an accuracy of $10^{-6}$.

In a satellite triangulation the directions of the lines connecting the ground stations with the satellites are determined directly by observation. Since the precise observations are made by comparing the satellite's apparent position to the background stars, the directions will be expressed in terms of right ascension and declination, referring to the mean equator and equinox of a certain date (depending on the catalogue from which the mean places of the stars were taken). They can be converted in the terrestrial (or geodetic) system by applying the needed corrections for precession, nutation and sidereal time, but they should be corrected also for the annual aberration (the effect of the diurnal aberration is negligible), for the effect of planetary aberration (or light travel time) and for parallactic refraction.

*The observations.* In a satellite triangulation angular quantities are primarily observed (directions as a general rule) to solve the space triangulation net. However, for satellite tracking other non-angular quantities, such as ranges and radial velocities, are measured as well. These quantities can be also used in geodetic studies.

Among the directional tracking systems the most accurate are the ones which photograph the satellite against the star background, and use the principles of photogrammetry and photographic astrometry to determine the direction. Measuring the coordinates of the satellite and of a number of reference stars on the photographic plate, we can derive the apparent position of the satellite.

One of the most versatile cameras, specially constructed for satellite tracking, is the Baker-Nunn Camera (a Super-Schmidt F/1, 50 cm focal length with a $5 \times 30°$ field) which follows the satellite during the exposure, to integrate the light and thus increase its ability to photograph faint satellites. An accuracy of $2''$ in the directions and 1 msec in time can be achieved. Fixed cameras (Ballistic Cameras) have been developed as well, attaining also a high degree of accuracy, but limited to bright satellites only.

Passive tracking (satellite illuminated by the Sun) limits visibility to a part of the orbit, but a flashing satellite will allow optical observations while the satellite is in the Earth's shadow and provide a simple mean for simultaneous observations. The same can be achieved by using laser beams at the ground stations to illuminate the satellite and retro-directive reflectors on board.

A precise timing system capable of providing the time of the observation to 1 msec accuracy is vital at a tracking station, due to the high velocities (of the order of 8 km/sec) with which satellites circle the Earth.

*Observations to satellites of known orbits.* If the geocentric position of the satellite at the time of the observation can be computed from the orbital information, the geocentric position of the observing station can be determined if at least two observations are made on two distinct positions of the satellite.

Let $S_1$ and $S_2$ (see Fig. 1) be the known positions of the satellite (not necessarily during the same pass) and Q the observing station. Let furthermore $I_1$ and $I_2$ be the observed directions expressed in the same refe-

rence system as the coordinates. Then the position of Q will be at the intersection of the two lines passing from $S_1$ and $S_2$ with directions $-I_1$ and $-I_2$ respectively. In practice the two lines will not intersect due to errors in the observations (and the computed coordinates of $S_1$ and $S_2$). Then the mid-point of the shortest distance between the two lines can be used as the most probable position for the station Q.

In practice, many more than two directions will be observed (actually they do not need all to be made on the same satellite), and an adjustment will have to be made.

Let then $x_i$ be the terrestrial (geocentric) coordinates of the $i$th satellite position and $I_i$ the observed direction (expressed as direction cosines). Let also $x_q$ be the (unknown) coordinates of the observing station. The vector relation between known, observed, and unknown quantities will be:

$$\mathbf{x}_i = \mathbf{x}_q + r_i \mathbf{I}_i \tag{1}$$

where $r_i$ is the (scalar) distance, or range, of the satellite. From the system of equations of the form (1) obtained from each observation, the unknown $\mathbf{x}_q$, as well as the auxiliary unknowns $r_i$ can be determined.

In practice it is more convenient to obtain the solution by computing corrections $d\mathbf{x}_q$ to approximate coordinates $\mathbf{x}_q$ of the station. The corrections will be of the order of 100 m and so $\dfrac{d\overline{\mathbf{x}}_i}{r}$ will be less than $10^{-3}$ and their squares can be omitted. Expressing the direction cosines in terms of right ascension $\alpha$, and declination $\delta$ (which are the observed quantities) and differentiating equation 1 we obtain for each observation the following matrix equation:

$$\mathbf{\Omega} \, d\mathbf{x}_q = \begin{pmatrix} d\delta \\ d\alpha & \cos\delta \end{pmatrix} \tag{2}$$

where:

$$d\delta = \delta - \overline{\delta} \quad \text{and} \quad d\alpha = \alpha - \overline{\alpha}$$

viz. the observed values minus those computed from the approximate coordinates and

$$\mathbf{\Omega} = \frac{1}{r} \begin{pmatrix} +\cos\omega\sin\delta & +\sin\omega\sin\delta & -\cos\delta \\ +\sin\omega & -\cos\omega & 0 \end{pmatrix} \tag{3}$$

$\omega = \alpha - \theta$, $\theta$ being the sidereal time. The elements of the matrix are computed with the approximate values $\overline{r}, \overline{\alpha}, \overline{\delta}$.

In equation 2 the factor $\cos\delta$ has been kept on the right-hand side, because $\cos\delta \, d\alpha$ and $d\delta$ have the same weight.

Each observation from a station will give rise to a matrix equation (2). All observations can be combined in a least-squares approximation for the determination of the best correction $d\mathbf{x}_q$ to the approximate coordinates $\overline{\mathbf{x}}_q$. There will be twice as many linear equations as observations, but only three unknowns. The observations can be weighted if their relative accuracies

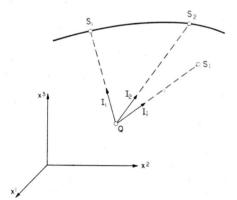

*Fig. 1.*

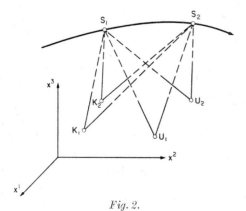

*Fig. 2.*

are known; and if the time of each observation has an important uncertainty, the two observed quantities $\alpha$ and $\delta$ constituting "one observation" should not be treated as independent, since as they both depend on time, they will possess a covariance.

If stations belonging to the same geodetic system (and thus to the same datum) participate in such an observational scheme, all observations can be treated as if made from the same station since it can be proved that the corrections to all stations are identical.

The accuracy of the coordinates for a station determined with this method will depend on the accuracy of the observations, and on the accuracy of the satellite coordinates used. The least-squares approximation will provide the complete variance-covariance matrix of the solution, but an order of magnitude for the variance can be obtained from the approximate expression, based on an ideal angular distribution of the observations:

$$\sigma_q^2 = \frac{\sigma_s^2 + \sigma^2 \tilde{r}^2}{n} \tag{4}$$

where $\sigma_s$ is the standard deviation (s.d.) of one satellite position, $\sigma$ the s.d. of one observation, $\tilde{r}^2$ the harmonic mean of the squares of the ranges, and $n$ the number of observations. Using $\sigma_s = +100\ m$, $\sigma = +4''$, $\tilde{r} = 3\ Mm$ and $n = 25$, the above expression gives $\sigma_q = \pm 23\ m$.

The effect of systematic errors in the observations and the satellite positions, and the effect of a non-perfect distribution of the observations, may increase the uncertainty to about $\pm 30$ m.

*Observations to satellites of unknown orbits.* In this case the positions of the unknown observing stations cannot be determined in a geocentric reference system. However, if both known and unknown stations have observed simultaneously a satellite, the observations made from the known stations will determine the coordinates of the satellite at the time of the observations. Those coordinates will then be used to determine the coordinates of the unknown stations with the previously described method. The positions of the satellite will refer to the same geodetic system as that for the known stations. Thus this method will provide a relative, rather than an absolute, solution since it will determine the coordinates of stations in a given geodetic datum.

The method of variation of coordinates can be used as previously. Every observation will be expressed in the form:

$$\boldsymbol{\Omega}\, \mathrm{d}\mathbf{x}_q - \boldsymbol{\Omega}\, \mathrm{d}\mathbf{x}_s = \begin{pmatrix} \mathrm{d}\delta \\ \mathrm{d}\alpha \cos\delta \end{pmatrix} \tag{5}$$

where $\mathrm{d}\mathbf{x}_s$ will be the corrections to the approximate coordinates $\bar{\mathbf{x}}_s$ of the satellite at the time of the observations. The observing stations will be either known (suffix $k$) or unknown (suffix $u$). Then the correction $\mathrm{d}\mathbf{x}_k$ will be zero and $\mathrm{d}\mathbf{x}_u$ will be the values to be computed, together with the auxiliary quantities $\mathrm{d}\mathbf{x}_s$.

The matrix of the observation equations will consist of submatrices $\boldsymbol{\Omega}$ and will be of the following form:

$$\begin{pmatrix} \boldsymbol{\Omega} & -\boldsymbol{\Omega} \\ (u, s) & (u, s) \\ \hline 0 & -\boldsymbol{\Omega} \\ & (k, s) \end{pmatrix}.$$

The upper part consists of observations made from unknown stations, the lower from the known stations; the left part corresponds to the unknon stations $(\mathrm{d}\mathbf{x}_u)$, the right part to the auxiliary unknowns $(\mathrm{d}\mathbf{x}_s)$. The method finally results in the solution of a linear system of equations with a great number of unknowns. All stations belonging to the same datum can be treated as one station, and thus eliminate a number of unknowns.

If there are only two geodetic datums involved, the one is taken as reference. The solution consisting of three unknowns can be obtained through a short-cut without much loss of rigorousness. The coordinates of the satellite are determined separately from observations from each datum and then the two sets of coordinates are compared.

When only two stations are observing simultaneously a satellite, the solution cannot provide the coordinates of the one station with respect to the other, but only the direction of the line connecting the two stations.

The accuracy of the computed coordinates will be determined after the least squares solution. The following approximate expression will give the standard deviation of the coordinates of the unknown stations (assuming an ideal net configuration with no errors in the coordinates of the known stations):

$$\sigma_u^2 = \frac{\sigma^2 \tilde{r}_k^2}{n \cdot m} + \frac{\sigma^2 \tilde{r}_u^2}{n}$$

using the same symbols as in (4) and $m$ being the number of the unknown stations. Assuming $\sigma = \pm 2''$, $\tilde{r}_u = \tilde{r}_k = 3\ Mm$, $n = 5$, $m = 5$ then $\sigma_u = \pm 15\ m$. The effects of systematic errors, an imperfect net configuration, and of the errors in the coordinates of the known stations may increase the total uncertainty to about $20$–$30\ m$.

*Satellite trilateration.* Just as in the case of surface triangulations, where the lengths of the sides of the triangles can be measured directly by an electronic distance measuring device and thus changing the "Triangulation" to a "Trilateration" it is also possible to measure directly the ranges in a space satellite net and thus perform a "satellite trilateration".

A trilateration space net can be computed—or rather adjusted—with the method of variation of coordinates as in a triangulation. The relation connecting $\mathrm{d}r = r - \tilde{r}$ with $\mathrm{d}\mathbf{x}_q$ and $\mathrm{d}\mathbf{x}_s$ is:

$$\boldsymbol{\Psi}\, \mathrm{d}\mathbf{x}_q - \boldsymbol{\Psi}\, \mathrm{d}\mathbf{x}_s = \mathrm{d}r \tag{6}$$

where $\boldsymbol{\Psi}$ is the row matrix:

$$\boldsymbol{\Psi} = (-\cos\omega\cos\delta \quad -\sin\omega\cos\delta \quad -\sin\delta) \qquad (7)$$

Three observed ranges to three known satellite positions will give the position of the unknown station. An adjustment will be required if more ranges are measured. In the case of simultaneous range observations to a satellite of unknown orbit, the net will be adjusted by including also the auxiliary unknown satellite coordinates.

If ranges are measured together with directions, a combined adjustment will be made, the matrix equation for each observation being of the form:

$$\begin{pmatrix} \boldsymbol{\Omega} \\ \boldsymbol{\Psi} \end{pmatrix} d\mathbf{x}_q - \begin{pmatrix} \boldsymbol{\Omega} \\ \boldsymbol{\Psi} \end{pmatrix} d\mathbf{x}_s = \begin{pmatrix} d\delta \\ d\alpha\cos\delta \\ dr \end{pmatrix}. \qquad (8)$$

*Bibliography*

GUIER (1963) Navigation Using Artificial Satellites, *The Use of Artificial Satellites for Geodesy*, Amsterdam: North-Holland.

HIROSE H. (1963) A Simple Method of Triangulation with the Use of Artificial Satellites, *The Use of Artificial Satellites for Geodesy*, Amsterdam: North-Holland.

KAULA W. H. (1962) Celestial Geodesy, *Adv. Geophys.* **9**, 191.

KUTUZOV I. A. (1964) Preliminary Results of Processing of Synchronous Observations of the Satellite ECHO, Academy of Sciences of the USSR.

McLELLAN C. D. (1963) *Trilateration*, radar in *Encyclopaedic Dictionary of Physics* (Ed. J. Thewlis), Oxford: Pergamon Press.

SCHMID H. (1961) Some Problems Connected with the Execution of Photogrammetric Multi-Station Triangulation, *Geodesy in Space Age*, Ohio: The University Press.

U.S.C. & G.S. (1962) *Satellite Triangulation*. U.S. Department of Commerce: Coast and Geodetic Survey.

VEIS G. (1960) Geodetic Uses of Artificial Satellites, *Smithsonian Contributions to Astrophysics* **3**, No. 9, 95.

VEIS G. and WHIPPLE F. L. (1961) Experience in Precision Optical Tracking of Satellites for Geodesy, *Space Research II*, Amsterdam: North-Holland.

G. VEIS

## SEA WAVES AND MICROSEISMS.
Microseisms (Greek: small shocks) are tiny vibrations in the ground with amplitudes up to 10 $\mu$. They cover a large range of frequencies from about 100 c/s downwards. The higher frequency ones are usually caused by traffic and other man-made agencies. Those in the range 0·5–0·1 c/s are, however, usually caused by natural means and have been shown to have a strong correlation with waves at sea and an increase of amplitude of these usually heralds the onset of strong swell off the coast and the arrival of a storm.

The association between sea waves and microseisms has been known of since the beginning of the century. It is only more recently that quantitative relations were established between their amplitude and frequency and a theoretical explanation was provided of the mechanism by which waves cause microseisms. The first quantitative relationship was obtained by Bernard in 1941 when he found that the microseisms frequency was double the wave frequency. This relationship was confirmed by Deacon (1947) when he compared the microseisms recorded at Kew observatory with the waves recorded at Perranporth, Cornwall. His results are shown in Fig. 1 and show clearly

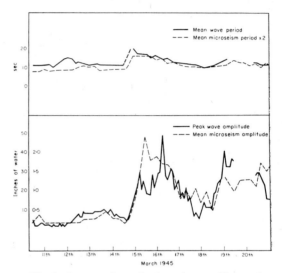

*Fig. 1. A comparison of microseisms at Kew and sea waves at Perranporth, Cornwall (after Deacon).*

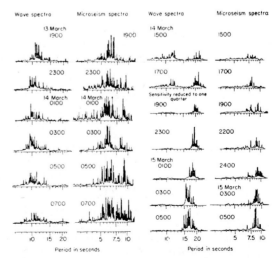

*Fig. 2. Comparison of wave and microseism spectra (after Darbyshire).*

the 2 to 1 relationship between the periods (or frequencies) and also how the amplitude of both sets of waves rise together. Darbyshire (1950) compared the wave spectra of waves recorded at Perranporth and the spectra of microseisms recorded at Kew and found that they were very similar when a 2 to 1 frequency ratio was allowed for (Fig. 2).

There remained the problem of finding the mechanism by which sea waves produce microseisms. This was a very difficult problem as the pressure due to ordinary progressive sea waves disappears at a depth greater than half a wave-length and so could not be felt at the sea bottom in the deep ocean. Longuet–Higgins (1950) following on some work by Miche showed that stationary waves have a second-order component which does not vanish with depth and this also applies when two progressive waves of different amplitude but the same period meet head-on. The pressure then becomes:

$$p \propto a_1 a_2 \sigma^2 \cos 2\sigma t .$$

This work was confirmed by experimental work in a testing tank by Cooper and Longuet-Higgins (1951). In deep water the compressibility of water has to be taken into account as the time for the pressure to be propagated to the bottom will be comparable to the sea wave period. In this case there is no uniform unattenuated pressure fluctuation but a compression wave whose plains of equal face are horizontal. Resonance occurs if the depth is $(\frac{1}{2} n + \frac{1}{4})$ times the length of the compression waves.

Trains of waves of the same period are likely to meet head-on near the coast where the incident waves get reflected and in the deep ocean in the area traversed by fast moving depression where the winds can occur in opposite directions in different parts of the ocean and so generating trains of waves moving in opposite directions. Darbyshire (1950) found examples of such *storms* producing microseisms.

In certain cases however, microseisms having the same frequency as sea waves are found. Bernard (1952) found some evidence for this and more recently Oliver (1962) found large 27 sec period microseisms in June 1961 which could be related to a peculiar onset of 27 sec sea waves which appeared at that time. Haubrich *et al.* (1963) compared sets of wave and microseism spectra for six days off the California coast and found the microseism spectra to show two peaks, a small one corresponding to that found on the wave spectra and a larger one at the double frequency. Darbyshire (1963) comparing wave and microseism spectra in South Africa, where coastal conditions are similar to California could only, however, find a peak at the double frequency. The effect at the same frequency can be ascribed to waves near the coast where the pressure at the sea bed is still appreciable. Hasselmann (1963) has considered various theories of microseism generation and concludes that they can only be formed when the exciting fields have a Fourier component which has the same phase velocity as the free modes of the elastic system. The Longuet–Higgins mechanism and the direct pressure of progressive waves on shallow water fulfil these conditions.

Microseisms have been used to determine the direction of the storm that produces them. Microseisms are surface ground waves of either the Rayleigh type where the particles move in ellipse in a vertical plane parallel to the direction of propagation or are Love waves where the particles move horizontally at right angles to the direction of propagation. If the motion was purely of the Rayleigh type, then the tangent and the angle of approach would be given by

$$\tan\theta = \bar{x}/\bar{y}$$

where $\bar{x}$ and $\bar{y}$ are the r.m.s. values of amplitude of the east-west and north-south components. In practice, however, there are usually Love waves present as well on the two horizontal components but not on the vertical. Consequently the direction of approach can be found by finding the coherence (maximum correlation allowing for phase lag) between the vertical and the two horizontals.

Then

$$\tan\theta = \bar{x}r_{xz}/\bar{y}r_{yz}$$

where $r_{xz}$ and $r_{yz}$ are the coherences.

Attempts on these lines to find the direction have been reasonably successful. Figure 3 (after Darbyshire and Hinde) shows such an attempt. The results are even more successful when the coherences are worked out for a band of periods of the spectra such as from 6 to 8 sec as was done in this case. Allowance has to be

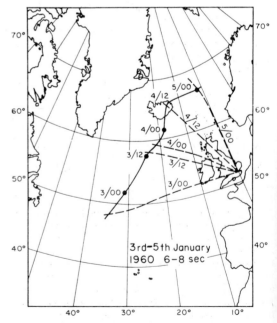

*Fig. 3. Calculated and observed bearings of a moving storm (after Darbyshire and Hinde).*

made for the refraction of microseisms. This effect not only changes the direction slightly but often causes natural converging and diverging lenses to be formed so the microseisms effects are diminished or increased under certain conditions (Darbyshire and Darbyshire 1957).

*Bibliography*

BERNARD P. (1941) *Etude sur l'agitation microseismique et ses variations, Ann. Inst. Globe.* **19**, 1.

BERNARD P. (1952) *Ser A. Trav. Sci. du Bur. Centr. Seismol. Fax.* **18**, 83.

COOPER R. I. B. and LONGUET-HIGGINS M. S. (1951) *An experimental study of the pressure variation in standing water waves, Proc. Roy. Soc. Lond.* A **206**, 424.

DARBYSHIRE J. (1950) *Sea waves and microseisms, Proc. Roy. Soc. Lond.* A **223**, 96.

DARBYSHIRE J. (1963) *A study of microseisms in South Africa, Geophys. J. Roy. Astron. Soc.* **8**, 165.

DARBYSHIRE J. and DARBYSHIRE M. (1957) *The refraction of microseisms on approaching the coast of the British Isles, Mon. Not. Roy. Astron. Soc. Geophys. Suppl.* **7**, 301.

DARBYSHIRE J. and HINDE B. J. (1961) *Microseisms, Research, Lond.* **14**, 8.

DEACON G. E. R. (1947) *Relation between sea waves and microseisms, Nature* **160**, 419.

HASSELMANN (1963) *A statistical analysis of the generation of microseisms, Rev. Geophys.* **1**, No. 2.

HAUBRICH R. A. *et al.* (1963) *Comparative spectra of microseisms and swell, Bull. Seis. Soc. Amer.* **53**, 27.

LONGUET-HIGGINS M. S. (1950) *A theory of the origin of microseisms, Phil. Trans. Roy. Soc. Lond.* A. 1–35.

OLIVER J. (1962) *A world-wide storm of microseisms with periods of about 27 seconds, Bull. Seis. Soc. Amer.* **52**, 507.

J. DARBYSHIRE

**SEMICONDUCTOR DEVICES.** The point-contact transistor, the first modern semiconductor device, was invented in 1948, shortly to be superseded by the junction transistor which was easier to manufacture and more reliable in operation. Since then a large number of specialized semiconductor devices have been produced, both to replace existing thermionic devices and to perform new functions. The following description of some of these devices will be dealt with under the headings Diodes, Transistors and Miscellaneous Devices.

### A) Diodes

*1. Point-contact and junction types.* Basically the operation of semiconductor diodes depends upon the rectifying property of either a point-contact between a metal and a semiconductor, or a junction between p- and n-type semiconductors. Both these types are finding wide application although under somewhat different operating conditions. The point-contact types are confined to low voltage–low current operation,

150 mA forward current and 100 V peak inverse voltage (PIV) being typical. The limits of the range of operating conditions for the junction types also include this range but extend to much higher ratings, the maximum being of the order of 100 A forward current and 500 V PIV. The use of silicon instead of germanium as the basic material has greatly increased the possible working range of temperature, 100–150°C ambient temperature being permitted for the silicon types. There are several special types of junction diode and these will be examined in detail.

*2. The Zener or avalanche diode.* When the junction diode is subjected to an inverse voltage which is equal to, or less than, the specified PIV, a very small inverse (leakage) current flows which may be several

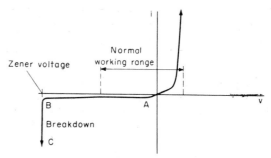

Fig. 1. *Characteristic of a junction diode.*

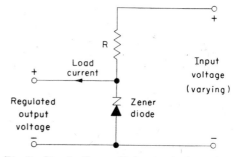

Fig. 2. *Simple Zener diode circuit for voltage stabilization.*

microamperes or only a fraction of a microampere, depending upon the type of diode (Fig. 1, Section AB). If the inverse voltage is gradually increased beyond the specified maximum, at some high value there comes a point at which breakdown takes place and a large inverse current flows (Fig. 1, Section BC). This current must be limited by external circuit resistance to prevent the diode from being destroyed. In the absence of any physical damage to the diode, the process is reversible. The physical process of the breakdown is not fully understood but is generally regarded as being a combination of the "Zener" and "Avalanche" Effects. When a large inverse voltage is applied to a *p-n* junction, the field strength in the depletion layer at the junction may be very high indeed. In a very thin depletion layer the field strength may be $10^6$ V/cm with an inverse voltage of only 5 V. C. Zener suggested from his work on dielectrics that with this order of field strength internal field emission may occur. This is the "Zener Effect" and its extension to semiconductors suggests that the covalent bonds break down releasing electrons and holes as charge carriers. If these carriers gain sufficient energy while still in the region of high field, they may produce other electron-hole pairs by collision. These carriers may produce others and the inverse current increases very rapidly by the "Avalanche Effect".

It is possible by judicious doping of semiconductors, to produce a diode that will break down at a specified inverse voltage. From Fig. 1, it may be seen that section BC of the characteristic corresponds to low resistance and it is within this region that the Zener Diode is used as a voltage regulator. A typical circuit is shown in Fig. 2. The output voltage will remain constant at the specified breakdown (or Zener) voltage of the diode, provided: (i) the input voltage is always greater than the Zener voltage, and (ii) the current taken by the load does not so starve the diode of current that its operating point moves into the high resistance region of the reverse characteristic. The value of the series resistance $R$ may be determined from the maximum input voltage minus the Zener voltage, divided by the maximum permissible diode inverse current. The maximum permissible load current may be calculated from the lowest input voltage minus the Zener voltage, divided by the series resistance. A short circuit across the load does not endanger the diode.

Zener diodes are available with breakdown voltages from 2 to 100 V, with a tolerance of $\pm 5$ per cent of their nominal value. Maximum current ratings range from 1 to 500 mA. Voltage regulating thermionic devices cannot be made for less than about 60 V.

*3. The controlled rectifier.* This device consists of two *p-n* junctions joined end-to-end to form a four-layer *pnpn* diode, connexions being made to the two end regions and to the second *p* region, as shown in Fig. 3. If the *N* region (Fig. 3) is made positive with respect to the *p-* region, the junctions *p-n* and *P-N* are reverse biased and *n-P* is forward biased. The leakage currents of the first two junctions are additive but as silicon is normally used, these currents are small. If the electrode *G* is open circuit and *p* is made positive with respect to *N*, the junctions *p-n* and *P-N* are forward biased but current is inhibited by the reverse-

biased junction *n-P*. As the forward bias across *p-N* is increased, a point is reached when breakdown occurs at the *n-P* junction in the same way as for the Zener diode. External series resistance is now required to limit the diode current. By variation of the magnitude and polarity of the bias applied to the gate-electrode *G*, the diode may be switched on at various forward voltages or prevented from switching. The characteristic for this device is shown in Fig. 4, where the section AB represents the switching from off to on and corresponds to a negative resistance.

Controlled rectifiers are available to switch current to a maximum of 20 A at hundreds of volts and are widely used in power rectification. Switching times

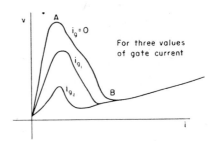

*Fig. 4. Typical family of characteristics for the controlled rectifier.*

are typically 1 μsec for "on" and 10 μsec for "off". The controlling signal at *G* works into a fairly low impedance so that it must supply power.

There is a distinct similarity between the operation of the semiconductor controlled rectifier and the gas-filled thyratron, but the former has the advantages of faster switching, low voltage drop, no filament and higher permissible working temperatures for the silicon types.

*4. The Esaki or tunnel diode.* This is fundamentally an abrupt *p-n* junction in which both sides are very heavily doped with impurities, doping levels being of the order of $10^{19}$ atom/cm³. With this concentration of impurities, the semiconductor is said to be "degenerate" which means that it is so heavily doped that it resembles a metal more than a semiconductor. There are a large number of electrons in the conduction band of the *n*-side and a large number of empty energy levels (holes) on the *p*-side. The consequence of such high doping is that the Fermi level on the *p*-side lies below the top of the valence band and on the *n*-side above the bottom of the conduction band (Fig. 5(a)). Moreover, the depletion layer is very narrow, of the order of 100 Å. The result of these factors is that for low values of forward bias electrons from the *n*-side cross the junction directly into the empty energy levels on the *p*-side. This process is called "tunnelling", and for small forward voltages the junction behaves as if it were ohmic. The condition for maximum current is reached when the bottom of the conduction band on

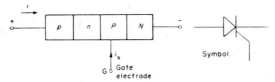

*Fig. 3. The four-layer triode or controlled rectifier.*

the $n$-side is opposite the Fermi Level on the $p$-side (Fig. 5(b)), since this is the condition of maximum overlap. Any further increase in voltage decreases the number of filled and empty energy levels which are opposite one another across the junction and the tunnelling current decreases (Fig. 5(c)). Eventually, when the bottom of the $n$-side conduction band is opposite the top of the $p$-side valence band, the tunnelling current is negligible, and further increase in bias simply produces normal diode diffusion current. The total characteristic is shown in Fig. 6. The negative resistance region AB has a slope of about 3 ohm;

this is the most useful part of the characteristic. For germanium, the values of the voltages at A and B are 50 mV and 350 mV respectively, and the currents at these points may be in the ratio of 15 : 1. The tunnel diode may be used as a bistable device by switching it between the points A and B. As an oscillator, the diode is capable of operation in the 1000 Mc/s (gigacycle/sec) range with switching times of 2 m μsec. Other possible uses include negative resistance amplifiers, pulse gates, voltage limiters and regulators. For reference devices, tunnel diodes are now available in

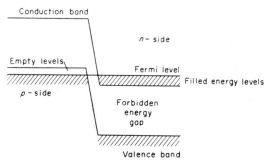

Fig. 5(a). Energy-level diagram for the tunnel
diode junction with zero bias.

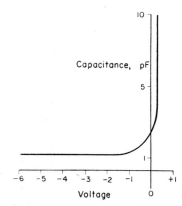

*Fig. 6. Total characteristic for the tunnel diode.*

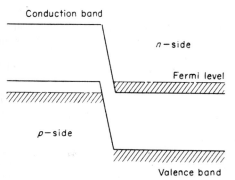

Fig. 5(b). Energy-level diagram for maximum
diode current.

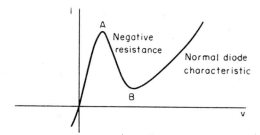

*Fig. 7. Typical varactor characteristic.*

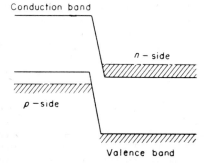

*Fig. 5(c). Energy-level diagram for high forward
bias showing decreased overlap.*

which the current at A in the characteristic is consistent to within 2 per cent of a nominal value, with a change of only 5 per cent over a temperature range of 150°C.

*5. The variable capacitance diode.* At a $p$-$n$ junction, an approximate capacitor is formed by the depletion layer (the dielectric) and the comparatively highly conductive $p$- and $n$-sides (the plates). The width of the depletion layer varies with the voltage across it, generally being very narrow for small forward bias and wide for reverse bias. Figure 7 shows a typical characteristic. This type of diode is also known as the "parametric diode", the "reactance diode", or the "*varactor*".

Graded junction non-linear variable capacitance diodes have been made which will operate up to 100 gigacycles/sec, and have therefore found application in microwave circuits. If a varactor is connected across a coaxial or waveguide transmission line, the impedance it presents can be varied by the bias applied to it. At small forward bias, the capacitance will be very high and will short-circuit the line, while at reverse bias the capacitance will be low and the diode will present a high impedance. Switching power of only a few milliwatts is sufficient to switch kilowatts of signal power.

Since the impedance of the diode is a function of the r.f. applied as well as the d.c. bias, it will function as an output power limiter because as the input power increases, there is a tendency for it to be reflected.

Another application is in harmonic generation, it being possible to make frequency doublers and triplers in the gigacycle/sec range at efficiencies of 50 and 30 per cent respectively.

One obvious use for the varactor is for automatic frequency correction in radio receivers. Here the d.c. component of the discriminator output voltage is used to bias the varactor and hence vary the reactance of the local oscillator tuning circuit.

### B) Transistors

*1. The field effect or unipolar transistor.* It is doubtful whether this device can be truly called a transistor but it is normally considered as such and is noteworthy as the semiconductor device whose operation most closely resembles that of the vacuum triode. The body of the device is a slab of $n$-type material with an ohmic contact at each end, and on each side (or around it) there is $p$-type material as shown in Fig. 8. The electrodes are called source, gate and drain rather than emitter, base and collector. If the source is made negative with respect to the drain, electrons move from source to drain. When the gate is biased negatively with respect to the source, the flow of electrons in the vicinity of the gate is limited by the transverse field, to a much narrower channel. Hence the resistance to electron flow is increased and the current is decreased. If now, the drain voltage is increased the electrons accelerate and there is a tendency for the current to increase. However, increasing the drain voltage also makes the body of the device more positive. This in turn increases the effective gate bias and the electron current is decreased. As a result of this the output

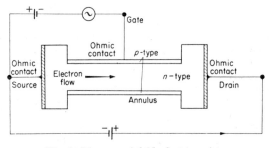

*Fig. 8. Diagram of field-effect transistor (cylindrical type).*

characteristic closely resembles that of the pentode valve. Since the signal input is across a reverse-biased $p$-$n$ junction, the input impedance is high, which is unusual for transistors. The action depends upon the majority carrier current and so the only frequency limitation is imposed by the time to traverse the body of the device. Available field-effect transistors operate up to about 50 Mc/s.

*2. The unijunction transistor (or double base diode).* Whether this device is a transistor or a diode, its main feature is that it has a negative resistance region in its characteristic. The unijunction transistor consists of a

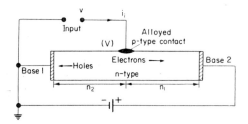

*Fig. 9. The unijunction transistor.*

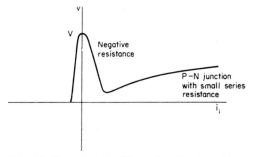

*Fig. 10. Characteristic of the unijunction transistor.*

bar of $n$-type material with connections at each end (base 1 and base 2) and a $p$-$n$ alloyed junction situated on the side of the bar more than half way from base 1 to base 2. (Fig. 9). The biasing polarities are shown on the diagram from which it may be appreciated that the voltage of the $n$-type material ($V$) in contact with the emitter is a proportion of that across the bases. As the input voltage $v$ is increased from zero, the $p$-$n$ junction remains reverse biased until $v$ is equal to $V$. If $v$ is further increased the $p$-type emitter injects holes into the $n$-type and these are swept to ground. These minority carriers provoke an equal number of majority carriers and the final result is to decrease the resistance of the $n_2$ section of the bar. This reduces the value of $V$ and further forward biases the $p$-$n$ junction. Now more current flows for less input voltage and this is expressed as the negative resistance region of the characteristic (Fig. 10). This process cannot continue indefinitely so there comes a point at which any further increase in voltage produces an increase in current and the characteristic becomes that of a *p-n* junction with small series resistance. The negative resistance region can be used for bistable operation although the upper frequency limit is low compared, for instance, with the tunnel diode.

*3. The tetrode transistor.* This type has the form of the normal *p-n-p* or *n-p-n* transistor with the addition of a second base connexion on the opposite side of the base region (Fig. 11). In the *p-n-p* type the second base is biased positively with respect to the first base, producing a transverse field across the base region. This alters the pattern of minority carrier flow, as shown in the diagram, restricting it to the region of the first base contact. The advantages of this are two-

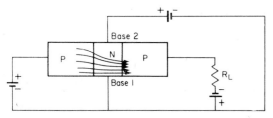

*Fig. 11. The tetrode transistor.*

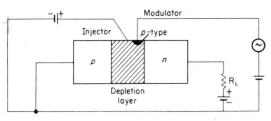

*Fig. 12. Diagram of the spacistor.*

fold. Firstly, it effectively reduces the collector junction area, thus reducing its capacitance. Secondly, it reduces the length of the path of the electrons which flow from base to emitter, hence decreasing the magnitude of the base resistance. Since collector capacitance and base resistance are two of the limiting factors in the high frequency operation of transistors, the tetrode transistor is especially useful at the higher frequencies.

*4. The PNIP transistor.* Another method of reducing collector junction capacitance in a junction transistor and hence increasing frequency response is to include a layer of nearly intrinsic material between the base and the collector. This form of transistor is used at collector voltages which ensure that the collector depletion layer includes the intrinsic region and is therefore very wide. This constitutes a decrease in collector capacitance. The field in the intrinsic layer is high by virtue of the high collector voltage, so that the additional time for the carriers to traverse it is not an important factor.

*5. The spacistor.* This is a single junction device (Fig. 12) and is operated at high reverse voltage so that there is a wide depletion layer. A metallic contact is made into the depletion layer in order to inject electrons. The injector voltage must be lower than the voltage that would exist at the injection point in the absence of the injector. The injected electrons move rapidly to the *n*-side and are collected. The significance of this device is that the electrons are injected *into* the depletion layer and not *across* it as in the case of the normal *p-n* junction; the net result being smaller transit time and hence higher-frequency operation. In fact, the improvement is not as great as would be expected, due to variation of the field in the depletion layer. A fourth, *p*-type, contact is added as shown, and is used to modulate the current.

*6. High-frequency transistors.* Many different forms of the junction transistor have resulted from the search for higher-frequency response. Some have already been described. The main objectives in the production of high-frequency transistors are a thin base region and highly conductive emitter and collector regions. Ingenious manufacturing methods have solved the problems of producing small adjacent regions of totally opposite properties. These methods have produced the surface barrier, mesa and epitaxial transistor. The basic operation of these types is fundamentally the same as for the normal low-frequency transistor, but the equivalent circuit may need some modification, in exact analysis, to account for some differences both in the relative positions of regions and in the pattern of carrier flow.

### C) Miscellaneous Devices

Silicon carbide is used to make fuses for the protection of equipment against excess voltages. The advantages of this semiconductor is that it does not obey Ohm's law, but has a resistance that decreases very rapidly with increasing voltage. This device is called a "*thyristor*".

"*Thermistors*" use the peculiar property of semiconductors that their resistance decreases with increasing temperature. Applications of the thermistor include protection of valve heater circuits against initial switch-on transients, temperature compensation of circuits and oscillator frequency control. Materials used include the oxides of manganese, cobalt, titanium and nickel.

The group of semiconductors called the "phosphors" include magnesium tungstate, cadmium sulphide, and the sulphate, oxide and silicate of zinc. These materials are capable of converting energy received from impinging particles into visible-light energy. Two applications are in scintillation counters and electroluminescent panels.

Strain gauges constructed of semiconductors are found to be fifty times more sensitive than the conventional ones using metals.

Semiconductors are used extensively in light-sensitive devices of the photoconductive and photo-voltaic types. In the former the most useful materials are selenium, cadmium sulphide, thallium sulphide, indium antimonide and several compounds of lead. The spectral response of these materials varies considerably and so the cell used must be defined by the application. Most important of the photovoltaic type is the *solar cell* which is finding considerable application in power supplies for telephone repeaters and satellites. The solar cell is a *p-n* junction in which the region of the junction is exposed to the light. The cell produces about 0·5 V at very small current, hundreds of units being necessary to obtain only 10 W of power.

The Hall effect in semiconductors is used in a range of devices for the measurement of flux density and radio-frequency power, current measurement by detection of the magnetic field around the conductor, for modulation down to very low frequencies and as multiplying elements in computers.

One of the most recent applications of semiconductor materials has been in lasers. *P-N* junctions of the semiconductor gallium arsenide are found to emit light of 9000 Å wave-length from the junction area during normal forward conduction. The wave-length of the light emitted may be varied by adjusting the operating temperature of the diode.

*Bibliography*

DEWITT D. and ROSSOFF A. L. (1957) *Transistor Electronics*, New York: McGraw-Hill.
JOYCE M. V. and CLARKE K. K. (1962) *Transistor Circuit Analysis*, New York: Addison-Wesley.
NANAVATI R. P. (1963) *An Introduction to Semiconductor Electronics*, New York: McGraw-Hill.
SIMS G. D. and STEPHENSON I. M. (1963) *Microwave Tubes and Semiconductor Devices*, London: Blackie.
SCROGGIE M. G. (1961) *Principles of Semiconductors*, London: Iliffe.
VALDES L. B. (1961) *The Physical Theory of Transistors*, New York: McGraw-Hill.

T. A. COEKIN

**SEMICONDUCTOR MATERIALS.** A great deal of attention has been given to semiconducting materials, especially since the invention of the transistor in 1948. The most commonly used semiconductors, germanium and silicon, have been thoroughly investigated, but in recent gears attention has been diverted to other elements and many compounds with semiconducting properties. The two main features of any semiconductor are the width of its forbidden energy gap and its carrier mobility. It has been found that a wide range of gaps and mobilities is available, particularly among the compound semiconductors. Many of these elements and compounds have been known to be semiconducting for many years but the need for better materials to produce special devices has accelerated the detail investigation of their properties. The theory of semi-conduction was more or less founded upon the study of germanium but modifications are now necessary to explain satisfactorily the phenomenon in other materials, particularly in some amorphous types.

*Elemental Semiconductors*

As well as germanium and silicon, at least some semiconducting properties are shown by diamond,

Table 1. Some properties of the elemental semiconductors

| Semiconductor | Energy gap eV | Mobilities cm² V⁻¹ sec⁻¹ | | Melting point °C | Remarks |
|---|---|---|---|---|---|
| | | Electron | Hole | | |
| Diamond | 5·3 | 1800 | 1300 | — | Semiconductivity due to impurities |
| Silicon | 1·21 | 1300 | 500 | 1420 | Difficult to make in pure form |
| Germanium | 0·79 | 4500 | 3500 | 940 | Most widely studied |
| Grey tin | 0·08 | 2500 | 2400 | — | Only stable below 13°C |
| Boron | ~ 1·7 | ~ 1 | ~ 1 | 2300 | Large reduction in resistivity with increasing temp. |
| Selenium | 1·5 | — | ~ 10 | 220 | Applications: photovoltaic cells and rectifiers |
| Tellurium | 0·34 | 1700 | 1200 | 451 | Hall effect inversion |
| Black phosphorus | 0·33 | 220 | 350 | — | Evaporated film |
| Grey arsenic | ~ 1·2 | ~ 65 | ~ 65 | — | Evaporated film |

graphite, grey tin, selenium, tellurium, boron and some forms of phosphorus, arsenic and antimony.

*Diamond and graphite.* Diamond and graphite are allotropic forms of carbon which are completely different microscopically and electrically. Diamond is hard, brittle and is almost an insulator, its resistivity being greater than $10^8$ ohm cm. The energy gap is very large, about $5 \cdot 2$ eV so its conducting properties are due almost entirely to impurities of which natural diamond contains many. Graphite, on the other hand, is soft, metallic and has a resistivity of $2 \times 10^{-4}$ ohm cm. The energy gap is zero at absolute zero and small at higher temperatures. The material thus displays a negative temperature coefficient just above absolute zero and a positive one at higher temperatures. Graphite is extremely anisotropic.

*Grey tin* has an energy gap of $0 \cdot 08$ eV and its intrinsic resistivity is $0 \cdot 0002$ ohm cm at room temperature. The addition of impurities from the fifth and third groups of the periodic table, produce $n$- and $p$-type material respectively. Antimony and aluminium are usually used.

*Selenium* has been used extensively in photocells and rectifiers but is still not fully understood. It has a gap width of $1 \cdot 5$ eV and only $p$-type has been studied. Selenium is a very useful semiconductor although in some ways it is unsatisfactory, for instance, in having a very small hole mobility.

*Tellurium* is metallic looking and has an energy gap of $0 \cdot 34$ eV. It is an intrinsic semiconductor at room temperature with a resistivity of about $0 \cdot 3$ ohm cm. The unique property of tellurium is that the sign of its Hall coefficient changes twice as the temperature is increased from a low to a high value.

*Boron* has been known as a semiconductor since the beginning of the century. It is a difficult material to prepare because of its very high melting point. Its resistivity at room temperature is about $2 \times 10^6$ ohm cm and this decreases by a factor of $10^5$ with an increase of about 500°C in temperature.

*Arsenic and antimony* are normally metallic and have only very slight semiconducting properties. If these materials are formed as evaporated films (grey arsenic and black antimony) which are amorphous, their semiconducting properties are more pronounced. This also applies to amorphous red phosphorous.

The properties of the elemental semiconductors are summarized in Table 1.

### Compound Semiconductors

It is possible to produce semiconducting compounds from pairs of elements from two different groups of the periodic table. Several classes of these semiconductors have been prepared, viz. using elements from the

*Table 2. The elements of groups II–VI of the periodic table*

| II | III | IV | V | VI |
|----|-----|----|----|----|
| Be | B | C | N | O |
| Mg | Al | Si | P | S |
| Zn | Ga | Ge | As | Se |
| Cd | In | Sn | Sb | Te |
| Hg | Tl | Pb | Bi | Po |

*Table 3. Properties of the III–V compound semiconductors*

| Semiconductor | Symbol | Energy gap eV | Mobilities cm² V⁻¹ sec⁻¹ | | Melting point °C |
|---|---|---|---|---|---|
| | | | Electron | Hole | |
| Indium antimonide | InSb | $0 \cdot 25$ | 65,000 | 700 | 523 |
| Indium arsenide | InAs | $0 \cdot 45$ | 23,000 | 100 | 936 |
| Indium phosphide | InP | $1 \cdot 3$ | 3400 | 650 | 1070 |
| Gallium antimonide | GaSb | $0 \cdot 8$ | 4000 | 700 | 720 |
| Gallium arsenide | GaAs | $1 \cdot 6$ | 4000 | 200 | 1240 |
| Gallium phosphide | GaP | $2 \cdot 4$ | – | – | – |
| Aluminium antimonide | AlSb | $1 \cdot 6$ | $\sim 100$ | 200 | 1080 |
| Aluminium arsenide | AlAs | – | – | – | – |
| Aluminium phosphide | AlP | $\sim 3$ | – | – | – |

groups II and IV, II and V, III and V and II and VI. Table 2 shows the relevant groups of the periodic table. The III–V compounds have received most attention although the others have certainly not been neglected.

The well known elements of group IV have the diamond type of crystal structure, the atoms joined by covalent bonds. In the III–V and II–VI compound the crystal structure is also of the diamond type but the bonding has a more ionic character. This produces a larger forbidden energy gap. Since the II–VI compounds have bonding of greater ionic character, they have larger energy gaps.

The properties of many of the III–V compounds are shown in Table 3. The compounds of this class will probably replace germanium and silicon in high-frequency devices such as microwave and switching diodes. Indium antimonide has a comparatively high electron mobility and this is now used for Hall effect devices since the power available is approximately 300 times greater than for germanium. Indium antimonide is also used in infra-red detectors.

The great variety of the compound semiconductors forbids any comprehensive survey. For further information about the III–V compounds and details of the other classes, the reader is referred to the bibliography. From the books listed there an exhaustive treatment of all types may be obtained.

*See also*: Semiconductor devices.

*Bibliography*

Dunlap W. C. (1957) *An Introduction to Semiconductors*, New York: Wiley.
Ioffe A. F. (1960) *Physics of Semiconductors*, London: Infosearch.
Putley E. M. (1960) *The Hall Effect and Related Phenomena*, London: Butterworths.
Shephard A. A. (1957) *An Introduction to the Theory and Practice of Semiconductors*, London: Constable.
Smith R. A. (1961) *Semiconductors*, Cambridge: The University Press.

J. A. Coekin

**SHOCKED METALS.** Studies of the effects of high intensity shock waves on metals date from quite recent times. The first systematic investigations were reported in the mid 1950's, and since that time an acceleration of interest is evident from the number of publications appearing in the open literature. For convenience, the work to date may be classed under two headings; that concerned with the strictly metallurgical effects of shock waves on metals, e.g. with changes in mechanical strength, dislocation substructure etc., as a consequence of shock loading, and that primarily concerned with the shock-induced phase transformation in iron.

The macroscopic physical situation in a material experiencing a high intensity pressure pulse is very different to that in material being deformed by any of the more conventional techniques. In the former, an element of the material, initially, say, at atmospheric pressure and temperature will feel the shock front as a pressure and temperature discontinuity with a rise time of about $10^{-10}$ sec, the pressure, which may be up to 500 times the static yield stress of the material, perhaps remaining constant for up to a few microseconds and then decaying to its original value at a rate considerably less than that in the shock front. During this isentropic expansion, the temperature of the element falls to a value less than that at the peak pressure but not to atmospheric temperature.

After the experience of such high pressure and extreme pressure gradients, it would be anticipated that the terminal properties and structure of shocked metals would be greatly different from those resulting from conventional deformation. Observations made to date indicate that in many ways the results of shock loading are not exceptional. However, the very short time characteristics of the shock wave result in the operation of deformation mechanisms which also occur rapidly. Thus deformation twinning is a commonly observed feature of shock loaded metals, not only in body-centred cubic and hexagonal close-packed metals and alloys where it is quite commonplace but also in face-centred cubic materials where it is normally only produced by conventional deformation at low temperatures.

From the earliest observations, it was evident that the passage of a high intensity shock wave results in considerable hardening of metals and alloys. This shock hardening, which has been found of commercial application, would imply the generation and multiplication of large numbers of dislocations in the material as the shock wave passes through. Observations in the electron microscope confirm this, as high densities of dislocations, usually arranged in subcell walls, are observed in shock loaded metals. However, high intensity shock waves are supersonic in velocity (at least with respect to the uncompressed state), whereas the edge and screw dislocations of conventional crystal defect theory are unable to travel faster than the transverse elastic sound velocity. Much debate has centred around the mechanisms whereby conventional dislocations can be generated and operate under the influence of a supersonic pressure pulse. As an alternative to the postulation of special supersonic dislocations, the idea that large numbers of dislocations are generated in the front, and then move small distances appears to be most in favour. However, tentative proposals have recently been presented, on the basis of observations of rapid disordering of an ordered alloy at a critical shock pressure, that deformation mechanisms other than those involving dislocations may arise at high shock pressures.

As appears most likely, mechanisms of deformation in a shock front involving conventional dislocations would require these latter to move at near-sonic velocities. It has been calculated that at such speeds, dislocations behave in quite unusual ways. The formation of superdislocations of multiple Burgers vector occurs and dislocations of opposite sign on the same slip plane are not self-annihilating. Also, it has been pointed out that point defect generation by jogged screw dislocations is much more efficient when the latter move at near-sonic velocities. In applying these ideas to metals after shock loading, it should be recalled that quite high transient temperatures are produced in the shock pulse and that the temperature does not return to its original value on expansion of the compressed material. The structure and properties found in terminal observations on shocked material will undoubtedly have been modified by this heating, and, at least partly, will depend on the efficiency of quenching of samples after shock loading.

The first experimental studies of the effects of shock waves on metals and alloys were fairly uncontrolled experiments employing non-planar shock waves and often involving complex reflections and interactions of these waves. The method of generation of the shock pulse usually involved direct explosive-sample contact, the pulses produced being of low pressure by present day standards. The experiments of Smith, reported in 1958, constituted the first metallurgical study of samples subjected to essentially planar waves where adequate precautions were taken to avoid undesired shock reflections and interactions within the samples. By the use of surrounding spall plates and momentum traps, and by generating the shock wave by impact of an explosively driven flat plate, planar shock waves of up to 600 kbar peak pressure were passed through various annealed metals and alloys, the samples being recovered undamaged by deceleration (and quenching) in water. Hardening of copper and iron samples more than that obtained by cold-rolling to 95 per cent reduction was observed, although the dimensions of the samples remained virtually unchanged and the grain structure was completely undistorted. Most of the subsequent studies in this field have also used planar shock waves generated by driver plate impact with precautions to avoid shock reflections in the samples. When these precautions are adequate, the observation of constant sample shape during plane shock wave loading has been repeatedly made. On the other hand, the subjection of samples to non-planar shock waves can yield confusing results. For example, foils of copper shock loaded to approx. 28 kbar by a divergent spherical wave have structures, both in terms of dislocation density and incidence of deformation twinning, akin to bulk samples subjected to planar waves of 150 kbar intensity. The exact way in which gross deformation by shock waves differs from planar wave compression is not at present understood.

In addition to studies of the hardening effects in shock loaded metals the characteristics of the deformation twins formed in such materials has received much attention. Thus, by the use of shock waves reflected obliquely at a free surface, it has been deduced that mechanical twins are formed on the rarefaction side of the shock wave in copper and brass, but on the compression side in iron. Further studies of single crystals of these and other metals subjected to planar shock waves have indicated that no general pattern exists for the twin formation event in the shock wave, even on the basis of the crystal structure of the metal.

The last five years have seen a variety of techniques applied to an increasing diversity of shocked materials. In general, the effects of shock waves have been compared with those of conventional deformation, and changes in substructure as observed by transmission electron microscopy and X-ray line broadening, in indentation hardness and tensile properties, in stored energy and in long range order in an ordered alloy have all been studied. It has been repeatedly observed that the overall action of shock waves, as evidenced by measurements and observations of these terminal properties is not exceptional—changes on shock loading may occur rather more spectacularly but are still only of the usual magnitude. However, it has recently become evident that the detailed results of shock loading are sometimes quite different to those of conventional deformation. Thus, tensile studies of copper and nickel sheets shocked in plane wave geometry have shown that the dislocation structure after shock loading is a highly interlocked one, which gives rise to discontinuous yield point effects in subsequent tests. Again, the limited annealing studies of shock loaded metals reported to date have shown that large numbers of unstable defects exist in the shocked material. On annealing, the presence of these is evidenced by considerable stored energy and hardness decreases before recrystallization. Further, the annealing textures of polycrystalline metal sheets may be very much modified by shock loading and subsequent recrystallization. Grain refinement and randomization of texture appear to be possible and much work is currently in progress to study these effects. The commercial aspects of this work may be important in the future.

In most of the studies reported to date, the only parameter of the shock wave given has been the peak pressure reached in the shock. Recent experiments on copper have shown however that this pressure is but one of the parameters which can influence the terminal properties of the material. For example, the time of application of the peak pressure appears to be of equal importance. On this basis it is evident that the details of the shock pulse must be controlled and measured for fundamental deductions regarding the shocked material to be possible.

Dating from the work of Bancroft and associates (1956) which revealed that the Hugoniot curve for iron exhibited a discontinuity at 130 kbar at ambient temperature, there has been much interest and speculation as to the origin of this effect. As a consequence of the discontinuity, a shock wave of magnitude of greater than 130 but less than about 330 kbar splits into two plastic waves of different velocities, the faster wave of 130 kbar being followed by a slower wave of the magnitude of the original pulse. (In addition, there is the small elastic precursor corresponding to the value of the dynamic yield stress.) The main concensus of opinion until recently has regarded the effect as probably arising from the familiar body-centred cubic to face-centred cubic phase transformation which occurs in iron at 910°C and one atmosphere pressure. Thus it was thought that this was suppressed in temperature by the pressure so that it occurred at about 330°C at 130 kbar. However as this temperature is still rather higher than calculations would forecast to occur in the shock front, much doubt surrounded the exact nature of the transformation. Optical metallography reveals that while mechanical twins or Neumann Bands only are seen at pressures less than 130 kbar, the structure becomes more complex, with greater density of twinning and the appearance of mottled markings between the twins at higher pressures. In addition a marked increase in the rate of hardening with shock pressure occurs at 130 kbars. The complexity of microstructure increases progressively in the pressure range of the two-wave region, but at pressures greater than about 310 kbar, simplifies again, though it is much finer in scale than at less than 130 kbar. Analyses of the crystallography of the twins formed at all pressures have shown them to be quite conventional, lying on {211} planes, as do the mottled markings produced at pressures in the two-wave region. The latter have been shown not to be twins as their formation is not inhibited, as that of twins is, by prior deformation.

In addition to work on iron of various purities, studies have been made of the response to shock waves of a wide variety of steels. The observations show that the effects arising from the pressure induced phase transformation are essentially dependent on the amount of $\alpha$-iron in the alloy, for the transformation takes place in this phase only. The twinning of this phase during shock loading is also unaffected in morphology by the other phases present, for mechanical twins are seen, for example, to propagate across the duplex $\alpha$-iron/iron carbide pearlite structure, apparently uninterrupted by the carbide platelets.

The examination of shock loaded irons by transmission electron microscopy has revealed that the mottled structure of metal shocked into the two-wave region consists of plates separated by high angle boundaries with high dislocation densities within them, suggestive of the product of a shear transformation. Indeed very similar structures are given by the conventional $\gamma \to \alpha$ transformation in carbon free alloy steels. These observations are not, of themselves, proof that the $\alpha$ iron has transformed to the face-centred cubic $\gamma$ phase during the passage of the shock wave.

Although from time to time alternatives to the $\alpha \to \gamma$ transformation have been proposed as responsible for the 130 kbar discontinuity in iron—for example, the Curie point transition has been suggested as a possibility—the experiments of Johnson and associates were the first to indicate that the phase transition at 130 kbar and ambient temperature was probably not the conventional body-centred cubic to face-centred cubic transformation. For by shock attenuation experiments at elevated temperatures, a temperature pressure diagram with a discontinuity at 115 kbar and 775°K was derived. On the basis of this and metallographic observations, it was concluded that below 775°K the high pressure phase was not face-centred cubic. Subsequently it has been shown that this phase is hexagonal-close-packed, designated epsilon. The transformation details have still to be determined.

Subsequent to the establishment of the existence of a shock induced phase transformation in iron, a polymorphic phenomenon on shock loading has been found in bismuth as well as in various non-metals. As yet, little is known of the metallurgy of this transformation.

Recently, shock waves have begun to be applied as a commercial tool, for example in metal forming and hardening operations, and as a method for the further study of properties or reactions in metals and alloys already known from more conventional experiments. Because of the ease of twin formation on shock loading, the morphology of the latter is readily amenable to study in shocked samples. Recently the mechanics of a commercially important hardening process, the ausforming process, have been elucidated by the use of shock waves.

Future work will no doubt involve similar applications of shock wave techniques to the study of phenomena in metal systems, in addition to further more comprehensive investigations of the structure and properties of metals and alloys after shock loading.

*See also*: Shock waves in solids.

*Bibliography*

Conference (1961) *Response of Metals to High Velocity Deformation*, New York: Interscience.
DIETER G. E. (1962) *Strengthening Mechanisms in Solids*, Amer. Soc. Metals.
APPLETON A. S. (1965) *Applied Materials Research*.

A. S. APPLETON

**SHOCK WAVES IN SOLIDS.** Solids in contact with detonating high explosives or involved in high velocity projectile impact have transient stresses of the order of several hundred kilobars induced within them. Work dating from the second world war has shown that since these stresses are very much greater than the shear strength of solids the stress system is effectively isotropic and thus equivalent to a hydrostatic pressure. Hence transient flow problems in solids may be described by the hydrodynamic treatments appropriate to fluids. In particular since the compressibility of solids decreases with increasing pressure a simple compressive stress wave builds up o a *shock wave* across which the dynamic and thermo-dynamic quantities comprising the flow variables change discontinuously. Similar considerations show hat a rarefaction or stress decreasing wave front preads out as it propagates and in general a negative r pressure decreasing shock cannot form.

A shock transition is described by the *Rankine–Hugoniot equations* relating the change in pressure $P$, pecific volume $V$ or density $\varrho = 1/V$, specific internal nergy $E$ and particle velocity $u$ across the shock front o the shock velocity $U$. If the variables ahead of the hock front are designated by subscript $o$ and velocities re measured relative to the material velocity ahead

From equation 4 it is seen that all $P$, $V$ states consistent with a given shock velocity $U$ lie on a straight line through $P_0$, $V_0$ of slope $-\varrho_0^2 U^2$. This is known as the Rayleigh line connecting the end states of a shock transition on the Hugoniot curve. The adiabat through $P$, $V$ passes between the Rayleigh line and the Hugoniot curve. The velocity of sound is $C = (\mathrm{d}P/\mathrm{d}\varrho)^{\frac{1}{2}}$ along the adiabat and at $P$, $V$ it follows that $C + u > U > C_0$ expressing the fact that a shock is supersonic with respect to the material ahead of it and subsonic with respect to the material behind it. From equation 3 the increase in internal energy in a shock transition is the area in the $P - V$ plane below

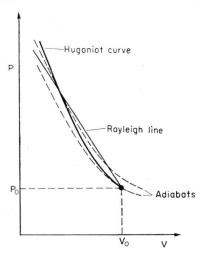

Fig. 1. *Form of Hugoniot curve and intersecting adiabats.*

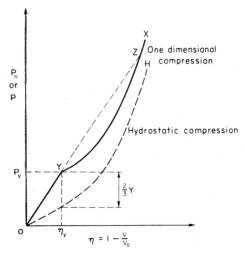

Fig. 2. *Comparison of one-dimensional and hydrostatic compressions of solids.*

of the shock then conservation of mass, momentum and energy yields,

$$\varrho_0 U = \varrho\,(U - u) \tag{1}$$

$$P - P_0 = \varrho_0 U u \tag{2}$$

$$E - E_0 = \tfrac{1}{2}(P + P_0)(V_0 - V) \tag{3}$$

Since in principle $E$ is a unique function of $P$ and $V$, equation 3 defines a unique curve in the $P$-$V$ plane, called the *Hugoniot curve*, which is the locus of all possible states attainable by a single shock transition from the given initial state $P_0$, $V_0$. The curve has the form shown in Fig. 1. It can be shown that the increase in entropy across a shock when expressed in terms of shock strength, $(P - P_0)$ or $(V_0 - V)$, is a function only of third and higher order terms. It then follows that the Hugoniot curve makes second order contact with the adiabat through $P_0$, $V_0$ and then diverges to regions of higher entropy. Equations 1 and 2 may be re-written as,

$$U = V_0[(P - P_0)/(V_0 - V)]^{\frac{1}{2}} \tag{4}$$

$$u = [(P - P_0)(V_0 - V)]^{\frac{1}{2}} \tag{5}$$

the Rayleigh line while from equation 5 the change in kinetic energy $\tfrac{1}{2}u^2$ is the area below the Rayleigh line down to the abscissa $P_0$. When $P_0$ is negligible the two energies are equal.

For the application of these concepts to plane stress waves in solids the Rankine–Hugoniot equations remain applicable if the hydrostatic pressure $P$ is replaced by the stress normal to the shock front $P_n$. The stress system is completely specified by including the transverse stress $P_t$ and the normal strain $\eta = 1 - V/V_0$. Since all displacements are normal to the wave front transverse strains are zero. The elastic response of the material, initially stress free so that $P_0 = 0$, to a one-dimensional compression is described by,

$$P_n = (K + 4\mu/3)\,\eta, \quad P_n - P_t = 2\mu\eta \tag{6}$$

where $K$ and $\mu$ are the bulk and shear moduli respectively. The hydrostatic pressure giving the same volume charge is,

$$P = K\eta \tag{7}$$

and since compressibility experiments show that $K$ increases slowly with $P$ this defines a convex downward curve in Fig. 2. Equation 6 holds so long as a yield criterion is not violated. For this system the *Tresca and von Mises criteria* give the same expression,

$$P_n - P_t \leq Y \quad \text{corresponding to} \quad P_n \leq P_y,$$

$$\eta \leq \eta_y \quad \text{or} \quad V \geq V_y$$

where $Y$ is the yield strength in simple tension. When the material is stressed beyond the yield point it is assumed that the anisotropic stress component $(P_n - P_t)$ remains constant and the shear modulus decreases with further strain such that for $\eta > \eta_y$, $2\mu\eta = Y$ and hence,

$$P_n = K\eta + 4\mu\eta/3 = P + 2Y/3. \tag{8}$$

Plotting this behaviour in Fig. 2 the material deforms elastically up to the point $Y$. Here the material yields, plastic flow begins and with further stress the material follows the path $YZX$ displaced upwards from the hydrostatic curve $OH$ by an amount $2Y/3$.

Stresses in the elastic range are propagated with a velocity obtained by substituting equation 6 in equation 4 giving the constant value $[(K + 4\mu/3)/\varrho_0]^{\frac{1}{2}}$ which is the usual expression for the velocity of longitudinal sound waves in solids. At the point $Y$ the compressibility of the material suffers a discontinuous increase which violates the conditions for shock stability so that stresses just above the yield point cannot be transmitted as a single shock step. The stress wave separates into two steps with an elastic precursor wave carrying the stress $P_y$ followed by a plastic shock moving with a velocity given by equation 4 with $P = P_n$, $P_0 = P_y$ and $V_0 = V_y$ relative to the material behind the elastic wave. At the point $Z$ where the Rayleigh line for the plastic shock is a continuation of the elastic line $OY$ the velocity of the plastic shock becomes equal to the elastic precursor velocity and stresses above the point $Z$ are carried as a single shock with a velocity given by equation 4 with $P = P_n$, $P_0 = 0$.

The values of $Y$ obtained from the measurements of elastic precursor wave amplitudes $P_y$ may differ by factors of two or three from static measurements, so the point $Y$ in Fig. 2 is often called the Hugoniot elastic limit to distinguish the type of situation to which it pertains. The order of magnitude of $Y$ under dynamic conditions ranges up to about 10 kb so that with normal stresses above about 100 kb, equation 8 shows that shear strength becomes unimportant. It is then permissible to regard $P_n$ as equivalent to a hydrostatic pressure and treat the solid as a fluid.

*Observations.* A shock wave moving into material in a specified state is defined by the variables $P, V, E, u$ and $U$ and a measurement of any pair is sufficient to determine the remainder through the Rankine–Hugoniot equations. The purely thermodynamic quantities are difficult to measure within their duration times of about a microsecond. Transducers are being developed for the direct measurement of pressure

in the several hundred kilobar range and though they perturb the system and suffer from calibration difficulties they are showing promise. High speed radiographic density determinations are inaccurate due to edge effects unavoidable in practical systems. The purely dynamic quantities $U$ and $u$ are the variables generally amenable to experimental determination. Shock velocities may be derived from observation of the time of arrival of a shock disturbance at preset positions using shock actuated probes which generate signals for display on time-calibrated high velocity oscillographs or by observing their arrival at interfaces using high-speed optical or radiographic methods.

The direct measurement of particle velocity immediately behind a shock front is not currently feasible but it is easily inferred in simple systems from interface velocities which may be obtained in a similar manner to shock velocities. The methods rely on the boundary conditions following a shock interaction at an interface which demand that the transmitted and reflected wave amplitudes are such that pressure and particle velocity are continuous across the interface. In the "deceleration" method a projectile plate is driven against a target plate of the same material. The collision generates identical transmitted and reflected shocks and symmetry demands that the particle velocity increment behind these shocks is half the impact velocity. Another approach is to measure the velocity of the free surface of a plate due to the reflection of a shock at normal incidence. At the surface the pressure is reduced to effectively zero by a reflected rarefaction wave and the free surface velocity is the sum of the incident particle velocity and the velocity increment in the same direction due to the rarefaction wave. As a consequence of the fact that the change in entropy through a shock is of third order in shock strength it can be shown that the particle velocities associated with shock and rarefaction waves of the same amplitude differ only in third and higher order terms which in many solids are not significant until the pressure is of the order of a megabar. It therefore follows that the two components of particle velocity are nearly equal and to a close approximation free surface velocity is twice the incident shock particle velocity. This is known as the "free surface velocity approximation". In a third method called the "impedance match method" a shock of known strength is transmitted across an interface from a material of known shock properties to the material whose shock properties are to be determined. In the $P-u$ plane the loci of possible reflected states in the known material may be plotted and a measurement of transmitted shock velocity used in conjunction with equation 2 is sufficient to determine the transmitted state satisfying the boundary conditions.

Shock data have been obtained in this way for a wide variety of solids including metals, alloys, ionic and molecular crystals, geological materials and plastics. Over a wide range of pressure within a given phase the data show an empirical linear relation,

$$U = U_0 + Su \quad (9)$$

where $U_0$ is the limiting plastic wave velocity $(K/\varrho_0)^{\frac{1}{2}}$ and $S$ is the slope, discussed below. The physical reason for this relation is not at present understood. Consequent upon the relation the Hugoniot curve takes the simple analytic form,

$$P = \frac{\varrho_0 U_0^2 \eta}{(1 - S\eta)^2}, \quad E - E_0 = \frac{1}{2}\left(\frac{U_0 \eta}{1 - S\eta}\right)^2. \quad (10)$$

Shock induced phase changes are generally indicated by a change in slope of the $U$–$u$ curve. If the transition is to a much more compressible phase then a situation analogous to the elastic-plastic transition exists and the shock separates into two steps. In addition the retransformation process in a subsequent rarefaction wave can result in the formation of a negative shock.

The application of shock wave data to a more complete thermodynamic description of high pressure states is made by assuming a form of equation of state and using the data to define unknown parameters. The equation of state usually used is the Mie–Gruneisen form,

$$P(V, E) = P_L(V) + (E - E_L(V))\gamma(V)/V \quad (11)$$

where suffix $L$ refers to the lattice contribution at $0°$K to the total energy and pressure and $P_L = -dE_L/dV$. $\gamma(V)$ is the Gruneisen ratio relating the thermal pressure to the thermal energy. It has the thermodynamic definition $\gamma(V) = V(\partial P/\partial E)_v = \alpha K/\varrho C_v$ where $\alpha$ is the volume coefficient of thermal expansion and $C_v$ is the specific heat at constant volume. This gives the initial value $\gamma(V_0)$. $\gamma(V)$ is also theoretically related to the variation of lattice frequencies with volume resulting in expressions for $\gamma$ of which the one due to Dugdale and MacDonald is slightly preferable,

$$\gamma(V) = -\frac{1}{2}V(P_L V^{2/3})''/(P_L V^{2/3})' - \frac{1}{3} \quad (12)$$

where the prime denotes differentiation with respect to $V$. Experimental values of $P$ and $E$ as functions of $V$ along the Hugoniot are used to find the functions $P_L$, $E_L$ and $\gamma$ consistent with equations 11 and 12. These functions can then be used to define the variation of $P$ with $E$ for all values of $V$ covered by the data.

The adiabats which intersect the Hugoniot curve and govern rarefaction wave behaviour are obtained by finding loci of states consistent with $P = -dE/dV$. Temperatures $T$ along the adiabats are derived by integrating $\gamma(V) = -d \ln T/d \ln V$ with initial conditions along $P = 0$ given by thermal expansion data. In this way temperatures of all relevant states may be determined including shock temperatures.

With the justification that at room temperature lattice effects are dominant, applying equation 12 to the Hugoniot curve equation 10 at $P = 0$ gives the

*Hugoniot curve constants and typical shock properties for some common metals* $(T_0 = 20°C)$

| Metal | Mg | Al | Sn | Cu | W |
|---|---|---|---|---|---|
| $\varrho_0$ (gm/cm³) | 1·73 | 2·70 | 7·28 | 8·90 | 19·17 |
| $U_0$ (mm/$\mu$s) | 4·48 | 5·38 | 2·64 | 3·96 | 4·01 |
| $S$ | 1·273 | 1·337 | 1·476 | 1·497 | 1·268 |
| $\gamma_0$ | 1·46 | 2·18 | 2·11 | 2·00 | 1·54 |
| $P$ (kb) | 251 | 317 | 376 | 425 | 497 |
| $V/V_0$ | 0·711 | 0·790 | 0·731 | 0·831 | 0·883 |
| $T$ (°C) | 690 | 440 | 840 | 340 | 200 |
| $U$ (mm/$\mu$s) | 7·09 | 7·48 | 4·38 | 5·31 | 4·71 |
| $u$ (mm/$\mu$s) | 2·05 | 1·57 | 1·18 | 0·90 | 0·55 |

interesting correlation $\gamma = 2S - 1$. For all the metals examined the experimental value of $S$ predicts $\gamma(V_0)$ to within 35 per cent with an average deviation of 15 per cent. Constants defining Hugoniot curves for some common metals are listed in the table together with parameters defining the shock induced in the metals in contact with a typical high explosive, composition B.

Techniques using strong shock waves have been used to subject solids to pressures up to 10 Mb, two orders of magnitude larger than those attainable by static methods. So far work has been mainly confined to measuring dynamic quantities and deriving thermodynamic properties. The measurement of other properties as a function of shock pressure awaits the development of suitable techniques adapted to the short observation times available and the destructive nature of the experiments. Topics such as shock induced phase transitions, dynamic fracture and metallurgical behaviour related to the high rates of strain involved are currently receiving attention. Problems associated with shock front thickness and microscopic deformations within shock fronts remain largely unresolved.

*Bibliography*

ALDER B. J. (1963) *Physics Experiments with Strong Pressure Pulses* in *Solids Under Pressure* (Eds. Paul W. and Warschauer D. M.), New York: McGraw-Hill.

DEAL W. E. (1962) *Dynamic High Pressure Techniques* in *Modern Very High Pressure Techniques* (Ed. Wentorf R. H.), London: Butterworths.

DUVALL G. E. (1961) *Properties and Application of Shock Waves* in *Response of Metals to High Velocity Deformation* (Eds. Shewmon P. G. and Zackay V. F.), New York: Interscience.

RICE M. H. *et al.* (1958) *Compression of Solids by Strong Shock Waves*, Solid State Physics **6**, 1.

I. C. SKIDMORE

**SMITH–PURCELL EFFECT.** The Smith–Purcell effect may be described as the interaction of a free, relativistically accelerated electron beam and a fine, conducting periodic structure such as a diffraction grating. The effect differs from similar microwave devices such as the Backward-Wave Oscillator in that radiation initially arises from charges induced on the surface of the periodic structure.

To demonstrate the effect an energetic, collimated electron beam is fired at grazing incidence over the surface of a metal diffraction grating with many hundreds of lines per inch rules in it perpendicular to the direction of the beam. The electrons passing over the grating, which is maintained at constant potential, induces positive charges which run along the undulating surface. These oscillating charges may be considered as moving Hertzian dipoles giving rise to electromagnetic radiation of the same frequency or frequencies as the Fourier modes of the oscillating charges, subject to a Doppler shift correction. A simple

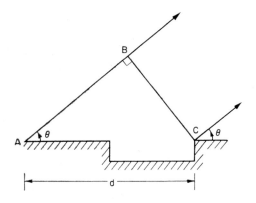

Huygens construction shows that a coherent wavefront such as BC in the figure arising from the passage of one electron from A to B must have a wave-length $\lambda$ given by

$$n\lambda = d(\beta^{-1} - \cos\theta), \qquad (1)$$

where $\beta$ is the velocity of the electrons expressed as a fraction of the velocity of light, and $n$ is the mode of the oscillation. ($n = 1$ for a grating with sinusoidal groove form.)

The above relation may also be derived by considering the relativistic Doppler shift of the frequency of oscillation of the induced charges when transformed to the laboratory frame of reference.

Equation 1 shows an interesting contrast with the fundamental relationship describing the Cerenkov effect, this being the hypothetical case of $n = 0$.

The output power in the first harmonic when a grating with a symmetrical squarewave groove form is used can be shown to be

$$50 \frac{iLa^2\beta^3}{d^4} \frac{(\cos\theta - \beta)^2}{(1 - \beta\cos\theta)^6} \mu \text{ watts sterad}^{-1}; \qquad (2)$$

where

  $L$ is the length of the grating in cm
  $a$ is the groove depth in cm
  $d$ is the grating pitch in cm
  $i$ is the effective inducing current in amp.

This shows the distribution of power in space to be a relativistic distortion of the familiar $\cos^2\theta$ lobe of any radiating dipole. As in the case of the Cerenkov effect this is a "shock wave" phenomenon and coherent wavefronts can only be produced in the forward direction.

Equation 1 shows that the spectral bandwidth of the emission will depend on the accuracy of the grating's rulings, the distribution of electron velocities and the magnitude of the cone within which radiation is accepted. There is an ultimate limit in spectral purity in that it is inherent in the nature of the effect that no emitted wave train arising from a given electron can be longer than $L/\beta$, i.e. the coherence time cannot be more than the time of passage of an electron over the rulings.

The output power is extremely low owing to the random arrival of the electrons at the grating (shot noise). This results in the separate wave trains from each induced charge adding incoherently. In the original experiment by Smith and Purcell, using a 350 keV electron beam and a 15,000 lines per in. grating, the estimated energy output was only 10 photons sterad$^{-1}$ for every mm of electron path. The visible radiation detected ($\sim 5000$ Å) was analysed by a second diffraction grating, and shown to obey equation 1 qualitatively. As would be expected, the radiation was strongly polarized with the electric vector perpendicular to the plane of the grating. Other experiments using lower voltages but finer gratings have demonstrated visible and near ultra-violet emission, and the radiation has successfully been amplitude and frequency modulated at source. While the prospect of a tunable, monochromatic source for the mid or far intra-red is attractive, equation 2 suggests that using manageable accelerating voltages the output power would be far less than that required for spectroscopy or communications, and this has been demonstrated by experiments.

It appears at the time of writing that the unmodified effect has found no useful applications; however, various proposals have been made to increase the power output. One suggestion is to use an electron beam which is sinusoidally modulated at infra-red frequencies. The arrival of electrons at the grating would then no longer be random, and the resulting coherence should provide much improved power outputs.

*Bibliography*

(1953) *Phys. Rev.* **92**, 1069.
N.P.L. Divisional Report, No. L1 8.
(1964) *Proc. I.E.E.E.* **52**, 4, 429.

P. B. CLAPHAM

**SNAP (SYSTEMS FOR NUCLEAR AUXILIARY POWER).** The SNAP programme was initiated in 1955 by the U.S. Atomic Energy Commision, to develop nuclear auxiliary power units for the American Space Research programme, using reactors and radioisotopes as an energy source with thermoelectric and turboelectric conversion systems. The programme was subsequently extended to include the design and construction of radioisotope fuelled generators for terrestrial use.

*Radioisotope generators.* The decay of a radioisotope, with the production of energetic charged particles and $\gamma$-radiation, results in the release of a considerable amount of energy. A variety of methods have been adopted to convert this energy into a more usuable form, but the most successful approach, and the one that has been adopted in the SNAP generators, is to allow the particles to dissipate their energy within the parent isotope to produce heat which is subsequently converted into electrical energy by thermoelements.

The efficiency of the conversion of the decay energy into thermal energy depends upon the absorption of the particles within the isotope. This is most efficient in the case of $\alpha$-emitters when all the $\alpha$ particles are stopped within the isotope and the can because of the small range of the $\alpha$ particle. $\beta$-emitters are a little less satisfactory because of the greater range of the $\beta$ particles, but their biggest drawback is the production of bremsstrahlung during the slowing down of the $\beta$ particles, which requires heavy shielding. Because of the low attenuation of $\gamma$-radiation, and hence the low density of energy deposited per unit volume, this form of decay energy is unsatisfactory for this type of energy conversion system.

Selection of a suitable isotope for a given generator involves the consideration of many aspects, the more important of which include the energy density (W/g), half-life, nature of activity, general availability, and cost and existence of a form or compound of the isotope that is suited to the proposed operating conditions. Out of the many hundreds of known radioisotopes, when the various criteria of selection are applied, only ten or less appear promising for use in generators. Table 1 gives a selection of these with some of the properties of interest.

These isotopes fall into two categories, the $\beta$- and $\beta$-$\gamma$-emitters which are products from the fission of uranium in reactors, and the $\alpha$-emitters which are obtained by specifically irradiating a starting material in a reactor. Because of their origin, the former are potentially available in greater quantities at less cost than the latter, although it is necessary to separate them from other fission products with less desirable characteristics.

The bremsstrahlung associated with the $\beta$-emitters requires shielding for the protection of personnel and radiation sensitive equipment, and accordingly this type of radioisotope is not well suited to space applications, where the shielding represents a weight penalty. Because of the cost and scarcity of $\alpha$-emitters, however, consideration is being given to the use of Sr 90 in SNAP 17 which is intended for use in a Communications Satellite. In orbit, the generator could be some distance from the satellite body and a relatively light "shadow shield" used. The $\beta$-emitters are much more suited to terrestrial application where the weight of necessary shielding can be tolerated.

Radioisotopes can represent a severe health hazard if not contained and shielded correctly. To ensure that no contamination or radiation hazard could arise in the event of a launch abort when a $Pu_{238}$ fuelled SNAP generator is involved, the generators have been subjected to rigorous testing under simulated conditions of all possible launch aborts.

*Conversion systems.* In the SNAP generators, the decay heat of the fuel is converted into electrical energy by means of semiconductor thermoelements in close contact with the fuel container, the remainder of which is well insulated to ensure maximum heat flow down the thermoelements.

The cold junctions of the thermoelements are provided with a heat rejection system suited to the particular environment.

*Table 1. Characteristics of isotopic heat sources*

| | Strontium 90 | Caesium 137 | Promethium 147 | Plutonium 238 | Curium 244 | Curium 242 | Polonium 210 | Cerium 144 | Cobalt 60 |
|---|---|---|---|---|---|---|---|---|---|
| Type of decay | Beta | Beta-Gamma | Beta | Alpha | Alpha | Alpha | Alpha | Beta-Gamma | Beta-Gamma |
| Half life, years | 28 | 30 | 2.7 | 89 | 18 | 0.45 | 0.38 | 0.78 | 5.3 |
| Specific power of isotope, watts (thermal)/g | 0.90 | 0.42 | 0.33 | 0.56 | 2.8 | 120 | 141 | 25.6 | 17.4 |
| Estimated isotopic purity, % | 50 | 35 | 95 | 80 | 98 | 90 | 95 | 18 | 10 |
| Typical fuel form | SrO | Glass | $Pm_2O_3$ | $PuO_2$ | $Cm_2O_3$ | $Cm_2O_3$ | Metal | $CeO_2$ | Metal |
| Active Isotope in Compound, % | 42 | 16 | 82 | 71 | 89 | 82 | 95 | 15 | 10 |
| Specific power of compound, watts (thermal)/g | 0.38 | 0.067 | 0.27 | 0.39 | 2.49 | 98 | 134 | 3.8 | 1.7 |
| Density of compound, gm/cc | 3.7 | 3.2 | 6.6 | 8.9 | 10.6 | 11.75 | 9.3 | 6.4 | 8.9 |
| Power density of compound, watts (thermal)/cm³ | 1.40 | 0.21 | 1.8 | 3.5 | 26.4 | 1150 | 1210 | 24.5 | 15.5 |
| Shielding requirement | Heavy | Heavy | Minor | Minor | Moderate | Minor | Minor | Heavy | Heavy |
| Emission requiring shielding | Bremst'ng | Gamma | | | Neutron | | | Gamma | Gamma |

Reference: Rohrmann, C. A., "Radioisotopie Heat Sources," IIW-76323 Rev. 1. October 15, 1963.

ex TID 200079

20*

*Table 2. Systems for nuclear auxiliary power—isotope power generators*

| Unit's designation | Use space applications | Power weight | | Isotope | Unit's design life | Status |
|---|---|---|---|---|---|---|
| | | W (electrical) | (lbs) | | | |
| SNAP-1 | Air force satellite | 500 | | Ce-144 | 6 mo | Programme cancelled |
| SNAP-1A | Air force Satellite | 125 | 175 | Ce-144 | 1 yr | Programme cancelled |
| SNAP-3 | Demonstration device | 2·5 | 4 | Po-210 | 90 days | Programme completed |
| Undesignated (modified SNAP-3) | Navy navigational satellites | 2·7 | 4·6 | Pu-238 | 5 yrs | 2 in space, 6/61 and 11/61, first in operation, second failed after 8 mo |
| SNAP-9A | DOD satellites | 25 | 27 | Pu-238 | 5 yr mission | 2 in space, 9/63 and 12/63, both in operation |
| SNAP-11 | Moon probe (surveyor) | 21–25 | 30 | Cm-242 | 90 day mission | Being tested† |
| SNAP-13 | Thermionic development demonstration device | 12 | 4 | Cm-242 | 90 day mission | Fabrication of fuelled unit Just initiated |
| SNAP-17 | Communication satellite | 25 | 28 | Sr-90 | 5 yrs | |
| SNAP-19 | IMP, Nimbus | 20 | 18 | Pu-238 | 5 yrs | Design study stage |
| Undesignated | Lightweight demonstration device | 6–10 | 3 | Pu-238/Sr-90 | > 1 yr | Being tested† |
| | Terrestrial applications | | | | | |
| Undesignated | Axel Heiberg weather station | 5 | 1680 | Sr-90 | 2 yrs minimum | Operating since 8/61 |
| SNAP-7A | Navigational buoy | 10 | 1870 | Sr-90 | 10 yrs | Operating since 1/64 |
| SNAP-7B | Fixed navigational light | 60 | 4600 | Sr-90 | 10 yrs | Operating since 11/63 |
| SNAP-7C | Weather station | 10 | 1870 | Sr-90 | 10 yrs | Operating since 2/62 |
| SNAP-7D | Floating weather station | 60 | 4600 | Sr-90 | 10-yrs | Operating since 1/64 |
| SNAP-7E | Ocean-bottom beacon | 6·5 | 6000 | Sr-90 | 10 yrs | Undergoing repairs |
| SNAP-15A/B | Nuclear weapons | 0·001 | 1 | Pu-238 | 4 yrs | Being tested† |
| SNAP-21 | Ocean bottom | 10 | — | Sr-90 | 5 yrs | Just initiated |
| Undesignated | Undersea seismograph | 5 | 500 | Cs-137 | > 1 yr | Terminated |
| Undesignated | Demonstration device | 5 | — | Mixed fission products | > 1 yr | Being tested† |

† Electrically heated generator.

In order to obtain maximum thermodynamic efficiency, the temperature difference between the hot and cold junctions should be as large as possible. In the environment of space, heat removal is solely by radiation which, owing to its temperature dependence favours a high rejection temperature. Since the hot junction temperatures is limited by the temperature at which deterioration of the thermoelement occurs, there is considerable incentive to develop materials capable of high temperature operation. For terrestrial applications, the incentive is not so great because of better cooling conditions available, particularly in Arctic and submarine regions.

Early space SNAP generators made use of *p*- and *n*- type lead telluride elements, the upper temperature of which was limited to about 450°C by sublimation. Such elements are still used in terrestrial SNAP generators. Recent developments in germanium silicon alloys now permit the use of hot junction temperature of 800°C.

The first radioisotope generator was built in 1958 by the Martin Company of Baltimore U.S.A. as a "proof of principle" model and designated SNAP 3. The heat from a 1760 curie polonium source was converted into electrical energy by means of 27 thermojunctions of *p*-type lead telluride doped with bismuth against *n*-type lead telluride doped with sodium. The output of each junction into a matched load was 93-25 mV when operating with a hot junction temperature of 1100°F and a cold junction temperature of 400°F. The full load voltage was 2·5 V with a power output of 3 watts which represents an overall efficiency of 4 per cent. This model was shown working to the President at the White House on January 13th 1959, nine months after the work had started.

A number of other generators have been built, which are listed in Table 2.

The SNAP reactor programme is aimed at developing a low power reactor that is capable of being started up and operated remotely, and which is simple, safe and reliable in its space environment. The initial concept is then used as a basis for successive units of higher power and improved performance. Units are identified by the title SNAP followed by an even number.

SNAP 2, with an output of 30 kW(electric) was the first reactor system to be initiated under the programme in 1956. In 1959, two other space reactor systems were introduced and designated SNAP 8 and SNAP 10. These had outputs of 30kW (electric) and 0·3 kW(electric) respectively. SNAP 10 was redesignated SNAP 10A in 1960, and its output increased to 0·5 kW(electric).

A terrestrial system, SNAP 4, using thermoelectric conversion to produce 1000–4000 kW (electric) was also initiated in 1960.

All systems are based on the same reactor concept of a combined fuelmoderator element of uranium–zirconium hydride and a beryllium reflector with modifications appropriate to the different power levels required.

In 1962, the SNAP 50 project of the U.S.A.E.C. and the *SPUR (Space Power Unit Reactor)* project of the U.S. Airforce were combined into the SNAP 50/SPUR project. This is to develop an advanced Rankine cycle/compact fact reactor system with an output of the order of 300–1000 kW(lectric) at an unshielded weight of 10–20 lbs/kW(electric) and a life of 10,000 hours.

The SNAP 2 reactor is a 50 kW (thermal) reactor using a homogeneous fuel-moderator of uranium–zirconium hydride with a beryllium reflector. Heat is extracted by liquid sodium potassium alloy (NaK) coolant operating into a heat exchanger with mercury as the working fluid in a turbine.

The core consists of a bundle of cylindrical fuel-moderator elements of uranium–zirconium hydrided to $6·5 \times 10^{22}$ hydrogen atoms per $cm^3$. Each element is clad in a thin walled steel tube with beryllium slugs at each end of the fuel elements to act as end reflectors; the steel tubes are coated internally to prevent hydrogen loss from the fuel elements. The core is contained in a vessel approximately $9''$ diameter and $20''$ long with an external beryllium reflector. Automatic control is effected by a thermo-mechanical link which rotates two semi-cylindrical beryllium drums and thus varies the effective thickness of the reflector.

Liquid NaK enters the core vessel at 1000°F and flows through the interstices between the fuel element tubes to exit at 1200°F; it then passes to a heat exchanger and vaporizes mercury which is expanded through a two-stage axial flow turbine driving an alternator; the mercury is condensed at 580°F in a radiation cooled condenser. The NaK pump, turbine and alternator are all mounted on a common shaft.

The weight of the entire system is about 600 lbs. with an output of 3 kW(electric) at 110 V, 2000 c/s, and is designed for 10,000 hr operation.

SNAP 8 was intended primarily as a power source for electric rockets, and used a reactor and conversion system based on the SNAP 2 concept, but operating at higher temperatures. The system was designed to deliver 60 kW(electric) for 10,000 hours using two 30 kW(electric) conversion units with the reactor operating at 600 kW(thermal). The same fuel-moderator as SNAP 2 is used but in smaller diameters ($0·6''$) to provide the higher power. The beryllium in the reflectors was replaced by beryllia since the operating temperatures were higher, i.e. 1350°K NaK outlet temperature.

SNAP 10 was designed to produce 300 W(electric), and used an entirely static thermoelectric conversion system in conjunction with SNAP 2 reactor operating at about 12 kW(thermal) and inlet and outlet temperatures of 900°F and 1000°F respectively.

The NaK coolant was circulated by an electro-magnetic pump which derives its current from a thermoelement operated by the temperature differential between the inlet and outlet pipes. $p$ and $n$ type lead telluride thermoelements convert the heat of the NaK into electrical energy.

The original SNAP reactor programme has undergone numerous changes as a result of changes in the associated space programmes, and the SNAP 2 system as a specific project has been cancelled but development is continuing.

The new systems objectives are shown in Table 3.

Development of SNAP 50 is in its early stages, and present thoughts are to use lithium coolant to transfer heat from uranium carbide (or perhaps uranium nitride) fuel elements at an outlet temperature of 2000°F into a boiling potassium turbogenerator. The use of high temperatures should reduce the size of radiator required with consequent saving of weight.

*Bibliography*

(Jan. 1964) *Systems for Nuclear Auxiliary Power—An Evaluation*, T.I.D. 20079.

J. L. CRASTON

Table 3.

| | SNAP 10A | SNAP 2 | SNAP 8 |
|---|---|---|---|
| Electric power | 0·5 kW | 3 kW | 35 kW min |
| Thermal power | 33·5 kW | 60 kW | 300–600 kW |
| Reactor outlet temp. | 1000°F | 1200°F | 1300°F |
| Reactor inlet temp. | 900°F | 1000°F | 1100°F |
| Life | 1 year | 1 year | 10,000 hrs |
| Power conversion | Si–Ge thermoelectric | Mercury Rankine cycle | 4 loop Rankine cycle |
| Specific weight | 1870 lb/kW (electric) | 490 lb/kW (electric) | c. 175 lb/kW (electric) |
| Fuel power density | 12·7 W/cm³ | 21·8 W/cm³ | 75–15 W/cm³ |

## SOLID CIRCUITS.

### 1. Introduction

Solid circuit is the name given to a complete electronic circuit, containing both passive and active components, fabricated within a small single crystal wafer of a semiconductor. It would probably avoid confusion with other fabrication techniques if the name Single Crystal Circuit (SCC) was used. As a direct result of this method of construction (outlined below) SCC's are very small, resulting in large component packing densities. For example, a wideband vol-

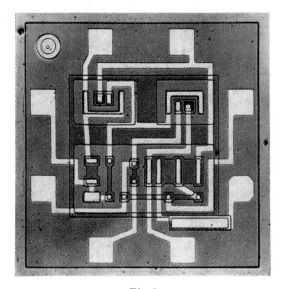

*Fig. 1.*

tage amplifier consisting of 6 transistors, 2 voltage reference diodes, and 14 resistors can be made in SCC form, in a wafer of silicon measuring 0·065″ × 0·065″ × 0·008″ (see Fig. 1).

The material most widely used in the construction of SCC's is in fact silicon, for the following reasons.

*a) Temperature limitations.* As it appears that the ultimate limitation on component packing density in a piece of electronics equipment will be set by the difficulty of removing the internally generated heat, silicon is preferred to germanium in view of the maximum operating temperatures for the respective devices (∼175°C for silicon, ∼80°C for germanium). This difference arises directly from the energy gaps of the two materials (1·1 eV for Si, 0·67 eV for Ge). In the future it is expected that gallium arsenide will be important in this context by virtue of its higher energy gap (1·35 eV) and low electron effective mass, both helping to extend the temperature range for devices up to ∼350°C.

*b) Reverse currents.* As we shall see below, reverse biased P-N junctions are frequently used in SCC's

to achieve electrical isolation between adjacent components. The lower energy gap (and hence higher intrinsic carrier concentration at room temperature) of germanium results in excessive leakage currents, compared with silicon.

*c) Device technology.* The method of preparing P-N junctions by diffusion, and the use of an oxide film on the surface as a barrier to unwanted diffusion, are the heart of SCC fabrication, and whereas these techniques work extremely well with silicon, using its thermal oxide ($SiO_2$) for diffusion masking, there is no correspondingly straightforward technique available with germanium.

*d) Surface passivation.* In order to achieve working SCC's it is obviously necessary that the fabrication technique is under sufficient control for ∼20 components on a single chip to be satisfactory simultaneously. The stabilizing of surface conditions on semiconductor devices by last-minute chemical treatment is not sufficiently reproducible to guarantee the high level of success required. On the other hand, it is found in the case of silicon that if the oxide film used for diffusion masking is left on the surface, it gives good device characteristics, with satisfactory stability. There is no comparable technique available for use with germanium.

For the above reasons, the following will be confined to a description of silicon SCC's from the points of how they are made, the design features, and their probable impact on the electronics industry.

### 2. The Fabrication of Single Crystal Circuits

*2.1. Mask making.* Given a circuit design, as shown in Fig. 2(a) the first stage is to design a layout of the required components on a small chip of silicon. This is shown in Fig. 2(b). This layout is then used to produce a sequence of patterns, each of which will be used at a particular stage in the process. These patterns are first of all drawn very carefully 250 times the final size. A photographic reduction of 10 : 1 is performed, followed by a further reduction of 25 : 1 at

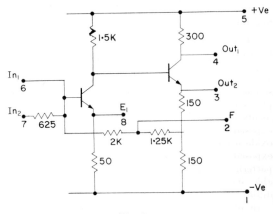

*Fig. 2 (a).*

which stage the final photographic plate is moved, after exposure, along X- and Y-directions in order to repeat the pattern some 150 times on a square $1'' \times 1''$. Contact prints are made from this master plate, and these are used in the processing of the silicon.

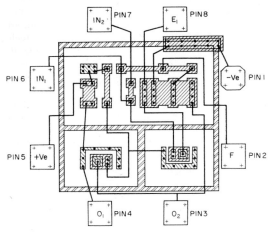

*Fig. 2 (b).*

*2.2. Material preparation.* There are several variations in the detailed technology of SCC fabrication, and only one of these is described below. Alternative techniques are described in the literature (see bibliography). A crystal of high resistivity ($\sim$100 ohm cm) P-type material is grown from the melt. This crystal is sliced, and the individual pieces are mechanically polished to give slices $\sim$1'' diameter and $\sim$0·010'' thick. These P-type slices are then used as the substrates for the iso-epitaxial deposition of a thin layer of N-type silicon ($\sim$10 microns thick, resistivity $\sim$0·5 ohm cm). All subsequent operations are performed on the N-type face of these slices.

*2.3. Oxidation.* Silicon slices are placed in a furnace at $\sim$1100°C. in an atmosphere of moist oxygen. Silica ($SiO_2$) films grown in this way (to a thickness of $\sim$1 micron) are suitable both for diffusion masking and for device passivation.

*2.4. Photoengraving.* The purpose of this operation is to transfer the pattern on the photomask (Section 2.1.) to the oxide film. This is achieved by applying a thin layer of photosensitive lacquer to the oxidized silicon, exposing this to ultra-violet light through the photomask, and then in successive chemical treatments the photo-lacquer is developed, and the non-exposed lacquer is removed. Holes are etched in the oxide using the exposed lacquer as a mask, and the exposed lacquer is finally removed. We now have a pattern of holes in the oxide film through which suitable impurities (e.g. Boron and Phosphorus) may diffuse.

In all but the first photoengraving step, it is necessary to achieve accurate alignment between the pattern already on the silicon slice and the new photomask.

*2.5. Diffusion.* In general, three diffusion steps are needed to make an SCC. The first stage is boron diffusion, at $\sim$1150°C using a boric oxide source. Boron (a P-type impurity) is diffused, in selected areas, through the thin N-type layer to meet the P-type substrate. This results in N-type "lands" separated electrically by P-type material (and hence by a high impedance PN junction). The next diffusion step also uses boron, but this time it is only diffused to a depth of $\sim$3 microns to form the base regions of NPN transistors, and

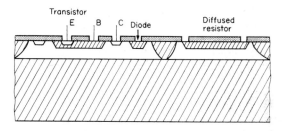

Key  Clear areas are N-type epitaxial deposit
       Dotted areas are thermal oxide
       Cross-hatched are P-type
N.B.  Not to scale

*Fig. 3.*

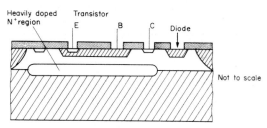

*Fig. 4.*

at the same time to form any resistors required in the circuit. Finally, phosphorus (an N-type impurity) is diffused into selected areas to form transistor emitter regions and low resistance $N^{++}N$ contacts.

A schematic section through a device at this stage is shown in Fig. 3.

When it is desirable to reduce the collector series resistance of the transistors, an N-type diffusant is used prior to epitaxial deposition, as shown in Fig. 4.

*2.6. Contacts.* Interconnexions between devices on the same chip are made using evaporated aluminium. External connexions are made by bonding aluminium wire between aluminium contact pads on the silicon and metal pins in the final package.

*2.7. Testing and packaging.* In all the processing described above, slices of silicon are handled, each of which contains potentially $\sim$150 SCC's. These circuits are tested electrically with an 8-probe tester so that faulty units can be marked as reject at that stage. The silicon slices are then scribed with a diamond and broken into ~150 individual chips which are then mounted in a suitable package (e.g. multi-lead T05 transistor package) ready for final testing.

### 3. Single Crystal Circuit Components

*3.1. Transistors.* All transistors made in SCC's are high frequency types with $fT$'s of typically 400–800 Mc/s. Storage times of <25 nanoseconds are achieved by gold doping. Typical voltage ratings are $BV_{CBO}$ ~12 volts, $BV_{EBO}$ ~5 volts. Common emitter current gains of 30–100 at collector currents of 10–50 mA are readily achieved.

*3.2. Diodes.* The emitter-base and collector-base junctions of transistors are used to provide the facility of fabricating diodes in a solid circuit. An emitter-base diode measuring $0.0015'' \times 0.0015''$ will give a forward current of ~10 mA at 1 volt. The capacitance of a similar area on 0.5 ohm cm N-type silicon is ~1.3 picofarad.

*3.3. Resistors.* Resistors in a SCC are normally diffused at the same time as transistor bases, and are therefore P-type. In general resistor values >1000 ohms are achieved using widths of $0.001''$. The present overall tolerance on a typical 10 K ohm resistor is ~±20 per cent, but resistor ratios can be readily made to about ±5 per cent of nominal value (these two factors seem likely to improve to ±10 per cent and ±2 per cent repsectively). The temperature variation of resistors is due to mobility variations (determined by lattice scattering). For a boron surface concentration of $5 \cdot 10^{18}$ atoms/cm$^3$ the sheet resistivity is ~160 ohms/square and the temperature coefficient is 0.25 per cent per °C from 0° to 150°C. Increasing the boron concentration reduces this value because of the onset of impurity scattering. At $2 \cdot 10^{19}$ atom/cm$^3$, $dR/dT$ is 0.1 per cent per °C.

The distributed capacitance associated with a typical resistor in a SCC is ~1 picofarad per 1000 ohm. This gives appreciable phase shift (~20°) at 400 Mc/s for a 100 ohm resistor.

*3.4. Capacitors.* Capacitors can be made in two distinct ways for a SCC. One method employs the capacity associated with a reverse-biassed P-N junction. Such capacitors are voltage dependent, with $C \propto V^{-n}$ where $0.33 < n < 0.5$. The area requirements limit the range of capacitance values at ~50 picofarads.

A second type of capacitor uses the silicon dioxide (as formed for diffusion masking) as a dielectric, with an evaporated aluminium counter-electrode. Such a capacitor measuring $0.020'' \times 0.020''$ gives a capacitance of 25 pF and a "Q" of 180 at 30 Mc/s.

*3.5. Inductors.* Achievement of inductance in a SCC is obviously very difficult. The use of particular devices (e.g. Field Effect Tetrode) to simulate inductance has been studied by some workers, but in general the approach is to avoid the problem by using SCC techniques to make broad band amplifiers and to use external frequency sensitive networks. The latter may be orthodox L-C circuits, evaporated L-C or evaporated R-C active filters.

*3.6. Parasitic components.* There are two troublesome sources of parasitic capacitance in SCC's. The first is the dsitributed capacitance associated with each diffused resistor, as mentioned in 3.3. above. With resistor widths of $0.001''$ this is no longer of prime importance, and when improved photoengraving permits widths of $0.0005''$ it will be further reduced. The more important capacitance is that associated with the isolating P-N junction. Using the technique described above it seems likely that circuits working at 100 Mc/s will be possible. Earlier isolation techniques limited useful operation at ~10 Mc/s.

### 4. Single Crystal Circuit Design Criteria

As we have already seen, SCC's are made on a piece of silicon ~1'' diameter, but the individual circuits are only ~0.070'' square. Furthermore, not all the 150 potential circuits from a given slice will operate. The causes of inoperation are random contamination at oxidation, diffusion, or photoengraving stages. As a result, whereas a useful quantity of 0.070'' square circuits can be achieved, the yield of circuits requiring 0.5'' square of silicon would be zero.

From a cost and reliability point of view it is important to minimize the number of connexions from the SCC to the outside world.

Combining the above two paragraphs, the most important aspect of design philosophy is "The achievement of the maximum functional complexity per unit of active area, with a minimum number of connexions."

A second rule of importance is to design circuits whose performance is a function of resistor ratios rather than absolute values. This helps overcome both the initial tolerance and the temperature coefficient, as discussed in 3.3.

### 5. Examples of Single Crystal Circuits
(all circuits are approx 1.5 mm²)

*5.1. Linear.* In Fig. 5 are shown the circuit and the SCC layout of the wideband voltage amplifier referred

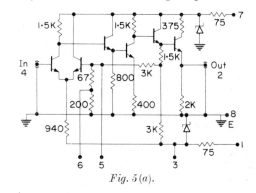

*Fig. 5 (a).*

to in section 1 above. This device has a typical voltage gain of 60, a maximum output voltage of 2 volts (R.M.S.) and a flat frequency response from DC to 20 Mc/s.

*5.2. Digital.* In Fig. 6 are shown the circuit and the SCC layout of a bistable circuit with integral "OR" gates. This device is used in counting circuits operating at 10 Mc/s.

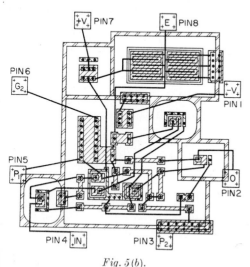

*Fig. 5 (b).*

## 6. Advantages of Single Crystal Circuits

There are 4 areas where SCC's appear likely to offer worthwhile advantages over other electronics techniques:

*6.1. Size.* In some applications (e.g. missiles) size and weight can be of prime importance. High speed computers require size reduction to minimize the effects of propagation delays (1 nanosecond per foot).

*6.2. Reliability.* There is a general requirement for electronics equipment to be made more reliable. SCC's constitute a useful step in this direction by virtue of the reduction in the number of soldered connexions; the possible standardization of packages and hence of interconnexions; the fact that only a few materials are used (e.g. silicon, silica, aluminium) and that all the processing is carried out in a controlled clean environment, thus facilitating adequate quality control.

*6.3. Circuit flexibility and performance.* When designing SCC's it is far from obviously true that a resistor is cheaper than a transistor. In fact the reverse is likely to be true (because of their relative area requirements). In view of this, a circuit designer can make free use of active elements and design for optimum performance.

In spite of the quite large temperature coefficient of resistors in SCC's it is found that because all the

components are within the one block of silicon (which has a good thermal conductivity) and the device characteristics are well matched, d.c. amplifiers with good drift characteristics can be made in SCC form. At the high-frequency end, the reduced wiring capacity and inductance—and the stability of stray C in the

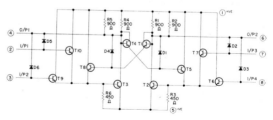

*Fig. 6 (a).*

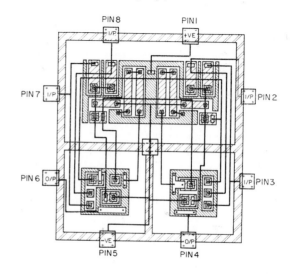

| | 'N' Substrate |
| --- | --- |
| | 'P' Isolation |
| | 'P' Base diffusion |
| | 'N'+ diffusion |
| | Aluminium contact |
| | Outer aluminium contact pads |
| — | Aluminium inter connections |

*Fig. 6 (b).*

evaporated interconnection pattern—are likely to give technical advantages to SCC's in the 100 Mc/s region.

*6.4. Cost.* In the long term there are reasons to expect that the techniques used to make SCC's, as discussed in section 2, will result in an overall reduction in the cost of equipment using SCC's compared with its orthodox equivalent.

*See also*: Microelectronics systems and techniques.

*Bibliography*

Bibliography on Microminiaturisation (1963) *Electronics Reliability and Microminiaturisation* **2**, 71, January-March.

DUMMER G. W. A. (Ed.) (1962) *Microminiaturisation, Proceedings of the AGARD Conference* Oslo 1961, Oxford: Pergamon Press.

KEONJIAN E. (Ed.) (1963) *Microelectronics*, New York: McGraw-Hill.

*Proceedings of National Conference on solid circuits and Microminiaturisation* West Ham 1963 to be published, Oxford: Pergamon Press.

D. H. ROBERTS

**SOLID, EQUATION OF STATE FOR.** The state of a solid is described by an equation relating the stress, the strain and the temperature. The derivation of the equation of state starts from the First and Second laws of thermodynamics, which express the energy change $dU$ of a system at temperature $T$ in a reversible process as the sum of the amount of heat $T \, dS$ absorbed from the external medium and of the work $dW$ done on the system by the external forces. Since the work done on a solid of volume $V$ under a homogeneous stress $\sigma_{ik}$ in changing its strain state $\varepsilon_{ik}$ by $d\varepsilon_{ik}$ is $dW = V \sum_{i,k} \sigma_{ik} d\varepsilon_{ik}$ whence the change in the Helmholtz free energy $F = U - TS$ is given by:

$$dF = -S \, dT + V \sum_{i,k} \sigma_{ik} \, d\varepsilon_{ik}. \tag{1}$$

The knowledge of the free energy as a function of temperature and strain permits the evaluation of all the thermodynamic properties. The derivative of the free energy with respect to the temperature at constant strain gives the entropy $S$, whence the heat capacity can be derived by an additional differentiation. On the other hand, the derivative of the free energy density with respect to the components of the strain tensor at constant temperature gives the equation of state:

$$\sigma_{ik} = \frac{1}{V} \left( \frac{\partial F}{\partial \varepsilon_{ik}} \right)_T. \tag{2}$$

The equation of state of a fluid and of a solid under hydrostatic pressure $P$,

$$P = - \left( \frac{\partial F}{\partial V} \right)_T, \tag{3}$$

is a special case of equation 2. The equation of state (2) can be looked upon as the equilibrium condition of the solid: for a given applied stress, and in particular for zero stress, it can be used to derive the parameters of the lattice cell of a crystalline solid as functions of temperature, and thus the thermal expansion. The bilinear term in the expansion of the free energy density of an elastically deformed body in powers of the elastic strain components gives the elastic coefficients of linear elasticity theory.

The evaluation of the free energy of a crystal requires the determination of the energy levels for the system of nuclei and electrons, and proceeds through a number of approximations. The motion of the nuclei is much slower than the motion of the electrons, because of the large mass ratio, and therefore that the nuclei during their motion essentially see the electron distribution as instantaneously adjusted to their own positions (*adiabatic approximation*; Born and Oppenheimer, 1927). One can therefore regard a solid as composed of essentially atomic particles (atoms, molecules or ionic cores, possibly embedded in a gas of valence electrons) and derive a simple equation of motion for these particles. The average positions of the nuclei in a homogeneously strained solid form a lattice, and the (strain dependent) energy of the system in the ground state is the sum of the lattice energy, the ground state energy that would be evaluated if the nuclei were taken to be at rest at their average positions, and of the zero-point energy, a consequence of the Heisenberg uncertainty principle which prevents complete localization of the nuclei. The zero-point energy is a small fraction of the lattice energy and can be neglected (*static lattice approximation*) in the majority of solids, a notable exception being the loosely bound solids of the rare-gas elements, especially the lighter ones. *A priori* calculations of the lattice energy have been attempted for a few, simple solids. However, it is convenient in much of the work on solids, and in particular in the evaluation of the equation of state, to adopt a plausible functional dependence of the lattice energy on the internuclear distances for each type of solid, with adjustable parameters which are usually determined from the observed equation of state in standard thermodynamic conditions. Thus, for an ionic or a molecular crystal, where the constituent atomic particles retain their identity to a good approximation up to moderately high density, the lattice energy is interpreted as the interaction energy of the constituent atomic particles; the main attraction is provided by the Coulomb interaction of the net ionic charges in the former type of solid and by the van der Waals interaction of the atoms or molecules in the latter, and is balanced by an empirical repulsion, a consequence of the Pauli exclusion principle which opposes the overlap of closed-shell configurations. Similarly, an expression for the lattice energy of a simple metal can be derived as a sum of the kinetic and exchange energy of the conduction electrons, the Coulomb energy of the system of conduction electrons and ionic cores, and an empirical repulsion between the ionic cores.

The stress-strain relationship of a solid in the static lattice approximation follows from the pertinent expression of the lattice energy as a function of the internuclear distances. The explicit temperature dependence of the thermodynamic properties is determined instead by the contributions of the excited states to the free energy. The excitations of interest are the thermal vibrations of the nuclei, thermal excitations in the electronic system and in the atomic and nuclear spin systems, molecular rotations, and so on. We discuss the first two contributions to the free energy.

(1) *Lattice vibrations.* The equation of motion of the nuclei is solved by expanding the potential energy in powers of the displacements of the nuclei from their average positions. Including up to bilinear terms in the expansion, a transformation to normal co-ordinates allows one to describe the nuclear motion as a superposition of harmonic-oscillator modes of motion. Therefore, in this approximation (*quasi-harmonic approximation*), the free energy of the solid is the sum of the lattice energy $U_L$ and of the free energy of an assembly of harmonic oscillators with strain-dependent frequencies $\nu_i$:

$$F(\epsilon, T) = U_L(\epsilon) + \tfrac{1}{2} \sum_i h\nu_i(\epsilon) +$$
$$+ kT \sum_i \ln(1 - e^{-h\nu_i(\epsilon)/kT}). \quad (4)$$

The second term on the right-hand side is the zero point energy. The contributions to the free energy due to the higher order terms in the potential energy (anharmonic terms) have been found to be very small in a number of solids at not too high temperatures. A complete evaluation of the stress-strain-temperature relationship from equation 4 requires the knowledge of the strain dependence of the vibrational frequencies, and has been attempted only for simple models of real solids.

The expression of the free energy given in equation 4 implies that the effect of temperature on the stress components and the elastic coefficients is simply additive at constant strain. We see also that the contribution of each mode of vibration to the free energy has the form of a product of the temperature times a function of $\nu_i(\epsilon)/T$. This form ensures that the pressure and its temperature derivative at constant volume, which is equal to the ratio between the coefficient of volume thermal expansion $\beta$ and the isothermal compressibility $K$, obey the following equations:

$$P = -\frac{dU_L}{dV} + \frac{1}{V} \sum_i \gamma_i U_i, \quad (5)$$

$$\frac{\beta}{K} = \frac{1}{V} \sum_i \gamma_i C_{V_i}. \quad (6)$$

Here, $\gamma_i = -d \ln \nu_i/d \ln V$ are quantities of the order of magnitude of unity, and $U_i$ and $C_{V_i}$ are the average energy and the heat capacity of the $i$-th mode of vibration. At classical temperatures, where $U_i = kT$ in accord with the principle of equipartition of energy, equations 5 and 6 combine to give the *Hildebrand equation of state*:

$$P = -\frac{dU_L}{dV} + \frac{T\beta}{K} \quad (7)$$

with $\beta/K$ constant at constant volume. On the other hand, the *Mie–Grüneisen equation of state* follows from equations 5 and 6 if one assumes that the quantities $\gamma_i$

are all equal, namely, that the vibrational frequency spectrum does not change shape upon a change of volume:

$$P = -\frac{dU_L}{dV} + \frac{\gamma U^*}{V}, \quad (8)$$

$$\gamma = \frac{V\beta}{C_V K}. \quad (9)$$

Here, $U^*$ includes strictly the contribution of the excited states to the internal energy and the zero-point energy, and the Grüneisen parameter $\gamma$, defined in equation 9 through measurable quantities, should be independent of temperature at constant volume. Thus the Mie–Grüneisen equation of state writes the pressure as the sum of a purely volume-dependent function and of a "thermal pressure" proportional to the thermal energy density through the Grüneisen parameter.

It should be noted that both the Mie–Grüneisen and the Hildebrand equation of state are, in principle, valid under less restrictive conditions than those used here in their derivation. In particular, a necessary and sufficient condition for the validity of the Mie–Grüneisen expression for the pressure is that the contribution of the excited states to the free energy satisfies a law of corresponding states. This holds, for instance, in the *Debye model* description of the lattice vibrations. Both experiment and theory indicate that the Grüneisen parameter is in fact constant in a number of solids down to temperatures as low as one-third of the Debye temperature.

(2) *Electronic excitations.* In a metal, thermal excitations of the conduction electrons give a strain- and temperature-dependent contribution to the free energy, which is easily observable as a contribution to the heat capacity at sufficiently low temperatures, usually below 4°K. The *free electron model of metals* schematizes the conduction electrons as a perfect gas of particles obeying the Fermi–Dirac statistics, and their contribution to the pressure–volume–temperature relationship is most easily obtained from the general equation of state of a non-relativistic perfect gas:

$$PV = \tfrac{2}{3} U. \quad (10)$$

In the ground state, all the one-particle energy levels below the Fermi energy $U_F$ are occupied, and all above are empty: the energy of the gas is easily found to be $\tfrac{3}{5} U_F$ per particle. At a temperature $T$, the energy available for thermal excitations is roughly $kT$ per particle, so that only those electrons whose energy is within an energy $kT$ from an empty level can be thermally excited. Hence, at temperatures $T \ll U_F/k$, only a fraction of the number $N$ of electrons in the gas, of the order of $kT/U_F$, are thermally excited: the thermal energy of each electron excited is $\tfrac{3}{2} kT$, and the thermal energy of the gas is proportional to $\left(\tfrac{3}{2} kT\right)\left(\frac{kT}{U_F} N\right)$. A detailed calculation gives $\pi^2/6$ as the proportionality factor. Therefore, the contributions of the thermal excitations of the conduction electrons to the pressure and to the thermal expansion of a metal are given by:

$$P_{\text{el}} = \frac{\pi^2}{6} \frac{(kT)^2}{U_F} \frac{N}{V}, \qquad (11)$$

$$\beta_{\text{el}} = \frac{2}{3} \frac{C_{V_{\text{el}}}K}{V}. \qquad (12)$$

The Fermi energy is proportional to $(N/V)^{2/3}$, so that $P_{el}$ varies as the 1/3 power of the density. The experimental information available on the thermal properties of metals near the absolute zero confirms the temperature dependence of the electronic contribution to the thermodynamic functions predicted by the simple free electron model, but yields numerical coefficients which are appreciably larger than predicted in a number of metals.

The forms of the equation of state for a solid discussed above presuppose the availability of a model to describe the lattice energy, which is of necessity different for different types of solid. The details of the interatomic forces become less and less important, however, as the density rises. At rather high densities, the Thomas–Fermi–Dirac statistical model of the atom provides an adequate description of the lattice energy and of the equation of state. The kinetic energy of the electrons becomes actually dominant over the potential energy at very high densities. This peculiar property of the electron gas, which implies that its "perfectness" increases as its density rises, permits a very simple qualitative description of the equation of state of matter at extremely high pressures: if the temperature is not too high, the pressure is given by equation 10 with $U = \frac{3}{5} N U_F$, where $N$ is now the total number of electrons in the substance. At still higher pressures, relativistic effects and nuclear reactions triggered by the capture of electrons by the nuclei come into play.

On the other hand, general forms of the pressure–volume–temperature relationship, of an empirical nature, have been proposed, which are useful to describe experimental data in an analytic form. An approach which is useful for rather small compressions expresses the volume as an expansion in powers of the pressure, or the pressure as an expansion in powers of the dilation, with temperature-dependent empirical parameters. A more fundamental approach uses instead the theory of finite strains to derive an empirical pressure–volume relationship which should be applicable to all substances. This equation of state fits fairly well the experimental data for a number of substances, including highly compressible substances such as the alkali metals, and has found an application in describing the state of matter in the Earth mantle.

*Bibliography*

BERNARDES N. and SWENSON C. A. (1963) *The Equation of State of Solids at Low Temperature*, in *Solids under Pressure* (PAUL W. and WARSCHAUER D. M. Eds.), New York: McGraw-Hill.

BORN M. and HUANG K. (1954) *Dynamical Theory of Crystal Lattices*, Oxford: The University Press.

FUMI F. G. and TOSI M. P. (1962) *On the Mie-Grüneisen and Hildebrand Approximations to the Equation of State of Cubic Solids*, J. Phys. Chem. Solids **23**, 395.

LANDAU L. D. and LIFSHITZ E. M. (1958) *Statistical Physics*, Oxford: Pergamon Press.

LEIBFRIED G. (1955) *Gittertheorie der mechanischen und thermischen Eigenschaften der Kristalle*, Handbuch der Phys. VII/1, Berlin: Springer.

SALPETER E. E. (1961) *Energy and Pressure of a Zero-Temperature Plasma*, Astrophys. J. **134**, 669.

M. TOSI

**SOLID STATE RADIATION DETECTORS.** Nuclear and ionizing radiations interact with matter to produce pairs of positive and negative charges, and the radiation can be detected by the current which results from the motion of these charge pairs in an electric field. In a gas, the charge pairs produced initially are positive ions and electrons; in a solid or condensed medium, radiation produces free electrons and holes, i.e. vacancies in the normally filled electronic levels which may pass from atom to atom behaving in effect as net positive charges. If the mobility and lifetime of the electrons and holes as free carriers is sufficient to allow them to move for a significant distance in an applied field, an electrical signal can be obtained from the electrodes.

First reports of the use of solid-state detection media appeared in 1945 when a description was given of single crystals of silver chloride to which electrodes had been applied which would give pulses from nuclear particles when cooled to 77°K. Later, many other materials were reported to have similar properties as bulk conduction counters; they included thallium brom-iodide, silver bromide and lithium-silver bromide, as well as solid and liquid argon. Other materials, notably certain diamonds, zinc sulphide and cadmium sulphide, could be operated at room temperature.

In any detection medium the number of charge pairs present due to thermal and other excitation must not be large compared with the number generated by the radiation to be detected. If the material contains impurities which are easily ionized, a reduction in free carrier density can be brought about by the presence of deep lying levels, in which carriers can be trapped. Cooling increases the effectiveness of the traps. In all of the early bulk conduction detectors this mechanism appears to operate. More recently deep levels due to copper and gold have been used to reduce free carrier densities in silicon and germanium to allow them to be used as bulk detectors at liquid nitrogen temperatures. Unfortunately the presence of deep traps also limits the free life of holes and electrons generated by the radiation, and this leads to a wide spread in pulse amplitudes and to polarization of the medium by the trapped charges so as to oppose the applied field after a period of operation. In materials such as cadmium sulphide and cadmium selenide, the dielectric relaxation time is sufficiently short to allow free carriers to enter the medium from the electrodes and to neutralize the trapped space charge (in case of these materials mainly due to trapped holes); the presence of these free negative carriers causes an increase in conductivity until the holes capture and recombine with the free electrons. This process results in many electrons passing between the electrodes for each electron pair generated by the radiation. Such detectors therefore have an effective current gain and are consequently attractive for measuring dose rates as direct current. However, electron trapping effects produce undesirably long response times at low dose-rates, e.g. response times of several minutes may be observed at dose-rates of the order of 1 r/hr.

A high electric field in a region of low carrier density is to be found in the "depletion layer" of a *p-n* semiconductor junction (see next paragraph). In 1948 the detection of alpha particles at point contact junctions in germanium was described (Ahearn 1948) but it was not until about 1956 that surface barrier junctions in germanium were developed for use as alpha particle spectrometers (Mayer and Gossick 1956). General interest grew in using *p-n* junctions as counting instruments in about 1960 largely with the development of silicon junction detectors which proved capable of measuring nuclear particle energies with a resolution greater than is obtained with any other ionization device. Silicon junctions were made having areas up to several square centimetres, and techniques developed to enable the sensitive depths to be increased to about a centimetre. The success of these methods in allowing junctions with thick depletion layers to be made in silicon, with atomic number 14, invited the application of similar techniques to germanium where the higher atomic number, 32, produces prominent photoelectric peaks in the distribution of pulse amplitudes from gamma rays.

A depletion region occurs in a semiconductor in the vicinity of a junction where the material changes abruptly from being *p*-type to *n*-type. Majority carriers diffuse across the junction, electrons from the *n*-type to ionize acceptors in the *p*-type material and holes to ionize donors in the *n*-type material. Equilibrium is established when the field produced by the ionized donors and acceptors at either side of the junction opposes further majority carrier transfer by diffusion. Such junctions are rectifiers, and a negative voltage applied to the *n*-type material reduces the effect of the positive space charge due to ionized donors and causes current to flow. A positive voltage applied to the *n*-type (reverse bias) prevents majority carriers from reaching the junction and widens the space charge layer. Thus a region is created from which majority carriers are excluded by the space charge field, and in which donors and acceptors are ionized and cannot readily generate free carriers. This carrier-free space-charge layer is known as the depletion region.

For a step junction, in which the density of donors $N_D$ on the *n* side is large compared with the density of acceptors $N_A$, the depletion region extends mainly into the *p*-type material and its width $w$ is given approximately by

$$w^2 = \frac{K(V - V_0)}{2\pi e N_A} \quad \text{c.g.s. units},$$

where $K$ is the dielectric constant, $e$ the electronic charge, $V_c$ the potential barrier at zero bias, $V$ the applied bias. This width constitutes the sensitive depth of the detector. Electron-hole pairs created by ionizing radiation in the depletion layer are swept out by the space charge field to opposite sides of the junction, and thus constitute a pulse of current flowing into the depletion layer capacitance. It can be shown that the capacitance, per unit area, $C$ is given by

$$C = \frac{K}{4\pi w} \quad \text{c.g.s. units}.$$

Vacuum floating-zone refining techniques are now used to produce high-purity single crystals of silicon in the form of ingots which are usually about 2 cm in diameter and several cm long; material with a resistivity in the region of several thousand ohm cm, indicating residual donor or acceptor concentrations of about $10^{12}/\text{cm}^3$, is now commercially available. *P-n* junctions can be made on this material by diffusing a donor into *p*-type silicon; for example, by heating a slice of *p*-type silicon at 900°C for about 20 minutes in the presence of a little phosphorus, a junction is formed at a depth of 2 microns. Boron may be diffused into *n*-type silicon in a similar way to form a junction. A second method of obtaining a *p-n* junction is simply to evaporate gold in vacuum on to the etched surface of a slice of *n*-type silicon and a junction forms on exposure to air, a satisfactory ohmic contact is formed when aluminium is evaporated on to the back of the *n*-type slice. A junction is similarly formed between a layer of aluminium evaporated on to *p*-type silicon; evaporated gold now makes a satisfactory non-injecting back contact.

The depth of the sensitive volume $w$ using diffused and surface barrier techniques approximates to

$$w = \frac{\sqrt{(\varrho V)}}{3} \times 10^{-4} \text{ cm on } p\text{-type silicon}.$$

$$w = \frac{\sqrt{(\varrho V)}}{2} \times 10^{-4} \text{ cm on } n\text{-type silicon}.$$

where $\varrho$ is the bulk resistivity of the silicon at room temperature.

If a sufficiently high voltage is applied the depletion layer may extend through silicon so that the sensitive volume completely fills the device. These are sometimes called *p-i-n* junctions. Using the best commercially available silicon, of the order of 10,000 ohm cm resistivity, and devoting great care to the elimination of electrical breakdown at the edges of the junction, it may be possible to apply up to 1000 volts reverse bias, giving a sensitive depth of a little over a millimetre.

To obtain depletion layers with much greater thickness, the junction field itself is used to distribute lithium, which is a donor in silicon, so as to compensate for an initial acceptor concentration. A slice of $p$-type silicon, on which lithium is deposited, is first heated to about 350°C for a few minutes so that a junction is formed about 100 μ below the surface. Reverse bias is applied to the junction and the temperature maintained in the neighbourhood of 200°C. Lithium ions are then sufficiently mobile to drift in the junction field, and move on to the $p$-type material effec-

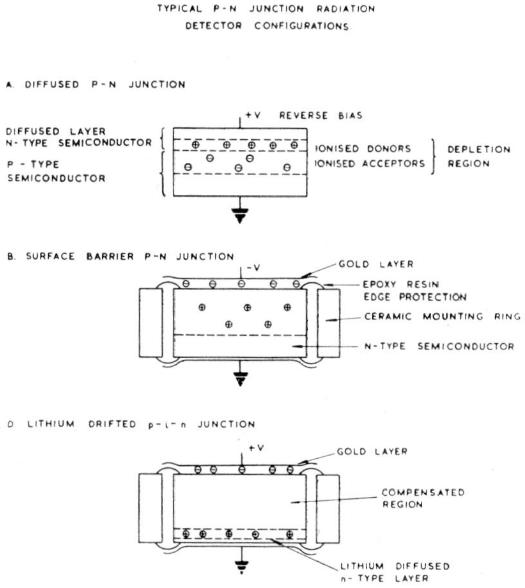

*Fig. 1.*

tively creating a layer of intrinsic semiconductor. Detectors with effective depths of over 10 mm can be obtained using this technique.

The band gap of silicon, 1·09 eV at 300°K, is sufficiently high to allow satisfactory operation of junction detectors at room temperature, but thermally generated leakage currents are doubled for each 10°C rise in temperature. There is often some advantage to be gained by cooling them a little below normal ambient temperatures.

The band gap of germanium 0·66 eV at 300°K requires that it be used well below room temperature as a nuclear particle detector. Cooling to at least dry ice temperature is necessary and more usually it is used at liquid nitrogen temperature. Junction detectors may be made by similar surface barrier or diffusion techniques to those employed in silicon. Thick germanium junctions are made by the method of lithium-ion drift, carried out at temperatures in the region of 60°C. Because of high mobility of lithium at room temperature, lithium ion drifted junctions in germanium must be stored at solid $CO_2$ temperatures

if gradual deterioration over a period of prolonged storage is to be avoided.

The high degree of pulse amplitude resolution obtained from semiconductor detectors when used as nuclear particle spectrometers arises from the low value of the work required to produce an electron hole pair. This is 3·50 eV in the case of silicon, and 2·8 eV in germanium (as in most semiconductors about three times the band gap). This may be compared with the values for gas detectors which require between 25 and 35 eV to produce a pair of ions. The relative spread in the amplitudes of the pulses produced by monoenergetic particles is related to the number $N$ of charge pairs produced and is proportional to $(F/N)^{\frac{1}{2}}$ where $F$ is a factor introduced by Fano to allow for the distribution of energy between ionization and excitation. In gaseous detectors values of $F$ as low as 0·09 have been achieved; in semiconductors $F$ is believed to be of order 0·5.

Other fluctuations in pulse amplitude are introduced by thermally generated leakage current in the detectors, and by shot noise in the input stages of the associated pulse amplifier. A low noise level equivalent to an r.m.s. fluctuation of 500 electronic charges at the input is achieved in a thermionic valve amplifier without much difficulty across an input capacitance of 10–20 pF. In exceptional cases, the noise level may be as low as 250 r.m.s. electron charges. Field effect transistor amplifiers have comparable or even lower noise levels when cooled to −100°C at low input capacitance, but junction transistor amplifiers usually have noise levels in the region of 2000 r.m.s. electron charges under similar conditions. Junction transistors can have lower noise than small thermionic valve amplifiers for an input capacitance in excess of 500 pF.

Silicon junction detectors of effective area 1 cm² used as alpha particle spectrometers have given line widths at room temperature on 5 MeV alpha particles down to 12 keV (full width at half amplitude). Used as beta-ray spectrometers resolutions of 5 keV have been reported on cooled specimens from 0·624 MeV beta particles. Germanium detectors can have comparable energy resolution, typically full widths of 5 keV have been reported on the photopeak from 0·661 MeV gamma rays.

A measure of the dose rate from nuclear radiation can be obtained by measuring the mean current from silicon junction detectors. Thin junctions, e.g. silicon solar cells generate 1 μA/cm² in gamma radiation dose rates of $10^4$–$10^5$ rad/hour. Dark currents from externally reversed biased silicon junctions at room temperature are usually of the order of 0·1–1 μA/cm².

Radiation produces damage in semiconductors by causing vacancies and interstitial atoms, which reduce carrier lifetimes. Typical levels at which radiation may impair performance are:

$10^{12}$–$10^{13}$    fast neutron/cm²
$10^{17}$        electrons with 1 MeV energy/cm²
$10^{10}$–$10^{12}$    alpha particles/cm²
$10^8$         gamma dose in rads.

The deterioration of carrier lifetime can be used to estimate the integrated radiation dose to the semiconductor, and silicon and silicon carbide junctions have been proposed for measuring fairly high levels of gamma radiation, e.g. > 100 rads.

Semiconductor counters have become established as a useful tool especially in low energy nuclear physics where the high energy resolution, small bulk and lack of windows make them especially useful as charged particle spectrometers. However, the carrier velocity at high fields approaches a limiting value ($10^7$ cm/sec in silicon) so that it is unlikely that they will ever compete with scintillation counters where large blocks of detector material are required for operation at speeds of the order of a few nanoseconds. The range of unexploited semiconductors, however, is large and other materials of special interest are those with high atomic number which could have application as radiation detectors if they become available at a purity comparable with that achieved at present in silicon and germanium.

*Bibliography*

AHEARN (1948) *Phys. Rev.* **73**, 524.
DEARNALEY G. and NORTHROP D. C. (1963) *Semiconductor Counters for Nuclear Radiation*, London: Spon.
MAYER and GOSSICK (1956) *Rev. Sci. Instrum.* **27**, 407.
TAYLOR J. M. (1963) *Semiconductor Particle Detectors*, London: Butterworths.

R. B. OWEN

**SOLIDS, EFFECTS OF GASES IN.** Under most circumstances solids are surrounded by gases. Very often these are just the gases of the atmosphere. Depending on the nature of the solids and the temperatures involved, the solids may absorb some of the gas atoms and subsequently the properties of the solids become changed. Atoms of elements which exist in the gas phase at normal temperatures are relatively small in size compared with other atoms and therefore they are usually rather mobile in the solid and reside in the interstices of the lattice structure rather than in vacancies created by absent atoms of the lattice. Hydrogen, oxygen, and nitrogen are absorbed to some extent by most solids. Although pure carbon is not a gas at room temperature, carbon forms the compounds carbon dioxide, carbon monoxide, and methane which are common gases at room temperature. This fact along with the small size of the carbon atom warrants its inclusion in this discussion. Carbon is absorbed to some extent by most solids. The inert gases such as helium, neon, and argon are only slightly absorbed by metals and semiconductors but are absorbed to a greater degree by glasses and polymers. The halogens, chlorine and bromine, react too strongly with metals to get beyond the surface layers, but are sometimes absorbed to a greater degree by other types of solid. Gaseous chemical compounds may be absorbed in their original molecular form in glasses and polymers but in crystalline solids, such as metals and semiconductors, the molecules decompose on the surface and the atoms enter the lattice structure separately. The discussion which follows will be concerned principally with gas atoms or molecules as impurities in solids and will neglect cases of nonstoichiometry such as that of oxygen in solid zinc oxide.

*Effects on mechanical properties.* When relatively ductile solids such as metals absorb gas atoms they become hardened and embrittled. Hardness is a measure of the ability of a solid to resist being scratched or indented, while brittleness is a measure of the degree to which the material can be plastically elongated before it breaks. Both of these properties involve the plastic deformation of the solid. An increase in hardness of a metal normally results in a corresponding increase in strength. The absorbed gas atoms harden and embrittle metals in several different ways. At low concentrations of carbon and nitrogen in iron the atoms are attracted to dislocations because of the distortion of the iron lattice in this region. The "clouds" or "atmospheres" of nitrogen or carbon atoms that form around the dislocations prevent them from moving and thus inhibit plastic deformation. However, when the applied stress becomes high enough, the dislocations suddenly break free from their nitrogen or carbon "clouds" and the strain rapidly increases in the form of a sudden yielding. If the gas atoms are removed from the metal the sharp "yield point" is no longer observed.

When iron containing higher concentrations of nitrogen or carbon is suddenly cooled from a high temperature, the gas atoms become trapped in interstices along one axis of the lattice structure. Under these circumstances the very hard tetragonal martensite crystal structure is formed rather than the normal cubic crystal structure of pure iron. Nitrogen or carbon can also combine chemically with the iron to form an iron nitride or carbide second phase which is brittle. The nitride or carbide precipitate particles furthermore interfere with dislocation motion in the iron and thus harden and strengthen the metal in that manner. Oxygen and hydrogen also produce precipitate particles in metals with which they react chemically.

A similar type of hardening occurs when gas atoms combine chemically with substitutional alloying elements in the solid. An example of this is the phenomenon known as *internal oxidation.* This effect occurs when a solvent metal contains an alloying element which has a greater chemical affinity for oxygen than does the solvent metal. The alloying element may then form oxide precipitates when the concentration of oxygen is too low to form oxide precipitates with the major constituent. When copper, which normally has a relatively high solubility for oxygen, contains small amounts of manganese or silicon, these elements form oxides which precipitate as finely dispersed particles. The precipitate particles obstruct the dislocation motion and thereby harden the copper. This process has been found useful as a means of strengthening very soft metals.

In cases where the normal solubility of the gas in the solid is small, atoms of diatomic gases may diffuse to internal voids or lattice discontinuities and combine to form the molecular species. If the dissociation reaction is slow compared to the recombination reaction, the gas molecules become trapped in the voids and accumulate until extremely great gas pressures are achieved. These precipitated gases are the cause of porosity or blowholes in metal castings. In the case of hydrogen in aluminium, small gas bubbles form on any type of lattice discontinuity such as grain boundaries and dislocation pile-ups as the hydrogen atoms come out of the lattice structure and combine to form hydrogen molecules. These arrays of gas bubbles obviously weaken the structure of the aluminium.

A long-standing enigma in metallurgy has been the embrittlement of steel by hydrogen. The high mobility of hydrogen in steel has made it difficult to obtain sufficient quantitative data to establish a detailed explanation for the effect. In mild or soft steel, this type of embrittlement, unlike other types apparently only occurs when the steel is strained very slowly. Thus it has been given the name, "*low-strain-rate embrittlement*". In general it is believed that the strain-rate dependence of this type of embrittlement indicates that it is controlled by the diffusion of hydrogen to certain localized areas. A popular theory is that the hydrogen atoms segregate to voids or openings in the lattice where they combine to form molecules of hydrogen gas. Enormous pressures are built up in the voids which cause the metal to fracture at stresses lower than normal. Another popular view is that the hydrogen adsorbs on internal surfaces in the steel, reducing the surface energy and causing the cracks to propagate at lower stresses. There are also indications that dislocations play some part in this unusual behaviour but their exact role is uncertain. Very hard and strong steels are subject to a phenomenon known as "*hydrogen-induced delayed brittle fracture*". When these steels contain sufficient amounts of hydrogen, they sustain their normal loads only for a period of time and then suddenly fracture catastrophically. Here again the period of delay before fracture is believed to be related to the time it takes the hydrogen to segregate to localized areas by diffusion.

Another similar type of embrittlement occurs when two different gas atoms combine chemically and precipitate out in voids in a solid. A typical example of this is the case in which both hydrogen and oxygen are present in copper or silver. The hydrogen and oxygen combine at grain boundaries or other lattice discontinuities to form water vapour which causes sufficient internal stress to rupture the solid. This effect usually occurs when the metal contains dissolve oxygen or occluded oxides and is then annealed in a hydrogen atmosphere. When the gases involved are hydrogen and oxygen this is some-times called "*water vapour embrittlement*".

*Electrical properties.* When gas atoms are absorbed by pure metals, the electrical resistivity is usually increased. The electrical resistivity of a metal is considered to be due to the scattering of the conduction electrons by lattice vibrations or by lattice imperfections such as impurities or structural defects (e.g. dislocations and vacancies). At high temperatures the effect of lattice vibrations makes an appreciable contribution to the electrical resistivity but at lower temperatures the contribution due to the impurity atoms and lattice discontinuities becomes the dominant part. At very low temperatures special conditions can exist in some metals which result in the disappearance of the electrical resistance almost completely, giving rise to the so called "*superconductivity*". This will be mentioned again later.

In metals, gas atoms such as nitrogen and oxygen in solid solution act as scattering centres for the electrons and thus increase the electrical resistivity of the metal. The resistivity is usually found to be directly proportional to the concentration of gas atoms in solid solution for low concentrations. For example the resistivity of niobium is increased by about $5.2 \ \mu\Omega$ cm per atomic per cent oxygen. When the gas atoms combine chemically with the metal atoms and precipitate as a second phase their contribution to the resistivity decreases considerably. Thus measurements of electrical resistivity have been used to study the rate of precipitation of the gas atoms out of solid solution. In some metals, hydrogen in solution also increases the resistivity of the metal. These metals are usually those with which hydrogen combines chemically at higher concentrations such as palladium and tantalum.

In the range of temperatures where *superconductivity* occurs it has been found that oxygen, nitrogen, and hydrogen in solid solution decrease the temperature at which the metal becomes superconducting in the hard superconductors such as vanadium, tantalum, and niobium. This decrease is approximately directly proportional to the concentration of the dissolved gas atoms at concentrations below the solubility limit. An example is that of oxygen in niobium where the transition temperature is decreased by $0.93°K$ per atomic per cent oxygen.

There are cases in which the effect of the dissolved gas on the resistivity of the metal is a secondary effect. As an example, recent studies of the effect of hydrogen on the resistivity of iron disclosed that hydrogen in solid solution had little effect on the resistivity. However, when the hydrogen atoms precipitated out as hydrogen molecules in small voids causing blistering and plastic deformation, the holes produced and the dislocations generated caused the resistance of the specimen to increase. The holes increase the current path and the dislocations act as scattering centres for the electrons. Thus the presence of the hydrogen in the iron caused the electrical resistance to increase but in a secondary manner. As another example, if the resistance of the metal decreases after it has been exposed to a gas atmosphere, it usually means that the absorbed gas atom has combined chemically with some other impurity in the solid and has caused the other impurity to be precipitated out of solid solution.

During the last few years some information has become available on the effect of dissolved gases on the properties of *semiconductors*. Since there is a gap between the valence band and the conduction band in semiconductors, the conduction of electricity is not only controlled by the scattering of the conduction electrons but is also controlled by the number of current carriers available (i.e. number of electrons raised into the conduction band or number of holes in valence band). Impurity atoms affect the electrical resistivity by acting as donors of electrons (to the conduction band) or acceptors of electrons (from the valence band leaving holes as current carriers). Hydrogen, helium, nitrogen and oxygen are known to be absorbed by the elemental semiconductors, germanium and silicon, when they are exposed to these gases at high temperatures. Hydrogen, helium and nitrogen apparently have little effect on the electrical resistivity of these semiconductors but there has been some indication that hydrogen may affect the lifetimes of minority carriers by producing recombination centres for holes and electrons. The absorption of oxygen in silicon and germanium during the growth of single crystals of these materials is a well known and important problem. The oxygen may be an impurity in the inert gas atmosphere or it may come from the fused silica crucibles used for growing silicon crystals. When it is absorbed by the semiconductor, it often forms clusters or complexes of $o_n$, where $n$ is the number of atoms in the group. The complex is usually neutral for $n$ less than 3 but when $n$ is 3 or 4 the complex becomes a donor and the resistivity of the semiconductor is decreased.

*Magnetic properties.* Gas atoms absorbed by solids may effect the magnetic properties of the solid in several ways. In soft ferromagnetic metals such as iron it has been found that a magnetic ordering of carbon and nitrogen atoms in solid solution occurs. These atoms reside in preferred interstitial sites which are dependent on the direction of the magnetization vector in each of the magnetic domains. Since the direction of the magnetization vector varies from one domain to another the motion of the domain wall requires re-ordering of the nitrogen or carbon atoms to new interstitial sites. The re-ordering is temperature dependent and therefore the presence of the interstitial atoms gives rise to a temperature dependent alternating current permeability.

When the concentration of nitrogen or oxygen is above the solubility limit in ferromagnetic metals, the result is the precipitated second phases discussed earlier. These precipitate particles impede the motion of the domain walls and thereby increase the coercive force. Since the precipitation process takes place very slowly at room temperature, the magnetic properties change slowly over a period of time and this is called "ageing". This is obviously an undesirable feature of magnetic materials and therefore they are usually heated to 100–200°C to hasten the ageing process and bring the system to equilibrium.

The paramagnetic susceptibility of palladium is decreased by the addition of hydrogen or deuterium up to a hydrogen to palladium ratio of about 0·6. The most widely accepted explanation is that the hydrogen contributes its electron to the $d$-shell of the palladium eventually filling the $d$-shell and removing this contribution to the paramagnetism. In a similar manner it has been observed that the saturation magnetization of nickel is reduced gradually to zero by the addition of hydrogen up to $H/Ni = 0·65$. The explanation for this effect is believed to be the same as that for the palladium–hydrogen system. A difference, however, is that the nickel hydride that is formed is unstable at room temperature and it has been observed only in thin films.

*Bibliography*

COTTERILL P. (1961) in *Progress in Material Science* **9,** 205, Oxford: Pergamon Press.
CUPP C. R. (1953) in *Progress in Metal Physics* 4, 105, Oxford: Pergamon Press.
VAN BUEREN H. G. (1961) *Imperfections in Crystals,* Amsterdam: North-Holland.

R. C. FRANK

**SONAR.** Sonar is similar in operation to the much better known radar. It is, however, an earlier kind of echo-location system as it dates from the First World War. The modern term Sonar is derived from Sound Navigation and Ranging; but the original term ASDIC is still frequently used in Britain (although its origins are obsure and still debated), and the term Echo Sounding is in current use for the particular application of Sonar to depth determination using a vertical beam.

The term Sonar is generally taken to mean an acoustic echo-location system operating in water—generally the sea. It is, however, applicable to similar systems operating in air, e.g. guidance aids for blind people. Sonar operating in air occurs naturally, as is well known, in many species of bat. Sonar in water occurs in nature in porpoises. It is generally believed that these naturally occurring sonar systems are more efficient in relation to their size and power than are any man-made systems.

The frequencies used in sonar systems are generally between 10 and 500 kc/s, with maximum detection ranges of a few miles down to perhaps 100 m at the higher frequencies.

The applications of underwater sonar are nowadays manifold. The original application to the detection of submarines is still of the first importance and there are many other military and naval applications. Civil applications include (a) the ubiquitous echo-sounder, which normally all ships nowadays use, and is an indispensible navigational aid; (b) fish location and classification systems for use both in fish catching and in fisheries research; (c) sonar equipment for whaling, a use which is closely akin to the hunting of submarines; (d) marine biological research; (e) hydrographic surveying and certain kinds of marine geological study; (f) civil engineering, e.g. the investigation of scour round bridge piers.

The majority of sonar systems, like most radar systems, use the transmission and reflection of a pulse of energy as their basis. The system is typified by the arrangement shown in Fig. 1. Individual systems show many variations, such as the use of a single electro-acoustic transducer for both transmission and reception, or the use of an oscillatory discharge from a capacitor into the transmitter so that no separate continuous oscillator and gating unit is required. The

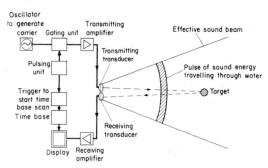

*Fig. 1. Schematic arrangement of a typical pulsed sonar system.*

pulses are always bursts of carrier frequency, although the extension of the term sonar to include marine seismic work using explosive pulses is becoming more common.

The distance from the transducers to a particular reflecting object (or "target") is indicated by the time elapsing between transmission of the pulse and reception of the echo. The display therefore always includes a time base, the traverse of which is initiated by the transmitted pulse. For simple systems of the type shown in Fig. 1 the commonest type of display is the *chemical recorder*, a remarkably efficient instrument which records the received information on sensitized paper, which may be a dry type such as Teledeltos, or a damp type in which the paper is impregnated with potassium iodide. In this instrument the time base is mechanical; the recording stylus is drawn across the paper. Since the mark is made when an echo is received, the position of the mark across the traverse indicates the range of the target. The paper is moved slowly in a direction perpendicular to the traverse of the stylus, so that successive traverses lie side by side.

If the range of the target from the transducers does not vary, the line is produced parallel to the direction of motion of the paper. It the range changes, e.g. due to the motion of the target, or of the ship on which the equipment is fitted, then the line is sloped relative to the paper motion.

Another kind of sonar system, which appears to correspond to that used by certain species of bat, is the "frequency-modulated" system which is typified by the block schematic of Fig. 2. Figure 3 shows the way

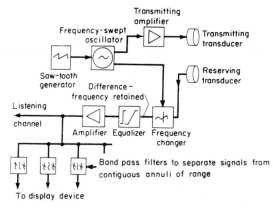

*Fig. 2. Schematic arrangement of a "frequency-modulated" sonar system.*

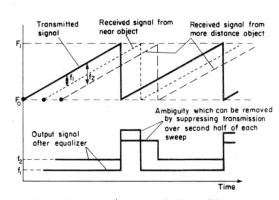

*Fig. 3. Frequency/time graph for a "frequency-modulated" sonar system.*

in which the frequency changes with time. The term frequency modulation is not used in quite the same way as in a frequency-modulation communication system because the variation of frequency is imposed on the transmitter by a saw-tooth generator and in general conveys no information. The sweep repetition period may conveniently be made twice that required for sound to travel from the transducer to the most distant range specified and back again. If the transmission is continuous as shown in Fig. 3, the difference-frequency output from the receiving frequency changer produced by the reflection from a fixed target is a steady frequency, (e.g. $f_1$ or $f_2$ in Fig. 3) from the time a signal is returned from the beginning of the new sweep until the end of the sweep at the transmitter. The actual frequency of the output is clearly a measure of the range of the target, and can either be detected by the ear, or measured using a bandpass filter system. There is an ambiguity, as indicated in Fig. 3, at the end of each transmitter sweep, but this can be reduced in importance by making the duration of sweep long compared with the target echo time, or by making the transmission semi-continuous.

In the simple systems outlined above a single beam is produced. To search a large sector, however, requires the transducer(s) to be swung mechanically so that successive transmissions, or groups of transmissions, "look" in different directions. Since the velocity of sound in water is only about one mile/sec, the rate of search is very slow in comparison with the corresponding rates of radar.

It is increasingly being found necessary to raise the rate of scan, and electronic sector-scanning systems have been developed with this object. In such a system the whole of the sector to be examined is insonified by a transmitted pulse which comes from a wide-beam transducer. A relatively narrow receiving beam is swung by electronic means across a sector within the time required for the pulse to travel its own length in the water. All directions within the sector are therefore effectively sampled on the one transmission. The block schematic diagram of the arrangement of one particular electronic scanning system is shown in Fig. 4. The receiving transducer is $n$ times the length of the transmitting transducer, and is divided into $n$

sections, where $n$ is the number of beam-widths (as measured between points where the power response has fallen to half) it is desired to contain in the scanned sector. If these $n$ sections were connected to $n$ corresponding, uniformly-spaced, taps on the delay line, then the beam would be deflected by an amount dependent on the phase-shift in the delay line. In the scanning system, frequency-changer equipment is inserted between the transducer sections and the delay line. The local oscillator which feeds all the frequency-changers is swept in frequency by the bearing time-base, so that the signal frequency received by the delay line varies over a range during every sweep of the bearing time-base. If then the delay line is made to have a phase-shift which varies over this frequency range from negative values to positive values, the beam will be swept from left to right during each sweep of the bearing time-base. The latter also deflects the spot on the cathode-ray tube from left to right, so that signals received on any particular bearing are recorded on that bearing on the display. The range time-base works in the usual way, so that the position of an echo-spot on the tube indicates the position of the echoing object on rectangular axes of bearing and range. (This kind of display is called the B-scan.) If the bearing scan is so rapid that it is completed within the duration of the pulse and is immediately repeated, no information is lost, and all directions in the sector are effectively looked at simultaneously, but with the angular resolution corresponding to the beam-width of the receiving transducer.

While the design of the electronic part of a sonar system is usually straightforward, the design of the *electro-acoustic transducers* is often quite difficult. Although there is a good deal of theory relating to transducers of various kinds, the limitations of practical materials are such that design has inevitably a very large empirical element. There are three main types of transducer in use for underwater applications. These are (a) magnetostrictive, (b) piezoelectric and (c) electrostrictive. In a magnetostrictive type, a magnetic field is applied to a piece of suitable magnetic material, causing the dimension of the piece to decrease along the axis parallel to the field. Thus when the field is alternating it is necessary to polarize the material if the acoustic wave is to have the same frequency as the electrical signal. The magnetostrictive effect also operates in reverse; received accoustic signals cause compression of the material which alters the magnetic field which in turn produces an e.m.f. in the electrical winding. The latter is generally a fairly thick wire with tough insulation so that the whole transducer can be directly immersed in the water. Such transducers are satisfactory for frequencies up to about 100 kc/s, and are easily constructed, although expensive.

Piezoelectric transducers use crystals in which the dimensions change according to the applied electric field. If the field is alternating the crystals vibrate and

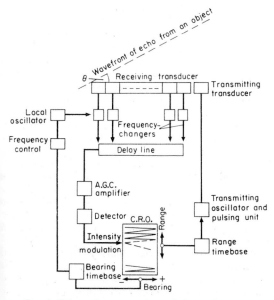

*Fig. 4. Schematic arrangement of electronic sector-scanning sonar system.*

give an acoustic radiation, conversely if the crystals are acted on by acoustic waves then they generate an electric field. Typical piezoelectric materials used for this purpose are quartz, ammonium dihydrogen phosphate, tourmaline and lithium sulphate. Whereas magnetostrictive transducers typically have low impedances of only a few ohms, piezoelectric transducers have high impedance typically of the order of 10,000 ohm and this requires high voltages for only moderate acoustic powers. They also have to be fitted in a water-tight container.

Electrostrictive transducers are becoming much more widely used and seem likely to displace the other types for the majority of applications. Examples of electrostrictive materials used for transducers are barium titanate and lead zirconate in ceramic form. The change of dimensions is dependent on the magnitude but not on the polarity of the applied electric field, and thus polarization is needed as with magnetostrictive transducers. Electrostrictive transducers generally have very convenient impedances of a few hundred ohms.

With all these types of transducer it is possible to achieve efficiencies of up to 50 per cent and $Q$-factors as low as about 5.

The beam-width of a transducer is inversely proportional to the dimensions of the transducer (measured in wave-lengths) on the appropriate axis. Thus at low frequencies the directional requirements necessitate very large transducers—or more accurately a large array of transducer elements. For this reason low-frequency echo-sounders, which normally operate at about 15 kc/s and obtain ranges of several thousand metres, operate with rather wide conical beams of perhaps 30°. High-resolution sonars, such as those used for studying the movement and behaviour of fish, need transducer arrays of tens or even hundreds of wave-lengths, and therefore operate on much higher frequencies such as 400 kc/s, with ranges of perhaps 100 m. The detailed design of arrays to have particular kinds of directional patterns is a rather specialized subject, but is closely analogous with the design of linear arrays for radio.

Although the design of a sonar system, as described above, involves many difficult problems and is the subject of much research, yet the greatest problems in the use of sonar undoubtedly arise in connexion with the propagation of acoustic signals in the sea. (For sonar systems operating in air, corresponding problems exist but are much less well understood.)

If the medium, i.e. the sea, were loss-free, infinite in all directions and uniform in all respects, then the spreading of the sound would be spherical; in other words the beam would expand uniformly in all directions perpendicular to its axis. Under these conditions the well known inverse-square law relates the sound intensity to the distance from the transmitter (sound intensity corresponds to power per unit area of cross-section). This relationship is usually expressed as a loss of 6 dB per doubling of range for

one-way transmission, or 12 dB per doubling of range for the echo-signal.

The sea is not, of course, infinite in all directions, but is bounded by the surface of the sea-bed. Thus in practice the spreading does not follow the inverse-square law, but the sound intensity, being to some extent canalized, falls off less rapidly than the inverse-square law. In the extreme case, called cylindrical spreading, the sound intensity is proportional to the reciprocal of the range, i.e. the loss is 3 dB per doubling of the range for one-way transmission.

In addition to this loss due to spreading, there is an additional loss due to absorption (and conversion into heat) of the sound energy by the water. This loss is, as would be expected, very variable. It is small at low frequencies but rises very rapidly with increase of frequency, and over the range of frequencies in common use at present, the loss expressed as dB/km is approximately proportional to the square of the frequency. At 50 kc/s the absorption loss varies from about 8 to about 16 dB/km.

Further propagation effects which are of great importance are caused by variations of the velocity of sound from one part of the sea to another. Under normal conditions the velocity of sound in the sea is about 1500 m/sec. The main causes of variation of this velocity are, in the usual order of importance: temperature, pressure due to depth, and salinity. The actual magnitudes of the effects can be indicated by various empirical formulae, of which the best known is that due to Wood, although later (and more complicated) formulae show it to be slightly in error. It applies in the temperature range 6–17°C:

$$V = 1410 + 4 \cdot 21t - 0 \cdot 037t^2 + 1 \cdot 14s + 0 \cdot 018d$$

where
$V$ = velocity in m/sec
$t$ = temperature in °C
$s$ = salinity in parts per thousand
$d$ = depth in m.

Perhaps the most serious effect of variations of velocity is the refraction of the sound beam when velocity changes with depth. This could cause a beam, perhaps normally horizontal, to be deflected to the sea bottom, where it will be reflected upwards only to be refracted down again, and so on. Along any straight line through the transducer, therefore, there are intervals of range where detection is impossible and others where it is possible. This kind of effect is, of course, more serious with low-frequency systems as they have the greatest nominal range.

Another effect of variations of velocity is the general scattering of the sound beam as it passes through turbulent regions; this leads to rapid fluctuation of received signal strength.

When the object of the sonar system is to detect small targets (as distinct for example from the sea bottom) the target signal has to be detected against a random background due to "reverberation"; this is the sum of all the numerous small echoes produced by back-scattering from sand and stone particles on the sea bottom, from minute air bubbles and other inhomogeneities in the water, from waves on the surface and so on. Although background noise in the sea, for example noise due to waves breaking, etc., may limit the maximum range of detection, reverberation on the other hand may be a limitation of performance at all ranges. This is because it arises from the signal transmission and at any time interval thereafter has a power level closely related to that of the signal. Thus for detection of small objects in areas where reverberation level is high, due for example to a shallow rough sea bottom, it is essential to use beams which are as narrow as possible in the relevant dimension. Because reverberation is normally the limiting factor in detection, an increase of transmitted power is not effective in improving detection over most of the range; consequently most sonar systems operate at acoustic powers of fairly low peak level, e.g. 1 W–1 kW.

*Bibliography*

ALBERS V. M. (1960) *Underwater Acoustics Handbook*, Pennsylvania: The University Press.
ALBERS V. M. (Ed.) (1963) *Underwater Acoustics*, New York: Plenum Press.
HORTON J. W. (1957) *Fundamentals of Sonar*, Annapolis: U.S. Naval Institute.
TUCKER D. G. (1956) *J. Brit. Inst. Radio Engrs.* **16**, 243.
WOOD A. B. (1930) *Textbook of Sound*, London: Bell.
Proceedings of Sonar Systems Symposium, (1962) Brit. Inst. Radio Engrs.

D. G. TUCKER

### SOUND PROPAGATION UNDER WATER. *Introduction.*

Sound is important under water because it propagates so well, whereas light and other electromagnetic radiations do not. The position is complicated because there are large variations with frequency, propagation path or mode, area and season. Practically, the frequency range from the order of 1 c/s to the order of 1 Mc/s is important; covering the interests of Geophysicists, Oceanographers, Marine Biologists, Fishermen, and Navies. This article will tend to concentrate on the audio-frequency range, and on results rather than methods. However, it should be mentioned that as sources, investigators have used both electrically-powered transducers, convenient for continuous waves and repetitive pulses; and underwater explosions, useful because of their high energy and wide bandwidth.

*Velocity in sea-water.* Sound velocity variations are significant because they control acoustic ray refraction in the sea. The velocity ($V$ in m/s) increases with the temperature ($T$ in °C), pressure (or depth $D$ in m), and salinity ($S$ in $^0/_{00}$), and these three parameters are arranged in the order of their practical importance. The older data were a few m/s in error, and for serious work the empirical 1960 Wilson formula should be used. To illustrate the behaviour near 0°C, zero depth and $35^0/_{00}$ salinity the expression is very approximately

$$V = 1449 + 4{\cdot}5T + 0{\cdot}017D + 1{\cdot}4(S - 35).$$

The full expression contains cross-product and higher-order terms. The non-linearity is only marked for $T$, the relation being shown in Fig. 1.

*Absorption in sea-water.* Sound is absorbed in pure water due to its viscosity, and there is a rather greater effect due to a molecular packing relaxation. Both absorptions are proportional to $f^2$, where $f$ is the frequency in kc/s. In sea-water the main additional absorption is due to a dissociation process of its magnesium sulphate content, despite the fact that this constitutes only 4.7 per cent of the total quantity of salts. This dissociation has a relaxation frequency $f_r$ about 140kc/s at 20°C, and around and below this frequency the absorption is mainly due to the magnesium sulphate. Schulkin and Marsh recommend an expression, written here to bring out the low-frequency $f^2$ dependence—

$$\alpha = \left( \frac{2{\cdot}34Sf_r}{f_r^2 + f^2} + \frac{3{\cdot}38}{f_r} \right) (1 - 6{\cdot}7 \times 10^{-5}D)\, 10^{-6}f^2$$

$$\text{neper/m}$$

where $f_r = 21{\cdot}9 \times 10^{\left(6 - \frac{1520}{T + 273}\right)}$ kc/s.

For 10 kc/s, 20°C, zero depth and $35^0/_{00}$ salinity $\alpha$ is $6 \times 10^{-3}$ neper/m or about 1 dB/nautical mile. The temperature dependence is important, note that at 0°C $f_r$ has fallen to less than half and correspondingly $\alpha$ has more than doubled. Depth or pressure dependence is also significant, at the deep ocean bottom, say 7000 m, $\alpha$ has fallen to about half.

It is difficult experimentally to separate true absorption (virtually direct conversion of acoustic energy to heat) from other causes of attenuation, such as body scattering and boundary losses. This is

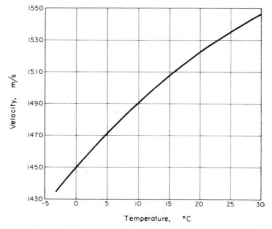

*Fig. 1. Velocity in sea-water (for $D = 0$, $S = 35$).*

especially so at low frequencies where $\alpha$ is small, but there is evidence that below 10 kc/s the $f^2$ formula gives too low a value. One alternative is the empirical 1957 Sheehy–Halley formula of $0.066f^{3/2}$ dB/nautical mile.

Without further consideration of propagation mechanisms the absorption gives a limit to the distance over which information may be transmitted. At 1 Mc/s this is a few hundred yards, at 10 kc/s a few tens of miles, but around 1 kc/s it is apparently the order of 1000 miles and often in practice determined by other factors. Note that compared to air the absorption is much less and the range possible is much greater.

*Shallow water.* Consider now what actually happens in the real ocean, starting with shallow water. In an ideal medium spherical spreading or 6 dB per distance-doubled would be predicted, but one of the first things a research worker in the subject learns is that this spreading law still holds in many situations where it has no right to. For example, in shallow water one might expect spherical spreading out to some range much less than the water depth, all the effective energy travelling along one ray path direct from source to receiver. At greater ranges there may be an additional path by way of a surface reflection, another by way

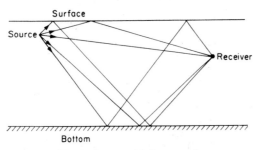

*Fig. 2. Some multiple ray paths.*

of a bottom bounce, and still more paths involving multiple reflections. Surface reflection is normally very good because of the enormous mismatch in impedance between water and air. Bottom reflection is good for grazing angles less than a critical angle, this critical angle exists because the velocity in the bottom material usually exceeds that in the water. We have now introduced the important idea of multiple rays, some of which are illustrated in Fig. 2. Another point is that the ray paths are in general curved rather than straight, as discussed later. At long ranges the wave nature of sound makes it convenient to abandon the idea of rays altogether, and to think in terms of modes which are similar in character to the modes in an electromagnetic waveguide. The sound energy is trapped between the surface and bottom, and tends to follow a cylindrical spreading or 3 dB per distance doubled law. Another complication arises because there is the extra absorption loss as the sound goes

through the medium, but even more important there are the cumulative losses due to imperfect reflections at the boundaries. The result is often close to the 6 dB law, which sometimes holds with surprising accuracy—even out to ranges of several tens of miles or perhaps 1000 times the water depth. This happens because of a cancelling out of the different effects, i.e. the gain due to the large number of possible rays is balanced by the losses on reflection.

*Variation with area.* Although the 6 dB law in shallow water is surprisingly common it must not be assumed that it always holds. One of the outstanding features of shallow-water propagation is how much it varies from place to place—and also from season to season and even from day to day. In fact this is a characteristic of water of any depth. Consider as an example some measurements of the propagation of 50 c/s sound to a range of only 5 miles, in two areas having water depths of a few tens of fathoms. In one the propagation was 18 dB better than spherical spreading, and in the other 39 dB worse. The difference is 57 dB, or half a million to one in intensity. As the main cause of the difference, in this and other cases, one can blame the bottom. A rocky bottom, with or without a thin layer of sediment on top, produces, surprisingly enough, a high reflection loss at low frequencies. There are a large number of subsidiary causes of propagation variation—water structure, bottom profile, sea surface, biological scattering layers etc.

*Shadow zones.* For shallow water there is this variation from area to area, but typically for a given area the mean loss increases reasonably regularly with

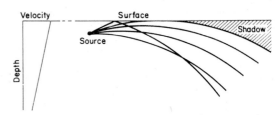

*Fig. 3. Shadow zone formation.*

range. This is not true for deep water, and some of the special features of deep-water transmission will now be looked at. One of the best known is the formation of a shadow zone at close ranges in downward re-fracting conditions. Typically these may occur in summer months in areas where there is little wind, plus continued sunshine to heat the surface waters of the sea. There may be a continuous decrease in temperature, with an accompanying decrease in sound velocity, as one goes deeper. This negative gradient has the effect of curving the sound rays downwards. Figure 3 shows that on simple ray theory no sound can reach an area near the sea surface beyond a certain range, even when surface reflected sound is taken into account. Note that in all the illustrations the ray angles are exaggerated, in fact for all propagation modes it is only the shallow-angle rays which are really effective. In practice shadow zones are an important feature, and may occur at ranges as small as one mile—or even less. Shadow zones have very low sound levels but are not completely dark acoustically, due to scattering, diffraction, and bottom reflections.

*Convergence zones.* As a sound ray penetrates deeper into the ocean it usually enters a region where the temperature is almost constant, but the pressure and therefore the sound velocity are increasing with depth. The ray is bent upwards and eventually returns to the surface. A certain number of rays can do this without being reflected at the bottom if the water is deep enough, as illustrated schematically in Fig. 4. Not only do these rays avoid any bottom reflection

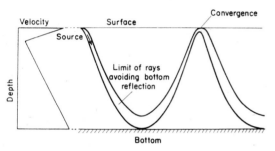

*Fig. 4. Convergence zone formation.*

loss, but it is found that they are focused by the medium—producing a caustic or convergence zone near the surface. Bottom reflected arrivals of the type shown in Fig. 2 are also present, but the magnified intensity in the convergence zone may be 100 times greater. There are further convergence zones occurring at equal range intervals determined by the water structure, about 30 miles for the Atlantic and Pacific and about 15 miles for the Mediterranean.

*Deep sound channels.* In shallow water it was shown that there was a large number of possible ray paths, which could alternatively be described as a channelling or concentration of the sound between the surface and bottom of the sea. Effective channelling is possible only because the bottom reflection loss is often very small for small grazing angles. In deep water there can also be channelling, but here even the small bottom losses may be avoided. For example consider near-horizontal rays leaving a source placed at the minimum of the sound velocity of curve of Fig. 5, i.e. at the axis of the sound channel. An upward-going ray is refracted downwards so that it eventually re-crosses the axis, and similarly a downward-going ray is refracted upwards. There need be no boundary reflections at all. Sound tends to spread out cylindrically and with very little attenuation, this being known as SOFAR propagation (from Sound Fixing and Ranging).

Transmission is so good at low frequencies that the limit is set by the distance over which one can find a continuous deep-water path. The longest possible path goes from the North Atlantic to the neighbourhood of Australia, half-way round the world, and relatively small charges fired near one end have been detected near the other end.

*Surface sound channels.* There are minor sound channels in deep water wherever there are minimal in the sound velocity profile, and these may have an important effect on propagation. The most important is the surface sound channel, with a velocity minimum at the surface. This occurs when the wind mixes up the surface layers of the ocean to produce homogeneous isothermal water. The only thing affecting sound velocity is the hydrostatic pressure, so that there is a positive velocity gradient. There is upward refraction

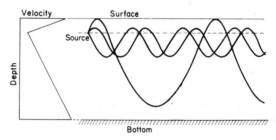

*Fig. 5. Sound channel trapping.*

and the channelled sound progresses by a series of surface reflections. One limitation is that surface channels (and other narrow channels) are not wide enough to trap the lower frequencies—typically those below a few hundred c/s.

*Fluctuations.* So far average transmission has been discussed, i.e. the transmission averaged over some indefinite period. If a pure tone is transmitted the level received may fluctuate wildly, though with a signal of appreciable bandwidth the fluctuations are reduced. There can be a large fluctuation from second to second, and the level averaged over several seconds may fluctuate from minute to minute and hour to hour. At a given range there may even be important level variations with periods ranging from fractions of a second to a year or more. Over several minutes the level may fluctuate through as much as 50 dB or 100,000 to 1 in intensity. There are many contributory causes, one of the most important being the diffuse reflection from a time-varying sea surface.

*Coherence.* There is great current interest in a range of topics which go under a variety of names—such as coherence and stability of wave fields. To give one example, consider two hydrophones at the same range from the source, but at different depths. Due to the multiplicity of ray paths, each of which is fluctuating, the total signal will not be the same at both hydrophones. These signals will again differ from the signal transmitted. These differences and fluctuations impose limits on some types of underwater signal processing.

*Conclusion.* In underwater acoustic experiments the parameter that can vary most is the transmission loss, which is the reason for its importance. Much progress in understanding it has been made, though there is still plenty to learn. The subjects where knowledge is most urgently needed are fluctuations and coherence, not to mention reverberations. In these areas and in transmission loss itself it is worth stressing the wide variety of problems encountered, one or more for each oceanographical condition.

*Bibliography*

EWING M. *et al.* (1948) *Geol. Soc. Amer. Memoir* **27**.
OFFICER C. B. (1958) *Introduction to the Theory of Sound Transmission, with Application to the Ocean,* New York: McGraw-Hill.
SCHULKIN M. and MARSH H. W. (1962) *J. Acoust. Soc. Amer.* **34**, 864.
STEWART R. W. (1962) in *Encyclopaedic Dictionary of Physics* (T. Thewlis Ed.) **5**, 192; Oxford: Pergamon Press.
WILSON W. D. (1960) *J. Acoust. Soc. Amer.* **32**, 1357.

D. E. WESTON

# SPACE MATERIALS, RADIATION EFFECTS ON.

## I. Introduction

Information on the radiation levels in space, although incomplete, together with results from the study of irradiation effects induced in nuclear reactors and accelerators can be used to infer the effects upon materials in space. However, the conditions in space are sufficiently different from those we are used to upon Earth that some of the inferences will be in error and some phenomena, for instance, the impact of gas atoms and micrometeorites at high speeds, are novel.

Until such time as nuclear reactors are used in space vehicles, the outer shell of the vehicle will suffer the most severe radiation, and protect the numerous complex components inside from the less penetrating radiations. It is particularly with the effects upon the outer shell that this article is concerned.

## II. The Interaction of Radiation with Matter

Radiation interacts with matter and transfers some of its energy. If the energy transfer is sufficient, either electrons or atoms, or both, can be ejected from their normal states, producing in the former an ionized atom and an electron, and in the latter a vacancy and a displaced atom in the crystal lattice. It is convenient to discuss these two displacement processes separately.

*1. Electron interaction.* Electromagnetic radiations in and below the infra-red range of energies (i.e., with photon energies less than 1 eV) interact with the atomic vibrations and heat the material they fall upon. In the visible range (∼1 eV) the electrons are raised to higher energy levels without escaping from the atom

and in falling back emit characteristic frequencies of radiation. Energies greater than 5 eV are able to eject electrons from the atom and ionize it, by the *photo-electric effect*; the higher is the incident energy, the greater is the kinetic energy of the ejected electron. At higher energies, electrons can be ejected by the *Compton effect*, where only a part of the initial energy of the photon is used; the photon being degraded to a lower frequency. At higher energies still (greater than 1·20 MeV) the photon has sufficient energy for *pair production*, in which an electron and a positron are generated. At even higher energies these scattering processes becomes less important and the incident photon is more likely to disintegrate the nucleus of the atom it hits. This can produce many particles of high energy.

In addition; when particles pass through matter the electrostatic forces may cause the excitation of the electrons and, if they are sufficient, eject the electron from its atom. This can only happen if the particle is travelling at a velocity higher than the electron in question. The critical energy for ionization thus depends upon the mass of the incident particle and upon the ionization potential of the atoms it excites. Heavy ions moving through their parent lattice are less likely to ionize than are the light ions, mainly because of the lower velocity of the former for the same energy, e.g. this critical energy is $1·7 \times 10^4$ eV when an aluminium ion moves through aluminium and $5 \times 10^4$ eV for copper ions in copper.

*2. Atomic displacements.* Energy in excess of about 25 eV must be transferred to the nucleus of an atom if it is to be ejected from its normal crystal position, when it becomes an interstitial atom in the crystal and leaves a vacancy behind. Conservation of momentum and energy determine the maximum energy $E_2$ which can be transferred to a stationary crystal atom, mass $m_2$, in a head-on collision with a moving particle, mass $m_1$ with energy $E_1$. This energy is given by the relation:

$$E_2 = E_1 \frac{4m_1 m_2}{(m_1 + m_2)^2}.$$

Glancing collisions will transfer less energy than this. All the initial energy $E_1$ is transferred when $m_1$ is equal to $m_2$, and decreasing amounts are transferred as the mass discrepancy increases. Table 1 gives the minimum energies various particles must have if they are to be able to transfer the 25 eV to displace an atom.

Electrons must have high energies before they can eject even atoms of low mass number, and so only if the electrons produced by the Compton effect, or pair production, have energies greater than these will they be able to displace atoms. Also, the cross-sections for these processes are relatively low so atoms are more usually ejected by direct collisions with impinging atoms or ions.

Although the energy of an electron is not likely to be sufficient to eject more than one or two atoms, an

Table 1. Threshold Energies, in eV, to Displace  n Atom
when the Displacement Energy is 25 eV  (  nchin
and Pease)

| Mass number of target | Electrons and gamma rays | Protons and neutrons | Alpha-particles | Fission fragments |
|---|---|---|---|---|
| 10 | $1 \cdot 0 \times 10^5$ | 76 | 31 | 85 |
| 50 | $4 \cdot 1 \times 10^5$ | 325 | 91 | 30 |
| 100 | $6 \cdot 8 \times 10^5$ | 638 | 169 | 25 |

energetic heavy atom can displace many atoms. For example, a proton with 10 MeV can transfer 0·77 MeV to an atom of mass number 50. This atom will collide with its neighbours and being of the same mass will lose its energy rapidly; enough energy being available to displace about $1 \cdot 5 \times 10^4$ other atoms. Thus, heavy particles produce local regions of very intense disturbance wherever they collide directly with an atom in the crystal.

### III. The Introduction of Impurities

The incident particle may come to rest in the bombarded material. If it is an electron this merely charges the material, but in other cases, an impurity element or ion is introduced. When free electrons are available, as in the case of a metal, protons become hydrogen atoms, and alpha-particles become helium atoms. A neutron either decays to a proton, emitting a beta particle of 0·782 MeV, or more likely, is captured by one of the atoms, making it a higher isotope which may decay to become a new element. Other particles can similarly transmute atoms, but with lower probabilities.

### IV. Range of Charged Particles

The energetic electromagnetic radiations penetrate matter deeply, and so do neutrons, the latter because of their inability to interact with electrons. However, charged particles, when they are travelling at high speed, lose energy by ionization and their range is short. Table 2 gives some of these ranges.

### V. Effects of Radiation

Ionization is of little consequence in metals, where the ejected electrons can quickly fall back to their lower energy states, the energy being lost locally as heat along the path of the ionizing particle. This makes metals rather insensitive to radiation. In semiconductors, conduction can be increased during the bombardment by increasing the number of carriers, and the conduction

Table 2. Range (in mm) of charged particles in aluminium

| Energy MeV | Electrons | Protons | Deuterons | Alpha-particles |
|---|---|---|---|---|
| 1 | 1·48 | 0·013 | 0·007 | — |
| 10 | 22 | 0·62 | 0·37 | 0·059 |
| 20 | — | 2·2 | 1·22 | 0·185 |
| 40 | — | 7·2 | 4·8 | 0·592 |

of insulators can be increased. These effects are proportional to the dose rate and are not permanent. Organic materials can be degraded and new compounds form due to the breaking of the chemical bonds by both ionization and atomic displacement. Elastomers are among the most susceptible to radiation damage; their bonds being broken. This causes either the long molecule to degrade, or neighbouring chains to cross-link, this latter producing a hardening and an increase in Young's modulus. Most of these elastomers are not satisfactory after a gamma-ray dose of $10^8$ ergs $g^{-1}$ or a neutron dose of $10^{15}$ cm$^{-2}$. However, man is much more sensitive, his maximum permissible dose being $\sim 10^5$ ergs $g^{-1}$.

The atomic displacements alter the properties of all materials except those where the atoms are free, as in a liquid or gas, or where the temperature is such that the atoms can quickly regain their normal positions by diffusion. These temperatures are normally above half the absolute melting temperature.

At very low temperatures the displaced atoms are unable to migrate once they have lost the kinetic energy acquired in the collision, and a concentration proportional to the dose builds up. These "point" defects alter the physical properties, for instance, the lattice parameter increases, the electrical resistivity of a metal increases, and ionic solids become opaque, due to the formation of colour centres.

At higher temperatures the displaced atoms diffuse and recombine with vacancies with the annihilation of both, or they may be lost at the surface or other sinks, or they may agglomerate to form clusters. These clusters, which may be three-dimensional, or two-dimensional plates one atom thick (when they are best described as dislocation loops), can affect the structure-sensitive properties of the crystal. For instance, the movement of dislocation lines in metals is impeded by them and many metals are hardened after neutron doses of $\sim 10^{16}$ cm$^{-2}$.

The bombarding particles or transmutation products entrapped will only affect the properties appreciably when their numbers exceed those of the impurities normally present. However, quite small concentrations of impurities which are normally gaseous can produce large changes. Hydrogen, for instance, can embrittle metals, and helium in sufficiently high concentration at a high temperature will precipitate as bubbles, increase the volume, and alter the physical properties.

### VI. Radiations in Space

The energy flux one astronomical unit ($1 \cdot 5 \times 10^{11}$ cm) from the Sun is 2 cal cm$^{-2}$ min$^{-1}$, and 99 per cent of this energy is composed of photons with energy less than 4 eV. Most of the energy is in the visible region and can be used for energizing solar cells. The infra-red radiation only heats the shell of the vehicle, whose temperature is determined by the surface emissivity.

Part of the radiation in space consists of high-energy cosmic rays ($10^2$–$10^{11}$ MeV), mainly protons; their maximum intensity on Earth is about 2 particles cm$^{-2}$ sec$^{-1}$. These particles will interact with the vehicle and energetic particles will result, but only if an intense source is approached will their intensity be important.

The inner Van Allen belt of radiation, which is a band of charged particles retained by the Earth's magnetic field as an approximate torus lying in the equatorial plane around the Earth with its centre at an altitude of 3600 km, sustains fluxes $10^{10}$ electrons cm$^{-2}$ sec$^{-1}$ of energies greater than 20 keV and $\sim 10^8$ electrons cm$^{-2}$ sec$^{-1}$ of energy greater than 600 keV. In addition there are fluxes $\sim 2 \times 10^4$ protons cm$^{-2}$ with energies greater than 40 MeV.

An outer belt at an altitude of 16,000 km contains higher electron fluxes ($\sim 10^{11}$ electrons cm$^{-2}$ sec$^{-1}$ greater than 20 keV and $\sim 10^8$ electrons cm$^{-2}$ sec$^{-1}$, greater than 200 keV) but lower proton fluxes ($\sim 10^2$ protons cm$^{-2}$ sec$^{-1}$ greater than 60 MeV).

Doses of $\sim 3 \cdot 6 \times 10^{14}$ electrons cm$^{-2}$ would be experience during one hour in the outer belt and this would produce more than $10^8$ ergs g$^{-1}$, enough to cause most elastomers to deteriorate. However, few electrons will penetrate an aluminium shell more than 1 mm thick. This makes it clear that materials sensitive to ionization should be protected in such regions. The fluxes of the higher-energy electrons are sufficient to cause $\sim 10^{16}$ displaced atoms cm$^{-2}$ in a year in the surface layer of a light element.

The protons will, however, penetrate more than 7 mm of aluminium, displacing atoms and introducing hydrogen. A satellite orbiting in the inner belt for a year would experience a dose of $6 \times 10^{11}$ protons cm$^{-2}$ and as each proton might displace about a hundred atoms along its path, about $10^{14}$ point defects cm$^{-3}$ (i.e., a concentration of about 1 in $10^9$ atoms) and of course $10^{12}$ hydrogen atoms cm$^{-3}$ would appear in a solid. Although the protons will be a hazard to any astronaut they are not likely to cause appreciable changes in any but the most sensitive materials, such as semiconductors, where $\sim 5 \times 10^{14}$ displacements cm$^{-3}$ can produce changes.

Thus the radiations in space are likely to prevent any radiation-sensitive material being used for a space vehicle shell. However, metals are not likely to be affected.

### VII. Surface Effects

The high speed of an Earth satellite ($\sim 10^6$ cm sec$^{-1}$) can result in appreciable bombardment from the atoms in space. At this speed a collision with a stationary atom is equivalent to bombardment with an energetic atom, e.g., a helium atom with 2eV or an iron atom with 30 eV. Space probes may travel considerably faster than this and as the energy is proportional to the square of the velocity these bombardments might erode the surface, whose emissivity normally determines the temperature of the vehicle.

The collision of atoms at low velocity can displace surface atoms *inwards* where they become interstitial atoms. Ions of energy as low as 75 eV cause damage

to the surface of a metal, doses of $\sim 10^{18}$ ions cm$^{-2}$ producing many dislocation loops by the aggregation of the interstitial atoms produced by the bombardment. These atoms only travel a few 100 Å beneath the surface and this is where the dislocations are seen. As the energies increase, so the surface atoms are given a greater momentum and can eject other atoms so that vacancies also are produced inside the metal and the number of point defects increases and also the penetration increases. The impinging atom can also be trapped in the skin and produce extra complications.

The collisions also *eject* atoms, which leave the surface. This latter process is known as *sputtering*. At 500 km above the Earth the pressure is $3 \times 10^{-8}$ mm of mercury, i.e. there are about $1 \cdot 5 \times 10^8$ atoms cm$^{-3}$. Their effective temperature is 2000°K, so their energy is 0·2 eV and they can be regarded as stationary. At a velocity of $10^6$ cm sec$^{-1}$ the satellite would sweep $4 \cdot 5 \times 10^{21}$ atoms cm$^{-3}$ year$^{-1}$. The number of atoms ejected per incident particle, i.e. the sputtering ratio, although not known at these low velocities, might be as high as $10^{-2}$ when $4 \cdot 5 \times 10^{19}$ atoms cm$^{-2}$ would be lost, i.e. $\sim 6$ microns of surface†. This effect could thus only destroy an optical surface after very long time.

Even at low vehicle velocities atoms of higher energy will bombard the shell. Solar winds have been discovered in which the average energies of the atoms are $\sim 50$ eV and their density is $\sim 10^3$ atoms cm$^{-3}$. The effect of such low densities is small, only $\sim 3 \times 10^{17}$ atoms cm$^{-2}$ impinging in one year in such a wind. Although the sputtering ratio is not well defined at these low energies, it might be $\sim 10^{-1}$, so that only $\sim 10$ atom layers would be eroded in one year. Even a vehicle travelling fast enough for the sputtering ratio to become $\sim 50$ (about the maximum attainable) the loss would not be excessive as the gas pressure in space is very low (equivalent to about one atom of hydrogen cm$^{-3}$).

Thus the surface erosion caused by colliding atoms would be most troublesome when the vehicle meets appreciable air friction, but in a normal orbit would probably not destroy an optical surface for a considerable time. In outer space even a vehicle travelling at very high velocities would not suffer serious erosion on account of the extremely low gas pressure.

### VIII. Micrometeorite Impact

Much more damaging than the impacts of single atoms on the surface of a vehicle will be those of the micrometeorites met in space. These range in size from less than a micron upwards; their numbers decreasing as their size increases. Their velocities relative to the Earth vary between $10^6$ and $10^7$ cm sec$^{-1}$ and so even Earth satellites will collide with considerable relative velocities. At a relative velocity of $10^6$ cm sec$^{-1}$ an iron

---

† As the sputtering ratio is higher for glancing incidence (a maximum at about 70° to the normal) the maximum effect will be found on an annulus subtending this angle with the direction of flight.

atom in one of these meteorites would have a kinetic energy of about 30 eV and at $10^7$ cm sec$^{-1}$ it would have 3000 eV. Although the average density of these micrometeorites is thought to be about $3.4$ g cm$^{-3}$, some may have densities as low as 0.05 g cm$^{-3}$. A micrometerorite of radius $10^{-4}$ cm might contain $10^{11}$ atoms, so the total energy involved in an impact could be $3 \times 10^{12}$ eV (i.e. $4.8$ ergs) at a relative velocity of $10^6$ cm sec$^{-1}$.

Recently, with the incentive that estimates of this sort provide, experiments on the penetration of small projectiles in metals have been performed. The velocities used have approached those of the slower micrometeorites and extrapolation of the results to high velocities is uncertain, but has been used to predict the probability of puncture of the outer shell of a vehicle.

This work has shown that whereas at low velocities the impinging particle punches a narrow, deep cavity, it loses its shape and disintegrates under the high pressures and temperatures involved at high velocities. A roughly hemispherical crater surrounded by a lip of displaced metal is then formed at the impact, but brittle materials tend to fragment around the crater. The volume of the crater ($V$) is proportional to the kinetic energy ($E$) of the projectile and satisfies the empirical relation

$$V = \frac{kE}{\varrho_t c_t^2}, \qquad (1)$$

where $\varrho_t$ is the density of the target, $c_t$ the velocity of sound in the target and $k$ a constant. Thus, if a projectile with a velocity equal to $c_t$ struck the surface of a plate of the same material, the crater volume would be $k/2$ times the volume of the projectile. For $\frac{1}{8}$ in. diameter spheres in the velocity range $(1 \text{ to } 5) \times 10^5$ cm sec$^{-1}$, $k$ was found to have the value of $25.4$ for steel, zinc, aluminium, magnesium and lead and in the range $(7.5 \text{ to } 22.5) \times 10^4$ cm sec$^{-1}$ a value of 82 for iron, copper, zinc, silver, aluminium, tin and lead.

The crater depth $P$ will depend upon the shape of the crater, but for a hemispherical one of radius $P$ equation 1 can be written:

$$\left(\frac{P}{D}\right)^3 = \frac{k\varrho_p}{8\varrho_t}\left(\frac{v_p}{c_t}\right)^2 \qquad (2)$$

where $D$, $\varrho_p$ and $v_p$ are, respectively, the diameter, density and the velocity of the spherical projectile. This relation suggests a linear relationship of the penetration $P$ with the projectile diameter, and experiments have verified this. The variation of penetration with velocity is vital if confident predictions are to be made at micrometeorite velocities. Here there is doubt. Relations of the form:

$$\frac{P}{D} \propto \left\{\frac{v_p}{c_i}\right\}^{1.4} \quad \text{and} \quad \frac{P}{D} \propto \frac{v_p}{c_t} \qquad (3)$$

have been suggested as respectively fitting the previously quoted results. For either of these relations (3)

to be compatible with equation 1 and a spherical crater, $k$ must be some function of the projectile velocity; the two values for $k$ already quoted would suggest this. If a value of 8 is arbitrarily assumed for $k$ at a velocity of $5 \times 10^6$ cm sec$^{-2}$ then from equation 2 the penetration of aluminium ($c_t = 10^5$ cm sec$^{-1}$) would be

$$P = 4.6D$$

(because the cube root of the ratio of the densities is approximately unity).

The flux of micrometeorites greater than $10^{-4}$ cm radius near the Earth is about 10 cm$^{-2}$ sec$^{-1}$. Using the above estimates each of these would produce a crater $9 \times 10^{-4}$ cm deep and $2.5 \times 10^{-6}$ cm$^2$ in area in aluminium. As $4 \times 10^5$ micrometeorites would hit each cm$^2$ in 10 hours, this would be the time taken for the surface to lose optical flatness. Particles of larger size have low fluxes and, for example, the flux of particles greater than $10^{-2}$ cm radius is $10^{-12}$ cm$^{-2}$ sec$^{-1}$. Using the above estimates, particles of this size would penetrate aluminium less than $9 \times 10^{-2}$ cm thick. Thus such an aluminium sphere with a radius of 1 metre would probably be punctured in about six months.

*Table 3. Values of $\varrho c^2$ for various materials*

| Metal | $\varrho c^2$, $10^{11}$ dynes cm$^{-2}$ | Metal | $\varrho c^2$, $10^{11}$ dynes cm$^{-2}$ |
|---|---|---|---|
| Tungsten | 35 | Copper | 12 |
| Molybdenum | 29 | Cast iron | 9·1 |
| Rhodium | 29 | Zinc | 9·0 |
| Beryllium | 29 | Gold | 7·8 |
| Chromium | 24 | Silver | 7·7 |
| Electrolytic | | Aluminium | 6·8 |
| iron | 20 | Iridium | 5·2 |
| Nickel | 20 | Tin | 4·6 |
| Tantalum | 19 | Magnesium | 4·2 |
| Platinum | 16 | Lead | 1·6 |

Equation 1 indicates that the minimum penetration will occur with materials with high values of $\varrho_t c_t^2$, i.e. those with high values of Young's modulus. Table 3 lists materials in the order of decreasing Young's modulus.

Of the low-density materials, beryllium has a much higher value of $\varrho c^2$ than aluminium, which itself is a little higher than magnesium.

### IX. Conclusions

Despite the varied electromagnetic radiations present in space their intensities are believed to be low and their only important effect is likely to be the surface heating due to the infra-red radiations. The most radiation-sensitive materials, e.g. semiconductors and elastomers are likely to be affected by the electrons in the Van Allen belts if directly exposed. Most of the sensitive devices will be shielded from the less penetrating radiations and then only the protons in these belts will be important and able eventually to impair semiconductors.

The outer shell of a vehicle should be made of a material insensitive to radiation, preferably a metal. Although this would suffer erosion by collision with gas atoms in space this is not likely to be appreciable except in the Earth's atmosphere. However, the bombardment of micrometeorites will pit the surface which will occasionally be punctured by the larger particles. This latter effect will determine the minimum thickness of shell which should be used for a given application and will restrict the materials which can be used.

*Bibliography*

BILLINGTON D. S. and CRAWFORD J. A. (1961) *Radiation Damage in Solids*, Oxford: University Press.
DUBIN M. *Scientific Uses of Earth Satellites* (edited by J. A. Van Allen) (1956) Ann. Arbor: University of Michigan Press; (1957) London: Chapman & Hall.
KINCHIN G. H. and PEASE R. S. (1955) *Rep. Prog. Phys.* **18**, 1.

R. S. BARNES

**SPALLING OR SCABBING.** Spalling or scabbing are terms used in the explosive metal working field to describe the process of fracturing due to the interaction between the tensile front of a reflected stress wave and its incident compression tail. The dynamics of such a process can be considered in the following manner.

When an impulsive load, such as that produced by the detonation of an explosive charge, is generated on one surface of a thick metal plate, a high-intensity, sharp-fronted compression wave is initiated which propagates outward through the metal. When this disturbance impinges against the free, opposite surface of the plate it is reflected as a wave of tension. If the wave is decaying in intensity, the superposition of the reflected tensile front on the incident compression tail produces a localized region of high tensile stress in the metal. When this tensile stress exceeds the critical dynamic fracture strength of the metal, a lateral fracture surface known as a spall fracture or scab fracture will be produced in the plate. Any metal piece removed from the parent body by this process is called a spall or scab. Should the incident stress wave be of sufficient amplitude, the spalling process may be repeated several times, with the remainder of the incident wave in each case impinging on a freshly created boundary surface; an effect termed *multiple spalling*.

Fractures of this type were first described by B. Hopkinson about 1912, and are sometimes referred to as Hopkinson fractures. Interest in such fractures has recently increased due to their occurrence in many areas of the newly developing explosive metal working industry.

J. PEARSON

**SPARK CHAMBERS IN HIGH ENERGY PHYSICS.** The spark chamber is an electronic device for locating the trajectory of energetic charged particles. An outgrowth of the spark counter, it originated in the observation by Keuffel (1949) that a parallel-plate spark counter discharge occurred at the point where the charged particle traverses the gap between the parallel plates.

*Properties of Spark Chambers*

A spark chamber system consists of an array of elements. A single element in the array usually consists of a pair of parallel plates; sometimes coaxial cylinders may be employed, or a plate may be replaced by an array of wires. In operation the gap between the plates is filled with a suitable spark-chamber gas, most often neon or a neon-helium mixture. Spark chamber detectors are generally operated as triggered arrays; that is to say, they are not continuously sensitive but are pulsed on for a short period upon the receipt of a suitable triggering signal. Consequently, auxiliary triggering apparatus must be used to furnish the required signal.

Upon receipt of the appropriate trigger a high voltage pulse is applied to the spark chamber elements. In the usual case of parallel plates separated by a centimetre or so, the voltage required in most gases

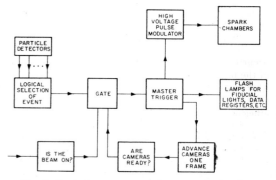

*Fig. 1. Logical arrangement for a spark chamber system with camera readout.*

is in the range 5–15 kilovolts. A suitable modulator is needed, which will apply this voltage with a rise time of $0.1\,\mu$sec or less and with a duration up to a half-microsecond or less. Usually a spark-gap modulator is used. In a properly operating spark-chamber system the application of the high voltage pulse will produce no breakdown between the plates unless a charged particle leaving an ionized trail behind it has passed through the gap just before the application of the high voltage pulse. If several particles have traversed the gap, discharges occur along or close to the path of each particle. Figure 1 illustrates the components of the typical spark chamber system.

Several different characteristic times describe the operation of the spark chamber; these may be defined as follows:

*1. Memory or clearing time.* The passage of an ionizing particle through the chamber leaves behind a trail of ion pairs which become the nucleus of the spark discharge. Such ion pairs will either recombine, diffuse to the walls, or be collected, by an electric field if the high voltage to produce a spark is not applied. The period following the passage of a particle, during which the high voltage pulse will produce a track, is defined as the memory time. In the absence of measures to shorten this time, it may be as long as 20–40 microseconds, depending on the gas medium. The memory time is usually minimized to permit the chamber to operate in a high intensity particle flux.

Two methods exist for decreasing the memory time. One is the application of an electric "clearing" field, generally up to 100–200 V/cm. This will produce clearing times, depending on the gap width, of perhaps 0·2–1·0 microsecond, the longer times applying to the wider gaps. An alternative method, independent of the gap width, is the use of chemical clearing or chemical recombination. Electronegative gases such as sulphur dioxide may be used to combine with the free electrons in the gas to produce negative ions. The spark discharge will not occur from heavy ions; free electrons are necessary for its initiation. Clearing times of about one or two microseconds can be obtained in this way. This method is most useful for very wide gap chambers (10 cm or more) in which electrical clearing times this short are not easy to obtain.

*2. High voltage delay time.* The interval between the occurrence of the triggering event and the appearance of the high voltage pulse on the spark chamber array. This time is usually from 0·1 to 0·5 μsec. The delay time must be less than the memory time, or loss of efficiency will occur.

*3. Dead time or recovery time.* The interval between the application of a high voltage pulse to produce a track and the time when the chamber is ready to record another event. This interval depends upon the time required to clear away every ion from the passage of a spark; it may vary over a considerable range. In bright sparks intended for photography the spark discharge may well produce $10^{12}$ or more ion pairs. These must all be removed by recombination and/or clearing before the chamber is again ready to be used.

Dead times from 0·25 to 10 milliseconds or more have been observed in spark chambers operated in various manners. The shorter times are associated with wire array chambers in which particle detection is achieved by electronic methods like core storage. The discharge current may then be much less than that necessary to produce bright sparks for photographs.

*4. Camera dead time.* In the case of chambers whose output is recorded by a camera on photographic film, the time necessary to advance the film in the camera to present a new unexposed frame to the chamber may be a limiting factor. The fastest currently available 35-mm pulsed film cameras have advance times in the range of 15–40 milliseconds and limit the speed at which successive events may be taken in visual chambers to a rate of the order of 40 frames per second.

## Triggering Systems

That the spark chamber is inherently a triggered device lends it one of its chief distinctions: its ability to record only events preselected by counter logic of considerable complexity. Such logic may include, for example, the accurate location and identification of the incoming particle, the assurance that the incoming particle has interacted in the target, and determination of the type of event selected. It may include anticoincidence counters to exclude events in which charged particles occur in undesired locations, and coincidence counters to determine the number, trajectory, and other properties of the charged particles it is desired to observe. Figure 2 shows a complex triggering system.

Compared with a scintillation counter, a spark chamber has a longer resolving time, and hence a relatively limited capability of tolerating high particle fluxes. The chamber will record all particles that occurred within an interval equal to the clearing time, ending at the instant of application of the high vol-

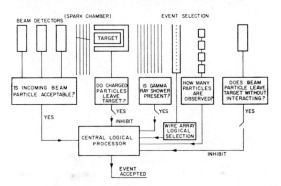

*Fig. 2. A trigger logic, with the anticoincidence counter surrounding the target to select events in which no charged particles leave the target, and the far downstream counter to exclude events in which the beam particle leaves the target without interacting.*

tage pulse. Thus, with a memory as short as 0·2 μsec, an average particle flux of one Mc/sec will produce, about 20 per cent of the time, a second unwanted track through accidental time coincidence. Consequently, it is possible to use not only scintillation counters as auxiliary trigger devices but also gas-filled detector like proportional counters, provided their delays are sufficiently short. A wire array has recently been used as a set of discharge counters for the logical selection of events for spark chamber triggering.

## Modes of Operation

Spark chambers may be classified either according to their mode of operation or according to the type of data which they yield.

The three modes of operation of spark chambers are known as the *sampling, projection* and *delineating* modes. In the *sampling* mode each element of a spark chamber samples the tracks of particles that traverse it and yields as data for each particle the coordinates of one point on each track. A spark chamber is an assembly of such elements, and the track is reconstructed from the ensemble of sample points. The conventional parallel-plate spark chamber with gaps of a centimetre or less is an example of this type of operation; each gap yields as data the coordinates of a sample point on the track of the particle traversing it. The optimum orientation of such chambers is at right angles to the trajectory of the particle traversing them. Figure 3 illustrates such a chamber.

When the path of the particle approaches parallelism to the plane of the spark chamber plates, the

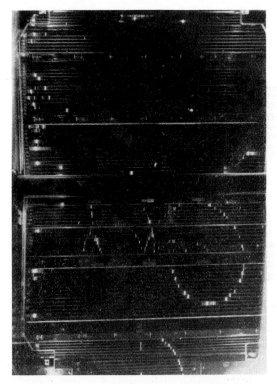

*Fig. 3. A set sampling chamber in a magnetic field. An incoming pion, at bottom, moving upward, strikes the target at above the lowest-chamber. One charged particle emerges backward, and three gamma rays are produced. They are converted into electron pairs in a lead sheet (not shown) between two chambers.*

operation of the spark chamber element transfers to the *projection* mode. (This mode of operation was originally termed a "discharge chamber" by Fukui and Miyamoto (1959) who first observed it.) In this mode of operation the particle moving more or less parallel to the gap produces a series of streamers or sparks along its path. The streamers follow the electric field

from one plate to the other. Their separation averages a few millimetres along the track in conventional gas mixtures at atmospheric pressure. In this mode the information obtained from the track is that obtained by projecting the trajectory on a plane parallel to the plates of the chamber and all information concerning its coordinate in the direction normal to the plane is lost (except for the beginning and end of the track). This mode of operation is advantageous for measurements of specific ionization since the density of the streamers along the track is a direct function of the primary ionization. Figure 4 shows a projection chamber track.

In the *delineating* mode the spark chamber is so operated that the observed track reproduces or delineates as closely as possible the actual trajectory of the particle through the system. In this case the operation resembles closely that of a cloud chamber, since it offers a gaseous medium in which the track of the particle becomes visible. Two types of delineating chamber operation are known at present. In one, the wide-gap chamber, the separation of the two parallel plates of a conventional chamber is increased to several centimetres or more, and a high-voltage pulse of short rise time applied. A particle traversing the gap at an angle not exceeding about 40° to the electric field will then produce a spark discharge which conforms to the trajectory of the particle (with the exception of small regions at the plates). As the angle of the trajectory with the electric field increases, the appearance of streamers that follow the electric field is noted, and finally at angles above about 45° the chamber goes into the projection mode. Figure 5 shows a wide gap delineating chamber.

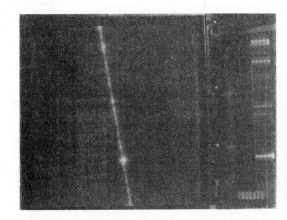

*Fig. 4(a). A cosmic-ray track in a projection chamber. At left the track is seen parallel to the electric field through the transparent gauze electrode; at right a 90° view shows the streamers crossing the gap. The gap appears to be subdivided by markers on the lucite window.*

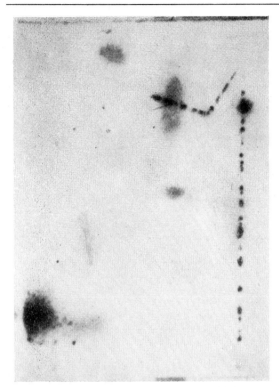

*Fig. 4(b) Projection chamber track scattered in the chamber gas. (Photo by G. Charpak.)*

The second type of delineating chamber is that in which the track of the particle is made luminous independently of the orientation of the track with respect to the chamber; it is a type of electrodeless discharge. To achieve this it is necessary to cause the primary track electrons to produce avalanches in the immediate vicinity of their point of production, with sufficient light intensity to be photographed, either by using a radio frequency high voltage pulse or by using pulses of very short duration. This mode of operation has the advantage that its sensitivity is completely isotropic; that is, it is independent of particle direction. Unfortunately, while such tracks can be seen even for minimum ionizing particles, they have not yet been produced with sufficient intensity to allow them to be photographed except with lenses of very wide aperture. In consequence, the depth of field is greatly restricted. Figure 6 shows a spectacular photograph in such a chamber (from Chekovani).

### Data acquisition

In sampling chambers there is a wide variety of procedures whereby data on the location of sparks may be recorded. The elementary datum in a sampling chamber is the position of a single point in space; it may be recorded directly in digital form, as well as in analogue form. In the conventional parallel-plate spark chamber, using camera and film recording, the locations of the sparks that sample the track are recorded on stereo photographs and may later be reconstructed in space by conventional stereo-reconstruction procedures.

Sampling chambers are subject to systematic errors; the track of primary ions produced by the particle may be shifted appreciably by the action of electric and magnetic fields before the high voltage that produces the spark is applied. The use of gaps in pairs, with oppositely directed electric fields, minimizes such errors.

At present most spark chamber experiments use parallel-plate sampling chambers with camera and film recording, with two or more stereo views. The photographs are then analysed by methods similar to those used in bubble chamber analysis.

*Fig. 5. (a) A pair of wide-gap delineating chambers above a six-gap sampling chamber, showing the same cosmic-ray track in both. A 90° view of the wide gaps appears on the right.*

*Fig. 5(b) Appearrance of the same chambers with a track making a large angle to the electric field. Note the streamer curtains.*

An alternative method of operation of spark chambers with visual data retrieval is provided by the use of television pickup tubes, whose resolution is, however, limited compared to that of a conventional camera and film. For sufficiently small chambers reasonably adequate resolution may be obtained, and the system offers the possibility of directly recording digitized data by using the temporary storage within the television pickup tube as a buffer out of which the data are read in digital form.

The *acoustic spark chamber* employs microphones and a clock to measure the elapsed time between the production of a spark and the arrival of the sound produced by the spark at several microphones situated at the perimeter of the chamber. A digital clock records the time of flight of the sound directly in digital form, and triangulation methods permit the location of a spark with considerable accuracy. The complexity of this method increases considerably when there is more

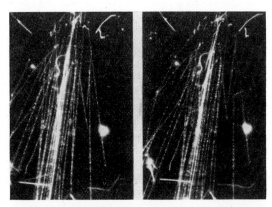

*Fig. 6. A comic-ray shower seen in small angle stereo in a wide-gap short-pulse delineating chamber one metre long (from Mikhailov et al. (1964)).*

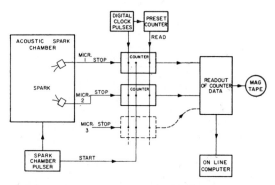

*Fig. 7. In the acoustic spark chamber the sound signals are used to stop a clock started by the high voltage pulse. Readings correspond to the time of flight of the sound from spark to microphone, and are later translated into coordinate data.*

than one spark in the chamber; the technical problems attendant upon multiple sparks have not yet all been solved. Figure 7 illustrates data collection in the acoustic chamber.

Finally, digitizing of the spark chamber data can be achieved by digitizing one of the two parallel plates, thus producing a wire array chamber. An array of parallel wires about a millimetre apart may be substituted for one of the plates of the chamber. Each wire is threaded through a small ferrite core of the type used in magnetic memories. The digital informa-

tion thus obtained locates a spark occurring on a given wire in the two dimensions of a plane perpendicular to the wire. It yields no information either about the position of the spark along the wire or about the intensity of the spark (in principle, methods are available for retaining both these types of information, but they have not been applied in practice).

Since the information obtained by a single wire array is incomplete it is necessary to use a second wire array at right angles to the first in order to obtain the missing third coordinate. Such a pair of wire arrays then gives a complete set of coordinates for a given sample point of particle trajectory. The pair of arrays is entirely analogous to the use of hodoscopes of geiger counters or scintillation counters and is subject to the same type of ambiguity in cases in which more than one particle is present. Such ambiguities may be largely removed by a third or redundant array. A wire chamber system then may consist of a series of elements, each element consisting of a set of three arrays, sampling the particle trajectories at a given plane. Figure 8 shows a wire array system.

The parallel plate geometry used for sampling chambers may be generalized to coaxial cylinders

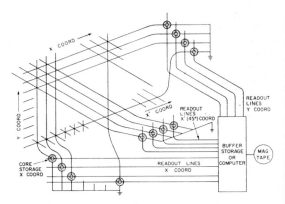

*Fig. 8. In a wire array system a spark to any wire sets a corresponding core. A set of these arrays, two at right angles, and one redundant array at 45°, is capable of recording digitally the coordinates of a large number of sparks.*

without affecting the operation; the cylindrical geometry offers advantages under certain circumstances, such as scattering from a point target. Combinations of parallel plate chambers can be used to devise systems in which particles can be adequately recorded no matter what their direction.

### Data Processing

The kind of data processing necessary with spark chambers is determined by the method used for recording the data and by the complexity of the event studied. Most experiments to date have used camera and film recording, and they have taken over, more or less unmodified, techniques previously developed for the analysis of data in bubble chambers. The photographs are scanned, measured on machines similar to those used for bubble-chamber measurements, and processed by computer programmes similar to, or identical with, those used for bubble chambers. A considerable degree of simplification may be possible, especially if all charged particles tracks are straight lines and the photographs are suitably composed. The measurement of tracks may then be simplified to operations with a ruler and protractor; computer operations may also be much simplified. For more complex events, or those in which magnets are used, the simplification is not so readily achieved.

*Automatic data processing.* The simplicity of the data in cases where particles move in straight lines has encouraged a movement toward automatic data processing. Spark chambers lend themselves to the application of automation both in scanning and measuring considerably more readily than do bubble chambers. This is because the photographs contain far less information (in the information theoretic sense) than do bubble chamber photographs. The useful data are consequently far less dilute and more readily recognized by mechanical means.

In general an automatic data processing system for spark chambers consists of a device which reads in coordinate data from the sparks, links them together into particle trajectories, then solves the kinematic equations of the problem and returns the answer to the experimenter in a usable form.

Most obviously adapted to automatic processing are the data taken with wire array chambers. They are, so to speak, born digitized, and cradled in magnetic core storage. Relatively few operations are necessary to convert the observed coordinates of points to tracks and to carry out the kinematics calculations characteristic of the event. Similarly, acoustic spark chamber data readily yield the digitized coordinates of a spark, at least when only one spark occurs in each gap. The same is true of spark chamber data registered optically either by television or by camera methods.

Significant progress in the automatic analysis of spark chamber data has been made for events in which there is only one track in each chamber and this track is a straight line. Successful "on-line" computation of the results of such experiments, that is to say during the operation of the experiment itself, has been achieved for television recording, for acoustic chambers and for wire arrays. For photographic recording using a camera and film, the operation is not on-line simply because the film has to be developed before it can be automatically scanned and measured. Were it not for this delay, this operation could be considered an on-line one as well, since the corresponding automatic analysis of photographs has also been achieved.

*Comparison with Bubble Chambers*

As a general purpose detector for visualizing complex events, i.e. events with more than one vertex, the spark chamber system (especially when using magnetic fields)

appears likely to offer serious competition to the bubble chamber in certain areas, notably those in which its ability to select a restricted class of events from all those occurring provides a major advantage. Since sampling chambers have limited resolution, they are mainly restricted to experiments in which high resolution plays no essential role; there are many of these. For systems using projection or delineating chambers, either alone or in combination with sampling chambers, this restriction may disappear. The spark chamber offers a particular advantage for events in which gamma-ray detection is necessary, in addition to the measurement of momentum with high precision; this is because the inhomogeneous construction of the spark-chamber system allows separation of these functions in separate detector volumes, each adapted to its function. High energy events in which neutral pions are produced exemplify this class.

The combination of triggering selectivity with high beam flux tolerance yields a rate of observation of events of a given class that reaches values two to three orders of magnitude greater than the bubble chamber. Thus, in a spark chamber with a hydrogen target, with a 1·1 BeV/c pion beam incident, associated production events in which both K-meson and lambda hyperon are observed may occur at the rate of two or three per pulse, as compared with one every few hundred pulses in a large hydrogen bubble chamber.

The spark chamber is at a disadvantage, weight for weight, in dealing with events of small cross-section produced by a diffuse beam, as of neutral particles such as the neutrino; a large bubble chamber combines high resolution and large target volume in a particularly advantageous manner.

*Bibliography*

ALIKHANIAN A. I. *et al.* (1963) *JETP* **17**, 522.
ARECCHI F. T. *et al.* (1961) *Energia Nucleare* 8, 213, 539; (1962) **9**, 713.
CHARPAK G. and MASSONET L. (1963) *Rev. Sci. Instr.* **34**, 664.
FUKUI S. and ZACHAROV B. (1963) *Nucl. Instr. & Meth.* **23**, 24.
MIKHAILOV V. A. *et al.* (1964) *JETP* **18**, 561.
*Proceedings of Informal Conference on Non-Film Spark Chamber Recording and Associated Computer Use,* March 3–6, 1964. CERN Report 64-30, Geneva: CERN.
ROBERTS A. (1963) *Spark Chambers: The State of Art, Proceedings of Int. Symp. on Nucl. Electronics,* 1, Paris; European Nuclear Energy Agency.
ROBERTS A. (1964) *Recent Progress in Spark Chamber Technology, Nucleonics* **22**, No 5, 68.
Sampling Chambers (1962) *Proceedings of the 1962 CERN Conference on Nuclear Instrumentation,* Geneva: CERN.
Spark Chambers Symposium (1961) *Rev. Sci. Instrum.* **32**, 480.

A. ROBERTS

**SPEECH MECHANISMS IN THE BRAIN.** *Receptor mechanisms*. Speech may be received in principle through the sense organs of any sense modality. The most important are touch (for the blind), vision (in reading), and hearing (normal spoken speech). The last of these is the paradigm of speech and language mechanisms. Man as a speaking animal antedates man as a reader or a writer by a very long period.

The receptor mechanisms therefore are in no way specialized for the handling of speech. The auditory and visual pathways are those which encode non-linguistic signals in general. The auditory pathway has a bandwidth from about 15 c/s to 15 kc/s, the upper end of which falls with advancing age. Much attention has been paid to the ear as a device for performing Fourier analysis on the incoming complex waves. An approximation to such an analysis seems to be performed, but this is best regarded as a coding function, rather than the primary function of the ear,

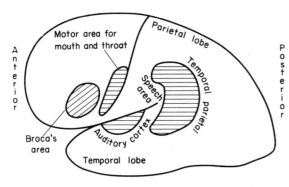

*Fig. 1. Left side of external surface of the brain.*

since the results of such an analysis, namely pure tones, are almost never heard by humans because of non-linearities in the auditory pathways, and are certainly never heard as components of speech. Considering the immense importance of speech and language to humans, it is hardly too much to say that the main function of the auditory pathway in man is to code speechwaves into a form which allows their analysis by the nervous system. The ear is best regarded therefore as a transducer rather than an analyser.

The sensitivity and bandwidth of the human ear do not greatly exceed those of many animals, and are inferior to several. The unique qualities of human speech cannot be assigned to any mechanism known to lie between the ear and the auditory cortex of the brain, which lies on the upper surface of the temporal lobe of the brain on either side (Fig. 1). In man sounds which enter one ear are represented on both sides of the brain at the cortex. The function and performance of the human auditory cortex seems to be similar to that of other animals, and does not include the handling of speech and language. Damage to this part of the brain will produce deafness: it does not cause specific loss of speech or language performance.

*Perception of speech.* Speech is remarkable for its resistance to distortion, transformation, interruption, masking by noise, etc. This is due to the great redundancy (in the Communication Theory sense of the word) which is found both in the structure of the speech wave itself and also in the words which it carries. The sequential dependencies between successive momentary values of the speech wave are very high, and there is a very great imbalance of frequency of use among the words of a language.

Most of the acoustic energy in spoken speech is provided by components of less than 1000 c/s, and is due to the vowel sounds. Most information (again in the Communication Theory sense) is provided by the higher frequency components which provide comparatively little energy and are associated with the consonants. The average spectral distribution of speech is given in Fig. 2 for a long term sample over many words. The relative importance of the high and low energy components is elegantly shown by the effects of centre-clipping or peak-clipping a speech wave. If the wave is clipped every time it crosses the time axis, and is then amplified, all amplitude modulation is lost, and what remains is a train of square waves of equal amplitude but varying wave-length. Such speech remains about 90 per cent intelligible. If, however, the centre of the wave is clipped out, and only peak values left, a rapid and catastrophic decline in intelligibility is found for even quite moderate degrees of clipping, since what is then removed are the informative, low-energy components.

Speech tends to be spoken at around 150 to 200 words per minute, but is intelligible at very much greater rates. Measurement is difficult, since at high rates so much is received per second, that items may be lost from short-term memory before they can be reported, due to overloading of the retention and retrieval, rather than receptive, mechanisms. But it seems that for short messages, using a small vocabulary such as spoken digits, ten words per second may still be fairly intelligible. Under normal speech loads, therefore, the system is operating well within its limits.

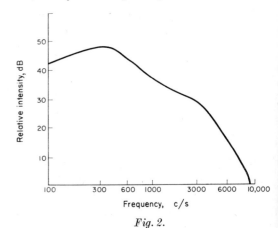

*Fig. 2.*

At loudness levels corresponding to normal "conversational" speech, speech is intelligible at about − 10 dB with respect to white noise, and is detectable as speech down to about − 18 dB. The intelligibility threshold can be altered greatly by making use of the semantic content of the signals. Telling a listener what classes of words he may hear lowers his threshold considerably, while giving him false information, or using a very large population of possible signals, raises it. The human brain has a tendency to hunt for meaningful sounds, and particularly for speech sounds, in ambiguous auditory situations. A classical example was found by Fromm during World War II. Danish listeners were told that the message to which they were to listen was English from a broadcast which had been "jammed" by the Germans. The material was

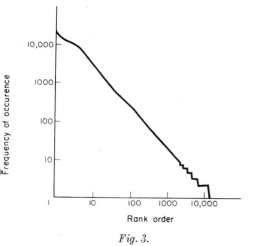

*Fig. 3.*

in fact Danish speech played backwards. The listeners were able to "identify" not merely words but whole phrases of English.

Several important mechanisms have been developed to aid the reception of speech in the presence of competing messages (the "cocktail party problem" as Cherry has called it). Foremost among these is auditory localization. Two sounds which are as little as ten degrees of angular separation apart are almost never mistaken one for another, and cause only very slight interference. Truly dichotic presentation gives an improvement of 100 per cent over listening to two messages presented to the same ear, measured by per cent correct reporting of messages received. Of lesser importance but still useful, are the cues provided by pitch and timbre (such as the difference between male and female speakers), and even loudness (it being easier to listen to the quieter of two messages when they are received over the same loudspeaker than one of two equally loud messages). The most difficult problem for the listener is the identification of the overall statistical properties of the message to be received. Once identification has been achieved, it is comparati-

vely easy to continue listening. The above mechanisms may be regarded as ways of detecting important cues for identification.

There is much evidence suggesting that a running comparison between simultaneous messages, and between messages now being received and those recently received is of primary importance in speech perception. Cherry has produced very precise predictions of auditory localization behaviour using auto- and cross-correlation equations to describe the relation between the signals at the two ears. Treisman has produced evidence that higher order analysis of speech works in a similar way. She carried out a series of experiments on "speech shadowing". If a subject shadows speech presented to one ear while to the other ear is presented the same message leading or lagging in time, then when the rejected message lags in time they are recognized as identical when they are as much as 6–10 sec apart in time. When the rejected message leads in time, then they are not recognized as identical until there is virtually zero time difference. This suggests that there is a comparison of new signals with memory traces of earlier ones.

Strategies such as those just described are powerful methods of coping with the problem of receiving one message under conditions of multiple input. Recent work suggests that what were thought to be limits on the input mechanisms of speech may often in fact be due to failures of storage and retrieval for output.

*Central mechanisms.* The regions of the cerebral cortex which are concerned with speech and language are fairly well defined, although their functional and anatomical organization are not well understood. In particular the way in which the comparatively slight change in brain structure conveys the enormously powerful ability to speak is entirely obscure. It is not clear from the functional anatomy of the brain why other species should not speak.

The regions concerned are shown in Fig. 1. They are peculiar in being localized usually only on one side of the brain, the corresponding area on the other side being used for a different function. Almost all other functions are bilaterally symmetrically represented in the brain. As a rough generalization, it may be said that if a person is right handed, then the speech areas lie on the left side of the brain. The strictness of the contralaterality is greater for right-handers than for left-handers, and cases have even been reported, though rarely, where the speech areas have been on the same side as the person's dominant hand.

Damage to Broca's Area results in quite severe speech deficit, which often seems to be mainly to do with speech output. The area is anatomically closely related to those parts of the brain which control the non-linguistic movements of the face and throat muscles. The effects of damage persist for some time but usually eventually pass off.

Damage to the parietal-temporal area results in adults in severe, permanent speech and language deficit. This often seems to be mainly of a perceptual, or cognitive nature, rather than to do with speech output. But these distinctions have recently been strongly challenged by Howes, in the CIBA symposium to which reference is made below. If damage occurs very soon after birth the other side of the brain may take over, but this ability to compensate for injury is lost very soon after birth.

Electrical stimulation of the speech areas may produce cries, vowel sounds, repetitions of words being spoken, or speech arrest. No coherent speech has been produced, merely the bricks, rather than the buildings, of language. In some patients with temporal lobe epilepsy, stimulation of the parietal-temporal region may cause them to "re-live" past experiences very vividly, hearing tunes, and seeing, smelling, etc. familar sights (see Penfield, below).

In examining the way in which speech is handled by these mechanisms we are looking at the central mechanisms of integration, analysis, memory, and synthesis, where information from all parts of the body, and from the external world, comes together with information stored in memory to interact with it, and to lead to the selection and initiation of responses. This is functionally probably the most complex part of the whole human body.

As an example of the interaction of information, we may note that very often when visual material such as letters of the alphabet are read, the confusions made between them are between those which sound alike rather than between those which look alike. As an example of the complexity of response we may note another experiment of Treisman's. She presented a message in English to one ear of a listener, who shadowed it. The other ear received the same message translated into French. Listeners who knew only little French found no interference from the rejected message; but bilinguals found it impossible to listen to one and ignore the other. The brain was responding to meaning regardless of the phonetic code in which it was carried.

If we wish to measure the rate at which information is handled by these mechanisms, we have a serious problem of measurement. Should we take as our unit phonemes, words, phrases, ideas, sentences, or what? It seems likely that the word is fairly appropriate, since it can be shown that normally phonemes are not processed singly. Moreover the rate at which people will read words in Western languages and Chinese ideograms is about the same, although meaning is coded into the two types of language rather differently.

At the level of the statistical properties of language is one of the most firmly established behavioural laws, *Zipf's law*. This states that if we take a sample of written or spoken language, and rank the words in terms of the number of times each occurs in the sample at our disposal, then

$$f(r) = k \cdot 1/R \qquad (1)$$

where $f(r)$ is the number of times a word of rank $R$ occurs, and $R$ is the rank position of the word, and $k$ is a constant. A typical plot of data giving Zipf's law is given in Fig. 3. It has been suggested that a more precise fit is given by the lognormal equation

$$f(r) = k(R + \alpha)^{-\beta} \qquad (2)$$

where $f(r)$, $R$, and $k$ are as above, and $\alpha$ and $\beta$ are parameters which are characteristic of the writer or speaker. The original form, however, gives a good approximation.

The form of the law is found in samples as small as 5000 words, and the source seems to have little effect. With slight adjustments of the parameters in the lognormal equation, for example, newsprint, normal adult speech, the prose of well-known novelists, children's speech, and even the speech of some brain-damaged speakers all fit the equation. The equation may be regarded as the canonical form of statistical description of language, since all other measures, such as type-token ratio, Shannon entropy, may be derived from it. It has been suggested recently that the effect of brain damage to the speech centres, which is usually described in terms such as a "loss of naming ability", "loss of syntax", "loss of colour words", etc. may better be regarded as a change in the mechanism which results in a shift in the parameters of equation 2, meaning that words of high rank number (low use) become still more rarely used. Thus the particular type of aphasia seen would be a reflection of the relative frequencies of different parts of speech or words in the normal speech of the damaged subject, rather than as separate clinical entities. The effects of brain damage might then be regarded as a failure of the retrieval system to cope with the problem of hunting through a store, organized in terms of probability rankings, for specific words which are hard to retrieve, leading to the use of synonyms, paraphrase, etc., as a time-saving device.

It will be apparent from the form of Zipf's law that there is gross redundancy (in the Communication Theory sense) within speech at the level of the emission of words. Some words are immensely more probable than others. The vocabulary of a verbally fluent adult may top 30,000 by quite a large margin, measured by his ability to recognize a word he comes across. The vocabulary of Basic English uses very roughly a tenth of that number. And recent work suggests strongly that in the middle of a passage of running speech or prose, the listener behaves as if the next word he must recognize is being picked from a subset of his vocabulary which is equivalent in Shannon information to about 2 Bits or less.

This reduction in the amount of information to be processed from moment to moment by an order of magnitude from 20 Bits to 2 Bits due to the great redundancy of speech explains why the brain is such an efficient handler of language. Since this kind of retrieval system, which selects moment-to-moment subsets of language has not yet been simulated, goes far to explain why computers, in which the events are several orders of magnitude faster than in the brain, and whose number of components is considerably smaller, are yet much slower at linguistic problems such as mechanical translation, and indeed many types of problem solving.

*Speech output.* Language which has been selected by the retrieval mechanisms of the brain is typically emitted as spoken speech. This may be regarded as the transforming of nerve impulses into soundwaves by the muscles of the larynx, lungs, throat, and mouth. A detailed treatment will be found in the article on speech sounds. There is now a large amount of information about what are the significant features of the sound waves, and several succesful speech synthesizers have been constructed.

The control of speech is particularly interesting as an example of a feedback-loop system. If a person is prevented from hearing his own voice, for example by playing loud white noise into his ears, then his speech alters. The pitch tends to rise, and the loudness increases, as if he were trying to "climb out of" the noise. His pronunciation usually deteriorates. More spectacularly, if he can only hear what he is saying through headphones, and a delay is introduced in the feedback, there is very marked deterioration in the quality of his speech, and when the delay is of the order of $\frac{1}{2}$ sec, he becomes unable to talk, unless he manages to judge on the basis of what the muscles of the throat and tongue feel like. Interfering with the feedback of stammerers may abolish stammering, and stammerers may be regarded perhaps as people who are unable to break out of the control of the feedback in order to emit a new sound instead of the one just emitted. It has recently been found that delayed visual feedback has a similar effect on writing.

*The role of speech among humans.* The importance of the difference between human and other languages is not often realized. There is no evidence for even remotely comparable languages in any other species. Logical analysis of the role of language in the handling of experience suggests that almost all "typically human" behaviour depends upon having a language with the grammatical, syntactic, and semantic structure which ours has. Without it, for example, planning for the future, promising, moral decisions, deception, altruism; scientific, philosophical, and religious theorizing; and broadly speaking what we call "mental life" in humans would all be impossible. The importance of the emergence of human language is comparable in importance, with respect to the qualitative changes in the life of the organism which it implies, with the development of many-celled from single celled organisms, or even, perhaps, with the emergence of life itself.

*Bibliography*

CHERRY C. (1957) *On Human Communication*, New York: Wiley.

CIBA Symposium on Disorders of Speech (in press).
MILLER G. A. (1951) *Language and Speech*, NewYork: McGraw-Hill.
PENFIELD W. and ROBERTS L. (1959) *Speech and Brain Mechanisms*, Oxford: The University Press. (1964), *British Medical Bulletin* 20, No.1, "Experimental Psychology".                    N. MORAY

**SPEECH SHADOWING.** The name speech shadowing is applied to the situation in which a listener repeats aloud a message that he hears while it is in progress. A practised person will generally be about a word or two behind the message he is hearing. Much use has recently been made of this technique in investigations into speech handling and attention mechanisms in the human brain.                    N. MORAY

**SPIN PACKET.** This name was coined by Portis to describe a homogeneous component of an electron paramagnetic resonance (EPR) line. If the EPR of a single paramagnetic ion could be observed it would have a certain linewidth $\delta\nu$ determined by the lifetime of its quantum states. This lifetime may be determined either by spin lattice, or spin-spin relaxation. In a specimen of paramagnetic material the internal magnetic fields vary from ion site to ion site because of the random orientation of the neighbouring magnetic moments. Hence the EPR frequencies of all the paramagnetic ions are slightly different, covering a range $\Delta\nu$. The component of the EPR line contributed by all of those ions whose frequency lies within the range of the line width $\delta\nu$ is called a spin packet. These ions can come to thermal equilibrium with one another rapidly because mutual spin flips can occur in which energy is conserved. Such processes are lower between one spin packet and another because energy is not conserved. Sometimes $\Delta\nu < \delta\nu$ in which case the whole EPR line is one single spin packet. The line is then said to be "homogeneous". If microwave power is absorbed at any frequency within the line width, which is $\delta\nu$, all of the paramagnetic ions contribute to the resonance. Alternatively $\Delta\nu$ may be greater than $\delta\nu$ when the EPR line of width $\Delta\nu$ comprises several spin packets. It is then said to be "inhomogeneous". The application of microwave power at any frequency within the linewidth $\Delta\nu$ only affects in the first instance those ions contributing to the spin packet at that frequency. If the "spin diffusion" of energy from one spin packet to the adjacent ones is sufficiently slow it is possible to "burn a hole" in the EPR line by saturating one spin packet. This hole can be observed by scanning across the whole EPR line rapidly after the saturation of the spin packet has occurred.                    J. M. BAKER

**SPORADIC-E IONIZATION.** The term sporadic E has its origin in the thirties when radio physicists observed that most reflections from the E-region were quite predictable in accordance with the theory of ionization by solar photons enunciated by Professor Sidney Chapman and others. E-region reflections which were not predictable were classed together as "Sporadic E", abbreviated $E_s$.

The behaviour of sporadic E is also quit edifferent from that of the regular reflections from the E-region. According to the magnetoionic theory originally developed by Sir Edward Appleton, reflections from the ionosphere, a birefringent medium in the presence of the Earth's magnetic field, take place for the two principal components, the ordinary and the extraordinary waves, at the levels where the refractive index for the component in question becomes zero. This happens at two frequencies separated by approximately one-half the Larmor frequency for free electrons. For each of these components, the transition from total reflection to total penetration of the ionosphere takes place over a very short span of frequencies (the order of 100 kc/s). In the case of sporadic-E reflections, however, not only is this transition frequently very gradual, but it also shows little if any measurable group retardation at the highest reflection frequency. In addition the highest frequency which can be reflected rises at times to as much as ten times the normal E-region critical frequency.

During the last three decades three possible structures for sporadic-E ionization have been considered. It is clear from the echoes that the layer must be thin, of the order of 1 km or so. Starting from this point, the reflections themselves could be caused by a thin layer, a ledge in the vertical ionization profile, or from a layer of scattering centres. It is now clear that to some extent all three of these processes take place. There has also been considerable question as to whether new ionization is produced or whether the existing ionization in the E-region is simply redistributed. Again it now seems clear that in some instances (in the auroral zone for instance) new ionization is indeed produced; while for temperate latitude and equatorial sporadic E, the reflections probably do not involve new ionization. Recent rocket measurements of the ionization profile in the E-region have shown both thin layers and ledges in the ionization profile. There has been a measure of agreement between these observations and the sporadic-E reflections observed on ionosondes. The distinction between the ledge reflection and the thin-layer reflection is that in the former the reflection is accounted for entirely by the gradient in the ionization profile in accordance with Fresnel's formulae. In the latter case, the maximum frequency of reflection is primarily accounted for by the maximum electron density which occurs in the thin layer, although of course to obtain this, one must also have sufficiently steep gradients to produce some reflection by the gradient mechanism. The horizontal dimensions of sporadic E vary considerably with latitude. In temperate latitudes, sporadic E frequently seems to behave as though it were a large flat platter several hundreds of kilometres in extent. In auroral latitudes, the physical dimensions frequently are much smaller,

the order of tens of kilometres, while at the equator a special type of sporadic E, which will be mentioned in more detail below, occurs as a large flat strip on the daylight side of the Earth.

The temporal characteristics of sporadic E are quite well defined geographically. In temperate latitudes, sporadic E occurs primarily between the months of May and August with quite a marked peak in the June-July period in the Northern Hemisphere. Also in the Northern Hemisphere, a secondary peak occurs in December and January. In the Southern Hemisphere the primary seasonal maximum is in the November to February period with the secondary peak in June-July. The diurnal maximum at temperate latitudes occurs in the late morning hours, and a small secondary maximum shows up in the evening hours. In the auroral zone the temperate latitude characteristics are largely wiped out and instead one finds the seasonal variation to be rather relatively minor, whereas the temporal occurrence of sporadic E is dominated by the diurnal variation which sees the effect occurring largely at night with a maximum at a time which seems to be dependent on local magnetic time to some extent. In the region near the magnetic equator (magnetic-field horizontal in the E-region), a special type of sporadic E associated with the equatorial electrojet occurs. This type, abbreviated $E_{sq}$, is observed on the daylight side of the Earth along a strip approximately 400 miles wide centred along the magnetic equator. Good agreement exists between the location of this effect and the flow of current along a narrow strip at a height usually just under a hundred kilometres which is thought to be driven by the dynamo potentials set up by tidal motions of the ionosphere.

There are also some interesting variations in the longitude of sporadic E. At temperature latitudes there is substantially more sporadic E observed in the Far East than in the Western Hemisphere or Europe; while in the auroral zone, there is substantially more sporadic E observed over Canada than in the part of the auroral zone which falls over Siberia.

The cause of sporadic E has been the subject of much speculation over the years. The most promising explanation at temperature latitudes is based on a suggestion originally made by J. W. Dungey. In the presence of horizontal wind shear taking place in the magnetic field, it is suggested that it is possible for a layer to be formed which results in a significant increase (order of magnitude) in the ambient electron densities through the redistribution of charge particles. In the auroral zone it is clearly possible for the influx of charge particles to create sporadic E-like layers. In equatorial regions the special peculiarities of sporadic E have been attributed quite successfully to plasma instabilities moving at acoustic velocities in the equatorial electrojet current stream. During the International Geophysical Year, scientists agreed to classify sporadic E into eight specific types in accordance with the appearance of the echo trace on the ionogram (record from an ionosonde). Considerable information is now available as to the occurrence and special peculiarities of these different "types" of sporadic E, and the reader is referred to the literature if his interest extends this far. Survey information on sporadic E may be found in the bibliography.

Sporadic E is important in radio communications primarily from the point of view of interference which may be caused by it. As $E_s$ is largely unpredictable, it is difficult to use it for regular service, but one can never be sure when an interfering station operating at frequencies too high to be propagated by the normal layers may suddenly be transmitted by sporadic E. This precise problem had much to do with the discontinuance of Channel 1 in the television band (44–50 Mc/s in the U.S.) and the movement of the FM band (42–50 Mc/s) up to 88–108 Mc/s. At present it is not unusual for those television stations operating at frequencies less than 80 Mc/s to be observed at distances up to 1500 miles away due to sporadic-E propagation.

*Bibliography*

SMITH E. K. and MATSUSHITA S. (Eds.) (1962) *Ionospheric Sporadic E*, Oxford: Pergamon Press.

THOMAS J. A. and SMITH E. K. (1959) *J. Atmosph. and Terr. Phys.* **13**, 295.

E. K. SMITH

**STRAIN GAUGES, SEMICONDUCTOR.** The most remarkable advance in strain gauge technology in recent years has been the development of semiconductor strain gauges. These strain gauges exhibit strain sensitivies up to almost two orders of magnitude greater than conventional metal ones. This high sensitivity removes the problems of handling the small signals produced by metal gauges but where high accuracy is required other problems are introduced involving the use of specially-designed circuitry.

In semiconductor strain gauges the effects of dimensional changes are negligible in comparison with the change in resistivity produced by the application of stress. Such gauges are made from single crystals whose properties (electrical and mechanical) are not isotropic. In such crystals the magnitude of the piezoresistance effect depends on the directions of the applied stresses and current flow with respect to the crystallographic axes of the crystal. Mathematically, the effect is described by a set of equations relating the three electric field components, three current density components, three direct stress components and three shear stress components. However, a considerable simplification can be made by considering the most useful case of a direct stress, the electric field component and the current density component all acting in the same ("longitudinal") direction. If $R_0$ is the unstressed longitudinal resistance of the crystal the piezoresistance ratio is defined as $\dfrac{\Delta R}{R_0}$ and is given by

$$\frac{\Delta R}{R_0} = \pi_\varrho \cdot S \qquad (1)$$

where $\Delta R$ is the change in resistance produced by the application of a longitudinal stress $S$.

$\pi_\varrho$ is the longitudinal piezoresistance coefficient. The gauge factor is then $G = Y_\varrho \cdot \pi_\varrho$.

Where $Y_\varrho$ is the longitudinal Young's modulus for the crystal.

The large piezoresistance effect in semiconductors can be discussed in terms of the many-valley type of energy band structure. In the many-valley model, a band consists of several equivalent energy minima, or valleys, located at symmetrical points in the Brillouin zone and occupied by equal number of charge carriers. The application of a uniaxial stress changes the lattice constant of the crystal slightly, and causes valleys whose constant energy surfaces are oriented, for example, along the direction of the stress to shift up or down in energy, and valleys oriented perpendicular to the stress to shift in the opposite direction. As a result of this energy shift the charge carriers are redistributed over the valleys, and the populations of the valleys are no longer the same. If the carriers have different mobilities in different valleys, which is generally the case, the average carrier mobility in the sample will then change, and a change in resistivity as a function of stress is observed. To obtain a large change in resistance, the direction of the applied stress must be such that all valleys are not acted upon in the same manner. For example, in n-type silicon the valleys are along the [100] reciprocal axes, so that a stress in the [111] direction is symmetrical to all of them and will not cause a large piezoresistance effect. Thus a large piezoresistance effect is observed only in certain crystallographic directions, with the direction of the large effect being determined by the symmetry of the valleys.

*Table 1. Gauge factors for some semiconducting materials*

| Material | Resistivity (ohm-cm) | Young's modulus (dyne/cm²) | Gauge factor | Crystal direction |
|---|---|---|---|---|
| Silicon p-type | 7·8 | $1\cdot87 \times 10^{12}$ | 175 | [111] |
| Silicon n-type | 11·7 | $1\cdot23 \times 10^{12}$ | $-133$ | [100] |
| Germanium p-type | 15·0 | $1\cdot55 \times 10^{12}$ | 102 | [111] |
| Germanium n-type | 16·6 | $1\cdot55 \times 10^{12}$ | $-157$ | [111] |
| Germanium n-type | 1·5 | $1\cdot55 \times 10^{12}$ | $-147$ | [111] |
| Indium Antimonide p-type | 0·54 | | $-45$ | [100] |
| Indium Antimonide p-type | 0·01 | $0\cdot745 \times 10^{12}$ | 30 | [111] |
| Indium Antimonide n-type | 0·013 | | $-74\cdot5$ | [100] |

For p-type silicon and similar semiconductors the mechanism of the piezoresistance effect is more complicated. However, a many-valley model with valleys along the [111] reciprocal axes may be used to visualize the symmetry of the effect in these materials.

Values of gauge factor for some materials of interest are given in Table 1.

Most, if not all, commercially available semiconductor strain gauges are made from silicon. The variation of strain sensitivity of silicon with crystal orientation and resistivity is shown in Fig. 1 and it will be seen that lightly-doped [111] p-type and [100] n-type are the most sensitive. These are the crystal types used

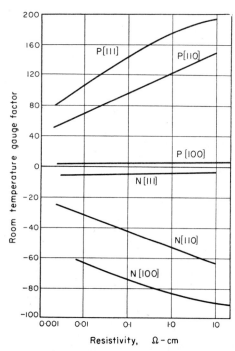

*Fig. 1. Variation of gauge factor of silicon with resistivity and crystal direction.*

in practice though they are generally doped down to about 0·1–0·2 ohm-cm for reasons which will be discussed below.

The gauges generally take the form of very fine filaments whose axes are in the appropriate crystal direction. The filaments are of the order 0·001 in. thick to make them flexible and are produced either by cutting and etching from crystals or by growing whiskers from the vapour phase. Recently gauges have been produced by diffusing the chosen impurity into a thin layer of suitable dimensions in the surface of a large piece of silicon of high resistivity or of opposite conductivity type to give p-n junction electrical isolation. If the large piece of material is a thin wafer it can be used as a strain gauge by bonding to the test structure in the usual way. More interesting appli-

cations of the diffusion technique are those in which the large piece is itself a structural member so that when load is applied to it it acts as a transducer. In this case four gauges can be diffused into appropriate positions on its surface allowing the formation of a full Wheatstone's bridge. By this technique the troublesome bonding procedure is eliminated and excellent temperature compensation assured. Transducers made in this way are now becoming commercially available.

There are several features of silicon strain gauges which make them quite different from wire gauges. The first of these is the non-linearity of the strain-resistance relation.

For typical strain gauge material ($\varrho \sim 0·1$ ohm-cm) the equations are:

$$\frac{\Delta R}{R_0} = 120S + 4000S^2 \qquad \text{(p-type)} \qquad (2)$$

$$\frac{\Delta R}{R_0} = -110S + 10,000S^2 \quad \text{(n-type)} \qquad (3)$$

(where $S$ = Strain).

From these equations it is clear that silicon gauges are inherently non-linear and that n-type silicon is inherently less linear than p-type. Further, equations 2 and 3 are parabolas and will cross the strain axis at another point besides zero strain. This point is $-30,000$ $\mu$strain in the case of p-type silicon and $+11,000$ $\mu$strain in the case of n-type showing that p-type strain gauges are more linear in tension than in compression, and n-type more linear in compression than tension.

Changes in temperature induce changes in resistance and in strain sensitivity. The temperature coefficient of resistance of silicon decreases as resistivity decreases and is about $100 \times 10^{-5}$ deg$^{-1}$C, for typical strain gauges. In practice the gauge is always bonded to a structure and in this case a second effect is introduced—the straining of the gauge due to differences in the coefficients of expansion of the gauge material and the structure. The zero drift of a bonded gauge is the sum of these two effects and is described by the equation

$$\frac{\Delta R}{R \cdot \Delta T} = a_g + G(e_m - e_g) \qquad (4)$$

where $a_g$ = temperature coefficient of resistance of strain gauge

$e_g$ = coefficient of thermal expansion of strain gauge

$e_m$ = coefficient of thermal expansion of structure material.

The coefficient of expansion of silicon is $3·2 \times 10^{-6}$ deg$^{-1}$C which is very much less than that of most structural materials so that the expression in brackets is always positive. For p-type gauges having a positive gauge factor the whole expression is positive and temperature drifts are high but, if expressed as apparent

*Table 2. Properties of strain gauges*

| | Gauge factor | Young's modulus (lb-in$^{-2}$) | Temperature coefficient of resistance (deg$^{-1}$-C) | Coefficient of expansion (deg$^{-1}$ C) | Apparent strain bonded to steel ($\mu$S deg$^{-1}$ C) | Temperature coefficient of gauge factor (deg$^{-1}$ C) | 1% Linear strain range | Breaking strain |
|---|---|---|---|---|---|---|---|---|
| p-type Silicon | 100 to 170 | $27 \times 10^6$ | 70 to $700 \times 10^{-5}$ | $3 \cdot 2 \times 10^{-6}$ | 15 to 50 | $-1 \cdot 4$ to $-4 \times 10^{-3}$ | $\pm 300$ $\mu$ strain | > 0·7% |
| n-type Silicon | $-100$ to $-140$ | $18 \times 10^6$ | | | Can be zero over small temperature range | $-1 \cdot 1$ to $-4 \cdot 5 \times 10^{-3}$ | $\pm 110$ $\mu$ strain | > 0·7% |
| Nichrome | 2·1 to 2·63 | $20 \times 10^6$ | $40 \times 10^{-5}$ | $17 \times 10^{-6}$ | 50 | Constant to at least 100°C | $\pm 1\%$ | > 2% |
| Advance | 2·04 to 2·12 | $24 \times 10^6$ | $\pm 2 \times 10^{-5}$ | $19 \times 10^{-6}$ | $-17$ | constant to at least 100°C | $\pm 1\%$ | > 2% |

strain, are still comparable with conventional metal gauges. However, for n-type gauges the two terms of equation 4 oppose and the drift is very much less. n-type gauges are available with specially matched properties to give very small drifts on particular materials.

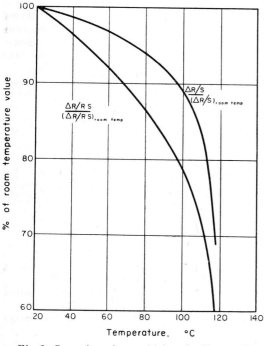

Fig. 2. *Loss of strain sensitivity of silicon with temper.* $\dfrac{\Delta R}{RS}$ = *gauge factor (constant voltage supply).* $\dfrac{\Delta R}{S}$ = *response of gauge with constant current supply.*

Another property of silicon gauges, which can have serious consequences if it is not allowed for, is the decrease in strain sensitivity with increase in temperature. A typical curve is shown in Fig. 2.

A table of the more important properties of silicon strain gauges is given in Table 2 with comparative values for conventional metal gauges .

*Circuits for semiconductor strain gauges.* Where semiconductor strain gauges are in use the resistance

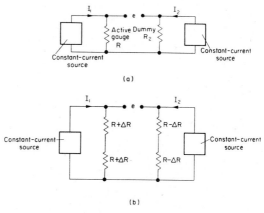

Fig. 3. *Constant current circuits.*
(a) *single gauge.* (b) *multi-gauge.*

changes may be large enough to seriously upset the linearity of a Wheatstone's bridge, e.g. a strain of 1000 $\mu$ strain will produce a $\dfrac{\Delta R}{R}$ of about 13 per cent (i.e. gauge factor = 130) and a non-linearity of over 6 per cent. Hence for any sort of precision work the equal-arm Wheatstone's Bridge is far from satisfactory as a reading circuit for a single gauge.

The only way to obtain true linearity of voltage output with resistance change is to supply the gauge with constant current and a suitable circuit for semiconductor strain gauges is shown in Fig. 3(a). The "dummy" resistance can be a second strain gauge used for temperature compensation. $I_2$ and $R_2$ are adjusted so that at zero strain $e = 0$ and then $\Delta e = I_1 \Delta R$.

This circuit is intended for use with a single active gauge and this is the most difficult case. It is often possible to use two or four active gauges in the circuit and if these can be made to give equal and opposite changes of resistance they can be arranged in the standard constant-voltage bridge to give a linear out-

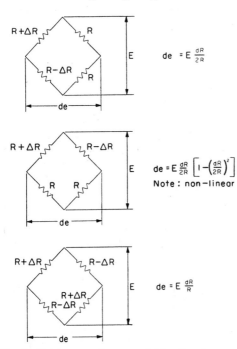

*Fig. 4. Multi-gauge wheatstone bridge circuits.*

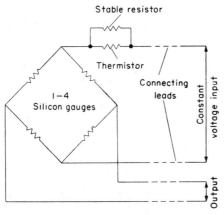

*Fig. 5. Compensation for gauge factor decrease with temperature.*

put (Fig. 4) or in a constant-current bridge (Fig. 3(b)). Such equal and opposite changes of resistance can be obtained from conventional gauges only where equal and opposite strains are available as in the measurement of bending or shear stresses. However, by the use of both p-type and n-type semiconductor gauges a fully active bridge can be employed for the measurement of strains of one sign. Not only do these multi-gauge circuits give output linear with total change of resistance in the circuit but the total change of resistance is linear with strain. Whether the bridge is composed of all p-type, all n-type or both p- and n-type gauges the gauge non-linearity arises from the (strain)$^2$ term in equations 2 and 3 which affects all the gauges to roughly the same extent and has no total effect.

The variation of gauge factor with temperature shown by these gauges can be compensated by the circuit shown in Fig. 5. A thermistor, exposed to the same temperatures as the strain gauge(s), is placed in one lead of the constant-voltage supply. The thermistor may then be shunted by a stable resistor of such a value that the bridge voltage increases with temperature at the same rate as the sensitivity decreases.

*See also*: Photoelastic strain-gauges and transducers. Strain-gauge transducers.

*Bibliography*

DEAN M. (Ed.) (1962) *Semiconductor and Conventional Strain Gauges*, New York: Academic Press.

G. R. HIGSON

**STRAIN-GAUGE TRANSDUCERS.** The resistance strain gauge is itself a transducer for the conversion of mechanical strain into electrical voltage and it is often used as the primary sensing element in transducers for the measurement of other mechanical quantities. The quantity in question is made to induce the deformation of a structure and this deformation actuates strain gauges. The structure is normally a high-strength material chosen for its linear and hysteresis-free stress-strain relation with wire, foil or semiconductor strain gauges bonded to its surfaces. It may alternatively be made from a semiconducting material (usually silicon) with strain-sensitive areas produced on its surfaces by diffusion techniques. A third form of manufacture is that in which the deforming structure itself is the strain-sensitive wires which are then known as unbounded strain-gauges.

Some of the requirements for an ideal transducer are the following: (1) Low input energy signal needed to actuate the transducer; (2) High output signal power level; (3) Good static accuracy; (4) High natural frequency, and (5) Small physical size. Strain-gauge transducers can satisfy all these requirements except no. (2) and those instrumented with semiconductor strain-gauges can meet this one also. Requirements (1), (4) and (5) are linked in so far as a reduction in the size of a transducer should result in a reduction in the input work required and an increase in the natural

*Output power and input work for strain-gauge transducers*

| Strain-gauge type (4-arm bridge) | Max. available output power (watts) | Input work (in. lb.) | Output power / Input work (watts/in. lb) |
|---|---|---|---|
| Bonded foil gauge<br>   gauge area $= 0\cdot1$ in. $\times$ $0\cdot1$ in.<br>   resistance $= 120$ ohms<br>   structure thickness $= 0\cdot002$ in. | $1\cdot6 \times 10^{-6}$ | $0\cdot01$ | $1\cdot6 \times 10^{-4}$ |
| Unbonded wire gauge<br>   resistance $= 350$ ohms<br>   wire dia. $= 0\cdot0006$ in. | $2\cdot4 \times 10^{-6}$ | $4 \times 10^{-4}$ | $6 \times 10^{-3}$ |
| Bonded silicon gauge<br>   gauge area $= 0\cdot25$ in. $\times$ $0\cdot006$ in.<br>   resistance $= 1000$ ohms | $12\cdot5 \times 10^{-3}$ | $2 \times 10^{-3}$ | $6\cdot25$ |

frequency. The input work is that needed to deform the gauge-carrying structure. Semiconductor strain gauges are superior to metal ones in this respect as they can be made to be very small while retaining a high resistance, but unbonded strain gauges require least input energy as all the work is done on the strain-sensitive wires themselves. The available output power is a function of the gauge factor, the maximum permissible strain and the heat dissipation ability of the strain-gauges. Semiconductor strain-gauges are best in this respect because of their high gauge factor. The ratio of output power to input work gives a good indication of the desirability of a transducer. Typical figures for the three types of strain gauge construction are given in the table.

Generally, a strain-gauge transducer is fitted with four strain gauges arranged so that when the transducer is actuated, two increase in resistance and two decrease. They can then be wired in a Wheatstone's Bridge circuit to give maximum unbalance. This simple circuit will be expected to show initial unbalance and will be liable to change both its zero and its sensitivity with temperature. A commercial transducer is compensated for these effects and its circuit is similar to that shown in Fig. 1. The auxiliary resistances added to the Wheatstone's Bridge circuit have the following functions:

$R_z$ is the resistance needed to bring the bridge into initial balance. It is shown here in series with the lowest resistance in the bridge but it could alternatively be arranged to shunt the highest resistance. This resistance should be wound from the same material as the strain gauges.

$R_T$ is a small quantity of material having a high temperature coefficient of resistance. It is placed in the appropriate arm of the bridge to cancel any zero drift. The amount of wire and its position in the circuit must be determined for each transducer individually from observations of the zero drift of the uncompensated bridge.

$R_E$ is a resistance placed in one of the voltage supply lines to maintain a constant sensitivity over a range of temperatures. Generally, the predominant effect on the sensitivity of the transducer is the decrease in modulus of elasticity of the strained member so that the transducer tends to increase its

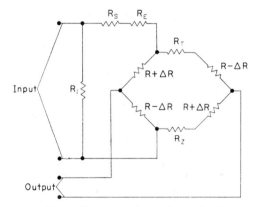

*Fig. 1. Compensated transducer circuit.*

sensitivity. $R_E$ is then a material of positive temperature coefficient of resistance which decreases the voltage applied to the strain-gauge bridge as the temperature increases to just oppose the natural increase in sensitivity. However, if semiconductor strain gauges are used the predominant temperature effect is the decrease in gauge factor of the strain-gauges so that the transducer sensitivity decreases with increase in temperature. In this case $R_E$ will be a thermistor which has a very high negative temperature coefficient of resistance. This thermistor needs to be shunted by a stable resistor whose value is chosen to allow the bridge voltage to increase with increase in temperature at a rate sufficient to just oppose the loss of gauge factor.

$R_s$ is a high-stability resistor which is then added to reduce the sensitivity of the transducer to a specified level. The transducer must, of course, be designed with excess sensitivity to permit this process.

$R_i$ may then be added to reduce the input impedance to a specified level.

Parameters commonly measured by strain-gauge transducers are: force; torque; pressure; acceleration and displacement.

*Force transducers.* The most familiar example is the load cell which measures compressive forces via the strain induced in a solid column or a cylinder (Fig. 2). Tensile forces may be measured by a similar arrangement of strain gauges on a rod, solid or hollow, in tension but often rings or ring variants, as shown in Fig. 3, are used. The strain-gauges are positioned to measure the maximum bending strains occurring in such shapes. The ring type may be used to measure tension or compression and is particularly useful where the force may change sign.

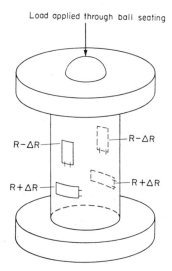

*Fig. 2. Bonded strain-gauge load cell.*

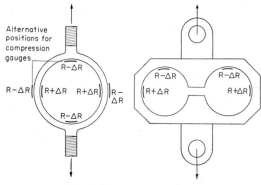

*Fig. 3. Ring-type transducers.*

*Torque transducers.* Torque may be measured by strain gauges bonded with their strain-sensitive axes aligned at $\pm 45°$ to the axis of the torque-transmitting shaft thus sensing the principal tensile and compressive strains. A transducer to be inserted between two separate shafts will normally have a fixed cover making contact via sliprings to the strain-gauges on the rotating member.

*Pressure transducers.* Many varieties of pressure transducer are available and only the most common ones can be described here. A simple principle is the measurement of the hoop strain on the outer surface of a thin-walled tube subjected to internal pressure. If this is a circular tube the sensitivity is low as the permissible stresses are low and only two active gauges can be used—temperature-compensating gauges are

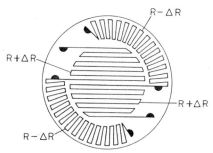

*Fig. 4. Bouffler-type diaphragm element (etched foil).*

stuck to an unpressurized section of the tube—and much greater sensitivity can be achieved by the use of a flattened or elliptical section tube. Such a tube will attempt to deform to a circular section under an internal pressure and strain-gauges can be attached to sense the resultant bending strains which will be tensile in the regions of least curvature and compressive in the regions of greatest curvature.

Another common scheme is the bonding of strain gauges to measure the strain in a diaphragm. A circular diaphragm with clamped edges undergoing deformations small in comparison with its thickness has its unpressurized surface in tension at the centre and compression at the edge so that four active gauges can be attached to one side of the diaphragm. Special strain-gauge forms have been designed to allow a full-bridge arrangement on one side of a diaphragm of minimum size (Fig. 4).

An alternative to measuring the strain in the diaphragm itself is the use of a very flexible diaphragm merely to separate the pressurized fluid from the transducer proper which may be an unbonded strain gauge arrangement similar to that shown in Fig. 5 or a bonded strain-gauge force sensor such as a beam or cantilever. A bellows may be used in place of a diaphragm.

*Acceleration transducers,* or *accelerometers* are often of the unbonded strain-gauge type as the high natural frequency associated with this form of construction is particularly important in this application. A typical arrangement is shown in Fig. 5. The mass M is attached to the case by the four groups of wires and any acceleration of the case is transmitted to the mass through these wires which are consequently strained

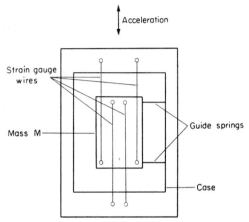

*Fig. 5. Unbonded strain-gauge acceleration transducer.*

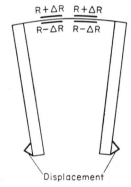

*Fig. 6. Bonded strain-gauge displacement transducer.*

in proportion to the magnitude of M and the acceleration. The movement can be conveniently damped by filling the case with liquid. The restraint of the mass to the case can be, of course, by some solid structural member, e.g. a cantilever, to which strain gauges are bonded but such arrangements are generally restricted to applications involving low frequencies and high values of acceleration.

*Displacement transducers.* The relative movement of two points will produce deformations of a structure, such as that shown in Fig. 6, which can be sensed by bonded strain-gauges. An important requirement of this type of transducer is that the actuating force should not be large enough to influence the quantity being measured. For small displacements an unbonded strain-gauge type, whose construction is basically similar to that shown in Fig. 5, is desirable.

*See also*: Photoelastic strain-gauges and transducers. Strain-gauges, semiconductor.

*Bibliography*

NEUBERT H. K. P. (1963) *Instrument Transducers*, Oxford: The University Press.
PASTAN H. L. (1964) *Electrical Compensation and Standardisation of Strain-Gauge Transducers*, Strain-Gauge Readings 6, No. 5.
LEE S. Y. and PASTAN H. L. (1961) *Design and Application of Semiconductor Strain-Gauge Transducers*, Instrument Society of America Paper No. 189-LA-61.

G. R. HIGSON

**SUPERANTIFERROMAGNETISM.** It has been known for a long time that finely divided antiferromagnetic solids, of which $Cr_2O_3$ is one, may exhibit magnetic susceptibilities considerably larger than normal. This is especially true at low temperatures. Matter in this state is variously referred to as colloidal as opposed to crystalline or, especially in applications to heterogeneous catalysis, as disperse as opposed to massive. Néel has recently (1962) attempted a theoretical interpretation of the effect and has suggested the name: superantiferromagnetism. The name is not widely used.

*See also*: Collective paramagnetism.

P. W. SELWOOD

**SUPERCONDUCTIVE ALLOYS, RECENT DEVELOPMENTS IN.** A significant proportion of recent research on superconductivity has been on alloys, and much of the stimulus for this has been due to the fact that it is possible to make very high field electromagnets using suitably prepared superconductive alloy wire. Magnets capable of producing fields in excess of 100 kilo-oersted are now available. These are used in research, but there is the prospect of constructing bigger magnets for use in nuclear machines and electrical power generators. The construction of superconductive transformers and even power transmission lines is being considered also.

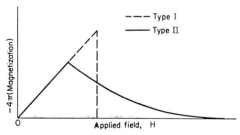

*Magnetization curves of type I and type II superconductors.*

The classification of superconductors into type I (soft) and type II (hard) is especially useful when considering superconductive alloys. The form of the magnetization curves for the two types is shown in the figure.

It is now generally agreed that the distinguishing feature between the two types is the sign of the surface energy between the superconducting and normal phase. The addition of alloying components to a metal shortens the electronic mean free path of the material in the resistive state. The shorter the electronic mean free path, the shorter the super-conductive coherence length $\xi_0$ and the longer the penetration depth $\lambda$. When, as a result of alloying, $\xi_0$ becomes smaller than $\lambda$ the surface energy between the superconducting and normal phases becomes negative and the material exhibits type II properties. Because of the negative surface energy, field penetration and superconduction become confined to filaments of the material, in agreement with the Mendelssohn sponge hypothesis. The flux filaments of the mixed state vary in size over a wide range from a single flux quantum ($2 \cdot 07 \times 10^{-7}$ G. cm$^2$) to areas large enough to be detected by magneto-optical techniques. The high critical fields and current carrying capacities of alloy materials used in the construction of super-conducting magnets are attributed to this *pinning* of flux filaments.

Alloys such as niobium–tin and niobium–zirconium have been specially developed which have a large number of extended defects such as dislocations and precipitation cells to pin flux. The properties of these alloys can be further improved by the introduction of strain by cold working.      P. R. STUART

## SUPERCONDUCTING SOLENOIDS.

### 1. Materials

The generation of usefully high magnetic fields by means of superconducting wires has recently become very attractive. Fields as high as 70 kgauss in small bores have been generated with Nb$_3$Sn (Kunzler 1962) and several magnets up to 60 kgauss (Hake 1962; Hulm 1962; Riemersma *et al.* 1962) have been made with Nb–Zr superconductors. Cold-worked niobium solenoids up to 10 kgauss (Autler 1960) and Mo–Rh solenoids to 15 kgauss (Kunzler *et al.* 1961) have been available for some time, but with the successful fabrication of Nb$_3$Sn with its critical field of 200 kgauss (Hart 1962) and Nb–Zr at 70 kgauss, interest has increased manyfold.

### 2. Design Procedures

*2.1. Field equation and volume.* The magnetic field in a uniform density solenoid can be written in terms of the current density in the wire:

$$H = j\lambda a_1 F(\alpha, \beta) \qquad (1)$$

where $j$, the current in the wire, is in amperes cm$^{-2}$, $\lambda$ = volume of wire/volume of coil, $a_1$ = inner radius

(cm) and

$$F(\alpha, \beta) = \frac{2\pi}{5} \varrho \ln \frac{\alpha + (\alpha^2 + \beta^2)^{\frac{1}{2}}}{1 + (1 + \beta^2)^{\frac{1}{2}}}.$$

The field can also be written in terms of the Fabry factor $G$, and the coil volume

$$H = j\lambda \left(\frac{1}{a_1}\right)^{\frac{1}{2}} G(V)^{\frac{1}{2}} \qquad (2)$$

where

$$V = a_1^3 2\pi\beta(\alpha^2 - 1). \qquad (3)$$

Having selected the necessary $F(\alpha, \beta)$ from equation 1 to generate the desired field at the allowable current density, nearly the minimum volume will be used if $\alpha$ and $\beta$ are selected from the crossing of the "minimum volume" line and the desired constant $F(\alpha, \beta)$ line. For cases where $\beta$ is greater than 2, this procedure will yield a result very close to the minimum volume; however, when $\beta$ is small, the ratio of the field at the conductor to the field at the centre becomes significant and the design procedure must be changed.

*2.2. Current-carrying capacity of wires.* It has been recognized since the first Nb–Zr solenoids were built that their current-carrying capacity was less than predicted on the basis of $(H_c, I)$ relations taken on short samples in applied fields. Montgomery (1962 a) has indicated that the discrepancy is due to persistent eddy-currents induced when the solenoid is energized. The eddy-currents are of sufficient magnitude that they tend to saturate the superconducting filaments. Indications are that the current-carrying capacity is dependent on the diameter and space factor as well as the field (Montgomery 1962 a). Large diameter coils appear to carry less current than small diameter coils at the same field.

Preliminary experiments indicate that the product of the current density in the wire and the coil space factor is roughly constant. It is therefore possible to spread out turns or layers and generate the same field with less material.

The current-carrying capacity can be increased to varying extents by several methods:

(i) If the wires are spread out, the current-carrying capacity of each wire goes up somewhat faster than the space factor decreases. Tests on coils at M.I.T. and Westinghouse (Hulm 1962) indicate that somewhat more field can be generated with less material by displacing the turns anywhere from $\frac{1}{2}$ a turn to 1 turn apart. This can be done by interwinding an insulating (or copper) filament between turns or between layers, or by building up the wires with metallic or insulating sleeves. This technique can represent a great saving in wire.

(ii) If concentric sections of the solenoid are energized in turn while all inner sections are held normal, higher currents can be made to flow. In coils run at M.I.T. all sections were thermally insulated by mylar barriers and copper cooling fins; the appropriate sections are held normal by about a watt cm$^{-3}$ of normal current joule heating while the outer sections are energized.

(iii) Gauster and Parker (1962) suggest running different currents through concentric windings, with the outer coils, where the field is low, carrying additional current. Riemersma et al. (1962) describe a solenoid where the wire is enriched with Zr in the outer sections. Zr-rich Nb–Zr exhibits a higher current-carrying capacity but a lower critical field.

Discrepancies between short sample test and solenoid have not been reported for niobium, MoRh or $Nb_3Sn$. However, the same magnitude of residual fields measured in Nb–Zr which indicate the presence of persistent eddy currents are also found in the other superconductors (Montgomery 1962 a). It is entirely possible that discrepancies will become evident when larger bore magnets are built with these other materials.

### 3. Fabrication Techniques

*3.1. Solenoids wound from the non-brittle super-conductors.* Nb–Zr, Nb, MoRh, etc., are not difficult to fabricate. Techniques for Nb and MoRh have been described (Autler 1960; Kunzler et al. 1961). Typical Nb–Zr construction uses 0·010 in. wire usually insulated with a 0·001 in. build of nylon, polyamide or formvar insulation. A 0·002 in. sheet of mylar is often used between layers to minimize short circuits.

Coils can be wound on insulated metallic or non-metallic bobbins. It is usually necessary to wind a coil from several lengths unless very long lengths are available. The sections can be conveniently spot-welded together outside the field (ambient field less than 10 kgauss) and power leads can be attached by spot welding platinum ribbons to the Nb–Zr and soldering the platinum to copper wire. Pressure contacts can also be used and are most successful when wires are twisted and strongly clamped between copper blocks or crimped in copper cylinders.

*3.2. Solenoids made with intermetallic compounds.* $Nb_3Sn$ solenoids are somewhat more difficult to fabricate, as the material as presently manufactured must be wound in place before final heat treatment at 950°C. Three coils have been described in the literature (Salter et al. 1962; Kunzler 1962; Betterton and Easton 1962), the first utilizing ceramic insulation on the wire, and the second a monel jacket and quartz cloth interlayer insulation. The interlayer quartz cloth is used to minimize the otherwise unwieldy time constant. Connexions to the $Nb_3Sn$ can be made by ordinary soldering techniques after copper plating, or with pressure contacts (Hanak 1962).

Short samples of $Nb_3Sn$ produced by diffusion of into Nb (Saur and Wurm 1962) have recently been fabricated and show great promise, especially as they can be wound in place after heat treatment.

### 4. Power Supplies and Electrical Operation

*4.1. Power supplies.* Superconducting solenoids need no continuous power, but voltage is required to change the field, and is usually required for control purposes. Standard current-regulated power supplies can be used, but the ripple content must be low as the solenoids have a rather low a.c. tolerance (Jones and Schenk, unpublished). Perfectly adequate operation can be obtained from a storage battery and a carbon-pile series resistor.

*4.2. Persistent current operation.* Once energized, a coil can be operated in the persistent current mode as described by Autler (1962) by introducing a superconducting short circuit around the coil. To keep the short circuit open during energization, a simple thermal switch using a $\frac{1}{4}$ watt carbon resistor can be used. The resistance of 0·010 in. diameter Nb–Zr wire at 18°K is about 1 ohm $ft^{-1}$.

*4.3. Quench protection.* Simple effective protection against excessive voltage build-up during a quench can be provided by using 1 ohm shorting resistors across each section of the magnet. When the coil goes normal, the parallel resistors prevent voltage build-up but the major part of the stored energy is dissipated in the coil as the normal resistance of the coil is much larger than that of the bypass resistors. Some large energy storage solenoids have utilized a layer of copper or aluminium between layers with the winding insulation "scratched" to contact the copper interlayer (Riemersma et al. 1962). Solenoids utilizing copper-jacketed wire are protected against violent quench characteristics since the currents switch to the copper jacket when the superconductor goes normal, and decay naturally. A study of a Nb–Zr solenoid during loss of superconductivity has been presented by Stekly (1962).

### 5. Energy Storage and Dissipation

The energy stored in a superconducting solenoid can be rather formidable even in small systems. For example, a solenoid of 0·5 in. inner diameter, 4 in. outer diameter, and 3 in. long, containing about 4 lb of wire and carrying 15 amperes of current, stores 1000 joules (watt-seconds) of energy. If this energy were all liberated in the helium bath it would vaporize 0·38 litres of liquid helium producing nearly 10 litres of gas. A very large magnet having a 12 in. inner diameter 14 in. outer diameter and 48 in. length would store 50 kjoules at a current of 1 ampere.

The energy storage can be conveniently calculated from $E = \frac{1}{2}LI^2$. Hulm et al. (1962) indicates that the measured value of $L$ approaches the calculated $L$ only for fields at the wire above 10 kgauss and may be only one-half the predicted value at lower fields.

When a coil goes from the superconducting to the normal state, much of the energy stored in the field is dissipated in the cooling bath, directly or indirectly.

Some of the energy released during a quench can be coupled into a copper winding and dissipated in an external load. The coupling must be good and an adequate cross-section of copper conductor provided to minimize the losses in the bath. Since spacing between turns is effective, the turns can be spaced with a bifilar copper winding which is coupled to an external load.

Energy dumped into the helium bath will go into vaporizing helium and then into raising the general temperature of the bath. Explosive conditions can result in large systems if precautions are not taken. It is necessary in large systems to confine the helium in small tubes to prevent rupture bursting stresses. Helium can be introduced into the general space for fast cool-down if necessary, but should be contained in small passages for maintenance of temperature under running conditions.

The effects of energy storage in large systems are discussed by Donadieu and Rose (1962) and Stekly (1962).

*Bibliography*

AUTLER S. H. (1960) *Rev. Sci. Instrum.* **31**, 369.

BETTERTON J. O. and EASTON D. S. (1962) in *High Magnetic Fields*, Cambridge, Mass.: M.I.T. Press.

DONADIEU L. and ROSE D. (1962) in *High Magnetic Fields*, Cambridge, Mass.: M.I.T. Press.

GAUSTER W. F. and PARKER C. E. (1962) in *High Magnetic Fields*, Cambridge, Mass.: M.I.T. Press.

HAKE R. R. (1962) in *High Magnetic Fields*, Cambridge, Mass.: M.I.T. Press.

HANAK J. J. (1962) *Bull. Amer. Phys. Soc.*, Series 2, **7**, 473.

HART H. R. (1962) in *High Magnetic Fields*, Cambridge, Mass.: M.I.T. Press.

HULM J. K. (1962) in *High Magnetic Fields*, Cambridge, Mass.: M.I.T. Press.

HULM J. K. *et al.* (1962) *Report* 62-108-273-P3, Pittsburgh, Penn.: Westinghouse Research Laboratories.

KUNZLER J. E. (1962) in *High Magnetic Fields*, Cambridge, Mass.: M.I.T. Press.

KUNLER J. E. (1961) *J. Appl. Phys.* **32**, 325.

MONTGOMERY D. B. (1962) *Report* AFOSR 3015, Cambridge, Mass.: M.I.T. National Magnet Laboratory.

RIEMERSMA H. *et al.* (1962) *Report* 62-108-281-P1, Pittsburgh, Penn.: Westinghouse Research Laboratories.

SALTER L. C. *et al.* (1962) in *High Magnet Fields*, Cambridge, Mass.: M.I.T. Press.

STEKLY Z. J. T. (1962) *Report* 135 *Contract* AF04 (694)-33, Everett, Mass.: Avio-Everett Research Laboratory.
                                    D. B. MONTGOMERY

## SUPERCONDUCTORS OF THE SECOND KIND.

Superconductors can be separated into two well-defined groups characterized by the value of the Ginzburg–Landau parameter $\varkappa$. Those (generally alloys) with a $\varkappa$ in excess of a critical value of $1/\sqrt{2}$ leading to a negative interfacial surface energy between normal and superconducting phases are generally termed "type II superconductors" or "superconductors of the second kind". They are easily recognized experimentally by their magnetic behaviour, flux penetration occurring gradually over a range of magnetic field instead of over a very narrow range as for an "ideal" or type I superconductor.

The first indications of type II behaviour were observed soon after Meissner's discovery of flux exclusion. Keeley and Mendelssohn repeated Meissner's measurements on a variety of pure metals and alloys and observed that for the alloys and some of the elements, flux penetration occurred over a range of magnetic field, and some flux always remained "frozen in" after the field was reduced to zero. Subsequent work showed that the electrical resistance remained zero up to fields considerably in excess of the value at which flux penetration began. Calorimetric measurements were compared with the well known Rutgers relation for the specific heat discontinuity:

$$4\pi\,\varDelta C = T_c \left(\frac{dH_c}{dT}\right)^2_{T_c}$$

and it was found that $\varDelta C$ was more nearly related to the slope of the critical field curve for flux penetration than the curve for restoration of the resistance.

These facts led to the "sponge model", in which the alloys were supposed to consist of a matrix of superconducting material with inclusions of material of higher critical temperature forming a multiply connected network. Flux trapping is easily explained by such a model because on reducing the field from high values the "sponge" would go superconducting before the matrix and flux in the interstices would be maintained constant by shielding currents. Also one would expect the calorimetric properties to be determined by the matrix while the resistive critical field would be determined by the filaments of the sponge; the high critical fields observed for many materials were believed due to the small dimensions of the filaments, since the London theory predicted that the critical field of a superconductor of dimensions less than the penetration depth should increase with decreasing specimen size.

Further work in the field was encouraged many years after the sponge theory by Kunzlers measurements of the current carrying capacity of several high-field superconductors, since his results suggested that the previously ill-fated idea of superconducting magnets might after all be a practical possibility. Mendelssohn's sponge model has been considerably elaborated and a more basic approach starting from the Ginzburg–Landau theory has been developed.

Abrikosov solved the Ginzburg–Landau equations for the case $\varkappa > 1/\sqrt{2}$ and was able to show that between a lower critical field $H_{c1}$ and an upper critical field $H_{c2}$ the superconductor exists in a stable "mixed state" in which quantized "threads" of flux (fluxoids) are arranged in a square array (except near $H_{c1}$, where the array is triangular). This is indicated in Fig. 1, taken from Abrikosov's original publication. Direct evidence of a regular fluxoid "lattice" has recently been obtained by neutron diffraction studies of the mixed state.

The upper critical field is given by

$$H_{c_2} = \varkappa \sqrt{2} H_c$$

where the thermodynamic critical field $H_c$ can be calculated from the area under the magnetization curve ($= H_c^2/8\pi$). Since $\varkappa$ is easily calculated if the normal state resistivity and specific heat are known, this equation is a useful check on the theory; in general observed and calculated values of $H_{c2}$ agree remarkably well. The lower critical field decreases with increasing $\varkappa$ in a rather complex manner but agreement of experiment and theory is again generally good.

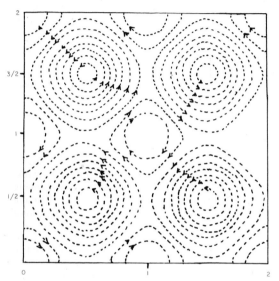

*Fig. 1. Microscopic structure of the internal field in the mixed state. The arrows indicate direction of current flow.*

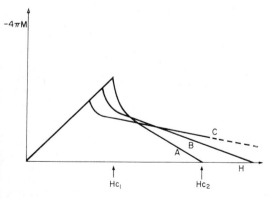

*Fig. 2. Diagrammatic magnetization curves for three alloy specimens A, B, C containing increasing quantities of the second component. The upper and lower critical fields are indicated for the alloy A. Note the vertical tangent at $H_{c1}$, and the linear approach to zero at high fields.*

Figure 2 indicates the manner in which the magnetization curves change with increasing $\varkappa$.

A calculation of the magnetization curve by Goodman, in which the mixed state consisted of alternating normal and superconducting laminae did not give the linear approach to zero of the magnetization at high fields observed experimentally, and although a later, more refined calculation rectified this the model seems to have been abandoned in favour of Abrikosov's.

Abrikosov type behaviour is followed quite closely by homogeneous strain-free alloys such as Ta–Nb solid solutions, but for metallurgically less tractable materials and specimens containing physical defects, dissolved gases etc., magnetic hysteresis, trapped flux and size effects occur, whereas none of these are to be expected on the Abrikosov theory. It is also found that such materials can support large transport currents in fields almost as large as $H_{c2}$ while the mixed state ideally is incapable of supporting transport current at all. The theory may be modified to include such effects since dislocations etc., can be shown to give rise to free energy barriers which "trap" or "pin" the fluxoids; alternatively the magnetization curve of irreversible specimens can be described quite well in terms of shielding currents induced in the threads of a "sponge" structure and similar behaviour has been observed in artificial "sponges" made by pressing mercury into porous glass.

Such shielding currents have been observed in the wall of a thin walled hollow cylinder of $Nb_3Sn$ by measuring the internal field for different values of an external field parallel to the axis of the cylinder. The shielding currents were observed to saturate at a critical value

$$J_c = \alpha/(B + B_0).$$

Where $B$ is the average induction in the cylinder wall and $\alpha$, $B_0$ are constants. A similar field dependence is found when the critical current of a wire is measured directly in transverse field. It may be explained by considering the "creeping" of bundles of one or more fluxoids over free energy barriers; the creep occurs by thermal activation aided by the Lorentz force $\mathbf{J} \wedge \mathbf{B}$. This "flux creep" implies that the shielding currents are in fact decaying with time, although since the decay decreased with the logarithm of the time it soon becomes very small and the current is then truly "persistent" for all practical purposes. It has been suggested on the above basis that the mixed state is to be regarded as a state of non-zero resistance, although the resistance is in all observed cases, vanishingly small.

This theory accounts for hysteresis, trapped flux and size effects and it has been suggested that it also explains the broad resistive transitions observed in transverse fields; the appearance of a voltage across the specimen is supposed due to an unbalanced e.m.f. due to flux creep which increases with increasing field as the flux creep rate increases. Broad transitions in longitudinal field are supposed due to the persistence of a thin superconducting sheath near the surface of the specimen up to a field $H_{c3} = 1.69H_{c2}$ This is a consequence of the Ginzburg–Landau equations for the case where the field is prallel to the specimen surface.

It is probable that the magnetic properties of most type II materials can best be explained by a combination of the Abrikosov theory with shielding current models, in which the field dependent shielding currents effectively describe the flux gradient inside the superconductor arising from the pinning of fluxoids by defects. The hysteresis in the magnetization curve is due entirely to these currents and can be expected to increase with cold work or other processes that enhance the critical current values. Similarly the hysteresis may be expected to decrease as the specimen size is reduced, as has been observed experimentally.

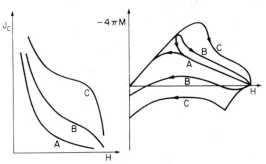

*Fig. 3. Increase of current carrying capacity $J_c$ and magnetic hysteresis for deformed specimen (schematic).*
*A annealed; B light cold work; C severe cold work.*

Investigations of the effect of various types of physical and chemical inhomogeneity on the current carrying capacity (and thus on the hysteresis) have been extensive, since the results are of interest with regard to superconducting magnet applications. Careful specimen preparation and prolonged annealing can result in almost zero current carrying capacity which can then be enhanced enormously by cold work (Fig. 3). The effect is probably greater if the dislocations are "decorated" by impurity atoms. Theoretical calculations indicate that dislocations may indeed "pin" the filaments with a pinning force of the same order of magnitude as the Lorentz force on a filament carrying its critical current (= bulk critical current density divided by dislocation density). Other calculations have shown that cavities (such as might be expected to occur in sintered and compacted materials such as $Nb_3Sn$) and composition fluctuations may also act as pinning centres. Precipitation of a second phase is found experimentally to lead to magnetic hysteresis, the extent of the hysteresis depending on the nature and size of the precipitate particles and on the matrix. Fast neutron irradiation leads to increased hysteresis in the intermetallic compounds $Nb_3Sn$ and $V_3Ga$ but not in homogeneous superconductors, and the hysteresis is thus probably due to disordering by exchange of atoms between lattice sites.

Hysteresis of a somewhat different nature can occur near to $H_{c1}$ where the specimen surface can act as a barrier to flux, the effectiveness depending on the degree of perfection of the surface.

Associated with magnetic hysteresis is the phenomenon of *flux jumping*; as the external field is increased the magnetization of the sample may change discontinuously, especially if the sample shows pronounced hysteresis and if the rate of change of applied field is large. Flux jumping is probably caused by local heating due to the motion of flux into the specimen, leading to a temporary transition to the normal state and allowing flux to penetrate suddenly; its occurrence is usually a limiting factor in the operation of superconducting solenoids.

*Bibliography*

ABRIKOSOV A. A. (1957) *Sov. Phys. J.E.T.P.* **5**, 1174.
BEAN C. P. (1962) *Phys. Rev. Letters* **8**, 250.
BERLINCOURT T. G. (1964) *Rev. Mod. Phys.*
GOODMAN B. B. (1962) *IBM Journal* **6**, 63.
KIM Y. B. *et al.* (1963) *Phys. Rev.* **131**, 2486.
KINSEL T. *et al.* (1962) *Physics Letters* **3**, 30.
KOLM H. *et al.* (Eds.) *High Magnetic Fields*, parts ID and IIIC, New York: Wiley.
KUNZLER J. E. (1961) *Rev. Mod. Phys.* **33**, 501.
LIVINGSTON J. D. (1963) *Phys. Rev.* **129**, 1943.
LYNTON E. A. *Superconductivity*, London: Methuen.
MENDELSSOHN K. (1963) *Cryogenics* **3**, 129.
ROSENBERG H. M. *Low Temperature Solid State Physics*, Oxford: Clarendon Press.
THEWLIS J. (Ed.) (1963) *Encyclopaedic Dictionary of Physics* **7**, 104, 110.

J. LOWELL

## TARDY AND SÉNARMONT COMPENSATION METHODS.

When a photoelastic model is observed in circularly polarized light, as will be the case for the conventional "dark field" circular polariscope arrangement illustrated in Fig. 1, then the isochromatic

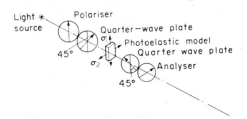

*Fig. 1.*

*Fig. 2.*

fringes observed will be whole order fringes given by the relationship $\delta = i\lambda$, where $i = 1, 2, 3 \ldots$

If we wish to determine the relative retardation $\delta$ at a point such as A in Fig. 2, then we can either introduce into the system an additional source of known birefringence in the form of an optical compensator of some type, or we may use the Tardy or Sénarmont methods of goniometric compensation.

The theory of these methods is adequately presented in most of the standard text-books on photoelasticity, and we shall therefore restrict ourselves to a consideration of the procedures to be followed when a system of the type illustrated in Fig. 1 is used.

*Experimental Procedures for Goniometric Compensation*

*Tardy method.* To determine the relative retardation $\delta$ at point A (Fig. 2):

a) With quarter-wave plates removed and polarizer and analyser crossed (plane polariscope system) rotate the crossed polarizer and analyser until an isoclinic covers point A. (In most polariscope systems polarizer and analyser may be mechanically linked so that they can be rotated together in the crossed position.) The transmission axes of polarizer and analyser are now aligned with the directions of $\sigma_1$ and $\sigma_2$ in the model, and the principal stress directions may thus be obtained from a calibrated scale.

b) Insert both quarter-wave plates with their axes crossed and at 45° to the transmission axes of polarizer and analyser (circular polariscope system—Fig. 1), thus removing the isoclinic from the field of view.

c) By inspection in white light determine the fringe order of the isochromatics in the vicinity of point A.

d) Rotate analyser clockwise. Depending on the relative directions of $\sigma_1$ and the polariscope components, either the higher or lower order fringe will move towards point A. Suppose the *lower* order fringe ($i = 2$ for the case illustrated in Fig. 2) moves towards point A. (We shall define such a rotation, in which the fringe order at a point is effectively *decreased*, as *positive*, irrespective of the rotation direction of the analyser.)

Continue rotating the analyser until the lower order fringe is coincident with point A.

e) Note the angular rotation $\theta$. The relative retardation at point A will now be given by: $\delta = \lambda(2 + \theta/180)$ if $\theta$ is measured in degrees.

f) Suppose we have pre-calibrated our model material so we know that, for example:

The first order fringe ($\delta = \lambda$) corresponds to $\varepsilon_1 - \varepsilon_2$ = 1000 micro-in./in.,
the 2nd order fringe ($\delta = 2\lambda$) corresponds to $\varepsilon_1 - \varepsilon_2$ = 2000 micro-in./in.,
the 3rd order fringe ($\delta = 3\lambda$) corresponds to $\varepsilon_1 - \varepsilon_2$ = 3000 micro-in./in.,—and so on.

Suppose further that $\theta = 60°$. Then the relative retardation at point A is given by $\delta = \lambda(2 + 60/180)$ = $2\frac{1}{3}\lambda$, and the shear strain $\varepsilon_1 - \varepsilon_2 = 1000 \times 2\frac{1}{3}$ = 2,333 micro-in./in.

g) As a check, rotate the analyser in the opposite (counter-clockwise) direction. The *higher* order fringe will now move towards point A. (The rotation will thus, by definition, be negative.)

h) Continue rotating until the fringe is coincident with point A.

Note the angle $\theta'$. The fringe order at point A will now be given by $\delta = \lambda(3 + \theta'/180)$, and if we have compensated correctly we will find that $\theta' = -120°$, giving $\delta = \lambda(3 - 120/180) = \lambda(3 - 2/3) = 2\frac{1}{3}\lambda$, and $\varepsilon_1 - \varepsilon_2 = 2,333$ micro-in./in. as before.

*Sénarmont method.* To determine the relative retardation $\delta$ at point A (Fig. 2):

a) As for the Tardy method.

b) Rotate polarizer and analyser 45° to remove the isoclinic from the vicinity of point A.

c) Insert analyser quarter-wave plate with its fast axis parallel to either polarizer or analyser transmission axis. (If the quarter-wave plate axes are not known, simply insert the plate and rotate it to obtain minimum background illumination, i.e. minimum intensity of light from the polariscope source that does not pass through the model.)

d) Compensate as described in sections (c) to (h) in the Tardy method.

G. S. HOLISTER

**TELEMETRY.** The word telemetry is a combination of Greek and Latin words signifying "measuring at a distance". Telemetry is concerned with the transmission of measured physical quantities such as temperature, velocity, displacement, humidity, blood pressure, pressure, acceleration, etc. to a convenient remote location in a form suitable for analysis and display. The link connecting the two locations may be an electric, pneumatic or hydraulic line, radio, modulated light beam, etc.

The first practical telemetry applications were made by the public utilities prior to World War I. However, the development of telemetry reached a high level with the advent of high-speed aircraft, missiles, and satellites. Simultaneously, the continuing development of industrial telemetry applications proceeded at a relatively slow speed until the experience of the military and governmental applications was applied to the industrial telemetry field.

Industrial telemetry takes advantage of new type transducers for measuring physical quantities to be telemetered, powerful miniature radio transmitters, long-life miniaturized self-contained power supplies (Fig. 1, IEC Straintel and battery) and better techniques of environmental protection. Industrial applications cover a broad segment of industry including utility, chemical, transportation, construction and machinery.

In addition, relatively recent applications in the biomedical field point to the three-fold trends in modern telemetry. Medical science is currently employing telemetry for use in experimental, clinical and diagnostic applications. Some of the particular body parameters telemetered include electrocardiac, electroencephalic, blood pressure, temperature, voice sounds, heart sounds, respiration sounds, muscle

tensions, etc. Similar studies are being pursued in the biological and psychological fields where more experimental latitude permits embedding of transmitters within living experimental animals. Advances in technology for specific applications in the military, industrial and medical field is proceeding post-haste.

The basic telemetry system for applications in any of the three fields consists essentially of three basic building blocks. These are the input transducers, transmitters and receiver station. Transducers convert

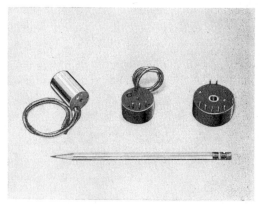

*Fig. 1. Miniature transmitters.*
*(Industrial Electronetics Corporation.)*

the measured physical quantity into a usable form for transmission. The conversion of the desired information into a form capable of being transmitted to the receiver control centre is a function of the type of transducer employed. Transducers convert the physical quantities to be measured into electrical, light, pneumatic, or hydraulic energy. The type of energy conversion is determined by the type of transmission desired.

The most common type of transducer generates electrical signals as a function of the changing physical quantity. There are many types of electrical transducers but one of the most common is the resistance wire strain gauge type. In resistance wire strain gauge type transducers, the ability of the wire to change its dimension as it is stressed causes a corresponding change in resistance to electricity. A decreasing diameter generally exhibits greater resistance to electricity. Similarly temperature sensitive material that will have changing electrical characteristics corresponding to the changes in temperature enables temperature measurement.

This is essentially what occurs in most transducers. The electrical output is varied as a function of changes in a physical parameter. Other type transducers perform essentially the same functions. These electrical changes can be transmitted by wire direct to a recording centre, data display area or to a data analysis section for evaluation. However, the apparent difficulties with use of wire in many applications have given rise to wireless transmission or telemetry.

In order to transmit transducer information out through the air, it is necessary to apply this information to a high-frequency electrical carrier as is commonly done in radio; application of the transducer information to a high-frequency carrier is commonly called modulation. High frequency or rapidly changing electricity has the capability of being propagated through space whereas low-frequency or battery, non-changing voltage does not possess this ability.

The technique used for applying or modulating the high-frequency carrier by the transducer output may involve any one of three different methods. It is possible to modulate a carrier by a change in amplitude, a change in frequency, or a change in the carrier phase. The last technique is similar to the modulation employed in transmitting T.V. colour information.

In colour T.V. the brightness signal is transmitted as amplitude modulation AM, the sound as frequency modulation FM and the colour as phase modulation or pulse coding. Pulse coding is used to modulate the

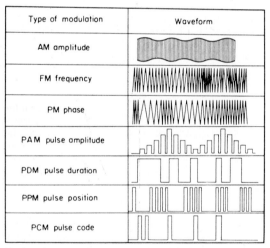

| Type of modulation | Waveform |
|---|---|
| AM amplitude | |
| FM frequency | |
| PM phase | |
| PAM pulse amplitude | |
| PDM pulse duration | |
| PPM pulse position | |
| PCM pulse code | |

*Fig. 2. Telemetry modulations.*
*(Industrial Electronetics Corporation.)*

radio-frequency carrier in either AM, FM, or PM. The various types of modulation that have been used for telemetry are shown in Fig. 2.

A common and extremely useful technique for increasing the information carrying capability of a single transmitting telemetry line is called multiplexing. When it is desirable to monitor different physical parameters, e.g. temperature and pressure, it may be wasteful to have duplicating telemetry transmission lines.

Multiplexing techniques can usually be considered to be of two basic types, frequency division multiplexing and time division multiplexing. In the frequency division multiplexing system different subcarrier frequencies are selectively modulated by their respective changing physical parameter; these sub-

carrier frequencies are then used to modulate the carrier frequency enabling the transmission of all desired channels of information by one carrier. At the receiver these subcarrier frequencies must be individually removed. This is accomplished by filters that allow any one of the respective subcarrier frequencies to pass. These subcarrier frequencies are then converted to voltage by discriminators. The discriminator voltages can be used to actuate recorders and similar devices. Time division telemetry systems employ pulse modulation or pulse code modulation. In these systems the information signal is applied in time sequence to modulate the radio carrier. Pulse modulation may be either amplitude modulation, frequency modulation or phase modulation. Current missile and space programmes employ PCM techniques since through coding techniques considerable increase in information can be transmitted. Various types of modulation that have been used in telemetry are shown in Fig. 2.

Telemetry began as a wire communication technique between two remotely located stations. As science extends its domains into the realm of space, telemetry will be the essential communications link between satellites, space ships, robots and other scientific devices yet to be designed.

C. H. HOEPPNER

# THERMAL EXPANSION, ANHARMONIC EFFECTS IN.

## 1. The Role of Anharmonicity in Thermal Expansion

*Coefficients of thermal expansion.* The coefficients of thermal expansion describe the temperature dependence of the size and shape of a solid body under constant stress. The number of independent coefficients depends on the symmetry. Under an isotropic pressure cubic crystals and isotropic solids (e.g. glass) possess only one coefficient of linear expansion, which may be defined by

$$\alpha = \frac{1}{l} \left( \frac{\partial l}{\partial T} \right)_p ; \qquad (1)$$

here $l$ refers to the distance between any two points of the solid. Hexagonal and some other crystals have two coefficients: $\alpha_\parallel$ for distances along the main crystal axis, $\alpha_\perp$ for distances perpendicular to this axis. A crystal of the lowest symmetry (triclinic) has the maximum number of independent coefficients, given by the temperature derivatives of the six independent components of the strain tensor.

For every homogeneous solid under isotropic pressure we can always define a unique coefficient of volumetric expansion, defined by

$$\beta = (\partial \ln V / \partial T)_p . \qquad (2)$$

$\beta$ is simply related to the coefficients of linear expansion: for cubic isotropic solids, $\beta = 3\alpha$; for hexagonal solids, $\beta = \alpha_\parallel + 2\alpha_\perp$. For ionic, metallic and valence solids, $\beta$ usually has a value between $10^{-4}$ and $10^{-5}$ deg$^{-1}$ at room temperatures. For less tightly bound crystals $\beta$ is usually considerably higher (for argon at $60°K$, $\beta = 6 \times 10^{-4}$ deg$^{-1}$).

*Derivation from the Helmholtz energy.* To see how $\beta$ depends on the Helmholtz energy $A$, and to relate it to molecular properties, we use the thermodynamic relations

$$\beta = \chi_T(\partial S/\partial V)_T, \quad S = -(\partial A/\partial T)_V; \quad (3)$$

here $\varkappa_T$ is the isothermal compressibility and $S$ is the entropy. If the Helmholtz energy can be expresse das the sum of contributions from systems which are only weakly coupled together (lattice vibrations, conduction electrons, magnetic spins, etc.), then also

$$S = S_l + S_e + S_m + \ldots, \quad (4)$$

$$(\beta/\chi_T) = (\partial S_l/\partial V)_T + (\partial S_e/\partial V)_T + \ldots \quad (5)$$

The contribution $(\partial S_l/\partial V)_T$ is present in all solids (although at some temperatures it may be much smaller than other contributions), and is a direct result of the anharmonicity of the lattice vibrations.

*Anharmonicity and thermal expansion.* We assume that the motion of the $\mathcal{N}$ atomic nuclei in the solid can be derived from a potential energy $\Phi$, and that $\Phi$ can be expanded as a Taylor series in powers of coordinates describing the displacement of the atoms from their equilibrium positions:

$$\boldsymbol{\Phi} = \boldsymbol{\Phi}_0 + \boldsymbol{\Phi}_2 + \boldsymbol{\Phi}_3 + \boldsymbol{\Phi}_4 \cdots \quad (6)$$

Here $\Phi_n$ contains all terms of the nth order in the displacements; $\Phi_0$ is the potential energy of the atoms in their equilibrium positions, and $\Phi_1$ vanishes because $\Phi$ is expanded about the equilibrium positions.

If there are no terms above the second order ($\Phi_3 = \Phi_4 = \ldots = 0$), the potential is said to be *harmonic*. The mechanics then reduces to a superposition of $3\mathcal{N}$ independent normal modes. Six of these have zero frequency, and refer to the translation and rotation of the solid as a whole; the remainder are vibrational modes, in each of which every atom vibrates harmonically with the same angular frequency $\omega_j$ ($j = 1, 2, \ldots 3\mathcal{N} - 6$). Increasing the temperature increases the mean amplitude of these vibrations, but the mean position of each nucleus is unaltered and there is no expansion. Thermal expansion can therefore arise from the lattice vibrations only if the potential is anharmonic. More stringently, symmetry demands that at least one odd order term in the expansion (6) should not vanish.

*Volume dependence of the frequencies.* A similar conclusion can also be reached from equation 3. The entropy of a harmonic lattice is given by the sum of contributions from the individual normal modes:

$$S_l = \sum_{j=1}^{3\mathcal{N}-6} s_j, \quad (7)$$

where

$$s_j = k\left\{[x_j/(e^{x_j} - 1)] - \ln(1 - e^{-x_j})\right\},$$

$$x_j = \hbar\omega_j/kT. \quad (8)$$

It follows from equation 3 that $\beta/\chi_T$ depends on the volume-dependence of the normal frequencies $\omega_j$. But for a purely harmonic lattice the normal frequencies are independent of strain (this can be proved by considering the effect of distorting the lattice by arbitrary external forces), and so the $(d\omega_j/dV)$ and hence $(\beta/\chi_T)$ are zero.

In a real anharmonic solid the atomic vibrations may still to a first approximation be expressed as the superposition of $3\mathcal{N} - 6$ normal modes, but the frequencies are now in general volume-dependent. The anharmonic contribution to the thermal expansion can therefore be regarded either as a consequence of the anharmonicity of the lattice vibrations or equivalently as a consequence of the volume-dependence of the normal frequencies. More generally, the strain-dependence of the frequencies of anisotropic crystals determines the temperature dependence of the strain under constant stress.

The strain-dependence of the frequencies is thus an essentially anharmonic property in terms of which we can conveniently discuss the role of anharmonicity in thermal expansion. A well-known example illustrating this is provided by a diatomic molecule with interatomic distance $r$ and potential energy $\varphi(r)$ between the atoms (Fig. 1). When $\varphi(r)$ is harmonic, the mean distance between the atoms remains constant at $r_0$

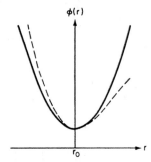

*Fig. 1. Potential energy of a diatomic molecule. ———, harmonic potential; ———, anharmonic potential.*

whatever the amplitude of the vibration; if the molecule is compressed or stretched, the potential about the new equilibrium distance remains harmonic with the same force constant and the vibrational frequency is unchanged. But for the asymmetric anharmonic potential the mean distance increases with increasing vibrational amplitude, while the frequency of vibration decreases with increasing mean distance. Conversely, an *increase* of frequency with mean distance would correspond to a negative thermal expansion.

## 2. Grüneisen Parameters and the Grüneisen Function

*Grüneisen parameters.* Although theoretical expressions for $\beta$ in terms of the potential $\varphi$ are rather.complicated, it follows from equations 3, 7 and 8 that $\beta/\chi_T$

depends only on $T$, the frequencies $\omega_j$, and their volume derivatives. We are here employing the quasi-harmonic approximation, in which we neglect the crystal anharmonicity when deriving normal frequencies $\omega_j$ at a given volume; the effect of anharmonicity enters only through the volume-dependence of $\Phi_2$ and hence of the $\omega_j$.

Because $s_j$ is a function only of $\hbar\,\omega_j/kT$, and the corresponding term in $C_V$ is simply

$$c_j = T(\partial s_j/\partial T)_V, \tag{9}$$

it follows from equation 3 that

$$\beta/\chi_T = \sum_{j=1}^{3\mathcal{N}-6} \frac{c_j}{\omega_j}\frac{d\omega_j}{dV}. \tag{10}$$

It is then convenient to define the dimensionless Grüneisen parameters $\gamma_j$ to describe the volume-dependence of each normal frequency:

$$\gamma_j = -d\ln\omega_j/d\ln V, \tag{11}$$

indicating that $\omega_j$ is varying as $V^{-\gamma_j}$. Equation 10 then becomes

$$(\beta/\chi_T) = \sum_{j=1}^{3\mathcal{N}-6} \gamma_j c_j/V. \tag{12}$$

The $\gamma_j$ are typically of order of magnitude unity, and can be positive or negative.

At high temperatures (greater than the Debye characteristic temperature $\Theta$) $c_j \approx k$, and then

$$(\beta/\chi_T) \cong \bar{\gamma}_j C_V/V, \tag{13}$$

where $\bar{\gamma}_j$ is the arithmetic mean of all the $\gamma_j$. In fact it appears from experimental data that equation 13 normally holds pretty closely for temperatures down to about $\Theta/3$.

*Grüneisen's rule.* If all the $\gamma_j$ had the same value, $\gamma$, equation 13 would then hold down to the lowest temperatures:

$$(\beta/\chi_T) = \gamma C_V/V. \tag{14}$$

Equation 14 is known as Grüneisen's rule. Since the temperature dependence of $\chi_T$ is small at low temperatures, it implies that $\beta$ varies with temperature in a rather similar way to $C_V$, with a $T^3$ dependence at very low temperatures.

In fact the rule is not generally valid, although it may be obeyed even when the $\gamma_j$ are not all equal. A necessary and sufficient condition is that the *shape* of the distribution of frequencies $\omega_j$ should not change with volume. Thermodynamically, the equivalent condition is that the entropy should be a function only of $T/\Psi(V)$, where $\Psi(V)$ is a characteristic temperature and $\gamma = -d\ln\Psi/d\ln V$.

Equation 13, shows that the rule is obeyed quite closely at high temperatures, with a value for $\gamma$ of

$$\gamma_\infty = \bar{\gamma}_j. \tag{15}$$

The rule is also obeyed at very low temperatures, since the entropy can be fitted to a Debye function

with characteristic temperature $\Theta_0$. The appropriate value of $\gamma$, $\gamma_0$, usually differs from the high temperature value; it is given by

$$\gamma_0 = -d\ln\Theta_0/d\ln V. \tag{16}$$

*The Grüneisen function* $\gamma(T, V)$. When Grüneisen's rule does not hold, equation 13 can be used to define a thermodynamic function

$$\gamma(T, V) = (\beta V/\chi_T C_V) = (\beta V/\chi_S C_P) \tag{17}$$

in terms of the experimental quantities $\beta$, $V$, $\chi_S$, $C_P$. If the thermal expansion is due only to the lattice vibrations, it follows from equation 12 that

$$\gamma = \sum \gamma_j c_j/\sum_j c_j. \tag{18}$$

The variation of $\gamma$ with temperature thus comes primarily from the temperature-dependent weighting factors $c_j$. There are also higher order anharmonic effects (due to the change of $\gamma_j$ with volume, and to the inadequacy of the quasi-harmonic approximation), but for most solids these are small, especially at moderately low temperatures ($T \lesssim \Theta/3$). We may therefore expect $\gamma$ to vary smoothly with temperature, flattening off at high temperatures to $\gamma_\infty$ and at low temperatures to $\gamma_0$. Since $\Theta_0$ depends only on the density and the elastic constants at $T = 0$, $\gamma_0$ can also be derived from the volume-dependence of the elastic constants at low temperatures. Calculations on specific models have indicated that $\gamma$ should be close to $\gamma_\infty$ above about $\Theta/3$, which is well below room temperatures for most solids.

$\gamma$ is of order of magnitude unity at all temperatures, and is therefore a convenient function both for plotting experimental data and for discussing its significance in terms of the volume-dependence of the frequencies.

*Non-vibrational contributions.* When there are separate contributions to the entropy, as in equation 4, we can define separate Grüneisen functions for each contribution:

$$\gamma_l = (\partial S_l/\partial \ln V)_T/C_l, \tag{19}$$

$$\gamma_e = (\partial S_e/\partial \ln V)_T/C_e, \quad \text{etc.,} \tag{20}$$

where $C_l$, $C_e$ are the corresponding contributions to $C_V$. Equation 17 then gives

$$\gamma = (\gamma_l C_l + \gamma_e C_e + \ldots)/(C_l + C_e + \ldots), \tag{21}$$

shewing explicitly that a given term is likely to contribute significantly to the thermal expansion only if it contributes significantly to $C_V$. When there is more than one contribution to $\beta/\chi_T$, equation 18 and the subsequent remarks apply not to $\gamma$ but to $\gamma_l$.

Several different types of contribution, and the methods of obtaining them by analysis of experimental data, are discussed in the review by Collins and White.

### 3. Behaviour of Real Solids

*Experimental methods.* In order to obtain $\gamma$ from experimental measurements, we need data for $\beta$, $C_P$ and $\chi_S$ over the required range of temperature. $C_P$ can be measured by standard calorimetric techniques down to very low temperatures, and $\chi_S$ (if single crystals are available) by the velocity of ultrasonic waves. However, $\beta$ falls off drastically at low temperatures, and very sensitive measurements are therefore required to obtain the variation of $\gamma$ in the important region below about $\Theta/3$.

A brief description of the different methods used to measure $\beta$ is given in the review by Collins and White. The most accurate equipment at present available will detect relative changes of length ($\delta l/l$) between $10^{-9}$ and $10^{-10}$, in principle giving the shape of the $\gamma$-$T$ plot down to temperatures of about $0 \cdot 01\,\Theta$. The length of a typical specimen is a few centimetres, and changes of length are detected to a limit of about a tenth of an Ångström.

*Van der Waals crystals.* Here the most important interactions between molecules are attractive dispersion forces and repulsive overlap forces. The binding is weak, the compressibility high, and the thermal expansion large compared with more tightly bound solids.

The simplest examples are the inert gas solids, which except for helium crystallise in the c.c.p. structure. A model with central forces between nearest neighbours gives only a small variation of $\gamma$ with temperature ($\gamma_\infty - \gamma_0 = 0 \cdot 3$), and the effect of interaction between further neighbours is to make the variation of $\gamma$ even less. The absolute value of $\gamma$ depends upon the particular interatomic potential chosen for the model; with various (6-$n$) potentials Horton and Leech have found values between 2·5 and 3·5.

Experimentally the crystals are difficult to prepare and handle, but preliminary $\gamma(T)$ curves have been determined for A and Kr. These indicate that at 60°K $\gamma \cong 2 \cdot 8$ (A), $2 \cdot 5$ (Kr), but are not sufficiently precise to cover the important range below $\Theta/3$. They reveal however a relatively large drop in $\gamma$ below 60°K, in the range where quasi-harmonic theory would expect $\gamma \cong \gamma_\infty$. If this drop is confirmed, it will indicate surprisingly large second-order anharmonic effects.

Measurements of $C_V$ for $^4$He at different volumes indicate that $C_V$ has the approximate form $f(T/\Psi(V))$, and hence that $\gamma$ has only a slight temperature-dependence at constant pressure. Solid helium is so anharmonic that the quasi-harmonic theory is not even approximately valid, and consequently this agreement with the predictions of a quasi-harmonic model requires further theoretical study.

*Metals.* Here the interatomic forces are due partly to the interaction between neighbouring metallic ions and partly to the conduction electrons. The dense packing in most metals suggests that central forces between the neighbouring ions are important, and the lattice model discussed above then suggests that the variation in $\gamma$ should be small. This behaviour has also been predicted by a continuum model, and is in fact usually observed, except at very low temperatures

where the electronic contribution $\gamma_e$ becomes important. An example is shewn in Fig. 2.

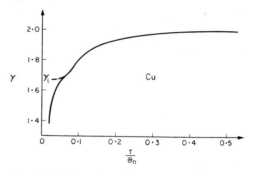

*Fig. 2. Variation of $\gamma$ with temperature for copper.*

*Ionic solids.* Here the most important interactions are the long-range Coulomb forces between ions and the overlap forces between nearest neighbours. Calculations on models for alkali halides reveal the possibility of a wide distribution of $\gamma_j$ values (typically ranging between $-1$ and $+3$), and a large variation

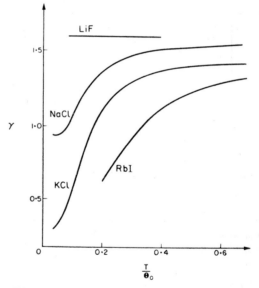

*Fig. 3. Variation of $\gamma$ with temperature for some alkali halides.*

of $\gamma$ with temperature. Marked variation of $\gamma$ with temperature is found experimentally in most alkali halides, especially in those of potassium and rubidium (Fig. 3). This behaviour appears theoretically to be made possible by the open structure of the NaCl lattice, and is particularly associated with very low values of the shear elastic constant $c_{44}$. In the lithium salts the cation is so small that overlap forces between anions increase $c_{44}$ appreciably, and the temperature variation in $\gamma$ is much smaller.

On this basis, Blackman (1958) has pointed out that the very open structure of zincblende should lead to still more striking effects, including a negative thermal expansion at sufficiently low temperatures. Similarly, the more dense packing of the CsCl structure would be expected to lead to a smaller variation of $\gamma$ than that observed in the rubidium salts. Unfortunately experimental data is lacking to test these predictions fully.

*Covalent crystals.* The simplest example of a covalent solid is provided by the elements of diamond structure, with cubic symmetry and tetrahedral bonding. In Si and Ge the open structure again is associated with a strong variation of $\gamma$ and with negative thermal expansion at low temperatures, although in the limit

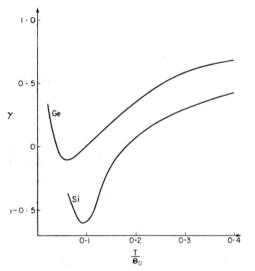

*Fig. 4. Variation of $\gamma$ with temperature for silicon and germanium.*

as $T \to 0$ it appears that $\beta$ becomes positive again (Fig. 4). A negative thermal expansion at low temperatures is also observed in normal hexagonal ice; this again is tetrahedrally linked, by hydrogen-bonding.

*Silica glass.* Although there is no long range order, the bonding of the Si atoms is predominantly tetrahedral through O atoms. Over most of the measured temperature range the variation of $\gamma$ is rather like that observed for many crystalline solids of open structure, with a negative expansion at low temperatures. Below about $0 \cdot 1\ \Theta$, however, this negative expansion appears to be greater than that observed in crystals; when $T = 0 \cdot 005\ \Theta$, $\gamma \sim -8$. At these temperatures the heat capacity is also anomalously large, in considerable excess of that predicted from the elastic constants. It thus appears (equation 5) that the corresponding excess entropy is strongly volume-dependent. The effect is presumed to be due to low

lying vibrational levels having their origin in localised configurations in the disorded lattice.

*Anisotroipc crystals.* Comparatively few measurements have been made at low temperatures, apart from a few hexagonal metals. Strong anisotropy is observed for Zn and Cd, with $\alpha_{\parallel}$ positive at all temperatures and $\alpha_{\perp}$ negative at low temperatures. Much less anisotropic behaviour is found for hexagonal metals with axial ratios close to the ideal value.

For the strongly anisotropic Cd (c/a $= 1 \cdot 89$) the values of $\beta$ calculated from $\alpha_{\parallel} + 2\alpha_{\perp}$ appear to show significant differences from those measured for polycrystalline Cd, and it has been suggested that this may be due to internal constraints or to preferred orientation in the polycrystal. If so, polycrystalline data may not be suitable for any simple interpretation and analysis.

### 4. Analysis of Experimental Data

The variation of $\gamma$ with temperature can be used to obtain information about the volume-dependence of the frequency distribution and properties dependent on it. Apart from the theoretical interest of this information, it enables us to make corrections for changing volume to a number of crystal properties.

With thermodynamic properties it is not always necessary to use the quasi-harmonic approximation and the concept of a frequency distribution. Thermodynamic relations can be used directly; in particular

$$(\partial C_P / \partial \ln V)_T = [\partial (\beta V) / \partial \ln T]_p / \chi_T, \quad (22)$$

$$[\partial (\beta V) / \partial \ln V]_T = [\partial (\chi_T V) / \partial T]_p / \chi_T. \quad (23)$$

These relations have the advantage that they are applicable even when there are appreciable non-vibrational contributions to the thermodynamic functions.

On the other hand, the analysis in terms of the quasi-harmonic theory shows that the volume-dependence of the thermodynamic functions at high temperatures, and of the zero-point energy, can to a fair approximation be estimated very simply from $\gamma_{\infty}$ alone. It also gives the volume-dependence of the Debye–Waller factor for simple crystals, and thus can aid the analysis of X-ray and neutron scattering data.

### 5. Higher Order Anharmonic Effects

At high temperatures the quasi-harmonic approximation may no longer be adequate fully to account for the anharmonic thermal expansion, and this may also happen when the effects of zero-point energy are large (as in argon). There have been several theoretical investigations of the breakdown of the quasi-harmonic theory, and some attempt to relate them to experimental data; but this is beyond the scope of the present article.

### Bibliography

BARRON T. H. K. *et al.* (1964) *Analysis of Thermal Expansion Measurements*, Proc. Roy. Soc. A **279**, 62.

BATCHELDER D. N. and SIMMONS R. O. (1964) *Lattice Constants and Thermal Expansivities of Silicon and of Calcium Chloride between 6 and 322° K, J. Chem. Phys.* **41**, 2324.

BLACKMAN M. (1958) *On Negative Volume Expansion Coefficients, Phil. Mag.* **3**, 831.

BLACKMAN M. (1959) *On the Lattice Theory of Expansion, Proc. Phys. Soc.* **74**, 175.

BORN M. (1923) *Atomtheorie des festen Zustandes, Encyklopädie der Math. Wissenschaften, Physik,* Vol. 3. Leipzig: Teubner.

COLLINS J. G. and WHITE G. K. (1964) *Thermal Expansion of Solids, Progr. Low Temp. Phys.* **4**, 450.

DAVIES R. O. (1952) *The Internal Pressure of Solids, Phil. Mag.* **43**, 473.

DOMB C. and DUGDALE J. S. (1957) *Solid Helium, Progr. Low Temp. Phys.* **2**, 338.

GRÜNEISEN E. (1926) *Zustand des festen Körpers, Handbuch der Physik* **10**, 1.

HORTON G. K. and LEECH J. W. (1963) *On the Statistical Mechanics of the Ideal Inert Gas Solids, Proc. Phys. Soc.* **82**, 816.

LEIBFRIED G. and LUDWIG W. (1961) *Theory of Anharmonic Effects in Crystals, Solid State Physics* **12**, 275.

SLATER J. C. (1939) *Introduction to Chemical Physics,* New York: McGraw-Hill.
                                                    T. H. K. BARRON

**THERMOVAPORIMETRIC ANALYSIS.** Thermovaporimetric analysis is an analytical technique in which the gaseous products of decomposition or reaction evolved on heating a substance are continually recorded as a function of temperature.

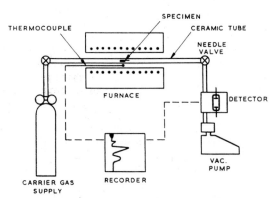

*Fig. 1. Apparatus for the thermovaporimetric analysis of ceramic materials using argon as a carrier gas and a discharge tube as detector.*

On heating a substance a number of changes can take place. Firstly there may be physical changes in the form of thermal expansion, sintering, crystallographic phase changes or desorption of gases. Secondly there may be chemical changes due to reaction of the components or decomposition in which one or more of the products is a gas.

Thermovaporimetric analysis (TVA) is concerned with the analysis of adsorbed gases and gaseous products of reaction, the rate at which these gases are evolved and the temperature of evolution. Figure 1 illustrates an experimental apparatus for TVA of solid materials using a discharge tube as detector and argon as a carrier gas, although other detectors and carrier gases may be used. The specimen is heated in a gas tight tube under a continuous flow of argon the purity of which is measured by the detector and recorded. As the temperature increases gaseous products of reaction are injected into the carrier gas. Under the conditions of constant gas flow the instrument behaves as a differential detector, the height of the peak traced out by the recorder (*thermovapourgram*) being proportional to the rate of reaction; the area under the curve being proportional to the volume of gas evolved. Provided the carrier gas flow rate is known, and the detector output has been pre-calibrated by injection of known gases and volumes into the carrier gas, the thermovapourgram will give a record of the rates of gas evolution at various temperatures.

The types of reaction most suitable for investigation by TVA are as follows.

(a) Simple thermal decomposition,

e.g.     $CaCO_3 \rightarrow CaO + CO_2 \uparrow$

(b) Stage dehydration and decomposition,

e.g.     $FeSO_4 . 7H_2O \rightarrow FeSO_4 . 3H_2O + 4H_2O \uparrow$

$\rightarrow FeSO_4 + 3H_2O \uparrow$

$\rightarrow FeS + 2O_2 \uparrow$

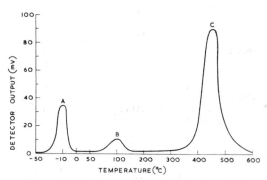

*Fig. 2. The thermovapourgram of the clay, Silklay at low temperatures and pressure. Peak A—absorbed water and gases; peak B—absorbed water; peak C—chemically bound water.*

(c) Gas adsorption and adsorption,

e.g. The clays $\rightarrow$ absorbed $H_2O$

$\rightarrow$ adsorbed $H_2O$

$\rightarrow$ chemical $H_2O$

illustrated in Fig. 2.

(d) Solid state reactions,

e.g. formation of carbides,

$$UO_2 + 3C \rightarrow UC + 2CO \uparrow$$

illustrated in Fig. 3.

(e) Diffusion of gases,

e.g. the diffusion of oxygen through a ceramic envelope at high temperatures, the diffused oxygen detected using an oxygen analyser.

(f) Gaseous reductions,

e.g. the reduction of oxides with hydrogen. Dry hydrogen used as carrier gas, the $H_2O$ formed on reduction detected on a humidity meter.

$$U_3O_8 + H_2 \rightarrow U_4O_9 + H_2O \uparrow \rightarrow UO_2 + H_2O \uparrow$$

(g) Gaseous oxidation,

e.g. the oxidation of carbonaceous matter as $CO_2$, the gas being detected using a katharometer.

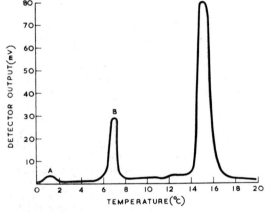

*Fig. 3. Thermovapourgram of the solid state reaction*
$$UO_{2+x} + C \rightarrow UO_2 + CO + \rightarrow UC + CO \uparrow$$
*Peak A absorbed gases; peak B the removal of hyperstoichiometric oxygen; peak C main carbide reaction.*

*Detectors.* The choice of a detector for use in TVA will depend upon the nature of the products of decomposition and how they differ in physical and chemical properties from those of the carrier gas. Detectors used in the field of gas chromatography can be used in TVA.

*Identification of decomposition products.* Using a detector which is sensitive to all impurities it is not possible to identify the various peaks on the vapourgram with known gases; however, by using a specific detector, a positive identification can be made of one gas. By using a *Hersch cell* oxygen can be positively identified and by using a moisture monitor water can be detected. Identification of a number of other gases

can be obtained by measuring the retention times on a gas chromatographic column.

*See also:* Hersch cell.

*Bibliography*

HERSCH P. (Oct. 1952) *The Chemical Age.*

KEIDEL F. A. (1956) *A Novel Inexpensive Instrument for Accurate Analysis of Traces of Water. The P₂O₅ Electrolysis Cell,* Conf. on Analytical Chem. and Applied Spectroscopy.

MACZEK A. O. S. and PHILLIPS C. S. G. (1960) *Retention Times and Molecular Shape,* in *Gas Chromatography,* (SCOTT R. P. W. Ed.) London: Butterworths.

RILEY B. (1960) in *Gas Chromatography* (SCOTT R. P. W. Ed.), London: Butterworths.

RUSSELL L. E. and HARRISON J. D. L. (1963) *"Sintering of U-C and (UPu)-C",* New Nuclear Materials including Non-metallic Fuels, Vol. 1 Vienna: International Atomic Energy Agency.

B. RILEY

**TIME STANDARDS.** A *unit* of measurement for any quantity is usually an abstraction which specifies the idealized concept underlying the realization of the unit. A *standard* on the other hand is a physical embodiment of the unit. Thus standards may occur at any level of use and accuracy. Here, however, we shall concern ourselves only with time standards at the highest level of accuracy and at the most inclusive level of use; namely, at the level of national and international standardization.

An operational definition of time is that time is a physical quantity which can be measured with a clock. A clock in turn is a device which generates a controlled, ordered, nearly continuous sequence of states or phases which can be identified and correlated, by observations, with other ordered sequences characteristic of some chosen phenomenon of interest, or which can associate a number with any single event in question. A unit of time is defined with reference to the particular clock in use. In practice, clocks are often based on nearly uniformly recurring phenomena, so that the unit is defined to be proportional to the period of the clock.

One may specify several requirements for the best possible time standard. It must possess continuity of operation, since time is unique in that it is not sufficient to establish the unit once and for all; one must continually generate and accumulate the units so that any arbitrary interval may be measured whenever desired. Possibly an acceptable substitute for continuity is renewability in the sense of the ability to re-establish the unit at will. This requirement will be acceptable provided that accurate enough other means exist for interpolating between successive determinations of the unit. Secondly, the standard must provide constancy in the size of the unit as time progresses, so far as can be told by comparison with other measures of time. In some cases one may accept a determinable and predictable variability in the period of the standard, so that corrections can be applied which lead to a constant unit. Thirdly, the standard must give an accuracy (with respect to the idealized concept of the unit) which is comparable to or greater than that of all alternate standards. This requirement also embraces, of course, the precision with which the standard is observable. Fourthly, the standard must be accessible to all who need it. Fifthly, the standard should possess a characteristic period of convenient size, because of the need to average over many cycles of the standard in order to improve its precision, to accumulate cycles easily in order to construct large intervals, and to subdivide cycles in order to construct small intervals. Lastly, the standard should be capable of continuously accumulating the unit so as to obtain epoch. Epoch means the state or phase of the standard referred to some arbitrarily selected initial state.

*Astronomical standards.* At this writing (1964), the internationally defined unit of time, the second, is the fraction 1/31, 556, 925.9747 of the tropical year for 1900 January 0 at 12 o'clock Ephemeris Time. The 1956 resolution of the International Committee of Weights and Measures defining this unit was ratified at the 1960 General Conference of Weights and Measures. This unit, identical with the Ephemeris Second, has supplanted the mean solar second (the second of Universal Time) because of a gain in precision of about one order of magnitude. The accuracy of realizability of this unit from astronomical observations is about 2 parts in $10^9$.

This definition reflects the special characteristics of astronomical standards which satisfy the requirements of continuity, constancy, accuracy, and accessibility. The size of the interval characteristic of the standard, that is the tropical year, is not too convenient. It is, perhaps, nearly optimum for precision of observation; but a single observation requires an inconveniently long time; and the tropical year interval must furthermore be subdivided by auxiliary means.

Astronomical standards provide epoch very well because of their continuity and their long characteristic periods, so that ambiguity of phase is easily avoided even with infrequent observation. Furthermore, astronomical standards have important auxiliary applications such as navigation and positional astronomy in which epoch, or state, of the astronomical standard has a use apart from the designation of the time of occurrence of single events.

*Atomic standards.* Atomic frequency standards are based on the frequency corresponding to a transition between two atomic energy levels. Such devices are now widely used as standards of time interval, that is, differences in time. They are also used to some extent as standards of time, that is, epoch (time as measured from some arbitrary initial instant according to the phase of the clock in use). Atomic standards have the advantage of renewability, although they are not automatically continuous for many thousands of years as are astronomical standards. Nevertheless, present engineering technology has achieved continuity for periods of several years. Constancy of determination of the size of the unit among various atomic standards may be strongly presumed to hold by the nature of atomic energy levels and indeed has been demonstrated to 1 part in $10^{11}$. Accuracy with respect to the value of the unperturbed atomic transition is typically about 1 part in $10^{11}$ now, or about two orders of magnitude better than the accuracy of determination of the Ephemeris Second. Precision of observation is high. Averaging times of only a few hours are long enough to show precision of observation of a few parts in $10^{13}$, in the sense of the standard deviation of the mean of several such observations. Atomic standards are universally available either by construction or by purchase. The size of the period characteristic of the standard is convenient, being the reciprocal of the atomic transition frequency. This frequency is typically in the gigacycle-per-second region. Thus it is easily averaged over many cycles in a matter of minutes, hours, or days. It is easily integrated by electronic means to construct large intervals of time. Small intervals, even down to the nanosecond region, are readily available because of the high frequency of the standard. The epoch of the atomic standard is useful for providing a uniform scale of time, common to all observers, upon which the occurrence of events may be placed. It does not, however, have presently known auxiliary usefulness such as in signifying when it is "noon" or when "spring" begins. Continuous atomic time scales have been constructed and maintained for several years by several laboratories. Techniques vary in detail, but all amount to the observation of the phase of a quartz oscillator with corrections as necessary for any fluctuations in its frequency with respect to an atomic standard. For example, we may define an arbitrary measure of time $\tau$ in terms of the elapsed phase $\Delta\varphi$ (that is, number of cycles) of an oscillator. If the oscillator is assigned a fixed nominal frequency $\omega_n$, then its indicated time is $\Delta\tau$ given by

$$\Delta\varphi = \omega_n \, \Delta\tau, \quad \text{or} \quad \Delta\tau = \Delta\varphi/\omega_n.$$

But if its actual frequency for that interval (assumed sufficiently short), is $\omega_a = \omega_s/r$ in terms of a fixed atomic standard frequency and a ratio $r$, we have the measure of atomic time $\Delta t$ given by

$$\Delta\varphi = \omega_a \, \Delta t = (\omega_s/r) \, \Delta t.$$

Note that these equations state the invariance of the oscillator phase as expressed in either system of time. These concepts define time in terms of a frequency standard; for if the oscillator is the frequency standard itself, then $\omega_a = \omega_s$ and $r = 1$, so that $\Delta t = \Delta\varphi/\omega_s$. In the general case, simple algebra gives

$$\Delta t = (\omega_n/\omega_s) \, r \, \Delta\tau$$

or, for long time intervals,

$$\Delta t = \frac{\omega_n}{\omega_s} \int_{\tau}^{\tau + \Delta \tau} r(\tau) \, \mathrm{d}\tau,$$

where $r(\tau)$ accounts for a non-uniformly running oscillator, expressed as a function of $\tau$. We may thus construct intervals of atomic time related to the two observables, the phase of an oscillator (that is, its indicated time $\tau$) and the atomically calibrated oscillator frequency $\omega_a$.

Since it is more practical to run quartz oscillators than atomic standards continuously, it would be desirable to correct for any predictable variations of quartz by programmed phase shifts at the output of the quartz oscillator thereby obtaining a measure of time always very close to atomic time. Mechanical phase shifters have been successfully employed recently to correct for frequency offsets and predicted frequency drift rates. Such systems have given measures of time within $\pm 0.3 \, \mu s$ per day of uniform atomic time. The magnitude of this absolute error increases as the square root of the number of observations; thus for 100 days the accumulated error would be expected to be $3 \, \mu s$. Since the measure of time itself progresses directly as the elapsed time, the relative error continually decreases and is limited only by the accuracy of the atomic standard.

$^{133}Caesium.$ The transition involved is between the two magnetic hyperfine levels in the ground state $(^2S_{1/2})$ arising from the coupling of the nuclear spin angular momentum $I$ and the electronic angular momentum $J$. In caesium, $J$ is due only to the spin of the valence electron. The transition is designated as

$$F = 3, \quad m_F = 0 \underset{\rightarrow}{\overset{\leftarrow}{\rightleftharpoons}} F = 4, \quad m_F = 0,$$

where $F$ takes on the integral values ranging from $I + J$ to $|I - J|$. For caesium, $I = 7/2$ and $J = 1/2$. The notation $m_F$ designates the magnetic quantum number for Zeeman splitting of the $F$ levels by an external magnetic field. It gives the component of $F$ along the field. The transition occurs at a frequency of 9,192,631,770 c/s.

The method of observation is by well known atomic beam techniques. Caesium metal heated in an oven vaporizes and atoms effuse from an aperture. These atoms first pass through a strong and inhomogeneous magnetic field. By virtue of the Zeeman splitting of the energy levels, a magnetic moment $\mu$ in the neighbourhood of 1 Bohr magneton is induced in the caesium atoms. The atoms are deflected from the instrument axis by the force due to the inhomogeneous field and of magnitude proportional to $(\mu \cdot \nabla) \, \boldsymbol{H}$. The atoms then pass into a drift space where a low, uniform magnetic field exists in order to prevent them from taking random spatial orientations. There an oscillating radio frequency interacts with the atomic dipole moment (which is associated largely with the electro-nic angular momentum) to induce either absorption or stimulated emission. In practice, the r.f. field is applied in two regions separated by a considerable distance. The separation is chosen variously from 0.5 to 5 m in order to achieve certain experimental advantages such as uniformity of the r.f. field and sharper linewidth for the atomic transition. If the radio frequency is not in resonance with the transition frequency, no transitions occur. The atoms then continue through a second strong deflecting field, are further deflected in the same sense, and miss the detector. If, on the other hand, the r.f. field is in resonance, transitions occur; and the effective dipole moment of the atoms in the strong deflecting fields is just reversed. A compensating deflexion occurs which focuses the atom onto a detecting target. The linewidth of the transition observed by this method is set by the uncertainty principle, $\Delta E \, \Delta t \cong \boldsymbol{h}$, or $\Delta v \Delta t \cong 1$, where $E$ is the energy separation of the atomic states, $\Delta t$ is the time of flight, $v$ is the transition frequency, and $\boldsymbol{h}$ is Planck's constant. Typically, an experimentally observed linewidth for a machine of effective length 160 cm is 120 c/s. Often the oscillator exciting the transition is locked to the resonance frequency by a servomechanism. This method has given results as accurate as manual setting of the frequency.

Accuracy is typically of the order of 1 part in $10^{11}$ with respect to the state separation in the unperturbed atom. Limitations on this accuracy are set by uncertainties in the magnitude and homogeneity of the uniform magnetic field in the drift space and by small but unknown phase shifts between the separated regions of the r.f. field.

*Hydrogen maser.* The transition involved is between the magnetic hyperfine levels in the ground state $(^2S_{1/2})$ of atomic hydrogen, arising from the same sort of interaction as described for caesium. The transition is designated as

$$F = 0, \quad m_F = 0 \underset{\rightarrow}{\overset{\leftarrow}{\rightleftharpoons}} F = 1, \quad m_F = 0.$$

Here $F$ takes on two values, 0 and 1, since $I = 1/2$ and $J = 1/2$. The transition occurs at 1,420,405,751·80 $\pm$ 0·03 c/s.

The transition is observed in a maser oscillator. The upper state of the transition is prepared by dissociation of hydrogen gas in an electrical discharge, followed by beam formation through mechanical collimation and subsequent magnetic focusing of the $F = 1$ state by the forces due to an inhomogeneous magnetic field on the magnetic moment. Atoms prepared in the upper state enter a storage bulb with a special wall coating such as polytetrafluoroethylene. The coating is chosen to have a weak interaction with the excited atoms to reduce perturbation of their energy levels by collision and to reduce the probability that atoms are lost by reaction at the wall. The bulb is contained in an electromagnetic cavity resonator. Any field present at the proper frequency causes stimulated emission of the atoms from the upper state to the lower state.

The radiation emitted is used to overcome internal power losses in the cavity and to provide power for external signals.

This method of operation has several features distinct from the atomic beam technique. The standard is an active oscillator which provides a signal as contrasted to a passive resonator which must be probed by an external signal. The natural linewidth is narrower due to the longer storage time, typically 1–3 sec. There is a high signal-to-noise ratio by virtue of the essentially noise-free process of amplification by stimulated emission. First-order Doppler shifts arising from the atomic motions are effectively cancelled because of the random directions of the atoms. The hydrogen Zeeman spectrum is simpler than that of caesium, consisting only of four levels rather than sixteen; thus overlapping of the tails of spectral lines is easier to avoid. The accuracy of the hydrogen maser with respect to the transition frequency of the unperturbed atom is estimated to be about 1 part in $10^{11}$. It is limited by shifts in frequency due to perturbation of the energy levels by collisions with the walls. These may be thought of as cumulative shifts of phase between the atomic dipole moment and the stimulating radiation due to perturbations of the energy levels during each wall collision. Another limitation has been due to temperature changes of the apparatus, which are mainly manifested as changes in cavity resonance frequency and consequent frequency pulling.

$^{205}Thallium$. The transition involved is in the ground state $(^2P_{1/2})$ of thallium, arising from the same sort of magnetic hyperfine structure as in caesium and hydrogen. The transition is

$$F = 0, \quad m_F = 0 \rightleftarrows F = 1, \quad m_F = 0.$$

Here $I = 1/2$ and $J = 1/2$. The transition occurs at $21,310,833,945\cdot9 \pm 0\cdot2$ c/s. The method of observation is by the use of an atomic beam. The detection of neutral thallium by surface ionization is harder than the detection of neutral caesium because of its higher ionization potential. It is also harder to deflect the thallium atoms because of a smaller effective magnetic moment in the strong deflecting field. Accuracy with respect to the transition frequency of the unperturbed atom is conservatively estimated at $2 \times 10^{-11}$. The main limitation is the phase shift of the r.f. radiation between the two separated field regions. There are some increased practical experimental difficulties with thallium, although in principle it offers certain advantages over caesium. These advantages are a lesser dependence of transition frequency on residual magnetic field, greater simplicity of the Zeeman spectrum, greater intensity of the beam signal (since a greater fraction of all the atoms in the ground state is in the $m_F = 0$ states used for the standard frequency transition), and a slightly longer time of flight for the same length of beam due to greater mass of the thallium atom.

*Other atomic frequency standards.* The ammonia beam maser formerly gave promise as an atomic frequency standard, but has been surpassed in accuracy by the above three. A rubidium device has been developed in which the upper magnetic hyperfine level is overpopulated by optical pumping, and transitions induced at the microwave resonance frequency are detected by changes in the absorption of the optical pumping radiation. This device is also of lesser accuracy than caesium, hydrogen or thallium. It does find usefulness, however, because it provides frequency stability of perhaps 1 part in $10^{11}$ over short intervals of the order of seconds in a simpler instrument than a caesium standard. Commercial models of caesium, hydrogen, and rubidium standards are available.

*Dissemination of time standards.* The very nature of time requires that its standards be immediately and continuously communicated to the place needed. A long-standing means of doing this has been by high frequency radio broadcasts from 2·5 to 25 Mc/s. These generally consist of a carrier of precisely controlled frequency and of time markers constructed by precisely modulating the carrier at intervals controlled by the carrier. The carrier generating the time signals is offset from the atomic frequency, and the time signals are jumped as necessary in order to give as good an approximation as possible to UT2. There can be some advantages in making atomic frequencies and time signals derived therefrom directly available by radio broadcast, and this feature may be added to some of the broadcasts in the future. High frequency standard frequency broadcasts cover roughly a continental reception range. The accuracy of frequency reception over a few hours is about 1 part in $10^7$ because of variable propagation effects associated with the motion of effective ionospheric height. The accuracy of time pulse reception is about 1 ms.

Very low frequency (VLF) radio broadcasts below 30 kc/s and low frequency (LF) broadcasts from 30 to 100 kc/s have provided a new method of time standard dissemination within the last decade. The range of VLF propagation has long been known to be large. Within the last decade the phase stability has been discovered to be excellent due to the special mode of propagation. This may be pictured roughly as a wave guided between the concentric spheres of the ionosphere and the Earth. It is well known that some guided modes have propagation constants which are insensitive to certain dimensions and boundary conditions of the wave guide. The diurnal rise and fall of the ionosphere does, however, give a small phase shift dependent on the direction and length of the propagation path. This shift is usually less than $2\pi$ rad at VLF even for intercontinental paths and usually is highly reproducible to within a few per cent from day to day.

The available bandwidths at 60 kc/s and 100 kc/s are enough to transmit effective time signals. In fact, the Loran C navigation system sends pulses of 100 kc/s radiation of rise time typically 60 µs, enabling time resolution between the arrival of the ground wave and the sky wave. Using special pulse reception techniques, timing accuracy is better than 1 µs and is due to the increased stability of the ground wave propagation over the sky wave. VLF broadcasts at 18 or 20 kc/s do not have the available bandwidth for resolution of time pulses to much greater than ±1 ms. Nevertheless, these broadcasts have proved highly useful for phase locking remote oscillators separated by continental distances and thus continuously transferring the time standard without slippage of more than a few microseconds. Once the time of the remote oscillator has been set with respect to the standard, it then remains synchronized and partakes of the full accuracy of the standard which drives the broadcasts.

Microwave pulses have been transmitted by communication satellite between continents for the synchronization of remote clocks. Results to 1 µs in the satellite link were obtained. This method largely avoids the influence of the ionosphere on the propagation because the microwave frequency used is less sensitive to ionospheric disturbance and because much of the path is outside the ionosphere.

Portable clocks capable of microsecond time resolution have recently been used to synchronize remotely located time standards. The clocks used have been both quartz and portable atomic standards. Uncertainties have been within a few microseconds for separations of continental scale and within a few tens of microseconds for separations of intercontinental scale. This technique provides a satisfactory way to disseminate time when used in conjunction with the VLF phase lock system described above.

*Bibliography*

BARNES J. A. and FEY R. L. (1963) *Synchronization of Two Remote Atomic Time Scales, Proc. I.E.E.E.* **51**, 1665.

BONANOMI J. (1962) *A Thallium Beam Frequency Standard, Trans. I.R.E.* I-11, 212.

CLEMENCE G. M. (1957) *Astronomical Time, Rev. Mod. Phys.* **29**, 2.

CRAMPTON S. B. *et al.* (1963) *Hyperfine Separation of Ground-State Atomic Hydrogen, Phys. Rev. Letters* **11**, 338.

ESSEN L., and PARRY J. V. L. (1957) *The Caesium Resonator as a Standard of Frequency and Time, Phil. Trans. Roy. Soc. London* A**250**, 45.

MOCKLER R. C. (1961) *Atomic Beam Frequency Standards*, in *Advances in Electronics and Electron Physics*, Vol. 15, New York: Academic Press.

J. M. RICHARDSON

**TIN CRY.** "Tin Cry" (cry of tin) are descriptive phrases referring to the crackling and snapping sounds heard when tin is plastically deformed. The origin of the phrases is unknown, however, knowledge of the

phenomenon undoubtedly dates back over 4000 years to the earliest known use of tin. The source of these acoustic pulses is the nucleation of mechanical twins induced during deformation. Such twins can form in a time as short as a few microseconds and a part of the energy associated with this sudden transformation occurs as a click or loud report (also referred to as acoustic emission). It follows that metals which twin readily exhibit the "cry" phenomenon to some degree. Metals in this class belong to hexagonal and tetragonal crystallographic structures, e.g. zinc, cadmium, magnesium, and tin. At cryogenic temperatures (4°K) the face-centred cubic metals copper, and gold–silver alloys which are not ordinarily considered to deform by mechanical twinning, have produced such twins accompanied by the "tin cry" response. Generally this acoustic effect is accompanied by a stepped or separated stress–strain curve. The generic term "twin cry" would appear more appropriate than the parochial "tin cry".

B. H. SCHOFIELD

**TRANSONIC FLOW IN NOZZLES.** When a gas flows through a convergent-divergent nozzle the velocity in the narrowest section (throat) is approximately sonic if the pressure ratio across the nozzle is greater than, for air, about 1·9. For pressure ratios less than this value the flow is everywhere subsonic with a maximum velocity in the throat region. At higher pressure ratios the velocity increases in the direction of flow becoming supersonic downstream of the throat. Assuming, as a first approximation, that the velocity $u$ is constant across any cross-section, the following differential relationship can be shown to hold

$$(1 - M^2) \frac{\mathrm{d}u}{u\,\mathrm{d}x} = - \frac{\mathrm{d}A}{A\,\mathrm{d}x} + \frac{\mathrm{d}X}{pa\,\mathrm{d}x} + \frac{\mathrm{d}Q}{\varrho h u A\,\mathrm{d}x} \quad (1)$$

where $M$ is the Mach number, $A$ the cross-sectional area, $X$ the shear stress on the fluid due to the retarding effect of the wall, $p$ the pressure, $Q$ the heat transfer to the fluid per unit time, $h$ the enthalpy per unit mass, and $\varrho$ the density.

If the shear stress and heat transfer can be neglected, $\mathrm{d}X = \mathrm{d}Q = 0$, it is easily seen that the velocity can only be sonic, Mach number equal to unity, when $\mathrm{d}A/\mathrm{d}x$ is zero. This is a minimum if the isentropic pressure–volume curve is concave upwards which is the case for all known gases, i.e. for isentropic, adiabatic flow through a nozzle sonic velocity is attained at the throat. The application of de l'Hospital's rule to equation 1 gives the velocity gradient at the throat, i.e.

$$\frac{\mathrm{d}u}{\mathrm{d}x} = \left[ \frac{1}{(\gamma + 1)A} \left( \frac{\mathrm{d}^2 A}{\mathrm{d}x^2} \right) \right]^{\frac{1}{2}}. \quad (2)$$

The assumption of constant velocity across any cross-section is satisfactory for a first approximation but it is easily seen that there will be a transverse velocity gradient due to the curvature of the streamlines. At the throat, the pressure must increase from the wall to balance the centrifugal acceleration and the velocity must decrease correspondingly. The transverse variation in velocity can be shown to be of order $R^{-1}$ where $R$ is the ratio of the wall radius of curvature at the throat to the throat half-height. Therefore, provided that $R$ is not too small, a better approximation to the flow in the throat region can be obtained as a small perturbation of the one-dimensional flow given by equation 1.

The first attempts to obtain the detailed flow field in the throat region of a symmetric nozzle assumed that the velocity potential $\varphi$ could be approximated by a double power series expansion terminated after the first few terms. Meyer, in 1908, assumed that the velocity increased linearly along the axis and obtained the coefficients $\varphi_{mn}$ of the power series

$$\varphi = \sum \sum \varphi_{mn}\, x^m y^n$$

up to and including the sixth order. This solution is illuminating in certain respects, the following features being brought out independently by Frankl and Lighthill. The sonic line is curved, concave upstream, with curvature directly proportional to the axial velocity gradient and, for two-dimensional flow, three times the curvature of the line joining the points where the flow is parallel to the axis. The sonic line must therefore intersect the nozzle wall upstream of the throat. A further deduction of considerable mathematical significance was the existence of a branch line in the hodograph plane whose trace in the physical plane was curved downstream with curvature equal to that of the sonic line. This hindrance to obtaining solutions in the hodograph plane was overcome by Cherry who was able to obtain an exact solution by means of an ingenious transformation. Unfortunately this solution is only applicable to two-dimensional nozzles.

A small perturbation approach seeking the velocity components as series expansions in $R^{-1}$ was suggested by Hall and has been found to be tractable for both two-dimensional and axially-symmetric nozzles. In this approach the nozzle shape is given and the flow field obtained whereas in most of the earlier theories the axial velocity distribution was given and the possible nozzle shapes deduced. Hall's method is also applicable to asymmetric two-dimensional nozzles and to axially-symmetric nozzles with centre-bodies. It has also been applied to swirling flows with interesting results. The reduction in choking mass flow due to swirl is well predicted.

Very little detailed study has been made of nozzle flows in which $dX$ and $dQ$ are non-zero. The one-dimensional approach has been rationalized by Shercliff and others showing, amongst other things, that in adiabatic flow with friction sonic speed occurs downstream of the throat. The effect is small, however, unless the nozzle is of almost constant cross-section. Recent studies of non-equilibrium flows in nozzles correspond in many ways to equilibrium flows with non-zero $dQ$. At present the available results are all restricted to the one-dimensional case.

*Bibliography*

OSWATITSCH K. (Ed.) (1964) Symposium Transsonicum. In particular the article *Transonic Flow in Ducts and Nozzles: a Survey* by I. M. HALL and E. P. SUTTON.

I. M. HALL

**TURBINE BLADES, VIBRATIONS OF.** In order to achieve a high efficiency of energy conversion in a turbine the blading profiles are thin. Although in practice mechanical strength considerations may dictate an increase in blade thickness from the ideal, the result is generally a relatively flexible member with low vibration natural frequencies.

The term "blade vibration" is a convenient one to describe certain modes of the whole turbine blade-wheel system where the maximum movement occurs in the blading. Other types of vibration, known as axial disk vibrations where the wheel vibrates with node loci along nodal-diameters also occur and are important in turbine design.

Although blade vibrations are vibrations where high alternating stresses are confined to the blading and the frequency of the system is chiefly controlled by the stiffness of the blading, this does not necessarily imply a true blade vibration since these conditions may apply equally well to a high nodal-diameter wheel vibration. For a stage where the blades are held together at the tips by a continuous shroud, the concept of blade vibration is best obtained by reference to the modes of vibration of an isolated cantilever bar which is firmly clamped at the root and supported at the tips. In such a bar the lower modes of vibration, illustrated in Fig. 1, are Fundamental flap, edge and torsional. In turbine blades the flexural modes are sometimes called tangential and axial to describe the motion relative to the rotor. Overtones of these vibrations also occur.

In modern turbines the blades are often of complex form being twisted and tapered to achieve greater aerodynamic efficiency and the vibration modes which occur are correspondingly complex, there being considerable coupling between flexural and torsional motion.

Further complication arises because of the way in which blades may be banded together for reasons of mechanical strength and vibration. Where blades are open-ended or connected together by a continuous banding at their tips the blading has a minimum number of natural modes. If the shrouding is not continuous, forming packets, or groups, consisting of a few blades only, other modes occur where the blade packet moves as a whole. The first three modes of such a packet are illustrated in Fig. 2. Where blades are connected together at intermediate positions along their length by lacing, or binding, wires, sometimes in complicated arrangements, the number and complexity of modes markedly increases.

Although open ended blades of overall length of 22 in. have been successfully used on the last stage of a steam turbine this is not general practice. Such

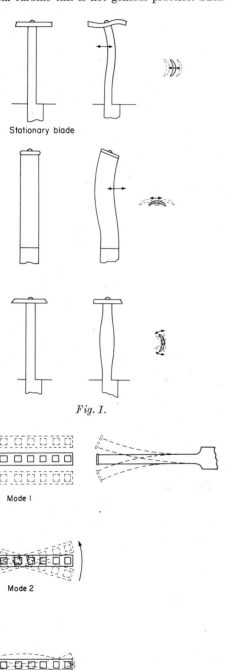

Stationary blade

*Fig. 1.*

Mode 1

Mode 2

Mode 3

*Fig. 2.*

a blade system has the simplest spectrum of frequencies and so give the most easily measured and controlled system. It is not used for long blade lengths mainly because the conditions for self-excitation of such flexible beams in a high-speed flow cannot be predicted with certainty and the possibility of unexpected self-excited vibrations exists. The attachment of a coverband or lacing wire effectively eliminates this possibility.

*Excitation of vibrations.* At first sight it would seem that, in turbines with full admission to the stages and freedom from reciprocating motion, the periodic forces tending to excite vibration will be slight and vibration troubles of minor significance. Any departure from perfect uniformity in flow velocity, pressure and temperature at outlet from the nozzles does, however, give rise to periodic forces which act on the blading. The magnitude of these forces may be small but should their frequency be the same as that of certain modes of natural vibration of the blades, an amplitude of vibration may be built up which is sufficient to cause failure by fatigue.

Not much is known about the magnitude of the various periodic impulses acting on the blades in a turbine except, perhaps, for the stage with partial admission, i.e. where the working fluid is admitted over only a section of the nozzle annulus. Here the blade is subject to a high bending load on passing one nozzle group and this load is completely removed before the blade meets the next. From a knowledge of the mean steam velocity, pressure and temperature at outlet from a group of nozzles, an estimate of this repeated loading and unloading may be obtained. Calculations made by Allen show that the forced alternating bending stress, due to repeated loading and unloading of the blade, may be three or four times higher than the steady mean stress. It is particularly important, in this arrangement, to have the lowest natural frequency of the blading substantially higher than the frequency of the impulsive force acting on the blades due to the wakes from the nozzle jets. Such exciting frequencies may be in the region of several thousand cycles per second so that, to satisfy this condition, a very short stiff blade is required. Fortunately, there is usually no great difficulty in fulfilling this requirement, since partial admission is normally used only in the first stages of steam turbines where the blade length is short.

With full admission each blade is subject to a much more uniform bending load but, because of the lack of uniformity in the steam flow round the nozzle annulus, and the action of inertia forces due to rotation of the rotor, various periodic forces act on the blades. For instance, the lack of circumferential symmetry that exists in the exhaust casing of a steam turbine, or the inequality in flow from the combustion chamber of a gas turbine are typical sources of periodic excitation which have components of low frequency at multiples of running speed of the machine.

Apart from the lack of circumferential symmetry which causes these low frequency disturbing forces, high frequency disturbances arise from the nozzles themselves and, superimposed on the mean steam or gas force on the blade, there is a ripple whose frequency relative to the moving blade is $NR$, where $N$ is the number of nozzle blades and $R$ is the rotational speed. For typical stages of steam turbines this frequency is between about 2000 and 6000 cycles per second at normal operating speed. Experience has shown that where the frequency of this ripple, which may conveniently be defined as the nozzle impulse frequency, is equal to the frequency of certain forms of blade vibration, early failures of the blading may be expected.

The possibility of blade vibrations being excited by the passage of the working fluid past the blades must always be considered. Such self-excited vibrations occur in axial compressor blading, for example by the mechanism of *rotating stall flutter*, but it is not accepted that similar vibration can occur in the relatively stronger turbine blading. Recent measurements made in a steam turbine have shown vibrations occur which could be caused by some such condition. These results are likely to lead to a greater emphasis on the measurement of the actual conditions of steam flow in the stages of operational machines.

Other exciting forces may be transmitted through the rotor from the driving end of the machine, arising from gearing or from the electrical forces in the generator.

*Design procedure.* Except for the very simplest types of blading, calculations do not allow an accurate prediction of natural frequencies to be made. This is not only because of the complex nature of the blade form but because the end fixings (and intermediate supports if present) cannot be approximated to reliably in the analysis.

In order to be sure that dangerous blade vibrations do not occur in service it is, however, necessary to be able to predict the natural frequencies that will occur. This is usually done by reference to the results of measurements on similar blading, using simple calculation techniques for small extrapolations, or by measuring statically on specially prepared blade groups or packets prior to stage construction. It is also necessary to know the effect that rotation will have on blade frequency. This information is obtained from tests made on production wheels, or specially made experimental wheels, run in a suitable test chamber.

This approach is less suited to gas turbine blade than steam turbine blading because of the greater difference between service and static conditions. Here it has been the practice for some years to measure the vibrational strains in the blading of running machines.

If it were necessary to completely avoid excitation of every one of the many modes which can occur in turbine blading, their design would be impracticable. The knowledge of which modes must be avoided is a matter of experience gained by observing the behaviour of past designs. The experience gained is likely to be restricted to a particular type of design.

*The effect of centrifugal force on blade frequency.* When centrifugal forces act on a turbine blade, its natural frequencies change because (i) there is a stiffening of the blade and (ii) the end fixtures alter.

It may be shown theoretically that, considering the blade stiffening alone, the relation between the frequency $F_R$ (c/s) at speed $R$ (revs/sec) and the static frequency $F_0$ (c/s) is given by the relation,

$$F_R = F_0^2 + BR^2$$

where $B$ is a coefficient, called the "*rotational coefficient*". Experiments have shown that this relation holds very closely for turbine blades where the changes in end fixings are insignificant such as, for example, on very long steam turbine blades. For such blades the value of $B$ may be in the range 6–10 for the lowest mode.

The changes which occur in the end fixings (or intermediate supports) are much more variable and depend upon the type of attachments used. Where these effects predominate the natural frequency may be lowered as speed increases, the change varying from mode to mode and following no simple rule. Eventually, of course, the frequency will rise again when the fixings have tightened and the blade stiffening predominates but this may not occur until well above the normal operating speed of the machine.

*Measurement of blade vibrations.* When turbine blades are stationary, the measurement of their natural frequencies is no more difficult than for other structural components. Care must be taken to ensure that the exciting device used does not modify the dynamic system and because of the small size of some blades and the need to inject moderate energy, a problem of experimental technique may arise.

To overcome this difficulty electromagnetic exciters were first employed. But not all blades are made of magnetic material and bowing, impact and electrostatic methods have also been used.

Detection and measurement of the excited vibration may be accomplished by using any one of a wide variety of proximity pick-ups using capacitance or inductance change or by using wire or crystal strain gauges.

On blades rotating in a test chamber the problem is more difficult. The early experimenters relied upon self excitation (i.e. by motion through air) and plucking by, for example, a stationary electromagnet. A sustained vibration may be excited in this way when the number of vibration cycles per revolution is an exact integer. The speed at which these conditions occur allows a curve of frequency against speed to be constructed, but the number of measured points is small. This method is generally only suitable for measurements on the gravest mode, and it has the disadvantage that many other modes are excited at the same time. Even on the simplest type of blading, i.e. with only root and tip fixings, the resonance curve measured on a blade is not simple and unless a controlled method of excitation is used, any one of a number of close vibrations may be excited.

More recently, piezoelectric crystals in the form of thin wafer units have been developed to overcome these difficulties. Here the reverse piezoelectric effect is used, a voltage applied across the crystal inducing a strain in it, which is transmitted to the surface to which it is attached. In this way close control of the exciting frequency is possible and precise measurements may be made. The amount of power that can be fed into one crystal is small and on large blades many crystals, which each measure about $1'' \times 1/4'' \times 0 \cdot 050''$ are used.

Detection of vibration on rotating blades running in a test chamber is no great problem and wire resistance or crystal type strain gauges, connected to receiving apparatus through slip rings may be used. Telemetering methods which involve no contact with the rotating parts are now available.

The measurement of blade vibrations from a running gas turbine were first made satisfactorily by Drew using a specially constructed wire resistance strain gauge and mercury slip-rings and later with a non-contacting slip-ring using R.F. coupling. A novel method of measuring the vibration of open-ended blades has been developed in which a tiny cylindrical magnet is inserted in the tip of the blade and a stationary grid of wires surrounds the wheel periphery. A frequency-modulated signal is received from the grid, which can be demodulated to show the blade motion.

Similar tests have now been made in steam turbines in normal operation using a radio telemetering method. Conventional wire resistance strain gauges are used in conjunction with miniaturized telemetering capsules attached to the rotating shaft.

*Bibliography*

Allen R. C. (1940) *Trans. A.S.M.E.* **62**, 689.

Campbell W. and Heckman W. C. (1925) *Trans. A.S.M.E.* **47**, 643.

Drew D. A. (1958) *Proc. Inst. Mech. Engrs.* **172**, No. 8.

Jones D. H. (1963) *Proc Inst. Telemetering Conference*, **67**.

Luck G. A. (1962) *G.E.C. Journal of Science and Technology* **29**, No. 2.

Reeman J. and Luck G. A. (1951) *G.E.C. Journal* **18**, No. 4, October.

G. A. Luck

**TYPE-TOKEN RATIO.** A statistic used in the description of written or spoken speech. It is the ratio of the number of different words occurring in a passage to the total number of words in the passage.

N. Moray

# U

## ULTRASONIC INTERFERENCE MICROMETER.

The ultrasonic interference micrometer was developed to give improved performance over the conventional ultrasonic pulse echo and resonance methods when used for thickness measurement of metal sheet or thin walled tube. Sections less than $\frac{1}{8}$ in. thick cannot easily be measured by the pulse echo method due to the finite rise time and width of the pulse, whilst the resonance method, although accurate on thin sections, requires the ultrasonic transducer to be in good physical contact with the part to be measured so that particularly with tubing, rapid scanning is difficult and measurement on a large scale becomes tedious.

The ultrasonic interference micrometer uses a combination of the two basic methods. A pulse system is employed, which allows a water coupling medium to be used between the transducer and the tube, and by making the pulse a 15 μsec long wave train which is short compared with the time of travel through a 2 in. long water column and long compared with the time of travel through the tube wall, it is possible to obtain a thickness measurement with an accuracy better than by the contact resonance method. The use of the water column allows rapid movement of the tube without disturbance of the ultrasonic coupling conditions and also allows focusing of the ultrasonic beam so that measurement is possible on confined areas down to 1 mm diameter.

In its simplest form the instrument consists of a pulsed variable frequency oscillator which drives a non-resonant transducer. The transducer sends out a train of ultrasonic waves at the oscillator frequency which travels through the water, is reflected from the tube surface and returns to the transducer. After amplification and rectification the received pulse is displayed on a cathode-ray oscilloscope. A block diagram of the instrument is shown in Fig. 1.

A sharply defined interference effect is observed within the reflected pulse when the oscillator frequency is tuned to the half-wave resonant frequency of the tube wall (i.e. $n\lambda/2$ where $n$ is an integer). This is a similar relationship to that obtaining for continuous wave resonance thickness measurement and, in the same way, thickness can be calculated from the measured frequency, if the velocity of sound in the particular material is known to the required accuracy. If this is not the case, the required value of velocity can be calculated from a measurement on a sample of

known thickness prepared from the same material. The appearance of the received pulse away from the half-wave frequency and exactly at the half-wave frequency is shown in Fig. 2(a) and (b) respectively. Figure 2(c) is a composite picture showing how the shape of the reflected pulse changes as the frequency

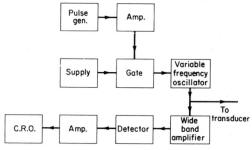

*Fig. 1. Block diagram of ultrasonic interference micrometer.*

Transmitted and reflected pulse

(a) Away from half wave frequency　(b) At half wave frequency

(c) Appearance of the reflected pulse at 10 kc/s intervals around the half wave frequency

*Fig. 2. Transmitted and reflected pulse. (a) Away from half-wave frequency. (b) At half-wave frequency. (c) Appearance of the reflected pulse at 10 kc/s intervals around the half-wave frequency.*

is varied in 10 kc/s steps through the resonant frequency of a 0·015 in. wall thickness stainless steel tube. Each step is equivalent to a thickness change of 25 μin.

To explain the observed interference effect it is necessary to consider both the steady state and transient conditions. At the half-wave frequency, the

360

distance travelled by a wave across the wall and back is such that, on emerging into the water on its return to the transducer, this wave will be 180° out of phase with the wave reflected from the front surface. This takes into account the 180° phase shift which occurs on reflection from the rear surface due to the acoustically "soft" boundary. In order that complete destructive interference can take place between the two waves they must also be of the same amplitude. Before this condition can occur a resonance must build up in the tube wall. At the beginning of the pulse, therefore, a reflection is observed which dies away as resonance builds up into a steady state condition. Once this condition has been reached there will be no reflection from the tube until the end of the pulse, provided the amplitude and frequency of the incident wave remain unaltered. When the pulse ends, however, there is no reflection from the front surface to produce destructive interference. A decaying wave, representing the release of the energy stored in the resonant tube wall, then returns to the transducer appearing as a second pulse on the oscilloscope trace.

With the simple instrument, hand tuning to the zero reflection point is necessary. This is not satisfactory for rapid measurement and some form of automatic sweeping and null detection is necessary, the ultimate speed of measurement being determined by the rate at which sweeping of the frequency can be carried out. By extending the length of the pulse to 60 μsec it is possible, within the period of a single pulse, to sweep over a frequency range which will cover a ±20 per cent change of thickness. A rate of 2000 measurements per second can then be achieved which for a full spiral scan of a typical 0·4 in. diameter tube gives a measuring speed of 30 ft/min.

*Bibliography*

AVEYARD S. and SHARPE R. S. (1963) *Application of Ultrasonic Pulse Interference*, Paper No. 20 4th Int. Conf. on Non-Destr. Test.

S. AVEYARD

**UNIVERSAL TELESCOPE (2-METRE REFLECTING).** The complete layout (Fig. 1) required even for a single large telescope in a major astronomical observatory is extremely extensive. Moreover since individual construction is involved in these installations, the total installation required becomes very costly. On economic grounds it has therefore to be considered how one can provide an installation which is suited to the expected future research problems, yet with the least possible call on financial and material resources. With all large telescopes this fundamental consideration has led to forms of construction in which the optical components can be joined in various combinations, so that the optical characteristics (focal length, aperture ratio, resolving power) can be suited to the requirements of particular domains of astronomical research. In all cases the variant forms of the telescope are based on the primary mirror. By combination of this primary mirror with correction systems, corrector plates and optically effective secondary mirrors, widely differing optical systems can be built up. Usually a parabolic primary mirror has been combined with correction systems and secondary mirrors, as for instance in the triple arrangement of a telescope as a parabolic reflecting telescope with a primary focus, as a Cassegrain telescope, and as a Coudé telescope.

For the investigations which can be carried out with this multiple telescope, the insertion of a large-field astrographic camera will generally be necessary. That was formerly effected in most cases by an astrograph with a lens system, but in recent decades Schmidt reflecting telescopes have been used.

A multiple telescope which is capable not only of the above-mentioned three variations but also of conversion to a Schmidt reflecting telescope was designed and built for the first time in the VEB Carl Zeiss Jena optical works, in the form of the 2-metre Universal reflecting telescope of the Karl Schwarzschild Observatory in Tautenberg (German Democratic Republic). This Universal reflecting telescope therefore combines practically all the possible optical arrangements of modern reflecting telescopes in one instrument. In this way only one building and one dome had to be constructed for the instrument.

The Schmidt system supplied to produce large-area stellar photographs, penetrating to the most remote regions, has a corrector-plate aperture of 1340 mm, a focal length of 4 m and a field of dimensions 24 × 24 cm, which because of the chosen difference between the diameters of the primary mirror and the corrector plate is photometrically extremely uniform right up to the edges. As the Schmidt system demands a structural length more than twice the focal length, this system determined the constructional design of the whole telescope.

In order to be able to utilize the full light intensity of the spherical primary mirror for the study of individual objects, especially of faint spiral nebulae, a correction system was designed which together with the spherical concave mirror yields an axial image quality identical with that of the focal image of a parabolic mirror. The focal length of this primary focus system is likewise 4 m and has accordingly an aperture ratio of approximately 1:2.

To arrive at an intermediate focal length the spherical mirror was combined with a second mirror of excess hyperbolic form to give a quasi-Cassegrain system, which has a focal length of 20 m and is used especially for spectral and photometric work on individual celestial objects.

In a fourth arrangement the above-mentioned mirror of excess hyperbolic form can be replaced with a similar one with greater curvature, so that an effective focal length of 92 m is created, which after reflection

*Fig. 1. Sketch of the general layout of the 2-metre Universal reflecting telescope. 1 Tube shutter; 2 Corrector plate; 3 Cassette introduction apparatus; 4 Cassette support cross with focussing device; 5 Aperture for Cassette introduction; 6 Support cross for Cassegrain and Coudé mirrors; 7 Compensation rods; 8 Spherical primary mirror, 2 m diameter; 9 Support bearing; 10 Positioning eyepiece-head of the guide telescope; 11 Operating control panel on the observation platform stage; 12 Course of the ray to the Coudé focus.*

at four plane mirrors and direction into the hour axis gives a focal image in the temperature-compensated basement rooms of the observatory.

It must be recognized that a greater effort is required in the design and construction of such a Universal telescope than in the construction of single instruments. Each of the four systems must ultimately show as good an optical performance as that of a separate telescope of the same order of size. The requirements that must be fulfilled by the design of the telescope can be collected into 12 main groups:

1. Simple convertability from one system to another within one day.

2. Guarantee of correct alignment after conversion has been carried out.

3. Automatic compensation of bending phenomena resulting from the force of gravity, particularly of those effects which produce a differential displace-

ment between the guide telescopes and the main optical system.

4. Automatic compensation of the focus shift caused by temperature changes.

5. Outstanding uniformity of the telescope drive compensating the Earth's diurnal rotation.

6. Devices to simplify the orientation of the telescope on to the desired constellation.

7. Easy possibility of checking that the drive is free from error.

8. The possibility of accurate adjustment of the telescope in its focussing and orienting motions.

9. Detailed precautions against maloperation.

10. Safety for the observer during work with the instrument at night.

11. Fully automatic drive of the observatory dome, to ensure that the instrument always has a clear view of the sky through the open slit.

12. Minimum maintenance work.

To enable the work of changing the optical system to be carried out easily and quickly, a winch has been provided in the telescope interior; this carries the optical components which have to be exchanged, and makes screwing and unscrewing possible without muscular effort. By means of calibrated prismatic contact surfaces perfect alignment after screwing up is always ensured.

Since it is not practical to construct such a large telescope tube so rigidly, that no disturbing deforma-

undergo a slight inclination which is made exactly equal to the displacement of the optic axis of the guide telescope. In this way all displacements between the optic axis of the guide telescope and the optic axis of the Schmidt reflecting telescope are eliminated.

The device serves simultaneously to compensate the thermal expansion of the telescope body. This, however, does not suffice to eliminate the effect of thermal distortion on the focussing. The primary mirror itself increases its focal length with rising

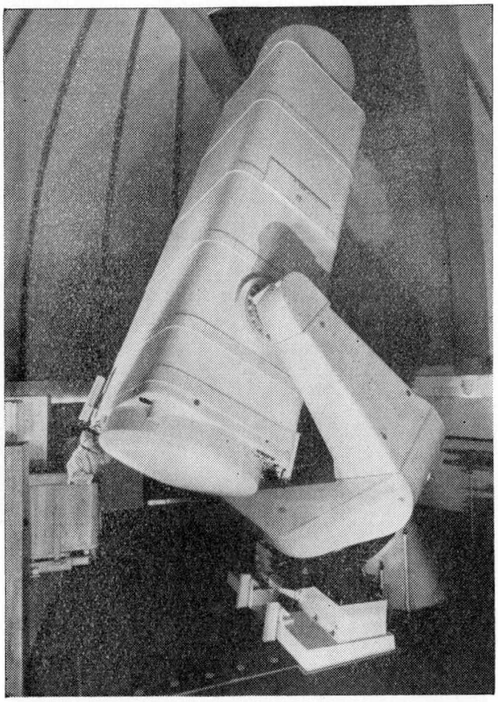

*Fig. 2. 2-metre Universal reflecting telescope working at night. Observer on the stage of the observation platform near the control at the positioning eyepiece head of one of the two guide telescopes.*

*Fig. 3. Putting in cassettes for Schmidt photographs with the 2 m Universal reflecting telescope.*

tions are caused by gravity, a compensating device had to be provided. For this purpose the floating suspension of the primary mirror was used. Adjustment can take place almost effortlessly, because the weight of the mirror is compensated. The adjustment is carried out by movable rods situated in the telescope body, the rods being supported on the wall of the tube at a predetermined cross-sectional plane. Further control rods extend from the same plane to the feet of support cross which carries the photographic equipment. By means of these rods the planes of the mirror and the support cross are always kept parallel to one another, and, corresponding to any bending of the telescope tube at the above-mentioned rod support, the planes

temperature. This effect too must be automatically compensated, but the compensation must operate at the same slow rate as that with which the glass takes up the temperature of its surroundings. Accordingly clamps of a synthetic material with a very high coefficient of expansion (about $100 \times 10^{-6}$) are ground on at the circumference of the mirror. The clamps are of such a size that their expansion has the same absolute magnitude as the corresponding focal variation of the mirror. The Invar steel rods for flexure compensation are connected to these clamps. The co-operation of both parts of the compensation system gives continuous adjustment of the photo-cassette position to the actual focal point.

In order to be able to cause such a telescope to follow the diurnal rotation of the celestial sphere with the requisite accuracy ($10^{-5}$), the bearing support of the hour axis (the principal rotation axis of the telescope which is orientated parallel to the Earth's axis) must have outstanding smoothness of operation and freedom from friction. For this a pressurized oil bearing is especially well suited. The whole mounting has its centre of gravity in the middle of a spherical zone of cast steel, which in its turn rests on two cushions of pressurized oil. Oil is forced into the supports at such a high pressure that the spherical zone rises from the bearing cushions and oil can escape freely all round (oil film thickness about 5/100 mm). Correspondingly high demands are also made on the driving mechanism itself. The transmission of the rotary motion through differential worm-gear to a large precision worm-wheel is considered as the best solution. Uniform speed of the driving motor offers no problem in the present state of regulator technique.

During the adjustment of the telescope it is very important to be able to read off its position easily, with conversion into the co-ordinate system of the celestial sphere already carried out. The telescope position is transmitted by electrical rotation recorders to the operation control panel, and is there continuously converted by means of electromechanical devices. For the positioning three graduated speeds are available: for rough positioning about 90°/min, for fine adjustment about 3° and 1°/min.

Even with the most precise mechanical construction, perfect following of the telescope is not guaranteed, since the light rays from the stars during their passage through the atmosphere undergo a refraction, dependent on the altitude of the constellation. This refraction of light is thus subjected to continuous variation and has the necessary consequence that the corresponding errors must be connected either by the observing astronomers or by a photoelectrically operating automatic device. The orders of magnitude of the maximum permissable errors with telescopes of this size are about 0·3–0·5 sec of arc. To make these corrections, very fine adjustments are available, which make careful operation possible with extraordinarily small speeds of 1 min of arc to 0·3 minute of arc per minute of time.

For checking the follow-on drive accuracy, two guide telescopes are available with the Schmidt system. They possess objectives of 300 mm diameter and 4700 mm focal length. For the same purpose when working with the Cassegrain or Coudé system, light is diverted from the beam just before the focus and led into a control eyepiece.

As the astronomer can only establish after the development of the photographic plate whether all the conditions for attainment of a good astronomical-photograph were fulfilled, control of correct operation must be guaranteed during observation. Those functions in particular which operate inside the telescope tube, invisible to the astronomer, are therefore connected to indicator lamps which allow the observer

to test at any time the momentary conditions in the telescope's interior. By locking devices, it is ensured that no exposure can be started unless the photographic cassette is in the right position and the mirror and telescope shutters have been opened.

During night work with the instrument, the observer must often be positioned at a considerable height above the floor of the dome. By a generous layout of staging it is made possible for the observer to carry out his control work in complete safety whilst seated comfortably. A one-man lift allows contact any time with the dome floor, without the observer's having to leave his place at the eyepiece.

In addition auxiliary instruments, such as high-tension equipment for comparison light and secondary-electron multipliers, or amplifiers and pen-recorders for automatic recording of measurements, can find a place near the observer on the observation platform.

*Fig. 4. 2-metre Universal reflecting telescope. On the observer stage of the observation platform. Control at the positioning eyepiece head.*

When the telescope is operated with the Coudé system, the observer is situated in the basement of the observatory, when the large spectographs for investigation of starlight are set up. These rooms are air-conditioned to ensure the highest optical quality of the very sensitive spectrographs. The observer thus has no free view of the sky and also he cannot verify the position of the dome. This is the reason for the construction of an automatic drive for the dome, so that it corresponds to the position of the telescope. The pathways of two infra-red light barriers are interrupted, if the dome is located correctly with respect to the telescope. If this is not the case, the path for one of the two light barriers becomes clear and thereby a rotation of the dome in the corresponding direction is switched on, until both light paths are again interrupted by the telescope tube.

With increasing structural mechanical and electronic expenditure for operating, controlling and supervision of the instruments functions, the necessary care and maintenance increase as well. Therefore modern methods of automation and control are applied only where they are necessary for carrying out the work, and bring an improvement in the observational data.

By careful supervision of the temperature in the dome space, climatic strains on the telescope are kept so small that the telescope is always in the best optical conditions.

A. JENSCH

# V

**VENUS, RECENT PHYSICAL DATA FOR.** The planet Venus was the subject of a formidable array of scientific investigations during the 1962 inferior conjunction, including the measurements during the successful flight of Mariner II. The results of these investigations have greatly increased our knowledge of our sister planet, but many questions remain unanswered.

The Mariner II microwave measurements showed unambiguously that limb darkening was present at 19 mm wave-length and therefore that the origin of the high thermal emission at cm wave-lengths is at or near the surface of the planet. After eliminating the limb darkening and dividing the resulting brightness temperature by an assumed surface emissivity derived from earth-based radar measurements of reflectivity, a fairly uniform surface temperature of about 700°K is obtained.

The surface origin of the high thermal emission was further supported by the discovery, prior to the Mariner encounter, that the low-frequency branches of the Venus bands of $CO_2$ had double maxima, with positions corresponding to 300 and 700°K. The appearance of the double maxima suggested that sunlight is reflected both by a cloud layer or layers with a definite base at an altitude corresponding to an ambient temperature somewhat in excess of 300°K, and by the surface. Further evidence for this interpretation was the discovery that the lines near the 700°K maximum were considerably broader than those near the 300°K maximum. The line-breadths corresponded to a pressure of several atmospheres, and indicated a surface pressure of the order of 10 atmospheres.

The Mariner II infra-red measurements also showed definite and substantial limb darkening in both the transparent ($8\cdot4\ \mu$) channel and the $CO_2$ ($10\cdot4\ \mu$) channel. Within calibration uncertainties, the brightness temperatures were the same for both wavelengths at each location on the planet, varying from about 240°K in the vertical to about 220°K near the limbs. The approximate equality of the brightness temperatures at both wave-lengths indicated that all temperatures read were cloud temperatures and that the cloud top was at a temperature below 220°K. The fact that one can "see" down to an average temperature of 240°K implies that the clouds were diffuse, and the fact that no radiation from the hot surface got

through implies that they were very thick, in agreement with the discussion above.

From the Regulus occultation data and the new surface and cloud top data, the radius of the solid planet can be estimated by assuming constant lapse-rates of temperature between the surface and cloud top and between the cloud top and occultation level. The result, using a $CO_2$ abundance of 10 per cent derived from spectroscopic measurements, is $6010 \pm 30$ km. This corresponds to a value of 900 cm/sec$^{-2}$ for surface gravity, consistent with the mass of the planet ($4\cdot87 \times 10^{27}$ g) as determined from the Mariner II trajectory measurements.

Rotation rates were obtained from independent and completely different active radar measurements using the Goldstone antenna. These included range-gated spectra of the reflection of modulated signals from isolated concentric zones on the planet, and the high resolution spectra of the reflection of a continuous wave signal. The continuous wave spectra showed a detail, suggesting a surface feature, which appeared to move slowly across the planetary disk. The results of both experiments indicated a sidereal rotation of about 250 days in the retrograde sense, with the axis nearly perpendicular to the orbit of Venus. The retrograde rotation was further supported by the observation of a phase lag in the results of passive radar measurements, consistent with the results obtained by the active measurements. The passive measurements showed an increase in brightness temperature with phase angle, indicating a higher bright-side temperature.

Other evidence for surface features was obtained from the observation of a cold area in the Mariner infra-red measurements and from measurements in the $8$–$14\ \mu$ window at Mt. Palomar. From radar reflectivity data, the reflectivity was estimated to be about 10 per cent and the dielectric about 4, suggesting that Venus has a dry, sandy, or rocky surface. From the same measurements, the r.m.s. slope is estimated to be between 4 and 7 degrees.

Water vapour has been recently detected with a high degree of certainty from balloon measurements during the dichotomy preceding the 1964 conjunction. The amount depends on the location of the effective reflecting layers, but is of the order of 10$^{-2}$ g/cm$^2$. There is also evidence that oxygen has been detected through measurements of doppler-shifted lines from the Crimean Observatory.

The presence of oxygen and water would eliminate several atomic possibilities for the cloud composition. The small amount of water eliminates it as a cloud component except at the very top layers and the cloud temperatures are high enough to definitely exclude carbon dioxide. The definite cloud base indicated by the double-maxima in the low frequency wings of Venus bands of $CO_2$ suggests that the clouds are condensation products, however. An attractive possibility is that the clouds are mostly composed of organic compounds, many of which have sufficiently high condensation or polymerization temperatures. About one metre STP of almost any gaseous compound containing a C—H bond would be sufficient to close a spectral window around $3·5 \mu$, which would otherwise allow more radiation to escape from the surface than can possibly be compensated by incoming sunlight. The remaining spectra are probably opaque enough through absorption by hot $CO_2$ and $N_2$ to account for the high surface temperature by a very effective greenhouse effect.

Most of the newly determined properties of Venus described above are listed in the following table.

*Newly determined properties of Venus*

| | |
|---|---|
| Radius (solid surface) | $6010 \pm 30$ km |
| Mass | $4·870 \times 10^{27}$ g |
| Density | $5·36$ g cm$^{-3}$ |
| Surface gravity | $900$ cm sec$^{-2}$ |
| Rotation period | $250 \pm 40$ days retrograde |
| Surface pressure | $10 \pm 3$ bars |
| Surface temperature | $700 \pm 150°$K |
| Cloud top temperature | $<200°$K |

Compositional estimates are not included in the table, as they are critically dependent on the relative reflectivity as a function of altitude, which is still very poorly understood.

*Bibliography*

Symposium on Radar and Radiometric Observations of Venus during the 1962 Conjunction (Feb. 1964), *Astron. J.* **69**, No. 1 (F. T. Barath, Ed.).

L. D. Kaplan

# VERTICAL TAKE-OFF AND LANDING PLANES.

With the increasing speeds of aeroplanes through the years there has come a demand for longer runways especially in the case of airliners. Thus have arisen larger and still larger aerodromes and airports. The natural reaction has been to consider how aeroplanes could be designed to take off with shorter runways, or in the limit, to take off vertically. At the same time the thrust available in the power plant has been steadily increasing. In general, a long range subsonic aeroplane with a lift/drag ratio of 16/1 requires about 6 per cent of its weight poundage as thrust and a high subsonic

speed aeroplane of the same type needs an engine thrust of about 8 per cent of its weight. Military aircraft of higher speeds reach the figure of 20—30 per cent so from this point of view a large thrust that could lift an aeroplane vertically off the ground is wasteful in power not required for ordinary flight. Nevertheless, the problem of the runway has become so acute that aeroplanes of various types have been designed and constructed in recent years which have sufficient power to take off and land vertically. Such aircraft either use the same powerplant for lift and for normal flight, in which case it is overpowered for the latter, or carry separate power units which are only used for vertical flight and are a dead weight in normal flight.

V.T.O.L. aircraft can be separated into the following classes: (1) tail-sitters which are propeller driven and may be jet-powered, (2) tilt-wing aircraft, (3) retractable-fan flat risers, (4) jet-powered flat risers and (5) vectored thrust aircraft. This excludes all forms of helicopters which are rotary wing aircraft that do take-off and land vertically, but which show up poorly in cruising flight, because the rotor is an inefficient source of lift and of propulsion with stringent limits on speed, though these limits are being gradually extended. There are also other aircraft known as convertiplanes which combine the features of both rotating and fixed planes and have not proved a practical success for a variety of reasons. These are also excluded from the present article.

The first category of tail-sitters contains aeroplanes which have been succesfully flown but their further development has been discontinued on account of practical considerations. Typical of such aeroplanes was the Convair XYF–I with a weight of 15,000 lb powered by Allison turboprob engines of 18,500 lb thrust. With a thrust greater than the weight the dual propellers with opposite rotation provided the power for the vertical take-off, with the aircraft standing vertically on its tail, and for hovering as well as for transitional flight to become horizontal. The hovering control was effected by using the propeller slipstream and the aircraft tail in combination with a pendulum function. The aircraft made its first flights in 1954 with standard engines and power plant installations, rotated in flight from its vertical position at take-off to the normal flying attitude of an aeroplane and then back again to land vertically.

The other V.T.O.L. categories mentioned above all start their flight with a horizontal fuselage so that the pilot's seat has not be to swivelled to keep him in a normal attitude with respect to gravity as was the case in the Convair aeroplane. The first of these latter, the tilt-wing aircraft, depends for its flight on the rotation of the wings through about 90° from a near vertical to the normal horizontal attitude, the engines being attached to, and rotating, with the wings. Aircraft of this type show considerable promise for future development and the best known is the winner of the 1961 design competition for a V/STOL transport of the

United States armed services. This aircraft, the Vought/Hiller/Ryan XC–142 A is powered by four 2850 s.h.p. General Electric T 64–6 turboprop engines driving shafting which links the propellers and a three- blade variable pitch tail rotor. During vertical take-off with the wings tilted vertically, roll control is by collective propeller pitch, yaw control by ailerons in the propeller slipstream and pitch control by the tail rotor. During transition a mixing linkage integrates this control with the normal aeroplane type of control. Similar in conception is a later study for aircraft with rotatable ducted propellers first for lift and then for propulsion, the wings being fixed: successful flights have been made.

Retractable fan-in-wing aircraft have been constructed in several countries. This type can be illustrated by the Ryan XV–5A V.T.O.L. research aircraft. It is a mid-wing monoplane of conventional lay-out with side-by-side seating for two crew. Two General Electric J85–GE–5 turbojets are mounted in a dorsal duct and are fitted with a system of valves by which the entire jet efflux can be diverted to drive a pair of 6½ ft diameter lift-fans in the wings and a smaller lift-fan in the nose of the fuselage. All three fans are driven by a ring of turbine blades around their periphery on which the jet efflux impinges. The air mass flow through the fans is several times greater than through the turbojets and is directed downwards by vanes to give lift. After take-off the pilot can deflect both sets of vanes from the wing fans rearward to obtain propulsive thrust for transition. The airflow through the nose-fan can be deflected upward or downward by curved doors for pitch control. When in horizontal flight the doors and louvres about and beneath the fans are closed to seal the wings and fuselage so that the aircraft flies like a normal type of aeroplane.

Jet-powered flat-risers depend on the direct lift from the jet thrust. The so-called Rolls-Royce "Flying Bedstead" consisted of a pair of Nene jet engines mounted on top of a large framework which ascended vertically, moved in any desired direction horizontally and was quite stable in flight above its air cushion. Speeds up to 15 knots were attained during 1954. This test vehicle was developed into the Short SC–I which had four Rolls-Royce RB–108 turbojets mounted vertically in a central engine bay, and a fifth RB–108 which exhausted horizontally at the tail to give thrust in forward flight. This craft (and the Bedstead) had "puff-pipe" control nozzles positioned at the nose and tail of the fuselage, and at the wing tips, which gave satisfactory control, the flow in the pipes being powered from the engine. The French V.T.O.L. tactical fighter Mirage III–V is a more powerful version of the same idea with four pairs of Rolls-Royce RB 162 turbojets giving a total of 35,200 lb jet lift solely for vertical flight. Its propulsion by a SNECMA TF–106 turbofan, developing 19,840 lb thrust with full afterburning will give a Mach 2 performance with V.T.O.L. capability.

The remaining method of V.T.O.L. flight is by turbojet vectored-thrust. The thrust of the engines can be either used in the normal way for forward propulsion or be deflected downwards to give vertical lift. This scheme has been developed successfully in the Hawker Siddeley P.1127 as a V/STOL strike reconnaissance aircraft with operational capability at speeds in excess of Mach 1 and has been flown in highly successful carrier trials on H.M.S. Ark Royal. The power plant is the 15,000–18,000 Bristol Siddeley Pegasus turbofan. A further development is the Hawker P.1154 with a Bristol Siddeley BS 100 vectored-thrust turbofan rated at over 30,000 lb thrust with plenum chamber burning to give a maximum speed greater than Mach 2. These last two aircraft are typical of the jet-lift types on which several countries are concentrating in 1965.

J. L. NAYLER

**VIDICON FOR X-RAY RADIOGRAPHY.** The X-ray-sensing vidicon camera tube permits direct conversion of X-ray images into video television signals without intermediate conversion into light. In combination with closed-circuit or other television amplifying, transmission, and display systems, the X-ray vidicon provides means for macroscopic electronic fluoroscopy and radiography. Typical systems provide image enlargements of the order of 30 diameters, with detail resolution better than 10 microns (390 microinches). Image contrast can be varied over wide limits, and can exceed the gradients possible with fine-grain, high-contrast radiographic films exposed and developed to film densities in the 2·0–4·0 H. and D. range. Images can be transmitted through extensive distances, and reproduced by one or several monitors, so that observers need not be subjected to radiation hazards. Test objects may be observed in motion at rates up to 18 in./min, or at higher speeds if objectionable output image blurring does not occur at frame rates of 30 images per second. With stationary test objects, X-ray integration can be performed in the vidicon tube to enhance signal levels and output image contrast. In this case, intermediate image-storage tubes can be employed to transform the intermittent output signals into continuous images on the monitor picture tubes.

*Design of X-ray sensing vidicon camera tubes.* With the exception of transmission window and target layers, X-ray-sensing vidicon camera tubes are similar to conventional light-sensing vidicon tubes (Fig. 1). Their cylindrical glass enclosures (typically about one inch in diameter and six inches in length) house an electron beam gun, accelerating electrodes, and intensity-control electrodes. Beam focusing and deflexion are provided by magnetic fields from external coil systems. The scanning electron beam passes through a fine-mesh screen (typically 750 wires to the inch) and impinges upon or is deflected from the semiconductor target layer, in accordance with its surface potential distributions. That portion of the scanning beam not utilized to discharge the target layer returns to ground through shields and collectors within the vidicon tube.

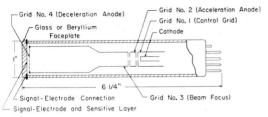

*Fig. 1. Structure of X-Ray Sensing Vidicon Camera Tube.*

The X-ray-sensing target layer of the vidicon tube consists of a suitable semiconductor, responsive to incident X rays by a change in electrical conductivity. Lead monoxide and other materials have been used in past developments. More recently, vapour-deposited amorphous selenium layers about 30 microns in thickness have been found preferable in terms of freedom from excessive lag in conductivity, following X-ray exposure. Lag effects in vidicon tubes contribute to excessive blurring of detail in continuous imaging of test objects in motion.

The semiconductor target layer is deposited upon an electrically-conducting layer that serves as the signal plate and is electrically connected to the signal ring or electrode near the face of the vidicon tube. With glass-window vidicon tubes, the signal layer can consist of a thin layer of tin oxide deposited upon the glass faceplate prior to deposition of the target layer. With beryllium-window X-ray vidicon tubes, substrates may be required to prevent deleterious interactions between beryllium and selenium target materials. If conducting, this barrier layer can also serve as the signal electrode.

Vidicon windows or faceplates must be designed to permit X-ray images to be projected onto the target layers without excessive attenuation or scattering within the window materials. Glass windows, even of the minimum feasible thickness to withstand external atmospheric pressures, can result in excessive attenuation of the X-ray beam, with low kilovoltage X-radiation. Beryllium faceplates, typically of about 0·030 in. thickness, have been found to be preferable, both in signal level and image contrast. In combination with beryllium-window X-ray sources, beryllium-window vidicon tubes permit extreme image contrast and detail resolution to be attained in X-ray television image enlargement systems.

*Operation of X-ray-sensing vidicon camera tubes.* Best performance is obtained from present types of X-ray vidicon tubes when very high X-radiation intensities (preferably 10–1000 r/min) are transmitted through the test object to the vidicon target. Continuous imaging at 1800 exposures per minute (corresponding to the 30-frame per second scan rate) thus provides

about 5 mr–0·5 r per image exposure. The seexposure speeds compare favorably with those of industrial X-ray films, which typically require from 0·1 to 2 r for optimum exposures. Beryllium-window, fractional-focus, constant-potential X-ray sources capable of outputs in the range of one-half million r/min, and operated at source-vidicon distances of 6–10 in., have been found to be preferable for use with vidicon imaging systems.

To sensitize the vidicon target layers to incident X-radiation, d.c. potentials are applied to the signal plate, and so to the surface of the target layer opposite that scanned by the electron beam. Output signals from the vidicon tube are direct functions of both the incident X-ray intensities, and of the target control voltage (typically from 10 to 50 V) applied to the signal ring (Fig. 2). During the X-ray exposure, the

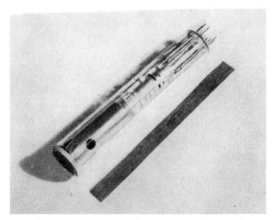

*Fig. 2. Photograph of X-Ray Sensing Vidicon Tube.*

target layer becomes locally conductive in response to the incident radiation intensity. Positive charge from the signal plate passes through conducting areas of the target layer to its inner face. When the scanning electron beam traverses such charged spots each 1/30 sec, electrons are deposited in sufficient numbers to neutralize these positive charges. (The balance of the scanning beam is repelled from the target surface, and is not utilized except in return-beam vidicon tubes.) At the instants of charge neutralization by the scanning beam, the target layer behaves primarily as a capacitive element. The capacitively-coupled signals (from charge neutralization) appear at the signal ring as output video signals from the vidicon camera tube. The output load resistors are commonly near 50,000 ohms. The X-ray video signals are of small magnitudes. Output currents are typically in the range from 0·01 to 0·1 microamperes.

Because of the relatively low output signal levels, it is particularly important that signal-to-noise ratios be extremely high at the input to the preamplifiers of the video chain. Typical signal-to-noise ratios for the X-ray-sensing vidicon are reported to be of the order of 200 dB. From the vidicon tube onward in the amplifier chains, good signal-to-noise ratios must be maintained. Amplification of noise signals, together with the image-carrying signals, cannot improve the output image quality, and noise can be a primary source of deterioration in output X-ray images. Sources of noise signals include not only the well-known contributions of resistors and other components of circuitry, and extraneous electrostatic or electromagnetic interference, but X-ray ionization effects as well. During exposures to very high X-radiation intensities, it is possible for portions of the vidicon camera enclosure to perform as ionization gauges. Air in these spaces can become highly ionized. Each exposed electrode of the camera circuitry may then collect ionization signals of potentially-large magnitudes, and in the case of grid circuits these may be injected as objectionable noise signals. For these reasons, pulsating X-ray sources such as high-frequency transformer type X-ray systems are usually undesirable for vidicon radiography. The higher-frequency components of radiation pulsations can produce signals in the frequency range transmitted by the video amplifier chain, often at levels that saturate amplifier components.

Vidicon output signal levels usually increase as the target control voltage is raised, whereas noise signals are relatively unaffected. Thus, the best performance is usually obtained high in the feasible range of target voltage. However, this feasible range is limited by the "cross-over" voltage of the vidicon tube. Above the cross-over limit, output images change from positive to negative, signal levels increase significantly, and image contrast may be reduced. Unfortunately, these apparently-desirable characteristics are off-set in this negative mode of operation by an excessive potential drop applied across the target layer, which is believed to lead to rapid deterioration of the target layer.

With proper operating conditions, X-ray-sensing vidicon tubes have performed excellently for periods of a year or more in daily service. However, amorphous selenium can be converted to an insensitive crystalline form at temperatures in the range of 139 F. Thus, for long life, the vidicon target must not be subjected to excessive temperatures, either from ambient sources, electron-beam heating, or excessive target potentials. Operation in the negative mode, with high target voltages, can apparently lead to target deterioration (often revealed by permanent white spots in the positive output images), and is presently not recommended by the tube manufacturers.

*X-ray vidicon image characteristics.* The image or raster area of the conventional one-inch diameter vidicon camera tube is 3/8 in. high by $\frac{1}{2}$ in. wide for the 4 by 3 aspect ratio standard. In the United States, the number of scanning lines in the image is usually set at 525, 675, 875, 945, or 1035 lines. These line numbers provide scan line densities of the order of 1400–2900 lines per inch. In well-designed television

chains, the horizontal resolution is comparable. The possible information density thus ranges from perhaps 2 to 8 million elements per square inch. The vidicon target layer, and the dimensions of the scanning electron beam, must be compatible with these resolution numbers if optimum output image detail is to be obtained. Considerable care is required in design and construction of television systems if this level of information transfer is to be maintained in the output images.

To date, even with 525-line television systems, X-ray vidicons can provide resolution better than 10 microns, provided that the incident X-ray image unsharpness is limited sufficiently, and high-quality television chains with very good signal-to-noise ratios are employed under proper conditions. It is routinely possible to resolve a human hair, 400–500 mesh wire screens, and minute biological specimens. For example, scientists at The Ohio State University have observed the birth of larvae from a common fly, and the feeding of mosquitos upon blood from mice, in detail with the vidicon X-ray image enlargement system.

Such image detail resolution would be of little value if the output images were not sufficiently enlarged to permit easy observation of fine detail. In typical television systems, the 5/8 in. diameter images from the X-ray vidicon can be readily displayed upon the screens of picture tubes with diagonal measurements ranging from 5 to 27 in. Corresponding image enlargements range from × 8 to more than × 40. Overscanning of the picture tube readily permits further enlargements up to 60 diameters or more. Much greater enlargements are feasible, as with projection television systems. However, enlargement without compatible increase in detail resolution offers no advantages provided that detail images are of sufficient size to be seen readily without further optical aids. Much of the basic work to date has been done at × 30 enlargement; in this case, a 1/32 in. detail appears almost an inch long in the monitor images.

Image quality, however, is a function not only of detail but also of image contrast. Conventional X-ray films used in industrial radiography provide film gradients in the range from 2 to 7, and so the processed films greatly amplify the contrast present in the X-ray beam image. The dynamic range of industrial X-ray films can approach values in the range of 10,000/1 (corresponding to the density range from 0 to 4 H. and D.). Thus, small differences in test object thickness or density can often be readily revealed by film radiography, and contrast sensitivities of 1 to 2 per cent are often attained in routine industrial inspection. Users have become accustomed to such performance levels, and X-ray image systems such as conventional fluoroscopy, which have nearly unity contrast transfer characteristics, have found little commercial acceptance in consequence.

Conventional electronic and television systems, on the other hand, have been most frequently designed for linear response characteristics, or even with dynamic compression. Gradients of 1·0, 0·7, and even 0·5 are common in closed-circuit television systems used with optical images. To be appropriate for use in X-ray inspection systems, overall television system contrast gradients or gamma values should be increased by factors of 5 to 10 or more. In addition, few electronic amplifier chains perform ideally over signal ranges of 10,000/1. Most are severely limited by signal-to-noise ratios at low signal levels, and by saturation of amplifiers at very high signal levels. These performance limitations should be overcome if X-ray television radiography is to equal or exceed film radiography in output image quality. It is generally far from sufficient merely to insert an X-ray-sensing vidicon into a conventional closed-circuit television chain, if an optimum X-ray television image system is desired. The limitations of such arrangements are most readily revealed by use of thicker test objects of dense materials, and critical X-ray image quality indicators such as penetrameters, in system evaluations.

As indicated by the slope of the transfer curves of Fig. 3, the gamma of the typical X-ray vidicon (analogous to the gradient of X-ray films) is characteristically less than unity. It ranges from 0·5 to above 1·0 in typical tubes, as a function of target voltage, target thickness, X-ray intensity and wave-length distributions, and the transparency of vidicon window to X rays. Thus, the enhancement of X-ray image contrast, so vital to attainment of high X-ray image sensitivity, is not obtained simply by use of an X-ray-sensing vidicon camera tube. Coordinated design of the overall system, from X-ray source to final picture tube, is usually required.

*Design of overall X-ray television image enlargement systems.* The primary consideration in design of X-ray television image systems utilizing X-ray vidicon tubes is the attainment of adequate image contrast. To match industrial inspection requirements commonly accepted in film radiography, the system should be capable of providing penetrameter sensitivities in the range from 1 to 2 per cent. Assuming adequate detail resolution (readily attained with vidicon systems and fractional-focus X-ray tubes), the contrast sensitivity must also be at least in the 1 to 2 per cent range. In practical terms, this means that a change of 1 to 2 per cent in test object thickness or material density should be readily observable. Per cent contrast can be defined by equation 1.

$$\%C = \%(\Delta x/x) \qquad (1\,\mathrm{a})$$

or

$$\%C = \%(\Delta\varrho/\varrho) \qquad (1\,\mathrm{b})$$

where $x$ is the test object thickness, $\Delta x$ is the just discernible increment in this thickness, or where $\varrho$ is the test material density, and $\Delta\varrho$ is the just discernible increment in this density. This incremental change in the test-object thickness or density is the *input signal* to the X-ray television inspection system.

The *output signal* of the entire television inspection system is the change in brightness observable in the image on the output picture tube. It might be conservatively assumed that a brightness difference of 5 per cent would be required in this output image to be readily observable. (The usual film density difference of 0·02 H. and D. units corresponds to about 4·72 per cent brightness difference when a radiographic film is viewed.) In this case, it would be necessary that:

$$\%(\Delta B/B) = 5\%. \qquad (2)$$

This value may be taken as a minimum performance requirement, in association with equation 1.

The relation between the output and input variables of a signal system, in so far as contrast is concerned, may be described by the slope or "gamma" of the transfer curve relating the logarithm of the output to the logarithm of the input variable (Fig. 4). For the case of the television X-ray image system, this gamma value is described by equation 3:

$$\gamma_s = [\%(\Delta B/B)]/[\%(\Delta x/x)] = (5\%/1\%) = 5 \qquad (3)$$

for the case of an X-ray image system providing 1 per cent contrast sensitivity.

The overall system gamma, $\gamma_s$, can be expressed as the product of the individual gammas of the sequential components of the image system. For the case of the television X-ray image system:

$$\gamma_s = \gamma_m\gamma_v\gamma_e\gamma_k \qquad (4)$$

where:

$\gamma_s = [\%(\Delta B/B)]/[\%(\Delta x/x)] = $ Overall system gamma.

$\gamma_m = [\%(\Delta I_x/I_x)]/[\%(\Delta x/x)] = $ Test material gamma, where $I_x$ is the radiation intensity transmitted through the material.

$\gamma_v = [\%(\Delta E_v/E_v)]/[\%(\Delta I_x/I_x)] = $ Vidicon tube gamma, where $E_v$ is the output video signal voltage from the vidicon.

$\gamma_e = [\%(\Delta E_a/E_a)]/[\%(\Delta E_v/E_v)] = $ Electronic amplifier chain gamma, where $E_a$ is the output signal voltage from the amplifier chain.

$\gamma_k = [\%(\Delta B/B)]/[\%(\Delta E_a/E_a)] = $ Kinescope picture tube gamma, where $B$ is the brightness of the kinescope screen image.

Typical gamma values measured for the experimental prototype television X-ray image system at The Ohio State University were:

$\gamma_v = $ 0·7 for glass-window vidicons; 0·8 for beryllium-window vidicons, variable from 0·4 to 1·2 with tube target design and target control voltage.

$\gamma_e = $ 1·0 for linear amplifier system, 2 to 10 for experimental gamma amplifier system optionally insertable into chain.

$\gamma_k = $ 2·5 for kinescope picture tubes, variable from 2 to 3.

In the absence of an absorbing test material in the X-ray beam, the television system gamma, with the linear amplifier chain:

$$\gamma_t = \gamma_v\gamma_e\gamma_k \qquad (5)$$

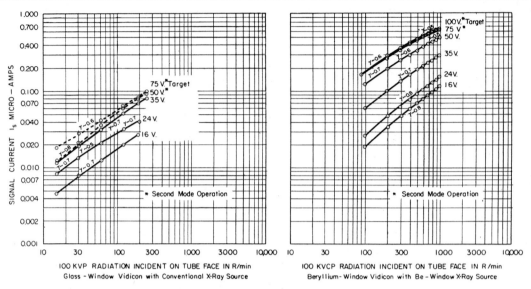

*Fig. 3. Transfer characteristics for experimental X-ray sensing vidicon camera tubes.*

ranged between 1·1 and 2.0 for glass-window vidicons (Fig. 4b), and between 4·0 and 10 for beryllium-window vidicons (Fig. 4c). In the latter case, far more soft (long wave-length) X-radiation from the X-ray generator passed through the beryllium-window structure of the vidicon tube than for glass-window vidicon tubes. Kinescope screen brightnesses were almost an order of magnitude greater (and signal-to-noise ratios correspondingly much improved) with the beryllium-window vidicon tubes. Such beryllium-window vidicon tubes show great response even to low-voltage X-radiation. Images have been obtained

of 400 mesh screens at 5 kV with a 10 W microradiographic X-ray source, with image integration periods of about two minutes. Consequently, it appears possible that the selenium-target vidicon tubes receive considerable signal strength from low-voltage (long-wave-length) components of the X-ray beams transmitted through test materials. Thus, performance is far better with beryllium-windows in both X-ray sources and vidicon tubes than when either element has a glass window or other source of inherent filtration of the X-ray beam.

The importance of test material gamma, $\gamma_m$, cannot be stressed too greatly. As shown in Fig. 5, test mate-

Slope or gamma values

*Fig. 4(a). Family of linear gamma value curves.*

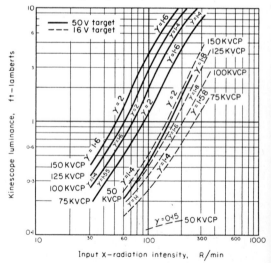

*Fig. 4(b). Transfer characteristic of X-ray television system with glass-window vidicon tube.*

rial gamma is highly responsive to the quality (wavelength distribution) of the incident X rays, and to the thickness and density of the test material. In general, for satisfactory X-ray image enlargements, the test material gamma should be at least unity. The greater the attainable test material gamma, with available radiation sources, and with adequate radiation intensities transmitted through the test material to the vidicon target, the better the output image contrast. Lowering the X-ray source kilovoltage increases test material gamma, but also lowers the total output radiation. Such reduction in input signal levels to the vidicon television chain also reduces the signal-to-noise ratio. A lower limit of perhaps 10 r/min of transmitted X-ray beam intensity is often required for good imag-

system, the resultant output image, following integration, can be displayed on the output monitors for periods of 1/2–4 min. Where integration periods are of the order of a few seconds to one minute, the electron beam scan can be activated at the end of each successive exposure period to provide continuous output images. Alternatively, it has proven feasible to enhance image contrast and signal levels by continuous scanning of the vidicon target, with integration occurring in the scan-conversion tube.

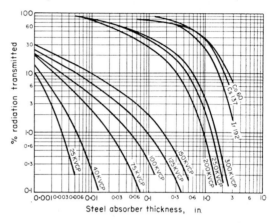

*Fig. 5(a). Transfer curves for steel in beams from various radiation sources.*

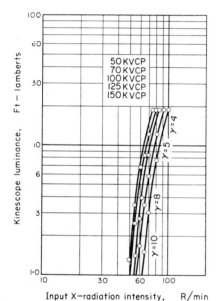

*Fig. 4(c). Transfer characteristic of X-ray television system with beryllium-window vidicon tube.*

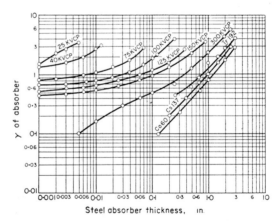

*Fig. 5(b). Gamma factors, $Y_m$, for steel absorbers and various radiation sources.*

ing. The optimum operating point is often determined experimentally, while adjusting X-ray source kilovoltage and television chain "black level" controls simultaneously for best image contrast and brightness.

*Operation with X-ray integration in vidicon tube.* With stationary test objects, it is feasible to discontinue scanning of the vidicon electron beam so as to integrate X-ray exposure and accumulate much larger positive charges on the target surface than with 30-frame-per-second scans. Signal levels increase with exposure time up to at least 15 min of signal integration on the vidicon target. In this way, excellent images have been obtained with X-ray input intensities lowered to 0·0005 r/min. This integration technique permits stationary X-ray radiography of thick or dense materials, or operation with very low X-ray source kilovoltages. With the addition of a scan-conversion

*Penetrameter sensitivities attainable.* The Ohio State University television image system was specifically designed for X-ray inspection of steel missile case materials and weldments of 0·1 in. thickness or thinner gauges. Its optimum performance is consequently attained in this range, with in-motion inspection. Image quality indicators, or penetrameters, are difficult to procure or manufacture for thicknesses corresponding to two per cent of material thickness in this range.

However, penetrameter tests have been made on steel and aluminium alloy specimens in the range from 1/8 in. of material thickness upward, for which U.S. penetrameters could be procured. When the 1-T hole is visible in a two per cent penetrameter of this type, it is certain that two per cent penetrameter sensitivity has been attained. In some cases, visibility of the 1-T hole is now taken to represent 1·4 per cent penetrameter sensitivity.

With steel and similar materials, the Ohio State University X-ray television image system has permitted visualization of the 1-T hole in two per cent penetrameters, through 1/4 in. of steel or one inch of aluminium alloy, during in-motion inspection. With integration or scan-converter modification of images of stationary objects, these limits for visualization of the 1-T hole in two per cent penetrameters have been raised to 1/2 in. of steel, and to 3 in. of 2219 aluminium alloy.

With very thin gauges of steel and lower-density materials, the apparent contrast sensitivity far exceeds that attained with the thicker specimens. However, no precision methods of evaluation, comparable to penetrameter tests, have been devised for this thickness range. Also, aluminium alloy spotweld images have been produced that are comparable to those attained previously with low-voltage, beryllium-window X-ray sources where the radiography was considered to be of the order of 0·2 per cent contrast sensitivity.

Tests made with the German DIN penetrameters (wires of various materials and sizes) have shown comparable performance, with the penetrameters located on the source side of the test materials.

*Bibliography*

HEIJNE L. *et al.* (1954) *Philips Technical Review* **16** (1), 23.

JACOBS J. E. and BERGER H. (1956) *Electrical Engineering* **72**, 158.

McMASTER R. C. and SATTLER F. J. (March 1962) *Gamma Characteristics of X-Ray Image Systems*, The Ohio State University, Dept. of Welding Engineering and the Engineering Experiment Station, p. 7.

R. C. McMASTER

## VIDICON FOR X-RAY RADIOGRAPHY, APPLICATIONS OF.

Three broad areas of potential application appear promising for the television X-ray image enlargement system described in the preceding article: (1) Inspection of critical aerospace materials, components, weldments, and assemblies, (2) Observation and control of welding and other critical processes, and (3) Use in scientific research and development.

*X-ray inspection applications.* Immediate application of the television X-ray image system is taking place in aerospace industries for inspection of thin-gage missile case materials, weldments, brazed joints, electronic components, printed circuit assemblies, and other small scale inspection operations. Studies are under way related to use of such systems for large-scale inspection of missile cases in increasing wall thicknesses.

*Fusion weldments.* The television X-ray image system has been proven effective as a low-cost, high-speed, high-resolution system for inspection of aerospace weldments, particularly in thin gauge materials. In electron-beam weldments in sheet materials, it has revealed unsuspected porosity accumulation along fusion sidewalls. This porosity, often with pore sizes smaller than 1/1000 in. in diameter, tends to lie in one plane along each wall of the fusion zone. In tungsten inert-gas (TIG) weldments, the enlarged images of tunsgten inclusions show that often many fine inclusions (again, often smaller than 1/1000 in. in size) can accompany the larger types of inclusions previously seen in film radiographs. Weldments produced by several forms of metal inert-gas (MIG) and carbon-dioxide gas-shielded metal arcs, particularly with automatic welding equipment, have been successfully inspected with the television system. Unusual sensitivity to transverse and longitudinal crack networks has been attained. Relative motion of test object and source, with television X-ray observations, has shown that many cracks revealed on film radiographs were actually much longer (often 3 to 5 times longer) than the film images indicated. The gross portions of the cracks continue in many cases into a fine network of subsurface cracks, some of which are visible only at certain narrow angles of incidence of the X-ray beam. Thus, it is evident that weld repairs, based upon gouging out and repair of crack areas revealed by film radiography, may often be inadequate since the crack extends far beyond the limits of the repair. In still other cases, extremely fine cracks have been found to extend from small random porosity indications, being subsurface entirely. Since random small porosity is sometimes acceptable for certain types of service, whereas all cracks are rejectable, misleading interpretations can be based upon film radiography alone.

To date, all fusion weldment discontinuities detectable by film radiography have been readily detectable by the television X-ray image system, in material thicknesses within its range of application. In many such cases, additional discontinuities, not detected by film radiography, have been detected quickly in the enlarged television images. In fusion arc weldments, cracks, shrinkage networks, porosity, slag inclusions, undercut, lack of fusion at sidewall and at the root of the weld, tear-drops or drip-through, suck-up or concavity at the root, and numerous other discontinuities have been observed. In fusion spotwelds in aluminium alloys with copper or zinc alloying elements, it has been possible to detect the dark ring outlining the nugget area at the interface, segregations of copper-rich eutectic, cracks, porosity, shrinkage defects, expulsion, and similar defects that X rays can reveal.

Weldments in miniature electronic components, fine wires, and sheets have also been examined with the television system. Electron tube assemblies, transistor details, connexions to leads of components such as resistors, capacitors, inductors, and transformers, and other components often only a few thousandths of an inch in diameter have been inspected.

With the scan converter system, faint indications of weld defects seen by in-motion inspection have been intensified and made clearly visible. This integration system now extends the capabilities for weld inspection to about 1/2 in. of steel and to about 3 in. of aluminium alloys.

*Brazed and soldered joints.* The television image enlargement system also provides unique detail sensitivity for inspection of brazed and soldered joints, particularly in thin gauge and small parts. In inspection of brazed stainless steel sandwich and honeycomb structures, the internal details of braze alloy voids and porosity, or loss of filleting or node flow, appear remarkably clear in the enlarged images. With honeycomb core wall only 1·5 mils in thickness, the individual fillets on either side of the core wall can be seen separately, Node flow, and in some cases even the depressions caused by welding the node flats together, can be seen. Lack of filleting, intermittent filleting (even for 1/1000 in. of length), cell repairs, cell wall perforations, crushed or distorted cells, and many other conditions are visible. In sandwich materials, or in edge members or close-outs of honeycomb panels, the distribution of braze alloy at the interface can be examined in detail. Often porosity of minute dimensions is found in the interfacial areas. Similar observations can be made of brazed joints in thin walled tubes and other brazed assemblies.

Of particular interest for aerospace electronic applications is the use of the television system to inspect miniaturized electronic components and printed circuit assemblies. With dip or hand soldered joints, it is often found that extensive porosity exists in the solder at terminals, or that terminals have been inadequately filled with solder to hold the leads. In many cases, it has been observed that thin, transparent (to X rays) films surround wires in soldered joints (observable only when the X-ray beam is exactly aligned with the sidewall of the soldered wire or part). These cases have been interpreted as indicative of cold-soldered joints. Other cases have been observed where a film of solder covers a broken lead wire, or where fine "whiskers" smaller than a mil in diameter extend partially or completely between printed circuit conductors.

*Foils, sheet and plate.* Limited observations have been made on thin gauge sheet and foils, and on steel and aluminium alloy plates of 1/20–1/2 in. in thickness. In several cases, striations in the direction of rolling are clearly revealed. Steel specimens, with fine cracks initiated by nascent hydrogen and stress, have been examined and all known cracks detected,

including some so fine that film radiography had generally failed to detect them. Some of these specimens also contained unexplained mottling as if films of non-metallic materials had been elongated by rolling. Weld cracks that extend outward into base material have also been seen in many specimens.

*Small castings.* Small castings of magnesium, aluminium and dense alloys have been inspected by the television system. In light alloy magnesium castings, individual grain structure was revealed, together with other discontinuities. In turbine blade alloys, structure is much more clearly evident, and often interferes with interpretation of the X-ray images. Small protrusions into internal cooling passages of blades could be seen clearly in precision investment castings. In some cases, observers thought they had detected the transient appearance of diffraction lines from coarse-grained castings (lines that moved many times faster than the images of structure).

*Graphite and non-metallic materials.* Extensive striations and other artifacts were visible in enlarged images of graphite specimens. Also visible were inclusions of denser materials, and apparent low-density areas, in some specimens. Fibre glass laminates have been examined extensively. With only a few layers, the individual strands or rovings often appear clearly, and the direction, lay, or density of fibre packing can be seen. With complete missile case walls, including liner materials, useful images were obtained that showed the principle fibre directions, void or low density areas, inclusions of dense materials, and other defects.

*Small mechanisms.* Small mechanical devices such as watches, fuses, relays, valves, and others can be examined readily, even in motion, with the X-ray image system. Movements of parts can be observed, and spacings or clearances measured to an accuracy of the order of 1/1000 in. from the enlarged images of hermetically-sealed assemblies. In one case, the movement of oil in a watch bearing could be seen clearly as the shaft rotated.

*Wires, tubes, and small rods.* Elongated objects such as wires, cables, coated welding rods, thin-walled tubes, range heater coils, and others are particularly well suited for inspection by means of the X-ray television system. If not more than 1/2 in. in diameter, they can be scanned longitudinally, often while rotating in the fixture. This permits many views, so that eccentricity of coatings or internal structures, and other irregularities can be seen from many angles. Longitudinal and transverse welds in tubes can be seen in many cases, even with tubes filled with other materials. Movements of fluids in tubes, or of bubbles in fluids, can be seen through the tube walls in some instances.

### Observation and Control of Welding Processes

Possibly of greater potential importance than post-mortem inspection applications is the use of the television image system to observe and control welding and brazing processes. To the extent that the formation of discontinuities or loss of control of welding processes can be observed during operation, and corrective measures applied, the number of rejectable conditions in finished weldments can possibly be greatly reduced. This can reduce the great costs of gouging out and repairing weldments (in structures that can be repaired), and may become essential in production of weldments than cannot later be repaired.

Preliminary studies have been made to determine the feasibility of direct X-ray observation and control of metal inert-gas (MIG) welding of 3/8 in. 2219 aluminum alloy sheets. Three major problem areas were encountered: (1) requirements for thermal shielding of the television camera from the heat of the welding arc and the molten and base metals, (2) modification of welding fixtures and equipment to permit introduction of the X-ray source and television camera unit at the point of welding, and (3) provision of X-ray shielding to protect the operator and environment from ionizing radiation exposures. Solution of these problems permitted direct observation of the X-ray images of (a) metal transfer, (b) the molten weld pool, and (c) the solidifying weld metal, according to the position of the camera unit along the line of welding. With the primitive initial set-up, images were not as clear as those obtained in weld inspection of finished welds, but the drops of metal transferring, the movements and escape of gas porosity bubbles from the melt, penetration of fusion to the root of the weld, and the formation of cracks in the solidifying weld metal were all seen clearly. Several new observations, such as the trapping of inclusions in the molten pool for long periods of time, crack growth in successive stages, and the dynamics of escape of gases from the melt, were of particular interest. Further studies are planned in the near future of submerged arc welding phenomena.

Considerable interest has been aroused by the possibilities of controlling welding processes directly by three-dimensional X-ray examination of the molten pool and adjacent areas. Such techniques could be applied to control of aerospace welding, particularly of large missile cases, if means could be found to introduce the X-ray source and television camera at the point of welding. An alternative procedure would be to insert the inspection system a few inches or feet down the weld line from the point of welding. If trends toward defects such as lack of fusion at the root, sideway drift of the arc tending to produce lack of fusion at a sidewall, trapping of porosity in the deposited metal, cracking during solidification or cooling, undercut or overlap, and others could be detected during welding, the process could be adjusted to avoid weld defects in the finished welds, or interrupted until corrections could be made. Potential cost, material, and time savings from such a development could be of great value to the aerospace industry. Rapid feedback of X-ray inspection information is the basis of such controls.

### Use in Scientific Research and Development

The television X-ray image system, like the light microscope and other instruments, can be used as an observation and measurement tool in many kinds of basic research, as well as in engineering development. Obviously, it could be of direct assistance in evaluation of new welding processes, materials, and procedures. X-ray examination of the process as operating conditions were varied could provide almost instantaneous information on the effects of welding variables. In many cases, the causes of weld defects might be identified. Studies of braze alloy flow and bonding might also be greatly expedited if tests were made while enlarged X-ray images were being observed. Of particular interest is the possibility of observing the propagation of defects and failures in materials with capabilities for plastic deformation before failure, during mechanical destructive testing. With electron beam welding, for example, the X rays generated by the electron beams themselves, upon absorption in the work material, might possibly serve as radiation sources internal to the melt, from which the point of impact, and the internal conditions, might be deduced.

Other possible research applications might include studies of the flow of metal in precision investment casting, of the successive deformation of inclusions or porosity during rolling or forging, and of the attack of corrosive fluids upon specimens at various temperatures. Still other possibilities include observation of enclosed structures and mechanisms during loading or operation, to determine stress distributions or deformations. Fluid flow problems might in some cases be expedited by studies of this type. And finally, biological studies on small animals, insects, and similar small forms of life might be made both in the laboratory and in outer space, to determine responses to various stimuli and environments.

*Bibliography*

McMaster R. C. *et al. X-Ray Image System for Non-destructive Testing of Solid Propellant Missile Cast Walls and Weldments,* Watertown Arsenal Contract DA 33-019-ORD-3385. Monthly Status Reports Nos. 1-26, Quarterly Reports Nos. 1-4, Interim Report No. 1, and Final Report in August 1962.

McMaster R. C. *et al. Nondestructive System for Inspection of Fiberglass Reinforced Plastic Missile Cases,* Watertown Arsenal Contract DA-33-019-ORD-3670, Monthly Status Reports Nos. 1-27, Quarterly Reports Nos. 1-6.

McMaster R. C. and Gericke O. R. (April 1962) *X-Ray Image Enlargement System for Inspection of Missile Case Walls and Weldments,* presented before Society for Nondestructive Testing, Cleveland, Ohio.

R. C. McMaster

# W

**WATER DESALTING.** *General.* Water desalting is the reduction or the complete elimination of dissolved inorganic salts from water. If the water is to be used for domestic purposes or for irrigation, a reduction of the total dissolved salts to about 500 ppm (parts per million) is sufficient. On the other hand, if the water is to be used in a high pressure steam boiler or in certain closely controlled manufacturing processes, the tolerable salt concentration may be only a few ppb (parts per billion). Raw water supplies vary in salt concentration from sea water with at least 35,000 ppm to surface and ground waters with as little as 50 ppm. The great solvent capacity of water results in the absorption of minerals from the earth, so that many rivers and wells contain water with dissolved salts in excess of 2000 ppm. Practical desalting plants have been applied to the reduction of salinity in water supplies containing from 1800 to 4000 ppm, as well as in sea water containing as much as 43,000 ppm.

Land-based desalting plants are presently found in remote locations where, alternatively, fresh water would need to be imported over considerable distances. Examples would include Kuwait and Bahrain on the Arabian Gulf, various islands in the Caribbean, Gibraltar, and the Isle of Guernsey. The latter location has the distinction of being the site of the only distillation plant installed primarily to meet an agricultural need. The Isle of Guernsey produces tomatoes for the early European market and commands premium prices for this crop. Ruinous droughts have occurred at intervals of about nine years. A distiller plant capable of offsetting the water shortage of a drought was installed, its costs of capital amortization and operation being justified as a form of crop insurance. The rest of the plants produce water for domestic and industrial purposes. The total installed capacity on a world-wide basis is of the order of 40 mgd (million gallons per day), with distillation accounting for at least three-fourths of the capacity and electrodialysis furnishing the remainder.

The chief obstacle to the more widespread use of desalted water is its high cost. This cost in 1964 is, in general, several times the prices charged for freshwater obtained from nearby wells or rivers. There are several reasons for this disparity, the chief one being the high capital investment and operating expense required for a desalting plant. Parenthetically, it may be noted that all of the processes used for desalting water are similar to processes used for the production or refinement of chemicals, all of which cost several dollars per ton as compared with the desired goal for water of only a few cents per ton. Other reasons for the difference between the accepted price of fresh water and the cost of desalted water would include the subsidies extended to fresh water production and the relatively small size of the desalting plants. Prices paid by the consumer for delivered fresh water from normal sources are of the order of 30–60 c per kgal. The lowest cost realized for distillation in 1964 is $ 1·00 per kgal at the production plant.

Schemes for desalting water are numerous and much experimental work is being done on the improvement and perfection of several of the schemes. All of the schemes may be grouped under four headings, according to the form of energy used. This grouping is shown in the following table:

Processes using heat energy—several forms of distillation.

Processes using mechanical energy—vapour compression distillation, reverse osmosis, freeze-separation, hydrate separation.

Processes using electrical energy—electrodialysis.

Processes using chemical energy—ion exchange, solvent extraction.

Of all the schemes listed, only distillation (in several forms) and electrodialysis are truly commercial, the others being in the pilot plant or experimental stages of development. The first production plant using freeze-separation was placed in service in 1963. Reverse osmosis was used in a pilot plant producing a few hundred gallons per day in 1964.

*Processes using heat energy.* Thermal distillation is the oldest process for desalting water, and its use dates back to the introduction of steam as a motive power for ships. It consists of the evaporation of pure water from saline raw water, the removal of this vapour from contact with the raw water, and its subsequent condensation. The equipment normally used is some form of shell-and-tube heat exchanger, in which a cylindrical shell surrounds a number of tubes which penetrate through two circular plates, closing the ends of the shell. Caps over the exposed ends of the tubes make possible the connexion of the tubes with a piping system.

Raw water passing through the tubes can be heated and partially evaporated by the heat from condensation of steam at a slightly higher temperature on the outside surface of the tube. A number of shell-and-tube heat exchangers con be placed in series so that a the vapour formed in the first one will condense in the second and produce additional vapour which flows into the shell of the third unit, and so on. The primary heat addition is only that in the first unit. This arrangement is called forward feed multiple-effect evaporation and is shown schematically in Fig. 1. This type of plant has the characteristic that the Gained Output Ratio (the ratio of the pure water produced to the steam used) is approximately proportional to the number of units in series.

Examples of the multiple-effect distiller may be found on most ocean-going ships and in all of the early units at Kuwait, Aruba, Curacao, and Gibraltar. The earliest large land-based plants had three effects and operated with a G.O.R. (Gained Output Ratio) of about $2 \cdot 3 : 1$. High fuel costs led to the use of six

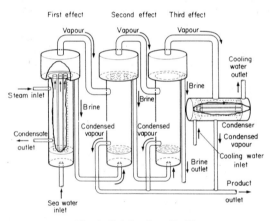

*Fig. 1. Triple-effect distiller.*

effects with a G.O.R. of $4 \cdot 6 : 1$. The largest single set of equipment of this type is the U.S.D.I. Demonstration Plant at Freeport, Texas, which produces one mgd with 12 effects and has a G.O.R. of about $9 \cdot 2 : 1$.

The multiple-stage flash distillation plant differs from the multiple-effect plant in that all interstage vapour piping is eliminated and, also, the G.O.R. is essentially independent of the number of stages or effects. The equipment consists of a long horizontal heat exchanger with condenser tubes running the full length of the heat-exchanger shell and located in the upper part of it. Bulkheads placed across the shell of the heat exchanger are pierced by the condenser tubes and serve to divide the shell into a number of stages. This general arrangement is shown in Fig. 2. Water flowing through the condenser tubes is heated by condensation of vapour on the outside of the tubes. Heated water leaving the condenser tubes is heated a few more degrees of temperature in an externally-

heated condenser and then is introduced into the bottom of the shell below the condenser tubes. An orifice at the entrance to the first stage slightly reduces the pressure of the water and causes the formation of vapour by flashing. The brine residue then passes through orifices in the successive bulkheads, producing increments of vapour in each stage and moving in a direction countercurrent to the flow through the condenser tubes. The G.O.R. for this equipment is dependent primarily upon the increment of temperature produced in the final heater and is, therefore, independent of the number of stages. Most of the distillation plants installed since 1959 are of this type, examples being found at Kuwait, the Isle of Guernsey, and the U.S.D.I. Demonstration Plant first installed at San Diego, California, and later move to Guantanamo Bay, Cuba.

Figure 3 shows temperature profiles characteristic of multi-effect distillation and multiple-stage flash distillation. Inspection of these charts will make clear the comments already made concerning the dependency of the G.O.R. on the number of stages or effects. Also, it will be noted that the temperature relationships in the multiple-stage flash distiller approximate those in a single-pass counterflow liquid-to-liquid heat exchanger. Increasing the number of stages would in-

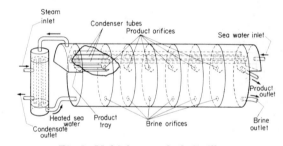

*Fig. 2. Multiple-stage flash distiller.*

crease the number of steps on the upper curve, thereby increasing the average temperature difference causing heat transfer. Reasoning of this type has led to the use of from 25 to 40 stages in multiple-stage flash distillers.

The practical problems encountered in the operation of thermal distillers of the above types are due to corrosion and salt deposition on the heat-transfer surfaces. When sea water is distilled, compounds of calcium and magnesium may become so concentrated at elevated temperatures that the brine will be supersaturated. If precipitation of these compounds takes place on the surface of the metallic heat-transfer tubes, a layer of scale will form and cause resistance to heat flow. Since the scale components found in many cases are calcium carbonate and magnesium hydroxide, it is possible to prevent their formation or to dissolve them by adding sulphuric acid to the incoming sea water. While this treatment effectively controls scale formation, it increases the hazard of corrosion. Other schemes for preventing scale formation generally involve an additive containing one of the polyphosphate salts. The effect of these additives is to cause the precipitated salts to form a fluffy mass suspended in the brine, which may be eliminated with the brine. All schemes for controlling scale deposition are less than 100 per cent effective and cost several cents per kgal of product.

The cost of fuel respresents a major share of the cost of distilled water, so that there has been great interest in the use of solar energy and other atmospheric effects to avoid the use of fuel. Solar distillers of the symmetrical greenhouse type have received a major share of attention but are still too costly for general

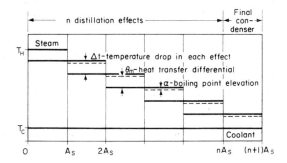

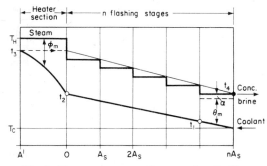

*Fig. 3. Temperature–area diagrams for distillers. (a) Multiple-effect distiller. (b) Multiple-stage flash distiller.*

large scale use. Dew collectors and dew ponds, which utilize nocturnal radiation to cause condensation of humidity, have likewise proven too costly, except where constructed by home labour for subsistence purposes.

*Processes using mechanical energy.* Turning now to processes using mechanical energy, the one commercial scheme is vapour compression distillation. In this scheme the vapour produced in a condenser-evaporator is compressed slightly and returned to the condenser side of the same heat exchanger at a pressure high enough that the condensing temperature will be greater than the evaporating temperature on the brine side. This process is the same as that of a heat pump

so that this process is more economical of energy than are the thermal distillation processes. It is widely used for small units, since these units can be mounted on a portable skid together with an internal combustion engine to furnish the mechanical power. The largest plant of this type is the U.S.D.I. Demonstration Plant at Roswell, New Mexico, which produces one mgd from well water containing 25,000 ppm of dissolved salts.

Freeze-separation also uses mechanical energy in the separatory process. When a saline solution is partially frozen at a temperature above the eutectic point, it is found that the crystals are those of pure water and all of the salts remain in the brine. By separating the ice from the brine and then melting the ice, the product would be pure water. Several arrangements have been tried, and some are in the pilot-plant stage. The one commercial plant uses the following cycle: Pre-cooled sea water is sprayed into a vacuum chamber. Part of the water freezes, a small amount evaporates, and the remainder becomes a concentrated brine. The ice crystals are mechanically removed from the vacuum chamber, drained, washed with a small amount of the pure product water, and delivered to the melting chamber. The vapour is pumped out of the freezing chamber, compressed slightly, and delivered to the melting chamber. As this compressed vapour comes in contact with the ice crystals, it condenses and melts the crystals in the process, so that the product leaving the melting chamber is composed of melted ice and condensed vapour.

The above process has the disadvantage of requiring very bulky equipment, due to the very large specific volume of water vapour at the low temperatures used—about 25°F. Other inventors have used butane as a refrigerant, mixing it as a liquid with the sea water before pre-cooling the mixture. After pre-cooling, the mixture is sprayed into the freezing chamber where the butane flashes into vapour and the ice crystals are formed. The butane vapour is compressed and pumped into the melting chamber to condense on the ice crystals and thereby melt them. Since butane and water are quite insoluble in each other, the separation of liquid butane from the product water is relatively simple. The major problems of this process, as with the one previously described, is the separation of the ice crystals from the brine. In all cases about 5 per cent of the product water is used in washing the ice.

The hydrate-separation process is very similar to the freeze-separation process. Under certain conditions of pressure and temperature it is found that propane and certain other materials will form solid crystals with water, in which one molecule of propane is associated with 17 molecules of water. Physically, it may be thought of as a crystalline structure in which the propane molecule serves as a keystone. The advantage of this process lies in the fact that the hydrating temperature is higher than the freezing temperature of saline water, and with some hydrating agents the hydrate crystals can be formed at temperatures only slightly less than 50°F. The details of the process are the same as for freeze-separation, the only difference being the substitution of the hydrate-forming chamber for the freezing chamber.

The last of the processes using mechanical energy is the reverse osmosis process. A thin plastic membrane is mechanically supported by a porous structure and saline water brought into contact with the exposed surface of the membrane. When the pressure of the saline water exceeds the osmotic pressure, the water will pass through the membrane, leaving the salts behind. Experimental developments up to 1964 include the casting and processing of cellulose-acetate membranes, which have good separatory properties and a useful life of about six months. The casting solution includes a small amount of magnesium perchlorate or certain other chemicals which is leached out after the membrane has been cast and seems to give the membrane the property of repelling the dissolved ions. Experimental plants have been built using an arrangement similar to a plate and frame filter press but the most successful seems to be the use of the membrane as a lining for the inside of small diameter tubes loosely woven from glass fibres. Saline water flowing through such tubes under high pressure causes pure water to pass through the tube wall and drip from the surface. The advantage of this process is that the osmotic pressure varies with the initial salinity of the water being purified, so that less energy is required for the purification of water of low salinity than for sea water. This differentiates this process from all of those previously described, since the energy input for any process using a phase change for separation (evaporation or freezing) is quite independent of the initial salinity of the water.

*Processes using electrical energy.* Electrodialysis is the only practical desalting process which directly uses electrical energy. It was applied to water desalting in the early 1950's when selective plastic membranes were invented. The first electrodialysis membranes were made of ion-exchange materials embedded in an inert matrix. Membranes containing cation exchangers would accept cations but repel anions, while those containing anion exchangers would accept anions but repel cations. The presence of an electric field would cause the accepted ions to move from one exchange site to another within the membrane and, thus, eventually pass through the membrane. Modern heterogeneous electrodialysis membranes are made in the same fashion and are widely used. Some homogeneous membranes are also in use, being made by the impregnation of materials such as cellulose acetate, cellophane and Kraft paper. The impregnating agents serve to modify the matrix structure and produce the equivalent of ion-exchange sites. These membranes are arranged in stacks, alternating cation and anion membranes, with a spacing between membranes of approximately one millimetre. Water flows through the spaces between membranes, and a direct-current electrical potential is imposed between electrodes placed against the outside membranes. The electrical potential causes the flow of current through the stack, with the ions serving as carriers of electricity. The potential causes the ions to migrate through the membranes and, due to the selectivity of the latter, half of the spaces between membranes will show a reduced ionic concentration and the other half, an increased concentration. Present designs are such that the length of flow path for the water is about seven feet and the reduction in salinity nearly 50 per cent.

Plants of this type are located in Bahrain, Libya, South Africa, and the U.S.A. The largest plant was built in South Africa in 1959 to process three mgd of mine drainage water with initial salinity between 3000 and 4000 ppm. This plant is no longer in operation since mining activity has been curtailed in this area. The largest plant in the U.S.A., located in Buckeye, Arizona, produces up to 650,000 gpd and reduces the salinity from 2200 ppm to 350 ppm. Water produced by this plant is said to cost 51c per kgal at a 48 per cent load factor.

Electrodialysis, like reverse osmosis, is sensitive in its energy requirement to the initial salinity of the raw water, being much less for well waters of a few thousand ppm than for sea water. All of the present plants treat water of less than 5000 ppm initial salinity, since the energy requirement for experimental plants processing sea water has been more than that of vapour compression plants.

*Processes using chemical energy.* Ion-exchange materials with hydrogen and hydroxide ions as the exchangeable ones have been used for the final desalting of water for laboratory purposes. However, the high cost of acid and alkali for regeneration has made this scheme infeasible for general use. Attempts to use recoverable regenerants have likewise proven too costly. For example, some experimental work was done using ammonium bicarbonate as the regenerant. The product water would then be heated to drive off ammonia gas and carbon dioxide, both of which would be dissolved and used for regeneration. The heat energy required proved to be nearly as great as that for efficient distillation to produce the same product water. Experimenters are still searching for a readily recoverable regenerant.

Solvent extraction has been investigated using materials such as diisopropylamine, methyldiethylamine, and triethylamine. These materials absorb water but reject salts and the solvent capacity for water varies considerably with temperature. Water is absorbed into the solvent at the temperature of maximum solubility and rejected at the temperature of minimum solubility. As of 1964 this process has been found more expensive than distillation.

E. D. Howe

**WEATHER, EFFECT OF MOON ON.** The idea that the Moon has an effect on rainfall is present in the mythology and folklore of many peoples, but until recently the idea had been rejected outright by geophysicists. This appears to have been due to two reasons. Firstly there were inconsistencies in the form of the effect. Just as many people favoured an influence after New Moon as favoured a similar effect after Full Moon. Secondly, the gravitational force of the Moon acting on the lower atmosphere is known to be small compared with that of the Sun and hence it would be expected to have a correspondingly smaller effect on the dynamics of the lower atmosphere.

A large amount of evidence has recently come forward from the U.S.A., Australia, New Zealand, India and South Africa that the distribution of rainfall is influenced by the phase of the Moon. In the U.S.A. and New Zealand, for instance, there appear to be two maxima, one between New Moon and First Quarter and the other between Full Moon and Last Quarter.

The pattern shifts with geography, i.e. the maxima do not occur at the same time of the lunar month in all places but there is sufficient uniformity in the pattern to suggest a world-wide lunar effect on rainfall.

The way in which the Moon produces these effects is still in question. Mechanisms which are being considered are:

(1) The gravitational effect of the Moon, although known to be small in the lower atmosphere, may still operate through some kind of trigger action.

(2) It has been suspected for some time that dust entering the top of the atmosphere—which could be determined by the Moon's position—has a controlling effect on rainfall. The dust might have this effect either by acting as an ice nucleus or indirectly through a disturbance of the ozone layer.

(3) It is known that the Moon modulates the solar corpuscular stream which in turn has an influence on ionospheric and certain geomagnetic phenomena. The effect of such disturbances on meteorological phenomena is unknown but has been suggested many times.

Evidence that the Moon has a definite effect on rainfall in most parts of the globe points to the need for much more knowledge on the behaviour of the atmosphere as a whole and, in particular, to an understanding of the mechanisms which lead to rain.

E. E. ADDERLEY

**WEATHER FORECASTING, LONG-RANGE.** Long-range weather forecasting, concerned with periods beyond 48 or 72 hours into the future, is still in its infancy as a science. The reasons for this are to be found in the tremendous complexity of the problem (almost equivalent to human behaviour), in the extensive data required in both space and time, and in the related fact that only a few top-notch scientists have yet been attracted to the subject.

Because of these difficulties long-range forecasters and researchers have had to supplement inadequate physical knowledge of atmospheric behaviour by statistics and synoptics (map portrayals). The basic data upon which present-day long-range forecasting is based is the same as for short-range prediction, but with particular emphasis on temperature, pressure, and wind data for various layers of the troposphere. The manner in which these data are processed, however, is different in long-range prediction inasmuch as no satisfactory method has been found to extrapolate wind and weather systems step by step out beyond two or at most three days. Not only do cyclone and anticyclone systems behave erratically during their lifetimes, but they die and become replaced by new systems. Thus long-range forecasting is primarily concerned not with the timing of individual systems but rather with analysing and predicting average or prevailing conditions over extended time intervals, usually in terms of their deviation from long-term climatological normal conditions. There are also many statistical techniques which draw upon the serial correlation of meteorological time series, upon quasi-periodic developments in the recent past, upon the tendency of developments to occur at certain calendar dates, and upon complex lag relationships between some characteristics of very large-scale atmospheric wind systems.

With high-speed computers, time-averaged mean patterns of pressures, temperature and circulation for the surface and other layers of the troposphere over most of the Northern Hemisphere are easily prepared —even the contouring and isobaric analysis is done by machine. When such patterns are constructed, let us say for weekly or monthly periods, the scale of meteorological systems is found to be quite different than one sees on daily weather maps. Instead of surface high and low pressure areas perhaps 500–1000 miles in diameter, cellular features about three to four times this size appear.

Thus, on one particular month's mean sea-level pressure map there may be a vast cyclonic whirl near Iceland, another in the Aleutians, perhaps one on the periphery of the Arctic Basin and a few large high pressure cells—perhaps one in the eastern Atlantic, another in the central Pacific, and a third over Siberia. The corresponding chart for mid-troposphere would show a meandering broad flow (westerlies) that resembles a roughly sinusoidal perturbation with dips (troughs) near and behind the surface cyclones and crests (ridges) behind the anticyclones. The latter features are the planetary waves of the hemispheric circulation; these together with the surface features are called "centres of action." The number, placement, and orientation of the centres of action in the hemisphere are by no means constant, so that the Januaries of different years will usually show different flow patterns and consequently different prevailing weather patterns. The central problem of long-range forecasting is to determine the evolution of these patterns in advance, for these are highly correlated not only with average weather for the period but also with the regions of birth of storms, their life histories, and ultimately with attendant weather over vast areas of the Earth.

Rigorous physical understanding of this class of meteorological phenomena is at present much less complete than that of the cyclonic and anticyclonic vortices and fronts on daily weather maps. During the past ten years dynamic methods using computers have been developed which greatly assist in predicting these short-period weather-map features by redistributing conservative quantities, particularly vorticity, in the basic upper-air currents. However, no satisfactory analogous methods have yet been worked out for predicting longer-range changes of the centres of action, consisting as they do of ensembles of many cyclones and anticyclones. This circumstances is largely due to the cumulative, interacting, and quantitatively uncertain effects of variable surface heating, by condensation, flow over mountain barriers, and friction.

Nevertheless, some skill in long-range prediction is possible because of a few fortunate statistical aspects of atmospheric behaviour. In the first place, the dense network of observations frequently indispensable in short-range work is not necessary because of the larger geographical scale of the systems. Secondly, in many aspects of their behaviour, cyclones, anticyclones, and planetary waves are persistently recurrent over weeks and sometimes months. Thus meteorological time series are serially correlated and average variations about a normal (for a month or season, for example) are much larger than would be expected if daily weather were randomly distributed. This statistical property implies that there are forces *external* to the atmosphere which force it again and again to repeat essentially the same series of weather developments. Aside from the geographically fixed Earth features, there must be other forcing functions which stabilize the long-period pattern. Otherwise, the observed persistence of abnormalities are difficult to account for.

Conceivably these external factors may be found in solar variations. A number of long-range forecasters have attempted to show this and apply the idea to practical forecasting. One method seeks out analogous periods in the past in which the mean circulation patterns resemble the present and which occurred in the same phase of the sunspot cycle. The forecast is then simply read off the past series, assuming, of course, that developments observed in the past will take place in the future. If there is some skill in this method, which has not yet been adequately documented, one can raise the question as to what part is accounted for by the chart analogue and what part by the solar analogue. It must be pointed out that no adequate physical theory as to how solar activity might affect long-period weather phenomena has been worked out.

Another possibility is that the external factors giving rise to persistent atmospheric flow pattern anomalies are close by—namely at the surface of the continents and oceans. These surfaces are indeed variable as heat reservoirs. For example, it is now known that temperatures in the upper 100 or 200 m of the ocean may sometimes reach several °F above or below normal over areas about half the size of the centres of actions. Over the continents, snow and ice cover is a major variable. Both these factors, arising from past weather regimes, may carry over effects to the atmosphere and influence its future behaviour because of the longevity, e.g. water motions and changes are quite sluggish compared to those in the atmosphere. Here again, methods for direct use in forecasting are only in the development stage. But assuming such factors operate, they are probably reflected in long-term meteorological records, and by this indirect route most long-range forecasts (e.g. a month) are made. Seasonal predictions of an experimental nature are also attempted.

There is, of course, the method of analogues in which synoptic experience plays an appreciable role because no two situations are ever found that are alike over the entire hemisphere. This is partly because available meteorological records are too short, so the selection of "the best analogue" must consider sequences and other items subjectively.

Another method involves keeping close track of a series of mean charts, applying (by computer) kinematic methods of extrapolation for the slow evolving singular features, together with empirically derived corrections associated principally with time of the year, place, and very importantly, the physical compatibility of different centres of action in terms of position and intensity.

Once the circulation patterns have been determined, the average anomalous weather patterns, cyclonic and anticyclonic tracks, etc., are derived using empirical methods and equations requiring computers for easy application. While of economic value, such predictions are still far from perfect. If desired, they may be expressed in terms of probabilistic success.

For medium-range forecasts up to five days, dynamical prognoses of circulation patterns form the basis of modern methods. Beyond two or three days the averaging technique described above is also employed, but with numerically predicted charts used as much as possible in the averaging, and with empirically derived lag effects between circulation and weather introduced whenever possible. The final product consists of forecasts both of average conditions for the 5-day period and of estimates of the day-to-day weather maps likely to develop. Obviously the detailed forecasts deteriorate rapidly in skill with time; generally the deterioration is of an exponential form.

*Bibliography*

World Meteorological Organization (1962) *The Present Status of Long-range Forecasting in the World*, Technical Note No. 48, WMO; No. 126, TP 56, WMO Secretariat, Geneva, Switzerland.

NAMIAS J. (1953) *Thirty-Day Forecasting; a Review of a Ten-Year Experiment*, Meteorological Monographs, Am. Met. Soc., Vol. 2, No. 6.

SAWYER J. S. (1963) in *Encyclopaedic Dictionary of Physics*, (J. Thewlis Ed.), 7, 167; Oxford: Pergamon Press.
                     J. NAMIAS

**WEATHER FORECASTING, NUMERICAL.** The basic laws of physics and dynamics which control large-scale motions in the atmosphere have long been known, but it is only since the Second World War that numerical calculation based upon the equations of motion and of thermodynamics has been applied directly to weather forecasting. Such procedures which are now beginning to take a place in weather forecasting practice have come to be known as "Numerical weather forecasting" or "Numerical weather prediction".

Three important factors have contributed to the development of practical methods of numerical weather forecasting in recent years. First, a network of observations of wind, temperature and pressure has become available over most of the northern hemisphere adequate to define the large-scale air motions both near the ground and up to above 10 km. This covers those regions of the atmosphere which primarily control weather developments over a few days. Second, ways have been found of specializing the equations so that they describe only the motions of primary significance in large-scale atmospheric developments. For example, the assumption of a pressure field corresponding to hydrostatic equilibrium eliminates sound-waves from the solutions. Third, electronic computers have become available and they can perform the necessary calculations (several man-years on a desk calculator) in time to provide a useful forecast.

The fundamental equations used as a basis for numerical weather forecasting are (a) three-component equations of motion, (b) the equation of continuity, (c) a thermodynamic equation relating the change of temperature of the moving air to the heat received, $Q$, by radiation and other non-adiabatic processes as well as to the change of pressure, and (d) the gas equation which is used to relate temperature, pressure and density. This set of equations provides six equations for the six unknowns, namely three components of motion, temperature, pressure and density. The initial values of the variables may be regarded as given throughout the volume of the atmosphere considered, and the mathematical problem is to determine their future values by calculation. Boundary values are needed at the lateral boundaries as functions of time to make the mathematical problem soluble, and some arbitrary assumptions have to be made in the expectation that, if they are well chosen, errors at the boundary will not penetrate too far into the volume during the forecast period.

In some meteorological disturbances, particularly in depressions and at fronts, significant amounts of heat are released from the latent heat of condensation of water vapour. The quantities depend on the water vapour content of the air, and to specify the problem completely a further equation is needed to specify the changes in the water vapour density resulting from air movements. However, considerable progress in numerical prediction over periods up to three days has been achieved while ignoring this and other contributions to the heat transfer, $Q$, i.e. the system has been treated as adiabatic. It has also generally been possible to treat the motion as non-viscous, but it has been found advantageous to make allowance for the frictional drag of the ground, and for the variation of the ground elevation. These factors can be readily incorporated in the lower boundary conditions.

The earliest successful methods of numerical weather forecasting eliminated pressure from the horizontal equations of motion to provide a relation between the vertical component of vorticity and the horizontal divergence; subsequently they treated the wind as related to the pressure field by the *geostrophic wind relation* or by a more general relation known as the "*balance equation*" which also had the effect of eliminating gravity waves from the solutions. Such mathematical representations of the motions are known as "*filtered models*". The simplest such model treats the atmosphere as a single layer of incompressible fluid, and the governing equation states that the fluid conserves its absolute vorticity about a vertical axis as it moves from point to point. (*Absolute vorticity is measured relative to axes not rotating with the Earth.*) This representation is known as the *barotropic model* and has provided forecasts of the winds at the 500 millibar level (approximately $5\frac{1}{2}$ km) almost, if not quite, as good as more refined models. Forecasts for other levels are, however, impossible without consideration of horizontal temperature gradients, i.e. a *baroclinic model*.

Recent work has been directed to the numerical integration of the basic first-order equations without the initial elimination of pressure. The basic equations have come to be known as the "primitive equation". Most of the difficulties of computational stability and truncation error arising in their solution have been overcome, but methods based on the primitive equations have not yet been introduced into operational practice.

At the present time numerical forecasts of wind and pressure in the atmosphere of temperate lattitudes from 1 to 3 days ahead give rather better results than subjective methods for levels from 3 to 10 km. For lower levels the two methods have similar accuracy.

In order to start a numerical forecast the initial pressure distribution must first be represented by numerical values at a grid of points at one or more levels in the atmosphere. The distance between the points is usually of the order of 300 km. For obvious reasons, observing stations are not situated at the corners of a regular grid, and the first step in a practical forecasting system is the interpolation between observations to provide the initial values at the grid-points. This procedure has come to be known as *"objective analysis"*, and is far from straightforward because the observation network is barely adequate to define the pressure field, and the observation soundings are subject to significant errors. The task for the electronic computer is comparable with that for the forecast, but the computation of a numerical forecast is able to proceed directly from the observations at upper-air sounding stations and surface observatories, and the whole task of preparing a numerical forecast may be completed in a period of 1–3 hours on a fast modern electronic computer.

*Bibliography*

KIBEL I. A. (1957) *An Introduction to the Hydrodynamical Methods of Short Period Forecasting*, Gostekhizdat, Moscow (Transl. edited by R. Baker), Oxford: Pergamon Press.

THOMPSON P. D. (1961) *Numerical Weather Analysis and Prediction*, New York: Macmillan.

J. S. SAWYER

## WEATHER SATELLITES.

*Definitions of some of the terms used in the article.*

Earth-oriented satellite. Satellite whose reference axis has a fixed orientation with respect to the Earth's surface.

Earth-synchronous satellite. Satellite whose orbit is such that it appears to hover over the satellite subpoint.

Satellite subpoint. Intersection of the local vertical passing through the satellite with the Earth's surface.

Space-oriented satellite. Satellite whose reference axis has a fixed orientation with respect to space.

Sun-synchronous satellite. Satellite whose orbital plane progresses 360° of longitude in 365 days. A satellite placed in such an orbit and inclined 100 deg to the equator would transit a given latitude equatorward of the 80th parallels at the same local times each day.

Vidicon. A photoconductive image pickup or television type tube. The image is focused on the vidicon screen by a lens, and the vidicon scanner transforms the image to an electrical signal which can be transmitted or recorded on magnetic tape.

Although the first successful meteorological satellite experiment was carried on Explorer VII in 1959, the first of the TIROS series of specifically meteorological satellites was placed in orbit in 1960, and the latest to the date of this writing, the eighth, was launched in 1963. All the TIROS satellites have carried two television cameras, and some in the series have had a radiation package aboard as well. All thus far have been placed in roughly circular orbits at a nominal altitude of 700 km; thus the orbital period is about 100 min. In each case the orbital plane has been inclined to the equator; the subsatellite point for

TIROS I through IV ranged between 48N and 48S, whereas that for TIROS V through VIII between 58N and 58S. Table 1 summarizes many characteristics of the TIROS series.

The meteorological satellite is still a relatively new observational tool, and the pictures are rather different in nature from conventional fixed-point observations taken at or near the Earth's surface. Consequently, the potential information content of the photographs has probably barely been exploited. The information available depends on many factors, among which are resolution, contrast and brightness. A lower limit to the size of features perceivable in the pictures is imposed by the resolution of the vidicon system, which is largely dependent upon the width of the individual television scan lines when projected to Earth. In TIROS wide-angle camera photographs resolution varies from about 2·5–5·5 km at the subpoint to several miles near the horizon. The overall average resolution is about 5 km; thus the smaller-scale cloud elements familiar to the ground observer are not distinguishable as such, and even areas that appear *clear* in TIROS pictures often actually contain fields of small cumulus clouds or optically thin cirrus clouds. The narrow-angle cameras provide resolution as high as 0·4 km over areas about 117 km on a side when the camera is pointed straight down, thus considerable cloud detail is possible over limited areas. Each picture taken by the medium and wide-angle cameras encompasses an area at least 1000 and 1340 km, respectively, on a side, so there is almost no upper limit to the size of visible features. Picture mosaics have given meteorologists a view of the Earth and cloud patterns on such a grand scale that even the largest cloud systems can often be seen in their entirety. Complete coverage of the Earth's surface, even of that part exclusive of the polar regions, is not possible with the present TIROS system because the satellite itself is space rather than Earth oriented, and because of the limitations imposed by power and readout station requirements. Several methods have devised for the geographical gridding of the satellite photographs. The most accurate gridding requires time-consuming graphical work performed by hand, whereas much faster methods employing high-speed electronic computers produce grids that are generally adequate for operational and may other purposes.

The cloud systems associated with extratropical cyclones and their attendant frontal systems were among the earliest and most striking synoptic-scale circulation features viewed by TIROS (Fig. 1a). Not all such cloud patterns correspond closely to the classical polar front, wave cyclone model, but general agreement is frequent enough to often enable the identification of cyclonic storm centres and their fronts from satellite pictures alone. Furthermore the cloud patterns often permit estimates of the stage of cyclone development and aspects of its thermal and vertical structure. Since large areas of the Earth, particularly oceanic regions and large portions of the

*Table 1. Information on TIROS meteorological satellites*

| Tiros No. | Launch date | Last regular pictures† | Camera angular openings (deg.)‡ | Usable pictures obtained | Radiometer channels* |
|---|---|---|---|---|---|
| I | April 1, 1960 | June 19, 1960 | 13 and 104 | 19,389 | — |
| II | Nov. 23, 1960 | Feb. 1, 1961 | 13 and 104 | 25,574 | 1, 2, 3, 4, 5; LRR |
| III | July 12, 1961 | Oct. 30, 1961 | 104 and 104 | 24,000 | 1, 2, 3, 4, 5; LRR; HR |
| IV | Feb. 8, 1962 | June 12, 1962 | 80 and 104 | 23,370 | 1, 2, 3, 5; LRR; HR |
| V | June 19, 1962 | May 5, 1963 | 80 and 104 | 48,562 | — |
| VI | Sept. 18, 1962 | Oct. 11, 1963 | 80 and 104 | 59,830 | — |
| VII | June 19, 1963 | Still operating‡‡ | 80 and 104 | (101,996)‡‡ | 1, 2, 3, 4, 5; HR |
| VIII | Dec. 21, 1963 | Still operating‡‡ | 104 and 108 (APT) | (73,156)‡‡ | — |
| IX | Jan. 22, 1965 | Still operating‡‡ | 104 and 104 | (57,081)[3] | — |
| X | July 2, 1965 | Still operating | 104 and 104 | (604)[3] | — |

† In some cases pictures were received sporadically for months after the indicated dates.

‡ Along diagonals of square picture format.

‡‡ As of July 7, 1965.

\* Medium-channel radiometer (nominal spectral ranges in microns).

Channel 1–6–6·5 $\mu$ (except 14·8–15·5 $\mu$ on TIROS VII).

Channel 2–8–12 $\mu$.

Channel 3–0·2–5·0 $\mu$.

Channel 4–7·5–30 $\mu$ (omitted on TIROS IV).

Channel 5–0·55–0·75 $\mu$.

Low-resolution radiometer (LRR).

Hemispherical radiometers (HR).

southern hemisphere, are poorly observed from a meteorological standpoint, TIROS photographs of such circulation features as major cyclonic systems can be of considerable aid to the forecasting services of many countries. The distinctive feature of well-developed mid-latitude cyclones is a spiral-like array of major cloud bands, whereas such a cloud pattern is seldom found in association with anticyclonic circulation centres. Anticyclones are characterized chiefly by relatively cloudfree areas over continents, and by large areas of low-level stratocumulus type clouds, often organized in cell-like configurations, over oceans. Frontal cloud bands are generally well marked, at least cold fronts, and it is often possible to delineate the active and inactive portions of a frontal zone from the appearance of the cloudiness in satellite pictures. The jet stream in middle and subtropical latitudes is often accompanied by bands or zones of thick cirrus to the right of the jet axis looking downstream, with little or no such cirriform cloudiness to the poleward side (Fig. 1a). The rather sharp poleward edge of the cirrus layer is often made detectable in the pictures by the shadow it casts on lower cloud decks.

In general the atmosphere over tropical regions is very poorly observed by conventional means, so knowledge and understanding of atmospheric processes in this part of the world have been gained slowly and with difficulty. The vast tropical oceans, however, breed many weather disturbances, some of which develop into dangerous storms. The orbital characteristics of the TIROS satellites have been such that coverage of the tropics has been generally superior to that of other regions. For the foregoing reasons, then, it has been the tropics that has received perhaps the greatest and most immediate benefits from meteorological satellite operations. Fully developed tropical cyclones exhibit a rather distinctive, if sometimes somewhat asymmetric, spiral cloud pattern in satellite photographs (Fig. 1b). Over the inner portions of mature storms there appears a canopy of dense cirrus that tends to partially or sometimes completely obscure the details of the inner bands of convective clouds. Preliminary studies have indicated that storm intensity is correlated with the size of the cirrus shield and the degree of organization of the interior banding. The hurricane *eye* is frequently discernible, as well as certain other characteristic features such as an annular ring of cloud-free air separating an outer convective band from the storm proper, and striations on the edge of the cirrus layer that are indicative of anticyclonic outflow aloft. Less well-organized tropical disturbances such as easterly waves and upper cold lows are sometimes detected in the satellite pictures also. Heavy convective cloudiness in connexion with the intertropical convergence zone (ITC) is viewed by TIROS satellites on nearly every pass over the equatorial regions. These observations indicate, however, that the ITC cloud band is never continuous across either the Atlantic or Pacific Oceans; rather, it

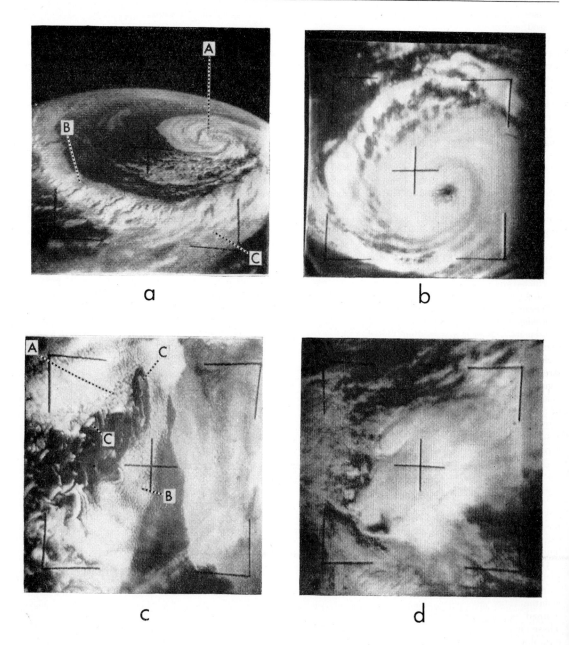

Fig. 1. Selected vidicon pictures from TIROS satellites: a.) Cyclone over North Atlantic Ocean near Nova Scotia on May 29, 1963; cyclone centre near A, frontal zone at B, jet stream clouds at C; TIROS VI, orbit 3691, 1005 GMT. b.) Typhoon Bess in North Pacific Ocean near Iwo Jima Island on Aug. 7, 1963; central pressure was 948 mb and maximum wind speeds were over 50 m sec$^{-1}$ near the time of this picture; TIROS VII, orbit 721, 0429 GMT. c.) Cellular and eddy cloud patterns off the coast of northwest Africa on April 24, 1963; cells at A, eddies at B, Canary Islands at C; TIROS V, orbit 4429, 1135 GMT. d.) Severe local storm cloud mass in middlewestern United States on April 30, 1962; damaging wind gusts of 30–55 m sec$^{-1}$, hailstones over 25 mm in diameter, and tornadoes were observed in Illinois and Indiana several hours after this picture; TIROS IV, orbit 1164, 1457 GMT.

consists of alternately active and inactive (or suppressed) segments of cloudiness.

Whereas TIROS pictures have been useful in the depiction of large-scale cloud patterns, they have revealed considerable cloud organization in the mesoscale also. Cellular arrays resembling the Bénard cells of laboratory experiments were first noted in TIROS I pictures (Fig. 1c). Although appearing in a variety of sizes, the cells are generally 30–80 km in diameter. Cloud masses 15–25 km in width form a ring, often only partially closed, at the periphery of the cell, leaving a cloud-free centre. The clouds themselves are masses of cumulus or stratocumulus, the latter being found beneath a low-level temperature inversion capped by dry air. Cellular cloud patterns have thus far appeared almost exclusively over oceans, usually in the tropics or subtropics. The physical processes governing their formation are incompletely understood, for although similar in some respects to the laboratory Bénard cells, they differ in others. The chief dissimilarity is in their width-to-height ratio of about 30, which is ten times that of Bénard cells.

Mesoscale eddy clouds are commonly found under conditions much like those associated with cellular clouds; in fact the two types often coexist in the same region (Fig. 1c). Eddy clouds occur in the immediate vicinity of island obstacles and to distances downstream indicating a persistence of up to some 20 hours. Again in this case the theory is in complete, but the evidence points to either a type of inertial oscillation or mechanical eddies in conjunction with eddy stresses somewhat different from those usually assumed. Another mesoscale cloud pattern seen in satellite pictures is wave clouds, mostly observed to occur in the lee of mountain ranges. The observed wave-lengths have been of the order of 5–15 km. Since the meteorological conditions commonly associated with their occurrence are well known, and the governing theory is fairly well developed, it may be possible to infer certain meteorological conditions by satellite observations of the presence and wavelength of mountain wave clouds.

Outbreaks of severe local storms appear in TIROS pictures as very bright, rather uniform masses of clouds some 150–350 km across, separated if not isolated from surrounding clouds, and having well-defined edges, particularly on the windward side. These cloud masses have been related chiefly to thick cumulonimbus anvil clouds that unite and spread long distance downstream from clusters or lines of the parent convective clouds. Ordinary thunderstorms appear as similar, but smaller, bright clusters ranging downward in size to isolated cumulonimbi about 10 km in diameter, each usually possessing a cirriform plume extending downstream in the direction of the vertical shear of the flow in the upper troposphere.

A five-channel, scanning radiometer of medium resolution and a low-resolution radiation sensor, together with Suomi-type hemispherical radiometers such as those on Explorer VII, were carried aboard various of the TIROS satellites (see Table 1). Each medium-resolution sensor measures intensity emitted in a particular direction over a selected portion of the visible and infra-red spectral range. The radiometer axis is mounted at 45° to the spin axis and its view is bi-directional, permitting some portion of the Earth to be scanned with each spin. Each channel observes intensity through a cone of view set nominally at 5°, and each is nominally aligned along parallel axes such that all channels view essentially the same area simultaneously. The resolution of this instrument is considerably less than that of the television cameras, and the resolution varies continuously through each spin. At TIROS altitudes any given radiation measurement represents an averaged value over an area of 700 mi² when the radiometer views directly below the satellite, but over about ten times that area when it views near the satellite's horizon. Calibration of the sensors was carried out before launch, but deterioration in the sensitivity of the instrument has raised doubts as to accuracy of the observations. The degradation rate of the various channels has been highly variable, furthermore.

The objectives of obtaining the "water-vapour window" (8–12 $\mu$) measurements were to depict areas of cloud cover, to make quantitative estimates of the altitude of cloud tops, and to measure the temperature of the Earth's surface in cloudfree regions. It has been found that these radiation data do define the large-scale cloud systems rather well, particularly the overcast middle and dense high cloud areas. With reasonable assumptions the radiation intensities can be translated into equivalent black-body temperatures at the cloud tops. Comparison of these temperatures with vertical temperature distributions derived from conventional radiosonde observations enables estimates of the cloud top heights. These estimates compare favourably with independent observations from aircraft and with radar cloud tops, and reasonable results have also been obtained in sparse-data areas using mean soundings. There are limitations to the use of this technique, however. Thin cirrus layers and areas of broken or scattered clouds, or areas where multiple cloud layers are viewed simultaneously can lead to difficulties in the interpretation of results. Within this same spectral range, moreover, water vapour and ozone are not totally transparent, which can cause measured intensities, especially those of the Earth's surface and of low overcast, to be too low. A corrective procedure that depends upon the vertical distribution of chiefly the water vapour has been developed, however.

Measurements in the 0·2–5 $\mu$ interval yield very nearly the total reflected solar energy, i.e. the albedo of a region, whereas the total infra-red radiation leaving the same region is determined from the 7–30 $\mu$ observations. The two intervals together, therefore, should show how much energy is returning to space and how much is absorbed by the Earth and its atmosphere. Meteorologists are making preliminary studies of the heat budget using these measurements, and eventually they hope to understand the relationship of the heat budget to the planetary circulation and its evolution. Incomplete coverage and the experimental nature of the instrumentation have thus far thwarted the achievement of these long-range goals, but TIROS radiation data have provided the means for initial probing into these problems. Figure 2 gives an interesting result from one of these studies. Using data from the 7–30 $\mu$ interval, mean outgoing long-wave radiation charts were prepared for 4–5 day intervals over the period from late November 1960 to early January 1961. In Fig. 2 latitudinal averages of the outgoing flux during this period are compared with previous estimates by other investigators.

Many geographic features are recognizable in satellite photographs. Ice-free oceans and seas, as well as large lakes and rivers, generally appear darker than land areas, therefore most coastlines are readily delineated (see Fig. 1c). Snow and ice fields are prominent because of their high reflectivity, but they may be difficult to distinguish from cloud. Mountain snow patterns often have a characteristic dendritic pattern formed by the contrast between the relatively barren and snow covered higher peaks and ridges and the often heavily forested and more often snowfree valleys. Sea ice and large floes have been observed frequently, especially in the Gulf of St. Lawrence area with the relatively high resolution, narrow-angle camera pictures from TIROS I and II. Even dust

storms and the dense smoke from large forest fires have occasionally been identified in satellite photographs.

Although to date the TIROS series has been largely experimental, the photographic detection and location of cyclonic storms over the data-poor regions of the world was an obvious benefit to daily weather analysis and forecasting from the very beginning. Meteorological research and technological improvements in the satellite system have been steadily contributing to increased operational utility of the satellite pictures. Communications problems and restricted geographical coverage are perhaps the chief reasons why the pictures are not even more useful. Nephanalyses portraying the cloud cover and major features of the cloud patterns are prepared from the pictures received at the command and data acquisition stations located in Virginia, California and Alaska. These nephanalyses are transmitted by facsimile to the National Weather Satellite Centre near Washington, D.C., then retransmitted to field forecast offices. Selected individual photographs are transmitted by photofacsimile also. Numerous word messages have been sent to the national meteorological services of other countries, including coded nephanalyses, advising them of the location of probable storm systems in their areas of concern. In a recent effort to make actual satellite pictures available to users outside of the United States of America, a special APT (automatic picture transmission) camera was carried aboard TIROS VIII. Any station in the world possessing relatively inexpensive ground receiving equipment can obtain

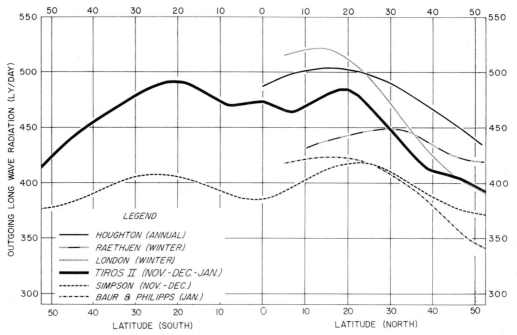

*Fig. 2. Latitudinal variation of outgoing long-wave radiation derived from TIROS data and compared with previous results (after Winston Rao, 1963).*

pictures covering a large surrounding area directly from the satellite when it is within visual range and has been programmed.

Many innovations in meteorological satellite design and instrumentation are in the planning or active development stage. Greatly improved coverage is one of the primary design aims in connexion with future satellites, this to be accomplished by placing it into a sun-synchronous near-polar orbit. One such earth-oriented satellite at the proper altitude could view every point on the Earth's surface twice a day, once in daylight and once at night. A single conventional TIROS satellite, in comparison, can observe only about 18 per cent of the world on any one day, and furthermore the area viewed by this type of satellite is continually changing. Under development concurrently is the complex data processing, distribution and archiving system that will be required to make rapid and efficient operational and research use of the enormous amount of different kinds of information to be continually received from one or more advanced type satellites. One experimental form of output from such a system is to be a computer-digitized and rectified mosaic of the cloud images received on each orbit. The Earth-synchronous meteorological satellite is another type under consideration. It would appear to hover above the satellite subpoint on the equator and, because of its great altitude, could continuously view the Earth's atmosphere over about 60° of latitude and longitude to either side of this point. New instrumentation is planned or under development also, among which are infra-red spectrometers, designed to obtain indirect temperature soundings, and a device to detect *sferics*, i.e. the radio frequency emissions by lightning. Perhaps the most far-reaching prospect is, however, the concept of using global weather satellites to interrogate remote, automatic weather stations or constant-level balloons, collect their data, locate them if necessary, and then transmit these observations to meteorological centres.

*Bibliography*

Kiss E. (1963) *Bibliography on Meteorological Satellites* (1952–1962), U.S. Weather Bureau, Washington, D.C.

National Weather Satellite Centre, U.S. Weather Bureau (1963): *Reduction and use of data obtained by TIROS meteorological satellites; WMO Tech. Note No. 49*, No. 131, T.P. 58.

Weather Satellite Systems (April 1963) *Astronautics and Aerospace Engineering* **1**, No. 3, 22.

Winston J. S. and Rao P. K. (1963) *Temporal and spatial variations in the planetary-scale outgoing long-wave radiation as derived from TIROS II measurements; Monthly Weather Review* **91**, 641.

E. P. McClain

**WIND TUNNEL, LOW-DENSITY.** Low-density wind tunnels are used to study rarefied gas dynamics or phenomena related to the molecular or non-continuum nature of a gas flow at low densities. The relative importance of such effects depends largely on the ratio of the mean free path to a characteristic dimension of the flow field, the Knudsen number. Low-density tunnels normally operate with Knudsen numbers in the range 0·01–10 and subsonic to hypersonic Mach numbers. The large Knudsen numbers are achieved by operating at test section pressures in the range 1–100 $\mu$Hg and it is this feature which distinguishes low-density tunnels from the more conventional wind tunnels.

Experimental work with low density supersonic tunnels started in 1947 at the University of California, Berkeley and the Ames Laboratory and since then many more similar tunnels have been built. The main features of a typical low density tunnel are shown in the figure. The test gas is fed through dryers and meters into a settling chamber or heater and is then expanded through a nozzle into the working section. The gas may then pass through a diffuser and is pumped out of the working section with vacuum

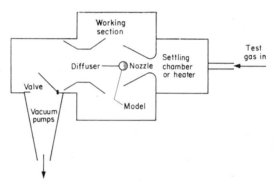

*Main components of a low-density wind tunnel.*

pumps. The gas is not normally re-circulated because of the contamination produced by the vacuum pumps but this is not a serious disadvantage since a typical flow rate is only of the order of 10 g/min. The basic components of a tunnel are considered in more detail below.

The settling chamber is used to allow the gas to obtain known uniform conditions upstream of the nozzle. For Mach numbers higher than about 6 or for hypervelocity conditions in the test section some form of heating is required and this is normally applied upstream of the nozzle though R.F. heating has been used to heat the gas in the nozzle. If the gas is to be heated merely to avoid the possibility of liquefaction in the working section then the heat requirements are comparatively modest and compact graphite resistance heaters operating with a few kilowatts have been used. If hypervelocity conditions are required in which the ratio of *stagnation temperature* to body wall temperature approximates that of free flight conditions then higher temperatures than can be supplied by electric resistance heaters are required and plasma jets or arc heaters have been used to obtain temperatures of the order of 5000°K. After passing through the arc heater the gas may then be expanded directly through a nozzle or may pass into an intermediate chamber, the plenum chamber, where the gas may approach equilibrium and its properties be more readily estimated.

Nozzles have been designed by using the method of characteristics to obtain the isentropic core and then adding the displacement thickness of the boundary layer along the nozzle wall. This method has proved to be satisfactory for axisymmetric supersonic nozzles up to about Mach 6 designed to operate with air at room temperatures. Higher Mach numbers have been achieved with conical nozzles but this type of nozzle has a disadvantage due to the axial Mach number gradient. Two-dimensional nozzles have been tried but are not now normally employed because of secondary flows which result in a less uniform test section flow than can be obtained with well designed axisymmetric nozzles. An inherent disadvantage with low density nozzles is that the boundary layer on the nozzle wall may occupy the major part of the test region leaving only a very small uniform core suitable for model testing. For example the uniform core of a nozzle with an exit diameter of 3″ designed to operate at Mach 4 may amount to only 3/4″ when operated at a static pressure of 40 $\mu$Hg and when the pressure is reduced to 25 $\mu$Hg the boundary layer will have extended to the centreline and the flow will become entirely non-isentropic. The boundary layer growth results in another disadvantage insofar as the effective area ratio changes with pressure and hence the Reynolds number cannot be varied independently of Mach number.

A number of techniques have been tried with limited success to overcome the difficulties associated with the boundary layer growth including cooling the nozzle wall with liquid nitrogen, use of boundary layer suction, freezing the test gas on the nozzle wall and the use of a free jet. The latter method utilizes the supersonic jet issuing from a sonic orifice which has the property that due to its associated shock structure, static pressures within the jet may be an order of magnitude less than the ambient static pressure in the tunnel. Thus for a given pumping capacity a free jet enables higher Mach numbers and lower pressures to be obtained than in a nozzle. However, there are severe axial gradients of flow properties in a free jet and so it is only suitable for certain types of experiments.

In a conventional supersonic tunnel a large proportion of the kinetic energy of the gas in the working section is normally recovered by allowing the gas to decelerate in a diffuser and attain a higher static pressure but the efficiency of a diffuser is reduced as the pressure is lowered and most low density tunnels have been designed to operate without the aid of a diffuser. There is now some evidence, however, to show that a properly designed diffuser can be used in a low density tunnel and result in a considerable saving in pumping capacity for given test section conditions.

Various pumping systems are employed to remove the gas from the test chamber. These are tabulated below together with the normal range of operating pressures.

| Pump | Pressure range |
|------|----------------|
| Steam ejectors | $P_\infty < 10 \ \mu$Hg |
| Oil booster diffusion | $1 < P_\infty < 50 \ \mu$Hg |
| Cryogenic | $P_\infty < 10 \ \mu$Hg |

A pump speed $\sim 10^4$ l/s is required to operate a tunnel with a nozzle exit $\sim 10$ cm diameter and this size is typical of many tunnels which have been built though tunnels utilizing cryopumping may have a pump speed $\sim 10^6$ l/s with a nozzle size of 100 cm. In addition to the continuous-operating type of tunnel a short duration hypersonic tunnel such as a gun tunnel or shock tunnel may be operated under low density conditions thereby enabling very large low density tunnels to be built without large vacuum pumps but at the expense of sacrificing the more simple instrumentation normally associated with a continuously-operating facility.

A considerable effort has been devoted to developing both conventional and novel forms of instrumentation to diagnose the properties of the gas in the test section and measure aerodynamic quantities on models and full details of many of these instruments are given in the bibliography.

*Bibliography*

DEVIENNE F. M. (Ed.) (1960) *Rarefied Gas Dynamics*, Oxford: Pergamon Press.

LAURMANN J. A. (Ed.) *Rarefied Gas Dynamics*, New York: Academic Press.

MASLACH G. J. (July 1959) *Slip Flow Testing Techniques*, AGARDograph 39.

TALBOT L. (Ed.) (1961) *Rarefied Gas Dynamics*, New York: Academic Press.

W. A. CLAYDEN

**"W" VALUE.** This is the mean energy expended forming an ion pair in irradiated gases; it includes the energy used both in ionization and excitation.

A. R. ANDERSON

# X, Z

**X-RAY DIFFRACTION TOPOGRAPHY.** X-ray diffraction topography is concerned with point-to-point variations in the directions and/or the intensities of X-rays that have been diffracted by crystals. From those variations the defect structure of the crystal may be examined.

Methods that mainly measure the local variations of the direction of the diffracted beam are useful for the detection of gross misorientations such as subgrains or grains (methods of Guinier and Tennevin (1949), Schulz (1954), Weissmann (1956)). Intensity mapping methods are chiefly concerned with individual defects such as dislocations, stacking faults, etc. In both groups there are experimental arrangements with both Laue case (transmission) and Bragg case (back reflection) geometry.

## I. Methods

*1. Schulz.* A bundle of white X rays divergent from a point source is diffracted by the crystal and recorded on a film (Fig. 1). The use of white radiation ensures

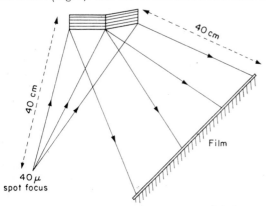

*Fig. 1. Typical arrangement for the Schulz technique.*

that there is no significant change of diffracted intensity due to different incident angles within the bundle. However, misorientations in the crystal cause gaps or overlap regions in the image. With dimensions shown (typical) rotations of 20 sec of arc can be detected.

Guinier and Tennevin used a similar technique in transmission. In a polycrystal or a heavily distorted single crystal many images from different grains

and/or different reflections may occur at the same time. Weissmann reduced the number of spots by using crystal monochromatized radiation. Furthermore, in order to identify the particular reflections which occur, he traced the direction of the diffracted beams with films at varying distances from the specimen.

*2. Berg–Barrett.* (Berg 1931; Barrett 1945; Honeycombe 1951) The crystal is set to Bragg-reflect the characteristic radiation from a line focus (Fig. 2). By placing the plate close to the crystal, geometric resolutions of $\approx 1\,\mu$ are realized. Across the diffracted

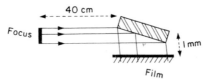

*Fig. 2. The Berg–Barrett arrangement.*

beam variations of the integral reflection power due to the imperfections in the crystal occur. Newkirk (1959) showed that single dislocations could be resolved by this technique and their Burgers vector experimentally determined.

The corresponding Laue-case arrangement has been developed by Barth and Hosemann (1958).

*3. Double-crystal.* (Bonse and Kappler 1958; Bonse 1958, 1962) X rays from a line focus are Bragg-reflected from a perfect reference crystal and then from the specimen, and finally are recorded on a film (Fig. 3). The reference crystal and specimen consist of the same kind of material so that exactly the same spacing of reflecting planes can be used in both crystals. As a result of this the shape of the rocking curve is principally independent of the spectral distribution $\Delta\lambda$ of the radiation used (dispersion is eliminated) and becomes 10–100 times narrower than any spectral line. This makes the method very suitable for measuring small tilts. When the specimen is slightly mis-set from the exact parallel position, i.e. on the flank of the rocking curve, tilts of only $\lesssim 0.1$ sec of arc and corresponding strains of $\left|\dfrac{\Delta d}{d}\right| \lesssim 10^{-8}$ to $10^{-9}$ result in detectable changes of reflected intensity. Deformations of this magnitude occur at distances of up to $50\,\mu$ or $100\,\mu$ from the cores of single dislocations, depending upon the material examined. The double crystal arrangement may also be used with the specimen set for transmission. When imaging larger tilts, for instance those that occur between subgrains or strained regions of a deformed crystal (>10 sec of arc), only

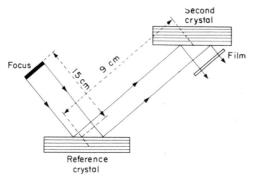

*Fig. 3. The double-crystal geometry.*

partial elimination of the dispersion by using a "monochromatizing" crystal with different spacing of reflecting planes or simply the use of characteristic radiation (Berg–Barrett technique) is usually sufficient for good contrast between regions with different orientations.

*4. Lang method.* (Lang 1958, 1959) In this method, a ribbon X-ray beam is collimated to a sufficiently small angular divergence that only one characteristic wave-length is diffracted by the crystal (Fig. 4). A stationary opaque screen allows only the diffracted beam to reach the photographic plate. A large area/

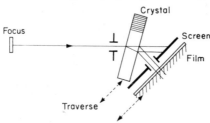

*Fig. 4. The Lang scanning method.*

volume of crystal can be surveyed by scanning both the crystal and photographic plate in synchronism past the incident beam. The line X-ray source in the method of Barth and Hosemann and the traversing movement in the Lang method achieve the same result, namely, a horizontally extended field of view. The drawback of the more complicated traverse apparatus, however, is compensated by the following advantages of the Lang technique: Lower background, less trouble with simultaneous reflections because a strongly collimated beam is used, and a higher resolution since the photographic plate may be placed closer to the specimen.

Many different types of lattice defects have been observed by this technique.

*5. Anomalous transmission.* (Borrman *et al.* 1958) In a perfect crystal incident and diffracted waves form the entity of the X-ray wavefield, which is a standing wave pattern fitting onto the set of reflecting planes. Depending on the position of the antinodes of this pattern the energy transport of the wave field occurs with anomalously high (antinodes coincident with planes) or anomalously low (antinodes between the planes, Borrmann effect) absorption. Therefore, a perfect crystal set at the precise Bragg angle is capable of transmitting interfering X rays at crystal thicknesses which would absorb almost all the energy of a non-interfering X-ray beam (Fig. 5). If $\mu_0 t \geq 20$ ($\mu_0$ normal

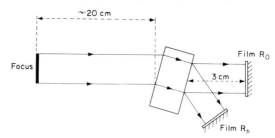

*Fig. 5. The Borrmann arrangement for anomalous transmission topography.*

absorption coefficient, $t$ thickness of specimen), both the transmitted diffracted beam $R_h$ and the transmitted direct beam $R_0$ have similar intensities and either can be used to obtain diffraction topographs. Lattice defects cause a local reduction in the anomalous transmission and are therefore seen as shadows.

## II. Causes of Contrast

With respect to the local amount of distortion produced by the lattice defects, the crystal may be divided into dynamical regions, where the primary wave $K_0$ and the diffracted wave $K_h$ are still coherently bound to each other as a wave field, and into kinematical regions where this coupling of $K_0$ and $K_h$ is destroyed because of too large lattice distortions. (Bonse 1964) Dynamical regions still behave locally like perfect crystals, though taken over longer distances, a continuous change of orientation and spacing of reflecting planes occurs. The wave field can adapt itself to these changes by a continuous variation of the direction of its energy flow $j$ and correspondingly of the ratio $I_h/I_0$ of the intensities of its components $K_h$ and $K_0$. (Penning and Polder 1961) Dynamical regions have locally the same narrow reflection curve as a perfect crystal (<10 sec of arc wide) and especially the same integral reflection power, i.e. the same area under the reflection curve.

*1. Homogeneous dilation and tilt contrast.* In reflection techniques, i.e. where the image is formed by the beam split off at the surface, the most straightforward cause for the contrast of dynamical regions is that these regions may simply be more or less mis-set from the diffraction peak given by the Bragg equation

$\lambda = 2d \sin \theta$. Mis-setting can be due to either a change of orientation $\theta$ or of lattice parameter $d$ or of both simultaneously. How large a mis-setting has to be before it causes a detectable intensity change depends on the "precision" of the diffraction condition, i.e. the width of the rocking curve (double crystal arrangement) or of the $\lambda$-spread (Berg-Barrett, etc.). In any case, the contrast should be proportional to

$$\frac{1}{W}\left(\frac{\Delta d}{d}\tan\theta \pm \Delta\theta\right),$$

($W$=width of rocking curve or $\lambda$-spread respectively). Owing to the extreme narrowness of the double crystal rocking curve this mis-setting is the main cause of contrast with the double crystal technique. It is worth noting that the sign of the contrast depends on the sign of the deformation.

*2. "Extinction contrast" or Direct image.* In kinematical regions the local reflection curve is generally lower and wider combined with an increase of integral (total) reflection power as compared with the perfect crystal. Cores of dislocations, up to ca. $2\mu$ diameter, are typical kinematical regions. What has been commonly called "extinction contrast" is exactly this increase of integral reflection power of kinematical regions. However, an increase of diffracted intensity will be observed only if the reflection curve is utilized too its full width, that is to say, only if the divergence of the incident beam is wide enough. This is in general the case for the Berg-Barrett, Barth and Hosemann and Lang methods, however, not with the double crystal technique. Therefore the former methods yield enhanced intensity from kinematical regions, and the latter decreased intensity (if the image of the kinematical region plays a role here at all).

The kinematical image contrast is also often referred to as the "direct" image. The sign of the contrast is independent of the sign of the deformation.

*3. Dynamical image.* The change which the entering wave field has experienced after traversing distorted dynamical regions also contributes to the image contrast, particularly in transmission methods. These effects are important for large values of $1/\chi_{rh}$ ($\chi_{rh}$ is the Fourier coefficient of order $h$ of electric susceptibility $\chi_r$, which is proportional to the structure factor $F_h$).

In reflection techniques, wave field beams can be bent back to the entrance surface where they add to the beam which is reflected at the surface itself (Bonse 1963).

In transmission methods, besides this bending of beam paths, the shift of energy between $K_0$ and $K_h$ is responsible for considerable contrast effects in the $K_0$ and $K_h$ waves when they leave the crystal at its exit surface (Hart 1963; Hart and Lang 1963). Moreover, the Borrmann contrast may also be counted in this category, since it is due to breaking up or continuous modification of wave field beams with anomalously low absorption into highly absorbed wavefields or single waves.

| Technique | Schulz, Guinier and Tennevin | Berg-Barrett | Double Crystal | Wide beam transmission (Barth and Hosemann) | | Scanning transmission (Lang) | |
|---|---|---|---|---|---|---|---|
| Apparatus | Simple | Simple | Complicated | $\mu_0 t > 10$ Simple | $\mu_0 t < 1$ Simple | $\mu_0 t \sim 3$ Complicated | $\mu_0 t < 1$ Complicated |
| Exposure Time | 10–25 hr | Short ($\sim$1 hr) | Short ($\sim$1 hr) | Long ($\sim$10 hr) | Short ($\sim$1 hr) | 10–30 hr | 2–10 hr |
| Defect for which technique is most suited with kind of contrast † | Grain mis-orientation, subgrains (1) | Subgrains (1) Dislocations (2) | Subgrains (1) Dislocations (1) Stacking faults | Dislocations (3) | Subgrains (1) Dislocations (2) Stacking faults | Dislocations (2) (3) | Subgrains (1) Dislocations (2) Stacking faults |
| Best geometric resolution | 50 μ | 1 μ | 1 μ | 1 μ | 1 μ | 1 μ | 1 μ |
| Sensitivity to deformations | Low | Low | High | High | Low | High | Low |
| Sensitive to the sense of deformations | Tilts: Yes Inhomogeneous deform.: No | Subgrains: Yes Disloc.: No | Yes | Yes | No | Yes | No |
| Thickness $t$ of specimen contributing to topograph ‡ | Schulz: $\leqq 5\,\mu$ G. + T. $50 \to 1000\,\mu$ | $\leqq 5\,\mu$ | $\leqq 5\,\mu$ (back refl.) $\leqq 300\,\mu$ (transm.) | $1 \to 5$ mm | $0 \to 2$ mm | $0{\cdot}1 \to 5$ mm | $0 \to 2$ mm |
| Dislocation image width †† | — | $1 \to 5\,\mu$ | Up to 150 μ | $\gtrsim 50\,\mu$ | $\sim 5\,\mu$ | Up to 150 μ | $1 \to 10\,\mu$ |
| Upper limit of dislocation density [lines/cm²] | — | $5 \times 10^6$ | $10^5$ | $5 \times 10^3$ | $5 \times 10^6$ | $5 \times 10^3$ | $5 \times 10^6$ |

† (1) homogeneous dilation and tilt contrast, (2) extinction contrast, (3) dynamical contrast.

‡ This is determined, in the Bragg case, by the extinction depth and, in the Laue case, by the value of $\mu_0$ (the absorption coefficient) for the material and the value of $t$ imposed by the technique.

†† Based on the assumption that this is determined by normal image overlap.

*4. Conclusion.* In the Berg-Barrett technique and in transmission techniques for $\mu_0 t \lesssim 1$ extinction contrast predominates. In the range $1 \lesssim \mu_0 t \leq 10$ both direct and dynamical images are visible. With $\mu_0 t \gtrsim 10$ practically only Borrmann contrast is observed.

Contrast in the double crystal arrangement is mainly due to homogeneous dilations and tilts, and, sometimes, to wave field beams which have been bent back to the surface.

The general condition for the visibility of a defect is that it produces a (sufficiently large) disturbance of the set of reflecting planes used. From this it follows, in the isotropic case, that dislocations can in principle vanish only if $\mathbf{g} \cdot \mathbf{b} = 0$ and $\mathbf{g} \cdot \mathbf{n} = 0$ ($\mathbf{g}$: diffraction vector, $\mathbf{b}$: Burgers vector, $\mathbf{n}$: vector normal to the slip plane). Since for mixed dislocations both equations cannot be satisfied simultaneously, mixed dislocations can never vanish exactly. In the anisotropic case dislocations can only vanish if their direction is normal to a mirror plane of elastic symmetry.

A particular cause of contrast not mentioned so far is due to the interaction of two wave fields (Pendellösung fringes) in wedge shaped portions of the crystal. Since stacking faults and dislocations may in effect generate wedges in crystals, they are often observed by these fringes.

### III. Comparison and Application of the Different Techniques

The Table gives a comparative survey of the different methods described above and their main fields of application. It has to be kept in mind, however, that the limits between different categories are somewhat diffuse, depending on the special experimental conditions under which a particular method is employed. Observations have been made of dislocations, stacking faults, low and high angle grain boundaries, twin boundaries, magnetic domains, precipitation and segregation of impurities. The materials so far studied by various workers include diamond, Si, Ge, InSb, NaCl, LiF, AgCl, SiC, MgO, $Al_2O_3$, BeO, Al, Cu, Fe:5% Si, calcite, quartz, and ice.

The methods are given below in the order of their decreasing sensitivity to lattice perturbation for the same thickness of crystal: Double crystal; anomalous transmission; Lang; Berg-Barrett, Barth-Hosemann; and Guinier-Tennevin, Schulz.

In some cases the transmission techniques may exceed in sensitivity the double crystal reflection method because a larger volume of crystal is sampled. There are, however, applications where high sensitivity is not desired such as sub-grain delineation, slip-band survey and any investigation of heavily strained regions. For these applications the less sensitive methods are more suitable. For stacking faults, dislocations, twins, and lamellar structures the transmission techniques are generally more informative than the Bragg-case techniques.

Webb (1962) and Azaroff (1964) discussed some of the aspects outlined here a few years ago. A fairly comprehensive list of literature up to 1961 may be found in those references.

*Acknowledgement*

This work was supported in part by the Advanced Research Projects Agency.

*Bibliography*

AZAROFF L. V. (1964) *Progr. Solid State Chemistry* 1, 347.
BARRETT C. S. (1945) *Trans. AIME* **161**, 15.
BARTH H. and HOSEMANN R. (1958) *Z. Naturf.* **13**a, 792.
BERG W. (1931) *Naturw.* **19**, 391.
BONSE U. (1958) *Z. Physik* **153**, 287.
BONSE U. (1962) *Direct Observation of Imperfections in Crystals*, New York: Interscience.
BONSE U. (1963) *Z. Physik.* **177**, 529.
BONSE U. (1964) *Z. Physik.* **177**, 543.
BONSE U. and KAPPLER E. (1958) *Z. Naturf.* **13**a, 348.
BORRMANN G. *et al.* (1958) *Z. Naturf.* **13**a, 423.
GUINIER A. and TENNEVIN J. (1949) *Acta, Cryst.* **2**, 133.
HART M. Ph. D. Thesis, Bristol, 1963.
HART M. and LANG A. R. (1963) *Sixth Intern. Cong. Cryst.*, Rome, Paper 12.4.
HONEYCOMBE R. W. K. (1951) *J. Inst. Metals* **80**, 39.
INTRETER J. and WEISSMANN S. (1954) *Acta Cryst.* **7**, 729.
KOHRA K. *et al.* (1962) *Direct Observation of Imperfections in Crystals*, New York: Interscience.
LANG A. R. (1958) *J. Appl. Phys.* **29**, 597; (1959) **30**, 1748.
NEWKIRK J. B. (1959) *Trans. AIME*, **215**, 483.
PENNING P. and POLDER D. (1961) *Philips Research Repts.* **16**, 419.
SCHULZ L. G. (1954) *J. Metals* **200**, 1082.
WEBB W. (1962) *Direct Observation of Imperfections in Crystals*, New York: Interscience.
WEISSMANN S. (1956) *J. Appl. Phys.* **27**, 389.

U. K. BONSE, M. HART and J. B. NEWKIRK

**ZERO SOUND.** A wave motion predicted by the Fermi liquid theory of L. D. Landau to exist in liquid $^3$He at very low temperatures. The velocity of zero sound is a little higher than that of ordinary sound. To observe zero sound one needs to excite liquid $^3$He at a high frequency so that the period of oscillation is much less than the time between collisions of the particles in the liquid. Under these circumstances ordinary sound cannot propagate. The conditions for observing zero sound can be attained if $^3$He at a very low temperature is excited ultrasonically at a very high frequency.

It has been suggested that zero sound should also be observable when electrons in a metal are suitably excited.

*See also*: Helium-3, propagation of sound in.

G. A. BROOKER

# INDEX

# ENCYCLOPAEDIC DICTIONARY OF PHYSICS

## *SCOPE OF THE DICTIONARY*

For convenience in planning, and to provide a framework on which the Dictionary could be erected, physics and its related subjects have been divided into upwards of sixty sections. The sections are listed below, but, as the Dictionary is arranged alphabetically, they do not appear as sections in the completed work.

Acoustics

Astronomy

Astrophysics

Atomic and molecular beams

Atomic and nuclear structure

Biophysics

Cathode rays

Chemical analysis

Chemical reactions, phenomena and processes

Chemical substances

Colloids

Cosmic rays

Counters and discharge tubes

Crystallography

Dielectrics

Elasticity and strength of materials

Electrical conduction and currents

Electrical discharges

Electrical measurements

Electrochemistry

Electromagnetism and electrodynamics

Electrostatics

Engineering metrology

General mechanics

Geodesy

Geomagnetism

Geophysics

Heat

Hospital and medical physics

Industrial processes

Ionization

Isotopes

Laboratory apparatus

Low-temperature physics

Magnetic effects

Magnetism

Mathematics

Mechanics of fluids

Machanics of gases

Mechanics of solids

Mesons

Meteorology

Molecular structure

Molecular theory of gases

Molecular theory of liquids

Neutron physics

Nuclear reactions

Optics

Particle accelerators

Phase equilibria

Photochemistry and radiation chemistry

Photography

Physical metallurgy

Physical metrology

Positive rays

Radar

Radiation

Radioactivity

Reactor physics

Rheology

Solid-state theory

Spectra

Structure of solids

Thermionics

Thermodynamics

Vacuum Physics

X rays